TECHNOLOGY AND HUMAN AFFAIRS

TECHNOLOGY AND HUMAN AFFAIRS

Edited by

LARRY HICKMAN

Department of Philosophy,
Texas A & M University,
College Station, Texas

AZIZAH AL-HIBRI

Department of Philosophy,
Texas A & M University,
College Station, Texas

Illustrated

The C. V. Mosby Company

ST. LOUIS • TORONTO • LONDON 1981

Printed in the United States of America

The C. V. Mosby Company
11830 Westline Industrial Drive, St. Louis, Missouri 63141

Library of Congress Cataloging in Publication Data

Main entry under title:

Technology and human affairs.

Bibliography: p.
1. Technology—Social aspects—Addresses, essays, lectures. 2. Technology—Philosophy—Addresses, essays, lectures. I. Hickman, Larry. II. al-Hibri, Azizah, 1943-
T14.5.T4414 303.4′83 80-18992
ISBN 0-8016-2164-X

GW/M/M 9 8 7 6 5 4 3 2 1 05/C/599

Preface

The point at which technology and the humanities intersect is the concern of this book of readings. It grew out of courses in the philosophy of technology offered at both the University of Texas at Austin and Texas A & M University over some seven years. While this book makes available a collection of material particularly suited to such courses, and thus fills a serious gap in available texts, it also offers the independent reader a foothold in this complex topic.

Two important omissions should be explained. We had hoped to include selections from the work of Martin Heidegger and Marshall McLuhan, both of whom have helped shape much current thinking about technology. But the late Dr. Heidegger's family holds the rights to his works, and to our misfortune it is not ready to see his work reprinted. The best we can do is refer the reader to the article we would have included—namely, an interview that appeared in *Der Spiegel* in 1976 and that was reprinted in English in *Philosophy Today,* Winter 1976, Vol. 20, No. 4/4.

In the case of Professor McLuhan the situation was more complex. Although permission was offered to reprint his essays "The Motorcar" and "Clocks," from *Understanding Media,* his required fee could not be met. Its payment either would have required the omission of other important essays or would have resulted in a substantially higher price for this book. We regret this unfortunate situation, and can only recommend that Professor McLuhan's important essays be sought out and read.

In this vein, we are happy to acknowledge a debt to the late Herbert Marcuse, who gave special permission to reprint a selection from his work *Eros and Civilization.*

While the selection and organization of the material included in this anthology involved to some extent a joint effort, the headnotes and introductions to Part One, "Technology and Everyday Affairs," and Part Three, "Contemporary Perspectives," are solely the work of Larry Hickman. The headnotes and introductions to Part Two, "Some Salient Views on Technology," and Part Four, "Technology and the Professions," are solely the work of Azizah al-Hibri.

In all introductions, those entries followed by an asterisk are included among the selections in this book.

There are of course many people whom we would like to thank for their help. These include John J. McDermott, Manuel Davenport, Robert Cohen, Richard Stadelmann, Terry Winant, and Connie Sharp. For the preparation of the manuscript, we thank Linda Kocman, Linda Hosea, and Linda Maybry.

Larry Hickman
Azizah al-Hibri

Contents

Part one
TECHNOLOGY AND EVERYDAY AFFAIRS

1 **Automobiles,** 13

Some meanings of automobiles, 13
DOUGLAS BROWNING

The automobile and American morality, 17
GLEN JEANSONNE

2 **Television,** 22

Editorial from *Radical Software,* 22
MICHAEL SHAMBERG and the RAINDANCE CORPORATION

From *Guerrila Television,* 24
MICHAEL SHAMBERG and the RAINDANCE CORPORATION

T.V., technology, and ideology, 37
DOUGLAS KELLNER

Television: the new state religion? 46
GEORGE GERBNER

Growing up on television, 50
ROGER ROSENBLATT

Toy, mirror and art: the metamorphosis of technological culture, 56
PAUL LEVINSON

3 **Clocks,** 66

The monastery and the clock, 66
LEWIS MUMFORD

Urban time, 69
JOHN J. McDERMOTT

Time and nonverbal communication, 71
THOMAS J. BRUNEAU

4 **Information and cybernation,** 77

Who will lead the way to the "information society"? 77
OLE ENGBERG

Policy problems of a data-rich civilization, 89
HAROLD D. LASSWELL

Technological models of the nervous system, 95
ANATOL RAPOPORT

The man-machine confrontation, 102
FRED R. BAHR

5 **Technology and the arts,** 108

Yes-no: art-technology, 108
JACOB LANDAU

The imminent alliance: new connections among art, science and technology, 115
JOSEPH W. MEEKER

6 **Technology and sex,** 123

From *Love and Will,* 123
ROLLO MAY

The technologies of man-made sex, 134
ROBERT T. FRANCOEUR

Hot and cool sex—closed and open marriage, 139
ROBERT T. and ANNA K. FRANCOEUR

7 **Technology assessment,** 146

Technology assessment in a sharp social focus, 146
HENRYK SKOLIMOWSKI

From *New Reformation,* 150
PAUL GOODMAN

Technology assessment: the benefits . . . the costs . . . the consequences, 156
JOSEPH COATES

Assessing technology assessment, 165
DAVID M. KIEFER

Part two
SOME SALIENT VIEWS ON TECHNOLOGY

8 Introductory statements on technology, 191

Toward a philosophy of technology, 191
HANS JONAS

The technological order, 205
JACQUES ELLUL

9 Four theories on the origins of technology and civilization, 215

The part played by labour in the transition from ape to man, 215
FREDERICK ENGELS

From *Civilization and Its Discontents,* 222
SIGMUND FREUD

From *The Instinct of Workmanship,* 224
THORSTEIN VEBLEN

Man the technician, 227
JOSÉ ORTEGA y GASSET

10 Prominent perspectives on technology and civilization, 231

HISTORICAL

The act of invention: causes, contexts, continuities and consequences, 231
LYNN WHITE, Jr.

The *vita activa* and the modern age: the reversal within the *vita activa* and the victory of *homo faber,* 241
HANNAH ARENDT

PSYCHOLOGICAL

Where are we now and where are we headed? 244
ERICH FROMM

The dialectic of civilization, 251
HERBERT MARCUSE

SOCIOLOGICAL

The three phases of the machine civilization, 254
LEWIS MUMFORD

From "Technology, Nature, and Society," 263
DANIEL BELL

POLITICAL-ECONOMIC

From "The Problem," 266
HENRY GEORGE

From "Technology, Planning and Organization," 269
JOHN KENNETH GALBRAITH

From "America in the Technetronic Age: New Questions of Our Time," 274
ZBIGNIEW BRZEZINSKI

RELIGIOUS

From "Technical Progress and Sin," 282
GABRIEL MARCEL

From "Man and Machine," 285
NICHOLAS BERDYAEV

EXISTENTIAL

From "My Fellowman," 290
JEAN-PAUL SARTRE

The stages of technology, 300
JOSÉ ORTEGA y GASSET

HEGELIAN-MARXIST

From *The Communist Manifesto,* 306
KARL MARX and FREDERICK ENGELS

From "Man and His Natural Surroundings," 311
MIHAILO MARKOVIĆ

Lordship and bondage, 316
G. W. F. HEGEL

PRAGMATIC-LIBERAL

From "Renascent Liberalism," 321
JOHN DEWEY

MEDIA-ORIENTED

A schoolman's guide to Marshall McLuhan, 326
JOHN M. CULKIN

The problem of responsibility in communications, 333
MARGARET MEAD

FEMINIST

The dialectic within cultural history, 338
SHULAMITH FIRESTONE

THIRD WORLD

Technology versus civilization, 347
DENIS GOULET

Contemporary Western man between the rim and the axis, 354
SEYYED HOSSEIN NASR

Man's nature, 358
RABINDRANATH TAGORE

Growth is not development, 362
SAMIR AMIN

Part three

CONTEMPORARY PERSPECTIVES

11 The autonomous technology debate, 381

The autonomy of technique, 381
JACQUES ELLUL

The house that *Homo sapiens* built, 389
ROBERT THEOBALD

Technology and human values, 393
MELVIN KRANZBERG

How technology will shape the future, 399
EMMANUEL G. MESTHENE

Do machines make history? 411
ROBERT HEILBRONER

The emerging superculture, 417
KENNETH BOULDING

12 The "small is beautiful" debate, 426

Technology with a human face, 426
E. F. SCHUMACHER

Machines that shield us, 432
JOHN LACHS

Small is dubious, 438
SAMUEL C. FLORMAN

13 The *homo faber* debate, 442

From *The Myth of the Machine: Technics and Human Development,* 442
LEWIS MUMFORD

Work and tools, 449
PETER F. DRUCKER

From *The Human Condition,* 454
HANNAH ARENDT

14 Technology and metaphor, 465

Technological metaphor, 465
D. O. EDGE

From *Zen and the Art of Motorcycle Maintenance,* 479
ROBERT PIRSIG

Part four

TECHNOLOGY AND THE PROFESSIONS

15 Technology and agriculture, 505

The energy we eat, 505
CAROL and JOHN STEINHART

Of mites and men, 519
WILLIAM TUCKER

Ergot: the taming of a medieval pestilence, 529
LEO VINING

16 Technology and the biomedical sciences, 536

The future impact of science and technology on medicine, 436
LEWIS THOMAS

On the dangers of genetic meddling, 546
ERWIN CHARGAFF

Recombinant DNA: fact and fiction, 548
STANLEY N. COHEN

Sex control, science and society, 554
AMITAI ETZIONI

17 Technology, business, and political economy, 563

The future world disorder: the structural context of crises, 563
DANIEL BELL

Bottle babies: death and business get their market, 575
LEAH MARGULIES

Industry's new frontier in space, 580
GENE BYLINSKY

18 Technology and engineering, 585

Are engineers responsible for the uses and effects of technology? 585
AARON ASHKINAZY

From *Destination Disaster: From the Tri-Motor to the DC-10, the Risk of Flying,* 588
PAUL EDDY, ELAINE POTTER, and BRUCE PAGE

Solar energy—a practical alternative to fossil and nuclear fuels, 597
HAROLD E. KETTERER and JOHN R. SCHMIDHAUSER

Bibliography, 609

Part one

TECHNOLOGY AND EVERYDAY AFFAIRS

An unavoidable fact presents itself to us as residents of the last quarter of the twentieth century: technology increasingly and inescapably shapes the structure of our day-to-day lives. Morning begins when sleep is interrupted by a technological artifact—the alarm. Toothpaste and toothbrush, coffeepot, clothes, and automobile are all parts of a cocoon which we as technological persons spin, and in which we live.

Some, such as the late Spanish philosopher José Ortega y Gasset,* have suggested that we technological human beings are like technologically naive forest- or jungle-dwelling human beings in at least one important respect. We are so involved with our environment, so enmeshed in its closeness, that we, like they, become unaware of it. Technology for us, just as nature for them, becomes an environment that is simply "given" and that we do not question.

Ortega, among others, suggested that when this situation occurs, when we begin to accept any system or environment, whether natural or technological or other, as just given to us, we have at that point lost our way. We have forfeited an important part of what it means to function actively as human beings. The technological man or woman who never ponders the meanings of automobiles, the person who has never considered the impact of integrated circuitry on his or her life, the person whose television viewing approaches the national average of six and one-half hours per day but who never pauses to consider the impact that television has had upon the world of his or her experience—all these individuals are adrift in an environment whose structure has become invisible to them.

For Ortega, as for the French philosopher and novelist Jean Paul Sartre,* this capitulation to and absorption in one's environment amounts to a terrible loss. For Ortega the uniquely human task is what he calls "autofabrication," the determination of one's own being. Sartre calls this kind of being "for-itself" as opposed to the merely "in-itself," indicating, in a vein similar to that explored by Ortega, that to be human is to choose oneself, to project oneself into the world, literally to make a *project of* oneself. To be human, then, in their sense of that word, is to be critical, to assess implications and alternatives, to weigh evidence, to ponder, to wonder—in short, to introduce ordered and effective mental activity into a world that is or has become unhuman or dehumanized.

We may carry out this program in many ways. Once we have taken that critical step back from our environment, be it natural or artificial, a variety of forms of analysis and expression is open to us. In our case, as men and women living in a technological environment, the conjunction of all these activities is an activity we call "technology assessment." Each of us has talents and interests that allow us to approach the problems of assessing our technological environment from our own unique perspective.

In the final section of Part One, we have collected several essays, each of which expresses a distinct viewpoint regarding the nature of technology assessment. These are the views of individuals who are working toward a definition of that inchoate discipline. They are attempting to identify the central task of technology assessment and to determine which branch of organized knowledge is most closely associated with that task.

But there is a wider sense of technology assessment, one that Ortega, Sartre, and others lead us to and that is reflected in the structure of this book. This wider sense, the one we present here, involves not only technologists and scientists, but historians, sociologists, philosophers, poets, psychologists, media makers, and in fact a whole spectrum of contributors to the discussion of what technology is and should be and what our response to it can and should be. In this wider sense, every individual living in a technological environment should be actively involved in the technological aspects of his or her everyday affairs, just as every individual living in a society should be concerned with the political process by means of which that society is governed.

There is, however, an important element in current technology that tends to militate against a careful, analytical approach to technology assessment. The speedup of life occasioned by electronic technology often leaves little time for contemplation. Computers now set the pace, as machines once did, for our daily lives. Computer designers now measure switching operations in terms of picoseconds (trillionths of a second) rather than the once familiar nanoseconds (billionths of a second). The movement of energy through the switches of computers approaches the speed of light. Decisions are made and often implemented long before the persons whom these decisions will affect the most are consulted.

This situation is, of course, not so much a reason to abandon our program of assessment as it is an argument to pursue it more vigorously. For only through continued struggle against apathy, ignorance, and blind acceptance either of change or of status quo as automatically good do we become human beings in the fullest sense of that term.

• • •

Consider, for example, the role that an adequate technology assessment could have placed at the time of the introduction of the automobile into American society.

Melvin Kranzberg* has reminded us that the automobile was uncritically hailed by early–twentieth-century New Yorkers as a wonderful solution to their problem of pollution. The transportation system then in use, carriages and trolleys drawn by horses, was creating a gigantic health problem because of the thousands of gallons of horse urine and the thousands of pounds of horse manure deposited daily on the streets of New York.

There were apparently very few individuals who considered the change in the moral climate of this country that would occur as a result of the introduction of the automobile. John Dos Passos, writing about America at mid–twentieth century, remarked that a whole generation of Americans was conceived, if indeed not born, in the backseats of Model T Fords. The mating rituals of American young people moved out of the closely monitored front porch swing into the private spaces within, or accessible only by means of, the automobile. As Glen Jeansonne* tells us in his essay "The Automobile and American Morality," a survey undertaken by sociologists at Southern Methodist University in 1936 produced the amazing results that 75% of the clientele of tourist courts in the Dallas area were not tourists at all but local couples.

The expectations and surprises registered by those who experienced the ingression of the automobile into American culture are to us today more than simply a source of amusement, although they are that. They give us clues about our own program of assessing the effects of technology, and they reveal significant aspects of what automobiles mean to us.

For example, Douglas Browning,* in his essay "Some Meanings of Automobiles," explores a distinctly human activity that he calls "personating" oneself—that is, playing at being oneself. He suggests that the automobile provides one the greatest opportunities to do so. In the automobile we are in a world that is both richly private and at the same time open to immediate and sudden intrusion. We become aware as no individual in nature could be of the importance of the category of the sudden. Being open to disclosure at any time, our automobiled lives become precarious, as when we pick up a hitchiker, make love in an automobile, or become aware that an accident is about to occur or has occurred.

These aspects of automobiles and many more constitute the fabric of an adequate assessment of the fact of automobiles. Technology assessment, in the sense in which we are thinking of it in this book, is thus an enormous undertaking. It is a task not unlike educating oneself to live in a technological environment.

• • •

Whereas at present no single academic discipline attempts to assess the impact of the automobile on the societies in which it has played an important role, the same cannot be said with respect to the impact of electronic technology, especially the electronic media. Taking their point of departure from the work of Norbert Wiener, the father of cybernetics, schools and departments of communications have now gone beyond his original sense of cybernetics as control (Wiener drew upon a Greek word for "helmsman" in coining the term) to explore media along historical, sociological, political, psychological, economic, and aesthetic lines, as well as combinations of these approaches.

Perhaps one of the most interesting questions arising from the assessment of electronic technology is its relation to print. This is of course more than simply a question of how we get our information, or what happens when we watch television as opposed to what happens when we read. The assessment of this relation provides the material for complex theories about the way in which technology has developed, and about its future.

Marshall McLuhan, who has probably done more than anyone else to explore the transition from print to electronic media as our dominant information source, has suggested that print is a "hot" or "saturated" medium, a "linear" one, with high definition. The print person, he argues, is an interior person (we normally read silently, alone). Print emphasizes one of our senses, the visual, at the expense of the others. It habituates us to thought that is linear and unidimensional in form. Television, on the other hand, is described as a "cool" or "unsaturated" medium, one that engages us on a social as opposed to an individual level, one that develops not only the eye but also the ear, and one that emphasizes the nonlinear, the multidimensional.

McLuhan's point may perhaps be clarified by attending to the way in which the printed page is digested, line by line, one paragraph following upon another, and, especially in the case of traditional literature, with the generation of a beginning, a middle, and an end. Print thinking brings with it a "conveyor-belt" perception of time, one that involves an orderly flow from past to future.

Television, on the other hand, offers time and space frames that are topsy-turvy in relation to those of print. Images are stacked on top of one another, and even dissolve into one another. There is often no beginning, middle, or end. The viewer is left to "saturate" the unsaturated material with his or her own contribution. It is not uncommon, for example, for a sixty-second commercial to have as many as thirty or forty "cuts"—that is, for the camera to change its angle, distance, or locale almost once per second. Yet we do not "see" these changes unless we are made aware of them: we have been conditioned to view the total "gestalt" or form of the commercial.

For McLuhan this technological "revolution," this move from print to electronic technology, calls to mind another technological revolution, one occasioned by the invention in the late fifteenth century of the printing press. He argues that our "post-literate" era has much in common with the "pre-literate" period. Each is more socially oriented (McLuhan speaks of the electronic "Global Village"), more emphatic in its recognition of the importance of a wider sensual perception of the world, and more open to nonstandard experience than was the world of print. At the same time, he suggests, our post-literate era will carry forward the critical attitude and interest in quantification generated in the age of literacy.

These revolutions do not occur without casualties. Enormous problems attend life lived in a period of transition between two overarching technological forms. How, for example, are print people to learn to assess electronic technology? The converse of this problem is of course worrying many educators today. How does one help students who are primarily oriented toward electronic media to gain access to the subtleties of print? Or put more straightforwardly, how can the children of the television era be taught to read critically?

There are those who have taken quite seriously McLuhan's suggestion that electronic technology has produced a new "super-system," a new cultural form, which supplants the images, myths, and metaphors of the old system with those of its own. Douglas Kellner,* for example, in his essay "T.V., Technology, and Ideology," argues that we are conditioned by television in ways that are by no means opaque to the viewer who cares to cast a critical eye. Television promotes a particular ideology, Kellner argues, one that is especially attractive because of the way in which it combines rational theory with images and slogans. George Gerbner* is worried that the elements of traditional religion that render the individual passive before a transcendent value-giver may be present in television. The myths and rituals associated with television, Gerbner claims, have "tremendous popular mobilizing power which holds the least informed and least educated most in its spell."

Whereas both Kellner and Gerbner see television as a source of information about social institutions, a kind of indicator of social and political trends, Jerry Mander, in his recent book *Four Arguments for the Abolition of Television* is appalled that television even continues to exist. Mander, a former advertising executive, worries about the kind of radiation that emanates from the television screen. He fears the de-education of the televiewing masses. He compares television to sensory deprivation, suggesting that it robs its viewers of a healthy sense of reality.

Perhaps no one has put the case against television more eloquently than has Frederick Exley. The main character of his vaguely autobiographical novel *A Fan's Notes* retreats to the fantasy world of television during bouts with mental illness. In a lucid period he writes:

> I watched—but there is no need to enumerate. Not once during those months did there emanate from the screen a genuine idea or emotion, and I came to understand the medium as subversive.
>
> . . . television undermines strength of character, saps vigor, and irreparably perverts notions of reality. But it is a tender, loving medium, and when it has done its savage job completely and reduced one to a prattling salivating infant, like a buxom mother it stands always poised to take one back to the shelter of its brown-nippled bosom.[1]

The assessment of electronic technology has also been undertaken by many nonacademic groups and individuals, many of whom are working in fields such as video art, video documentary, and film, as well as syntheses of these areas. Especially significant has been the Raindance Corporation and its production arm, Top Value Television (TVTV).

While academic and scholarly evaluations of television, such as those of Kellner, Gerbner, Mander, and Exley, are usually addressed to a constituency deeply involved in print, TVTV offers a critique of television from within that very medium. (Persons working in such alternative ventures as TVTV often differentiate between television and video, with the former term referring to the broadcast medium and the latter to a kind of meta-television that quite often provides a sharp critique of the accepted assumptions and conventions of broadcast television.)

The coverage provided by TVTV of the 1972 Republican convention in Miami, for example, aptly titled "Four More Years," not only provided insights into the nature of television news reportage but also cleverly assessed the nature of television documentaries within what was itself a television documentary.

An important aspect of TVTV's productions is that its equipment, which makes use of half and three-quarter inch videotape, is available to virtually anyone who can raise the price of a good stereo system or a late-model automobile. Portable videotape has thus been shown to be a truly democratic, decentralized, and popular form of electronic technology. Since the early work of TVTV, this fact has become even more obvious. Japan now exports over 100,000 video cassette recorders to the United States each month, and with them the capability to make one's own television programs.

Individuals and groups, such as TVTV, who have sought to democratize electronic media have suffered setbacks, to be sure. Federal courts have rejected the claim of the Federal Communications Commission that each cable company over a certain size has the obligation to provide free access to a "citizens" channel. But the trend overall is toward a breakup of the traditional monopoly of CBS-NBC-ABC-PBS over television in America. Satellite transmission of video material now threatens to make network programming obsolete. Cable television in the 1980's will offer thirty or more channels, splintering "mass" audiences into interest groups. Information programs will be transformed from lowest-common-denominator banality to in-depth analyses of previously esoteric but nonetheless important issues. Fred Silverman, President of NBC, is aware of the decline of the great networks as we now know them and is intent on changing the orientation of his company to meet the challenges of the next decade. "I look at the 80's not with any trepidation," he says. "There will be enormous opportunities. By the 1990's we could be feeding six networks simultaneously. There could be a concert network, a Spanish network, there could be a children's service from sign-on to sign-off, there could be a news service, there could be all sorts of things."[2]

One of the elements precipitating this change in the attitudes of network executives and their clients, the corporate advertisers, is what is called "time-shift" viewing. Advertising rates are based on the number of people the network can deliver at a given hour. But with the increased use of video cassette recorders, the decision of when to watch a broadcast shifts from network to viewer.

Arthur C. Clarke predicted the rise of satellite broadcast television as early as 1945 in an article written for *The Wireless World*. Later, in *Profiles of the Future,* he predicted that the first country to begin sophisticated satellite television transmission would see its language become accepted worldwide. Already in countries such as India satellite broadcasting is being used to tie the central government to remote areas, where generator-operated television sets are necessary to receive programming.

Cable and satellite television bring with them enormous possibilities for liberation from ignorance, boredom, and isolation. Imagine, for example, a cable channel oriented toward the interests and viewing habits of bedridden senior citizens, a group whose major activity in any case is television viewing.

• • •

A third technological artifact, fully as important as the automobile or television to the structure of our everyday affairs, is the clock. Serious questions need to be asked in assessing the role of clocks in our lives. What does the clock measure? Are there varieties of duration for which the clock is not an adequate measure? What part did the clock play in the formation of western industrialized man? Are there significant differences between the time of the analogue clock culture and the time associated with the emerging digital clock?

Officially, a second is defined as 9,192,631,700 cycles of the frequency associated with the transition between the two energy levels of the isotope cesium 133. But the clock certainly does not *measure* this cycle; it is only correlated to it. Neither does the clock measure human biological or psychological time. We now know that biological rhythms "speed up" and "slow down" and that uniformity must be imposed upon, not read out of, the organism. If clock time measured psychological time, we could never take "time out" for psychological recreation, to heal our psychological disorders by, as it were, stepping outside time for a time. A view that is probably more appropriate is that the time of the clock is artificial time, not something that actually existed before the advent of the clock.

This is not to say that there are no regularities within the universe that we can correlate to clock time. It is rather to say that the clock—especially the analogue clock, with hour and minute hands—is a machine whose product is uniformity, a way of artificially representing the passage of events in a spatial dimension.

Lewis Mumford,* in his essay "The Monastery and the Clock," argues that the medieval monk was the first "modern" man because he was the first to organize his day by means of the clock. He has also argued that the clock, no less than the printing press, laid the conceptual foundations for the machine age, with its emphasis on uniform, repeatable products manufactured by "assembly line" methods.

In his important book *The Medieval Machine,* Jean Gimpel presents a thesis counter to that of Mumford. The first mechanical clocks, he argues, "did not primarily tell time, but rather were built to forecast movements of the sun, the moon and the planets. Remarkably enough, the time telling clock which dominates our lives was only a by-product of the astronomical clock."[3] Earlier clocks, prior to the fourteenth century, he suggests, indicated unequal or "temporal" hours. "Egyptian, Greek, Roman, Byzantine and Islamic water clocks had indicated unequal, or temporal hours—in all these civilizations the day was divided into hours of light and hours of darkness, generally periods of twelve hours each."[4] Of course the result of dividing day and night into periods of twelve hours each was to produce "hours" that varied; at the latitude of London the "hour" varied from 38 minutes to 82 minutes!

Marshall McLuhan, following Mumford's lead, compares the clock to the blast furnace, which, he writes, "speeded the melting of materials and the rise of a smooth conformity in the contours of social life."[5]

Both Mumford and McLuhan argue that perceptions of time vary radically from one technological age or stage to another. Mumford points out that whereas the machine age, the age he calls the "Paleotechnic" period, emphasized cheap labor, the age that preceded it, the medieval wind, water, and agricultural age he calls the "Eotechnic" period, emphasized "cheap time." The medieval peasant worked fewer hours than did his counterpart in the English factory of the nineteenth century. The agricultural worker of the medieval period lived in an environment in which time was flexible and fluid. His means of measuring time were closely associated with the rhythms of his own organism. But during the time of Paleotechnic technology the worker became a part of a nonorganic structure in which the only important time was the machine time of the clock. His workday actually expanded to the point that fourteen to sixteen hours a day, six days a week, year round, were spent in service to the machine. This compression of organic human beings into mechanically uniform time structures, Mumford points out, led to the situation in which the average life span of the worker during the depths of the Paleotechnic or industrial age was some 20 years shorter than that of his middle-class counterpart.

Charlie Chaplin parodied this situation in his film *Modern Times,* in which he portrayed an assembly line worker whose work was speeded up to the point that he suffered a mental breakdown. The comical havoc he then wrought left the factory in shambles. He saw the foreman's nose and the buttons on the skirt of the plant secretary as just more nuts and bolts to be tightened.

When Mumford calls the "Neotechnic," or post-industrial, age has seen a move away from this tyranny of abstract clock time. Enlightened corporations are now beginning to use "flex-time" as part of their employee relations programs. In companies in which this system is applicable, the worker is given a time span of twelve or fourteen hours in which to accomplish eight hours of work. Flex-time allows the worker to align the workday to his or her own biological and psychological needs and preferences. In most cases it increases productivity and allows the worker flexibility in meeting social, familial, and personal needs which are outside the sphere of the workplace, and it may have the added benefit of relieving overburdened transportation facilities.

Among the persons who have assumed the task of assessing technology are those who seek to understand the replacement of analogue watches with digital ones. Joseph Coates,* for example, has argued that digital watches amount to a retrogression in time keeping. Even persons who are numerically illiterate can tell time by means of analogue clocks—especially those used in many Third World countries, which have twenty-four instead of twelve divisions. While this criticism is perhaps irrelevant in highly industrialized and literate western societies, the important difference between these means of keeping time nevertheless bears examination.

It is interesting to note that the move from mainspring to integrated circuit is probably the most important advance in time keeping since the thirteenth century. If we take seriously the technological "phyla" propsed by Anatol Rappoport*—namely, (1) tools, (2) clockworks, (3) heat engines, and (4) information processors—the clock has undergone the major shift from (2) to (4).

Digital clocks abstract time from space in totally new ways. The sundial, the hourglass, the weighted clock, the pendulum clock, the clockwork or mainspring clock, even the quartz watch, all represented time relative to space. But the digital clock represents time abstractly, as the relation between digits, lines, and dots displayed on a surface.

The comedian George Carlin expressed a marvelous glimpse into the nature of this shift when he noted that with the analogue, mechanical clock it is much more difficult to work during the last half of an hour than the first half, since gravity itself appears to oppose the regular progression of time, its forces seeming to retard the upward movement of the minute hand.

A new degree of precision is delivered by the clock with integrated circuitry. If one is wearing an analogue watch, the correct response to "What time is it?" has been to round off to the nearest five-minute mark. To give 9:37 as an answer is to be thought a bit overly interested in detail. But when one's watch actually says 9:37:54, a new degree of precision has been introduced that is hard to overlook or to deny.

• • •

It is by now a commonplace that we are in the midst of a technological revolution more extensive and complete than any since the development of the machine in the eighteenth century. It is now no longer machines that offer our civilization its primary models and metaphors, but information processing devices as unlike the machine as the central nervous system of the human being is unlike the system of muscles.

Marshall McLuhan has suggested that new forms of technology do not do away with older forms, but encapsulate them and raise them to the level of an art form, presenting them in a new light and establishing new relations among their parts. It is in precisely this sense that the new electronic technology does not cause the extinction of the machine, but rather gives it new meaning. The very name "cybernetics," with which this technological revolution has often been tagged, expresses such a relation between a working mechanism and its control device.

An indication of the extent to which this technological revolution has progressed may be obtained from Figure 1 in the article by Ole Engberg,* "Who Will Head the Way to the Information Society." His graph depicts the way in which the United States since World War II has deployed an ever smaller percentage of its total work force in industry, agriculture, and services, while the percentage of workers involved with information has steadily risen. A look at the nature of the three largest American corporations to emerge during the post–World War II era tells the same tale: Xerox, IBM, and Polaroid are all basically information-oriented industries.

How are we to assess a technological change of such vast dimensions? Alvin Toffler, the author of *Future Shock* and other studies of our new "information society," has suggested that in fact many individuals never succeed in making this assessment. The future now intrudes into the present so rapidly that adequate assessment is for many impossible.

But there is probably a more optimistic answer to this question. Human beings can assess a rapidly changing environment of information technology in the same way they have always assessed objects and events—namely, through careful, critical, and imaginative probes into the substructure of their environment.

Assessing the structure of information technology involves an important problem that is unique to the field of technological history: it is not that there is too little information, but quite often the contrary. Increasingly, so much information is available that new means of classification and analysis must be developed. A double danger exists here: not only is there a tendency to recoil before information overloads, but there is also quite frequently the reverse tendency—to become so "familiar" with vast amounts of information that all information becomes trivial.

The issues emerging from the cybernetic or information revolution are complex ones, but they resonate with problems that are traditional. What is to be the role of the individual *vis à vis* institutions that are strengthened by sophisticated information gathering and storage systems? Is democracy compatible with information-centered societies? Will the computer-controlled machine replace man in undesirable ways? How does one define (or redefine) privacy in an information society?

Aldous Huxley's *Brave New World* was published in 1932 and George Orwell's *Nineteen Eighty-Four* in 1949. What these works have in common is an anti-utopian view of technology, one in which the future is portrayed as being dangerous because of the control that institutions will have over individuals as a result of the institutions' superior information gathering power.

As we approach 1984, there are many who argue that quite the opposite has occurred. The information revolution, they suggest, has produced a situation in which men and women control their own lives and activities in ways that were previously unknown. The counterculture movement of the 1960s, for example, detailed so well by Theodore Roszak in his widely read book *The Making of a Counterculture,* was undeniably an important phase of the information revolution. Elements of this movement included a renaissance of interest in non-Western world views and life-styles and greater attention to "inner," nonquantified information often regarded as unimportant to machine-technology man. In an important sense the participants in this movement were not so much "flower children" as "electronic children." The first truly popular electronic music emerged from this movement, as well as a host of metaphors that still strike many as important. (The West Coast musical group Country Joe and the Fish appropriately entitled an album *Electric Music For the Mind and Body.*) In the struggle between the individuals of the movement and the institutional powers that attempted to contain and control them, it was the most often the latter that assumed the politics and aesthetics of the former.

As information exploded, citizens seemed to demand increasing amounts of it. Far from it being the case that governing institutions had a monopoly on information, it was often they who were under attack from citizens' groups newly empowered and emboldened by fresh information concerning the working of their government. Some have even argued that the fall of Richard Nixon from power resulted from a kind of *hubris* that was perhaps acceptable in the old machine era but totally inappropriate to the age of information. According to this view, the seeds of his destruction lay in his assumption that he alone could control information, an assumption evidenced, for example, by his practice of making tapes and by his wiretaps to find "leaks" such as the source of the "Pentagon Papers."

The rise of citizens-band radios, the increasing number of special interest groups organized by means of computers, the renaissance of the magazine (this time as a means of addressing groups of individuals with similar interests rather than a nationwide audience), and even the current interest in personal home computers and videotape players are all symptomatic of a society in which initiative has been taken by individuals and groups of individuals who see the information revolution both as a challenge and as an opportunity to democratize their institutions. Referring to another chart included in Ole Engberg's essay, Figure 2, one can see that, given current trends, the cost of information storage will continue to decrease during the coming decades. This will make possible the continued involvement of the individual in his political and economic institutions.

But the fact that the cost of information is decreasing is no guarantee that the present trends toward a decentralized, participatory democracy will continue. There are those who argue that the mushrooming of information systems may well distract individuals from their true interests, that electronically abstracted information is at best entertainment and at worst a source of false information about the world.

Apart from this fear that information technology may turn out to be yet another opiate of the people, there is the fear, registered by many, that privacy will decline and that individuals will be open to new forms of abuse. In a recent study Professor David Linower of the University of Illinois surveyed seventy-four large corporations on the subject of how they protected the privacy of their workers. He found that forty-one percent had no policy regarding when to divulge information, that eighty-five percent released personnel material to creditors, and that forty-nine percent released data to landlords. A full seventy percent did not inform workers about the kinds of records kept about them.

There are also those who fear that cybernetic technology will displace human workers on a scale that will eventually cause suffering. In answer to these contemporary Luddites† it has been said that workers have in many situations been the first to welcome the cybernated or computerized machine to their factory or workshop, since machines do not mind the dirty and dangerous work that must often be done there. Fred Bahr,* for example, looks to the future of man-machine relations with hope, suggesting that as long as we can manage to see technology as the property of all mankind we have hope for a bright technological future, and that under such circumstances we need not fear the expansion of the role of the cybernated machine in our society.

• • •

The classical Greeks did not distinguish between the fine arts, such as painting and sculpture, and the practical, or what we today would call the technological, arts. Their word "techne," obviously the root word of our term "technology," meant "art," "craft," "skill," "method," or "work of art."

In his essay in Part Two of this book, Jacques Ellul* characterizes technology or "technique" as an autonomous force that displaces the natural order of things, an order involving

†Ned Lud led a late–eighteenth-century revolt against machines. His name was given to those who subsequently expressed an anti-machine or anti-technology viewpoint.

the ends or goals of things, with an artificial order in which instruments and means are the only consideration.

Now it may be that Ellul is overly pessimistic, and even that he has put matters a bit extremely. His contention will be discussed in the introductions to Parts Two and Three. But for the most part, those who write about technology accept a characterization of it that involves elements of control, artificiality, instrumentality, and more or less autonomy.

Compare this with Rudolph Arnheim's characterization of art. In his view art is a process of ''revitalizing'' the world. An artist is a person who is able ''to take the crust of familiarity off everyday objects and make us see them again.''[6] For Arnheim, art reveals ''the nature of things,'' in that the artist re-presents our ordinary world to us in alternative and often unfamiliar ways. The artist provides another perspective from which to view the objects and events of our world.

There is an obvious contrast between, on the one hand, an abstract force that attempts to bring everything into conformity with its own laws and, on the other hand, a method of providing fresh insights and of taking a step outside whatever system prevails by means of the introduction of the novel and the unexpected into human experience. Some therefore have argued that the arts may be just the antidote for the more somber aspects of technology. Ezra Pound called the artist ''the antennae of the race,'' meaning that the artist collects the ''messages'' being broadcast from a civilization, messages to which the artist is particularly sensitive. Kurt Vonnegut has called this view the ''canary-bird-in-the-coal-mine theory of the arts,'' a reference to the practice of taking birds into mines because they were more sensitive than humans to the presence of noxious gasses. A bird's sickness or death indicated to the miner that it was time to leave the mine.

Of course artists have not always responded negatively to technology. Artistic creations relating to this subject have been as varied as the artists themselves. Giacomo Balla and Umberto Boccioni, part of the Italian Futurism of the early twentieth century, saw in the rise of technology the means of overcoming traditions that rendered men and women servile. But there have also been artists such as Giorgio de Chirico, the Italian painter in whose work technological artifacts are often presented quite menacingly. Others, such as Robert Rauschenberg, Claes Oldenburg, and Mel Ramos, have treated technological subjects in a playful manner. So artists, like everyone else, can be divided into those who are optimistic about the prospects of technology and those who are pessimistic.

Artists can be classified in another, and perhaps more interesting, way. Those who have addressed technological subjects in media that were traditional, or that antedated the technological developments that were the subject of those works, can be distinguished from those who have embraced new technological instruments and materials in order to discover the parameters of those developments.

Electronic technology has elicited more of this second type of artistic response than did machine technology. While it is true that there have been those such as Morton Livingston Schamberg, who incorporated modern plumbing pipes into his works of art, and those such as Laslo Moholy-Nagy, who constructed playful sculptures of new alloys and plastics, these artists probably did not exert any great influence upon designers of alloys and plastics (or upon manufacturers of plumbing pipes). In the case of electronic technology and the arts that arise out of it and interact with it, the situation has been quite different. New types of musical instruments have been developed by and for persons experimenting with electronic music. New types of video synthesizers and special effects generators have been developed by and for persons who work with video art.

Jacob Landau* has argued that it is important in this time of the ingression of technique into every portion of our lives that the artist be able to say ''no'' as well as ''yes'' to technology. His worry is that, since the artist has traditionally needed a patron to support him, he will become technology's yes-man. There is certainly the danger that many artists will succumb to the demands of the technological system, and perhaps even more danger that they will succumb to the crude commercialism that often attends it. But at the same time there are artists who are working to provide us with new insights into technology and who see technology not as a vast inhuman system but as something arising out of what it means to be human and to interact creatively with the world of objects and events.

• • •

If McLuhan, Mumford, and others are correct in asserting that we are in the midst of a technological revolution more far-reaching than any since the fifteenth century, we can expect considerable changes in sexual attitudes to accompany this revolution. It is true that individuals and societies have responded differently to the importance of sex during different technological periods.

According to Marshall McLuhan, for example, we are now emerging from a period dominated by the eye, a period of linearity, whose models were the railroad, the telegraph, the assembly line, and the line of print. McLuhan associates with this period a kind of meager, unidimensional approach to all of life, and in particular to human sexuality. This eye-dominated, linear sexuality had as its early manifestation—during the rise of industrialism in England, for example—the development of the attitude that sexual activity was unnecessary at best and disgusting or downright unhealthy at worst. Sexual impulses and feelings did not fit in with this mechanized view of man—man as an extension of, as ancillary to, a great mechanistic society modeled upon the factory. The realities of life, the tangibles, as Lewis Mumford has pointed out, were thought to be debits and credits and capital and profits, and the worker was treated as a part of the machinery he served. The intangibles were health, fresh air, and physical and emotional well-being.

Siegfried Giedion has included in his important book *Mechanization Takes Command* a study of the bath and bathing practices from earliest recorded times until the 1940's. In this study he documents the decline of interest in personal hygiene and the rise of the notion that the body is corrupt and sinful; all of this he associated with the rise of industrial civilization and the decline of sex.

In its later phase, the paleotechnic mentality gave rise to the unidimensional, visual sex of the Playboy bunny. "The fold out playmate in Playboy magazine," McLuhan suggests, "—she of outsize breast and buttocks, pictured in sharp detail—signals the death throes of a departing age."[7]

The point McLuhan makes here is that when the mechanical, linear, visual, unidimensional technology gives way to the electronic, nonlinear, oral-aural, multidimensional technology, our total value system changes accordingly. The Playboy bunny said "Look but don't touch." The centerfold lady had freckles and moles airbrushed out of existence so as to appear almost synthetic. But the electronic age is one in which the total environment becomes sensually alive. By "multidimensional" it is meant that a whole range of qualities becomes important in sexual activities, not just how he or she "looks."

It is also true that in the age of electronic technology sexual practices become more varied, alternative life-styles become more acceptable, and new worlds of experience become available to the individual because of the very nature of the electronic visual media. Groups that may have been fragmented, such as women or gays, begin to coalesce on a national or international level once channels of communication begin to allow feedback between formerly fragmented groups, and even between those new coalitions and the wider society.

Electronic media have a mass appeal, allowing unthought-of alternatives to become available to a wide audience. But electronic media are also specific in the sense that smaller interest groups can present points of view, or even gain access to channels devoted to communication within groups holding similar values. At the present time, for example, the cable program "Midnight Blue" serves just such a specialized audience in Manhattan. A program specifically dedicated to pornography, it attempts to select its audience only from those who share the general value structure of its producers. A stern warning is issued prior to each segment asking anyone who objects to sexually explicit material to "switch to another channel, or better yet, turn the damn thing off."

There are those, such as Rollo May,* who have argued that we today run a very special risk, that of making sex subservient to "technique." Copulation without eros, he claims, dehumanizes. It destroys feelings, undermines passion, and blots out individuality.

Others are more hopeful. Anna and Robert Francoeur,* for example, see an era of "cool" sex in the wings, an era of joyful exploration, of attempts to plumb the richness of a full range of human sexual experience, a time when one's sexual partner will no longer be regarded as "property."

• • •

The final section of Part One comprises four views of the nature of technology assessment. Joseph Coates* argues that technology assessment should be an affair in which technology itself assesses its secondary and tertiary effects, and he gives examples of how this kind of work can be done. David M. Kiefer* calls for what he terms "corporate Cassandras," people working within technologically oriented corporations who will act as irritants to otherwise smooth decision-making processes. Henryk Skolimowski,* on the other hand, expressing a view that is closer to the one we present in this introduction, calls for a technology assessment that takes "sociomoral" elements into account and that is more a part of philosophy than of technology. Paul Goodman,* too, writes that technology assessment is, through and through, a moral problem.

All of these activities probably fall within the legitimate range of technology assessment. It is our argument that the assessment of technology, like the assessment of the political acts and processes that govern the society in which one lives, is the business of every person who is awake to his or her environment.

The models for the assessment of technology come from literature, from philosophy, from the sciences, and from technology itself. It is now more than ever essential that technological societies have among their citizens individuals who have gone beyond the narrow technical education now offered by many of our universities. These will be men and women who take as their task increasing our understanding of the technological world about us, a world that includes not only clocks, automobiles, television, computers, and art objects, but practically every object with which we come into contact.

NOTES

1. Frederick Exley, *A Fan's Notes,* (New York: Simon and Schuster, Inc., 1977), pp. 180-181.
2. Desmond Smith, "Television Enters the 80's," *New York Times Magazine,* 19 Aug. 1979, p. 66.
3. Jean Gimpel, *The Medieval Machine,* (New York: Penguin Books, 1977), p. 153.
4. Gimpel, pp. 165-168.
5. Marshall McLuhan, *Understanding Media,* (New York: McGraw-Hill Book Co., 1965), p. 155.
6. "Eyes Have They, But They See Not" (A conversation between James R. Peterson and Rudolf Arnheim), *Psyogy Today,* June 1972, p. 92.
7. Marshall McLuhan and George B. Leonard, "The Future of Sex," in *The Future of Sexual Relations,* ed. Robert T. Francoeur and Anna K. Francoeur (Englewood Cliffs, N.J.: Prentice Hall, Inc., 1974), p. 20.

1

Automobiles

Some meanings of automobiles

DOUGLAS BROWNING

■ Douglas Browning *is a professor of philosophy at the University of Texas at Austin. In addition to his interest in the philosophical meanings of the automobile, evinced here, his publications reflect a concern with ethics, metaphysics, and philosophical anthropology. He has also published a book of poems.*

As in the case of his other work, Browning here articulates a perception that is uniquely human in its scale. It is the interaction of the automobile with the individual human person, as opposed to its relation to society as a whole, that fascinates him. To this end he provides a new meaning for that seldom used verb "to personate"—namely, to "play at being oneself." The automobile allows such "personation" because it absorbs our bodies, extends them, and allows us to enjoy that larger function.

The automobile also introduces us to the category of the sudden. This is not the "suddenly happened," which is an old form of human experience, but the "suddenly happening." Finally, the automobile affords us what Browning calls "anonymous privacy," a concept he illustrates with great sensitivity in three mobile vignettes: the hitchhiker, loveplay in the automobile, and death by highway accident.

Of course there are many meanings of automobiles. I am interested in just three interdependent meanings which I take to be philosophically instructive yet seldom if ever explicitly formulated. These three are the enlargement of the body, the introduction of the category of the sudden, and the deepening of the concept of anonymous privacy. To sum up these points, the distinctive meaning of automobiles with which I am concerned herein is self-enrichment through the anonymity of speed. With it is involved a new understanding of the human self.

Let me state emphatically that I am not concerned with the meanings of automobiles *to* the human being. If meanings come about only though human beings, still there is the living fact of the automobiled human being, just as there are the facts of the economic human being or the enfamilied human being. Meaning is a fact about the concrete fact of organism in environmental transaction and not a fact about a mind separate and distinct from environment.

The personal automobile, especially one with clutch and gearshift, serves to enlarge the body beyond the skin. This first point is the most obvious. Such enlargement begins with the simple fact of the instrumentality of the automobile as a means of conveyance. The automobile is a vehicle for satisfying demands, and is as such a tool like a hammer. But whereas hammers extend only our hands and arms, automobiles extend our legs, hands, fingers, arms, eyes, ears, and so on. The automobile is a thoroughgoing tool. Moreover, though all tools in a sense enlarge the body, some tools become more than simple means when they are chosen for the enjoyment of their functioning. Among those so chosen, some become especially personated. By the verb "to personate" I mean here the proper correlative to the verb "to impersonate." To impersonate is to play at being another; to personate is to play at being oneself. The automobile in the lives of many is a thoroughgoing tool within which the skinned body is absorbed and enjoyed for its functioning and in terms of which one plays at being a self.

□ This essay appears for the first time in this volume.

Too, an automobile is like a suit of clothes. It is our traffic habit, our highway wear. Clothes serve the ends of modesty and decoration; the automobile gives us anonymity and status. Both protect us from the elements, natural and social. But as everyone is well aware, some garments come to be identified with one's character. They merge into the personality. They are stuck by some secret glue to one's privacies. They are personated.

The richness of this meaning of the automobile is not exhausted by the consideration of it as a personated, thoroughgoing tool and garment. In order for the automobile to enlarge the body, it must be demachined, for the automobile is a machine indeed while man is not. Many grave men once believed that all of nature was a vast machine and that the human physiological plant was one very marvelous machine within the larger one. Men seldom believe this today. But suppose it were true. It is obvious, is it not, that the physiological machine is amazingly personated, organic, pervaded with feeling, intimate. This could only be by some sleight-of-hand, some arch-trickery, whereby what is a machine is discovered to be functioning as though it were not a machine. This is what I mean by demachining. There is the science-fiction story of the humanoid robot that comes to serve as a personal friend of the hero. There is the companion story of the robot who asserts his rights to life, liberty, and property. There is more here than mere personification, for in each case the machine comes to function as what is no longer a machine. The automobile is not only personated, it is necessarily demachined in the process.

Contrast the power of automobiles as the grand machine of the turnpikes, still incompletely demachined, with the personation of the fine frame, which is so one with the protoplasm as to be uncommanded and unmastered. The swell machines are fun to drive, for we set ourselves up as bronco-busters to master their idiosyncratic powers. In fact any unfamiliar automobile is such a machine. It is the autonomous other for our mastery. However, the demachined automobile is not an automobile under control, as though it were another we have mastered; it functions as uncontrolled as our hands and feet. Who thinks of his hand as a thing to be mastered and controlled? And if one did, would it not thereby be meaningfully other than the body?

From what has been said so far it must be clear that not everyone who drives is an automobiled human being. Some of you may not know what I am talking about. The automobile remains a thing *for* you, merely a vehicle, an object of fear or astonishment, or perhaps a friend like a favorite dog, a child, or a mistress. The meanings of which I speak are, as yet, unrealized potentialities in your experience. You must not be disconcerted, for it is a simple enough truth that the meaning of things often eludes us. If you have never been a party to the kind of fact of which I speak, yet someday you may, and then perhaps you shall find this meaning within the fact.

I have said that the three meanings of automobiles with which I am concerned are interrelated. The initial point, the enlargement of the body, cannot be made adequately without anticipating the next two. Let me make two transitional comments.

First, the identification of automobile with self, its personation and demachining, seems to be at a peak at high speeds. This is, I suspect, analogous to the fact that one's identity with the flesh-and-bone body is experienced most profoundly in those periods of its peak yet effortless functioning. The demachining of the automobile requires, however, an added inducement, which is satisfied in the transformation of landscape at highway speeds into a uniquely alien environment characterized by anonymity and suddenness. Thus the dualism of self and the world is radical, and clearly the automobile itself exists at the still center which "I am." This experience of peak identification I call "the phenomenon of the Texas highway."

The second comment is closely related to the first. The automobile has an inside and an outside. This is a psychological as well as a spatial structure. What passes as ordinary visual and auditory sensations of externals in walking life becomes, in driving life, split between sensations of landscape and sensations of the automobile interior. The latter sensations take on an introspective aura, especially at high speeds, and the automobile interior becomes a nonpublic place of privacies. The character of one's internal life becomes enriched, and a larger self is realized. The dualism of self and world is compounded. This aspect of personation, whereby interiors of sensation become introspections, is the other side of the processes of certain psychoses in which the internals of phantasy are read wholeheartedly into an external world. I call this aspect introspectation. As an illustration consider the role of the automobile radio. Like our eyes and ears, it links us to the world, for through it we sense the happenings of the day; yet the sounds made are not themselves the link from dashboard to the outside, but the interiorized links from receptor to consciousness and as such fall within the body.

I now pass to the second meaning of automobiles, the introduction of the category of the sudden. I am referring to a new way of seeing things, a fresh principle that has arisen into public consciousness since the coming to be of the automobile and according to which the world takes on character and meaning of which human consciousness was not previously aware. I will not say that the basic structure of the universe has itself changed, for I know nothing of the universe, but I suggest that the world as the object of human consciousness has altered since the coming of the automobile. And the alteration that has come to be is simply this: the world presents itself as a place of the sudden, as an arena wherein the rapidly developing lurks to spring across the background of the slow, methodical passage of days. The basic category of processional growth, the natural way of the coming to be of man, trees, and love, is in no way replaced. But now the promise and fact of the sudden cuts across nature as the rapidly shifting shimmer of moonlight plays over the surface of the waves of a patiently toiling sea. Birth, life, love, and death, spring, summer, autumn, and winter are there. The ripening of friendship, the rewards of long labor, digestion, pregnancy, creation are there. But also the scream of tires, the yellow light, the narrowly avoided accident.

I would like for you to make some efforts to understand me. I am not maintaining that the sudden is a fact of automobile experiencing alone. It is, no doubt, more apparent when one drives and clearest when one drives in traffic at high speeds. My suggestion is much more radical than this. I am suggesting that the sudden has become a category of all of our experience. I am also suggesting that the automobiled human being is the source and sanction of the publicity of this category and that this is one important meaning of automobiles. I suggest that contemporary man is the first man to live in a world of suddenness, that daily life, art, religion, philosophy, politics, conversation, education, and indeed all human concerns are infected with experience of and concern for crisis, and that the automobiled human being of traffic and highway is the breeder, infector, living reinforcement, and purest exemplification of the flash happening.

Now is the time to rectify one misunderstanding. Throughout this discussion you have suspected perhaps that the perspective of the sudden was neither new nor concerned with automobiles. What you have done is to confuse the sudden with the abrupt, the suddenly-happening with the suddenly-happened, rapidity with discontinuity. These are by no means the same. The category of the suddenly-happened has been native to thought and experience as long as men have lived in an environment in which important things suddenly happened. And man always has. One suddenly dies, one suddenly wakes up screaming, tigers suddenly spring from trees. But the suddenly-happening is not the over and done with, the finished, the beginning abruptly or the instantly closed out. It is something going on, transpiring, but at the very limit of human capacities for adjustment or control. It directs our attention to a new facet of man, his reaction time. Man becomes, as does his age, crisis oriented. Perhaps this is not totally new. Maybe horse soldiers and gladiators participated in some such world as ours, but the fact is more pervasive now. It is public domain. It is common property. It is categorical.

A new phenomenon requires a new attitude. The attitude one takes up in the automobile is controlled by its adequacy to the landscape as a speedily shifting anonymous field of lurking suddenness, violence, and crisis. The attitude is an orientation to the appearance of the sudden, though the sudden does not constantly appear. When it does appear, the sudden moves a special object of concern across a background of speed and anonymity, as though it were a vividly red flash on a dull greyish ground. The attitude is directed to the occurrence of such flashes, but not in order to react to them, but in order to develop with and control them. This means that the attitude must be a constant tension which is both patient with long stretches of routines and capable of instant transformation into rapid and precise business. The attitude is not one of contemplation of the road, for the absorption in the object of attention precludes action. Nor is this attitude an ordinary tension or anxiety, for these allow neither long inactivity nor coolly perfect execution. In fact there is nothing quite like it. It is best understood by exemplification. Consider driving long distances at high speeds. What happens: There is an immense coolness, almost trance-life, in the performance of routine affairs. The time comes when the automobile is no longer handled; it handles itself. You, the driver, sit and await the exigencies. Now suppose you stop for a sandwich and a cup of coffee. Notice how difficult it is to make the transition. Your head rings: because your eyes are too wide to take in the slow subtle movements of human expressions, you have to give your order twice: you have to take a moment to get your bearings. You have to reorient yourself to a world of greater viscosity.

The category of the sudden is only understood by a glance backward at the meaning of the automobile as an extension of the body. You can see now that the new body serves to increase the scope of the manageable far beyond that in any previous age. At night on the highway the headlights pick out an anonymous landscape that extends approximately the length of control. The automobile moves into a world of secrets, any one of which may emerge at the extreme of vision as a suddenly unfolding drama of collision. Such things may be managed. The mere punch of a foot serves to brake. The slight twitch of a hand swerves an enormous mass into another lane. The scope of the manageable is increased in distance, speed, mass, and momentum. The sudden as the advent of such management is precisely the initial impetus for the demachining of the automobile.

And now let me make a brief comment in transition to a discussion of the last meaning of automobiles to be considered. The new attitude is controlled by the scope of the manageable and posits a world of exigencies, of potential explosiveness, which is radically other than the automobiled human being himself. The human self, enlarged by the automobile, becomes a fact of mobile privacy. We are now in a position to understand this contemporary man.

Privacy, we sometimes feel, is a function of anonymity. The anonymity of the automobile, its faceless commonness, is mobile, and hence it is a constant of human traffic, a mask for all occasions. It does not reveal us as do our face and speech, nor must the automobile assume the frightening fixity of assumed joviality, wittiness, or sternness. The automobiled person is from the outside the only typical specimen of the mass man. On the highway one always travels incognito.

In proportion to the perfection of anonymity in the automobile privacy is deepened. The automobile is spacious, like a spacious soul, and contains room for the greatest and the smallest of intimacies. We cannot sing loudly and with proper flourishes even in our homes, but the roar of engine and the passage through anonymous landscapes allow us such expressions in the automobile. The relaxation of the strains of holding steadily to our pedestrian masks effects release in the private car. These expressions no longer have the character of overt behavior but become as internal, demachined, personated, and introspectated as the automobile itself.

There is thus an inside and an ouside to the automobiled human being. From the outside there is anonymity. From the inside there is privacy. The windshield is the frontier between the person and the world. The world is the place of exigencies, the coiled potentialities of violence, the possibilities of the sudden. The person is the autonomous master manipulator with the enriched intimacies of a mobile hideout.

But there is an ambiguity involved. Such utter facelessness on the outside and such complete privacy on the inside are only the typical facts of the flow of traffic; these facts are strained when an intrusion occurs. There are sometimes riders, sometimes love, sometimes death. These are intruders in the privacy and the anonymity, brutally contradicting both. If intrusion could destroy anonymity without destroying privacy, if intrusion could cancel privacy without cancelling anonymity, then there would be no ambiguity. But intrusion actually serves only for the destruction of the simple inside/outside structure of anonymity/privacy. The bringing of outsides into insides, the world into the person, gives the intruders a peculiar quality of intimacy not otherwise encountered. The peculiarity is due to the fact that the intruders are ambiguous—ultimate strangers in a private land yet ultimately personated interiors from a public landscape. Love, death, and riders are internal-externals, which if not contradictions are at least riddles. But the driver of the automobile is a party to the intrusion also. Without him intrusions would be only entries. The automobiled man in his moment of intrusion is an ambiguity.

I will illustrate with three vignettes of intrusion.

Vignette one: the hitchiker. I step into the automobile bidden and yet, as an outsider in his domain, an intruder. I am welcome; I should make myself at home. I am not welcome; I should take up as little of the privacy as possible. I am a question from the start. He does not know me. He does not trust me. I am bringing into his life the unpredictable, the unassimilated, the faceless, the strange. But now he speaks to me. He is running around on his wife; he hates his job; he fears death by cancer. How privileged I am to be the object of such confidence. I am in his deepest soul, a sharer in his hideout, yet surely no one has ever been so external to his person. Interloper and confidant, I am irremediably ambivalent to him. I become, even to myself, a pun.

Vignette two: love-play. This is a mobile rendevous. You enter myself as something stolen from the world. The secret is deep indeed. We may not linger, for the anonymity of our place is subject to immediate discovery. Wariness is mixed with the recklessness of intimacy. The outside peers into the inside. The ambiguity of sex is unique and exciting. By invitation the alien world admits to my trespass. A piece of the landscape transgresses my soul. I am myself the one who is most naked and vulnerable. I am at the same time the anonymous manipulator. Desperate passion is one with infinite reserve. The suddenness of the world wraps into our privacies, as though the automobile were turned inside out.

Vignette three: death. The traffic fatality is statistical. There is in every highway death the symbolic anonymity of a figure in a column of figures, an announcement on the radio. But the suddenly perceived inevitability of final intrusion unfolds the eternally personal moment. You will participate in your death. There in the stark aloneness of the quiet center of suddenness, you will be totally aware. The very enlargement of the body underlines the vulnerability, the contingency of your automobile. The death is public, for it is an object of traffic and the openly observable recovery of a swollen body by the anonymity of landscape. The automobile is still, no longer personated, but a simple inert piece of the environment. It is a personal death; it is a common end.

These vignettes of intrusion take us beyond the meanings of automobiles and introduce us to the meanings of contemporary men. Therefore they may be labeled metaphysical escapades.

The automobile and American morality

GLEN JEANSONNE

■ Glen Jeansonne *is Professor of History at the University of Southwestern Louisiana, Lafayette, Louisana.*

In our time of the ubiquitous automobile we tend to forget that its introduction to the American scene involved a considerable displacement of values that were certainly traditional and that were regarded by some at the time as being immutable.

In this essay Professor Jeansonne has afforded us a remarkable insight into the moral foment and cultural confusion that accompanied the introduction of the automobile to American culture. Were the automobile to be supplanted in our own time by another form of technology, would there be corresponding difficulties?

The American automobile has traveled the whole circuit from hero to villain. Once enshrined as a liberating and democratizing agent, it is now condemned as a major cause of pollution and congestion. Ever since the early automobiles clanked along dusty roads Americans have both loved and hated their cars.

During the pioneering stages of automobiling, Tennessee law required that a motorist advertise his intention of going upon the road one week in advance.[1] Vermont enforced an ordinance compelling motor car drivers to hire a person to walk one-eighth of a mile ahead of the car, bearing a red flag.[2] A law introduced in the Illinois Legislature stated:

> On approaching a corner where he cannot command a view of the road ahead, the automobilist must stop not less than one hundred yards from the turn, toot his horn, fire a revolver, halloo, and send up three bombs at intervals of five minutes.[3]

The pioneering era of the automobile had ended by the twenties. There had been 8,000 cars registered by 1900. As the twenties opened, there were 8,000,000. At the close of the decade there were almost 23,000,000.[4]

Not a phase of American life was untouched by the automobile. Americans no longer measured distance in miles, but in minutes. Isolated wasteland became choice property when a major highway cut through it. The automobile made elopements easier, increased the income of parsons who specialized in quick matrimo-

☐ From *Journal of Popular Culture,* 8(1974), 125-131.

nials, and swelled the duties and the fees of village constables. It made bootlegging profitable and prohibition impractical. It enriched the American vocabulary with such words as "flivver," "skid," and "jaywalker." It captured the hearts, imaginations, and pocketbooks of Americans.[5]

Several factors contributed to the automobile's impact on society. It provided flexible, time-saving transportation. Engineering improvements and a great increase in service stations made it a reliable means of transportation. Highway expansion and improvements opened new areas to travelers. A great variety of both economical and luxury cars became available. Marketing techniques were improved and credit sales extended. Advertising benefited from the growth of movies and radio. General prosperity provided abundant purchasing power. Wartime standardization and pent-up demend helped spur production. Favorable government policies toward business and public adulation made the enterprise highly satisfying for those who could stand the competition. Finally, the spirit of the decade was characterized by a distate for moral crusades and a readiness for adventure.

As automotive producers experienced prosperity or suffered decline, they pulled other industries along with them. In 1929 the automobile industry consumed 85% of the rubber imported, 19% of the iron and steel made in the United States, 67% of the plate glass, 18% of the hardwood lumber, 27% of the lead, and 80% of the gasoline. On person in every ten worked for the automotive industry or affiliated companies.[6]

Psychological effects were equally poignant, if less measurable. The manufacture of automobiles demanded streamlined mass production and dealt with a product which lent itself readily to standardization.[7] Little skill was needed to operate the automotive machines. The ability to maintain a methodical pace, eliminate wasted motion and follow instructions precisely constituted the requirements demanded of a machine operator. Social intercourse was limited by noise, fear of falling behind the cycle and the realization that failure to pay close attention could create safety hazards. Pride in achievement was lost in the endless drone of the assembly line. One worker complained:

> There is nothing more discouraging than having a barrel beside you with 10,000 bolts in it and using them all up. Then you get a barrel with another 10,000 bolts, and you know every one of those bolts has to be picked up and put in exactly the same place as the last 10,000 bolts.[8]

High wages and abundant jobs partially compensated for the drabness of automotive work. Entire cities, such as Cork, Michigan, were revitalized by the automobile industry. When Henry Ford built his Fordson tractor works there in 1917, the community economy was languishing. By 1929 workers at the Fordson plant were receiving five million dollars yearly in wages. Salaries paid by the Ford organization, which operated on a five-day week, were higher than the best wages paid by plants operating on a six-day week—fifty percent higher in some cases.[9]

Henry Ford himself was considered, if not an outright deity, certainly an agent of God. Ford made money by mass-producing cars and selling them for less than his competitors. In 1914 the price of the Model "T" was about $450. By 1924 the price had been reduced to $290. A poll of college students that year voted Henry Ford the third greatest figure of all time—Jesus Christ winning first place and Napoleon running second. Upon being told this, Ford's only comment was that he personally would have voted Napoleon first, rather than Christ. "Napoleon," he said, "had real get-up-and-go."[10]

While stimulating the economy of many cities, the automobile helped to alleviate some of the isolation and loneliness which had made rural life bleak. The automobile provided the farmer with access to the city and threshers, tractors, reapers and combines afforded him the leisure to utilize his new-found opportunity.[11]

The genesis of the automobile also meant the death of the one-room school. In Montgomery County, Pennsylvania, for example, sixty one-room schools had been closed through the organization of eighteen consolidated schools by 1923. New York City employed 180 motor buses to transport students.[12]

Although the automobile centralized rural activity, it led to a decentralization of the city. The twenties saw the growth of a new urban phenomenon, the suburb. Previously, cities had by necessity been compact. Now they could sprawl beyond the shadows of factories. In New York City as early as 1920, 420,000 people traveled back and forth daily by automobile. A survey taken among Michigan industrial workers several years later showed that 650,000 of 850,000 polled depended entirely on their cars to get to and from work.[13]

In 1922 the National Department Store in St. Louis inaugurated a novel trend by locating a branch store three miles from the center of town. The day of the branch location had arrived, soon to be followed by the development of the shopping center.[14]

As dwellings mushroomed in the suburbs, office buildings rose to the sky in the city's core. With space in the city's business district at a premium, the automobile became a substantial liability, for it was a voracious space-consumer. The automobile, which had promised to provide workers with the luxury of the suburb, demanded severe recompense by accelerating the blight of the city.[15]

To escape city monotony or the strain of daily work, millions of Americans went vacationing. Everyone now had a chance to be as free as a tramp and to do it respectably. By 1923 motor touring was being termed "the great American summer sport." For the rich there were elegant and majestic hotels; for the moderate, boarding houses, modest lodges and private homes. For the poor and the adventuresome there was camping. Anyone who could afford a car could afford a motor camping trip. No longer were vacation resorts the monopoly of the rich. Frank Brimmer, auto camping editor of *Field and Stream,* termed motor camping "the finest melting pot of our American democracy."[16]

The lure of the road proved too great a temptation for many churchgoers. Motor trips began to rival church socials as leisure activities in small communities. The rural church experienced the same fate as the small rural school. Churches were forced to follow the highways. Those distant from good roads declined and disbanded. Those near well-traveled highways grew and prospered.[17]

Frustrated theologians and parents attributed much of the moral laxity as well as religious neglect to the advent of the auto. The motor car was amoral if not immoral. It was undeniably iconoclastic and irreverent. As one observer remarked: "A load of booze is no heavier to an auto than an equal weight of deacons."[18]

Some uses of the automobile especially shocked defenders of sexual purity. The car displaced the parlor in the scheme of courtship. Young people traveled to neighboring towns where they could cut loose from parental supervision and gossiping whispers. As a perceptive critic noted: "A bulwark of American morality had always been the difficulty of finding a suitable place for misconduct."[19]

The judge of a juvenile court called the automobile "a house of prostitution on wheels." He explained that of all the girls brought before him for sex crimes in the past year, one-third had committed their offense in cars.[20]

In 1927 a police officer in New York City arrested a pair of newlyweds for petting in the rear seat of their car. The case was dismissed when it came before court, but the arresting officer expressed dismay at the trend of public morals. He said: ". . . conditions have been getting worse, with all sorts of couples kissing and hugging in automobiles and the girls openly smoking cigarettes."[21]

American morality received a further loosening with the growth of roadside tourist courts. These made it easier for families to travel late and avoid having to dress formally. They also made it convenient for unmarried couples with other things in mind. F.B.I. Director J. Edgar Hoover called the tourist courts "little more than camouflaged brothels." A survey conducted by sociologists from Southern Methodist University in 1936 found that 75% of the patrons of tourist camps in the Dallas area were not tourists at all, but local couples. At one camp a succession of couples had rented a certain cabin sixteen times in one night.[22]

The automobile was also on trial in the academic world. It was banned at Penn State and the University of Illinois, severely restricted at Yale, and condemned at Princeton. In 1926 Princeton banned the use of automobiles by students holding scholarships. The Secretary of the University explained: "Boys who can afford to drive a car in Princeton are not justified in asking financial aid from the university." The decision was buttressed by the Secretary's opinion that ownership of an automobile was "detrimental to the academic career of a student."[23]

Women as well as young people felt themselves freed from ancient taboos by the motor car. In her automobile the lady could smoke in private or sample bootlegged liquor. She could vacation alone, view movies, shop in city fashion centers, and temporarily escape the frustrations of raising children.

Gasoline cars were at first considered an improper concern for the lady who wished to remain a lady. Ladies were not supposed to possess the muscle or technical ability to drive or even appreciate cars. Motoring entailed wearing rough clothes and getting dirty, which was neither very fashionable nor very comfortable.[24]

Both attitudes and automobiles were soon forced to change to accommodate the ladies. Women demanded that cars be handsome as well as sturdy. Advertisements began to appeal not only to the pocketbook but to the aesthetic sense of the buyer. Bright colors become stylish.[25]

The automobile helped emancipate women and young people, but it also offered unprecedented opportunities to criminals. It facilitated bank robberies, bootlegging and insurance fraud. Car theft itself became a major problem.

Thieves were able to strike any part of a city and flee quickly. State lines were obliterated as the auto enabled criminals to greatly expand their escape routes and increase their refuges.[26]

Henry Ford received many congratulatory letters after he introduced his V-8 model in 1932. One letter, handwritten, was post-marked Tulsa, Oklahoma, April 10th, 1934:

> Mr. Henry Ford, Detroit Michigan
>
> Dear Sir:
>
> While I still have got breath in my lungs I will tell you what a dandy car you make. I have drove Fords exclusively when I could get away with one. For sustained speed and freedom from trouble the Ford has got every other car skinned and even if my business hasn't been strictly legal it don't hurt anything to tell you what a fine car you got in the V-8.
>
> Yours truly,
> Clyde Barrow[27]

Although it aided criminals, the auto also helped law enforcement. Patrolmen were able to cover wider areas and render emergency aid faster. More equipment could be carried in patrol cars and they could be dispatched quickly from headquarters to the scene of a crime.[28]

The automobile wrought subtler changes in the attitudes of honest Americans. This had long been a nation of frugal workers, but the motor car helped alter the tradition which linked indebtedness with immorality. Consumption, not hoarding, was the new virtue. In 1926, 75% of all motor vehicles purchased were brought on the installment plan. Automobile purchases involved a hierarchical structure in which the buyer borrowed from the dealer, the dealer from a finance company, and the finance company from a commercial bank.[29]

The car became a leading status symbol. By 1925 there were 20% more cars than telephones. Only 12% of farm families had running water, but 60% had cars. When asked by a Department of Agriculture investigator why her family owned a car but not a bathtub, a farm wife replied: "Why, you can't go to town in a bathtub."[30]

Across the nation, the automobile helped to build a more fluid society. The frontier which historian Frederick Jackson Turner said had closed in 1890 now opened again. Dissatisfied wage earners could tramp the highways in search of better fortunes. Probably they did not often find them, but the mere opportunity made contemplation of the future more hopeful.

The automobile did not make Americans a mobile people. The pioneers who traversed the Atlantic and the "Forty-Niners" who crossed the continent to California were hardly sedentary. But the automobile did make a naturally mobile people more mobile. Like most technological advances, the automobile was not so much a social innovator as a catalyst to innovation. America would have become more democratic, materialistic, urban, industrial, and less strait-laced even without the automobile. But the auto stimulated change at an almost revolutionary rate. It accelerated the pace of American life. It enabled more people to do more things in less time at distant places. The things that they did were both admirable and mischievous. That depended on the person, not the car.

The most striking image generated by the auto was that of speed. The automobile was exciting and frightening, challenging and frustrating. It brought new solutions and multiplied old problems. It increased freedom but demanded responsibility. It freed workers to travel while it chained them to the assembly line. It ended the farmer's isolation but increased his sense of inferiority by revealing the affluence of his urban neighbors. It brought efficiency to industry, new opportunity to recreation, and anarchy to traditional morality.

When prosperity ended in the early thirties the frustrated demand for the automobile and the things it made possible increased tension. Those who could still afford to drive automobiles became objects of bitter jealousy. But the desire to own and operate one's personal car remained deeply entrenched. It had become a part of the American dream.

NOTES

[1]Dwight Lowell Dumond, *America in Our Time* (New York, 1947), 42.

[2]William E. Leuchtenburg, *The Perils of Prosperity* (Chicago, 1958), 185.

[3]"Motor Laws that Make Lawbreakers," *Literary Digest,* LXXVII (May 12, 1923), 58.

[4]John Bell Rae, *The American Automobile: A Brief History* (Chicago, 1965), 87; 238.

[5]"The Automobile: Its Province and Problems," *Annals of the American Academy of Political and Social Science,* CXVI (November, 1924), vii; New York *Times,* November 25, 1923.

[6]"Automobiles and Prosperity," *New Republic,* LXI (January 29, 1930), 263-64.

[7]Charles Reitell, "Machinery and its effect upon the workers in the Automotive Industry," *Annals of the American Academy,* CXVI (November, 1924), 38-39.

[8]Charles Rumford Walker and Robert H. Guest, *The Man on the Assembly Line* (Cambridge, 1952), 11;69.

[9]Detroit *News,* September 15, 1929.

[10]Keith Sward, *The Legend of Henry Ford* (New York, 1948), 279.

[11]John M. McKee, "The Automobile and American Agriculture," *Annals of the American Academy,* CXVI (November, 1924), 14; J. C. Long, "The Motor Car as the Missing Link between Country and Town," *Country Life,* XLIII (February, 1923), 112; Ralph Cecil Epstein, *The Automobile Industry: Its Economic and Commercial Development* (Chicago, 1928), 7; Lloyd R. Morris, *Not So Long Ago* (New York, 1949), 394-95.

[12]LeRoy A. King, "Consolidation of Schools and Public Transportation," *Annals of the American Academy,* CXVI (November, 1924), 74-77.

[13]New York *Times,* November 7, 1920.

[14]Morris, *Not So Long Ago,* 385.

[15]John Ihlden, "The Automobile and Community Planning," *Annals of the American Academy,* CXVI (November, 1924), 200-204.

[16]Frank E. Brimmer, "Vacationing on Wheels," *American Magazine,* CII (July, 1926), 176.

[17]Warren H. Wilson, "What the Automobile has done to and for the Country Church," *Annals of the American Academy,* CXVI (November, 1924), 84; "Where the Car has helped the Church," *Literary Digest,* LXX (July 16, 1921), 53.

[18]Joseph K. Hart, "The Automobile in the Middle Ages," *Survey,* LIV (August 1, 1926), 494.

[19]Frank Robert Donavan, *Wheels for a Nation* (New York, 1965), 164.

[20]Morris, *Not So Long Ago,* 382.

[21]New York *Times,* September 28, 1927.

[22]David Lewis Cohn, *Combustion on Wheels: An Informal History of the Automobile Age* (Boston, 1944), 225.

[23]New York *Times,* November 14, 1926; October 19, 1926.

[24]Morris, *Not So Long Ago,* 387.

[25]New York *Times,* January 6, 1928.

[26]Arch Mandel, "The Automobile and the Police," *Annals of the American Academy,* CXVI (November 24, 1924), 191-94; New York *Times,* January 7, 1923.

[27]*Time,* May 2, 1934, p. 45.

[28]Mandel, "The Automobile and the Police," 191-94; New York *Times,* January 7, 1923.

[29]Cohn, *Combustion on Wheels,* 259; Epstein, *The Automobile Industry,* 116-19.

[30]Morris, *Not So Long Ago,* 381.

2

Television

Editorial from *Radical Software*

MICHAEL SHAMBERG AND THE RAINDANCE CORPORATION

■ Michael Shamberg *is one of the founders of the Raindance Corporation and Top Value Television (TVTV). The essay reprinted here is a statement of the editorial purpose of the journal* Radical Software. *Of particular interest is the introduction of* Ⓧ, *a sign that is said to represent the antithesis of the copyright.*

As problem solvers we are a nation of hardware freaks. Some are into seizing property or destroying it. Others believe in protecting property at any cost—including life—or at least guarding it against spontaneous use. Meanwhile, unseen systems shape our lives.

Power is no longer measured in land, labor, or capital, but by access to information and the means to disseminate it. As long as the most powerful tools (not weapons) are in the hands of those who would hoard them, no alternative cultural vision can succeed. **Unless we design and implement alternate information structures which transcend and reconfigure the existing ones, other alternate systems and life styles will be no more than products of the existing process.**

Fortunately, new tools suggest new uses, especially to those who are dissatisfied with the uses to which old tools are being put. We are not a computerized version of some corrupted ideal culture of the early 1900's, but a whole new society because we are computerized. Television is not merely a better way to transmit the old culture, but an element in the foundation of a new one.

Our species will survive neither by totally rejecting nor unconditionally embracing technology—but by humanizing it; by allowing people access to the informational tools they need to shape and reassert control over their lives. There is no reason to expect technology to be disproportionately bad or good relative to other realms of natural selection. The automobile as a species, for example, was once a good thing. But it has now overrun its ecological niche and upset our balance or optimum living. Only by treating technology as ecology can we cure the split between ourselves and our extensions. We need to get good tools into good hands—not reject all tools because they have been misused to benefit only the few.

Even life styles as diverse as the urban political and the rural communal require complex technological support systems which create their own realities, realities which will either have to be considered as *part of the problem, or, better, part of the solution,* but which cannot be ignored.

Coming of age in America means electronic imprinting which has already conditioned many millions of us to a process, global awareness. And we intuitively know that there is too much centralization and too little feedback designed into our culture's current systems.

The only pieces of public technology, for example, which are responsive to human choice are electric-eye doors and self-service elevators. Street-use patterns and building designs completely structure our experience rather than vice-versa. *(The people belong to the streets).* When you get into mass communications systems other than the telephone not only is control cen-

☐ From *Radical Software,* Summer, 1970.

tralized, but decision-making is an institutional rather than a people process.

• • •

Fortunately, however, the trend of all technology is towards greater access through decreased size and cost. Low-cost, easy-to-use, portable videotape systems, may seem like "Polaroid home movies" to the technical perfectionists who broadcast "situation" comedies and "talk" shows, but to those of us with as few preconceptions as possible they are the seeds of a responsive, useful communications system.

• • •

Videotape can be to television what writing is to language. And television, in turn, has subsumed written language as the globe's dominant communications medium. **Soon, accessible VTR* systems and video cassettes (even before CATV† opens up) will make alternate networks a reality.**

Those of us making our own television know that the medium can be much more than "a radio with a screen" as it is still being used by the networks as they reinforce product oriented and outdated notions of fixed focal point, point of view, subject matter, topic, asserting their own passivity, and ours, giving us feedback of feedback of information rather than asserting the implicit immediacy of video, immunizing us to the impact of information by asking us to anticipate what already can be anticipated—the nightly dinnertime Vietnam reports to serialized single format shows. If information is our environment, why isn't our environment considered information?

So six months ago some of us who have been working in videotape got the idea for an information source which would bring together people who were already making their own television, attempt to turn on others to the idea as a means of social change and exchange, and serve as an introduction to an evolving handbook of technology.

Our working title was *The Video Newsletter* and the information herein was gathered mainly from people who responded to the questionaire at right. While some of the resulting contents may seem unnecessarily hardware-oriented or even esoteric, we felt that thrusting into the public space the concept of practical software design as social tool could not wait.

In future issues we plan to continue incorporating reader feedback to make this a process rather than a product publication. We especially hope to turn the interest and efforts of the second and third television generations on college campuses, whose enormous energies are often wasted by the traditional university way of structuring knowledge, towards the creation of their own alternate information centers. (We are of the first television generation ourselves.)

To encourage dissemination of the information in *Radical Software* we have created our own symbol of an *x* within a circle: ⓧ. This is a Xerox mark, the antithesis of copyright, which means *DO* copy. (The only copyrighted contents in this issue are excerpted from published or soon-to-be published books and articles which are already copyrighted.)

The individuals and groups listed here are committed to the process of expanding television. It is our hope that what is printed here will help create exchanges and interconnections necessary to expedite this process.

* Videotape recorder (ed. note).

†Cable television (ed. note).

From *Guerrilla Television*

MICHAEL SHAMBERG AND THE RAINDANCE CORPORATION

The following is an excerpt from a book designed as a kind of primer on the nature of what Shamberg calls "Media-America." Power, it is claimed, has passed from capital, and even from the usual political mechanism, to those who control the means of disseminating information.

Rich in metaphor ("print is television's Thorazine"), popular in its approach, and radical in its proposals, this assessment of television has been shared by many who have enlarged the boundaries of that medium.

Americans are information junkies.

Almost every American home has a television set which is turned on an average of five-and-a-half hours a day as part of the environment, regardless of what's playing.

"All-news-all-the-time" radio stations recycle information every fifteen minutes but nonetheless captivate people for hours as a sort of information-Muzak background to any activity.

Homes and apartments are decorated with magazines like dentists' offices because they feel strangely sterile without them. Some people can't even handle the solitude of sitting on a toilet unless they have a toke of print to keep themselves occupied.

Instamatic and Polaroid cameras are travelers' tools because people know places as photographs and photograph them so they'll seem real. Home movies are a kind of surrogate sperm which ensure biological continuity on an information level. Taking pictures, regardless of content, has become an end in itself.

Organisms have always needed a minimum of information or novelty to stay alive and alert and ever-evolving, but Media-America has made that minimum a staple right behind food, clothing, and housing. Electronic media have become looped-in to our neural networks. We need a minimum of information flow not only for physical survival, but also for psychological balance because electronic media are as omnipresent as light.

My own addiction began when I was five, the year my family got its first TV set. As somebody later calculated for those of us of the first television generation, we spent more time with TV than we did with our parents.

I believe this dependency represents human evolution: to the degree to which it stimulates us, it enhances our survival. When it numbs us, it threatens our ability to adapt. In either case, the information environment is an inexorable part of our ecology.

But just as the consequences of disturbing natural ecology were ignored until we were surrounded by omni-pollution; so too is a media-ecology an alien sensibility to the people who control change in America.

The 1960s were a Pearl Harbor of the senses. Whole new technologies conditioned us from birth to relate to a world which was not that of our parents' childhood. It came as a sneak attack because print-man, impervious to his own bias, was unable to perceive that any time there is a radical shift in the dominant communications medium of a culture, there's going to be a radical shift in that culture.

What perils us now is that in electronic environments consequence is simultaneous with action. About the longest lag between us and ecological suicide is fifteen minutes, the time it takes the missiles to get to Moscow. Yet, there is no sanctioned study and understanding of media-ecology.

This is especially reflected in our schools and universities which, as I got older (I am now twenty-seven), wasted more and more of my time. As a media freak, *my* "homework" usually consisted of watching TV and when that was forbidden because I had to "study," I holed up in my room reading any and all magazines I could get my hands on.

Slowly, cumulatively, I came to realize that the school environment was wildly out-of-sync with an electronic environment it refused to acknowledge. It wasn't until I graduated from a so-called college, however, that I developed confidence in my intuition that *I* was right, and *they* were wrong. My information processing modes, conditioned by electronic media, were better reality models than theirs, which were and still are based on print.

In short, I had to learn how to survive in Media-America despite the very institutions whose job it is to teach survival. And that's heavy. Because a culture with sanctioned education processes which are out-of-phase with the life-process can't last very long. It's just such a culture that fears and even hates the most perceptive of its children.

For the demand of media-children that schooling be relevant is not, as the print-men mistakenly believe, a philosophical plea to be debated with print-based modes of perception called "rationality." Rather, it is an intuitive biological response.

Most of the sanctioned information models in Media-America are irrelevant *biologically*. They will not allow us to survive and adapt.

Unfortunately it's not just pre-Media-Americans who don't track what's going on. Ironically, many children of the first and subsequent television generations who have no choice but to rebel against the way Media-America is being managed nonetheless refuse to see that using the media as tools in-and-of themselves (not for political propaganda) can lever enormous social change. I have talked to college students who seriously wanted to blow up their school's computer instead of demanding access to it. Where's that at?

The same people who might be naturals with new information tools precisely because they're dissatisfied with the uses to which old tools are being put are still trapped in anachronistic political models.

The media have their own bias, just like guns. You shoot a gun and you kill somebody. Their being dead is a consequence of technology, not ideology. But not all technologies have the same consequences. Some enhance life more than others. If we can understand how to orchestrate these technologies, we can work directly on the level where Media-America is shaped.

This is the software or design level. How many political decisions of twenty-five years ago, for example, have as much influence today as do those criteria which went into the technological structuring of broadcast-TV? Aren't more American communities a reflection of highway design than of political philosophy? Would our culture be so rigid today if early school designers had been hip to living spaces like domes and inflatables?

While you can argue forever that the latter are reflections of political bias, in Media-America real power is generated by information tools, not by opinion. The information environment is inherently post-political.

This Meta-Manual is here to lay out why the information environment is a good and verifiable reality model; why we must perceive media structures biologically (media-ecology); and why videotape, particularly portable video systems, can enhance survival and generate power in Media-America.

The moon landing killed technology. Far from being the ultimate technological art, it demonstrated on a rather elegant scale that our hardware can do anything we want once we figure out the software. In fact, no major hardware breakthrough was needed to get us to the moon. The Chinese were on to rockets a few thousand years ago and computers are about as exceptional as automobiles. Supersoftware in control of super-complex hardware put us on the moon.

We learned that it's only a matter of time before developments will come about once we've scheduled them as software, which is why NASA* can blandly tell us when men will be on Mars and other planets. Similarly precise predictions made only a few decades ago were treated as science fiction, at best.

The death of hardware is the ultimate transformation of America to Media-America. It embodies our total shift from a product- to a process-based culture. It's much like the difference between renting a car and owning one: you pay for the service of using it (process), not for the value of ownership (product).

Not for nothing are the people who are hung-up on preserving products as a measure of worth the same ones who are most bewildered and alienated by Media-America. (Reagan called the National Guard to clear People's Park in Berkeley because it was private property, not because he objected to the way it was being used.) And not for nothing are property freaks called "reactionary." Because what they object to is essentially human evolution.

The dominant technology of a culture determines its character. Agricultural societies, for example, were spawned by breakthroughs in farming technology. Farming overloaded into trade and generated the great shipbuilding/

*It's ironic that NASA, probably the greatest government agency produced by America, has killed patriotism. National boundaries are simply not a motivating image when we have photos of the Whole Earth.

ocean-going societies whose ultimate act was to get the Pilgrims to America.

Over the next couple of hundred years America got itself together, laid out a government, and prepared for "take-off" in the nineteenth century. Then the great patriarchal fortunes (which still haunt us, especially in New York) were scored in basic tool-up and energy industries like railroads, coal, steel, and so on.

They overloaded into mass production and the automobile, which has had more effect on life in America than all the native philosophy we learned about in our high school "civics" classes.

The savvy gained from automobile production, essentially the orchestration of the production line, got us through the so-called "Second World War."

Computer technology, the next step of man, was developed almost as a by-product of World War II by Cybernetic Superstars like Norbert Wiener and John Von Neumann (how many of you learned *those* names in school?) who were doing weaponry research for the military.

The commercial synergy of that stream culminated in the first real computer, Eckert and Mauchly's ENIAC, at the Moore School of Electrical Engineering in Philadelphia in 1948. At that time, broadcast television was just a few years old.

None of the technologies which are dominant today and make America Media-America were in the public space until after World War II. (Yet World War II still dominates accepted political and cultural thinking.)

All of the post-war corporate successes, however, have to do with radical new technologies; technologies which do not produce things but which *process* information: the IBMs, Xeroxes, and Polaroids, their suppliers and competitors.

Now, less than fifteen years after they got going, these industries are practically public utilities and the real entrepreneurs are scoring in software and leasing companies which have no vested equity in any particular hardware system.

Meanwhile, production lines are automated and controlled by processing lines of blue-collar information workers like clerks and secretaries and computer-programmers, none of whom work anywhere near the actual manufacturing plant. Instead, they and their supervisors—executives who spend their time having meetings and managing memos and computer print-outs—work in urban office buildings which are like huge, on-line filing cabinets.

Concomitant with this is the ascendance of the super-psychological marketplace where psychic benefits replace physical ones and we're exhorted to buy moods and services, or processes instead of objects.

You just can't track what's going down in Media-America with product models. Communications and computers are our central shaping technologies, and they have absolutely no object value. (What good is a TV set that won't turn on?) Only as process are they worth anything to us.

Now the upshot of all this is that history is practically worthless as a survival model. Not only is there no precedent for television and radio and airplane travel and so on prior to this century at the outside, but even those phenomena are evolving within a fraction of a lifetime.

I got my first transistor radio when I was twelve. It cost $75. Now you can play with them in your crib and they cost less than $7.50. When they break down you throw them away. Similarly, the carcasses of television sets are common sights in city garbage cans. Twenty-five years ago they were high technology. And jet travel, which to me was a big deal, is no more phenomenal to a kid growing up in Media-America than the existence of the sky itself.

It's not just that history lacks any of the forces operating today which makes it a burnt-out medium. It is also useless because it is product-based. Historical evidence like books, buildings, painting, and sculpture, is what's survived time, not necessarily what was crucial about being alive in the past.

The closest we can come to decoding the process of a past culture (outside of re-creating it through historical movies or flaky Walt Disney pageantry) is by reading about it. That means a print-grid has been laid on our entire past. When the schools try to interface that bias with kids who have been electronically imprinted, the results range from boredom to hostility.

After all, we read mostly for entertainment, not survival, and we throw away our (paperback) books instead of keeping them under glass like Gutenberg Bibles; buildings aren't built to last; and art ranges from pure process to artifacts like Warhol silkscreens and Rauschenberg lithographs which are an enhancement of process.

The new art media are mechanical processes capable of an infinite number of copies. Only the object-value mentality of scarcity economics keeps us from mass-producing art. The result is an artificially supported seller's market which has grossly inflated the prices of "one-of-a-kind" paintings simply because there aren't that many more to go around. Even fairly recent paintings are going for up to $100,000 apiece.*

Nonetheless, the most far-out art object of our time—a box with an electronic canvas capable of monitoring this whole planet and others, *i.e.*, TV—is readily available to everyone. High art is a myth promulgated by pre-Media-America.

There simply are no process artifacts from time past which can compete with even the worst television show as a kinetic experience. What will pass as historical artifacts from our culture are storage media like computer tape and videotape which are essentially process retrieval systems.

Evolution is parlay, or we'd still be back in the sea. If cells just combined one-on-one, organisms would be larger but not necessarily more complex. The best we could hope for would be a race of giant amoebas. Instead, we are enormously complex creatures relative to our size and development time.

As some scientist once charted out, if we take all of known evolution and equate it to one hour, then man comes in around fifty-seven minutes on the clock of life. In other words, the most complex organism to come along so far took but a fraction of the evolutionary process to emerge, which is a characteristic of parlayed complexity.

There is clearly some sort of velocity of evolution which is accelerating and parlaying back into itself so that the rate of acceleration is itself accelerating. Changes which used to take place over generations now must be accommodated in fractions of a lifetime. Nothing is extraordinary if it happened before you were born. Television and heart transplants and space travel are all *a priori* experiences to children whose parents still consider them phenomenal. Even the age at which girls first begin to menstruate has dropped over the last few decades, according to *The New York Times*.

But our bodies cannot keep up with what evolution via our minds would have us do. So we are evolving through our technology.

Computers come in generations, just like people. The moon landing module was a collective foetus given birth by a synergy of mind which could not propel its physical body alone into space. We send men into space instead of machines because the state-of-the-art of human beings is more advanced than that of our extensions. By the year 2000, however, computers may be pound-for-pound dollar-for-dollar more advanced. Then we'll start sending them.

This acceleration of succession simply can't be accommodated by the old model which calls it "change" and then generates distress because "change" is disorienting. Just because Richard Nixon's President doesn't mean that all of the 1970s are the 50s in drag, a sort of amphetamine replay of everything that's happened since World War II. "Future shock" is a condition of trying to lay yesterday on today. It's a bias of generations. Nobody who's hip to Media-America suffers from it.

Instead of being in the advanced stages of an old culture we are on the threshold of a new one. Human evolution has done a flip-flop. Like the lines on a logarithm graph which bunch together and then space out again we have crossed over from super-swift technology to the crude, nascent stages of controlling our own evolution. Already we can countermand nature's aging with biological engineering like heart pacemakers, synthetic arteries, transplants, and so on. Next, genetic engineering will enable us to control human development from conception on.

The consequence of this crossover is that we can no longer differentiate between man and machine. The ecology of America is its technology.

We can best understand and manage technology in a biological context. (For example,

*A truer use of past artwork which is process-based is what David Smith calls "image ecology," or the re-cycling of images. Similarly, Ant Farm keeps an "image bank" of old photos and drawings clipped from back issues of magazines. Interspersed throughout Ant Farm's graphics, these images of how we saw technology (especially television) in the Forties, Fifties, and Sixties gives context to our current visions and makes them seem more transient, less immutable.

Indeed, history will only survive as a medium if it is generally re-cyclable. Studies of ancient civilizations which have no connect to current contexts will be yawned off. However, where information from the past has contemporary application then it has contemporary value.

Warren Brodey suggests that we research the past for forgotten skills (*e.g.*, alchemy) which were lost when mechanization took command but which are needed again now as we revert to ecologically-valid life styles.

consider automobiles as a species which has simply overrun its niche. Indeed, all technologies evolve like organisms; towards doing more with less. Machines get smaller, cheaper, and easier to use. Compare transistor radios to antique crystal sets; portable videotape units to television studios, or desk-top mini-computers to room-filling units.)

This realization is only now being selected-in to our survival grids even though it was summed up twenty-five years ago in the subtitle of Norbert Wiener's book, *Cybernetics,* i.e., *Control and Communication in the Animal and Machine.* What Wiener passed off as an elegant scientific breakthrough is also a major conceptual re-structuring.

Before then man and machine were similar only when human beings were pictured as mechanical extensions of the assembly line. In the early part of this century studies were actually done to see how man could better mimic mechanical motion. Wiener reversed all that. His studies showed that machines could be understood in the context of animal or biological processes, not vice-versa.

Technologies are embodiments of mind. Or, as Buckminster Fuller says, anything nature will let us do is natural. Technology is thus neither all good nor all bad because nature itself is not such a binary system. Forms on their way up co-exist with ones on their way down—which enhances the diversity essential to evolution. Once dinosaurs made it. Now they would be out of place, to say the least. And so on.

There is no possible realm which has an independent technology on one side and us on the other. While Luddites may have a comprehensive program, they are ultimately spiting themselves. Nor can we run back to Walden Pond; it has to share its oxygen with the Con Ed plants in New York City.

Yet destroying technology or running away from it are our only accepted technological strategies other than unchecked growth, and they are especially prevalent among my contemporaries who should know better.

A better answer is to stay around and begin to structure and relate to technology. It should be remembered that pollution is not a product of man in general, but of men who are already into killing other men.

Media and man evolve together. The bias of each new medium is that it seems more congruent with mental process than the one it supplanted. This happens in two stages.

First a new medium seems fabulously "real" and excites people no end. Perspective painting, for example, was once such a turn-on. Today you can still find people who think that film is a mainline into the brain.

But after a while a medium seems so natural that its effects are taken for granted and its bias is given the status of "objectivity." This is what happened with print, which put its hooks into thought process for five hundred years. Out of that experience came all sorts of biases, like the Cartesian mind-body dichotomy, which couldn't have made more sense at the time.

In videospace it's impossible to know where your insides end and your outsides begin. However, print-people in television think they are being "objective" by using TV as a radio with a screen. In other words, they wrap "subjective" images in objective words.

But kids who have never known a world without television don't make the value judgment that personal contact is real and TV is unreal.

My own experience straddles print and television. TV is "where the action is," but when I want to cool out I read a while because it slows me down. Print is television's Thorazine. Reading is an experience I select to fit my mood, not a natural process which transcends any mood.

Like techno-evolution, media evolution is accelerating. Within less than a century man has developed three major media: radio, film, and television. Each is an overload of a previous one, and was used first to do a better job than an old medium. Film first imitated theater. Television combined both film and theater. Or offset printing, for example, which was a natural response to an overload on the letterpress system, now has its own indigenous forms: paperback books and the underground press.

Then, the information diversity generated by offset printing overloaded into the Xerox machine which was initially a system for individual control of existing information. Now it is a new publishing medium. It can be used as a vanity press or to make every man his own anthologizer. In either case it completely destroys the notion of information as property to be copyrighted. Like videotape, which can be duplicated the same way as audiotape, the

Xerox system is biased toward a process rather than a product use of information.*

High information density within a medium will also generate new soft media, or software patterns. For example, the great breakthroughs in biological research over the past decade have all come as a result of its being a high information field which demanded that the process of information-use itself be understood before discoveries could be made.

Information comes together and diversifies exactly like the evolution of organisms. In fact, it's impossible to draw a line between the evolution of man and that of his media. They feed each other. The more sophisticated an organism's media are, the more complex the organism.

Ants, for example, are limited in what they can pass on from generation to generation because they have no external storage media. Because man does, he can mutate himself within a generation or less. Instinct is but one of man's many media. But it's all an ant's got.

Man's media processes are cultural DNA; the assimilation of them we call education. For a medium to function like DNA, its genetic analogue, it must have three modes: record, storage, and playback.

Print meets those criteria. We record with writing, store on paper, and play back through reading. Film has never supplanted print because its three modes are expensive and demand an intolerable lag time for processing. Moreover, film technology, especially in playback, demands a fetishist's attention to equipment and environment, which is why film is a cult medium.

Film is the evolutionary link between print and videotape. Like reading, seeing a movie is essentially a solitary experience. Unlike print, film is highly kinetic.

Media also exist in symbiotic or hybrid forms. Film and TV combine into movies on television. The symbiosis of radio and TV is Walter Cronkite. And so on.

But each medium also demands its own context. Until the development of videotape it was possible to view TV as a hybrid. With videotape, however, television becomes a total system and succeeds print as our cultural DNA. Recording on videotape is analagous to writing, the tape itself is equivalent to paper, and playback through a TV set is video read-out. Only by pushing film to its limit can it match the ease of operation at which videotape begins. Videotape as a process medium frees film to become an art form.

A failure to understand which medium is cultural DNA at any point in time is counterevolutionary. Because American education, which is only now getting into *film,* refuses to verify the assimilation of video literacy it has become antisurvival. In that context, rebellion is a biological response.

Evolution is essentially a process of information storage and retrieval. That's what genes are all about. Resisting the neurophysiological congruence of television and brain is schizophrenic. It may be that there will be no clear-cut new medium to succeed television, only symbioses of video, lasers, computers, and beyond. But cultural DNA is sure to ascend to new hybrid forms. Already people find holograms phenomenally "real."

Just as techno-evolution is gradually phasing out our bodies with increasingly sophisticated support technologies like heart machines, synthetic arteries, and so on, so too is media evolution transforming us into whatever technology can best record and retrieve information.

Some scientists for example are using a computer transfer system where one keys in his latest theory for feedback by others. The result, coupled with the amazing amplification of thought process that the computer already offers, is a whole new process of mind, which supplants "human" relationships.

Right now, the human brain in symbiosis with computers is the best thing going. But if some fabulous computer can process intelligence better than man all by itself, then at that point the computer may be man. . . .

A system is defined by the character of its information flow. Totalitarian societies, for example, are maintained from a centralized source which tolerates little feedback. Democracies, on the other hand, respect two-way information channels which have many sources.

*Information can be owned only if it is in product form. Thus, copying technologies like Xerox, audio and videotape recorders destroy the notion of copyright because they allow for unrestricted processing of information. The upshot will be that markets will develop not for product information, as such, but access to tailored assemblages of information which reflect the user's taste and needs.

When a culture has only crude communications technology information flow is reflected in social ritual. But in an electronic culture like Media-America the communications systems themselves, not philosophy, are what shape social structure. Similarly, it is the structure of bureaucracy, not the decisions which are pumped through it, that determines government policy.

Because we are in an information environment, no social change can take place without new designs in information architecture. Redesign at any sublevel will only generate frustration. Many of Media-America's problems can be understood as a clash of information structures.

Print information, for example, is biased toward hierarchy and control because it fosters linearity and detachment. Electronic media are the opposite. They are everywhere all-at-once. Schools, which are based on print and centralized control of information, can no longer contain students who can be their own authorities simply by turning on a TV or a transistor radio. It's the very structure of TV that undermines the nature of school administrations, regardless of what the programming is.

But the structure of broadcast television contains its own schizoprenic contradictions. We get too much news to accept authority based on restriction of information flow. Yet pre-Media-Americans are conditioned to trust authority because "the President knows more than we do." Nonetheless our video sense of death in Vietnam is no less vivid then the President's.

Agnew's attacks on television are successful with pre-Media-Americans who are anxious because they know too much and yet believe that authority is based on someone knowing more than they do. While Media-Americans ask the government to get in sync with the information environment, Agnew demands the opposite.

Agnew is right about broadcast television being a system which minimizes diversity (although he makes it sound as if he wants his viewpoint in place of others, not alongside them). This is inherent in the technology which has no capacity for feedback.

Television sets, for example, are also called "receivers," a one-way term for a system which conditions passivity. So on the one hand we're given information to respond to; on the other there are no sanctioned channels of response. This results in spontaneous attempts at feedback like the pro-war construction workers who stormed around on Wall Street and then rushed home to see themselves on the evening TV news shows.

It may be that unless we re-design our television structure our own capacity to survive as a species may be diminished. For if the character of our culture is defined by its dominant communications medium, and that medium is an overly-centralized, low-variety system, then we will succumb to those biologically unviable characteristics. Fortunately techno-evolution has spawned new video modes like portable videotape, cable television, and video-cassettes which promise to restore a media-ecological balance to TV.

And we're going to need similar technologies to save our cities. Modern urban design is largely a function of homogeneous information-processing structures. The result is an ahuman "international" style of curtain-wall architecture coming from hack construction company designers as well as from fashionable architects like Philip Johnson.

Instead of buildings which stress their structural elements and give a sense of tactility, print-men give us visual masterpieces which are to be looked at but not experienced. Where there used to be diversity of ground level shops, we now get homogeneous forty-foot-high granite slabs studded with elevator doors.

Park Avenue in New York, the quintessence of all that, has absolutely no human scale. Similarly they're killing off the Avenue of the Americas with buildings which are designed to expedite, not engage, pedestrian traffic. The only exception is the Saarinen-designed CBS building with its triangular columns which shift their perspective as you walk by. Everything else is unresponsive hard cybernetic technology which controls you and not vice-versa. Except perhaps for self-service elevators. But they're linear.

Corporations, which are still structured around centralized information flow, demand that their subsidiaries be responsible to a "home" office rather than to the cities which house them. Urban centers thus function as support systems for metanational corporations which do not feed back to any local community.

Executives are either transferred frequently or use their offices as locker rooms for when they're not traveling. Travel itself is reduced to a problem in information processing. Either computers shuffle planes and passengers around or you can stay at home and teleport yourself with telephones and telefax.

Most of the men who control our cultural decision processes live in suburbs which bleed off deteriorating cities* while offering none of the diversity of a total country or a total urban environment. Suburbs may be ideal rest areas for executives who can feed on the variety of a city during the day, but as full-time whole environments they rob kids of any survival experience. Not for nothing can dope be found in almost every American high school.

Only the more powerless urban classes have a vested interest in the livability of cities, and they are informationally indigent.

Man has this planet by the balls not just because he knows more than other organisms, but because he knows that he knows. (I think that I think, therefore I am.) This process of perception of perception is called apperception, and it's a measure of our evolutionary complexity.

Because we are apperceptive creatures we're not preoccupied with stimulus and response, but with the mechanisms of stimulus and response; or not with action, but with the metalevels which control action. Not until this century, however, has that understanding been embodied in technology.

The first Industrial Revolution was essentially in technologies which imitated the work of the human body, only faster and better. The radical technologies of the Cybernetic Revolution mechanize control systems and thus extend the brain, which is exclusively a control system. The image of the brain trying to do its own physical labor—beating your head against the wall—symbolizes futility.

Information is the energy of control systems. Just as electricity magnifies physical action, information amplifies control processes. With the information output of just one finger, for example, I can control a multi-billion dollar network and relay information anywhere on earth. That's called "making a phone call."

Information, like other forms of energy, also has its basic unit, which is called a "bit." One bit of information is the least amount needed to make a binary, yes-or-no response. Combinations of binary decisions generate more complex controls. Only five brain cells or computer circuits, for example, can generate thirty-two different combinations. As the actual capacities of both brain and computer technologies range from millions to billions of circuits, their potentials are immense.

But the concept of a bit is essentially a sop to the product mentality. (Computers are sold or rented for prices which relate to their bit capacity.) For unlike other forms of energy, information has no product or potential mode. It is inherently process, which is why it's the energy of evolution.

We can calculate how much of other types of energy is available energy, like knowing the life of a battery. Thus they have value in a money/product economy. But information simply is

*This is largely because taxation is product based, *i.e.*, you are taxed where you physically live instead of for the services which also sustain life. Thus, executives use the municipal services of cities, which are subsidized by the less affluent who live there, and then retreat to the suburbs.

Similarly, the American political process does not reflect the global context in which the American government must operate. As U.S. citizens we had no say in the central issue of the 1960's, the Vietnam war, because participatory democracy represents only hard territory.

We do not elect officials who deal with foreign policy even though their decisions affect us enormously. This was tenable before an electronic world of simultaneous cause and effect when the media were unable to tell us what was happening, and the cost of fighting a foreign war was offset by economic exploitation.

Now, however, we are all global citizens. And while we should not have the right to force our decisions upon alien cultures, we must be able to control our government in its global actions.

Congress had no voice on the Vietnam war because its constituency is local. This creates a dangerous imbalance wherein voters are expected to surrender collective decisions to the Executive branch of government, which for the most part is non-elective, and at best asks us to trust them.

It would have been unthinkable, much less a gross political liability, for Nixon to have tried to suppress a classified government report on domestic events as he did with the New York Times' publication of the Pentagon's study of the Vietnam War.

Even more chilling was the response of Maxwell Taylor, a former general and Ambassador to Saigon, to a question about whether people are entitled to know about those kind of decisions:

"I don't believe in that as a general principle . . . What is a citizen going to do after reading these documents that he wouldn't have done otherwise? A citizen should know those things he needs to know to be a good citizen and discharge his function." (In other words, to be a good citizen there are things we must not know!)

That we are asked to sacrifice elective control over global affairs is, at best, to have a government which is a benign dictatorship. We need to elect executives like Secretary of State, Defense Secretary and so on, or the power of the executive will be awesomely imbalanced when we begin to settle on other planets.

not information unless it's applied, or processed.

The more process-oriented an information medium is, the sillier it seems as a product. That's why we approve of book collecting, indulge people who collect old magazines, and lock up people who hoard old newspapers.

An unapplied bit is at best a fact. In context, it can become information. Here's an example: It is a fact there there is a chair in my room. When that chair is out of the way or I have no use for it, it remains a fact, a piece of data. But if I need the chair or it's in my way as I move around the room, then it enters into a feedback loop and becomes information.

Very little of what we experience is information, although everything is data. As you read this page only the words and paper are being processed, yet your eyes are also taking in the book binding, maybe the table it's resting on, and so on. Not to mention what your other senses are doing. Meanwhile inside your head previously assimilated data may be flipping into new contexts and thus becoming information independent of any external input.

Try this. Imagine the difference between yelling "fire" in the proverbial crowded theater and in your bathroom. Clearly the value of the information can be measured only in process. Information out-of-context is merely data or if particularly misplaced: dada.

Gregory Bateson thus defines information as "a difference that makes a difference." And Norbert Wiener has pointed out that clichés, because they are so familiar, have practically no information value. In other words, information is not information unless it reveals something new.

This means that unlike other forms of energy, which can ultimately be neither created nor lost, only re-distributed, information is inherently regenerative, or negentropic. (Negentropy is the reverse of entropy. Both words are from the lexicon of cybernetics. I acquired these elegant tools of thought only after I completed my formal education. Unless maybe you were a physics or an engineering student they were probably left out of your education too.)

Unless information regenerates from a past state there are no differences to transform. This book is a synergy of books I've read and experiences I've had. If everyone who reads it thinks it's a lot of crap then the transforms will remain personal. But if it makes sense to others then its range of regeneration will be greater and its function will become one of expanded consciousness, which is the drift of evolution.

Feedback is the key concept of the Cybernetic Revolution for without it there would be no control technologies; only machines which run willynilly onward until they burn out or are stopped externally.

When Wiener discovered that control and communication in the animal and machine are the same, he meant that both respond to feedback process. The initial application of feedback concepts was in gunnery where to target a missile it was necessary to consider each successive state as stemming from the last rather than assuming an even velocity over the range of firing.

Before feedback came along people believed that the universe was a big machine wherein each action was initiated independent of the preceding one. And God was the mechanic who kept the thing running.

That was groovy for a long time because with a seemingly inexhaustible supply of resources we didn't have to worry about consequences. That's how pollution happened. Pollution is empirical evidence of feedback.

Evolution proceeds through feedback. Nature selects-in those creatures which are going to make it by the environment feeding back to animals and vice-versa. When the two are congruent, as with say an aardvark's long nose and enough ants to go around, then survival is enhanced. When the characteristics of an organism deviate from what the environment can feed back, then it perishes. Without feedback a creature exists in a vacuum and that's impossible.

Man, thinker that he is, experiences not just physical feedback, but psychological as well. In fact, feedback is a prerequisite for the verification of experience. People who get no feedback or who refuse it become autistic or catatonic.

In Media-America, our information structures are so designed as to minimize feedback. There is no feeding back to broadcast television; you can call up a radio talk show but the announcer usually works you over; and there's only so many times you can write a "letter-to-the-editor."

This makes for incredible cultural tension because on the one hand people cannot ignore media evolution, while on the other they require feedback for psychological balance. The result was the 1960s: every conceivable special interest group, which was informationally disenfranchised, indulged in a sort of "mass media therapy" where they created events to get coverage, and then rushed home to see the verification of their experience on TV.

Mass media therapy, however, is at best an *ad hoc* remedy for social problems because it demands abnormal behavior which cannot be integrated into normal living patterns.

The now legendary 1968 Democratic convention was energizing for people who were on the streets of Chicago because it was extraordinary in a superficial way that life is not: demonstrations and combat, staying up all night listening to music and smoking dope, a clear-cut enemy, and so on. That's exhilarating stuff, but totally unapplicable to an ongoing life style. The streets may belong to the people, but they're a crummy place to live.

But if our information structures are so designed as to minimize feedback and verify only what is essentially abnormal behavior, the psychological survival of Media-America is threatened. And mass media therapy will continue.

Moreover, if people are unable to believe that their collective will has a collective effect on the physical environment, they retreat from their feelings of impotency into conspiracy theories of social action.

Such a lack of feedback is exactly the opposite of democracy in America as de Tocqueville saw it: decentralized, self-governing units of people who could see that their decisions were being carried out.

It's nostalgia to think that that type of balance can be restored politically when politics are a function of Media-America, not vice-versa. Only through a radical re-design of its information structures to incorporate two-way, decentralized inputs can Media-America optimize the feedback it needs to come back to its senses.

Media-America is on information overload. The proliferation of information technology from techno-evolution and media evolution has revealed a sort of Parkinson's Law of Media: "Information expands so as to fill the channels available for its dissemination."

The result is that everyone feels they have to know more, instead of knowing differently. Thus people sign up for speed reading courses but refuse to try smoking dope. Or they subscribe to news magazines which haven't changed their formats very much in the last thirty years instead of finding re-organized print resources like the *Whole Earth Catalog*.

But no system can survive continual overload unless it's re-structured. When electrical systems keep burning out or dams break we don't re-design them the same way. Similarly, the way to respond to "all-information-all-the-time" isn't to try to pump more into the same old head, but to treat it as a new medium, and expand your head.

It's unlikely there will be a significant, discrete new medium beyond television. Instead we're going to have symbioses of media. Things like Xerox machines giving a print-out of a televised copy of the daily newspaper; three media in one; or model testing by community groups using computer terminals to cure problems. (All electronic information can be transmitted as the same type of binary pulse. Thus the cable in cable television can supply any type of end-terminal from TV to computers to holographic chambers.)

"All-information-all-the-time" is thus an amalgam of media which transcends any specific hardware configuration. It is in essence a soft medium and requires new software patterns instead of beefing-up the old models. (It is also verification of the ultimate ascension of media evolution to a purely process condition in which mind will succeed mind independent of gross matter.)

Only reasonably adequate criteria of perceptual relevance permit a species to survive. If output—behavioral, genetic, or otherwise—doesn't sensibly correspond to input the species courts extinction. Almost any model can, of course, be verified by the user if he tries hard enough (that's what paranoia's all about), but some enhance survival more than others.

If I believed, for example, that there are no automobiles on city streets during rush hours, unless I modified that model very quickly, I'd be wiped out.

The breakdown of old models and the inability of authority to legitimize new ones by understanding that Media-America is discontinuous with the past is central to the crisis in our culture.

A recent study discovered that kids who think their parents are sources of useful information are the ones who get along best with mom and dad.

Formal education, *i.e.*, school, is exclusively a survival process. Creativity is a high form of survival because it optimizes flexibility. Kids who rebel at the inflexibility of schools are merely responding biologically to being subjected to an anti-survival mode.

The main problem is that in an age of process, authority is still using product models, things that you can see. Yet, as Buckminster Fuller points out, we went off the visual standard fifty years ago: from wire to wireless, track to trackless, and so on.

Nonetheless, the news media, because they have been unable to develop a process vocabulary, zoom after visually-oriented or product information every chance they get.

Each front page of *The New York Daily News* exploits some visual event as a headline (rape, murder, explosions, labor stoppages instead of wage negotiations, and so on) atop a huge photograph which is really a mini-TV screen. Often the picture has nothing to do with anything else; it just looks good, like a shot of two paraplegic sisters tearfully embracing after not having seen each other for fifty years when one was mistaken for a cow when they landed on Ellis Island, that sort of thing.

In such a context, resentment against the young is embodied in a visual symbol: long hair.

(Of course the other extreme is the baroque sophistication of a paper like *The New York Times,* especially the Sunday edition, fully half of whose content is information about information: studies, reports, policy discussions, with headlines like ''Wider Study on Humans Urged,'' as Peter De Vries once wrote.)

But all the old anthropomorphic models have been destroyed by techno-evolution. Our machines do things which are in no way representative of physical extensions. A moon landing module, for example, has absolutely no traditional, visual aerodynamic characteristics. And computers, even if they extend our brains, have no human analogue because you simply can't ''see'' the brain think.

''Reality'' is now represented by information media like brain wave monitors or computer print-outs which are not visual reproductions but whole languages other than sight. The upshot is that man must now self-reference in an information space which has no product markers, only process.

The final supplanting by information territory of visual space came with the space program, ironically. Rockets are launched one place, controlled another, tracked in foreign countries, and retrieved in the ocean. Those flag decals which read ''Good old USA, first on the moon,'' are reactionary precisely because they're so nostalgic.

The nostalgia is for a product-based view of social change which sees what's going on as the content of stable institutions, rather than realizing that the very nature of the institutions themselves is re-configured through events. (Governments always build their offices to last and be seen. The bias of architecture and construction in general is toward tearing down whole structures and rebuilding them in waves instead of incorporating time into their design. It's as if cities trade in their old buildings for new models every fifty years with no adaptation in between. Thus transition is discontinuous and disorienting. There is no respect for the diversity offered by old buildings coexisting alongside new ones. Nor do landlords budget space for expansion and play, or use materials which can be changed.)

The result is a discontinuity between the powerless who have to experience an institution as process (cops really do beat you over the head when you live in a ghetto) and those who have no personal contact with it and thus still buy the government product.

The ultimate shortcoming of the product mentality is that it makes for palliative rather than remedial solutions. We are forever treating problems after they become visible rather than redesigning whole systems. We're always reacting instead of acting. The result is that government is geared towards crisis management, not anticipatory response. (In fact, it's almost as if the government doesn't know what to do unless there's a crisis or something stridently visible to manage. It's not just because he needed a catchy title that the one book Nixon ever wrote was called *Six Crises.* His whole style of governing is one which demands visual confrontations which are ''resolved'' by his going on television so he can be seen doing his agonizing. The ultimate result of crisis government is a demoralized citizenry, however, because people are misled into thinking that things are out of their control and that only authority knows better.)

Cars, for example, are products; something you can see. And each American is hyped-up to buy at least one of his own instead of demanding a more collective form of transportation. (Cars are essentially tools of the mythology of the individual generated by a book culture. Media-Americans favor buses because they are communal.)

The result of all that, around New York City at least, is that the normal state of "expressways" during daytime hours is a traffic jam: row after row of automobiles with a capacity of up to six carrying but one or two people.

The real problem, of course, is one of transportation, which is a process word.* Nonetheless, the history of American transportation is one of our subservience to the production line. Each step always seemed a better palliative and no comprehensive remedy or symbiotic plan was developed. The government funded the Interstate highway system which ultimately undercut the railroads. Then when Penn Central goes bankrupt the government is asked to bail it out. Resuscitating the railroads means undercutting the airlines. And so on.

The reason the automakers are all of a sudden being such good citizens and vowing to clean up their cars is because that keeps us from seeing that the real problem is mass transportation in general and none is imaginative enough to undertake the necessary re-tooling.

The ultimate worth of a survival model depends on whether or not it is ecologically enhancing. There is an inherent consciousness in media evolution which thrusts forward survival models into the information environment when they are needed. For example, the image of man as a meta-ecological super-creature has been selected-out by the ascendence of information models, or "best-sellers," like the ape books by Desmond Morris, René Dubos, and Robert Ardrey which portray us as just smart apes who are not immune to ecological laws.

Concomitant with this is a feeling that there is an information territorial imperative that we respond to media ecologically. That's why people don't like the fact that data banks possess them by infiltrating their privacy with often erroneous credit and surveillance information.

Consumerism is a similar response to information pollution. Savvy customers, rather than demanding fragmentary data about all the wonderful things a product will do for them, are now demanding useful information about what it will do to them.

This generates an overload on our basic financial modeling system. Money costs used to be the only constituents of financial models. But that wasted our environment. So now social costs are being figured in. In other words, basic survival re-adjustment must take place at the software level before the operation of hardware can be re-channeled.

This means we need a continual understanding of technology, as a model in the public space. But the only on-line models we get are political or financial, like stock market tables and analysis, and political columnists. There isn't one newspaper or TV station which has a daily technology or media column.

We're just now understanding the consequences of hardware, but nobody's anywhere near publicizing potential software fall-out. Twenty-five years too late people are concerned about the effects of broadcast-TV. But no one is looking into the potential of the technologies which will replace it. How are kids going to respond to cable television and videocassettes? Nobody asks, and as a result bad design decision are being made.

CBS, for example, has developed a videocassette system called EVR. Its dominant characteristic is that it has no record mode. To do that CBS had to design a technologically reactionary piece of hardware. Rather than use videotape, which is indigenous to television, CBS chose a film medium because it won't allow you to do your own recording.

That was a deliberate design decision. It was probably motivated by men who think of information as property and thus wanted to minimize copying. The software ramifications are that people can't generate their own information with the system.

Now CBS is pushing very hard to sell EVR to school systems. That means that educational retooling money for those schools which buy EVR will be tied up for years in a system which minimizes student participation, other than to let them choose from a pre-recorded library over which they have no control. Interaction will be minimized when most educators agree that that's precisely the opposite of what is needed. Yet, because they have no sense of software they're going to frustrate their own wishes.

Software fall-out is a consequence of the media's failure to develop a grammar of process. Other than to anthropomorphize it, the popular press has developed no language to personalize the effects of technology. We do not

*Bucky Fuller says I Seem To Be a Verb because that's a process state, one of becoming. A media-background vocabulary stresses similar process words like "happening" and "trashing."

learn about scientists and technologists in the same colorful way we learn about politicians or even athletes.

Yet the bias of technology, not ideology, is where the real power lies in Media-America. Our having no on-line technological analysis is an anti-survival mode.

In that context, Nixon can get away with indignation at the vote against more funds for the SST with the stated reasoning: 1. we've already put lots of money into it, and 2. without it we'll become a "second-rate power."

An overload of attention was paid to the political consequences of his logic, but no one called his bluff on the real, technological importance, the perpetuation of old myth-models: 1. if you've spent money on something, that validates it, so spend more, and 2. technology is basically a power tool.

We have no myth-modeling system for using the future (feedforward). Thus we can't adjust feedback models to attain desirable future states. Without a mythology of the future, "future shock" prevails.

Future shock is nothing more than the experience of product man lost in an age of process because he expects the future to be just a bigger past, not different. When it's not the same as they remember, people get upset.

Automation was one of the great triumphs of the future shock myth. Everyone mistakenly assumed that an age of automation would be exactly the same as it was, only some people would be put out of work by machines. What really happened was that automation created a new industrial climate and with it new, but different jobs. Anyone who's into systems could have told you that when you change a variable, you change the system.

Take a similar example of something that's not yet here but is coming, teleportation, or the physical beaming of matter, which is a potential of the future. When it happens, product man will be amazed. Yet teleportation as process is already here, the future is in the present: I can get my physical body from, say, New York to California in only six hours in teleportation chambers which are within the state-of-the-art, *i.e.,* airplanes. Actual teleportation may take place at the speed of light, and that too is available now via telephone and television and telefax, which can transmit facsimiles of documents over phone wires.

Our lack of feedforward estranges us from ourselves. Important new materials, for example, are produced in a rear-view mirror: soybean food supplements are made to taste and *feel* like chicken or beef; plastic formica is given wood-grain patterns rather than an enhancement of its own material qualities; naugahyde replaces leather but looks the same (have you ever seen a live nauga?); and television sets are sold in French Provincial cabinets, which is very very heavy.

By embodying a nostalgia for the past in new materials we retard an acceptance of the present and by extension, the future, because the future is in the present. Instead we use the future as a place to re-*produce* the present, at best without its fuck-ups.

T.V., technology, and ideology

DOUGLAS KELLNER

▪ Douglas Kellner *is Associate Professor of Philosophy at the University of Texas at Austin. His publications have included critical studies of the Frankfurt School and insightful studies of electronic media and film, articulated from a radical standpoint.*

In this essay he argues that television, far from being a "vast wasteland," is instead a medium that "teems with images conveying . . . values, ideologies, and messages." These images, he suggests, "are coded into narratives that provide a set of American morality plays." Although he draws on an intellectual tradition that includes the works of Marshall McLuhan, Herbert Marcuse, T. W. Adorno, Jurgen Habermas, and others, his essay bristles with concrete examples from the medium itself.

The central cultural role of the broadcast media in advanced capitalist society has changed the very nature and social function of ideology. In this essay, I shall explore some of the contradictions in ideology under the impact of the communications revolution and the broadcast industries. My analysis suggests that the site of ideological transmission for the majority of people in advanced capitalism is shifting from print media to electronic media, especially television. In an earlier paper on ideology, I suggested that when "ism-ideologies" like liberalism and Marxism were institutionalized in capitalist and socialist societies, there was a decline of rationality, and ideology became increasingly fragmented, mythic, and imagistic.[1] Although hegemonic ideology tends to be conservative in that it legitimates dominant institutions, values, and ways of life, it is not devoid of contradictions, as some theories of a monolithic dominant ideology would claim. Instead, in advanced capitalist societies, hegemonic ideology tends to be fractured into various ideological regions, with ideologies of the economy, politics, society, and culture often being contradictory. Hence, in advanced capitalism, there is no one unifying, comprehensive "bourgeois ideology," and hegemonic ideology is consequently saturated with contradictions.

Many radical theories of ideology have neglected the role of mass media images and messages in the transmission of ideology. Although Alvin Gouldner, for instance, is aware of the importance of television and devotes many interesting pages in *The Dialectic of Ideology and Technology* to analyzing both print and electronic media, he does not want to include the images and messages broadcast by the electronic media in the domain of ideology.[2] Louis Althusser highlights the role of the educational system as the primary vehicle of ideology, while ignoring the mass media.[3] These and many other discussions of ideology rely too heavily on an overly linguistic paradigm of ideology that cannot account for recent developments in the electronic media which have produced a new configuration of ideology in advanced capitalism.

To overcome the deficiencies of earlier theories of ideology, I propose that we view ideology as a synthesis of concepts, images, theories, stories, and myths that can take rational, systematic form (in Adam Smith, Locke, Marx, Lenin, and so on) or imagistic, symbolic, and mythical form (for example, in religion or the culture industries). Ideology is often conveyed through images of country and race, class and clan, virginity and chastity, salvation and redemption, individuality and solidarity. Images of success through hard work or lucky breaks, or of happiness through wealth or domestic life, may be effective symbolic devices to propagate ideologies of possessive individualism, rising mobility, or the family. Ideologies are further disseminated in myths of redemption or revolution, which are often embodied in both political rhetoric and popular culture.[4] It is the very combination of rational theory with images and slogans that makes ideology a compelling and powerful social force. Ideology roots its myths in theories while using its theories to generate myths and to supply a rationale for social domination (if the ideology attains hegemony).[5] Hence ideologies have both rational and irrational appeal; they combine rhetoric and logic, concepts and symbols, clear argumentation and manipulative propaganda techniques. They not only employ metaphors (like "master race," "iron curtain," or "new deal") but also use theoretical constructs that lend them rational force.[6]

☐ A longer version of this paper, entitled "T.V. Ideology and Emancipatory Popular Culture," appeared in *Socialist Review*, No. 45(1979), 13-54.

Past theories of ideology have failed to analyze properly the ideological apparatus that transmits ideology.[7] Previously, ideology was transmitted through an elaborate ideological apparatus and set of rituals: military and patriotic pomp and parades; judicial ceremonies and trappings; religious rites; university lecture halls that invested professors with a priest-like aura when they delivered lectures/sermons; political conventions, speeches, and campaigns; etc., etc. After being formulated, usually in print media, ideologies were transmitted through ceremonies and rituals that attractively presented and conveyed them through an ideological apparatus. In this way, ideology in bourgeois society has always been bound up with mythologies and rituals. However, the central role of the broadcast media in advanced capitalism has endowed television and popular culture with the function of ritualistically transmitting mythologies and hegemonic ideology. Hence, there has been an increase in the imagistic, symbolic, and mythical components of ideology in advanced capitalism, and a decrease in rationality from earlier print media forms of ideology, requiring a new theory of ideology and ideological transmission which accounts for these changes.

T.V. AND HEGEMONIC IDEOLOGY

It is part of conventional wisdom that television and the electronic media have provided a new kind of cultural experience and a new symbolic environment that increases the importance of images and decreases the importance of words. Many argue that the television experience is more passive and receptive than reading print. It is claimed that American television has produced a form of cultural experience in which people passively receive ideology that legitimates and naturalizes American society and everyday life. Such a strategy of image production-consumption and cultural domination follows the logic of advanced capitalism as a system of commodity production, manipulated consumption, administration, and social conformity.[8] In the words of Susan Sontag:

> A capitalist society requires a culture based on images. It needs to furnish vast amounts of entertainment in order to stimulate buying and anaesthetize the injuries of class, race, and sex. And it needs to gather unlimited amounts of information, the better to exploit natural resources, increase productivity, keep order, make war, give jobs to bureaucrats. The camera's twin capacities, to subjectivize reality and to objectify it, ideally serve these needs and strengthen them. Cameras define reality in the two ways essential to the workings of an advanced industrial society: as a spectacle (for masses) and as an object of surveillance (for rulers). The production of images also furnishes a ruling ideology. Social change is replaced by a change in images. The freedom to consume a plurality of images and goods is equated with freedom itself. The narrowing of free political choice to free economic consumption requires the unlimited production and consumption of images.[9]

Undoubtedly, American television plays an important role as an instrument of enculturation and social control. What is not yet clear is *how* television conveys hegemonic ideology and induces consent to advanced capitalism. In the following analyses I shall suggest how television images, narrative codes, and mythologies convey hegemonic ideology and legitimate American society. But I am also concerned to show how the images and narratives of American television contain contradictory messages, reproducing the contradictions of advanced capitalist society and ideology. Against leftist manipulation theories, which solely stress television's role as a purveyor of bourgeois ideology, I shall argue that the images and messages of American television contain ideological contradictions and have contradictory social effects. Accordingly, after discussing how television functions as a vehicle of hegemonic ideology, some exploratory analyses of what forms emancipatory popular culture might take in advanced capitalist societies will be proposed.

TELEVISION IMAGES, SYMBOLS, AND PALEOSYMBOLIC SCENES

In addition to photographic images of people, cars, cities, and natural objects, television contains a wealth of symbolic imagery.[10] For example, although the image of the sea has a plurality of meanings in the cultural tradition, the crashing waves after every episode of *Hawaii Five-O* punctuate the torrent of passion unleashed in the story and symbolize wild, chaotic human energies that must be brought under control. Television builds on and uses traditional symbolism, but it also creates its own symbols: Jack Webb's stacatto interrogation procedures, authoritarian personality, and crisp recitation of the facts form a symbol of law and order; the immaculate middle-class homes of the situation comedies or soap operas become symbols of bourgeois domesticity; Ironside, Ben Cartwright, and Walter Cronkite become father symbols; Mary Tyler Moore provides a symbol of the independent working woman; the soap operas generate symbols of stoic endurance through suffering; the game shows exhibit symbols of greed, commodity fetishism, and money-worship; and commercials attempt to create symbols that will manipulate consumer behavior.

Symbolic images endow certain characters or actions with positive moral features and other characters or actions with negative, immoral features, providing positive and negative models of identification. When television symbols become familiar and accepted, they become effective agents of enculturation.[11] All symbols have a historically specific cultural content and serve as instruments of social control. For instance, *Kojak* symbolized triumphant authority, law and order, and a stable set of values in an era of political upheaval, cultural conflict, and confusion. His forceful advocacy of traditional values invested him with symbolic significance; his features crystallized into a symbolic structure, linking his macho personality, bourgeois values, and authoritarian views on law and order.

It is a curious fact of our society that today television is the dominant producer of cultural symbolism. Previously, symbols were often fused into a hierarchy of significance by a well-defined code such as Christianity or Communism, or by a great poet or writer who created an organically consistent symbolic world, but television symbolism is fragmented and contradictory. Nonetheless, it can be argued that television symbolism currently provides the most significant model of behavior and norms of action. Television's imagery is thus prescriptive as well as descriptive, prescribing the proper attitudes toward the police, the property system, and consumer society. Television images assume a normative status: they not only reveal what is happening in the society but also show how one adjusts to the social order. Furthermore, television demonstrates the pain and punishment that result from not adjusting, as well as the machinations of power and authority. Television imagery thus contains a picture of the world and an ethics. It shows us that we should drive a car, have a nice house, wear fashionable clothes, drink, smoke, keep a pretty smile, avoid body odor, and stay in line. The endless repetition of the same images produces a television world where the conventional is the norm and conformity the rule.

There is a class of television symbols that have powerful effects on consciousness and behavior but that are not always readily identifiable or conventionally defined. Building upon Freud's notion of scenic understanding and the concept of paleosymbolism proposed by Habermas and Gouldner, I call these sets of imagery *paleosymbolic*.[12] The word "paleosymbolism" means a sort of "before symbolism" or "underneath symbolism" (in the archaeological sense). Paleosymbols are tied to particular scenes that are charged with drama and emotion. The paleosymbol does not provide or integrate holistic constructs such as the cross, the hammer and sickle, or an aesthetic image, in each of which a wealth of meaning and significance are crystallized in one image; rather, the paleosymbol requires a whole scene in which a positive or negative situation occurs. Freud found that certain scenic images, such as a child being beaten for masturbation, have profound impact on subsequent behavior. The images of these scenes remain as paleosymbols that control behavior—for instance, prohibiting masturbation or producing guilt and perhaps sexual inhibition. Paleosymbols are not subject to conscious scrutiny or control; they are often repressed or closed off from reflection, and they can produce compulsive behavior. Thus Freud believed that scenic understanding was necessary to master scenic images and that, in turn, this mastery could help one to understand what the scenic images signified (resymbolization) and how they influenced behavior.[13]

It is possible that paleosymbolic scenes on television function analogously to the sort of scenic drama described by Freud. Television scenes are charged with emotion, and the empathetic viewer becomes heavily involved in the actions presented. An episode on a television adaptation of Arthur Hailey's novel *The Moneychangers* may illustrate this point. An up-and-coming junior executive is appealingly portrayed by Timothy Bottoms. It is easy to identify with this charming and seemingly honest and courageous figure; he is shown, for example, vigorously defending a Puerto Rican woman accused of embezzling money. It turns out, however, that the young man stole the money himself to support a life-style, including gambling, that far outstripped his income. He is apprehended by a tough black security officer, tries to get away, and is caught and beaten by the black. There are repeated episodes in which we may identify with the young man trying to escape, and then feel pain and defeat when he is caught and beaten. Some of the escape scenes take place at night in alleys, recreating primal scenes of terror and pursuit. Furthermore, to white viewers the fact that the pursuer is black may add to the power of the imagery, building on socially inculcated fear of blacks and reinforcing racism. Hence, the paleosymbolic scene is multidimensional and multifunctional. The paleosymbolism in this example carries the message that crime does not pay and that one should not transcend the bounds of one's income or position. What happens if one transgresses these bounds is dramatically shown, perhaps frightening potential white-collar criminals into remaining within the boundaries of law and order. In case the moral does not sink in during the pursuit scenes, our young embezzler is brutally raped on his first day in prison during a scene remarkable for its explicitness. Paleosymbols on television are especially effective because although one may forget the story, or even the experience of having watched the program, a paleosymbolic image may remain. After watching *The Moneychangers,* one may have a psychological block against stealing or gambling, or a fear of blacks or of going to jail, because of paleosymbolic images in the program, which are multiplied, repeated, and reinforced by other programs and thus serve as powerful vehicles for transmitting morality and ideologies.

Paleosymbolic scenes may play an important role in promoting sexism and racism, in that images of blacks and minorities can be shaped by media images. In the first decades of film, blacks were portrayed stereotypically as comical, eye-rolling, foot-shuffling, drawling imbeciles, usually in the role of servant or clown—precisely the image fitting the white power structure's fantasy that blacks should be kept in their place. Then, during the intense struggles over civil rights, blacks began appearing both as cultural neuters who were integrated into the system and as evil, violent criminals. The latter, negative stereotype was prevalent in television crime dramas that featured black dope dealers, prostitutes, pimps, militants, and vicious killers. These negative images were presented in particularly dramatic scenes that conveyed paleosymbolic images of blacks as evil and dangerous—a subtle mode of manipulation that might have served to shape viewers' images of blacks and to reinforce racism. The viewer was more likely to have a stronger paleosymbolic image of the black junkie shooting up dope and killing a white person to feed his habit than of the good black cop who finally apprehends him, since the paleosymbolic scenes involving the evil black were more charged with emotion and dramatic intensity.

Television has communicated stereotypical views of women as foolish housewives, evil schemers, or voluptuous sex objects. Images of women as scatterbrained clowns (Lucy, Gracie Allen, Edith Bunker) or as adultresses, destroyers, or greedy egotists (in soaps, crime dramas, melodramas, and so on) have been multiplied through paleosymbolic scenes. Although negative images of women certainly have been countered by more positive images (but never by truly emancipatory ones?), the negative images seems to remain most forcefully in the viewer's mind because of its place in the scenic narrative.

In a crime drama, for example, the dramatic climax disclosing the murderess' evil act makes the strongest impression on the viewer. In a soap opera, the paleosymbolic image of adultery and subsequent distress endows the characters and their actions with moral opprobrium that may evoke active dislike. Endlessly multiplied paleosymbolic scenic images of women committing adultery and wreaking havoc through their sexuality help create sexist stereotypes of woman-as-evil, building on and reinforcing mythical images ranging from Eve in the Bible to the seductive schemer on *Days of Our Lives*. These paleosymbolic images may overpower more positive soap opera images of women and create a stereotypical consciousness of women per se as evil (or at least provide a strong negative image of ''evil woman''). Likewise, in a situation comedy, although the mothers/women often manifest admirable traits, these traits are frequently overshadowed by slapstick crescendos in which the woman star (say Lucy, Rhoda, or Alice) is involved in a particularly ludicrous situation, thus promoting a stereoype of woman-as-scatter-brained-dingbat while the paleosymbolic image lingers on.

Television commercials, too, utilize paleosymbolic scenes. They involve scenes that associate desirable objects or situations with the product being huckstered. For instance, one commercial shows Farrah Fawcett with an automobile: Catherine Deneuve purrs and seductively caresses an auto in another paleosymbolic extravaganza. The cars are associated with sexuality, beauty, and highly positive scenic qualities. Other commercials create negative paleosymbolic scenes, with ring-around-the-collar, bad breath, an upset stomach, a headache, or tired blood producing situations of anxiety or pain—which, of course, can be relieved through the miraculous products offered. Television commercials contain in an extremely compressed form a paleosymbolic scenic drama that attempts to invest images with negative or positive qualities in order to influence behavior. Since people often are on guard when they view commercial messages, the makers of commercials find the attempt to manipulate consumer behavior particularly challenging. In fact, perhaps one of the reasons why television entertainment has become an effective socializer is that people are not aware that they are being indoctrinated; they tend to relax their guard and immerse themselves in the program. This makes television images and scenes an increasingly important and powerful instrument of socialization.

The inhabitant of T.V. world lives in a cultural domain of symbolic images and paleosymbolic scenes. The television world does not consist of "pap," nor is it a "vast wasteland"; rather, it teems with images conveying an "impression of reality," as well as values, ideologies, and messages. These images are coded into narratives that provide a set of American morality plays.

TELEVISION SITUATION COMEDIES AND MELODRAMAS AS AMERICAN MORALITY PLAYS

A television situation comedy contains a conflict or problem that is resolved neatly within a preconceived time period. This conflict/resolution model shows that all problems can be solved within the existing society, thus serving to reinforce the dominant morality and institutions. For example, a 1976 episode of *Happy Days* saw the teen hero Fonzie out with an attractive older woman. He learns she is married, and a set of jokes punctuates his moral dilemma. Finally, he sits down with the woman and tells her he has heard she is married. When she says, "Yes, but it's an open marriage," he responds: "No dice. I've got my rules I live by. My values. And they don't include taking what's not mine. You're married. You're someone else's." He gets up, shakes her away, and is immediately surrounded by a flock of attractive (unmarried) girls—a typical comical resolution of an everyday moral conflict that reinforces conventional morality. In a 1977 episode, dealing with the high school graduation of the series' main characters, Fonzie moralizes, "It's not cool to drop out of school," thus serving as a mouthpiece for the dominant ideology of education. In a 1978 episode, when his friend Richie is seriously injured in a motorcycle wreck, the Fonz "reveals his compassion in an emotional prayer for his friend" (*TV Guide* description). With eyes to heaven, he prays, "Hey Sir. He's my best friend. . . . Listen, you help him out and I'll owe you one." Here ideologies of religion and exchange reinforce each other and in turn promote ideologies of middle-class utopia. Television thus becomes not only the opiate of the people but also their active instructor and educator.

Interestingly, the working-class character Fonzie here is used as the spokesman for middle-class morality. The Fonz represents a domestication of the James Dean/Marlon Brando 1950s rebel. Whereas Brando and his gang in *The Wild One* terrorized a small town, and Dean in movies like *Rebel Without a Cause* was a hopeless misfit who often exploded with rage against the stifling conformity and insensitivity of those around him, Fonzie quits his gang the Falcons and comes to live in the garage apartment of the middle-class Cunninghams. In this way, the Other, the Deviant, serves to legitimate the middle-class point of view by submission to the middle-class way of life and espousal of middle-class values. Fonzie's defense of the dominant morality creates a melting pot effect whereby it is implied that all good people share similar values and conform to the same institutions, ideology, and way of life. Hence, *Happy Days* provides a replay of *Ozzie and Harriet* and earlier television family morality plays, with Richie Cunningham starring as David Nelson, the all-American good boy, and Fonzie as the irrepressible Ricky Nelson, whose "hipness" made him an effective salesman for the middle-class outlook.

Television melodrama also serves to propagate conservative ideologies. The world of the melodrama is full of conflict, suffering, and evil.[14] Not only is there an intense conflict between good and evil, but there are also clearly defined ways to depict good and evil. In most television series, the regular characters are good and intruders are evil; fear of the outsider and paranoic adhesion to the tribe are thus promoted. Moral Manicheanism is the metaphysic of narrative melodrama, and teaches that conventional morality is good and its transgression is evil. After a highly emotional conflict, Good triumphs and order is restored to the universe. The heated discussion of television violence and sex fail to note that these elements are the very staple of melodrama, since they heighten emotional impact and dramatize the moralities portrayed. Moralistic opponents of television sex and violence fail to note that it is precisely these features that help to reinforce the moral codes they themselves subscribe to, for transgressors of the established norms are always punished. Television melodrama creates a fantasy of the moral order of the universe and endows its protectors with heroism and its subverters with odium, thus providing a repertoire of morality plays.

For instance, the miniseries *Loose Change* portrayed the fates of three women who went to Berkeley in the 1960s trying to "make it" and "be free" in the 1970s. It reduced the explosive politics of the 1960s to television melodrama, emphasizing the pain and punishment inflicted for not conforming and the rewards for adjusting to the existing order. It presented the 1960s as a disorderly, chaotic period to be eschewed for the order and stability of the present. Television melodrama—like earlier forms of bourgeois melodrama—invariably punishes those who refuse to conform, and the suffering imposed on the characters reinforces the message that only through conformity to the existing society will one avoid pain and tragedy.

Television morality plays present *rituals* that transmit hegemonic ideology.[15] Soap operas and situation comedies celebrate the values of family life and domesticity. The soap operas ritualize the suffering brought about by transgressions of social norms. Situation comedies celebrate the triumphs of the norms, values, and good will that enable one to successfully resolve conflicts. Each program has its own formulas and conventions that are ritualistically repeated in every show. For example, the comedy hit *Three's Company* (which was often number one in the ratings war during the seasons from 1977 to 1979) celebrates the sexual attractions of two single women and a single man who live together. Every episode deals with suggestions of sexual temptation among the three, or their dates, and involves eventual frustration and renunciation. Every episode portrays the sexual advances and frustrations of the landlord's wife and her husband's lack of sexual interest in her (often interpreted as his impotence). The young man pretends to be a homosexual in order to placate the moralistic landlords, and the show contains repeated episodes of feigned homosexuality. These rituals enable the audience to play out fantasies of taboo sexual desire—and of renunciation of such illicit desires. Hence, despite the sexy facade, the program conveys traditional, puritanical ideologies of sexuality.[16]

Other situation comedies allow the audience to experience dramatizations of their own problems with interpersonal relations, work, the family, sexuality, and conflicts of values; they offer opportunities to experience ritualistic solutions to everyday problems that take the form of rites of submission to one's lot and resignation (*Laverne and Shirley, Rhoda,* and *Alice*) or rites of problem-solving through correct activity and change or adjustment (*Happy Days, All in the Family, Maude,* and most Norman Lear sitcoms). Rites of submission often involve the use of a facade of individual self-assertion to promote ideologies of conformity and resignation. For instance, *Laverne and Shirley* provides narratives that inculcate acceptance of miserable working-class labor and social conditions. Although Laverne and Shirley may occasionally rebel and assert themselves against bosses and men, they generally adjust to their life conditions and try to pull through with humor and good-natured resignation. The *Laverne and Shirley* theme song boasts, "We'll do it our way . . . we'll make all our dreams come true . . . we're going to make it anyway," but poor Laverne and Shirley simply espouse middle-class values and dreams, and do it the system's way. They usually fail to realize these dreams, and every episode ends with acceptance of their jobs and social lives. The series tries, however, to make working-class dreams as appealing as possible for Laverne and Shirley and the audience, thus helping working-class people in similar situations to accept their fate with a smile and good cheer.

Action-adventure series provide ideologies of law and order and idealize authority figures. They often champion the individual hero who is dedicated to the preservation of the social order (the cowboy, cop, soldier, detective, and so on), and they legitimate violence used to eliminate evil and protect the established society. These morality tales contain television mythologies.

TELEVISION MYTHOLOGY

Television images and stories produce new mythologies that are operational for the exigencies and problems of everyday life. Stripped of theological mystification, myths are simply stories that instruct and that explain and justify practices and institutions for people in a given society. They have a powerful effect on consciousness and are lived in daily life, shaping and influencing in often unperceived ways our thought and action. Myths deal with the most significant phenomena in human life: they enable people to come to terms with death, violence, love, sex, labor, and social conflict. Myths link together symbols, formula, plot, and characters in a pattern that is conventional, appealing, and gratifying. Joseph Campbell has shown how mythologies all over the world reproduce similar patterns, linking the tale of a hero's journey, quest, and triumphant return with rites of initiation into maturity.[17] It is a mistake to ascribe myth solely to a primitive form of thought that has supposedly been superseded by science (this was the Enlightenment view), for the symbols, thematic patterns, and social functions of myth persist in our society, and they are especially visible in television culture.

Jewett and Lawrence, in their fine book on American popular culture, *The American Monomyth,* have described a recurrent pattern of an "American monomyth." The pattern begins with an Edenic idyll, which is then interrupted by trouble or evil (Indians, rustlers, gangsters, war, monsters, communists, aliens, things from another world).[18] The community, the common people, is powerless to deal with this threat and relies on a hero with superhuman powers (the Westerner, Superman, Supercop, Superscientist, the Bionic Man, or Wonderwoman) to resolve the problem. In the ensuing battle between good and evil, the hero wins, often through macho violence (although redemption can take place through a character of moral purity like Heidi, Shirley Temple, or Mary Tyler Moore; a person of homey wisdom like the Wise Father, Dr. Welby, or Mary Worth; or even a magical animal like Lassie, Old Yeller, or Dumbo). Myths are tales of redemption that show how conflicts are resolved, evil is eliminated, and order is restored. The myth requires the Happy Ending.

Far from having abandoned or transcended mythical culture, Jewett and Lawrence show in convincing detail that much American popular culture is structured by mythical patterns and heroes. Their examinations of *Star Trek, Little House on the Prairie,* Westerns, Walt Disney productions, *Jaws* and other disaster movies, and various superheroes demonstrate that mythical patterns and themes (retribution through blood violence, salvation through superheroes, redemption through mythic powers, and so on) are operative in many major works of popular culture and that these themes are used to purvey American ideology and submission to social authority. *The American Monomyth* traces the historical rise of the myth of the American hero, shows its many manifestations in contemporary popular culture, and suggests that television provides a repertoire of superheroes and mythologies.

At stake here is seeing the role of imagery, symbols, and narratives as vehicles that transmit ideology. Roland Barthes' example in *Mythologies* of the picture of the black soldier saluting the French flag on the cover of *Paris Match* may illustrate this point.[19] Barthes explains how this single image conveys ideologies of the French empire, the integration of blacks, and the honor of the military. Likewise, television commercials propagate ideologies of consumerism; the images of America that appear on television electron-coverage programs propagate ideologies of democracy, freedom, and competition/winning; the images of Carter in the 1976 Presidential campaign communicated ideologies of the country, the small town, the family, and religion; and television sports coverage transmits ideologies of macho heroism, competition, the triumph of winning, and the agony of defeat. Every night on network television the endless stream of crime and violence contains Hobbesian ideologies of human nature and law and order. The quick resolution of problems and conflict in the situation comedies and in other shows promotes an ideology of conflict resolution within the present system.

For instance, the cop show *Starsky and Hutch* deals with the fundamental American conflict between the need for team conformity and the need for individual initiative, between working in a corporate hierarchy and being an individual. Starsky and Hutch are at once both conventional and hip individuals; they do police work and yet wear flashy clothes, drive around in a colorful car, and have lots of good times. They show that it is possible to fit into society and not lose one's individuality. The series thus mythically resolves conflicts between the work ethic and the pleasure ethic, between work or duty and having a good time. Television mythology cuts through the Gordian knot of social contradiction and ties together antitheses into a neat package by speciously resolving conflicts and by enabling individuals to adjust to society and its exigencies.

CONTRADICTORY TELEVISION IMAGES AND MESSAGES

My analysis suggests that the forms of television narratives and codes tend to be conservative and ideological and that the content of American television is saturated with ideology. American television is rigidly divided into well-defined genres that have dominant conventions, formulas, and codes. Situation comedy, melodrama, and action-adventure series utilize conflict-resolution models that show that all problems can be resolved within the current society and way of life. These programs transmit ideologies of power and authority, the family and sexuality, law and order, professionalism and technocracy, and myriad other ideologies that legitimate the established institutions, values, and way of life. But like all ideology in advanced capitalism, hegemonic television ideology is full of contradictions. Television conveys conflicting views of, for example, the family and sexuality, power and authority, that reflect ideological conflicts, contradictions, and social changes in advanced capitalism. For instance, in the 1950s a rather coherent hegemonic ideology of family life dominated television situation comedies like *Father Knows Best, Leave it to Beaver,* and *Ozzie and Harriet.* The middle-class family unit was idealized as the proper locus of sexuality, socialization, domesticity, and authority. A single dominant ideology of the family thus reigned triumphant in the early television world. In the 1960s and 1970s, however, one-parent families began to become more commonplace, as did broken families, and there was an increase in the number of people living alone, such as divorcées and independent working women. All these changes reflected the breakdown of the traditional family and the fracturing of the dominant ideology of the family in American society. Contradictory portrayals of the family and sexuality began to appear on television as conflicting ideologies developed.

Likewise, in the violent world of the television crime drama, the previous "iron fist," authoritarian, law-and-order ideologies of such classics as *Dragnet, The Untouchables,* and *The FBI* were challenged by liberal morality plays like *Mod Squad, Dan August,* and *The Streets of San Francisco,* which featured more liberal, tolerant, "velvet glove" ideologies of the police and power. In the 1970s, new ideologies of the police and authority appeared in the macho individualism of *Starsky and Hutch, Baretta, Serpico,* and other series that featured passionately individualist cops who battled corrupt and inefficient authority and power figures. In contrast to these programs, series like *Ironside, The Rookies,* and *S.W.A.T.* stressed teamwork and the submission of the individual to corporate conformity and hierarchy. Hence, the ideological region of power and authority became saturated with contradictions within hegemonic ideologies of the police and authority in the television cop show.

Moreover, the fact of individual decoding and processing of television images and narratives contains the possibility of the production of contradictory messages and social effects. Individual television viewers are not passive receivers of encoded television messages, as some manipulation theorists posit; rather, individuals process television images according to their interests, life situations, and cultural experiences (of which social class is the determinant factor). Hence, Gerbner and Gross's analysis of television violence[20] indicates that middle-class viewers of such violence tend to be scared into social conformity and paranoic fear of crime, making them suspectible to law-and-order political ideologies, whereas ghetto children, criminals, and persons prone to violent behavior may act out violent or criminal fantasies nurtured by heavy television watching. Likewise, although *Three's Company* and *Charlie's Angels* are encoded as vehicles of puritan sexual morality, they may be decoded as stimulants to promiscuity or sexual fantasy. Although *Laverne and Shirley, Rhoda,* and many situation comedies are encoded as rituals of resignation and acceptance of the status quo, individual images or programs may be processed to promote dissatisfaction or rebellion. Even the most blatantly conservative-hegemonic images and messages may have contradictory social effects. For example, images of consumerism, money, and commodity happiness on commercials, game shows, and other programs may cause rising expectations of happiness through affluence which, if frustrated by social conditions, may breed discontent. Even though news programs and documentaries on the whole attempt to legitimate the political-economic-social system, their images and messages may help lead the viewer to critical views of business, government, or the society. As long as individuals in advanced capitalistic societies are more than totally manipulable robots, they can process television images and messages in ways that may contradict the ideological encoding of the "mind managers."

Furthermore, more progressive content may subvert the rather conservative effects of television codes. The introduction of more topical and controversial content into the form of situation comedy by Norman Lear and his associates helped produce a new type of popular television, as did the introduction of the miniseries and docudrama forms. Even within some of the most conservative television forms, like the crime drama, paleosymbolic scenes and images may contain subversive messages. For instance, *Baretta* and *Starsky and Hutch,* often criticized as among the most macho shows on television, often contain paleosymbolic scenes that are anti-authoritarian, and they and other cop shows frequently have broadcast attacks on the FBI and the CIA in recent years. Hence, although paleosymbolic scenes often convey hegemonic ideologies and promote sexism and racism, they frequently are also double-edged, contradictory, and full of conflicting meanings.

NOTES

[1]See my article "Ideology, Marxism, and Advanced Capitalism," which appeared in *Socialist Review,* 42, November-December 1978, pp. 37-66. Again, I am indebted to the editors of *SR* and the Austin Television Group for criticisms and comments on earlier drafts, especially to Jack Schierenbeck, who suggested many ideas, formulations, and structural-stylistic changes that were incorporated into this essay. I would also like to thank Carolyn Appleton and Marc Silberman for helpful comments and aid in preparing the manuscript.

[2]Alvin W. Gouldner, *The Dialectic of Ideology and Technology* (New York: The Seabury Press, Inc., 1976). Gouldner argues that initially the major symbolic vehicle for ideology was print technology, which was primarily conceptual and relatively rational (p. 167 ff). He perceives that, "in contrast to the conventional printed objects central to ideologies, the modern communication media have greatly intensified the nonlinguistic and iconic component and hence the *multimodal* character of public communication" (p. 168). I reject here Gouldner's disjunction between print media/ideology and electronic media/symbolic (non-ideological) imagery, for in my view the electronic communications revolution has provided powerful new instruments of ideological transmission and social hegemony. Moreover, Gouldner tends to equate print media with rational discourse, and electronic media with relatively irrational symbolic imagery. This notion tends to exaggerate the rationality of print media, which have frequently been, and still are, vehicles of sensationalistic, irrational propaganda, as well as blatantly conservative ideology. Gouldner also fails to discern the relative rationality of the ideology transmitted by television. He notes that the shift from a "newspaper to a television-centered system of communications" leads to "altogether differently structured symbol systems: of analogic rather than digital, of synthetic rather than analytic systems, of occult belief systems, new religious myths" (p. 170). He fails, however, to draw appropriate conclusions, arguing: "In this, however, there is no 'end' to ideology, for it continues among some groups, in some sites, and at some semiotic level, but it ceases to be as important a mode of consciousness of masses; remaining a dominant form of consciousness among *some* elites, ideology loses ground among the masses and lower strata" (p. 179). Against this position, I am arguing that ideology has had a remarkable new impact on individuals in advanced capitalist societies through the technology of the communications revolution, which has transformed the very substance of culture and ideology.

[3]Louis Althusser, "Ideology and Ideological State Apparatuses," in *Lenin and Philosophy* (New York: Monthly Review Press, 1971), pp. 158 ff.

[4]On the relation between ideology, myth, and revolution, see Georges Sorel, *Reflections on Violence* (New York: The Free Press, 1950) and Lewis S. Feuer, *Ideology and the Ideologists* (New York: Harper & Row, Publishers, Inc., 1975). Feuer has written probably the worst book on ideology in recent history. Why? See the next footnote.

[5]I do not agree with Lewis Feuer that ideology is essentially mythical. Against Feuer, Gouldner's emphasis on the relative rationality of ideological discourse is clearly correct. Feuer's strategy is to claim that all ideological discourse is a form of cognitive pathology in order to debunk, above all, Marxism. Feuer completely neglects to discuss ideology as hegemony, and fails to see that the "science" that he counterposes to ideology itself takes ideological forms. (See Lewis S. Feuer, *Ideology and the Ideologists.*) Feuer needs a good lecture on historical specificity to discern the differences between the Old Testament, Marx, fascism, and the New Left, which he lumps together in one amorphous category of ideology that he assimilates into mythology. Surely this kind of historical storytelling is itself a mythology that is fully ideological.

[6]In a fine analysis, "The Metaphoricality of Marxism" (*Theory and Society,* 1, No. 1 [1974], pp. 387-414), Alvin Gouldner suggests that much of Marxism's appeal, power, and success lies in the attractiveness of its metaphors: socialism and the proletariat, bondage and revolt, alienation and its overcoming, class struggle and community. Extending this line of analysis, one could show that all ideologies owe much of their appeal to their metaphoricality, their symbols and images.

[7]Louis Althusser, "Ideology and Ideological State Apparatuses." Althusser really does not analyze the "ideological apparatus" here and falsely assumes a monolithic "state ideological apparatus." In fact the ideological apparatuses are not all state controlled and are fractured into various agents of socialization, social practice, and ritual that are full of contradictions. For a critique of Althusser's analysis of ideology see "Ideology, Marxism, and Advanced Capitalism."

[8]On advanced capitalism see Herbert Marcuse, *One-Dimensional Man* (Boston: Beacon Press, 1964); Henri

Lefebvre, *Everyday Life in the Modern World* (New York: Harper & Row, Publishers, Inc., 1971); Jürgen Habermas, *Legitimation Crisis* (Boston: Beacon Press, 1975); and Ernest Mandel, *Late Capitalism* (London: New Left Books, 1975).

[9]Susan Sontag, *On Photography* (New York: Farrar, Straus & Giroux, Inc., 1977), pp. 178-9.

[10]For a more detailed articulation of my theory of television images, see my forthcoming paper "Television Images, Codes, and Messages."

[11]For a discussion of the theory of socialization assumed here see my forthcoming article "Network and American Capitalism," which will appear in *Theory and Society* in 1980.

[12]On the concept of paleosymbolism see Jürgen Habermas, "Toward a Theory of Communicative Competence," in *Recent Sociology*, ed. Hans Dreitzel (New York: Macmillan, Inc., 1970), and Alvin Gouldner, *The Dialectic of Ideology and Technology* (New York: The Seabury Press, Inc., 1976). Habermas and Gouldner claim that the concept of paleosymbolism derives from Freud, but they provide no source references, and I have not been able to find it in Freud's writings. In any case, the concept is rooted in Freud's notion of "scenic understanding" (see the Habermas source) and is consistent with Freud's use of archaeological metaphors for the topological structures of the mind. See, for example, Freud's *Civilization and Its Discontents* (New York: W. W. Norton & Co., Inc., 1962), pp. 16 ff. The term "paleosymbol" thus refers in my usage to scenic imagery that remains in the viewer's mind but that is not mediated by concepts or interpretive understanding (in other words, these images are prelinguistic or nonconceptual).

[13]See Habermas, "Toward a Theory of Communicative Competence;" Freud, S., *The Interpretation of Dreams*, trans. and ed. James Strackey (New York: Basic Books, Inc., Publishers, 1955); and Alfred Lorenzer, "Symbol and Stereotypes," in *Critical Sociology*, ed. Paul Connerton (New York: Penguin Books, 1976).

[14]On the historical background of the concept of melodrama, see James L. Smith, *Melodrama* (London: Methuen & Co. Ltd., 1973).

[15]On television ritual see my more detailed study in "Network and American Capitalism" (forthcoming, 1980, *Theory and Society*); on hegemonic ideology, see "Ideology, Marxism, and Advanced Capitalism."

[16]Even more curious is how the new sexploitation series promote puritan morality. The stars of *Charlie's Angels* rarely, if ever, have lovers or erotic relationships, and the women on the new "t and a" series *(Flying High, The The American Girls)* make a point of articulating their old-fashioned morality. The fact is that despite the increasing sexual references, jokes, and innuendoes, there is no real eroticism on television, and television comedy and melodrama still, for the most part, push the old, obsolete puritan morality.

[17]Joseph Campbell, *Hero with a Thousand Faces* (New York: The New American Library, Inc., 1956).

[18]Robert Jewett and John Lawrence, *The American Monomyth* (Garden City, N.Y.: Anchor Books, 1977).

[19]Roland Barthes, *Mythologies* (New York: Hill & Wang, 1972), pp. 109 ff.

[20]George Gerbner and Larry Gross, "Living with Television: The Violence Profile," *Journal of Communications*, 26, No. 2 (Spring, 1976).

Television: the new state religion?

GEORGE GERBNER

■ George Gerbner *is Dean of the Annenberg School of Communications at the University of Pennsylvania and a well-known and highly regarded communications theorist.*

Professor Gerbner here argues that the changes which have occured in the structure of mass media as a result of the transition from print technology to electronic technology have begun to alter our social organizations. Whereas print media fragmented, electronic media, especially television, assembles heterogeneous audiences by means of imagery and ritual, a phenomenon not unlike what has traditionally been known as religion.

Both classical electoral and classical Marxist theories of government are based on assumptions rooted in cultural developments of the eighteenth and nineteenth centuries. These developments also gave rise to mass communications and eventually to research on mass communications. In the past few decades, however, rapidly accumulating changes brought about a profound transformation of the cultural conditions on which modern theories of government and of mass communications rest. That change presents an historic challenge to these theories and to scientific workers concerned with these theories. I would like to sketch the nature of that challenge and to make a few tentative suggestions about the tasks ahead.

Human consciousness seems to differ from that of other animals chiefly in that humans experience reality in a symbolic context. Human consciousness is a fabric of images and messages drawn from those towering symbolic structures of a culture that express and regulate the relationships of a social system. When those

□ Reprinted from *Et cetera* Vol. XXXIV, No. 2 (1977) by permission of the International Society for General Semantics.

relationships change, sooner or later the cultural patterns also change to express and maintain the new social order.

For most of humankind's existence, these systems of society and culture changed very slowly and usually under the impact of a collapse or invasion. The long-enduring, face-to-face, pre-industrial, pre-literate cultural patterns, relatively isolated from each other, encompassed most of the story-telling, and the rituals, art, science, statecraft, and celebrations of the tribe or larger community. They explained over and over again the nature of the universe and the meaning of life. Their repetitive patterns, memorized incantations, popular sayings, and stories demonstrated the values, roles, productive tasks, and power relationships of society. Children were born into them, old men and women died to their ministrations, and both rulers and the ruled acted out their respective roles according to their tenets. These organically integrated symbolic patterns permeated the life space of every member of the community. Non-selective participation of all in the same symbolic world generated mistrust of strangers, the quest for security through protection by the powerful, and a sense of apprehension of and resistance to change. Conflicts of interest were submerged and dissent suppressed in the interests of what to most people seemed to be the only possible design for life.

All of this changed when the industrial revolution altered the contours of power and the structure of society. The extension of mass production into symbol-making correspondingly altered the symbolic context of consciousness and created cultural conditions necessary for the rise of modern theories of government.

One of the first industrial products was the printed book. Printing made it possible to relieve memory of its formula-bound burdens and opened the way to the endless accumulation of information and innovation. "Packaged knowledge" (the Book) could be given directly to individuals, bypassing its previously all-powerful dispensers, and could cross the old boundaries of status and community. Images and messages could now be used *selectively*. They could be chosen to express and advance individual and group interests. Printed stories—broadsides, crime and news, mercantile intelligence, romantic novels—could now speak selectively to different groups in the population and explain the newly differentiated social relationships which emerged from the industrial revolution. Print made it possible for the newly differentiated consciousness to spread beyond the limiting confines of face-to-face communication. Selectivity of symbolic participation was the prerequisite to the differentiation of consciousness among class and other interest groups within large and heterogeneous societies.

Publics are created and maintained through *publication*. Electoral theories of government are predicated upon the assumption of cultural conditions in which each public can produce and select information suited to the advancement of its own interests. Representatives of those interests are then supposed to formulate laws and administer policies that orchestrate different group interests on behalf of society as a whole.

Marxist-Leninist theories of government similarly (albeit more implicitly) assume cultural conditions that permit selectivity of symbolic production and participation, and thus differentiation of consciousness along class lines. Lenin's characterization of the press as collective organizer and mobilizer assumes (not unlike advertisers do in capitalist countries) that the major mass media are the cultural organs of the groups that own and operate (or sponsor) them. Only in that way could working class organizations (or business corporations) produce ideologically coherent and autonomous symbol systems for their publics.

Before these theories of government came to full fruition, the cultural conditions upon which they were explicitly or implicitly based began to change. Private corporate organizations grew to the size and power of many governments. The increasingly massive mass media became their cultural arms and the First Amendment their shield. Commercial pressures made the service of many small, poor, or dissenting publics impractical. Public relations replaced the autonomous aggregation of many publics. Public opinion became the published opinions of cross-sections of atomized individuals rather than a differentiated mosaic reflecting the composite of organized publics, each conscious of its own interest.

In the young socialist countries and People's Democracies, mass media became centralized organs of revolutionary establishments. Their governing responsibilities made it difficult to cultivate a distinctly working class consciousness and to institutionalize the critical functions of the press.

These problems and difficulties arose under

essentially print-based cultural conditions. But in the past few decades even those conditions began to change.

The harbinger of that change is television. The special characteristics of television set it apart from other mass media to such an extent that it is misleading to think of it in the same terms or to research it in the same terms. Furthermore, these special characteristics are only the forerunners of the prospect of an all-electronic organically composed and orchestrated total symbolic environment.

What are these special characteristics of television? My observations are based primarily on our research and experience in the United States. We do not yet know to what extent they are applicable to other countries. (That, I think, should be an early task for communications research to discover.)

1. Television consumes more time and attention of more people than all other media and leisure time activities combined. The television set is on for six hours and fifteen minutes a day in the average American home, and its sounds and images now fill the living space and symbolic world of most Americans.
2. Unlike the other media, you do not have to wait for, plan for, go out to, or seek out television. It comes to you directly at home and is there all the time. It has become a member of the family, telling its stories patiently, compellingly, untiringly. Few parents, teachers, or priests can compete with its vivid demonstrations of what people of all kinds are like and how society works.
3. Just as television requires no mobility, it requires no literacy. In fact, it shows and tells about the world to the less educated and the non-reader—those who have never before shared the culture of the literate—with special authority and force. Television now informs most people in the United States—many of its viewers simply do not read—and much of its information comes from what is called entertainment. As in ancient times of great rituals, festivals, and circuses, the information-poor are again royally entertained by the organic symbolic patterns informing those who do not seek information.
4. These organic patterns have to be seen—and analyzed—as total systems. The content differentiations of the print era, where there were sharp distinctions between information (news) and entertainment (drama, etc.) or fiction and documentary or other genres, no longer apply. Besides, viewers typically select not programs but hours of the day and watch whatever is on during those hours. Unlike books, newspapers, magazines, or movies, television's content and effects do not depend on individually crafted and selected works, stories, etc. Assembly-line production fills total programming formulas whose structure encompasses all groups but serves one overall perspective. Story-telling (drama and legendary) is at the heart of this—as of any other—symbol system. "Real-life" demonstrations of the same value structure, as in television news, provide verisimilitude and "documentary" confirmation to the mythological world of television. All types of programming within the program structure complement and reinforce one another. It makes no sense to study the content or impact of one type of program in isolation from the others. The same viewers watch them all; the total system as a whole is absorbed into the mainstream of common consciousness.
5. For the first time since the pre-industrial age, or perhaps in all of history, there is little age-grading or separation of the symbolic materials that socialize members into the community. Television is truly a cradle-to-grave experience. Infants are born into a television home and learn from its sounds and images before they can speak, let alone read. By the time they reach school age they will have spent more hours with television than they would spend in a college classroom. At the far end of the life cycle, old people, and most institutionalized populations, are almost totally dependent on television for regular "human" contact and engagement in the larger world. Only a minority of children and older age groups watch the few programs (none in "prime time") especially designed for them. Unlike other media, television tells its stories to children, parents, and grandparents, all at the same time.
6. Television is essentially in the business of assembling heterogeneous audiences and selling their time to advertisers or other institutional sponsors. The audiences include all age, sex, ethnic, racial, and other interest groups. They are all exposed to the same repetitive messages conveying the largest common denominator of values and conduct

in society. Minority groups see their own image shaped by the dominant interests of the larger culture. This means the dissolution of the concept of autonomous publics and of any authentic group or class consciousness. Television provides an organically related synthetic symbolic structure which once again presents a total world of meanings for all. It is related to the State as only the church was in ancient times.

All this adds up to a non-selectively used cultural pattern which can no longer serve the tasks of cultivating selective and differentiated group, class, or other public consciousness. The pattern is formula-bound, ritualistic, repetitive. It thrives on novelty but is resistant to change, and it cultivates resistance to change. In that, too, television's social symbolic functions resemble pre-industrial religions more than they do the media that preceded it. The process has tremendous popular mobilizing power which holds the least informed and least educated most in its spell. Results of our research (reported under the title "Living With Television: The Violence Profile" in the Spring 1976 issue of the *Journal of Communication*) indicate that television viewing tends to cultivate its own particular outlook on social reality even among the well educated and traditionally "elite" groups.

Heavy viewers of television are more apprehensive, anxious, and mistrustful of others than light viewers in the same age, sex, and educational groups. The fear that viewing American television seems to generate, the consequent quest for security and protection by the authorities, the effective dissolution of autonomous publics, and the ease with which credible threats and scares can be used (or provoked) to justify almost any policy create a fundamentally new cultural situation. The new conditions of synthetic consciousness-making pose new problems, difficulties, and challenges for those who wish to realistically analyze or guide public understanding of society.

Researchers and scholars of communication and culture should now devote major attention to long-range cross-cultural comparative media studies that investigate the policies, processes, and consequences of the mass-production of major symbol systems in light of the respective structures and aims of different social systems. Do media really do what they are designed to do according to the theories governing (or used to explain) the societies in which they exist? What are the differences and similarities among them? What are the cultural and human consequences of the international exchange of media materials? What are the effects of changing cultural, technological, and institutional conditions upon the social functions of media, particularly television? What are the new organizational, professional, artistic, and educational requirements for the effective fulfillment of societal goals in different cultural and social systems? And, finally, how can liberation from the age-old bonds of humankind lead to cultural conditions that enrich rather than limit visions of further options and possibilities?

These are broad and difficult tasks but we can at least begin to tackle them. Much depends on the success of the effort.

Growing up on television

ROGER ROSENBLATT

■ Roger Rosenblatt *is literary editor of the* New Republic *and the author of* Black Fiction *and numerous articles. In this entertaining and insightful essay Rosenblatt argues that it is not in the interest of commercial television as a medium to address or to portray the reasonable, responsible, experienced, mature individuals whom we normally call "adults." Rather, the medium presents individuals who are whimsical, who are forced to make decisions in the absence of adequate reasons, and who soar in realms of fantasy. In short, television presents as models people who have never grown up.*

Adulthood on American television is represented most often and most clearly on family shows. These continuing stories about the adventures of families became popular in the fifties and have grown more so since (I can count thirty-five). Most have been light comedies *(The Stu Erwin Show, The Life of Reilly, Life with Father, Make Room for Daddy, Father Knows Best, Ozzie and Harriet)* or farces *(The Munsters, The Addams Family).* A few have been gentle melodramas *(One Man's Family, The Waltons, Little House on the Prairie).* Fewer still, such as *All in the Family* and *The Jeffersons*—comedies essentially—have reached for a deeper nature by dealing with real conflicts. The titles of these shows have generally shifted from a father-centered conception of the family to a community operation, but that, I think, represents social appeasement more than genuine change. In fact, the television family has been an unusually consistent institution, far more stable than its real-life counterpart.

One reason for this stability is that families on television are not families with special coherences; they consist of interchangeable parts. Family members share the same surnames, live in the same house or apartment, or, more specifically, the same combination living room and kitchen, and they hang around together and recognize each other. Ordinarily families consist of one father, one mother, and some children who confront problems of the magnitude of surprise parties, garbled telegrams, overcooked chickens, high-school proms, and driving lessons. We could trade the mother of the Andersons on *Father Knows Best* for the mother of the Waltons without ruffling the chickens.

Since the family arrangement is so basic to television we could even exchange certain family members with characters on non-family shows. The relationships among the detectives on *Hawaii Five-O* or the paramedics on *Emergency* are no less familial than those among the *Partridge Family* members or those in *Family Affair.* If Kojak were to marry Mary Tyler Moore and raise the Brady Bunch, we would still get the same conception of problems and solutions as each holds alone: every week Kojak, Mary, and the children would rid Minneapolis of ethnic abuses.

That conception of problems and solutions follows the same formula for all family shows: a problem is made evident within the first three minutes, usually as a result of some new direction or decision of the family (a picnic, a new car); one by one, the family becomes aware of the problem so that each member can adopt the characteristic stance which he or she adopts for all occasions; the problem then serves as a catalyst for the individual performances that occupy the rest of the show; the problem is "solved," gotten rid of, at the wire by means of a telephone call or some other *deus ex machina*. The show ends with everybody laughing. (Detective shows use much the same formula, incidentally, substituting murders for picnics and stool pigeons for the *deus.*)

In so standardized a situation it would be hard to pick out the grown-ups were it not for certain allegorical assignments. When a problem strikes *Father Knows Best,* for example, the family arranges itself around it, responding as humors: Jim is Recalcitrant Wisdom, sought only (yet always sought eventually) as a last resort; his wife Margaret is Anxiety; elder daughter Betty is Panic or Extreme Emotion; her brother Bud, Dullness; and little Cathy is Childlike Sensitivity, providing an unspoiled account and "special" perception of the problem as it develops. On *All in the Family* the Bunkers' childishness and panic reside in the

□ Reprinted by permission of *Daedalus,* Journal of the American Academy of Arts and Sciences, Boston, Massachusetts. Fall 1976, *American Civilization: New Perspectives.*

father; therefore, when a problem (abortion, death, racism) hits the Bunkers, the mental process of dealing with it begins out of control in the one member of the family who is supposed to represent order and authority. Thanks to Edith, whose constitutional bewilderment serves as Right Instinct, things eventually calm down on *All in the Family,* but are rarely resolved or set straight. A bunker is a bulwark, a fortification; problems attack the Bunkers at a terrific rate, but ricochet just as quickly, denting nothing.

Jim Anderson and Archie Bunker seem very different kinds of adults, and in more than stylistic ways they are. It is better to dwell in the palace of wisdom than panic, and so better to see Anderson as a model of adulthood than Bunker. Curiously, however, neither father is necessarily identifiable as an adult. In fact nobody in the allegorical arrangements of TV families is identifiable as a child or adult except by physical size, dress, and age. Experience, one thing that might effect such identification is not called upon. Anderson's wisdom usually does not derive from experience but from inborn perspicacity. Bunker's terrors rarely result from experience, but burst upon a problem as if one like it had never been seen before. There are more wise than panicky grown-ups on family shows, but they simply function as the wisest minds in groups of contemporaries. Their wisdom is separable from their adulthood, and so says nothing for adulthood generally, suggests no particular advantage to growing up.

This, I believe, is one of the ideas that television has about adulthood—that adulthood is, may be, perhaps should be, unconnected to experience, and exists as an admirable state of mind (when shown as admirable) insofar as it exhibits the most general virtues. Experience is not evil, merely unnecessary for growing up, and thus memory, the medium of experience, is unusable. If this proves true, then what appears to be an expression of approval or encouragement on television's part, in usually assigning wisdom to grown-ups, may in fact be the opposite; because to praise something in terms applied to many things is to exaggerate the possibility that those same terms fit other things more appropriately, leaving the original object peculiarly vulnerable by comparison. The question ought to be raised here, however, as to whether there *are* such things as ideas in television; or are the notions we pick out merely ephemera tied to commercial interests, which vary with those interests?

Because it emerges from a single history, because nearly all its parts, at least on commercial television, operate on the same standards of success (sponsors, ratings), because its technical methods are the same for all stations and what variety exists within broadcast companies is still the same variety for all such companies; because it appeals to and reaches the same audience in relatively similar situations, and because its opinion of that audience is continuous and stable, television must, I think, be regarded as a world *in toto*. That is, whatever differences in tone or invention there are among situation comedies, mysteries, quiz shows, talk shows, et al., none of these "genres" is basically separable from the others. Among all are the samenesses of the medium fed by every show's active awareness of every other show. They function alertly in William James's systemic universe: cooperate, reinforce themselves, create, sustain, and eventually proselytize for a single vision of most things.

Adulthood, then, as one of the ideas set forth by television, is not presented solely in one type of show, but in all types, even—perhaps particularly—in shows that have nothing to say about grown-ups, but simply show them in action. Quiz shows, for example, usually have grown-ups playing what appear to be the simplest kinds of children's games, which range from pie-tossing *(People Are Funny)* and bedroom olympics *(Beat the Clock)* to versions of spelling and math bees on shows that test contestants on information. *(It Pays to Be Ignorant,* both a radio and TV show, was an interesting perversion of these information quizzes.) Yet adults are not, in fact, behaving as children on these shows, nor are the shows trying to bring out the child in them. The quiz show takes grown-ups as they are, and takes them seriously. Having detected certain weaknesses in our conception of adulthood, which sometimes means our self-esteem, it goes to work on them.

Let's Make A Deal is the most popular American quiz show, and probably the most sinister. Its audience is divided into two sections. There is the regular audience, and in front of it, in a roped off portion, is the participating audience, would-be contestants who have come dressed as animals or in other outlandish costumes and who forcefully vie for the attention of the master of ceremonies. The master of ceremonies patrols the aisle, choosing players at random.

Only a few can participate, so every person in costume continually screams and flails his arms in order to attract the emcee as he makes his selections. Those whom he chooses to "deal" with can barely contain their excitement and have to be forcibly quieted before the show can continue.

What the master of ceremonies offers these people is a choice between unknowns. He tells somebody dressed as Mother Goose, for instance, that he may have whatever is in this box (which a professionally delighted assistant produces) or whatever is behind that curtain (to be drawn apart by a long-legged girl). The contestant is baffled, but he chooses the box. Before he has a chance to open it, however, the master of ceremonies says, I'll give you five hundred dollars for whatever is in that box. The contestant hesitates. Six hundred. The regular audience shouts, "keep it!," "sell!" Seven-fifty. The contestant decides to keep the box. When the lid is lifted there may be jewelry on display worth two thousand dollars, and the contestant howls and whoops. If the contestant is a woman, she flings her arms around the emcee's neck and kisses him powerfully. A man usually jumps up and down like a great cartoon frog. Or, there may be a sandwich in the box, whereupon everybody guffaws and the contestant collapses in disappointment. Sometimes there is a live animal behind one of the curtains, and, as the audience roars, a look of genuine terror comes into the contestant's eyes as he or she not only deals with the despair of losing, but with the possibility of taking home a pig or a mule.

The excitement of the show derives not from the price or size of the prizes available, but from the act of depriving people of their ability to make informed decisions, in other words, to reason. To deprive them of their sanity at the outset, the show's producers insist that the contestants disguise their actual appearances in order to qualify for losing their reason. When the "deals" are presented, there is nothing for these people to go on but bare intuition, tortured and prodded by the rest of the audience shouting "the curtain," "the box," "the money." In the center of the mayhem, controlling it, is the master of ceremonies, offering people money for things which they cannot know the value of, things which they cannot see, distributing punishments and rewards as capriciously as the devil he is.

What *Let's Make A Deal* does for, or to, the reasoning process, a new quiz show, *The Neighbors,* does for civility. *The Neighbors* enlists five participants, all women, who live in the same neighborhood and know each other very well. These women are seated in rows, three over two. The two are the principal contestants, but all five get into the act. They are positioned on a white fake-filigreed porch. The floor is green, to suggest a village sward, and hedges and potted geraniums are placed about to suggest a quiet suburban neighborhood.

As with *Let's Make A Deal,* the audience is in constant frenzy, here not goaded by the emcee who affects the studied quiet of a sermonette preacher (and in fact plays the role of a trouble-making parson), but by the questions he puts. We asked your neighbors, he tells the two principals, which one of you wears short shorts in the neighborhood to attract other women's husbands? A collective gasp and false hilarity are followed by each of the pair guessing that the majority of the three neighbors picked herself or the other one, justifying her guesses either in terms of the character deficiences of the other principal or the maliciousness of the three neighbors. If a woman guesses herself and is right, she "wins" and is mortified. In the process, at least one of the principals will insult at least one of the three in the upper row. The game continues.

In the second stage of the show, the emcee tells the two principals that one of their neighbors has said something vicious about them, and they must try to guess who. Each of the three neighbors now competes to convince first one and then the other of the principals that it was *she* who made the slur. They do this by revealing secrets or confidences which until that moment on national television were shared only with the contestant in question. Each neighbor is encouraged by the fifty dollars she will receive if one of the principals picks her incorrectly. At various intervals in the program, prizes have been mentioned—such as large supplies of La Choy Chinese food and four gallons of paint—that will go to all the contestants. But what the emcee calls "the fantastic grand prize" will not be revealed until the end. As in the first stage of the show, the principals justify their guesses of one of the three neighbors by emphasizing that particular one's unsavoriness, but they will also cover their bets by generalizing unfavorably on the trio as a whole.

Stage three of the show has the emcee tell

the principals that all three of the neighbors are unanimous in the opinion that they are . . . and here the comments incorporate vanity, spite, noseyness, selfishness, ungratefulness, corpulence, snobbery, and so forth. Here finally is the full weight of their neighbors' judgments. The two principals guess themselves or each other as the insults are enumerated, the prize money increasing with each question. Eventually one wins the grand prize—a kitchen—and the five neighbors, having destroyed their neighborhood, meet for a good hard hug as the show ends.

Civility, which is a grown-up attribute, is deliberately undermined on *The Neighbors,* as indeed it is on other quiz shows and other shows generally on television. Reason is undermined on *Let's Make A Deal* and elsewhere. Memory and experience are undermined on family shows. And the total effect of all such underminings is the undermining of authority, which may be the central adult attribute. Oddly, the only voices of authority undisputed within television are the newsmen—oddly, because they are often the voices least trusted by us. Yet even on the news there is a certain undermining of adulthood, not by overt derision, but by the general cultural theater in which almost all news programs participate.

There are no more grown-up-looking or -sounding people on television than newscasters. John Chancellor, David Brinkley, Walter Cronkite, Howard K. Smith, Harry Reasoner—all are about as adult, in the theatrical sense, as one can get: clear and forceful in presentation; well barbered and tailored in appearance; deliberate and settled in manner; non-panicky in tone; emotionally stable almost, but not quite, to a fault. In these terms, if Eric Sevareid were any more grown-up, he would have to deliver his speculations from Shangri-la, and he may yet. These people have nothing quite so much in common with anything as with each other. Differences among them are detectable, to be sure, but television does not seek to point out their differences, certainly not to elaborate on them, because it has decided that we do not simply expect the news of the news, but the world of news as well, a nightly *Front Page*.

We have two distinct yet cooperative images of newsmen in our culture, images established and reinforced in the movies more than anywhere else. One is the adventurous reporter—hatted, trench-coated, either hard-drinking and -loving (to be played by James Cagney) or belligerently innocent, awkward yet cocky, seeking only the whole truth (to be played by Joel McCrea—or now is it Robert Redford?); contemptuous of money, pure of motive, defender of underdogs—the discoverer, a man who dares to go where no one else has been permitted, or has thought, to go before. His editor is the other image: snorting, stomping, wild yet established, the authority (hirer and firer) on which the reporter depends, at once romantic, envisioning scoops and stopping the presses, and suspicious or afraid of the truth as well—unlike his colleagues, he has come to believe in City Hall and political realities. In the end, nevertheless, he is as one with his reporter-antagonist as they collaborate to set history right for a moment. Afterwards, we trust, they revert to their separate barkings and ravings.

In television these two images combine into a single character—both Perry White and Superman—who is a sort of polished version of each of his contributive actors. Often he is posisitioned in a simulated newsroom lined with desks at which simulated reporters sit, busy with simulated copy. As he talks, people walk on and off the set carrying messages and bulletins. Typewriters clatter in the background. The newscaster appears to have just looked up from his desk in order to tell the latest story. Or, as with Chancellor, one only senses the background activity. The newscaster is trusted by his reporter-colleagues—they report in before our eyes, courteous and efficient. We feel that they feel that the newscaster is one of them, the grounded former ace. He is also the editor, directing the sequence of presentations. No antagonism here; no hysterics or passion either. The newscaster has been neutralized into a model of both decorum and sensibility (to be played by Frederick March).

The effect of this deliberate theatricality is to separate the news from the character who speaks it. Whatever we may think of the content or order of the newscaster's presentations, he remains in the clear, which of course allows him longevity in his position. Yet the clear in which he remains, bright and mellow as it is, is not adulthood in real or ideal terms, because adulthood is not a form of play-acting in which a man is dissociated from his words—quite the opposite. So even here, with newscasters looking and sounding as grown up as can be, adulthood is misrepresented as a pose, a set of trappings through which information may be conveyed without concern for integrity or emotional and intellectual responsibility. (This is not to say that Chancellor, Cronkite, et al., are without thoughts or feelings about the news; only that the mold they fill makes their minds impertinent.) The result, curiously, is that television newsmen have come to represent not a celebration of stature, another adult attribute, but a mockery of it.

More curiously, the only place on television where some of the attributes of adulthood, or ideal adulthood, are realized—helpfulness, guidance, gentleness, self-sacrifice—are on children's shows. Saturday-morning cartoons generally recreate the old chase-and-miss mayhem of movie cartoons; even today Tweety-Bird and Road Runner still watch their pursuers rolled like dough under boulders. And there are other "adventure" cartoons that merely make cartoons of the real shows on television (a gratuitous art). But *Sesame Street, Captain Kangaroo,* and *Mr. Rogers* do in fact show adults behaving thoughtfully and compassionately, often creatively. The trouble, of course, is that the adults on these shows are depicted either as wise oversize children or as those living their lives solely *for* children, or both. No child measures his parents by the adults on these programs because no parent spends his day as Mr. Rogers does, except Mr. Rogers.

Where television invents its characters and situations, adulthood is chided, scoffed at, ignored, by-passed, and occasionally obliterated. What could be the purpose (assuming there is one) of these underminings? Not satire, certainly—the destruction is too scattergun, and there is no moral position or corrective imagination behind it. In many ways the depiction of adulthood seems merely the revenge of a peevish child, albeit a Gargantuan one: a free-wheeling diminution of adult stature for a laugh or for the hell of it. If there is a scheme here, it need not be intentional. As one instrument of popular culture, television contains the properties of popular culture as a whole, and its collective attitude toward anything may be grounded in the general ways in which popular culture works.

Conventional wisdom has it that contemporary life changes so rapidly that we are unable both to see its shapes and to apply standards of value to the whole or its parts. Since popular culture carries the signs of the times, people who exclaim over the transitoriness or shakiness of modern man ordinarily use elements of popular culture as referents. Look closely, however, and the signs read differently. Rather than disowning the past (history and traditions), popular culture persistently resurrects and reinforces it. But it works by sleight of hand. It operates on two levels of tradition simultaneously, and it uses one to cover the other and, in many ways and for some important reasons, to disguise its existence.

The more obvious level of tradition in popular culture is the one it creates for itself. This is tradition born of mounting conventions. In television, for example, it takes shape by building program upon program, format upon format, character role upon character role, situation upon situation down to things as small as lines of dialogue and gestures—all continually repeating themselves as soon as a certain receptivity on our part has been perceived. All forms of "newness" and "change" rely upon these conventions both as the standard against which their apparent novelty may be measured and as the future repository for our quick assimilations. The "new" becomes the "old" in a flash, as it also becomes part of, and strengthens, this level of tradition.

This process of assimilation, seen by some observers as genuine change, is like fast-falling rain which creates its own water surface on the ground. It is *not* the ground but a *surface on* the ground. We, in turn, move like cars in a rainstorm on that surface, at a high speed often determined by the rain itself. The surface tradition built by popular culture, because it is ephemeral in its parts, only exists as a total body by means of the rapidity of additional elements, reinforcements, to it. This tradition is formed not necessarily by merit on objective standards of excellence (although it may be so), but by repetition.

Advertising on television uses this sort of repetition not only to sell products, but also to sell people selling products. Actors who perform in commercials very often perform in regular programs as well, and in similar roles—the know-it-all mother on the situation comedy *Rhoda* pushes Bounty paper towels; the authoritative "professional dishwasher" who advocates Ajax liquid once had his own detective series. The effect of these interchanges is to turn the actor himself into a product that becomes as familiar as the thing he hawks. The mere sight of him in any context elicits visions of clean counters and shiny dishes and, naturally, sustains a terrifyingly smooth transition between advertising and programming.

While this is happening, we are told that it is not. Instead, we are assured that everything presented to us by various media is new and startling, to which announcements we willingly suspend disbelief. The reason we are told that everything is new is simple: those formats or character roles of which we recently grew tired may easily be refurbished, and we, seeing them again and again, may have our powers of discrimination worn down accordingly. Why we suspend disbelief, however, is a much deeper problem. Part of the answer, I believe, lies in the fact that popular culture makes us just as happy as we wish to be, no more, no less. The other part is our abhorrence, perhaps fear, of making connections generally, which allows us to go along with, and in some instances to become, those who tell us "there's a new you coming every day."

This abhorrence or, at the least, avoidance is at the heart of the second and darker level at which popular culture operates. This level of tradition is real tradition—those elements of American factual and intellectual history that encouraged and permitted our start as a nation and that have both dogged and inspired us since. Our history—the significant part of it—resides in popular culture, often confused and jumbled, often hiding like the purloined letter, nevertheless coming through with the inevitability of fate in the classics. It comes through with particular clarity on television and with particularly particular clarity on the subject of adulthood.

I believe the belittling of adulthood on television is no different at base from Emerson's decision to shuck the courtly muses of Europe. It derives from a wish for improbable freedom in the name of some higher, if indeterminable, virtue. "Abstract liberty, like other mere abstractions, is not to be found," said Burke. But Burke did not imagine a world so dominated by the expression of abstractions that the abstractions could develop their own symbolic logic, and rapidly become the definitions of themselves.

To be without memory, reason, civility, stature, and authority sounds like a wish for savagery. In fact, it may turn out to be so, though it is unlikely that television consciously promotes such a wish. There is, however, a state of mind that falls short of savagery, which simulates a dream state, where all freedoms associated with savagery flourish without histrionics. We have no name for this state, for it did not exist before television and still does not exist outside it. But, whatever its name, it is a state of freedom—apolitical, though it admits politics, asocial, though it depends on social life—a celebration of pure irresponsibility.

In many ways, television is the medium of irresponsibility, which is why the idea of adulthood within television is a contradiction in terms. The medium itself allows freedoms that no other medium will; it doesn't hold us like the theater or movies: we can place it where we choose, we can eat, do push-ups, answer the door. We are free to spin the channels, free to take or leave it. In turn, it shows us freedoms never won before—tapes and repeats that play havoc with time and sequential actions, with order and the idea of order. Of course these freedoms are illusory, but that doesn't seem to lessen our interest in them. Television is so far the only medium to take us out of history, making paltry such nuisances as original sin.

"Grow up!" as an imperative means "behave and control yourself": understand your limitations and be reasonable and civil accordingly. One does not grow up on television. It is not in television's commercial interests to have one do so, because a free-floating mind is more apt to buy large quantities of La Choy Chinese food. But we are complicit in this as well, having found and tacitly urged on television a strange answer to our wildest dream. The question, how free can you be?, which bestrides the democracy as does no other, is in television rhetorical.

You *can* yell fire in a crowded television set. You can do anything you please. Adults are interchangeable and lack the virtues of adulthood, and soap operas rely on perpetually changing characters and circumstances. New series every season, new logos for networks. All this for us, as we sit alone in our separate houses dazzled by the freedoms within the medium and the ferocious self-reliance *of* the medium. Our memories go, too, for the experience of television is itself unmemorable.

As for the ever-recurrent question of influence, it is hard to tell what these images do. I seriously doubt that watching panicky or wise adults on television will make children grow one way or the other, or that seeing adults forfeit sense and manners will cause children to do likewise. But what of these fierce freedoms: the message continually sent, dot by dot, that a person needs no one but himself in this world and no other person needs him? What of the message of the box? Every night in America the doors lock, the screens glow bright, and man sits down to see how free he can be. Nothing will disturb him, if he can help it. He is a grown-up, after all, and has earned his independence.

Toy, mirror and art: the metamorphosis of technological culture

PAUL LEVINSON

▪ Paul Levinson *is Assistant Professor of Communications at Fairleigh Dickinson University.*

"Toy, Mirror and Art" argues a broad thesis, under which it is possible to gain important new insights into television. Technological artifacts, it is claimed, undergo a life of three stages. In the first stage there is the gimmick, the toy, the product of the tinkerer. Although many technological artifacts never progress to a second stage, those which do so begin to "mirror reality," to take as their content the social and objective, and to acquire a mass audience. In their third stage a new phase of non-reality is developed, that imaginative reorganization of reality which we call art.

The varied produce of our technological media—the amalgam of television shows, movies, books, recordings, etc., known collectively as "mass," "popular," or "technological" culture—has been the subject of considerable recent study and controversy. Arguing primarily from aesthetic and sociological perspectives, champions and critics of popular culture have alternately praised and condemned it as aesthetically democratizing and degrading, socially stabilizing and stultifying, and so forth.(1) Curiously missing from such discussions, however, is any serious analysis of the technological basis of popular culture. While theorists usually acknowledge that it is technology that makes most popular culture possible, they have apparently been content to view the connection as axiomatic and undeserving of further research.(2) Thus, an otherwise comprehensive summary of "Theories and Methodologies of Popular Culture" in a recent *Journal of Popular Culture* issue discussed everything from structuralism to cultural geography and popular culture, with barely a mention of technological underpinnings.(3) The omission is even more remarkable when one considers that Marshall McLuhan, one of the first to write seriously of popular culture, was also one of the first to point out that technological media are much more than passive conveyors of information and content.(4)

It is perhaps understandable that subtleties in the technological shaping of popular culture have gone unexplored, since the broad outlines of the relationship are so obvious. There seems little profit in pursuing, for example, the fact that without the invention of the motion picture camera there would be no film industry, and with the technological achievement of the phonograph, no popular recording culture. Yet upon closer examination, such simple connections begin to display an increasing number of complications. Why, to stay with the same example, did film attain a cultural prominence forty years before music recording, when the motion picture camera was in fact perfected shortly *after* the phonograph? While differences in societal receptivity, economics, and the like were no doubt in part responsible, it seems plausible that certain elements in the very mechanics of film and record production may have stimulated the first and inhibited the second as they arose in cultural impact and esteem.

Film, as the first product of the nineteenth-century electro-chemical revolution to achieve artistic notice in the twentieth, might be a good place to begin an inquiry into the technological determination of technological culture. In tracing the changing usages and perceptions of film

☐ Reprinted from *Et cetera* Vol. XXXIV, No. 2 (1977) by permission of the International Society for General Semantics.

from its first appearance in society, it may be possible to discern a relationship between the level of technological sophistication in film and the type of popular culture each technological level engendered. To the extent that such observations are generalizable, they may suggest a series of principles that describe a step-by-step development, common to all technological culture, from new medium to widespread influence. Such principles may also have some bearing on the aesthetic controversies about culture, helping to explain how and why some technologies facilitate more "artistic" creations than others. The inquiry may also have some implications for the evolution and appliance of more "practical" technologies, and elucidate the distinctions and similarities between technologies used primarily for work, and those used for entertainment.

STAGE ONE: TECHNOLOGY AS TOY

Writing of the inception of film technology in *A Short History of the Movies,* Gerald Mast describes an interesting pattern:

> The first film makers were not artists but tinkerers. . . . Their goal in making a movie was not to create beauty but to display a scientific curiosity. The invention of the first cameras and projectors set a trend that was to repeat itself with the introduction of every new movie invention: the invention was first exploited as a novelty in itself. . . .(5)

A survey of early "talkies" like *The Jazz Singer,* first efforts in animation such as Disney's "Laugh-O-Gram" cartoons, and indeed the supposed debut of the motion picture itself in *Fred Ott's Sneeze,* supports Mast's observation of technology's supremacy in the beginning stages of technological culture. In each instance elements of plot, characterization, and what little content there is, play a subservient role to the exposition of the new gimmick, and perform in effect as low-key vehicles for a highly visible technique. The enjoyment in these primal forms lies in a fascination with the process—not the product of the process, but the process itself—in seeing and hearing a man sing on film, for example, rather than caring *what* the man sings on film. Thus, in the medium of film, at least, new technologies have made their entrances like the brash new kid on the block, in a flexing of muscle and raw technique that transcends and for all purposes *becomes* the content. In a sense, then, the most important content—or popular culture—of a new medium is the medium itself.

McLuhan has explored the concept of technology-as-the-content-of-technology, suggesting that *out-moded* or post-functional media often serve as the content for newer media. (Plays and books, for example, become the content of the new medium, film, and film becomes, in turn, the content of the newer medium, television.)(6) McLuhan goes as far as to say that usually invisible technologies only become visible when no longer in use—a proposition that at first seems to contradict Mast's contention, but may in fact serve to complement it. For incipient media are as out of the mainstream as obsolescent forms—it's the familiar equation of childhood and old age—and as such occupy equivalent if opposing positions in the medium development cycle. Thus, McLuhan's model may be reduced to a basic expectation that the discernibility or observable impact of any medium will vary *inversely* with the usage or functioning of that medium in the overall society; that is, the workings of technology, like the blades of whirring fans, are most visible both before and after they reach the peak of their function, and the triviality and trickery of pre-pubescent media are as nonfunctional and hence ostentatious as the ritual pomp and funeral exhibition of media in demise.

The role of neonate medium as societal plaything is perhaps best documented in the history of film. But it is also readily apparent in the trajectories of most other communications media, and indeed in the invention and implementation of many technologies, used for entertainment and otherwise. William Orton's celebrated refusal to inexpensively buy up Bell's early telephone patents on grounds that the new device would never be more than an "electrical toy";(7) the corporate decision to initially promote the phonograph as a "novelty" music box *in spite of* Edison's early assertions that his new invention could perform more practical tasks;(8) the amateur crystal set radio fad of the 1920's and the gawking at televisions in department store windows in the 1940's; the continuing popularity, in our own time, of computer "games,"(9) as well as the propensity of programmers to couch computer terminology and print-outs in cute phrases and configurations; and, most recently, the Citizen's Band or "CB" radio "craze"(10)—all testify to the tenacity with which the novel medium is perceived and tends to be employed as a toy. Moreover, examples of technologies not specifically concerned with communication, but nonetheless at first utilized for peripheral amusement, are even more varied and abundant. The Chinese discovery of gunpowder and the principles of rocketry, and their use solely as holiday and children's entertainment; the initial application of Newtonian mechanics to devise intricate dolls or "automatons" in the eighteenth century;(11) the debut of ether as a giddy party drug well before its medical properties were exploited(12)—all suggest that the "toy principle" may far exceed the province of popular culture and communications media. It is even tempting to suggest that *all* new technologies may gain first admittance into society as court jesters and Trojan horses, with their physical presence clearly visible, but their potentialities poorly understood.(13)

But if practical technologies indeed begin as playthings, what forces are needed to transform the playthings into practical technologies? That this transformation is by no means inevitable is documented by various curiosities of history, such as the failure of the Aztec civilization to use the wheel other than in children's toys, and the Chinese confinement of their gunpowder, rocket, and printing inventions to use on only ceremonial occasions. Abbott Payson Usher, the technological historian, sees this problem as central to an understanding of technology. "The history of invention," he writes, "is a study of the circumstances that have converted the simple but relatively inefficient mechanisms of early periods into the complex and more effective mechanisms of today."(14) Usher views economic needs and perception of technological potentials as the most important of these circumstances, but numerous other factors have been linked by theorists to the development, and nondevelopment, of specific technologies. Victor von Hagen, for example, points out that any Aztec attempt to use the wheel in more practical ways would have been foiled by Mexico's steep-walled landscape,(15) or the lack of a physical environment conducive to the technology of the wheel. McLuhan emphasizes the importance of compatible *media* environments, proposing that the Chinese ideograph was the main impediment to Chinese use of the printing press for mass communication, on grounds that ideographic writing is not well suited for reproduction on mass, movable type.(16) And Friedrich Hayek contends that the massive application of invention to industrial tasks in the nineteenth century—aptly characterized by Alfred North Whitehead as "the invention of the method of invention" (17)—was made possible by the uniquely invigorating climate of free capitalism.(18) It is thus fairly plain that the ignition of technological growth has often come from outside the specific technology itself, in a combination of supportive social, economic, media, and even physical conditions. It also follows that while the toy phase may be prerequisite to subsequent technological development, its existence by no means *guarantees* that development: lacking the proper environment, the technological toy may long endure in a case of "arrested" development. (In this sense, the confinement of "ESP" phenomena to largely show-business and "magic" roles in our own society may constitute a failure to exploit a potentially useful "mental" technology due to lack of a proper attitude on our part. Future historians may well regard this failure in the same way we regard the Aztec "failure" with the wheel.)(19)

In the case of the popular culture technologies of the past hundred years, however, the emergence of new communication toys has almost always led to their more extensive use as practical media and/or mass art. The lesson of the toy principle for popular culture, then, is that mass art forms don't spring full-blown from the head of new technologies, but rather pass through a series of developmental stages beginning with a naive, raw, almost "contentless" flexing of hardware. That there is little "mass" about these incipient mass media is obvious in the bygone kinetoscope parlors of primitive film and the ear-horns of early victrolas—devices that doled out entertainment on a purely personal, one-to-one, fragmentary basis, characteristic more of the individual experience of toys that the mass experience of popular media. And yet with surprising regularity, these primordial media evolved into technologies and cultures of universal impact.

STAGE TWO: TECHNOLOGY AS MIRROR OF REALITY

In the history of film, it wasn't long before Edison's kinetoscopic oddities faced stiff and ultimately overwhelming competition from the Lumieres' presentation of "actualities" on the screen. Rather than photographing sneezes and what amounted to other filmic gag-lines, the Lumieres pointed their cameras at real-life events—workers leaving a factory, a baby's meal, and the famous train entering the station. Although the novelty of movie technology undoubtedly played a major role in the appreciation of these early "documentaries," the cries and jolts of audiences upon viewing *L'Arrive d'un train en gare* in 1895 (approximately six years after Ott's nasal acrobatics) clearly indicated a new focusing on content—in this case, a real train chugging into a real station, at an angle such that the audience could almost believe the train was chugging right in at *them*. The superficial amusement and curiosity characteristic of the earlier gimmick films were replaced with the deeper emotions of fright, sorrow, and so on—emotions that one would expect in a replication of a real-world interaction. In effect, the adoption of reality as film content distracted from the technology and artificiality of the film experience, directing attention to the non-technological content—the events depicted on the screen—and in turn enhancing the believability of the content, i.e., belief that the events on the film were "really" happening. The co-option of reality in media thus becomes a self-fulfilling loop, in which the very mirroring of actuality tends to disguise the mirroring process and promote the actuality. It is perhaps the spiraling power of this media loop that accounts for the riveting impact of media technology once it left the infant toy stage and evolved into the succeeding reality/mirror phase.

If, as Whitehead said, the most important invention of the nineteenth century was invention itself, the most important development of the twentieth century was, as Bertrand Russell saw, the suspension of disbelief.(20) The public's willingness to respond to an electronic transcription of a voice as if it were a *live* voice, and to a photo-chemical likeness of a face as if it were a *real* face, soon enabled communications technology to effectively recapture or substitute for the real world on a massive scale. And it was this mirroring of the real world, with the attendant prominence of content and invisibility of technique (qualities which media theorists from McLuhan to Jacques Ellul have long seen as the defining traits of mature technologies)(21), that became the *modus operandi* of communications media. The telephone, of course, shortly confounded William Orton's assessment and became a major artery for both business and personal conversation. And while the phonograph's transition from toy to reality-transcriber may have been less obvious than that of the telephone, it was nonetheless profound. By 1893, the attempt to sell the phonograph as a "novelty" had run its course, and Edison, regaining control of his invention, planned for the introduction of popular music records.(22) This signalled a shift in phonographic emphasis from technique to content, the content in this case being the reality of a past musical performance retrieved and captured on the record. Television and radio, after even briefer tenures in the toy stage, began functioning as rather mature transcribers of reality—where they for the most part continue today. For the broadcasting of film, video tapes, or recordings, television and radio perform as much of a reality transmission as when actual events and live performances are broadcast. The non-reality component, if it exists in the film or recording broadcast, is in the original film or recording, *not* in its transmission on television or radio. In fact, film and recordings on television and radio satisfy all the requirements of the stage two "reality" mirror: the film is the reality/content and the film/content is paramount in the audience's awareness, even as the underlying transmitting technology of television goes unnoticed. The audience suspends its disbelief and pretends it's seeing *a film* rather than a televised *broadcast* of a film.

Conveyance or interaction with reality is apparently the second and terminal phase of development, not only for most communications technology, but for most other applications of technology as well. When such curios as rocketry, electricity, and phosphorescence were finally harnessed for practical purposes—when the little toys finally grew up—they extended our physical control of the real world and, as Fuller, Hall, McLuhan, and many others have pointed out, in effect acted and continue to act as surrogates for our arms, legs, hands, and bodies.(23) There thus appears to be some merit to the proposition that communications media differ from other technologies only in specific application and content, sharing essentially the same developmental patterns and dynamics.

There does appear, however, to be at least one interesting difference in the development of film and most other technologies. Whereas, as indicated before, the transformation of most techniques from side-show toys to mainstream appliances seems to have been sparked primarily by forces and attitudes *outside* of technology, the growth of film from gimmick to replicator was apparently in large part dependent upon a new technological component. As described earlier, the "toy" film played to individuals who peeked into individual kinetoscopes; but the "reality" film reached out to mass audiences, who viewed the reality-surrogate in group theaters. The connection between mass audience and reality simulation, moreover, was no accident. Unlike the perception of novelties, which is inherently subjective and individualized, reality perception is a fundamentally objective, group process—tested in the social consensus, as George Herbert Mead and social psychologists have long stressed—and as such is strengthened and even predicated upon mass experience. The creation of a group audience for simultaneous and reinforcing perception of the reality-surrogate film required that a new technology of film *projection* had to be devised and hooked into the existing communication chain. This suggests that technological determinism may have played a greater role in the development of film culture, and perhaps of all popular culture by extension, than it did in the appliance of more practical technologies.

In addition, film (and recording) seem to be distinguishable from most other media and technologies in one other significant respect: as implied earlier, it is film and recording that now provide the wellspring of imagination and originality for television and radio broadcasting. This nonreality is not the pre-reality of technological toys, but the post-reality of media that have mastered the straight transcription of the real world and have gone on to something beyond—the rearrangement of the real world to create fantasy, eloquence, and art. To accomplish this feat, a technology must evolve to yet a third phase—a phase that can copy reality, dissect it, and put it back together again in new and intriguing ways.(24)

STAGE THREE: TECHNOLOGY AS MIDWIFE TO ART

The present discussion began with an inquiry into the relationship between technology and popular art, and a specific question as to why film and recording, which were first introduced into the culture at approximately the same time, developed into popular art forms at such different rates of speed. Thus far, however, the discussion has had little to do with popular art—talking on the phone certainly doesn't constitute a popular art (at least, not for the average speaker), and, as indicated, what aesthetic content there is in radio and television derives from the content of the film, play, record, script, and not from its broadcast. The connection between technological process and technological art, then, remains yet to be defined.

Once again, the history of film might prove instructive. An oft-told story has it that George Melies, another French film pioneer, was shooting his camera at pedestrians and vehicles on the Place de l'Opera in Paris one fine spring day in 1898 (in the "actuality" fashion of the Lumieres), when his camera jammed. Thinking his film ruined, Melies nevertheless cleared the aperture gate, reset the film, and started shooting again—taking the film home for development just for the amusement of it (apparently one of the essential ingredients of many great discoveries). When the print came out, and Melies projected it, he received a little surprise: there, at the spot in the action where the camera had stopped and then started, was a magical transformation—men turning into women, children into adults, and a passing bus instantly materializing into a hearse! The film, in other words, gave no discernible indication that the camera had stopped and started with several seconds elapsing; all that was apparent was the continuous "reality" of a bus suddenly changing into a hearse. Melies had inadvertently hit upon the potent intervention of editing.(25)

The edit of course proved to be the key in the transformation of film from reality transcription to popular art: the discovery that disparate pieces of film, shot at different times and places and reflective of different realities, could be spliced together so as to project what would be accepted as one continuous *new* reality, freed film from dependence upon literal reality.(26) Film no longer need be wedded to the natural rhythms of time and space to create a natural, flowing experience; the editing room could create its own rhythms, which were equally "natural" and palatable to the perceiver. Within less than twenty years, Griffith and other exploited this opportunity to mold, bend, shape, fracture, and reconstruct realities to the dictates only of the writer/director/editor's imagination. Film now had a life of its own.

From Altamiran cave paintings to Victorian

literature, the ability not only to retell but refashion reality in the retelling has been a hallmark of "art."(27) It is not surprising, then, that film's transcendence of reality touched off its explosion as a popular art form, which by the 1920's had become both global and golden. Conventional film history, of course, traces the employment of editing and its artistic vistas to the chance tinkering of Melies and his followers; but the serendipity of film editing can perhaps be better understood as not so much the personal fortune of Melies, as the serendipity of film's original *technology*. The mechanics of film were never *intended* to allow for an alteration of reality—celluloid was used for its flexibility in projection, not for its amenability to splicing—and the "chance" discovery of editing was thus a chance uncovering of a hidden capacity already present in the technology of the medium. In this specific sense, then, the popular art of film can be seen as a direct outgrowth of its peculiar technology.

The development of technological art thus appears dependent upon the special capacity of a technology, first designed as a toy and second used as a reality-substitute, to transcend reality and make new ones. This toy/mirror/art or pre-reality/reality/post-reality dialectic of technological development bears some interesting resemblances to several well-known models of human development, including Piaget's sensorimotor, concrete, and formal (abstract) stages of intellectual growth;(28) McLuhan's oral, written, and electronic eras of communication;(29) Freud's oral, anal, and genital stages of psychosexual development (which Walter Ong has intriguingly compared to McLuhan's stages of communication, *e.g.*, written and anal are retentive and reality-oriented)(30); and Arthur Koestler's Jester, Sage, and Artist as the three unfolding expressions of human creativity.(31) Note how technology as toy displays the subjectivity of the oral, the nonseriousness of the joke, the flexing of muscle for its own sake characteristic of sensorimotor activity, and the emphasis upon technique or delivery common to humor, oral communication, and sensorimotor behavior. Technology as mirror stresses accuracy, objectivity, and prominence of content or "knowledge" as befits both the sage and the scribe, as well as the literal transaction with reality basic to concrete operations. And technology as art, combining elements of the previous two stages, is both serious *and* subjective, capable of the emotional intensity of the genital stage (the multi-dimensionality of electronic communication) and the abstraction and restructuring of reality—the triumph of form over content—of the formal stage of intellectual functioning.

Most technologies, however, perhaps too well-suited to the second stage mirror task, simply lack the ability to make the artistic jump. Thus the telephone is purely a medium of reality communication, and still-life photography, for all its aesthetic aspirations, remains essentially a medium of literal replication.(32) Other technologies, such as radio and television possess the ability to restructure reality and create art, but are limited by convention and economic pressures to simple reality-mirroring of previously created filmic, theatrical, or musical art, as discussed earlier. (The quick switching of video cameras and perspectives on talk shows like "The Tonight Show" often create ambiences that don't exist on the set, and thus may constitute a bona-fide "art-form" *produced*—rather than merely transmitted—by television. Experimentation with video editing, computer character generation, and so forth may also be a source of potential art.)

The case of phonograph/recording technology is even more unusual. Initially, the hardware used to record was hopelessly reality-bound—sound was stored and reproduced first on electric wires and then on discs, neither of which allowed for splicing, alteration, or rearrangment once the recording was made. Thus lacking the hidden potential of celluloid film, rubber records continued for better than forty years as a glorified Xerox operation for musical performances. It wasn't until the addition of a completely new tape technology in the 1940's—a spliceable medium which was initially introduced for remedial purposes, so as to make more accurate replications—that recording attained a faculty for reality alteration or art. Magnetic sound tape not only allowed for easy editing, but for overdubbing, multi-track sel-syncing, and a general reshuffling of recorded sounds to make for imaginative new combinations. Thus, within twenty years of the introduction of tape recording, artist/producer/songwriters like Phil Spector, the Beatles, and others turned recording into the popular art-form of the generation—a two billion dollar industry that has at times surpassed even film and television in combined sales and cultural impact. (Note that the twenty years from inception of tape technology to Beatles parallels the twenty years from Melies to Griffith in film.)

The differential in the rise of film and recording as popular art-forms can thus now be explained as follows: both were initially conceived as toys, and both were quickly adopted for reality transcription; the same technology that enabled film to adequately replicate the real world enabled film to reconstruct the real world, so film soon evolved into a popular art; but the technology that enabled recordings to adequately replicate its real world contained no such double advantage, so recording remained a simple transcription device until the addition of a new mechanism capable of reality alteration. It is thus apparent that, as suggested at the outset of the present discussion, the relationship between technology and technological art is no simple cause-and-effect matter.

Indeed, the addition of a new component to an already-productive technology cannot even always be depended upon to enhance the medium's capacity for art: as suggested in the earlier reference to the first "talkie" films, the introduction of sound technology to the silent film in effect *reduced* the medium to the state of a toy—setting the whole technology back to stage one by creating a new medium, as it were. In this regard, some critics insist that to the present day, film has never recovered from the introduction of sound—that speech and dialogue have been used at worst as a toy and at best as an unimaginative, literal exposition of plot (some of the work of Orson Welles and perhaps Robert Altman might be an exception), to the detriment and even destruction of lofty artistic styles developed during the silent era.(33)

Moreover, the connection between technology and popular culture is further complicated by the tendency of various technologies to operate, not singly or in isolation, but in conjunction, often cross-influencing one another. Edison, it is said, invented the phonograph to perfect the telephone, and a motion picture process to enhance the phonograph.(34) The role of phonetic writing as a prerequisite to the mass usage of the printing press has already been alluded to (see Note 16, above). And the rise of music recording culture, though clearly a product of its own technology, was augmented by, of all media, the television: when television co-opted radio as a medium of live entertainment in the early 1950's, radio was forced to rely much more extensively on recorded music to attract its listeners—and thus provided a sustaining forum for a recording technology already ripe for popular art. (In a similar fashion, the FCC's decision in 1965 that all FM radio stations must broadcast programs different from their AM radio affiliates hastened the development of the LP record as an art-form—for many FM stations turned to what was previously considered "noncommercial" album music, thus giving the LP a much needed public forum.) This type of technological interaction of course demands an eventual analysis of the economic and social factors that mediate the technologies.

A complete discussion of technology and art inevitably invites some consideration of aesthetics—yet the complexity of the technology-to-culture equation makes an aesthetic of technological art and culture rather difficult. Criticisms of technological art have often been insensitive to gradations in technological process, and have been frequently directed at immature media that are physically incapable of, and make no pretense to, any type of technological "art." Thus, Jose Ortega Y Gasset, for example, condemns "modern art" as, among other things, "play and nothing else," and "of no transcending consequence"(35)—qualities that, in the perspective of the present discussion, can be seen as more properly belonging to the technological toy than to technologic art. This confusion of technological stages—an error of premature judgment and "mistaken identity" born of an inability to see technology as an evolving, developmental process—was recognized by Susanne Langer, whose assessment of filmic evolution aptly complements Mast's observations on novelties that served as the springboard for the present analysis. Langer writes:

> With every new invention—montage, the sound track, Technicolor—its [film's] devotees have raised a cry of fear that now its 'art' must be lost. Since every such novelty is, of course, promptly exploited before it is even technically perfected, and flaunted in its rawest state, as a popular sensation, . . . there is usually a tidal wave of particularly bad rubbish in association with every important advance. But the art goes on.(36)

The problem, of course, is that while stages of technological culture are indeed distinctive and successive, they by no means are mutually exclusive—the current popularity of "Sensurround" gimmickry and "wildlife sage" reality-mimicry in the movies suggests that, having attained the *capacity* for technological art, film need not necessarily always *produce* art. Instead, technologic evolution, like its biological model, allows for the co-existence of earlier and later designs, with an assortment of aesthetic ramifications. But if an awareness of media evolution cannot provide a definitive aesthetic for popular culture, it can at least offer a useful yardstick for making such judgments. Moreover, it perhaps at last reveals a common ground between the critics and champions of popular culture—the first looking at the caterpillar, the second at the butterfly, of the same technological process.

SUMMARY AND CONCLUSIONS

The ways in which technologies engender and encourage mass culture and art are complex and multi-faceted, yet have often been oversimplified or taken for granted. Although a physical invention must lie at the root of every technological art, very rarely if ever do inventions have immediate mass cultural impact or flowering. Instead, new technologies usually make their first appearance in the culture as novelties, gadgets, gimmicks, and toys. The content here is dominated by, and an exposition for, the new technique; the perceptual experience is personal, subjective, and highly individual rather than "mass"; and the toys usually perform on the sidelines of the overall society. Due more often than not to shifts in societal attitudes rather than developments in technology, the novelty item eventually (though not always) becomes a more practical device, used for various types of literal transactions with reality. The content in this phase attains a high prominence while the visible technology recedes; the perceptual experience is markedly social, objective, and "mass," as the entity of "audience" comes into play for the first time; and such transcribers of reality usually occupy significant and often central positions in the society. At this point, the evolutions of practical and artistic technologies are virtually indistinguishable—the difference being that practical technologies remain at the reality level, while artistic media must evolve to yet a third stage. To achieve this artistic jump, a medium must have the capacity not only to replicate reality, but to rearrange it in imaginative ways. Performance at this level entails a blending of features from the previous two stages: technological art is nonfunctional and subjective like the toy, yet nontrivial and (in most cases) group oriented and content dominated like the reality-surrogate. In the case of film, the suprarealism ability was inherent in the original realism technology, so the development of a popular film art was relatively swift. In the case of music recording, the capacity for reality alteration came only with the nonpurposeful addition of a new technological component, so the rise of a popular recording culture was correspondingly delayed. In yet another case, radio and especially television have the capacities for technological art, yet function as aesthetic parasites in relying upon other media for creativity and art. It is thus apparent that the technology for transcending reality, and its two antecedent stages, are necessary but not sufficient conditions for the fostering of mass culture and art; the remaining conditions probably lie in the interaction of various technologies both among themselves and with more abstract, non-technological elements of society.

NOTES AND REFERENCES

1. Herbert J. Gans, *Popular Culture and High Culture* (New York: Basic, 1975), summarizes many of the extant criticisms and defenses of popular culture.
2. A few theorists have argued that technology is *not* the necessary basis of popular culture, citing such non-technological cultures as the oral folk music tradition; see Ray B. Brown, "Popular Culture: Notes Towards a Definition," in *Side Saddle on the Golden Calf*, George H. Lewis, ed. (Pacific Palisades, Cal.: Goodyear, 1972), pp. 5-11; and Bruce A. Lohof, "Popular Culture: The *Journal* and the State of the Study," *J. of Popular Culture*, 6, No. 3 (1972): 438-455. Since even folk music, however, has become increasingly dependent upon the technology of electronic instruments and recording, it seems fair to say that such cases represent a diminishing series of exceptions to the technological rule.
3. Vol. 9, No. 2 (1975): 353-508.
4. Principally in *The Gutenberg Galaxy* (Toronto: University of Toronto, 1962), and *Understanding Media*, 2nd ed. (New York: Mentor, 1964).
5. (New York: Pegasus, 1971), p. 15. Unless otherwise indicated, examples of film history to be cited in the ensuing discussion come from Mast's account.
6. In the McLuhan schema, technology once liberated from function often becomes not only "content" but "art." "The machine turned Nature into an art form," McLuhan writes, by making humans nondependent upon natural technologies for survival. (See *Understanding Media*, p. *ix*.) Note that the art here is not the direct product of a technology at work—as is film montage, for example, from the technology of editing—but rather the peculiar result of a technology *not* at work. A fine recent example of this unusual type of art genesis appeared in a *New York Times* travel piece that seriously described New York City's increasingly nonfunctional subway system as a "delightfully elevated *tour de force*," and "a scenic de-

light," featuring "track-wheel music." (See Stan Fischler and Richard Friedman, "Subways," 23 May 1976, Section 10, pp. 1, 22.) With ridership diminishing and service reduced, the mechanics of the subway system are now apparently capable of being appreciated not for what they do (or don't do), but for what they "are."

7. Matthew Josephson, *Edison* (New York: McGraw-Hill, 1959), p. 141. Moreover, Orton, as president of the Western Union Telegraph Company, was apparently steadfast in his low regard for the "talking telegraph." According to an amusing little article aptly entitled "Three Great Mistakes" by S. H. Hogarth in the November, 1926 issue of *Blue Bell,* Orton coursneled his hapless friend Chauncey M. Depew to pass up a chance to purchase one-sixth of the new Bell telephone enterprise for a mere $10,000, because, in Orton's view, "the invention was a toy" with no "commerical possibilities." Meanwhile, John Brooks relates that use of the telephone in England was delayed for at least a decade due to the conviction that it was only a "scientific toy." (See *Telephone: The First Hundred Years* (New York: Harper & Row, 1976), p. 92.)
8. Josephson, p. 172. Among the more practical applications of the phonograph envisioned by Edison but long unimplemented were recording of letters and books for the blind, preservation of lectures and public addresses, and permanent transcription of telephone conversations.
9. A recent *New York Post* article reports that sales of home computers are "spreading like wildfire," and mostly to "techno-fetishists" who play a variety of visual and intellectual games with computers. David Ahl, editor of *Creative Computing* who was interviewed in the *Post* story, sees the current computer phase as consistent with a more general pattern of media development: "When the principle of radio was first discovered," Ahl explains, "it was the amateurs who developed the first sets. . . . it's the same here [with computers]." Peter Keepnews, "The Latest Do-It-Yourself Fetish: Computers," *New York Post,* 9 June 1976, p. 47.
10. Although C.B. was first introduced by the F.C.C. in 1958, it was relatively unknown by the general public until the recent fanfare. Predictably, C.B.'s first burst into public awareness has been accompanied by a jargon of code-names and pass-words, both accoutrements of gimmick usage. And perhaps most significant is C.B. enthusiast and writer Michael Harwood's assertion that the messages relayed on Citizen's Band "are often inconsequential"—clear evidence, again, of new technology overpowering content, or being operated just for the fun of it. See Michael Harwood, "America With its Ears On," *The New York Times Magazine,* 25 April 1976, pp. 28, 60, ff.
11. Siegfried Giedeon provides a colorful account of "invention in the service of the miracle" from Alexandrian religious plays to mechanical ducks that defecated in the 18th and 19th century courts of Europe, in *Mechanization Takes Command* (New York: Norton, 1948), pp. 32-35. See also Robert S. Brumbaugh, *Ancient Greek Gadgets and Machines* (New York: Thomas Crowell, 1966), who points out that "the Greeks invented the steam engine, but to them it was only a toy" (p. 4). Brumbaugh then documents the Greek invention of numerous other mechanical devices used primarily to amuse and amaze.
12. Rene Fulop-Miller, *Triumph Over Pain,* trans. by Eden Paul and Cedar Paul (New York: Literary Guild of America, 1938), pp. 95-97.
13. Cyril Stanley Smith argues along similar lines in "On Arts, Invention, and Technology," *Technology Review* 78, No. 7 (June 1976): 36-41, pointing out that practical metallurgy began with the making of ornamental necklaces, wheels first appeared on toys, lathes were used to carve snuff boxes a century before their use in heavy industry, and metal casting was first perfected for making bells rather than cannon—all of which suggests to Smith that technology may originate more from playful and aesthetic impulses than practical need.
14. *The History of Mechanical Inventions,* 2nd ed. (Cambridge, Mass: Harvard University, 1954), p. 117.
15. "The wheel, had the Mexicans had it," von Hagen writes, "would have done them no good as all is up and down, and high valley is walled from high valley almost throughout the length and breadth of the land; its heights are only passable to foot traffic." *The Aztec: Man and Tribe* (New York: Mentor, 1961), p. 18.
16. *The Gutenberg Galaxy,* p. 185. In a similar way, Roger Burlingame explains the failure to actualize most of Leonardo's inventions as due to a lack of "collateral" technology. (See "The Hardware of Culture," *Technology and Culture* 1, No. 1 (1959), p. 16.)
17. *Science and the Modern World* (New York: Macmillan, 1925), p. 136.
18. "That the inventive faculty of man had been no less in earlier periods," Hayek explains, "is shown by the many highly ingenious automatic toys and mechanical devices constructed. . . . But the few attempts towards a more extended industrial use of mechanical inventions, some extraordinarily advanced, were promptly suppressed . . . the beliefs of the great majority of what was right and proper were allowed to bar the way of the individual innovator. Only since industrial freedom opened the path to the free use of new knowledge . . . has science made the great strides which in the last hundred and fifty years have changed the face of the world." *The Road to Serfdom* (Chicago: University of Chicago, 1944), pp. 15-16.
19. Of course, not everyone agrees that application of technology to practical tasks is socially desirable. Lewis Mumford, for example, sees pre-industrial, "esthetic" inventions as more fundamentally human, and more beneficial to society, than "utilitarian" appliances; he thus views pre-19th century incipient technologies not as "arrested" development at all, but as the finest expressions of the human inventive impulse, and laments their absence in our modern culture. See *The Myth of the Machine,* vol. 1: *Technics and Human Development* (New York: Harcourt Brace Jovanovich, 1966), pp. 252-253.
20. As cited by McLuhan in *Understanding Media,* p. 68. It was Samuel Taylor Coleridge who first identified "that willing suspension of disbelief for the moment, which constitutes poetic faith," *Biographia Literaria,* ed. J. Shawcross (1817; reprint ed: London: Oxford University, 1907), vol. 2, p. 6.
21. See William Kuhns, *The Post-Industrial Prophets* (New York: Harper Collophon, 1971), for a comparison of the work of McLuhan, Ellul, and other media theorists. For more on the narcotic capacity of media mirrors to disguise their operation, see McLuhan *Understanding Media,* "The Gadget-Lover," pp. 51-56.
22. Josephson, pp. 330-333. Edison's interest in the popu-

lar record was apparently due not only to foresight, but to the pressure of rival inventors—most notably Emile Berliner, whose "flat-disc" record in the 1890's took a sizeable bite out of Edison's "novelty" phonograph market.

23. See, for example, R. Buckminster Fuller, *Nine Chains to the Moon* (Carbondale, Ill.: Southern Illinois University, 1938), pp. 38-39; Edward Hall, *The Silent Language* (New York: Fawcett, 1959), p. 60; and McLuhan, *The Gutenberg Galaxy* and *Understanding Media: The Extensions of Man*.
24. This distinction between pre-reality and post-reality technology will perhaps run contrary to conceptual frameworks that distinguish primarily between reality and non-reality, and are thus prone to view both toys and art as a same technological expression belonging to a single non-reality or non-practical class. Thus Mumford, as suggested earlier, views pre-industrial technologies as the source of both games *and* the most genuine art (see Note 19 above), and indeed later argues that art attempted by post-industrial technologies is in effect a contradiction of terms, or an "anti-art." (See *The Myth of the Machine,* vol. 2: *The Pentagon of Power* (New York: Harcourt Brace Jovanovich, 1970), pp. 361-368, *et passim.*) From a rather different perspective, Freud has equated the artist, child, primitive, and psychotic, in his "Relation of the Poet to Day-Dreaming," reprinted in *On Creativity and the Unconscious* (New York: Harper & Row, 1958), pp. 44-54, and *Totem and Taboo,* trans. A. A. Brill (New York: Vintage, 1918). And McLuhan's description of post-operative technologies as art, and their similarity to the pre-functional technological toy (see Note 6 above), suggests at least one type of art which may be analogous to the toy. For the most part, however, the evidence of popular culture as well as intuition points to keep divergences between technological toys and art. While both *Fred Ott's Sneeze* and the movie *Chinatown,* for example, are indeed non-practical, the first merely distracts from the real world through razzle-dazzle, whereas the second subtly restructures reality through imperceptible technique; the sneezing nose in the first is diversion, the bandaged nose in the second is commentary and symbol. Thus, the discussion which follows will accept Susanne Langer's observation that while both games and art are nonutilitarian, art is serious and games are not. *Philosophy in a New Key,* 2nd ed. (New York: Mentor, 1951), p. 42.
25. There is apparently a sliver of suspicion that the Melies anecdote may be apocryphal. Eric Rhode, for example, in his recent *History of the Cinema,* (New York: Hill and Wang, 1976), p. 34, prefaces his recounting of the episode with a weighty "it is alleged that. . . ." On the other hand, Maurice Bardeche and Robert Brasillach, writing much closer to the source in *The History of Motion Pictures* (New York: Norton, 1938), p. 11—as well as Lewis Jacobs' "George Melies: Artificially Arranged Scenes," first published in 1939 and reprinted in *The Emergence of Film Art,* ed. Lewis Jacobs (New York: Hopkinson and Blake, 1969), p. 11—present the Melies story without qualification. In any event, the specific manner in which the editing principle was discovered, whether accidental or other, is not as important as the fact of discovery itself—which, as the discussion will shortly emphasize, was an all but inevitable if unintended consequence of the particular technology of film.
26. It is Edwin S. Porter, not Melies, who is usually credited with being the first to *physically* splice the film for story construction, as in Porter's *The Great Train Robbery* made in 1903. (See Jacobs, pp. 20-21.) Melies' technique of stopping the camera, rearranging the scene, and starting to shoot again was a cruder method of rearranging or "editing" reality.
27. Langer refers to this supra-reality quality as "semblance," and defines the "artist's task" as follows: "to produce and sustain the essential illusion, set it off clearly from the surrounding world of actuality, and articulate its form to the point where it coincides with forms of feeling and living," *Feeling and Form* (New York: Scribner's, 1953), p. 68. See Sergei Eisenstein, *Film Form,* trans. and ed. Jay Leyda (New York: Harcourt, Brace and World: 1949), for a discussion of film art as montage or creation of new realities through editing.
28. See Howard Gardner, *The Quest for Mind* (New York: Knopf, 1973), pp. 51-110, for a summary of Piaget's model.
29. See both *The Gutenberg Galaxy* and *Understanding Media*.
30. Walter Ong, *The Presence of the Word* (New Haven: Yale University, 1967), pp. 92-110.
31. *The Act of Creation* (New York: Dell, 1964).
32. As Stanley Milgram has recently pointed out, "The English language is blunt about the nature of photography. A photographer *takes* a picture. He does not *create* it." "The Image-Freezing Machine," *Psychology Today,* 10, No. 8 (January 1977), p. 52.
33. See, for example, Rudolf Arnheim, *Film as Art* (Berkeley, Cal: University of California, 1968), foreword and pp. 229-230; and Francois Truffaut, *Hitchcock* (New York: Simon and Schuster, 1966).
34. Josephson, pp. 161, 385; also Mast, pp. 25-26.
35. *The Dehumanization of Art* (1925 reprint ed.: Princeton: Princeton University, 1968), p. 14.
36. *Feeling and Form,* p. 412.

3

Clocks

The monastery and the clock

LEWIS MUMFORD

▪ Lewis Mumford *is one of the most prolific of all those who write technology's history. Among his many books are* Technics and Civilization *and* The Pentagon of Power.

Mr. Mumford here argues that the first "modern" man was the medieval monk, because of the way in which his life was regulated by the clock. He sees in the clock the first machine, in this case a machine for dividing time into uniform segments. "The modern industrial regime, he concludes, "could do without coal and iron and steam easier than it could do without the clock."

Where did the machine first take form in modern civilization? There was plainly more than one point of origin. Our mechnical civilization represents the convergence of numerous habits, ideas, and modes of living, as well as technical instruments; and some of these were, in the beginning, directly opposed to the civilization they helped to create. But the first manifestation of the new order took place in the general picture of the world: during the first seven centuries of the machine's existence the categories of time and space underwent an extraordinary change, and no aspect of life was left untouched by this transformation. The application of quantitative methods of thought to the study of nature had its first manifestation in the regular measurement of time; and the new mechanical conception of time arose in part out of the routine of the monastery. Alfred Whitehead has emphasized the importance of the scholastic belief in a universe ordered by God as one of the foundations of modern physics: but behind that belief was the presence of order in the institutions of the Church itself.

The technics of the ancient world were still carried on from Constantinopole and Baghdad to Sicily and Cordova: hence the early lead taken by Salerno in the scientific and medical advances of the Middle Age. It was, however, in the monasteries of the West that the desire for order and power, other than that expressed in the military domination of weaker men, first manifested itself after the long uncertainty and bloody confusion that attended the breakdown of the Roman Empire. Within the walls of the monastery was sanctuary: under the rule of the order surprise and doubt and caprice and irregularity were put at bay. Opposed to the erratic fluctuations and pulsations of the worldly life was the iron discipline of the rule. Benedict added a seventh period to the devotions of the day, and in the seventh century, by a bull of Pope Sabinianus, it was decreed that the bells of the monastery be rung seven times in the twenty-four hours. These punctuation marks in the day were known as the canonical hours, and some means of keeping count of them and ensuring their regular repetition became necessary.

According to a new discredited legend, the first modern mechanical clock, worked by falling weights, was invented by the monk named Gerbert who afterwards became Pope Sylvester II near the close of the tenth century. This clock was probably only a water clock, one of those bequests of the ancient world either left over directly from the days of the Romans, like

☐ From *Technics and Civilization* (New York: Harcourt, Brace and World, Inc., 1963) pp. 12-18. © in U.K. Routledge and Kegan Paul, Ltd.

the water-wheel itself, or coming back again into the West through the Arabs. But the legend, as so often happens, is accurate in its implications if not in its facts. The monastery was the seat of a regular life, and an instrument for striking the hours at intervals or for reminding the bell-ringer that it was time to strike the bells, was an almost inevitable product of this life. If the mechanical clock did not appear until the cities of the thirteenth century demanded an orderely routine, the habit of order itself and the earnest regulation of time-sequences had become almost second nature in the monastery. Coulton agrees with Sombart in looking upon the Benedictines, the great working order, as perhaps the original founders of modern capitalism: their rule certainly took the curse off work and their vigorous engineering enterprises may even have robbed warfare of some of its glamor. So one is not straining the facts when one suggests that the monasteries—at one time there were 40,000 under the Benedictine rule—helped to give human enterprise the regular collective beat and rhythm of the machine; for the clock is not merely a means of keeping track of the hours, but of synchronizing the actions of men.

Was it by reason of the collective Christian desire to provide for the welfare of souls in eternity by regular prayers and devotions that time-keeping and the habits of temporal order took hold of men's minds: habits that capitalist civilization presently turned to good account? One must perhaps accept the irony of this paradox. At all events, by the thirteenth century there are definite records of mechanical clocks, and by 1370 a well-designed "modern" clock had been built by Heinrich von Wyck at Paris. Meanwhile, bell towers had come into existence, and the new clocks, if they did not have, till the fourteenth century, a dial and a hand that translated the movement of time into a movement through space, at all events struck the hours. The clouds that could paralyze the sundial, the freezing that could stop the water clock on a winter night, were no longer obstacles to time-keeping: summer or winter, day or night, one was aware of the measured clank of the clock. The instrument presently spread outside the monastery; and the regular striking of the bells brought a new regularity into the life of the workman and the merchant. The bells of the clock tower almost defined urban existence. Time-keeping passed into time-serving and time-accounting and time-rationing. As this took place, Eternity ceased gradually to serve as the measure and focus of human actions.

The clock, not the steam-engine, is the key-machine of the modern industrial age. For every phase of its development the clock is both the outstanding fact and the typical symbol of the machine: even today no other machine is so ubiquitous. Here, at the very beginning of modern technics, appeared prophetically the accurate automatic machine which, only after centuries of further effort, was also to prove the final consummation of this technics in every department of industrial activity. There had been power-machines, such as the water-mill, before the clock; and there had also been various kinds of automata, to awaken the wonder of the populace in the temple, or to please the idle fancy of some Moslem caliph: machines one finds illustrated in Hero and Al-Jazari. But here was a new kind of power-machine, in which the source of power and the transmission were of such a nature as to ensure the even flow of energy throughout the works and to make possible regular production and a standardized product. In its relationship to determinable quantities of energy, to standardization, to automatic action, and finally to its own special product, accurate timing, the clock has been the foremost machine in modern technics: and at each period it has remained in the lead: it marks a perfection toward which other machines aspire. The clock, moreover, served as a model for many other kinds of mechanical works, and the analysis of motion that accompanied the perfection of the clock, with the various types of gearing and transmission that were elaborated, contributed to the success of quite different kinds of machine. Smiths could have hammered thousands of suits of armor or thousands of iron cannon, wheelwrights could have shaped thousands of great water-wheels or crude gears, without inventing any of the special types of movement developed in clockwork, and without any of the accuracy of measurement and fineness of articulation that finally produced the accurate eighteenth century chronometer.

The clock, moreover, is a piece of power-machinery whose "product" is seconds and minutes: by its essential nature it dissociated time from human events and helped create the belief in an independent world of mathematically measurable sequences: the special world of science. There is relatively little foundation for this belief in common human experience: throughout the year the days are of uneven duration, and not merely does the relation between day and night steadily change, but a slight journey from East to West alters astronomical time by a certain number of minutes. In terms of the human organism itself, mechanical time is even more foreign: while human life has regularities of its own, the beat of the pulse, the breathing of the lungs, these change from hour to hour with mood and action, and in the longer span of days, time is measured not by the calendar but by the events that occupy it. The shepherd measures from the time the ewes lambed; the farmer measures back to the day of sowing or forward to the harvest: if growth has its own duration and regularities, behind it are not simply matter and motion but the facts of development: in short, history. And while mechanical time is strung out in a succession of mathematically isolated instant, organic time—what Bergson calls duration—is cumulative in its effects. Though mechanical time can, in a sense, be speeded up or run backward, like the hands of a clock or the images of a moving picture, organic time moves in only one direction—through the cycle of birth, growth, development, decay, and death—and the past that is already dead remains present in the future that has still to be born.

Around 1345, according to Thorndike, the division of hours into sixty minutes and of minutes into sixty seconds became common: it was this abstract framework of divided time that became more and more the point of reference for both action and thought, and in the effort to arrive at accuracy in this department, the astronomical exploration of the sky focussed attention further upon the regular, implacable movements of the heavenly bodies through space. Early in the sixteenth century a young Nuremberg mechanic, Peter Henlein, is supposed to have created "many-wheeled watches out of small bits of iron" and by the end of the century the small domestic clock had been introduced in England and Holland. As with the motor car and the airplane, the richer classes first took over the new mechanism and popularized it: partly because they alone could afford it, partly because the new bourgeoisie were the first to discover that, as Franklin later put it, "time is money." To become "as regular as clockwork" was the bourgeois ideal, and to own a watch was for long a definite symbol of success. The increasing tempo of civilization led to a demand for greater power: and in turn power quickened the tempo.

Now, the orderly punctual life that first took shape in the monasteries is not native to mankind, although by now Western peoples are so thoroughly regimented by the clock that it is "second nature" and they look upon its observance as a fact of nature. Many Eastern civilizations have flourished on a loose basis in time: the Hindus have in fact been so indifferent to time that they lack even an authentic chronology of the years. Only yesterday, in the midst of the industralizations of Soviet Russia, did a society come into existence to further the carrying of watches there and to propagandize the benefits of punctuality. The popularization of time-keeping, which followed the production of the cheap standardized watch, first in Geneva, then in America around the middle of the last century, was essential to a well-articulated system of transportation and production.

To keep time was once a peculiar attribute of music: it gave industrial value to the workshop song or the tattoo or the chantey of the sailors tugging at a rope. But the effort of the mechanical clock is more pervasive and strict: it presides over the day from the hour of rising to the hour of rest. When one thinks of the day as an abstract span of time, one does not go to bed with the chickens on a winter's night: one invents wicks, chimneys, lamps, gaslights, electric lamps, so as to use all the hours belonging to the day. When one thinks of time, not as a sequence of experiences, but as a collection of hours, minutes, and seconds, the habits of adding time and saving time come into existence. Time took on the character of an enclosed space: it could be divided, it could be filled up, it could even be expanded by the invention of labor-saving instruments.

Abstract time became the new medium of existence. Organic functions themselves were regulated by it: one ate, not upon feeling hungry, but when prompted by the clock: one slept, not when one was tired, but when the clock sanctioned it. A generalized time-consciousness accompanied the wider use of clocks: dissociating time from organic sequences, it became easier for the men of the Renascence to indulge the fantasy of reviving the classic past or of reliving the splendors of antique Roman civilization: the cult of history, appearing first in daily ritual, finally abstracted itself as a special discipline. In the seventeenth century journalism and periodic literature made their appearance: even in dress, following the lead of Venice as fashion-center, people altered styles every year rather than every generation.

The gain in mechanical efficiency through co-ordination and through the close articulation of the day's events cannot be overestimated: while this increase cannot be measured in mere horsepower, one has only to imagine its absence today to foresee the speedy disruption and eventual collapse of our entire society. The modern industrial régime could do without coal and iron and steam easier than it could do without the clock.

Urban time

JOHN J. McDERMOTT

▪ John McDermott *is Professor of Philosophy at Texas A & M University. A noted authority on the American philosophers William James, John Dewey, and Josiah Royce, he has also published essays that assess the American city.*

Professor McDermott here brings to our attention ways in which urban time is unlike rural time, and in an extended and felicitous exploration of time's metaphors he encourages us to rethink the impact of clock time on our experience.

Having considered in some detail the experiencing of urban space and artifact as a context for human life, we turn now to a beginning analysis of some characteristics of urban time. Once again we must forgo a full discussion of the contrasting experience of nature time, offering only some contentions gleaned from previous consideration. Nature time can be described as fat time, running in seasons, even in years and decades. To a young city boy the pronostications of the *Farmer's Almanac* were as strange and alien as if they were made for a millennium hence. Nature times gives room to regroup, to reassess, featuring a pace and a rhythm turned to the long-standing, even ancient responsive habits of our bodies. Feeding off the confidence in the regenerative powers of nature, time is regarded as realizing, liberating, a source of growth.[1] The rhythm of nature time share with nature space a sense of expansiveness such that in nature we believe that we "have time" and that "in time" we too shall be regenerated. As exemplified by the extraordinary journey of the nineteenth-century Mormons, their "trek," a distinctive nature phenomenon, points to salvation "in time." Space is the context for "in time," both taken as a long period of time, walking from Nauvoo, Illinois, to the Great Salt Lake, Utah, and being saved "in the nick of time." In nature time, undergoing, doing, and reflection function simultaneously, thereby providing little need for "high culture," chunks of reflection taken out of the flow of experience. Perhaps the most obvious way to describe nature time is to call it baseball time, referring to a game in which the clock technically plays no role and which conceivably could last to infinity, tied to the end.

By contrast, urban time is thin time, tense, transparent, yielding no place to hide. Urban time is clock time, jagged, self-announcing time, bearing in on us from a variety of mediated sources, so often omnipresent and obtrusive that many people refuse to wear watches in an effort to ward off its domination. Why are clocks when worn on our bodies called watches? The first meaning of "watch" was to go without sleep, that is, to beat nature. Is the urban "watch" to "watch" time passing or is it to make sure that no one steals our time, as when we hoard time by saying that we have no time?

Clock time, like clock games, carries with it the threat of sudden death, an increasing urban phenomenon. But sudden death can be averted with but "seconds" to go. Urban time is "second" time, which may very well mean second chance or surprise time. We now have clocks which tell time in hundredths and even thousandths of seconds, reminding us how much faster we must go if we are not to be obsoleted, left behind, for cities have little patience with the past. Some people in the urban environment like to think that they live by nature time, but this assumption is self-deceptive, for

□ From "Space, Time and Touch: Philosophical Dimensions of Urban Consciousness," *Soundings*, 57(1974), 268-271.

such claims are relative only to the frantic urban pace. Clock time, after all, overrides nature time, as, for example, when at one second after such claims are relative only to the frantic urban pace. Clock time, after all, overrides nature time, as, for example, when at one second after midnight, in the pitch dark, your radio announcer says good morning and describes the events of your life that day as having happened yesterday.[2]

Beneath this somewhat anecdotal discussion of urban time there reside some significant implications. The rapid pace of urban time radically transforms the experience of our bodies, which often seem to lag behind. The network of communication media, which blankets a city like a giant octopus, constantly tunes us in to sensorially multiple experience, even if vicariously undergone. In a braod sense, when the setting is urban we experience less identity in spatial terms than proximity to events: rather than having a place, we identify ourselves relative to events taking place. In a city, when giving directions, the question as to "how long it will take" is not answered by the spatial distance traversed; rather the allotted time is a function of potential interventions, for the time of city-space is activity measured. Our imagination, fed at all times by the messaging of electronic intrusions,[3] races far ahead of our body, which we often claim to drag around. Yet, despite the pace, the apparent garrulousness and noise, urban life is extraordinarily introspective, enabling us to carve out inner redoubts of personal space. The urban person must protect himself against the rampaging activity of time, which dismantles our environment with alarming speed. It is a cliché that you can't go home again, for the spiritual and psychological experiences of childhood are unrepeatable; but further, for one seeking his urban childhood there is added the almost inevitable burden of having his physical environment obliterated. The urban past is notoriously unstable, so that in urban experience we often outlive our environment.

Just as the prepossessing verticality of urban space encourages us to endow body-scale places as experiential landmarks, so too in the rush of urban time is it necessary to further endow such loci with the ability to act as functional clots in the flow of time, in short, to stop time for a time. In urban argot we call this "hanging out," and we might reflect on how it differs from a new version, oriented toward nature and called "dropping out."

The verticality of urban space turns vision inward, and the speed of urban time revs up our capacity for multiple experiences, thereby intensifying the need for inner personal space to play out the experiences subsequently in our own "good" time. More often than is supposed, urban man does attain management of his inner personal space, and, contrary to the offensive cliché of antiurban critics, anonymity is *not* a major urban problem. As a matter of fact, urban life is crisscrossed with rich interpersonal relations, brought about by the extraordinary shortcuts to interpersonal intimacy which flourish in the type of situation we have been describing, namely, a welter of experiences, rapidly undergone, yet transacted by the ability of the person to impact some of them in both space and time and convert them to sources of emotional nutrition. Having cut to a bare minimum the time-span required to forge urban interpersonal relationships, in general one does not expect that longevity be a significant quality of these relationships. The pace of urban time, coupled with the people density of urban space, churns up tremendous possibilities for interpersonal life, for the multiplicity of relations widens considerably the range and quality of the intersections and transactions operative in our daily lives. As we see it, then, the major problem in urban life centers not in the relations between persons but rather in the relations of persons to the urban environment and in the studied institutional insensitivity to the aesthetic qualities germane to the processes of urban space and urban time.[4]

If we come full circle and remind ourselves of the need for urban consciousness-raising, a warning is in order. The healing and amelioration of the contemporary American city will be stymied if our efforts betray an ignorance or insensitivity to the experiential demands of the original qualities of being human in urban space and in urban time. As in most human situations, the caution of William James is a relevant here:

> Woe to him whose beliefs play fast and loose with the order which realities follow in his experience. They will lead him nowhere or else make false connexions.[5]

NOTES

(1.) Nature time is not always kind and regenerative, as the Dakota sod-farmers of the nineteenth century and the Okies of the twentieth century discovered. A recent and extraordinarily powerful and original version of the systematic madness often found in nature time is the photographic essay by Michael Lesy, *Wisconsin Death Trip* (New York: Pantheon, 1973). The setting is rural Wisconsin from 1895 to 1900 and the common experience is laced with misery and affliction.

(2.) Cf. Robert Sommer, *Design Awareness* (San Francisco: Rinehart Press, 1972), p. 66. In his chapter on "Space-Time," Sommer tells us that "a San Francisco radio station announces the exact time 932 times a week."

(3.) Surveying big city newspapers, one finds that the screeching headlines of the first edition frequently do not merit even a paragraph in the last editions. Are these instances references to pseudo-events, or is it the pace?

(4.) We refer here not only to the erosion of aesthetic quality in the urban environment, symbolized by the faceless projects of the days of urban renewal, but to the more subtle and equally important fact that we fail to articulate, let alone sanction, the still existing aesthetically rich experiences of city life. The development of such an articulation is equivalent to an urban pedagogy. For an important step in this direction, cf. Jonathan Freedman, *Crowding and Behavior* (New York: Viking Press, 1975).

(5.) James, *Writings*, p. 205.

Time and nonverbal communication

THOMAS J. BRUNEAU

■ Thomas Bruneau *is Assistant Professor of Speech-Communication at Eastern Michigan University.*

Professor Burneau argues that clock time, the western industrial time already discussed by Lewis Mumford in this section, influences our behavior in areas below our perceptual and conceptual thresholds. The antidote to this unconscious acquiescence to uniform, spatialized time is "research" into alternative time structures, or what he calls "time shifts riding on the back of a constant time culture."

What, then, is time?—if nobody asks me, I know; but if I try to explain it to one who asks me, I do not know. Augustine, *Confessions*.

Only man hopes and only man has a notion of time, and death. Though time remains for him a *memento mori,* creative man builds his own time.[1]

It is impossible to mediate on time and the creative passage of nature without an overwhelming emotion at the limitation of human intelligence.[2]

This essay will attempt to focus the reader on cultural, social, and subjective aspects of human time-experiencing. Due to "space" considerations, I will be able only to hint at a number of interesting aspects of human time-experiencing, thus limiting the supports for my claims.

The hyphen between *time* and *experiencing* indicates the dynamic relationship of experiencing and time *variations*. The subject of "time" is complex and profound—and, yet, appears to be simple and obvious. In this essay, I will discuss: "Clock-time and Insanity," "Common Reactions to Clock Insanity," "The Meaning of *Now,*" and "Current Movements into the Moment" (popular journeys into different kinds of mental "time-shifts" or "time-wraps").

Put your watches away for a "while." (A *while* is a strange word we often use to justify our flights from interesting experiences, e.g., "I can stay only for a 'little while,' or, I just can't sit around here talking and 'whiling away' the hours!") It is my hope that the reader will while, while he or she reads about whiles.

CLOCK-TIME AND INSANITY

Both Marshall McLuhan and his "father," Lewis Mumford, have said interesting things about the message of the clock as a medium. They have, however, ignored any elaboration about the magnitude of the clock's impact or its widespread influence on the behavior of people. We should preface our inquiry by saying that clocks do serve some important purposes related to productivity, efficiency, and order. The clock, however, influences people to act with bizarre mannerisms. Clock control and influence are often below the cognitive and perceptual thresholds of most people. Time control has even evaded the deep study of "nonverbal communication theorists." In "passing," I would like to say that clock-time and, especially, time-experiencing outside of clock-time *are the bases of all nonverbal behaviors*. This should become more apparent to even the most "spatial" reader as we proceed.

The click-clunks of many variations of punch-in and punch-out clocks resound off the rigid walls and borders of this country. The RPMs in crepe paper engines pound against the insides of crepe paper machines. Angry people with unshifting, intent eyes begin the "narrow" trip down the freeway or subway. Traveling on

□ From *Journal of Popular Culture,* 8(1974), 658-666.

their linear "time-ribbons," with blank, empty stares, the Americans burrow into a little tunnel in the back of their heads. The tunnel leads directly to the safety of "home," being "on-time" for dinner, and another time-ribbon freeway called the "six o'clock news." "Regularity" in evening TV viewing is becoming increasingly a patterned behavior. Walking on worn sidewalks and pathways at work every day, Americans seldom wonder why millions of miles of sidewalks in their neighborhoods are seldom used. Structuring their space to structure their every movement, they take pride in building their geometrical prisons. As the clowns of the Western world, they find deep security in traveling, in like-fashion, mental time-ribbons.

Locked into their TV puppet shows, they cry out for slow-motion time "shots" of bodies crunching, motorcycles and rockets flung through death-defying air—to the querulous glances of the Third World. With greedy eyes, they gather up these "violent times" with much applause. They shatter the quiet air with their high-speed, clamorous machines. As Thomas Merton would have it, "They [those who love their own noises] bore through silent nature in every direction with their machines, for fear that the calm world might accuse them of their own emptiness."[3]

Ennui is little understood as a function of time-experiencing. However, one thing seems fairly certain to me: mass boredom results from uniformity of time conception and uniformity in responding to clock-like, temporal and rhythmic pacers. These pacers often extend and expand from habitual conceptions of clock-time. We utilize many different points and activities in our waking days to "*keep*-time." We are discussing here the "patterning of time" in American culture.

Driven by super clocks which count off milliseconds, nanoseconds, and picoseconds—we "keep-time" with Greenwich—at an accuracy of one-hundred-thousandths of a second. The signal from Greenwich is clear, intense, and powerful. This semiotic monster is obeyed by Americans on a very large scale—as if the signal were generated from some "heaven."

Despite ulcers, neuroses, psychoses, schizophrenias, and a whole lot of other "oses" and "phrenias," we dive into our projects and plans without understanding that there is often more wisdom in a snail than a tight schedule.

Despite the notion that words become tasty and palatable, expanding mental-time when we chew them well, large numbers of people with speed-reading orientations ignore the possibility that they have not "gotten the word." Some tinker-toy babies with erector-set minds become adults, computer experts, who collect "permanencies" they little understand into codes, input these into machines they do understand, and gleefully claim output designed to help you travel along even more narrow time-ribbons. Erecting electrifying castles, they choke-out their explanations about the "meaning of zero." Dividing clock-time as if it were the only mathematical operation worthy of pursuit, they split the second into involuted infinities, all the while hoping for more precise and faster technology.

Our machines, institutions, and organizations direct schemata and schedules at a tempo which is often increasingly below conceptual and perceptual thresholds. (Clock-time directs the machines, institutions, and organizations). We catch ourselves running and, "sometimes," find ourselves saying, "Where-the-hell-am-I-going—what-the-hell-am-I-doing—what-the-hell-am-I-running-to-or-from?" Then we run even faster to avoid being "left behind."

COMMON REACTIONS TO CLOCK-INSANITY

Two common reactions to clock-like *impositions* of temporality on mental-time are: quests for permanency and punctuality.

The clock as a message extends to help us reify, thingify, and object-ify our spaces (mental spaces as well as those we imagine to be completely external to us). We will discuss in another section of this essay notions of mental time.

Our quests for permanence in the fluxion of imposed temporality are taking on absurd-looking robes. World militaries march in temporal precision like rosy-faced puppets in the eyes of people whose cultures have been badly damaged by border-makers and "watchmen." Our silent cities of the dead support monuments which are very weighty, indeed. The graveyard is a depository for stylized monuments built to remind us of the "last ditch" efforts to achieve some security and durational stability. The concept of "home" in large sectors of the society is coming to mean: a stable "resting" place, decorated to prevent any intrusions on personal clock-time (i.e., impersonal, personal time).

We gather up and place around us our mementos and treasures of "good times." Collectors all, we try to achieve some form of tangible permanence. Some collect objects from nature, some collect artistic expressions, and many collect "antiques" and family "keepsakes." Some collect and save money as their "rock against aging." Some collect unchanging dogmas and ideologies to gather dust on their frontal lobes. Some people even gather nouns to defy the verbs of the world. Some people expend their energies creating and recreating the events of memory (the PAST). One can feel a sense of permanency when identifying with histories—especially those histories of family and ethnic origin which some are fond of retelling. Concerning the permanencies of historicity: *Thens* are, at best, assumed permanencies of fading memories.

As living becomes a process of adapting to greater and greater speeds, the quest for permanent-like stabilities and concrete-like, momentary experiences increases. In short, obsession with permanency in the face of electronic speeds leads to a counter-reaction—an obsession with expanding the present moment. (We will discuss this more "later"). As part of the reaction to increasing speeds, we may also note a growing "repetitive compulsion" in American culture.

Punctuality in an odd sort of behavior. It seems to be pushing at us behind every wall and time-ribbon. It is difficult to imagine why someone would enjoy being "right on the dot." There certainly is not a great deal of elbowroom on a dot and I have found the mobility there to be quite circular and restricted.

Adherence to punctuality and periodicity has the effect of an admission of compliance to those who control temporality and, consequently movements. This admission of compliance is especially prominent in "gathering" situations where one does not wish to be in attendance in the first place and, in the second place, in situations where an authoritarian cloud hangs over the scene. Punctual behaviors can and often are manipulated to initiate, develop, and preserve order, power, and position. Snub a few "mandatory" meetings by coming late or missing them—if you question my last statement. Of course, some punctual behavior are responses to important events, respected events, or respected persons.

There is much to be learned about punctuality, a great deal of study due behaviors of punctuality. The person with "little revolutions" in his or her head should understand tha punctual and periodic behaviors are rooted in the creation, maintenance, *and* dissolution of all types of groups. Punctual behaviors provide one of the major bases and prerequisites for "grouping." This notion does not seem to take on significance until a called meeting takes place and ends with a "leader" sitting alone in an empty room.

Punctual behaviors are not simple matters of "being-on-time." There are other "times" which we form-ulate and which are punctual in the sense that they are periodic or pointed (e.g., at this "point in time"). Some "times" are categorical. All of these times can be viewed as extensions of punctuality. These times "place" the flow of time into discrete, categorical units. In this sense, then, punctuality and periodicity can be viewed as quests for permanency. In short, permanency quests are often extensions of clock-like habituations extending from clock-time habituations. As the degree of punctuality of persons increases, those persons can be forced in their communicative exchanges toward two-dimensional semantics and bi-polar thinking. Punctuality and the packaging of "times" seem to provide resistance to slow, durational inquiry.

When people usually talk about "saving time," "making time," "setting aside a little time," "taking time," or "wasting time," they are customarily and unwittingly converting time into spatial representations of the flow of time. The flow of time in the spatial view becomes a commodity, e.g., "time is money." Active involvement with such "times" can cause anxiety and can also prevent deliberate, probabilistic (futuristic) inquiry using slow-time.

"In passing," it seems appropriate here to speculate about periodic, punctual people. J. T. Fraser has remarked that ". . . a person's view of time is a method of discerning his personality. . . . We may almost say: Tell me what you think of time and I shall know what to think of you."[4] It does seem very reasonable that a person's conceptions, expressions, and behaviors relating to time-experiencing may provide valuable cues about his person-ality. Having been one who watches people watch their watches and, all the while, wondering what they are *watching out* for, I have an opinon to offer.

It seems to me that people who run between

"points in space" on connecting time-ribbons, people who frequently exhibit that odd, wrist-turning gesture, and those who frequently glance at clocks on walls are also those persons who have a *certain, dead-sure* attitude about life and living. Should we not begin to deeply study the relationships between highly punctual behavior or rigid temporality and closedmindedness? It seems to me that highly punctual people are those who are "deathly" afraid. When one often views time as "running out," one begins to view time as a devouring, consuming force with some mysterious intent to render one "asunder." Thus, one should avoid "being late," "at rest," "stopped," or "stilled." Could it be that being late and passing on beyond this world are related feelings? What I am saying is that there seem to be some remarkable relationships between rigid temporal responses and fears of authority and death.

THE MEANING OF *NOW*

The characteristics of the English language help to confuse our discussions about human time-experiencing. When we strive to remove spatial and static conceptions and semantic predispositions about the PAST, the PRESENT, and the FUTURE, we discover the momentary mergings and reciprocity of human temporality. When we analytically examine the meaning of "now," we begin to wonder about tense systems and begin to realize that representations of nouns are verbs. When we view language as a psycholinguistic-temporal process, the dynamics of memory (PAST), sensation-perception-conception (PRESENT), and anticipatory cognition (FUTURE) blend into the mental moment. We live in moments of mental tensives. While it appears to us that "time passes" into some great warehouse called the past, the past is only a function of the processes of remembering during the moment. In this sense, then, the past is dynamic and undergoing fluxion. In this view, the past no longer becomes some supermarket holding memories, the past is shaped and reshaped by the physiological and psychological constraints of the moment. Histories and futures are momentary. The difficulty in discussing time-experiencing arises when we use words such as "is" and "it." While we use such words and many similar words for convenience, such words should be viewed as quests for permanencies by verbal minds which move *constantly*. (We will deal with minds which move *variably,* too).

Not only does the present moment function as a complex and dynamic relationship between mnemonics, sensory based psycho-physiological-electro-processes), and the processes of probabilistic thought, anticipations and expectancies—the now expands and contracts.

Time, then, does as the human brain does. The *now* is what the human brain does *while* it is doing what it does. Physicists and mathematicians are fond of externalizing time in spatial representations and positing "relative" observers in expanding space. What of the relativity of the *mental* time and space of these observers? It is my position that time distortions, time-warps, time-shifts, and temporal relativity are but variations of the psycho-physiological and chemo-biological processes of *developing* human brains—what else? The reader is referred elsewhere for some exciting readings which deal with these matters with sophistication and detail.[5]

If time *does* according to variations in neurological process—what then does this kind of time function *on* or *in?* A "reasonable" question. The backdrop for "time" is an interesting challenge. It is like asking, what is outside the entire universe. The backdrop for subjective or psychological time-experiencing is little understood. We could say that the backdrop is "the basis" for life and motion, but one does come to recognize that the backdrop of time moves, *too*. The study of "interior distances" of psycho-physiological and bio-chemical time-space is just beginning to have contractions and expansions. I do, however, find excitement in the idea that the "nothings," "unknowns," and "empty spaces" in the functioning of the brains of humans are the frontiers of time theorists. The future of the study of time should tend toward deeper probes into the functioning of the *intercepts* between psychophysiological-biochemical processes and those processes which we often refer to as conscious and unconscious thought. Simply stated, the future of time study is in the intercepts of brain and mind. The meaning of "now" will be further described below.

CURRENT MOVEMENTS INTO THE MOMENT

Depending on the level of neural functioning and psychological functioning to which we refer, the "now" can be viewed differently.

For instance, concentrating our attentive vigilance on some object or event changes time-experiencing. We are becoming a highly visual culture. When we attempt to *become* the object or event we behold, we alter time-experiencing. This process can be viewed as a quest for permanency. Through such efforts, we can become children of the moment. Walt Whitman said it well: There was a child went forth every day, And the first object he look'd upon, that object he became, And that object became part of him for the day or a certain part of the day, Or for many years or stretching cycles of years.

It is well known that meditational states achieve an expansion of the moment by minimizing external stimuli. Such popular activities as yoga, transcendental meditation, and the like are examples of movements into the moment.

The visual explosion is making tremendous demands on our attention. We are not merely molders of our own time-experiencing—the intensity, duration, and frequencies of exterior sources of stimuli are increasing. The image expands the now. Image management functions to expand the now. When we willfully cut out or are forced to cut out environmental sources of stimuli because of the intensity, duration, and frequency of input stimuli, we expand the now. The image and slow-motion techniques seem to encourage the ignorance of peripheral stimuli. Through sensory shocking, the images burrow into our heads.

In a clamor, one finds many devices to retreat from the enticements of the rapid-fire visual and auditory bombardments of American culture. Strivings for solitude and silence are especially popular reactions against imposed temporalities and taking part in stimuli gluttony. I have elsewhere attempted to relate time-experiencing to the silencing of our own minds for thoughtful purposes and the role of silence and silencing in interpersonal relations.[6] We often achieve silence by daydreaming, thinking in the abstract, and taking other forms of mental vacations. We can find silence in a noisy crowd —we can will this silence as well as be driven to silence by information overloads. When we activate the wrinkles of inquiry and puzzlement in our faces, we activate *interrobangs* of time-expansion. An interrobang can be graphically represented by the merging of the exclamation and interrogation marks:

This mark can be viewed as a mark of mental-time variation from some state of equilibrium. This mark is a mark of "questioning-uncertainty." The blank stares associated with some of our mental journeys seem to coincide with mind-time expansions of "now." Bergson has called such silences—durée. Maslow has discussed certain of these silences as "peak experiences." Jung has referred to such journeys as somewhat "timeless" probes into the unconscious. Freud has remarked that "dreams" are timeless.

The retreats to quiet and restful places by large numbers of people are increasing—there are many indices of this increase. These retreats may be matters of "finding a place to think." Solitude and silence help to expand the now. Our quest for that little cottage on the unknown lake, or that little island where others dare not tread (our study?), or that far-off place in the mountains may be simple reactions against imposed temporality. However, such quests can also be responses to an increasing patterning of mental-time in this culture. As to student escapism, behavioral objectifiers and curricular planners are creating narrow time-ribbons and many educators have somehow lost the notion that students need time (as defined) to expand time (as defined).

I do not wish to be an "alarmist." Yet, time-distortions, as a result of the ingestion of every sort of chemical and metabolic change agent, are becoming *very* widespread. How many drug-addicts, alcoholics, and "soft drug" people can dance on the head of a pin? (On the pin of a head?), I can not ascertain. It is known that many hallucinogens alter time-experiencing by altering body temperature, neural speed, metabolic rates, and the rates at which chemical processes happen. The agents which millions are ingesting induce catalytic (and sometimes catatonic) reactions in the human brain which are deceptively complex. The reactions to the ingesting of psychotomimetic stimulants such as LSD, mescaline, etc. (as well as depressants and tranquilizers) are being enjoyed by millions of youth. (And let us not forget or disregard the massive influence of the pharmaceutical people.) Such reactions are well known to scientists studying time-distortions; what is not so well known, however, is how time-distortions affect human relations and communication and, especially, *intra*-personal communications. (See note 5 for information about time-distortions and drugs.)

One can feel "communal" or "unified" in a group of people exhibiting timeless-like trances without any personal interaction with others. One can become a lonely floater in mental spaces he little understands. Tripping over one's ego through the annihilation and ignoring of the communication of others can spawn large numbers of self-proclaimed geniuses who have given up on both books and listening. Drugs which change mind-time can result in burnt-out, frightened or zombie-like youngsters who could "care less" about others—and care more about such time-associated abstractions as Gods, Devils, Spirits, and other forms of deity. The problem and the danger is expressed well by Paul Fraisse: ". . . there are moments when we try to escape completely from the pressure of [social time] but it is a dangerous freedom when it becomes a habit, for it can lead to insanity, which is precisely a break between the individual and [others]. . . . Human equilibrium is too precarious to do *without* [italics mine] fixed positions in space and regular cues in time."[7]

I do believe we must begin to discuss the seriousness of this matter without cheap references to valleys of dolls and blowing our minds. Diving into zero with Alice and sipping tea at the Mad-hatter's party or sitting with Thomas Mann on magic mountains are positive time-shifts. Changing the basic neuro-chemo-electro processes of a large segment of an entire generation, however, is another matter. Ego-soothing "trips" can result in lonely crowds and mind-cave burials. I see a possible development of an entire generation with drug-induced time-shifts riding on the back of a constant-time culture.

The "time-shifts" induced by retreats into silence and solitude may be healthy reactions to impositions of a quickening temporality. We may also consider time-shifts of fantasy, day-dreaming, and many forms of mental stance as necessary and capable of producing many forms of positive, creative expression. Metaphor and artistic movement, for instance, are examples of such positive shifts externalized. However, time-shifts and warps where a person has not the slightest notion of where the journey began or where the person "came out" may be just exhibitions of intellectual and behavioral finger-painting.

In this essay, I have tried to develop several ideas which are important aspects of time-experiencing: clock-time influences people in many ways; imposed temporalities are major forms of persuasion; spatial configurations induce time-experiencing variations; there is a cultural movement into the "now" and a desire to expand this now; this reactive movement can result in the traveling of "worthwhile" mental pathways or, on the contrary, the traveling of pathways which lead to being hopelessly lost on circular, space-time ribbons. One of my hopes in writing this essay is that people will come to consider the study of time-experiencing as basic to the study of human communication. Time is the basis of nonverbal experience and communication. Perhaps one of the astronauts, when on the moon, said it better than I.

While in deep, silent inquiry over a rock formation, the astronauts were behind schedule. Despite pleas for moving on by Earth's mission control, they were caught-up in the analytic silence of the "now." Finally (?), apologizing for the delay, one astronaut without understanding the significance of his statement, blurted out: "Gee! Maybe time is different in space!"

NOTES

[1]J. A. M. Meerloo, "The Time Sense in Psychiatry," in *The Voices of Time,* ed. J. T. Fraser (New York: George Braziller, 1966), p. 252.

[2]A. N. Whitehead, *Concept of Nature* (Cambridge University Press, 1920), p. 73.

[3]T. Merton, *No Man Is an Island* (New York: Harcourt, Brace and World, 1955), p. 257.

[4]J. T. Fraser (ed.), *The Voices of Time: A Cooperative Survey of Time as Expressed by the Sciences and by the Humanities* (New York: George Braziller, 1966), p. xix.

[5]The reader interested in temporal insanity, time distortion, and temporal aspects of mentation is directed to J. A. M. Meerloo, *Along the Fourth Dimension* (New York: John Day, 1970); J. Cohen, *Psychological Time in Health and Disease* (Springfield, Ill.: Thomas, 1967); H. Yaker, et al. *(eds.) The Future of Time* (Garden City, N.Y.: Doubleday, 1971); A. Hoffer and H. Osmond, *The Hallucinogines* (New York: Academic Press, 1967); R. Fischer, "Biological Time," in *The Voices of Time* (see above).

[6]T. J. Bruneau, "Communicative Silences: Forms and Functions," *The Journal of Communication,* 23 (March 1973), pp. 17-46.

[7]P. Fraisse, *The Psychology of Time,* trans. J. Leith (New York: Harper and Row, 1963), pp. 289-290.

4

Information and cybernation

Who will lead the way to the "information society"?

OLE ENGBERG

▪ Ole Engberg *has worked in industrial production engineering and production management and for a Danish computer manufacturer.*

In this article Engberg argues that the present trend toward an information-based society is inevitable and that such a society will, by means of greater automation, provide increasing options for the individual. He also argues that unless the present competitive materialism is supplanted by new attitudes toward work and consumption, the result will be a totalitarian society.

Increasing use will be made of automated production and information communication techniques in the years to come. The result will inevitably increase the ranks of the unemployed. We must therefore put more energy than we have done hitherto into changing our present work ethic. It is no good extolling the ennobling qualities of work if, at the same time, we are cutting down jobs through automation.

A long-term and innovatory project, beginning in primary school, is what is needed. And in the short-term, we must concentrate our efforts on ensuring continued free access to information.

THE NEGLECTED INFORMATION SECTOR

The information sector employs half the labour force

In 1975, the Organization for Economic Co-operation and Development (OECD) published a report by the American, Edwin B. Parker [1],[1] in which, with Marc Porat, he analyses a number of problems that merit discussion. The report includes a graphical history of the evolution of the labour force in the United States since about 1850 (Fig. 1). The curve indicated by the solid line represents the information sector, one which has trebled over the past thirty years. From employing 15 per cent of the country's labour force in 1945, it has risen to well over half.

The information sector is made up of all those in the three traditional occupational sectors—agriculture, industry and services—who work with information. An appendix to Parker's report, containing figures from the official statistics for 1970, indicates the labour force in 400 separate groups. Of these groups, 170 belong to the information sector, the largest being officials and managerial staff, teachers, secretaries, accountants and engineers. Authors, postmen, printers and archivists are also included. The smallest group is 'clerical assistants, social welfare'.

Even if this classification can be questioned —as Parker himself does—the general trend is highly convincing. A similar survey in other industrialized countries would very probably produce a similar curve, with the traditional two- to four-year time-lag in relation to the United States.

Automation of information processing

Just as the drop in employment in agriculture can be ascribed to mechanization and the

1. Figures in brackets correspond to the references at the end of the article.

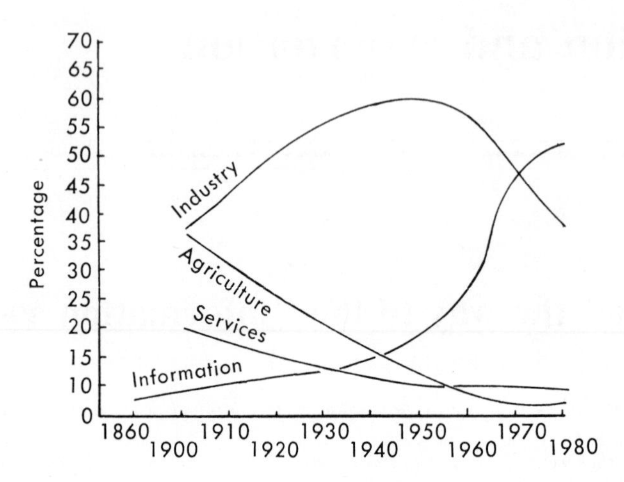

Fig. 1. United States labour force: four sector aggregation. *Source:* Edwin Parker and Marc Porat for OECD, 1975.

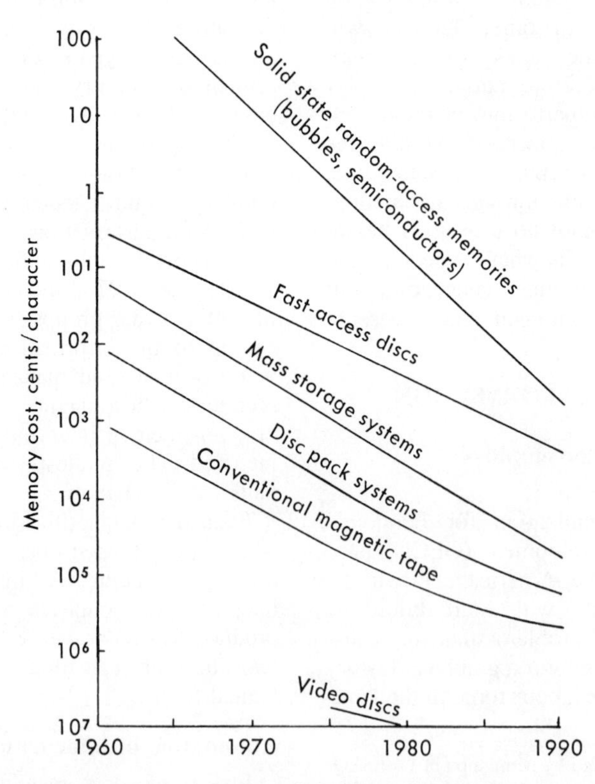

Fig. 2. Memory cost trends. *Source:* Report of the Privacy Protection Study Commission. Appendix 5: Technology and Privacy. (Stock No. 052-003-00425-9.)

drop in industrial employment to automation, increasing use of electronic data processing (EDP) techniques can be expected to result in significant structural changes in the information sector.

Figure 2 comes from a report recently published in the United States on the subject of technology and privacy [2].

This report contains corresponding curves for the actual calculation units, etc. A fall in prices can be expected for many years to come. New areas in which automatic techniques can profitably be applied are constantly being opened up.

Parker considers that there are already several areas within the information sector which are neglected from the investment angle with the result that full use is not made of possible applications: (a) expansion of the educational system so as to make the range of instruction offered independent of time or place, namely use of video tapes and programmed courses, to make instruction essentially automated and personalized and thus create an entirely new market (a new category of student) in addition to the present one; (b) automation of the payment system (despite the various risks involved, both as regards privacy and employment); (c) speeding up of business transactions through automation, especially of international paperwork; (d) improvement of consumer information as regards price, quality, speed of delivery, etc., by means, for instance, of setting up terminals, e.g. in libraries; (e) more effective public administration, especially at the service level; (f) extension and improvement of the public telephone system through video, teleconferencing, etc.; (g) improvement of the public health service. It is hard to contest this approach and Parker gives [an] interesting explanation of why the necessary investments have not been made and points to the danger that this field may be taken over by commercial interests.

Future employment trends

Printers are well aware of the implications of automation. Secretaries, middle-grade managerial staff and several other categories are now about to be hit in their turn. This is clear from the sales prognoses carried out by EDP suppliers in respect of text processing equipment. Such equipment is sold not only on the basis of promises of better service. It is financed partly by the savings made on salaries. In large companies with the necessary financial resources, staff dismissals can be reduced or avoided completely through a reasonable personnel policy, i.e. natural retirement. In smaller companies, such equipment is often a matter of life or death and EDP therefore ensures that the whole enterprise is kept going.

The "fortunate", those who keep their jobs, cling to their carefully acquired rights and, for example, through the intermediary of the tax authorities, share the amount saved with those who are put out of work: unemployment compensation.

In short, local, national and international competition makes it necessary to step up productivity. The claim of the instigators of this drive that more work is created by automation "if one looks at the economy as a whole" is understandable, though not necessarily correct. Nor indeed does this claim get much support except from the economists, even when it is made with great conviction, as was the case, for instance, when the EDP experts included this subject [3] on the programme of their own world congress (that of the International Federation of Information Processing), in 1977.

One of the most recent attempts to clarify this problem was made by the Swedish Central Bureau of Statistics [4], which notes, with many reservations, that EDP has taken over 60,000 to 90,000 jobs and created 30,000 new jobs. This amounts to a net decrease of approximately 1 per cent of the total labour force. The report estimates that much more could be done to exploit the possibilities for savings of labour.

Figure 1, from Parker, not only reveals the scale of the information sector but conceals a problem. The curve shows only the relative employment situation, and then only up to 1980. Figure 3 is based on Parker's four curves, as well as certain ILO employment statistics [5] relating to the total population of the United States in various years. A rough estimate is also made of working hours, age groups, etc. The trends indicated, rather than the precision of the information, are of interest.

Occupation in the information sector (curve I) is expected to drop. Investments will probably be made in the seven areas mentioned by Parker (items (a) to (g), [above]. The result of these investments will naturally be to slow the drop in employment somewhat, since development and installation require extra manpower—but afterwards there may, on the other hand, be a considerable drive to make such investments pay. Presuming, of course, that by then we have not revised our approach to work and money.

The other three sectors will also become increasingly automated, with the result that Parker's relative statistics in Figure 1 may be said to apply until the year 2000. If, however, one looks at the total employment figures for all four sectors as a whole, the drop in the total employed (curve T) will be quite significant. This forecast corresponds fairly closely to a Swedish prognosis [6] carried out in 1972, which anticipates that working hours will be halved. That was when there was a considerable shortage of manpower, and before anyone thought of the present unemployment crisis.

The outlook is therefore not particularly promising. Figure 3 shows the various possibilities open to us if unemployment is to be kept to a minimum:

We have probably exhausted our possibilities of increasing production through exporting goods to the less-developed countries. They are frightened that our essentially competition-orientated culture will accompany such exports; they prefer to keep control over development in their own hands. The only exception is arms—and here the industrialized countries exploit most production and export possibilities to the full.

We can also try to persuade our fellow-citizens to move from the available work-force category (curve W) into the top group: longer education, early retirement, with all their attendant problems.

Finally, W can be brought down even closer to T by expanding free time.

There would also seem to be no grounds for believing that our own consumption can continue to grow. Gunnar Adler-Karlsson has written, in a paper [7] for the Swedish Secretariat for Futurology *(Svenska sekretariatet för framtidsstudier),* "To maintain full employment at a rate of eight hours work 48 weeks a year for 40 years, it is essential to commercialize free time to the full by making 80,000 Kr. worth of plastic boat an entirely natural material human need, calling for State subsidies." This sentence figures in one of the chapter headings, and the paper as a whole is entitled "NO to Full Employment, YES to Basic Material Security" and outlines some of the policy aspects as well as a model solution. This model specifies limits for the size of houses, the number of boats, and private aeroplanes, which may be produced by the automated factories. Both the shortage of resources and environmental considerations are increasingly imposing restrictions. The underlying

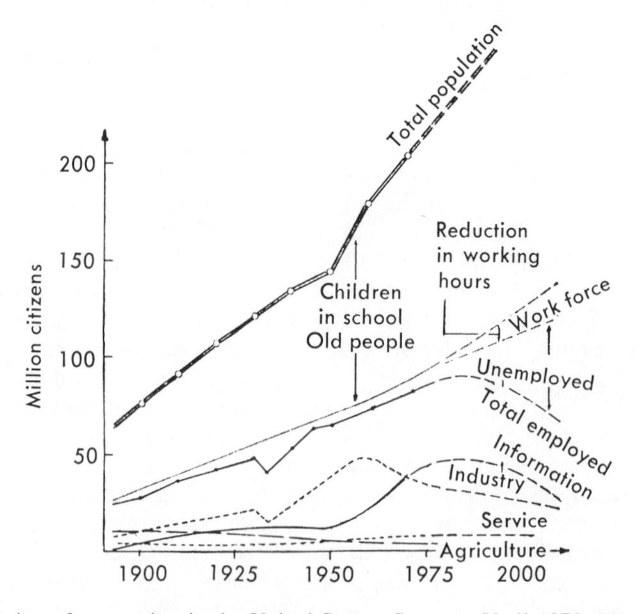

Fig. 3. Distribution of occupation in the United States. *Sources:* Until 1970: *Historical Statistics of the United States, From Colonial Times until 1970,* Bureau of Statistics, Washington, D.C., 1975. (Part. I, p. 8 and 139.) Information sector by Edwin Parker for OECD in 1975. After 1970: author's estimate, as of February 1978.

reason is the need to start to think differently so that the information society can be run according to human, not only economic, conditions.

TOWARDS AN INFORMATION SOCIETY

What is the information society?

Parker does not deal with this matter in detail. He merely points out that we already live in an information society, since more than 50 per cent of the labour force works with the communication of information, even if we have not awakened to these facts and their consequences.

A detailed description of the information society has been provided by Japan, *The Plan for an Information Society–a National Goal Towards Year 2000* [8]. This plan, which incorporates a Japanese programme of action, is based on a theory advanced ten years ago by Kenichi Koyama [9].

The Japanese plan was prompted by several of the problems which are easier to detect in other countries than in one's own: steadily expanding towns, which become more and more difficult to run effectively, and increasing automation of production machinery, with the resulting unemployment and other national and international consequences—which do not need to be developed in greater detail in this article. Reference is also made to environmental problems and the shortage of resources.

Masuda [10] reproduces the main points of this plan in a short article. The above trend can be counteracted if we (a) make full use of EDP for individual selection of information (so far we have only been scraping the surface); and (b) create system-oriented information, i.e. automatic selection, combination and, if need be, concentration of information previously scattered. Clearly these applications should be accessible to the public and not be reserved for private owners or public officials.

Table 1. Comparison of information society with agricultural and industrial society*

	Agricultural society	Industrial society	Information society
Production power structure			
Production power form	Land production power (farmland)	Production power of motive power (steam engine)	Information production power (computer)
	Material productivity	Material productivity	Knowledge productivity
Character of production power	Effective reproduction of natural phenomenon	Effective change of natural phenomenon and amplification	Systemization of various natural and social function
	Increase of plant reproduction	Substitution and amplification for physical labour	Substitution of brain labour
Product form	Increase of agricultural product and handiwork	Industrial goods, transportation and energy	Information, function and system
	Agriculture and handicraft	Manufacturing and service industry	Information industry, knowledge industry and systems industry
Social structure			
Production and human relations	Tying humans to land	Restricting man to production place	Restricting man to social system
	Compulsory labour	Hired labour	Contract labour
Special character of social form	Closed village society	Concentrated urbanized society	Dispersed network society
	Permanent and traditional society	Dynamic and free competitive society	Creative and optimum society
	Paternalistic status society	Social welfare type controlled society	Social development type multifunctional society
Value outlook			
Value standard	Natural law	Materialistic satisfaction	Knowledge creation
	Maintenance of life	Satisfaction of sensual and emotional desires	Pursuit of multiple social desires
Thought standard	God-centred thought (religion)	Human-centred thought (natural science)	Mankind-centred thought (extreme science)
	Ecclesiastical principle	Free democracy	Functional democracy
Ethical standard	Law of God	Basic human rights, ownership rights	Sense of mission and self-control

*Source: Yoneji Masuda.

Masuda goes on to describe how as a result of these developments the individual's dependency on time and place will gradually decrease. At the same time the need for grouping together in factories, schools and offices will also disappear. This in turn will affect forms of employment, bringing about a transition from the usual, permanent job to short work contracts, thus affecting forms of leadership and political standards. With good reason, Masuda devotes the last part of his article to the radical behavioural changes which life in the new society will require. A comprehensive survey of the adjustment process is perhaps shown most clearly in Table 1, taken from Masuda's article [10].

Bound for collision

The basic and most essential message of the picture of the information society presented by Parker, and more particularly by Masuda, is of a society where automation has eliminated the industrial society's competition for material possessions, rooted in status symbols or a craving for power.

Such a change of behaviour in the industrialized countries will cause many problems. Competition has gained a secure hold in our culture and the manner in which we have organized the leadership of both private enterprises and public authorities. Most people are, not without reason, fairly satisfied with this system which has given the industrialized countries a very high material standard of living and created a society where no one needs to be destitute. We are so pleased with the results that we overlook the fact that the industrial society is in the process of being replaced by the information society. Those of us who are engaged in activities influencing public opinion will be hit particularly hard. Our work makes it difficult for us to question the ethics underlying the present power structure, which guarantees us our influence.

None the less, we are being forced to it by the inevitable process of automation. Fewer and fewer people are able to place their trust in us as the discrepancy between what we say and what we do becomes more and more obvious:

We promise decentralization and the right of participation, while at the same time building up administrative EDP systems on a national scale to ensure total equality and improved statistics for the experts' use.

In a competitive society, the combination of liberalism and statistics creates the belief that everybody is to be above average. Those who can manipulate the letter of the law to their own advantage enjoy increasing respect. The politicians have no choice. They have continually to enact new provisions and see that these are put into effect. "Millimetre justice" may forever secure employment for EDP specialists.

We say that the administrative use of EDP can ensure the necessary "comprehensive picture" which has so far been lacking, yet at the same time we standardize and simplify descriptions of the private individual without explaining how this comprehensive picture is composed or what it tells us.

Once again this conflict may be explained by the need for effective administration. Fear of the personal identity number is an unfortunate expression of the resulting mistrust. Characteristically, hardly anyone has tried to explain that this number makes it technically possible to group together the various details that sum up an individual and thus obtain a more complete description as well as reducing errors. This will make it possible to show consideration for the individual, even in public administration and policy-making.

We say that job satisfaction increases when people are given sufficient time to carry out their work properly, yet at the same time we step up the tempo for reasons of competition.

The well-educated, well-paid employees in permanent jobs have found it necessary to set up their own trade unions to ensure that their particular group is not adversely affected. Such is the pressure from the less well-off. Once again, carefully gained rights are in danger.

We extol the ennobling qualities of work, the cornerstone of our industrial ethic, yet at the same time, in the name of competition, we are automating many jobs out of existence.

This conflicting attitude is so firmly entrenched that only the unemployed react by asking, "Whose interests does this work ethic actually serve?"

In the well-off, industrialized countries, it is probably necessary to have only slightly higher unemployment or to risk slightly greater with the poor countries, for some of these conflicts to become obvious. We must therefore revise our approach.

WHO WILL LEAD THE WAY TO THE INFORMATION SOCIETY?
Our generation is hardly trained for the task

The purpose of the preceding sections was to emphasize three truisms:

Continued automation of production and information is inevitable. Very few people would wish to stop it, nor is the optimist at all interested in doing so. Almost everyone is concerned with having his basic material needs satisfied as quickly, easily and generously as possible.

Such automation can be a dynamic force in the information society by providing increasing options for the individual. This is desirable.

If we do not, meanwhile, manage to get rid of the present competitive mentality (focused on material possessions) and turn instead to individual creative use of information, automation may lead to a totalitarian society.

For many reasons the present generation is not prepared to make full use of the information society's potential. We are far too influenced by what we learned as children and our loyalty for our employers, customers, pupils and colleagues.

The trouble is that we really have learned to regard work as ennobling. At home and at school, as well as in our jobs. In the very first years of school the indoctrination starts with the idea of the recreation break, which was—or was intended to be—desirable. Anyone who thought otherwise was peculiar, an ugly duckling.

On the other hand, for most people to learn something, to work, to do something useful became a special activity, which had to be remunerated. This remuneration corresponded to the effort, which in its turn was measured in terms of hours. Unpaid work hours came to be looked on as lack of solidarity.

This expression sums up all the good old attitudes. The sacred cows of the competitive society and of the class war are at stake: "It was necessary to fight to bring the working week down from 60 to 48 hours, and from 45 to 40. Similarly the 2-3-4 weeks summer holiday was no free gift! The capitalists have given away nothing. All this has been achieved through solidarity." Gunnar Adler-Karlsson has also written [7]: "On the consumer market the individual has unlimited freedom of choice. On the labour market he has exceedingly limited freedom of choice as regards his working time." This struggle has been going on for so long that its very target—working hours—has disappeared from sight. The wear and tear of traditional work is vanishing with the advent of automation.

This distinction between work and free time would seem to be the motivating force of the competitive society. If this distinction can gradually be done away with, the democratic information society, as opposed to the authoritarian or even totalitarian information society, can be achieved.

Long-term work: schools

School attitudes will have to be changed even more. This will take a long time. The educational system would seem to be the heaviest and most inflexible element of society. It takes twenty-five years to change course.

Schools are at last getting round to preparing our successors to be good citizens—for the growth society of the 1960s! We are urged to be acquisitive, for competitive reasons. The only mistake is that today's pupils will be taking their place in society when the next decade is well under way.

Many, perhaps most, teachers are aware of this fact and do their best to instil in their pupils a critical attitude to the causes and effects of the competitive mentality, even if curricula, salary regulations, class quotas and other requirements as regards the teachers' own work make it impossible for them to provide a good example themselves. Their impact is therefore limited.

There are, however, examples of how it is possible to run a school without having every forty-five minutes to demarcate the difference between work and free time. One example is to be found at Tvindskolerne (name of locality) [11]. Here schools have pupils who are above the compulsory schooling age and operate under a special law governing private schools. Under this law, large subsidies are allocated (up to 85 per cent) to cover the operational costs and the staff have considerable freedom in planning the teaching programme. A number of particularly interesting features may be mentioned: (a) school hours can be spread over all twenty-four hours as the pupils themselves decide, since these schools are often boarding schools; (b) the pupils join in any tasks that may be at hand, including cooking, fishing, handwork, farming, etc.; (c) the teachers' salaries are pooled, and allocated by the collective teaching staff; (d) good economic management and growing student interest have allowed considerable expansion; and (e) the schools take a strong interest in developments abroad. In several cases the schools' programme includes cheap, rough-and-ready expeditions (in their own thirty-year-old buses), lasting several months, to other parts of the world.

The next step that can be hoped for is that the ordinary State secondary schools will open the way for experiments designed to free them from the tyranny of the school timetable and show that one is either busy or sleeping. The demarcation between working hours and free time is artificial and implies an employer-employee relationship which is replaced by very different bonds in the true information society.

A box with electronics

Once this difficult barrier—the rigid school timetable—has been broken down, the educational system can begin to train pupils in the technology they will have to live with and use in the information society.

What is needed is a large box with electronic components, some books on electronics, a roof over one's head, and a small measure of constructive interest. These experiments then become a local assignment, requiring help from local electricians, engineers, mechanics, radio enthusiasts and others who can read a blueprint and use a few tools. One also needs a boiler which has an automatic device to open and shut the heating system—as well as whistle —"in case".

Alternatively, the local anglers' club may say: "If you can make an automatic gauge which shows whether there is poison in the river, we will undoubtedly find a use for it." Or an ordinary family may volunteer to act as guinea pigs for any invention.

To derive enjoyment from testing ideas (creativity) on one another, it is not necessary to be an organization. It is highly stimulating and educational when no specialist is in charge and there is full acknowledgement of mutual dependency. Such activities are so creative that most productive enterprises are likely to become interested. This applies both to the producers of electronic components and to those whose responsibility it is to control an industrial process, large or small. All that is needed is a constructive frame of mind.

The electronic box idea may be carried further: electronic components are the mechanical building materials of the information society. In my youth, they were called Meccano or Erector; today, they are known as Lego building blocks. The cost, manipulation and combination possibilities of the electronic box are on a higher level than those of the Meccano set and, thus, emphasize the distinction between the industrial society and the information society and the respective demands that each makes.

The most important and characteristic difference is that the Meccano set can usually only be used to make very rough miniature models of physical reality. The electronic construction is itself the real thing, since it works with information rather than with materials. The difference can be illustrated by a simple example. With Meccano I can build a lift. With electronic components I can build a machine to control the lift. It does not matter whether my electronic machinery is controlling a model or the real thing. The important thing is that this control machinery has a memory. Once it has that, it can be programmed to obey instructions in the order in which they are received or to change the sequence of the execution. Most readers will have had the opportunity to reflect why such an automatic device has not been installed as they observe an empty container passing in the right direction.

The exciting, challenging, stimulating aspect to all this is that what is important is not the machinery or the electronic component but the programmer's ability to look into the future and decide what is going to happen, "if". Only imagination and local conditions set the limits. Memory and ability to calculate are no longer stumbling blocks. What matters is the ability to anticipate what might happen, "if". Here creative abilities and correct analysis of local conditions have their chance. Practical limitations vanish. In the expert's professional terminology, this is called modelling.

The most important contribution that can be made by our generation is—as I tried to make clear—to urge that schools continue to experiment so as to find educational forms that make it possible for our successors to feel the necessity of living in a democratic information society. If our generation makes no such demands, an autocratic system may gain the upper hand.

Christiania—the free town

Alongside such school experiments, there is a need to learn more about societies which have not combined industrialization and free competition in the same way as our own. There are a number of examples. We have read about societies which are treated as developing countries even if their present script dates back several thousands of years. There are also religions which do not place emphasis on the collection of material possessions.

However, such groups can be found in our own pressured industrial society, where those who have jobs envy those who have the freedom to enjoy a sunny day, or to complete what they are doing or let an interesting conversation draw to a natural close. And where the unemployed are ashamed at being out of the race.

In Denmark the best known society of this kind is Christiania [12]. This is a former military establishment, 0.3 kilometres square, 9 minutes' walk from the centre of Copenhagen. It comprises 150 buildings: workshops, dormitories, the commander's residence, ramparts, trees, moats, etc., all dating from the seventeenth century, when Christian IV was king. Hence the name.

When this area was abandoned in 1971 by the army, squatters broke in and within a short time there were at least 1,000 inhabitants who declared this district "the Free Town of Christiania"—free from outside society as regards its police, courts, specialists and professional guardians. The inhabitants, to a large extent, consist of society's outcasts and individuals wishing to protest against the majority's way of living. Some no longer have the energy or ability to manage by themselves outside Christiania. Others have so much surplus energy that they can disregard traditional standards and norms.

The inhabitants of Christiania live off unemployment benefits, social welfare, local crafts, and a number of environmental tasks such as the upkeep of the area. Some have permanent jobs in the city. The main feature of the Free Town would seem to be a lack of respect for steady work. In Christiania one only works in order to live, or when it is amusing. The inhabitants make few demands as regards basic comforts.

The Free Town lives beyond the pale of law. Fencing has been knocked down and over a period Christiania has infringed on property rights and thus undermined a form of society which, with the help of industry, has achieved such a profusion of material goods that no one needs any longer to starve or freeze. In the Free Town people live the whole year round in a way which most of us can only envisage during the summer holidays, when we ourselves choose how we want to use our time.

Opinions about Christiania are sharply divided. Some wish the area to be levelled to the ground—if necessary with the support of the police—"this dangerous irresponsibility must be crushed". Others hope that the plague will spread.

Almost everyone is agreed that these divergent views and wishes are dictated by what people think is best for those who come after them. Very few people believe that their own existence will be changed significantly by what happens to the Free Town. Yet the fight by the inhabitants of Christiania for the right to decide themselves how to use their time is a matter of keen interest to the courts, the government and the outside population. It can be hoped that this interest will continue for many years to come, though it may of course result in the Free Town's being razed with bulldozers. Alternatively it may just wither away with the good wishes of an unconcerned outside world.

Unfortunately no EDP firm has yet had the courage to install an operation in Christiania. The far-sightedness of these firms cannot be doubted. Perhaps, however, the inhabitants of Christiania will not give their permission—unlike our higher educational institutions when they received a similar offer fifteen years ago.

It would have been interesting to be able to say what use the Free Town had made of EDP, and for what purpose. It would have been equally interesting to see how the outside world would have reacted to requests, e.g. for official data or transfrontier data transfer [13]. These questions will be dealt with in the following paragraphs.

Automation on the user's terms

A characteristic feature of almost all EDP systems is that they have been established on the administrator's terms. Their purpose is to enable the administrator to achieve greater efficiency, save money, and provide better service.

Hardly any EDP system is set up on the basis of what the user requires. User participation experiments are currently being carried out in the private sector but so far little progress has been made. This is due not to lack of goodwill but to lack of training and experience. In public administration, it will take a long time before the politicians, who ought to act as the citizens' representatives, acquire an overall view of these problems.

In hardly any case has the question been asked from the start: "What human needs, apart from the traditional administrative functions, can be served by EDP?" Little has been written about experiments of this kind. One of the best illustrated examples is a Canadian tourist service system, a project which provides data on tourist attractions, eating places, open hours, which was progressively expanded. In 1974 it already covered several other needs, including the local Chamber of Commerce directory and information on conditions for getting children into day nurseries [14]. A similar system exists in California. This system is non-commercial, which may be the reason why it has been difficult to get exact details of it.

It is easy to imagine how such systems are likely to develop. For example, terminals could be installed in libraries where anyone (preferably anonymously) could go in and ask, for instance, about their social rights. It is possible to programme such a system so that the terminal prints the necessary questions and the person seeking information only needs to answer "Yes" or "No" or give commonplace information so as to obtain the following results: "If the questions have been answered correctly, you are entitled to . . ." "You may apply for . . . at the office (address), to which you should present the following papers . . ."

EDP on the user's terms is an undertaking for which we should prepare ourselves already at this stage. This type of system is becoming an obvious consequence of our adjustment to the information society. This adjustment is expressed in the fact that it is considered natural and reasonable to let machines carry out work which people would prefer to avoid. Our entire present-day work approach will therefore be changed so that people spend their time and abilities on tasks that machines cannot perform. It will, however, take a long time to get the information society to function.

Until then we must, as experts in automatic techniques, expect there to be less and less enthusiasm among a large section of society over our motives and our work, and its results [15]. This can be good cause for trying to introduce systems which operate on the user's terms and not just concentrating on getting the experts' closed world to function better.

FREE ACCESS TO INFORMATION

Is money to determine access to information?

Parker starts by asking: "Who is interested in investing in information?" He himself answers this question: Given the present economic theories and ethical standards. no one is interested. First, the private investor has to have guarantees against illegal copying. At best, information can be invaluable when it is secret. Once it is made public, there is no reason for paying for anything other than retrieval, transmission and presentation.

In the old days (only a few years ago), this process could be guided through distribution of the paper on which this information was printed and the price asked. It was considered a human right that the information made public (reproduced) should be accessible to everyone through the libraries.

Today, in the space of seconds, EDP makes it possible to retrieve, transmit, present or copy almost all printed information. This information is increasingly available in machine-readable form. Information has thus been released from the confines of its traditional packaging (paper). There is a general call for new rules of conduct. Patents, copyright, legislation on public records, satellite concessions, radio legislation, trans-national data flow [13], opposition to the charging of fees for library documentation services and payment for EDP aid in education and research are all relevant to the discussion. If commercial interests gain control over even one of the essential components of the information flow, they are clearly better placed to assess the value of particular information and carry out investment calculations. With good reason one can share Parker's fear for "1984 situations", as depicted in Orwell's novel forecasting the future [16].

A task for specialists: freedom of information

The above remarks are a gentle challenge to the competitive society, namely a challenge to the decision-making authorities to give the next generation a reasonable chance to take charge of their own lives. Despite the fact that many committed experts [17] are already studying this subject, there is little interest on the part of the electors and consequently the politicians. Not until the problem becomes political can any stand be expected.

The problem is free access to information,

which today is regarded as self-evident. There is no censorship. On the contrary, we have libraries serving everyone free of charge. If a library cannot satisfy anyone's wishes, it will get hold of the particular book or a facsimile. One of the greatest and cleverest industrial tycoons in the world was among the first persons to champion the idea that use of a library must be free. Most of Andrew Carnegie's fortune was used to set up free public libraries while only a small part went on recognizing heroic deeds.

This idea—free access to published information—is now threatened. Not because anyone wishes to charge for borrowing a book or a periodical. That can continue as before.

Payment may, however, become the determining factor for access to information when it comes to retrieving information in a steadily increasing information jungle. Anderla [18] has studied this matter on OECD's behalf.

From a slightly longer-term angle, there is the risk that the help provided by public libraries in tracing documentation will become dependent on private commercial documentation centres since the library services themselves cannot provide sufficiently effective service.

The publisher's obligation to supply libraries

In most countries publishers are required by law to supply one or more copies of all printed material published, whether books or newspapers. Copies are supplied to the national libraries, which are responsible for registration and for producing catalogues showing what the country has produced in the way of printed material as well as ensuring that such material is available to the public.

These catalogues, which require a great amount of work, are developed on an international scale and previously provided a sound basis for the help libraries gave in tracing documentation and obtaining relevant material.

For several reasons, private (commercial) documentation centres are starting to offer a service on which researchers are dependent and which the libraries are finding it increasingly difficult to provide: (a) the volume of documentation has grown to such an extent that the individual research centre can no longer keep up on its own with what is happening in the international field; (b) the specialized documentation centres can find potential users all over the globe; (c) they can use EDP for both communication and especially tracing purposes. These tracing systems have over the years become unbelievably sophisticated [17, Appendix XI] and time-saving. The libraries therefore will eventually have to install such systems and pay for them. This, however, is not the whole story.

Researchers and libraries will gradually become increasingly dependent on these systems, without having any real influence over their content. Research can be directed in an undesirable manner if the researcher's actual access to information is determined by the policies of documentation centres in other countries. Operation of a documentation centre can be compared to operation of a publishing company or editing a newspaper. It is impossible to include everything. Someone has to lay down guidelines for what is to be omitted. Which interests are going to decide what we, as borrowers, can get hold of in an information society?

Libraries have taken an interest in these problems on an international scale for many years. Most significant are the efforts made to ensure smooth exchanges of the traditional catalogue data in a machine-readable form, so as to avoid unnecessary duplication in connection with cataloguing. This work is, however, only the first step in this direction. The really exciting and significant stage is the exploitation of magnetic tapes, which are increasingly being used for the editing of texts and preparation for printing—photo-setting. These tapes can, after "typesetting", be used for automatic analysis of the entire text and can therefore be of considerable objective assistance in the classification and selection of keywords, the two most expensive and demanding tasks when it comes to use of the sophisticated documentation tracing systems.

If society wishes the library service to be able in the longer term to ensure research and business circles and the ordinary citizen access to documentation uninfluenced by private interests, it is necessary (a) to enact laws requiring publishers to supply libraries with magnetic tapes; (b) to make major efforts to ensure standardization; (c) to extend international co-operation between libraries as regards the use of documentation systems; and (d) to ensure increased political understanding for this work, which will inevitably require considerable resources, in particular library staff and EDP equipment.

Communication of information [19] is becoming one of the most important activities in the information society. One can only hope that those who work with libraries will guide future developments. The EDP experts are trained first and foremost to respect efficiency and think in quantitative terms. Their influence therefore often operates in the wrong direction. Those who are in charge of the libraries must ensure that quality is also guaranteed. They must therefore learn something about EDP which can best be achieved by working with it.

SUMMARY

"On the consumer market the individual has unlimited freedom of choice. On the labour market he has extremely limited freedom of choice as regards working time." This statement, which I have already cited on page 83, points to perhaps the greatest weakness of our present society: admiration for the natural sciences in particular has become so great that electors and consequently politicians have been induced to discuss future social relationships on the experts' terms. There is no reason to believe that this will work to the advantage of a democratic social system.

This situation leads us to the question: "What interests are we serving through cementing respect for a work-oriented ethic which is incompatible with the automated society—the information society which we have already created? The interests of our own children?"

REFERENCES

1. Parker, E. Background Report, *Proc. Comf. on Computer/Telecommunications Policy*. Paris, Organization for Economic Co-operation and Development, 1975.
2. Privacy Protection Commission, *Technology and Privacy*. Appendix 5, Washington, D.C., Government Printing Office, 1977. (Order No. 052-003-00425-9.)
3. *Computer World*, 22 August 1977, p. 1.
4. *ADB och Arbetskraften*. Stockholm, National Central Bureau of Statistics, 1977.
5. *Statistical Yearbook*. New York, N.Y., United Nations. See various recent editions; Henderson, J. Changes in the Industrial Distribution of Employment, 1919-59, *University of Illinois Bulletin*, Vol. 59, No. 3, August 1961.
6. *SACO Tidningen*, No. 11, 1972, p. 27.
7. Adler-Karlsson, G. NEJ til Beskaefrigelse, JA til Materiel Grundtryghed. Copenhagen, Erling Olsens Forlag, 1977.
8. *The Plan for an Information Society—A National Goal Towards Year 2000*. Tokyo, Japan Computer Usage Development Institute, 1972.
9. Koyama, K. Introduction to Information Society Theory. *Chuo Koron*, Winter 1968.
10. Masuda, Y. Social Impact of Computerization—An Application of the Pattern Model for Industrial Society. *Proc. International Future Research Conf.* 1972. (Held in Kyoto, Japan.)
11. Petersen, K. *Ungdomsuddannelser i International Belysning*. Copenhagen, Statens Trykningscentral, 1977.
12. *Christiana, et Samfund i Storbyen*. Copenhagen, Nationalmuseet, 1977.
13. *Trans-National Data Regulation*. Available from On-Line, Uxbridge (United Kingdom). (Papers from a conference held in Brussels in February 1978.)
14. Carroll, J.; Tague, J. The People's Computer, An Aid to Participatory Democracy. In: E. Mumford and R. Sackman (eds.), *Human Choice and Computers*. Amsterdam, North Holland, 1975.
15. Engberg, O. What Systems Do We Need in the Future? In: R. Buckingham (ed.), *Education and Large Information Systems*. Amsterdam, North Holland, 1977.
16. Orwell, G. *1984*. London, Martin Secker & Warburg, 1949.
17. *Forskningsbibliotekernes Målsaaetning*. Copenhagen, Beteenkning af Forsknings-bibliotekernes Fellerad, 1977.
18. Anderla, G. *Information in 1985. A Forecasting Study of Information Needs and Resources*. Paris, OECD, 1973.
19. *Vetenskaplig och Teknisk Informations-försörjning* (Statens Offentliga Utredningar SOU 1977/71). Stockholm, Liberforlag, 1977.

Policy problems of a data-rich civilization

HAROLD D. LASSWELL

▪ Harold Lasswell *is one of America's leading political scientists. His numerous books include* The Future of Political Science *and* In Defense of the Public Order.

Although there are those who fear that the current information explosion will overload available social and political structures and precipitate chaos, Professor Lasswell is more optimistic. He suggests ways in which the greater information handling capacity of our civilization may be used to limit the power struggle among the large nations, to recover the dignity of the individual, and to re-democratize our institutions through increasingly shared power.

No doubt we can anticipate at least some of the policy problems of a civilization whose stocks of promptly retrievable data are vast and growing. If we were to carry on this exploration in a rigorously systematic fashion it would be necessary to canvass the world social process as a whole, and to give full consideration to trends, conditions, and projections for every territorial component and for the principal value-institution sectors of every society. In this way attention would be directed to the potential impact of the shaping and sharing of information on the shaping and sharing of every other valued outcome in human affairs.

We could use as a comprehensive model of the human social process the following: *Human beings* seeking to optimize *values* (preferred outcomes) through *institutions* utilizing *resources*. The giving and receiving of information is a preferred outcome (the *enlightenment* value); and if we examine the aggregate flow of information gathering, storing, retrieving, processing, and dissemination, we might draw up an annual balance sheet of the *gross enlightenment outcome* of the world (or any lesser unit). If we subtract the information losses (from destruction of information), it is possible to arrive at a *net* outcome figure. The *accumulation* (investment) total would be the amount of newly stored information in data-banks, libraries, and other documentary collections, plus the increment of expert knowledge memorized by the scientific and scholarly community. The *enjoyment* would be the volume of information used by decision- and choice-makers during the year. (Note that, unlike some valued assets, an information re-use does not necessarily deplete the asset.)

☐ From International Federation for Documentation, *31st Meeting and Congress, Proceedings of the 1965 Congress, Washington, D.C., October 7-16, 1965* (Washington, D.C.: National Academy of Sciences).

As our civilization of science and technology moves toward universality, the rapid enlarging of its capital stock of available knowledge introduces repercussions in every value-institution sector. Problems arise whenever changes occur in the expectation map in reference to which past value demands have been partially stabilized by individual and group participants in the social process. By revising the map of expectation, the act of accumulating information introduces an accelerating tendency toward further innovation and possible instability. We might take the various value-institution sectors in sequence and attempt to foretell emerging problems. This would make it necessary to examine government, politics, and law; and also to explore the network of institutions specialized to enlightenment (mass media, research agencies, storage and retrieval facilities, for example). We would assess economic institutions (production, distribution, investment, consumption of wealth); institutions of well-being (concerned with safety, health, and comfort); institutions specialized to vocational, professional, and artistic skill; affection (family, friendship, loyalty); respect (social classes); rectitude (ethics, religion).

As a selective device we focus on three large policy areas that cut across the entire social context: world security, individuality, democracy.

IMPLICATIONS FOR WORLD SECURITY

As the world's stock of information increases, will the prospects of world security improve? The key point to be made in this connection is that big-scale information retrieval easily favors the making of rapid, highly centralized decisions. However, the decisions that most vitally affect world security are not made by a globally inclusive agency of public order.

On the contrary; whatever the words in which international obligations are phrased, the effective decision-makers of Moscow, Peking, Washington, and other capitals do *not* expect to be able to rely on an efficient system that maintains at least minimum public order. Hence there is not prompt and open access to vital information. The nation-states will continue to seek to monopolize rather than share information which is supposed to bear immediately on national security. Great ingenuity will continue to be exercised in attempts to misinform the intelligence services of rival states.

Perhaps surprise will become more difficult as world social and political data are more inclusive, and cross checks are improved by means of better simulation models. On the positive side it will be possible in a data-rich world to spell out in far greater detail than hitherto the net gains of worldwide cooperation.

We must make full allowance for the strength of forces that favor monopolies that block the dissemination of information to the entire world community. In a world divided by the expectation of war and other forms of collective violence, the control of information falls easily in the hands of governmental, party, or other private monopolies. In small countries of high economic, social, and cultural levels of activity, parochialism is diluted by rather high per capita frequencies of foreign travel, message exchange, and transnational publications. In world security terms, however, this is not the most significant trend. The modern communications revolution has been unable to universalize the outlook of mankind. The new instruments have been utilized by the managers of the information media to overcome localism. However, *the chief gainer from reduced localism has been, not a common world perspective, but intermediate attitudes of a more parochial character.* The great continental units—like the U.S.A., Russia, and mainland China—absorb the focus of attention of the overwhelming percentage of their population. National self-references rise more sharply than do more inclusive references. The flow of information is controlled to perpetuate the patterns of segregated access that correspond to the value-institution structure of a divided world arena.

It has been suggested that heroic policies must be adopted if the parochializing and divisive consequences of the present control of information are to be overcome. Drastic proposals include: give a World Communication Network legal access to all inhabitants of the globe (and if a regime in power is in opposition, engage in subversive circulation of tiny receivers); give a World Board of Education control of an hour a day of the attention of all school children; arrange for at least a year's education abroad before adolescence for all children, and another year between adolescence and young adulthood. It is perhaps unnecessary to point out that these proposals are at present unwelcome almost everywhere. More immediate policy suggestions favor the self-interest of scientists and scholars by proposing that facilities be greatly multiplied to allow foreign travel, study, and information services for at least ten percent of all professionally trained people each year.

Unfortunately it cannot be asserted with confidence that scientists and scholars will be more effective than the public at large in breaking through the divisive structure of world affairs. Top flight scientists and scholars have long distinguished themselves for their humane and enlightened approach to the problems of man. However, most of their less successful colleagues and competitors have shrewdly adapted themselves to the professional opportunities of a divided and nationalistic world arena. The inference would appear to be that *so long as those who specialize in the advancement of knowledge depend for support on sources that seem to benefit from a divided globe, the vested and sentimental interest in division will in fact continue.*

IMPLICATIONS FOR INDIVIDUALITY

We turn now to a brief examination of the implications of a data-rich civilization for individuality. It would be congenial to most scholars, at least, to be able to demonstrate that individuality varies directly with society's stock of information. Does not a fuller map of past trends and conditions, and of future probabilities, multiply the bases of inference for the creative imagination of mankind, and thereby provide more direct and indirect opportunities for human capabilities to find expression in socially acceptable diversity?

Unfortunately we cannot propose an affirmative answer without assessing the weight of counteracting factors. (1) New information may be employed to concentrate the management of motivation in the hands of a few, thereby reducing the flow of spontaneous creativity in the entire society. (2) Individuality (and creative expression) depends on more than simple exposure to information. An essential requisite is a motivation that welcomes and does not turn away from perceptions of possible change. (3) There is the question whether individuality can withstand the invasions of privacy that seem inextricably interwoven with a data-rich civilization. Let us consider each of these points in turn.

(1) It is to be assumed that the study of factors that condition human response will become progressively more refined, and therefore more available for manipulators of political, economic, and indeed all collective responses. We do not doubt, for example, that in popularly governed states rival parties and candidates will take full advantage of the current stock of knowledge about the influence of culture, class, interest, personality, and crisis level factors on response. Pre-tests will provide information about audience predisposition toward contingent choices of issue and appeal. Rival political manipulators will continue to uncover new conditioning factors that will enable them to exert a relatively strong impact on the target (at least the first time around). While it must be conceded that the chief protection of public judgment from monopoly management is *competition,* a further safeguard is *open disclosure of knowledge about the predispositions of voters, legislators, administrators, or judges.* If popular government depends on an informed opinion, a category of particularly important information for everyone is knowledge of how he can be manipulated and thereby deprived of the degree of choice that one might have.

I note that *disclosing information about how they are manipulated to those who are manipulated* is a policy that can be applied far beyond the arena of politics. The policy may include permission to technicians and managers to use new knowledge of motivation for a short time prior to disclosing it to the target audience. The advantage of initial use is that it provides an inducement to research and to the cultivation of knowledge. Political parties can be required to turn over their data and simulation models after elections. Business enterprises, whether private or public, can be required to disclose their merchandising information regularly in order to put the consumer on guard. Charitable fund raising and missionary data can also be made public.

True, one does not necessarily alter one's opinion because he becomes aware of the factors that usually shape it. But if one's factor-determiners are continually brought to attention, the likelihood is improved that an individual will ask himself whether his response is, after all, satisfactory when reviewed in the light of all the information at his disposal. Consider Mr. A, who is reminded that his party, candidate, and issue responses can be attributed to his Republican or Democratic family tradition; to frequent association with neighbors and business associates of one partisan coloration; to selective viewing, listening, and reading of one-sided media; to local favors granted by the political machine (and the like). Mr. A may recognize his determiners and go on responding to them. Then again, he may begin to wonder, and thereby become more sensitive to other factors.

Although it is usually said that information about courts, commissions, legislatures, or electorates is important for purposes of prediction, this is not necessarily the chief objective. It may be argued that the principal social purpose served by information about motivating factors is insight; that is, making available to the chooser knowledge about his dispositions which, when recognized, may lead him to reflect critically on the appropriateness of his behavior as judge, official, or leader.

(2) We have said that information may be misperceived; and that acquired mechanisms of misperception can reflect a deep motivation to turn away from innovation and to abandon any attempt at creativeness. Both the advancement and the effective application of knowledge in a process of decision are conditioned by automatic modes of perception and selection. Clinical psychology, in particular, emphasizes the point that perceptions and choices are often made rigid by defense mechanisms that inhibit the individual from taking note of new events or of entertaining new options of thought. It is not to be overlooked that these defense mechanisms have positive as well as negative functions, since they protect the person from anxieties aroused when earlier conflicting drives are mobilized. The cost of new perceptions or of novel options is thereby made too high in sectors that are taboo.

Is it to be taken for granted that the children and youth of tomorrow will be socialized (educated) for the encouragement of novelty and creativity? Many of us would be skeptical of the educational programs of totalitarian societies. But the future of relatively open "mass societies'' cannot be viewed with unqualified optimism. As the dangers of ''mass society'' become more obvious, is it not probable that influential elements in any state will seek to reduce erratic crowd responses by aiming education at more perfect indoctrination, especially in reference to the key symbols, doctrines, and rites of public policy?

Already we see the outline of techniques that, if stringently employed for indoctrination purposes, can circumscribe both the acquisition of selected categories of information and imaginative new uses of knowledge. Combinations of hypnosis with drug-induced states and the use of learning machines can evidently specialize the stores of information put at the command of individuals, and circumscribe their potentialities for retrieval. Sheer quantity of information does not guarantee wide-ranging use if either retrieval or processing is subject to crippling automatic mechanisms of regulation.

(3) Another factor whose impact on individuality (especially creativity) is worth considering is individual privacy. Although the issues are not in principle new, a data-rich world can have individual records of unheard-of detail promptly available. Shall we attempt to bar the gathering and storage of specified classes of information? Shall limits be set on who shall have access to personal data, for what purposes, and by what procedures?

It is conceivable that human beings will undergo general revulsion against the invasions of privacy that become more common as man's auxiliary brains are put to more and more detailed use. A few years ago when modern psychiatry was first applied to the study of politicians, the London *Times* editorialized in a sardonic mood that in the future campaign biographers would feel constrained to demonstrate that even in the cradle the candidate had lived a blameless life. Will the implacable demand for data about the decades, years, months, weeks, days, hours, and minutes of career development require, in addition to gynecological and obstetrical data, a record of everybody's feeding schedule, toilet behavior, and test performances at frequent intervals? Are we going to take literally the maxim that whatever concerns man is necessarily of concern to mankind's storers of information?

I suspect that there is no limit to the data that may be regarded by some scientist as of research importance. There seems to be no clear way to prescribe a limit; hence the mighty man-machine tide of information will probably swell and swell. Only think of all the psychologists, sociologists, physicians, nurses, welfare workers, child specialists, educators, tax collectors, police officers, salesmen, creditor agents, personnel administrators (and so on and on) who will continue to collect and collect and collect.

Reflection suggests that the most likely areas of policy regulation are not information obtaining or storage, but access. Consider briefly a few past policies of restricted dissemination.

It is generally agreed that great damage can be done to children and young people if certain kinds of information are made public, even to restricted audiences. The stigma of illegitimacy is an example, or of interracial or intercast parentage where these matters are taken seriously. Protection is given to family medical records, especially when a traditionally low-respect blemish is involved (idiocy), or when a low-rectitude involvement is implied (syphilis). It is common to destroy or seal records of early delinquency or failure.

Adults also obtain the protection of non-disclosure, though the policy is less stringent than with children. In a society that affirms its commitment to human dignity, whatever causes anguish without significant social gain tends to be avoided. One historic policy is designed to obtain for the community the benefit of allowing people to make a new start, unencumbered by ever-present reminders of the past. In countries where great new resources lay idle until labor and capital were added, the tendency was to overlook a man's past in the informality of everyday life and also in official records. People were not asked to identify the eye color of their grandparents before they could get a job or a driver's license. In a world whose established systems were widely viewed as immoral, the refugee might have a seemingly crooked or contemptible past; but it was felt that in a land of opportunity he might very well make good. In all societies some form of reprive or statute of limitations seems to operate as a means of giving hope to the deprived classes. Thus prisons may be cleared at the birth of a royal heir, and political prisoners released on the coronation of a new monarch.

As the data-rich civilization closes in, these public policies are almost certain to become obsolete, as indeed they are nearly everywhere today. If we begin to open up new zones of settlement in outer space on natural or artificial satellites, these perspectives and policies may revive. But the trend is more likely to be in the opposite direction, emphasizing the area of careful selection of personnel on the basis of exhaustive information. Space explorers and colonists will presumably be carefully selected, owing to the cost of transfer.

It may be that the coming years will witness a drastically new approach to the problem of protecting self-esteem from the crippling effect of unnecessary deprivation, and of obtaining for society the advantages of a buoyant, self-confident, congenial contributor to human relations and to the cultivation of resources. Instead of adopting the strategy of privacy, even secrecy, which in any case seems most applicable to a rather isolated frontier community, a contrasting strategy of insight may win acceptance.

By the *strategy of insight* several things are meant. A data-rich society can provide information that reduces the false pride of any individual, family, or community. The distorting effect of facts such as illegitimacy, a history of mental disorder or of secret vice comes from the ego's sense of isolation and vulnerability. It seems quite possible to change such an outlook by enlarging the scope of accessible information about man and society. As we push our data back through the years (while also carrying it forward), the objective, scientific point of view toward psychology and culture will become more widely shared. Like old families, large community groups will discover the existence of closets full of skeletons—of criminals, mentally disordered, defective, and illegitimate—as well as men and women of eminence. *If the educational system takes care to train the individual to see his enormous potentialities, and to discover the many strategies by which a specific negative characteristic can be overcome or compensated, the resulting personality systems can be expected to do without much privacy.*

IMPLICATIONS FOR DEMOCRACY

Although our discussion of security and individuality has had many side-implications for democracy, the problem has not been given the direct treatment that is warranted by its importance.

We are told that a vast system of documentation that relies on automatic methods is almost certain to possess built in biases that affect public policy in the direction of centralization, concentration, monopoly, regimentation, and monocracy. These biases are described as consequences of the seeming advantages gained by operating the technicalities of storage and retrieval in ways that obtain economies of scale.

The analysis suggests further that reliance on accumulated documentation establishes an attitude that, when over-emphasized—as is bound to happen in a data-rich civilization—is dangerous to genuine power sharing. The reference is to an *attitude of inquiry, exaggerated at the expense of commitment, conviction, and argumentation*. The attitude of inquiry includes the tendency to translate statements of preference or determination, when challenged, into a demand for further factual investigations. Hence the style of problem solving seems factual, objective, and emotionally calm. It becomes bad form to insist on arguing at length in support of a goal value, a policy objective, or a strategy. Hence the apparent philosophic composure of such a decision process. It operates with a built-in short circuit toward fact-form inquiry or manipulation.

However—the analysis proceeds—surface composure of this kind actually works in favor of the established system of public order. In totalitarian regimes this is obvious, since the devices of documentation are employed as instruments of planning to keep the top elites in place. In non-totalitarian regimes the consequences are less evident. But the most sophisticated programs are nevertheless likely to be initiated and run by and for the benefit of top elites in politics, business, and other positions. The analysis suggests that the procedure is consciously or unconsciously slanted for the perpetuation of the oligarchical or monocratic advantage of those on whom the technical personnel must depend for safety, income, and other value inputs.

Are there strategies at hand that may neutralize or even reverse tendencies of the kind? The inference is that there must be *explicit training in commitment*. Willingness to come to a decision and act as a responsible leader must be rewarded in the educational process and not treated as an indication of an unsophisticated perspective. Similarly, *rewards must be forthcoming in the policy-forming and -executing processes* throughout society if temporizing and acquiescence are to be met by countereffective forces. This implies genuine participation in power sharing, even in situations where unexamined tradition allows important decisions to be the prerogative of a few.

Another way to formulate what has been said here is to say that modern documentation technique strengthens the tendencies in every large scale society toward bureaucratism, and that *realistic awareness of the pervasiveness of the danger* is itself a prerequisite to the adoption of strategies to nullify the threat.

The case for the cultivation of decisiveness is totally insufficient as a strategy of democracy unless it is joined with command over comprehensive information. In a relatively open society it is conceivable that control of the basic stocks of information is effectively pluralized, and is not allowed to gravitate into oligarchic or monocratic hands. A *policy of pluralism implies that the same stocks of data are differently used and interpreted by influential voices in the decision process.* Military experts must be open to challenge, not only by military experts but by civilian officials; executive authority by legislative authority; government office holders must be open to equally informed criticism by party experts; party experts must be open to adequately grounded criticism put forward by the non-power sectors of society—economic, scientific, medical, ecclesiastical, educational, journalistic.

Issues of preference and volition, not of information, rise at many places in a problem-solving policy process. There are the questions that come up in the *clarification of goal:* shall we postulate moving toward a world of human dignity or a world of indignity, of caste supremacy? Some questions arise in connection with the *examination of historical trends:* how far back and how detailed shall our inquiries go, and how complete shall our attempts to be empathize with past generations? Questions occur in the scientific *quest to explain* the social and physical process: how willing shall we be to sacrifice life in experiments and in field expeditions? Questions also relate to the *projection* of future developments, and the *invention, evaluation, and selection of alternatives:* how much risk of loss shall we take in the hope of how much gain?

If an information-rich civilization is to share with its members a common image of past, present, and future, it must find effective ways of providing *comprehensive, selective, and realistic experiences of the whole map of knowledge and informed conjecture.* As a symbol of what is required I have sometimes spoken of a social planetarium adapted to the orientation needs of every region, nation, province, city, locality, and pluralistic component of the world community. Such a series of rooms, buildings, presentations would provide alternative interpretations of the past of civilization and of potential goals and strategies of future development.

We are suggesting that, *if the many pluralistic elements in modern industrial nations seize promptly on modern methods of documentation, it will be possible to sustain political initiatives in the decision process of sufficient intensity to support a public order in which power is genuinely shared.* Shared data means shared power; a monopoly of data means a monopoly of power.

Our exploration of some of the policy problems of a data-rich civilization suggests the wisdom of anticipating as early as possible the implication of these emergents for such problem areas as world security, individuality, and democracy. Once more we confirm and apply the maxim: knowledge is power.

Technological models of the nervous system

ANATOL RAPOPORT

■ Anatol Rapoport *is Professor of Psychology and Mathematics at the University of Toronto, and a leading spokesman in the field of general semantics. His books include* Operational Philosophy *and* Game Theory as Theory of Conflict and Resolution.

In what remains one of the best introductory pieces on computers and the human nervous system, Professor Rapoport recounts the progression leading to the development of the computer by describing four "technological phyla": tools, clockworks, heat engines, and information machines. In this progression, he argues, human ingenuity has come ever closer to reproducing in the non-natural world the properties that characterize living beings.

It is not often that a book written in another age suddenly acquires an astonishing up-to-date-ness. This does happen when some prophecy suddenly passes from the realm of the fantastic to the realm of the imminent. Such a prophecy was contained in the book *Erewhon* by Samuel Butler, written in 1872. The prophecy has to do with the evolution of machines, particularly machines endowed with a property which has seldom been attributed to machines—intelligence. As stated by Butler in rather poetic terms, the prophecy envisages a world in which the machine becomes the dominant system of organization (in the way living things are systems of organization). Like living things, the machines of the future metabolize, reproduce, maintain themselves, and in general seem to have an aim in life. The one frightening thing about the *genus machina* is its parasitic dependence on the *genus homo*. The mechanism of natural selection is supposed to function in such a way on that form of "life" as to select those variations which are especially capable of catering to the compulsions of human beings—namely, their compulsions of caring for machines. Gradually, what had started as a symbiosis between man and machine passes into parasitism, so that finally man becomes domesticated by the machine.

Almost the same prophecy is stated in more realistic terms by N. Wiener in his *Cybernetics* (New York, 1949). Wiener envisages the Second Industrial Revolution ushered in by machines able to perform tasks requiring an average intelligence with the resulting dislocations and crises similar to those which followed the First Industrial Revolution, when the "stupid" machines first appeared on the scene of history.

Our purpose here is not to discuss the merits or the limitations of these prophecies, but rather to point out that the sudden dramatic revival of the "intelligent machine" idea (be it a metaphor or a myth or a profound insight) is indicative of a really significant historical event—a major intellectual revolution.

Like the second Industrial Revolution, of which Wiener writes, this intellectual revolution is also the second in recent times. The first one occurred in the seventeenth century with the creation of mathematical physics. Perhaps I should make clear what I mean by an intellectual revolution. I think of such revolutions metaphorically as crystallizations of thinking around new, powerful concepts. In the seventeenth century these central concepts were those of mechanics—force, momentum, particularly energy. They became the central concepts of classical physics and of technology which came into being during the First Industrial Revolution.

The second intellectual revolution, now occurring, brought forward another powerful new concept, that of "quantity of organization," a concept of high degree of sophistication and bearing within it the seeds of extremely far-reaching consequences. It is this concept, also called negative entropy and "amount of information," which makes the anthropomorphic conception of the machine especially intriguing, particularly because through it the common features of "intelligent" or "purposeful" behavior of the higher animals and "automatic" behavior of "higher machines" are made apparent.

Now the personification of machines and "mechanization" of organism are not new. The former has mythological roots in the medieval legends of the Golem and the Homunculus. The latter appears, for example, in the writings of Descartes.

□ From *The Modelling of Mind*, ed. Kenneth M. Sayre and Frederick Crosson (Notre Dame, Ind.: University of Notre Dame Press, 1963), pp. 25-39.

The question "Are living beings machines?" has long been treated as a metaphysical question, presumably answerable on metaphysical grounds. Since metaphysics is more or less a lost art, we must learn to look at that question somewhat more critically, that is, with semantic awareness. We must translate it into other questions, such as, "To what shall the name 'living thing' be applied?" and "To what shall the name 'machine' be applied?" "Is there an overlap among the referents of the two terms?"

Putting the question this way, we see that the answers to the first are relatively clear, while the answers to the second are not nearly so clear. Barring certain borderline cases (viruses, etc.) we have no difficulty recognizing the class of objects to which the name "living thing" can be unambiguously applied. Not so with machines. This is so because living things are "given." They have remained about the same for as long as we can remember. But machines have evolved rapidly within the span of human history. We realize keenly that there are machines today which our grandfathers could not have dreamed of, and, by extrapolation, we feel that we can't really say what the limits of the world of machines may be. If we think about the matter a little more, we realize that machines in their evolution undergo "mutations" of tremendous magnitudes. Where it takes eons for a new biological species to develop, a new technological "phylum" has on occasion come into being within a generation.

By a technological phylum I mean something similar to a biological phylum. If the latter is defined by a very general plan of organization in a wide class of living things, the latter is defined in terms of a *principle of operation*. We can, if we wish, distinguish four technological "phyla," which came into being successively.

The first phylum we could call *Tools*. Tools appear functionally as extensions of our limbs and they serve primarily for transmitting forces which originate in our own muscles. In the transmission of force, sometimes a mechanical advantage is gained, as in the crowbar, a screw, or a pulley. However, the work done by a machine of this sort is actually work done by our own muscles. Therefore a machine of this kind, a tool, does not give the impression of "independent" action, and so it did not occur to anyone to compare tools to living things.[1]

With the second phylum it is a somewhat different story. This second phylum of machines we could call *Clockworks*. In a clockwork a new principle of operation is at work, namely, the *storing* of mechanical energy. A typical clockwork is wound up, that is, potential mechanical energy is stored in it, which may be released at an arbitrary later time and/or over a prolonged period of time. A clockwork does give the impression of autonomous activity, and doubtless this crude resemblance of a clockwork to a living thing (residing in its quasi-autonomous activity) gave the craftsmen of the late Middle Ages and of the Renaissance ideas of constructing mechanical dolls and animals. Perhaps the first ideas of automata sprang from the same sources. Characteristically, Descartes speculated on the possibility that animals were elaborate clockworks and, equally characteristically of his age, excluded humans from this class, as possessors of "souls."

We may observe in passing that the bow and the catapult are also clockworks by our definition, since mechanical energy is stored in these machines to be released later (in the case of the crossbow, it may be released much later). However, the "autonomous" action of these machines is so brief that they do not give even the appearance of being "alive."

The first comparison of living things to machines, therefore, was made with regard to clockworks. It is not surprising that this metaphor was not particularly fruitful for the understanding of the living process. We know now, of course, that *energy* is stored in living things, but this energy is not stored in the form in which it is stored in clockworks (mechanical stress) and so was not recognized as such. Living things are not wound up to keep going, and this absence of the most essential characteristics of a clockwork in living things made the early mechanical intepretation of life a sterile one.

This comparison get a new lease on life with the appearance of the third phylum of machines. This phylum includes primarily the *Heat Engines*. Again an entirely new principle enters into their operation. As with tools and clockworks, the output of the heat engine is an output of energy which had been put into it. But whereas the energy put into the earlier classes of machines was in the form of mechanical stress, which is obviously associated with our own muscular effort, the energy put into a heat engine is contained in a *fuel*.

Consider the vital difference between the two situations. It is obvious even to a child that the

[1]However, personification of weapons does occur. Note also the legend of the Sorcerer's Apprentice.

tool is not autonomous, because the tool is geared at all times to muscular effort. A child or a very primitive person may believe that a clockwork is autonomous, but it is still easy to convince him that it is not, because the winding up is still a result of someone else's muscular effort. No such effort is apparent in the fuel. Fuel is "fed" to the heat engine. The analogy to living things (which also need to be fed in order to operate) becomes ever stronger.

The comparison between heat engines and organism passed beyond the metaphorical stage and bore real scientific fruits. It became apparent that fuel is in a very real sense the food of the engine and equally apparent that food eaten by organisms likewise functions as "fuel." The principle of energy conservation was shown to hold in living things—a serious blow to the contentions of the vitalists, which sent them on their long and torturous retreat. Biochemistry was born. More and more processes characteristic of life were shown to be instances of processes reproducible in a chemical laboratory. An analogous revolution was occurring in technology. In fact, it would be not inaccurate to say that the First Industrial Revolution occurred when it became apparent that machines could be constructed which did not need to be "pushed" but only "fed" in order to do the work.

Driven out of physiology, the vitalists took refuge in psychology. Here, in the realm of thought and purpose, of emotion and insight they felt they would remain safe from the onslaught of the mechanists, materialists, determinists, and reductionists. The label "nothing-but-ism" was derisively pinned on the philosophical outlook of those who believed that even the most complex manifestations of the living process, including the intricacies of men's psyche, could somehow be described in terms of analyzable behavioral components which, in turn, could be related to observable events in space and time.

And so the focus of the battle between the vitalists and the physicalists shifted to psychology, where it remains at this time. The line between the two camps is, of course, not sharply drawn. Like the political spectrum ranging from extreme left to extreme right, the range of convictions concerning the nature of mental processes stretches from extreme behaviorism to vitalism or mysticism. The gestaltists can, perhaps, be assigned intermediate positions.

I am sure you all know the main outlines of the controversy. The opening offensive was undertaken by the behaviorists, the champions of what in some circles bears the unattractive name of S—R psychology. The method has a strong physiological bias. Technological analogies are frequently invoked. The earliest of these was the "telephone switchboard" model of the central nervous system. The environment was supposed to act on the organism by a series of stimulus configurations, which activated combinations of receptors, which initiated impulses, which traveled along nerve fibers, passed through the central nervous system to other nerve fibers and into the effectors, whose activity accounted for the overt behavior of the organism, which was proclaimed to be a sole legitimate object of study in psychology. Behavior was viewed as a grand collection of units called reflexes.

The model was seen to be inadequate from the start. If to every configuration of stimuli there corresponded a definite set of responses, how was learning (the acquisition of new responses to the same stimuli) possible? However, this seemingly embarrassing question proved a blessing in disguise, for the discovery of the conditioned stimulus by physiological means strengthened the reflex theory of behavior. It was shown that the paths of the impulses could be *systematically* changed. The switchboard model was shown to be still useful. Learning was accounted for by the "switchings" of the connections.

Hot on the heels of the behaviorists' successes, however, came a more serious critique called *gestaltism*. Gestaltism deserves serious attention, because its ideas were the direct precursors of a new approach to the theory of the nervous system, which is the subject of the present discussion. The gestaltist critique was not simply a reiteration of the vitalist faith and did not confine itself to derisive labels like "nothing-but-ism" directed against behaviorism. It was much more specific and constructive and was based on at least two clearly identifiable characteristics of behavior, which did not seem to fit into the behaviorist scheme, namely, the recognition of "universals" and the equi-finality of response.

The recognition of universals means the following. Suppose an organism learns to respond to the sight of a particular square in a certain way and to a particular circle in a different way (say open a box marked with a square but not with a circle). The phenomenon is clearly an instance of conditioning. A strict behaviorist (telephone switchboard) explanation would have to rest on the assumption that the stimuli originating from the receptors activated by the sight of the square are "switched" by the conditioning process to paths leading to the proper effectors for opening the box. However, it is known that the conditioned stimulus can be varied considerably *after* the conditioning has been established and still elicit the response. For example, if the original conditioning was to a white square on a black background, it can be subsequently changed to a black square on a white background, which, at least in the retina, excites the *complementary* receptors, i.e., precisely those which were *not* involved in the conditioning process. Roughly speaking, the organism responds to the square as a "square," regardless of the receptors involved. Hence the emphasis on the term *gestalt* (the configuration perceived as a whole, rather than a complex of elementary stimuli). The gestaltists maintained that the behaviorists' emphasis on the stimulus response pathways detracted from the importance of "universals" or abstractions in the act of perception.

(If a counter-argument is offered to the effect that in the perception of a geometric figure only the receptors affected by the edges of the figure are involved, it can be countered by other interesting evidence, such as the well-known phenomenon where familiar maps are not recognized if the continents appear in blue and oceans in yellow, or the still more baffling phenomenon that the shapes of objects can be recognized regardless of position, size, or orientation.)

The equi-finality of response argument is even more powerful. It has been observed that once an animal has learned to perform a task (say to run a maze to a reward) it will perform that task with whatever means are available to it. If its legs are amputated, it will roll through the maze. Clearly, such behavior cannot be explained in terms of a series of reflexes, each setting off the next, since the performance may involve totally different effectors each time.

This equi-finality of response naturally leads one to talk of *purposeful* behavior, in which only the *goal* is relevant and not the particular configuration of neural events which come into play. This seeming inevitability of invoking teleological notions opens the door to more vitalist arguments. The notion of "purpose" seems to resurrect the ancient classification of causes into "efficient" and "final" and to give new life to the ailing idea that the behavior of living and non-living things could not possibly be governed by the same set of laws.

It is at this point that concepts associated with the fourth phylum of machines become exceedingly important. We recall that the first phylum (tools) operated primarily as force transmitters; the second phylum (clockworks) as storages of energy resulting from mechanical stress; the third phylum (heat engines) as transformers of different forms of energy into mechanical energy. Now the fourth phylum of machines operates on the principle of storing and transmitting something called *information*.

Already the telephone switchboard model of the nervous system employs a technological analogy with a communication device rather than a conventional engine. The primary concern of psychology is not so much with "what makes the organism active?" as "how does it know what to do?" Not the source, the transformation, or the utilization of energy by the organism is of prime significance but its *organized* disposition. What the psychologist actually studies is not how much activity has been performed but the sequence of specifically directed acts, which when organized one way may give one set of results and organized in another way (or randomly performed) may give an entirely different set, even though the amount of energy expended remains the same. To give a homely example, consider the difference between closing the door and then turning the key and turning the key and then closing the door. The machines of our fourth phylum are primarily concerned with *systematizing* operations in which utilization of energy is involved. The amount of energy used is not important. The "power" of these machines is not "muscular" power but "mental." The giants among them are capable of receiving, transmitting, and storing complex sets of directions, i.e., large amounts of "information." This is why technological analogies with these machines are of particular interest in psychology. These machines simulate not muscular effort (like their ancestors did) but human intelligence.

Just as the concept of energy and its transformations was able to explain the "activeness" of organisms, which could not be explained on the basis of externally applied stress (as tools are activated) or by internally applied stress (as clockworks are activated), so the concept of "information" promises to do the same for a much larger area of the living process, namely, the "intelligent" and "purposeful" aspects of living behavior.

What is this thing called information? There is now a wealth of literature on the subject and it is not within the scope of this presentation to develop the ideas of this literature. I think, however, that a reasonably good idea of the nature of "information" can be given by a few examples. I will not attempt to make these ideas precise. I will try to appeal to intuitive understanding, even at the risk of being vague.

Information bears a similar relation to energy as organization to effort. One can best see this in an example where the inadequacy of a theory based on energetic considerations alone is obvious. Consider the automobile traffic in a large city. Suppose the proverbial man from Mars decided to study this traffic. He might measure the rate of flow of cars along the city's arteries. He would correctly relate that flow to the speed with which the cars traveled and, being a good physicist, he would relate the speed to the power of the engines. And so he would be satisfied, perhaps, in explaining the rate of flow by energetic considerations.

Next suppose that all the traffic lights failed. Certainly the speed of the cars and thus the rate of flow of traffic would be reduced. Suppose our Martian stuck to his conceptualization in terms of energetics. He would then have to ascribe the reduced flow (or speed) to some failure of the automobile engines, and he would be wrong. The failure is not of the engines but of the traffic lights. True, it takes energy to activate the traffic lights, but it is negligible compared with the energy it takes to move the cars. Energy has therefore little to do with the traffic problem under consideration. The key concept is not that of energy but of *directions for the utilization of energy* (commands "stop" and "go" properly patterned), i.e., a matter of *information*. If the traffic lights are not functioning, the driver of a car does not know what to expect at each intersection and, playing safe, he slows down. The accumulated slow-downs of all cars at all intersections turn out to have a greater effect on the over-all slowing of traffic than the occasional full stops at the red lights. In the case of regulated traffic lights, set for certain speeds, the flow of traffic is most efficient. The cars are, in effect, "organized" or bunched up along the roads in such a way that the bunches on one system mesh with the empty spaces on the system perpendicular to it, and the flow is continuous without stops.

Examples can be multiplied at will. Children well-trained in fire drills leave a burning building in a surprisingly short time, while a disorganized mob may never leave it. The success of a military action depends both on fire power and on proper coordination of the units. Fire power is measurable in terms of energy units, but coordination is measurable in terms of something else: the rate of flow of information and the precision of timing in carrying out the sequence of necessary steps. Productivity of an industry depends on the amount of power available (energetics) but to no less extent on the skill of the workers (coordination of activity within the individual) and the skill of management (coordination of activity of the several workers). While it was traditionally assumed that these coordinating functions must be performed by "reasoning beings," i.e., men, it became gradually apparent that a great many of them could be performed automatically (by traffic lights instead of policemen, IBM machines instead of filing clerks, automatic steering mechanisms instead of helmsmen, electronic computers instead of human calculators). There arose then the intriguing idea that there may be a general "psychology" applicable both to the behavior of these devices and at least to certain aspects of human behavior.

Now let us pause for a moment and take stock of what we have said. Historically the technological analogies purporting to explain the behavior of living things have been geared to prevailing technological concepts. As technology became more involved, the analogies could be extended to more facets of behavior. We are now entering a new technological era—the era of "intelligent machines," called automata and servo-mechanisms. The understanding of the principles on which these machines are constructed and operate promises to extend our understanding of the living process still further.

We must, however, if we are to say something significant, indicate more specifically where that promise lies. We have two pieces of evidence in support of our rather optimistic view. The first is the tremendous stride forward in the understanding of the living process, which resulted from the previous discovery of just one far-reaching principle—that of transformation of energy. The second is the progress being currently made in the analysis of the vague teleological and vitalistic notions of "purpose" and "intelligence."

Let us recall, at the risk of becoming repetitious, why the understanding of the living process presents difficulties. Living things seem to differ from nonliving in three fundamental respects (immediately apparent to the naive observer).

1. They seem to be "autonomously" active (i.e., the motive power seems to come from the inside rather than be impressed from the outside as in the case of moving inanimate objects).
2. They seem to be guided by purpose and intelligence.
3. They maintain their integrity, grow, and reproduce.

The first technological analogy (with the clockwork) attempted to explain only the first of these characteristics and it did so very poorly. A clockwork is, to be sure, activated from the "inside" for a while, but there is no question about what the source of this activation is. The clockwork simply gives a *delayed* response to a stress (a push) impressed on it.

It is different with a heat engine. There is no obvious push there. The engine is *fed* in a very real sense and is activated by the food it "eats." The analogy to a living organism is in the case of a heat engine far from superficial. But the "muscular effort" of the engine is still externally directed. The locomotive is guided by the rails; the boat by the rudder. A simple engine is "told what to do" at every step of the process. Here the analogy with the living organism fails.

Now it is clear why the development of automata and servos naturally extends the analogy. The mechanisms of *control* are now built into the machine. We now want machines to behave "purposefully" and intelligently, and since we have to design the machines, we have to analyze the notions of purposefulness and intelligence into component parts.

Really no sharp distinction can be drawn between intelligence and purposefulness. Any definition of one is sure to involve the other. Let me therefore describe very roughly the present status of "intelligence" and/or "purposefulness" in our machines which will then naturally lead me to the concluding remarks on the modern ideas of the nervous system. I view "intelligent" machines as consisting of two kinds, automata and servo-mechanisms. The only distinction I make between them is that the automaton is guided by a program of discrete steps or directions fed into it, while the servo-mechanism is guided by observing the effects of its action on the outside world. Thus a juke box which plays a number of selections in the order selected by the customer (in response to the buttons pushed) is an automaton, and so is an electronic computer. A target-seeking torpedo, on the other hand, or a gyroscope, I would call a servo-mechanism. Both exhibit "purposefulness" and "intelligence," although if we adhere to the intuitive popular meanings of these terms, the servo-mechanism seems to specialize in purposefulness and the automaton in intelligence. This seems so, because the automaton seems to be able to follow *explicit* directions, "When so and so, then so, unless so or so, in which case so . . . " (the program), while the servo seems to be guided by a goal. This difference is only apparent. To an outsider, the automaton may well seem to be guided by a "goal" ("Find the solution of this equation") while someone intimately familiar with the operation of a servo can describe its operation in terms of a program.

This equivalence of "program" and "goal" is the principal idea of the modern theories of the nervous system. One point must be kept in mind, however. Program and goal may be logically equivalent, but it does not by any means mean that a description of an operation of an organism or a machine is equally convenient in terms of one or the other.

Let us take a trivial example. We wish a ball in a cup to "seek" to come to rest at a particular point. Here the desired behavior of the ball is described in terms of a "goal," and nothing is simpler than to design a device which will exhibit just such behavior. Take a cup of any convex shape and place it so that the desired point is the lowest. To describe the same kind of behavior in terms of a "program" would necessitate an infinite number of statements, each of which tells which way the ball is supposed to move if it finds itself in a particular position. Such a description in terms of discrete statements (an explicit program) is, of course, out of the question. A description of "intermediate complexity," however, can be given, namely, as a set of differential equations of motion which imply a stable equilibrium at the desired point.

Of the three descriptions clearly the first, "stating the goal" of the ball, is the simplest. What enables us to realize this "goal" by a mechanical device is our ability to see the problem as a whole. Similar considerations apply, I believe, to the theory of the nervous system.

The first attempt to account for gestalt phenomena in strictly behaviorist terms was made by McCulloch and Pitts in 1943. They showed that any pattern of behavior which could be described by a *program* was realizable in an automaton of a specified construction and (herein lies the importance of their idea) they gave an "algorithm" for the construction of the automaton based on the program. Automata, of course, operate on the same principle. The limitations of this approach, however, are immediately evident. The *whole difficulty* is to describe the action of the nervous system in terms of a program of discrete elementary steps.

The task looks more hopeful if "goal-seeking" steps are allowed in the description of the program. If, for example, the construction of a mechanism for keeping a certain muscle-tone constant is known, one of the directions in the program may read, "Plug in that mechanism." Thus with one stroke an immense number of elementary steps is "described."

The value of information-theory in this approach to the nervous system is now apparent. The McCulloch-Pitts picture represents behavior in terms of firing patterns of individual neurons. With 10^{10} neurons in the human body, there are $2^{10^{10}}$ such possible patterns at each instant of quantized time. This number is utterly unthinkable. Nothing whatsoever can be said of a system with that many distinguishable states where nothing is known about how the states are to be classified.

To put it in another way, the amount of information per unit-time needed to describe such a system is 10^{10} bits (the amount of information coming over an ordinary telegraph wire is considerably less than 30 bits per unit-time).[2] It is quite another matter, however, when sub-systems are "organized" to work in prescribed ways, when touched off by proper signals. The *amount of organization* of such subassemblies *reduces* the amount of information necessary to transmit over the channels.

This consideration leads to two complementary conclusions. Rigidity of behavior in organisms requires smaller capacities of channels over which information flows. Contrariwise, greater channel capacities allow for greater flexibility of behavior.

We are thus led to ideas of the nervous system which involve not minute blueprint structures (a hopeless approach because of the tremendous complexity of the nervous system) but which involve overall statistical concepts such as channel capacity, storage capacity and other parameters familiar to the modern communication engineer, such as redundancy, signal-to-noise ratios, etc.

They are concepts analogous to the over-all concepts in terms of which the operation of the "muscle engines" is understood: power, efficiency, compression ratios, etc.

It is the development of the corresponding *over-all* concepts of communication and complexity which made intelligent machines possible and which gives us promise of future understanding of living behavior, particularly of the functions of the nervous system.

I need not, I hope, emphasize that none of these considerations are relevant to the question of whether thinking machines "really think." I admit I do not understand the question. The really pertinent question is whether similar *abstractions* can be utilized in both the theory of intelligent machines and in the theory of living behavior, particularly that governed by the nervous system.

We know that both organisms and machines receive, transmit, store, and utilize information. The question of how information is "utilized" is particularly interesting. We now know what food is used for: three things, namely, as a source of heat, a source of locomotive and chemical energy, and a source of materials for growth and restoring worn-out tissues. All these elements are being constantly dissipated by the organism: heat by conduction and radiation, energy by motion, materials through breakdown and excretion.

Can it be that besides energy in the form of food and sunlight, organisms also feed on something called "information," which serves to *restore the order,* which is constantly being dissipated in accordance with the Second Law of Thermodynamics?

The formal mathematical equivalence between entropy (the measure of disorder in a physical system) and information (as defined mathematically) was commented on by Shannon, Wiener, MacKay, and others. Can it be that this is no mere formal mathematical equivalence, such as obtains between an oscillating mechanical system and the analogous electrical one, but a more fundamental equivalence such as that between heat and energy or between energy and matter? Can it be that 1.98 calories per degree mole (the difference in entropy between two moles of two separated perfect gases and two moles of their mixture in equilibrium) is actually equivalent to 6.06×10^{23} bits of information—the amount it would

[2](Editor's note [Sayre and Crosson]: in information theory, the unit measure of conveyed information is the "bit" (binary digit). If transmission of either of two symbols is equally probable the actual transmission of either would convey one "bit" of information).

take to separate the mixture into the constituent parts in terms of yes-no decisions?

If there is such a conversion factor, how do the information receiving, information transmitting, and information storing organs operate to convert information into negative entropy or its concomitant "free energy" and, perhaps, vice versa?

The intriguing nature of these questions has stimulated some of us to undertake the study of communication nets from the information-theoretical point of view. This approach necessitates the description of such nets not in terms of detailed structure but rather in terms of gross statistical parameters. The flexibility and far-reaching adaptability of the behavior of higher organisms almost demands this sort of approach. Perhaps the most fundamental characteristic of living behavior as distinguished from that of man-designed machines is in the *sacrifice of precision* for safety. It is not important that a response be precise but rather that an equivalent response be given under a great variety of conditions or handicaps. It is more important to be "roughly" correct in practically every case than be "precisely" correct in every case but one and altogether wrong in that one. It is necessary to relate totally new situations *approximately* to situations already experienced, and it is necessary to leave certain portions of the nervous system "uncommitted," so that new behavior patterns to meet new situations can be organized. When machines are built possessing these characteristics, we may expect an even closer analogy to the workings of actual nervous systems. That the day is not far off can be inferred from the fact that mathematicians like von Neumann already do not shirk from theoretical investigations aimed at throwing light on the most typical of life processes—reproduction. I am referring to his recent calculations on the number of elements required in an automaton which can not only perform specific tasks assigned to it but also is able to reproduce itself, given a mixed-up aggregate of its elementary constituents. An actual materialization of such a machine would, of course, give startling reality to the prophecy in Butler's *Erewhon*.

Such is the state of the present studies, which are extensions of the technological analogies of the living process, particularly of the integrating functions of the nervous system. It is hoped that these studies are now approaching a level sufficiently sophisticated to yield enlightening and lasting results.

The man-machine confrontation

FRED R. BAHR

■ Fred R. Bahr *is a member of the faculty of the Department of Management, Graduate School of Business Administration, University of Southern California.*

Professor Bahr calls for a "humanistic" attitude toward technology; that is, he asks that we see technology as the product of human effort enlisted in the service of all mankind. He argues against those Luddites who fear that automation will cause widespread unemployment, and he attacks those such as Jacques Ellul who fear that technology has somehow become autonomous.

He suggests that we look to the future of man-machine relations with hope, and with the confidence that has been characteristic of mankind during the periods of his greatest achievement.

Popular fears about "thinking" machines have been both long-standing and persistent. While man has remained the same throughout the ages of civilization, he has increasingly surrounded himself with machines of ever greater technological capacity and complexity. Of these, computers, which in certain respects simulate human thought processes, have met strong resistance both from organizations and from the people in them. Fears of unemployment and loss of identity in automated systems are largely chimeras, however, for when considered humanistically, computers are merely highly useful tools for extending man's capabilities.

The psychological capabilities of man have grown at a much slower pace than the technological advancements which surround him. This fact underscores the need to see both man and machines in their true productive relationship.

□ From *Business Horizons,* 15 (October 1972), 81-86.

THE UMBRELLA OF UNCERTAINTY

Few myths have been more persistent, more persuasive, or more unrealistic than those concerning technological change and automation. The machine in general and the computer in particular seem to have touched some raw nerve, some irrational and deep-rooted fear—a fear reflected in legends of man-made robots who turn on their creators.[1] Even science fiction writers have largely adopted Isaac Asimov's three laws of robotics which prohibit robots from injuring or disobeying man.

Along with the other stress-producing developments such as the harnessing of nuclear energy, a rapidly increasing world population, and continued expansion of transportation and communication capabilities, computers, and some of the proposals of what to do with them, are making George Orwell's *1984* seem less and less like social science fiction and more and more like reasonable prediction.

Almost from the beginning, the scientific revolution has produced a number of threats to man's egocentric conception of himself. Copernicus showed that our world, the earth, is not the center of the universe, Darwin demonstrated that man is part of the same evolutionary stream as other higher forms of life, and Freud suggested that man is not fully the master of his own mind.[2] The electronic computer now emerges as another challenge to man's concept of self.

There has been a good deal of speculation as to the nature of this challenge. At first glance it does not appear to be a reaction to a purely physical phenomenon—that is, the concern does not originate in the material properties of the computer. We are surrounded with machines and other artifacts of far larger and more impressive proportions. But if the confrontation is not physical in nature, then it must somehow be linked to the computer's functional characteristics or its powers to manipulate data. If this is true, the threat must be perceived as being intellectual in nature.

Somehow the computer seems to have taken on the image of an automous entity—a kind of superbrain which thinks the way humans do. It appears to formulate instant solutions to highly complicated problems which the "ordinary man" cannot even begin to understand. "This anthropomorphic notion of a machine which can possibly out-think man is not easy to assimilate or live with. It suggests that man is less unique than he once thought and therefore somehow less important."[3]

MAN AND HIS UNIQUENESS

While it has been maintained that the physical evolution of man's body and mind came to a near standstill some 20,000 to 30,000 years ago,[4] socio-cultural influences have continued to modify human life and at an accelerated rate. The physiological needs and drives of modern man, his responses to environmental stimuli, his potentialities and limitations, are still determined by the interplay of the 20,000 genes that governed potential responses in the paleolithic age. "The fact is that man is partly geared to the past by physiological processes over which he has normally no control."[5]

The result is the emergence of a kind of biological Proustianism: the responses of the individual to the conditions of the present are always being influenced by biological remembrances of things past. Because of this hereditary limitation, it is often difficult for him to think things through before acting. To develop long range plans and to abide by them requires consummate and dedicated effort.

Still, man alone has been able deliberately to change the environment to suit himself. This capability dramatizes and draws attention to the central mystery of man—his persistent restlessness. Native aggressiveness, reinforced by prolonged periods of infancy and youth, probably account for it. It is the force behind discontent, the search for novelty and diversity of experience. Considered in concert with the human need for variety in sensory stimuli, it accounts also for man's inability to concentrate on specific tasks for extended period[s] of time.

As a source of intellectual productivity, man shows marked contrast in his capabilities. As just noted, he has limited native competence for the development and execution of long range plans and an inability to concentrate be-

1. Charles E. Silberman, *The Myths of Automation* (New York: Harper & Row, Publishers, 1966), p. vii.
2. William D. Orr, ed., *Conversational Computers* (New York: John Wiley & Sons, Inc., 1968), p. 221.

3. *Ibid.*
4. Ellsworth Huntington, *Mainstreams of Civilization* (New York: Mentor Books, 1959), p. 43.
5. John E. Pfeiffer, *The Emergence of Man* (New York: Harper & Row, Publishers, 1969), p. 421.

yond short periods of time, yet he also possesses an enormous and often unappreciated capacity for the assimilation and retention of information.[6] This potential has been assessed as being several thousand times greater than the storage capacity of the most powerful computer in existence today.

In addition, man has enormous powers of selective perception, self-organization, learning, and finally adaptability. He is capable of arranging an immense amount of impinging sense stimuli into patterns which simulate the environment. He also has learned to reject sense stimuli of low interest and to admit to consciousness only those which are important to him as an individual.

Most individuals can distinguish meaningful sequences of cause and effect in extremely complex situations. When faced with the necessity of making decisions of a probabilistic character on the basis of insufficient information, they can produce excellent statistical estimates involving many interacting factors. This particular capacity for self-determination had led some scholars to confer on man the unique distinction of being a "thinking" or "rational" animal. Recent experiments with other forms of animal life demonstrate rather conclusively, however, that they too must be credited with limited measures of rationality. A sense of awareness and the ability to form abstract ideas also do not seem to be completely human prerogatives.

There is one trait which is distinctively if not uniquely human: it is the making of even more complex tools. From the dawn of civilization, implements have been indispensable for man's survival and growth. They have been the catalyst in bringing about the dominance of Homo Sapiens as a competing species. They have also given a special character to the nature of modern society.[7]

Moreover, as the "facilitators of increased productivity," tools have become indispensable extensions of man's own physical capabilities. The ever-growing demands of contemporary social structure have created a special kind of dependence on their continued presence and effective use. The computer represents a new and significant plateau in that continuing relationship.

THE COMPUTER IN CONTRAST TO MAN

Only a very small segment of the population have had any substantial contact with computers, yet these machines have provoked an enormous amount of interest. There seems to be a sort of "grass-roots" consensus that the computer can do anything, solve any problem. In contrast, there is little popular understanding of the necessary role that people play in computer-assisted problem solving. What the population sees is the machine, an independent source of data-manipulating power. This has led to an unhealthy sense of idolatry "for the machine as all-powerful, as incomprehensible, and as independent of man's will."[8]

This characterization is both inaccurate and unfortunate. Computers are, by simplistic definition, "machines for recording numbers, operating with numbers, and giving the result in numerical form."[9] They have been employed to carry out long and tedious, but conceptually simple, manipulations. Because the design of these machines endows them with extraordinary ability to perform mathematical and supporting logical calculations in amounts and at speeds that completely overshadow man's comparable capabilities, people have tended to generalize from such demonstrations of computational prowess. They have used these demonstrations as the basis for ascribing the characteristics of infallibility to the computer. This value judgment does have some degree of surface validity. Both man and computer:

1. receive information to provide the basis of action;
2. possess memory mechanisms;
3. utilize the feedback of information from "outside" their own system as a means of comparing actual with intended performance;
4. possess central decision organs which determine what is to be done on the basis of that information feedback.

Such similarities have only served to reinforce the notion of a machine which can possibly "out-think" man.

6. George A. W. Boehm, "That Wonderful Machine, the Brain," in Stanley Young, ed., *Management: a Decision-Making Approach* (Belmont, Calif.: Dickenson Publishing Company, 1968), pp. 35-39.

7. Norbert Wiener, *The Human Use of Human Beings* (Garden City, N.Y.: Doubleday and Co., 1954), p. 16.

8. Orr, *Conversational Computers*, p. 220.

9. Norbert Wiener, *Cybernetics* (Cambridge, Mass.: The M.I.T. Press, 1961), p. 116.

AN INITIAL ASSESSMENT

The guardedness with which the nonscientific community has greeted the introduction of the computer is a measure of our narrowness in understanding the man-machine interface. Here is a tool devoted to making precise, speedy, and virtually infallible mathematically-based decisions. It is a conditioned reflex, a device which extends the ability of man to associate numerical and logical concepts to produce still others.[10]

To be sure, computers have been responsible for introducing a unique and new dimension of discipline into the on-going man/machine activities. They impose procedural imperatives which require order, precision, and consistency. Regardless of the size and speed of the machine, all three of these demands must be met. In effect, the computer obligates us to sort vague ideas and feelings out from clearly formulated expressions of value if we wish to manipulate ideas and information by machine. Once this has been accomplished, the computer is capable of executing long and detailed programs of data manipulation otherwise beyond human capability. In many cases, the logic operations performed by machines permit a rationality in decision making and a precision of control hitherto unattainable.

As previously suggested, the operation of these machines is closely parallel to the behavior of humans in that the results of output or demonstrated behavior are "reported back" and used to modify subsequent actions. There is, however, one crucial difference in the nature of the output.

> The source of the mechanical brain's stunning potentialities is that the "reporting back" is flawlessly truthful. In human beings a variety of factors—prejuduce, external pressures, neurosis—can falsify the "reporting back" and make behavior unresponsive to the evidence.[11]

This issue of communicative integrity further sharpens the need to understand the nature of the man/machine relationship. The computer with all of its manipulative capability has even less native intelligence than a newborn child. It has no voluntary reflexes, only an amazing capacity to do what it is programmed to do. The slightest error in instructions, one bit of information out of millions, will trigger incorrect manipulation of data by the computer or no response at all. The potentialities of the computer for flexible and adaptive responses to assigned tasks is therefore no narrower and no wider than the intellectual and emotional potential of man himself.[12]

THE SCOPE OF CONFRONTATION

There are approximately 70,000 computers currently functioning within the United States. Business firms and related activities had already spent between 20 and 25 billion dollars on operating computer systems. This represents an investment of approximately two percent of the current annual gross national product. The Federal Government, as the largest single user of computers, presently uses more than three billion dollars worth of data-processing equipment.[13]

The search for new and expanded applications of computer technology continues. More than half a million workers direct their creative talents to that end. Within the next two or three years, the number of computer applications will move past the 100,000 plateau. The growth anticipated in computer usage has only underlined a number of unanswered questions associated with proper utilization of this new productive resource. These questions usually take the following form:

1. What kind of information should be collected and stored?
2. How and to whom should its output be made available?
3. What will the price of this service be in terms of dollars, time, and other resources?
4. How should these costs be allocated to users?
5. What will be the effect of this computer activity on existing, more conventional man/machine relations?

In a broader sense, these questions represent real anxiety over the impact that the computer will have on the future shape and structure of society as well as on its goals and purposes.

DETERRENTS TO ACCOMMODATION

This measure of concern becomes further aggravated when consideration is given to the influence exerted by conventional organiza-

10. *Ibid.*, pp. 117-120.
11. Wiener, *Human Use of Human Beings*, p. 107.
12. Peter F. Drucker, *Technology, Management & Society* (New York: Harper & Row, Publishers, 1970), p. 174-75.
13. "The Trillion Dollar Economy," *Business Week*, October, 1970, pp. 60-61.

tions. These existing entities, in themselves, can be formidable barriers to any proposed change not considered to be in their interests. The continuing demands survival imposes have encouraged institutions to develop sophisticated defense mechanisms; protective considerations permeate the fabric of their formal goals and strategies. They can and often do represent static constraints on progress. Efforts to circumvent their barriers introduce a whole new dimension of complexity to any process of change.

Individual members, in responding to the pressures on them, often replicate the organization phenomenon: they react according to self-interest. Collective efforts can significantly modify the announced patterns of organizational activity. Formal statements of institutional purpose can be subtly reshaped by this dual "filtering" process.[14] These two sources of influence, organizations and their membership, often work in concert to thwart efforts to introduce technological change, even when the results are clearly in the general public interest.

The result can be a muddled matrix of organizational and individual activity. Management often becomes the art of improvision. The manager responds by employing organizational methods and processes "to solve problems without knowing how he solves them."[15] Operations become defined only indirectly or are not explicitly defined at all. This has prompted one expert to observe that "the biggest problem in bringing an information system to fruition is blending it into an operating organization that is not accustomed to the discipline required."[16]

The other principal deterrent to the widespread use of the computer within organizational activity is more of a sociological nature. It involves the purported hardships imposed on the general work force by the introduction of automation. Apprehension over this issue has been heightened by the number of writers who have described this new phase of industrialization as imposing a set of imperatives—insistent pressures for urbanization, professionalization of management, and the rapid growth of professional and technical employment—on the working rank and file. Jacques Ellul has gone so far as to announce that the technical take-over has already occurred and that it is irreversible.[17]

The anxiety such statements arouse has manifested itself in two ways. First, it is evident in the public's pre-occupation with the issues of job displacement, loss of job security, and even intellectual inadequacy. The second is apprehension over the loss of personal identity through the assignment of productive tasks to devices for automated execution.

Silberman has attempted to refute these myths of automation in his book of the same title. His research shows that:

1. Automation is not a significant cause of unemployment—in large part because there is not a great deal of true automation.
2. New technology is exerting far less impact than had been assumed on the kinds of work men do and the amount of education and skill they need to do it.
3. Man is not losing control of his instruments; technology is not taking over, nor is it effacing human will.[18]

He further shows that automation is not structuring employment opportunities; instead, this new technology is progressively enlarging the sphere of human action and choice by contributing to the general growth in industrial productivity and by eliminating tedious and low-skill productive tasks.[19]

A LOOK TO THE FUTURE

Today we are developing computers of even higher order of capability. These machines will be an intricate part of our intellectual and physical equipment as they contribute directly to our capacity for thought and action. This advance could not have come at a more propitious time; man has already reached the limits of his personal productive capabilities. If he is to continue to respond to his own growing needs, man is forced to find new artificial extensions to his muscles and senses. The computer epitomizes the shape and scope of that effort.

The evolution of still more elaborate and

14. This is an imposing stumbling block to the proponents of increased automated activity within the organization. Processes of this nature feature closed-loop feedback control where the human element is not required in the monitoring sequence.

15. Allen Newell and Herbert A. Simon, "Heuristic Problem Solving," *Operations Research,* VI (January-February, 1958), pp. 1-10.

16. "Trillion Dollar Economy," p. 66.

17. Jacques Ellul, *The Technological Society* (New York: Vintage Books, 1964), pp. 89-94.

18. Silberman, *Myths of Automation,* p. x.

19. *Ibid.* pp. 19-22.

complicated man/machine systems will present some new and intriguing problems in accommodation. The most immediate hurdle to be cleared is technical in nature. It concerns the resolution of differences in communicative techniques employed by man and machine. The descriptive languages currently used by man in the processes of thinking and reasoning do not lend themselves to computer manipulation. At the same time, the manipulations that man performs on these linguistic elements are usually more subtle and complex than routine mathematical calculations. This suggests a major distinction in future man/machine relations: the computer of tomorrow will have to have the capability to respond to words in the form of verbal abstractions as well as numerical data.

The more important distinction in this evolving relationship concerns the promise it brings to the task of resolving the growing problems of organizational complexity. Once we have established linguistic communication with the machine, we will have opened up a whole new vista of responses to the most intricate processes of thought and action. To capitalize on these advancements in man/machine relations will require that the computer be thought of in humanistic terms. It is the creation of man, the outgrowth of technology. As such, it can and should be made to serve the needs of man. One renowned scientist characterized this challenge when he said:

> My plea is for a humanistic attitude toward technology. By this I mean that we recognize it as a product of human effort, a product serving no other purpose than to benefit man—man in general, not merely some men: man in the totality of his humanity, encompassing all his manifold interests and needs, not merely some one particular concern of his. Humanistically viewed, technology is not an end in itself but a means to an end, the end being determined by man.[20]

20. Hyman G. Rickover, "A Humanistic Technology," *The American Behavioral Scientist* (January, 1965), p. 3.

5

Technology and the arts

Yes-no: art-technology

JACOB LANDAU

■ Jacob Landau *is chairman of the Department of Graphic Arts at Brooklyn Pratt Institute.*

Mr. Landau describes the split between the world of artists and the world of technological reality. The insightful artist, the honest artist, he argues, has given way to the artist as technician or supersalesman. The true artist is now afraid to create because his creation can be so easily co-opted. When this happens he becomes "a smug facet of the collective dream," a phenomenon which Landau thinks is endemic to present technology.

He concludes his essay with a call for a "syncretic" man, a type of artist capable of being an individual, cutting through the systems of patronage that industrial technology espouses, and thereby humanizing technology instead of becoming its "yes-man."

I have often overheard young people say after a pregnant pause in a conversation about almost any subject from art to baseball, "What terrifies me is. . . ." All themes somehow lead into a no-man's-land for everyman's fear. I am not ashamed to confess that I too am afraid.

What terrifies me as a working artist is what appears to be a split in the thinking of modern man, a deep and perhaps fatal cleavage between illusion and reality. This is perhaps more serious than C. P. Snow's two cultures' breach, or the East-West polarization. Although these splits are interrelated, both science and technology are deeply implicated, and have, so far, failed to eliminate the road towards reintegration. The tendency to split reality may indeed be inherent in man's structure, rooted in the emergence of his forebrain and consciousness. Mind, as the latest emergence in nature's hierarchy of innovations, is also the highest—it subsumes and seeks to dominate the rest. It stands at one pole in the ego-world field, and perceives the world as "otherness." It can endeavor to sink to the level of structure, to drown in simple *being* when it suspects itself and learns to fear its *becoming*. Or it can seek to rise above its rootedness and aspire towards a human or divine unity. In the ego-world field, as in all of nature's fields or systems, a conflict of opposing forces appearing as vectors or tendencies aimed at poles or thresholds occurs. At the center of a field, where the vectoral tendencies are very nearly equal, an uncertainty zone exists, giving rise to an oscillating wave or pattern of alternating polarity. Minds faced with uncertain choices tend to undergo flip-flop, yes-no inversions—life-death, hope-despair, freedom-security, pride-humility, angel-devil, goodguy-badguy, play ball-dropout, join-cop out. Yet, increasingly, the middle of a field, whether it be in mind, between mind and society, or in society, is where everyone wants to be, safe but anxious, encapsulated in the centrism of mass society, the consensus of political action, the narrow parameters of discourse in any sub-culture, the status quo-ism of institutions, the play-the-game of individuals. And, increasingly, the urge to explode violently outward towards opposite poles is created by the unbearable tension of uncertainty in the middle. As centrism is a feature of modern society, avant gardism is its alter ego, flying outward to the fringes of a field, pushing against its outer limits towards break-throughs and the estab-

☐ From *Dialogue on Technology*, ed. Jacob Landau (Indianapolis: The Bobbs Merrill Co., Inc., 1967), pp. 91-106.

lishment of new systems, escalating towards danger in search of perspective, certainty or meaning. Paradox seems to rule our lives, and multiplicity, despite McLuhan, is exceedingly hard to come by or live with.

Our consciousness is but a frail, thin thread, a one-dimensional line. Simultaneity is a function of the dialogue between consciousness and the lower levels of the psyche. The dialogue itself seems to be structured as an alternating rhythm; as a waving above and below the threshold of consciousness. Each distorts the other. Man cannot be rational without distorting the emotions, nor can he feel without distorting his thought. A rebel literally cannot see his opponent. If he does, he risks becoming a reformer. Yet most great minds were synthesizers, and the synthesizing tendency is an important in man's affairs as is the polarizing tendency. If an uncertainty principle seems to underlie the interplay of contraries, if riding one vector seems to involve penalties with respect to its twin, if wave-like behavior is characteristic of creatures and societies as they seek an all too transient equilibrium all too quickly undermined by events, then the individual should know more about field dynamics in order to avoid being dominated by unconscious or unknown forces, being led to accidental or controlled choices, being forced into catastrophic crises or a stifling avoidance of change and growth. He should, but it seems, he cannot.

The individual man has never felt more impotent than he does now. He feels as if he counts for nothing. He is incorporated in groups and collectives, each of which is hierarchically subordinated to the great power-blocs, and each of which is totalized by the new communications technology. Power is all, and "establishment" is the word for an ultimate global technological collective that seems to dominate all other sub-collectives enveloped by it. A man wants to feel competent to choose his destiny, yet he is chosen. He dreams of importance beyond his space and time; yet he is in the main forced into time-bound mediocrity and inconsequence. Not only does man feel incapable of influencing the social vectors except in quantitative terms, but he is also afraid to step out of line: one man, one vote, a poor man's immortality. Man can function technically, but not in a human way; partially, but not as a whole. I believe that much modern anxiety flows from this fundamental split between meaning and behavior. All meaning vectors tend towards universality; but all pathways are dead-ended and turned back on the self. The individual cannot bear the dichotomy between desiring to function in a meaningful and successful social way, and being obliged to operate self-preservationally within the group. He wishes to belong, but can only function as an attacker against all comers, as a particle among particles. His modes of belonging and functioning become more attenuated, and he is obliged to conceal his anger and frustration.

Two points are relevant here. First, despite the fact that you cannot taste, weigh, measure, accumulate, escalate, fabricate or distribute morality, it is an inevitable outgrowth of human emergence. Freedom of choice is a human privilege; its inevitable byproducts are good and evil, right or wrong. Life and death choices at the edge of life's field are relatively easy to assess as good or evil—if you are alone. If you make the wrong move, you lose. But Sartre reminds us, if you choose, you choose for others too, and your error may be the next man's fate instead of your own. Second, it is fashionable to view such individual problems as subordinate to group behavior, to see morality as relative to structure or pattern. Some structural thinkers are inclined to say: it does not matter whether you are capitalist or socialist, blue or violet, technology will lead you wherever it wishes to go. Or, it does not matter what you say on television; the medium is the message. Or, it makes no difference what you paint or write about, it is the "significant form" that counts.

Have we forgotten the word "content"? A man can kill to preserve life, his own in self-defense, or the lives of others in mutual defense, or he can kill for the sake of killing, or to protect one man's power over others, or to destroy a people or race. In this example, we can distinguish a content difference between structurally similar acts. I tend to see morality as a content difference. It is easier, however, to polarize, to say something is absolutely right or absolutely wrong, or to say that morality does not count at all. Or, as is often the case in group dynamics, it is safe to cling to the center, at the heart of ambiguity and darkness, to mill around in a consensus culture and avoid rocking the boat. That way lies madness, the bomb, or the maturing technological crisis. The artist in his own intuitive, complex, symbolic, indirect, bumbling and often contradictory fashion, has been heralding the coming of the crisis for some two centuries.

Since the coming of the machine, the artist felt himself to be an outsider, if not always and inevitably a Luddite, as C. P. Snow would have us think. Many artists of the nineteenth century, from Goethe on, believed in science, but many also had mixed feelings about technology. And all artists were affected, pro and con. William Blake reacted strongly to the coming of industry. Early in the century, he bore witness to its tendency to despoil the land, deform the people, and bind man to the machine in the image of a fiery crucifixion. Blake wrote:

> Bind him down, sisters, bind him down . . . .
> on Ebal, mount of cursing.
> Malah come forth from Lebanon, and Hoglah
> from Mount Sinai,
> Come circumscribe the tongue of sweets,
> and with a screw of iron
> Fasten this Ear into the Rock. Milcah,
> the task is thine.
> Weep not so, sisters, weep not so;
> our life depends on this.

For Blake, Locke, Newton and the machine age were abstract, pitiless mechanistic agencies of human enslavement:

> I turn my eyes to the schools and universities
> of Europe
> And there behold the Loom of Locke, whose
> Woof rages dire,
> Wash'd by the Water-wheels of Newton: black
> the cloth
> In heavy wreathes folds over every Nation:
> cruel Works
> Of many Wheels I view, wheel without wheel,
> with cogs tyrannic . . .

The prophetic Blake saw clearly that a new environment had replaced nature as an envelope of promise and punishment for man. Most of his contemporaries, seeing only the promise, failed to understand the psychic consequences of industrialism. Seen from the perspective of a New York penthouse, the road traversed since the Enclosure Acts is all onward and upward. But seen from Blake's perspective, as he stood at the threshold of the new age and Janus-faced looked both ways, the price paid in human suffering and waste seemed far too high for the promised rewards. With Blake stood a few lonely giants: Beethoven, Goethe, Goya, and Balzac. They saw the old and the new simultaneously, the good and the bad in each. They were the last to see life as a whole, to find a meaning structure which hung together, and to unite the contraries of heart and mind, man and society, form and content, part and whole. Their questions and answers differed, but to me they loomed high above the nineteenth century plain, and cast mighty shadows ahead.

The artists who followed were split a thousand ways. In making such a judgment, I do not say they were inferior as artists. Art is what it is: unique, priceless, unarguable, incomparable. We cannot easily question it for what it is not. Yet the artist, too, has paid a price for his liberation from patronage, his involvement in a free market, his need to identify himself through exaggeration and sell himself by all means open to his ingenuity. The new structure of technology has ruptured his primary relation with reality, and all else followed. John Dewey has pointed out that, "When the linkage of the self with its world is broken, then all the various ways in which the self interacts with the world cease to have a unitary connection with one another. They fall into separate fragments of sense, feeling, desire, purpose, knowing, volition."

In consequence, art moved inward towards the centers of feeling and inspiration, toward the psychic levels beneath consciousness. It simultaneously moved outward towards an attempt to see purely and without subjective bias or distortion. This was dichotomy, and although it led on to new adventures, it prompted instability and tremendous psychic disturbance, which, like an unbearable itch, drove the artist to frantic efforts in search of release. Both Appolonyian and Dionysian, scientific and artistic temperaments among artists were freed from the rational controls of the enlightenment, and the rational-technical orientation of their own society to pursue *personal vectors*. The artist was the human part of industrial man crying out for the dehumanized part which was appendaged, iron-faced, to the machine.

In the dynamic of naturalism, a vector towards the objective pole leads to an ultimate threshold between imitating and replicating reality. Since replication is impossible, since imitation in any one art is limited by formal and technical boundaries, the painter could not quite succeed in eliminating himself from the picture. The objectivism which impelled him towards the edge of a representation-distortion polarity compressed his image to a fragment in time, an instant of appearance. The uncertainty which arises from the space-time, appearance-motion polarities caused an appearance vector to freeze the image in space, to distort its motion in time.

Conversely, in the subjectivism of the romantic impulse, a vector towards subjective thresholds leads towards distortion of appearance in favor of motion. Although distortion of either-or is a built-in feature of polar tension and uncertainty, the nineteenth century began the process of seeing them not as contraries which may be synthesized, but as mutually exclusive opposites. The resultant dualistic instability between the warring halves caused them to undergo flip-flop inversions. Romanticism was the underlying idea for the entire century; yet it had no style of its own. Naturalism is by definition styleless, since only the addition of an artist's personality to raw reality can make a style. Romantic painters like Gericault and Delacroix, and even traditionalists like Ingres, look surprisingly naturalistic in photographic reproduction. Their form approach had all the love of particularity we find in Coubet. Proto-scientific, impartial, rational, analytic, pragmatic naturalism favored matter over motion: esthetic, subrational, impulsive, impractical romanticism, following Blake's dictum that "Energy is eternal delight," favored motion over matter. *As subjectivism was the essential content of the century, so objectivism became its form*—a manifestation within the ranks of art, of the art-science split.

THE COMING OF CÉZANNE

As the century wore on, other splits became manifest. The subject of a work of art declined in importance at both poles. The objective artist was turning into an *eye,* and what he saw mattered less than the arrangement and the manner of seeing. Impressionism transformed the objective moment into a subjective moment in the perceptual field. Ordering of parts, according to visual-esthetic criteria, prefigures the coming of Cézanne, and of Cubism. On the other hand, the subjective artist was turning into a *nerve ending,* and whether he wept over an old shoe or a crucifixion mattered less than the fact that he wept. Art for art's sake emerged as the reigning trend in the 1870's, and since each division in the field of art produced a sense of liberation from complex entanglements, the latest and last of the major splits gives birth to the ideal of a pure art divorced from human concerns, and serves at first to rejuvenate the artist, rescuing him from *fin-de-siècle* decadence.

It is interesting to note that an art of pure subject matter and appearance became the approved art of the academy-salon system; while an exciting art of esthetic-expressive involvement moving away from the subject-appearance pole *in a new way* arose outside of its sphere of influence. The new way consisted of an altered mix of subjective components; imagination and fantasy began to replace the mere seeing or feeling to which academic art had degenerated, and the unique blend of idea and form (form as idea or idea as form) began to appear. Academic painters like Lematte, Ferrier, Besnard, Wencker, Bramtot, Fournier, Lebayle, Mitrecey, Lavergne, Sabateé, among others, were Prix de Rome winners from 1870 to 1900. The list of artists who failed to win, constituting an honor roll of great innovators includes: Degas, Sisley, Pizarro, Cézanne, Seurat, Signac, Lautrec, Bonnard, Matisse, Rouault and Dufy. These artists were but meagerly supported by the rising (and still current today) dealer-client-museum system. As a true avant-garde, they paid a heavy price for their rebellion.

It is ironic to note that democracy, in the long run, spelled the end of security and social utility for the artist. Almost none, from Blake on, wanted to serve society. However, almost none were free of anxiety about their role, and many compensated by viewing themselves as seers or scientists. Like subjectivism, universalism, or an impulse to speak to and for multitudes, was an essential part of the artist's dream. Yet he feared adulteration and loss of identity in the market place, and integrity was equally in his dream. The illustrator, popular painter or designer came to produce artless functions; while the "fine" artist clutched to his breast, protestingly at first and then proudly, his image as the alienated creator of functionless arts. His new direction proved fruitful. He followed its vector towards the immense new field of structural innovation and found that the problem of meaning, already tied to isolated moments of beauty, of sensation or emotion, became more and more bound up with the problem of representation, with what to represent, how to represent it, and ultimately, whether to represent at all. The invention of the camera earlier in the century merely helped to precipitate the maturing crisis of representation—it did not create it.

In this sketch of nineteenth century tendencies, I do not wish to imply that no other shadings existed. On the contrary, nuance and variety abounded, change was rapid, paralleling the quickened pace of technology, and the ideal of purity was but a potential. In calling attention to the interplay of forces in the social and psychic fields, I wish to stress only what appear to be major tendencies, hopefully of significance in understanding our own time and its creation of flip-flop man.

THE INVENTION OF PICASSO AND MATISSE

From the beginning of the twentieth century, the artist has pursued with ever increasing intensity the vector towards unconsciousness. Picasso and Matisse, two giants who stood at the threshold of the new art of structure and dream, exploded in a burst of invention which dazzled the world. Picasso summed up the notion that consciousness was henceforth to be downgraded when he said: "I do not seek, I find."

The host of painters within the three principal movements of Cubism, Surrealism, and Expressionism, who first explored the unknown land of the psyche, were drunk with the joy of discovery. They were also persuaded that at last they could speak to multitudes, that they had invented a universal language of the senses, that the new art which was based on what one critic called "significant form" would provide a "direct linking of man to man." The scientific idea of structure as a universal component of what we experience as quality, of simple fundamental patterns which create the diversity of nature's forms, of growth and developments in all fields as possessing an underlying harmony and uniformity of motion, was a breathtaking glimpse of the universal. It served, however, to downgrade mind and consciousness while promoting the release of sensory-affective potentials.

The new technology was equally reductive in nature. According to Jacques Ellul, it reduced the multiplicity of means available to "the one best means." It thus created a network of means, and a network of technicians committed to the creation of means, to innovation. It based all of its calculations on number, on the probable behavior of groups, of aggregates rather than individuals. It possessed inherent drives towards totality. Wave after wave of technique and its echoes spread outward to global limits. Each man in the technical dynamic was alternately particle and wave in a unified dynamic field.

The mass culture which arose at the beginning of this epoch, is likewise quantitative and reductive. The electric media, while promoting sensory integration, do so at a price. The media give, man receives. Feminized, man becomes means for the development of other means; he is a consumer who keeps the "Satanic Wheels" in motion. The media reduce all ideas to least common denominators. They aim for saturation. They create a uniform network of surface excitement masking the violence beneath. Their content is progress, their form is entertainment. They homogenize all that enters their domain, rendering it uniformly imbecilic. They invent and disperse a fictitious image of reality which destroys man's dialogue with the real world. The "good news" of the ads, which is the "good news" of technology, is the content of its mass culture. The "bad news" of violence, war, race riots, murder and all the rest, is the *content of* technology, even as progress is its form. The violence flows from its built-in totalitarianism, its vectors towards maximum dominion, its reductionism, its promotion of the war of all against all, its sacralization of technique and its neutralization of all moral codes, its schizoid split between high purpose and low means, and last, its division of man into a progress-violence, yes-no polarity.

The growth of public relations, managed news and staged government leads to the emergence of a "show biz" methodology—if things are not going well, if problems arise, if your opponent shows you up, never mind, just make it look good. The fake excitement on the surface of mass culture is designed to convey an equally fake image of fun, success, and completeness. The results burst on the public from time to time as shocking, unforeseen and apparently uncaused events. Yet if we seem to prefer the "bad news" to the "good news," it is not for textural reasons as McLuhan contends, but because it is *more real* than the "good." The disparity between the false and the real is guilt and fear-provoking. Those who can, run to the couch in search of a tragic flaw. Most, however, become "fixed" in Ellul's sense of being "adapted to the degree that they have become inert, unable to take risks. . . ." William Burroughs' image of the mass media as "junk" or narcosis leaps to mind (even the word "fix" is suggestive), as they act to drug the masses and, as expressed in William Morris' remarkably prophetic utterance of 1896, to make "all men contented together though the pleasure of the eye was gone from the world. . . ." In such a totalizing environment, the artist-dissenter became a culture-hero. He was, in fact, almost the only heroic human around in a setting given over to the creation of technical, adaptable, inauthentic, cowardly people.

Yet no man can be absent from his culture. Despite his strong inner resources, his tradition of heroic individualism, the artist reflected the technology in all that he did. His hatred of consciousness was provoked by the failures of technical man, and by the suppression of man's senses and instincts in the technical environment; yet it paralleled the wholesale flight from awareness which had everywhere taken place. His search for purity, expressed in repeated attempts to eliminate all contraries from the image and arrive at an essence, parodied the prevailing drive towards polarization, towards the eliminating of contradiction because it provokes ambiguity and pain.

The artist's enhancement of pictorial excitement, of visual and tactile intensity, of shock almost for its own sake, repeated the inherent tendency of mass society towards promotion and aggrandisement, towards the entertaining visual and visceral commotion on the surface of its culture. His restless experimentation with new ultimate handwriting can only result in a perfect sign, an identifying quirk, squiggle or tic. The artist achieves an ultimate identity, which in an exploitative market situation has the survival value of a corporate image, at the cost of his freedom and maneuverability.

TRAPPED IN A DEAD END

A perfect sign is by definition incapable of further development—and so is the artist. He must either repeat himself endlessly, obsessively, or overthrow his own identity by inventing a new image. He is trapped in a dead end. He would not have been trapped had he not already been consumed by the boom-bust cycle of the new art game called show and sell. The dream of over a century, that the artist would someday experience a great return to the bosom of man, has at last been realized. Pop, Op, with the media mixes hard on their heels, are the first art styles of modern times to enter the mass-mediated environment and be parlayed into fashions. The new crop of artists are either technicians or sophisticated salesmen, exploiting the current glorification of creativity. Tip sheets for potential investors warn of painters whose prices have reached "cyclical tops" and finger others slated for stardom. The artist-hero, who skirted the abyss of martyrdom, becomes a celebrity in the last act, and in happy-ending style, winds up as copy for gossip columnists and ladies' journals.

Nothing could be more tragicomic than this return to usefulness. From the standpoint of society, it has been long overdue. Yet a technological system, which cannot provide better highways, equality of opportunity for minorities, adequate hospitals, clean rivers, unpolluted air, better-than-average schools, a decent standard of living for the "underdeveloped" people living in outmoded technologies, or peace on earth, is hardly to be denounced for failing to treat art as nothing more than a tolerable if somewhat suspicious activity. From the standpoint of the artist, a moment of good living in the sun of fashion is worthwhile; but not at the price of the very authenticity for which he is being celebrated. The media may be suspected of doing what they always do, betting on something which is either no longer dangerous, or which can now be transmuted into a pseudo-happening, a charade of meaning.

In the last act, the habit and pose of revolution itself becomes marketable, and the explorer, turned exploiter, consumes himself and his traditions. In desperation he espouses total experimentation, the antiart of happenings, of anomalous, improvized occurrences which cannot be codified or reproduced. This represents the breaking up of all structure and the prevention of closure either for the artist or the client. The artist is now *afraid to create,* lest his "difference" be programmed into subsidized systems of discontent and turned into one more smug facet of the collective dream. A painting which can be read on the run in thirty seconds, or less, is already part of the background of life. It looks good on orangerie walls overlooking a swimming pool. In its heyday, field art was a background in search of a figure. The figurative art which then arrived followed the tradition of treating man as an object of idle sensation, as a plastic form, as anything but a human.

Instant art has one unrecognized virtue. It is the most democratic art form every developed, precisely because of its almost total return to sensation and spontaneity. Because of its process nature, it is most valuable to the artist as creative experience rather than as artifact. The mass needs spontaneity and creation more than it needs a second bathroom. It can use a sensory-esthetic awakening as an antidote to the rational poisons transmitted by all modern means of communication.

It would, as an ultimate inversion, counteract all the totalizing vectors by setting up a resistance pattern. Action painting, found-object art, collage, assemblage, some branches of Op and Pop, and a dozen other styles are capable of imitation and reproduction. They can be packaged, varied infinitely, sold over counters or produced on Sunday between chores. They look alike because the vector of personalization, having crossed a threshold, has inverted individuality of expression to collectivity, to the birth of an "international style."

Today's elite art means reductionism and quantification of method; it is death for art as we know it. Paradoxically, it also means a potential quickening of life for everyone, a revival of sensory and erotic awareness, a return to the chthonic, to doing all things beautifully, to discovery, to acquiring a value and meaning for life beyond mere passivity and purchasing. The elite artist, curator, critic and art school, so long as they remain trapped in a decayed avant-gardism, will continue to play out their charades of pseudo-significance, unaware that they have been devoured by mass culture, and converted to personnel in the communications establishment. Art intended as entertainment or decoration is at best a minor art. It is one more means in the chain—for the artist a means to celebrity and affluence, for the masses a means to forgetfulness or titillation.

THE ALIENATED WEB

Two models for human behavior exist at the threshold. One is the corporate model. It is a major tendency in both capitalist and socialist technologies. The corporation or corporate state is an overriding necessity for the single man. In it he has no true individuality. Corporations are totally unfree as institutions, and channels of communication up and down the chain of command are normally blocked by a protocol system. The individual corporation says, "If you don't like it here, you can go elsewhere," a choice which disappears in the corporate state. The problem of reform is the universal problem of "How do you fight City Hall?" Paul Goodman points out that the corporation does not recognize policy errors. Though only the corporate elite has individuality and can set or alter policy, it is not responsible before the public. The corporation is equally not responsible. Then who is? The reformer who fails to operate according to the rules of the game, the technician on whom an error of policy can be blamed, are expelled as offending particles. The rest of the corporation closes ranks against such individuals along technical lines. They are always held guilty of rule violations or accused of technical inadequacy.

The corporation views with suspicion *the man who cares,* the one who breaks through the alienated web of corporate relations to the *content* of a problem, to the good or bad of a given policy, and who wishes to do something about it. Corporate mentality does not like critical feedback, and corporate morality is acceptance of the "medium as message," or form as meaning, of the network or system as perfect and complete except for small details. It is characteristic of such morality that it permits choice in the realm of detail, in styling cars for example, but not in social control of the automobile.

The corporation or bureaucracy suppresses the natural, the biological in man. It aims to program individual talents towards technical utility. It promotes an educational system which is built around facts rather than inquiry, which in teachings and books provides our children with an impossibly sweetened view of reality and a flood of informational know-how. Its vector is pointed towards absolute control, towards the ant society. The social insects institutionalize the social good by eliminating random individual behavior, creating a total environment.

Such societies degenerate through overspecialization, that is, through overdoing the good of society. It is in the cards that the next stage of technique will arrange for the "direct linking of man to man," not through art but electronically. Perhaps we will thus be able to tune in on other people's minds directly. The consequences of such a medium extension are not at this time even remotely foreseeable. It could, however, lead to maximum entropy, to the last dance of the particles before another big bang and a fresh start.

A second model of behavior is the breakthrough to authentic individuality, of which the artist, even in his decadent phase, has been one of the stardard-bearers. The Negro in America and the "underdeveloped" in other lands have brought another kind of authenticity and courage to the media-market of ideas. The avant-garde among the young have moved beyond range of the built-in safeguards we have provided for them, to a more free but more dangerous realm of exploratory behavior. In reply to the technology's proffer of things in place of love, they have chosen love in place of things. They have abandoned the expectations of simplicity which impel corporate man to yes-no thinking; they have repudiated the old fashioned categories of good and bad radicalism, totalized capitalist or socialist panaceas, and have emerged with a more integrated sensitive, esthetic and flexible view of the world.

The degeneration of this outlook occurs when an invisible threshold beyond transindividual behavior is crossed and pure individualism is indulged. Hallucinogenic release from reality is as large a danger as institutionalized forms of junk. They are sedatives to ease the pain.

Yes-no man is committed to consensus, to drift and automatism. He is McLuhan's man who "acts without reacting," Lancelot Law Whyte's "dissociated man," Gerald Sykes' puritanical and pragmatic man. He is the one who says: "I did not know, I only obeyed orders, I am not responsible." He is Erich Fromm's narcissistic man who, unable to create, wishes only to destroy. Flip-flop awareness, already reduced to the border of unconsciousness, fears awakening more than extinction.

The new syncretic man is a growing force in modern life. He is neither pro-science nor anti-science, pro-technology nor anti-technology. He is a blend in which art and the esthetic are primary forces, only because art has been and remains closer to the human. His function, and the function of the artistic vector, is to humanize the technology. He demonstrates the meaning of pain as a symptom of disease, the awareness of pain as a guarantee of survival, the worth of art as a disturbing activity designed to promote awareness, and the increase of awareness-in-depth as an antidote to fear.

The imminent alliance: new connections among art, science and technology

JOSEPH W. MEEKER

■ Joseph W. Meeker *is Interdisciplinary Professor at Athabasca University, Edmonton, Alberta.*

The original alliance between science and art was sundered, suggests Professor Meeker, by the rise of sixteenth-century science, with its emphasis on the quantitative and its deprecation of the qualitative.

But new alliances are beginning to take place. Science has relinquished its goal of the conquest of nature and its illusive search for an immutable "truth" about nature; new emphasis has been placed on that area where science and art are tangent–namely, technology.

Meeker does not argue that this new alliance will solve the world's ills, nor that it will attack such problems as overpopulation, the energy crisis, or pollution of the environment. Instead, he contends, it may well serve to render the world more intelligible, and will enable us to confront our problems as well-balanced organisms.

Science and art were united at the dawn of human consciousness some 50,000 or 60,000 years ago. With that perspective, the past four centuries of separation between them seem relatively minor. On the whole, they have enjoyed a great and durable companionship, rooted in the nature of things and necessary to the well-being of mankind. There are indications now that science and art are about to be reunited in a stronger bond than they have known before. Their separation has taught them how much they need one another. A third party, technology, is near at hand to counsel and to guide them in the ways of the real world, which both art and science sometimes forget while pursuing their attractive fantasies.

Science and art have drifted apart in recent times because they were persuaded that they had incompatible interests. Francis Bacon and many others in the 16th century counseled scientists to avoid the "delicate learning" of the arts and to concentrate upon proper scientific goals like the conquest of nature.[1] Artists, uninterested in conquering nature, withdrew into

☐ From *Technology and Culture,* 19(1978), 187-198.

[1]Francis Bacon, *Novum Organum,* bk. 1, aphorism 71.

the isolation of their own sensibilities and looked disdainfully upon mundane science fiddling with its mirrors and test tubes. There both sides have stayed for the past few centuries: art keeping its soul pure and its hands clean while it searched among the clouds for something to do, while science and technology proceeded to rearrange the earth with little sense of form to guide their activities. Despite occasional attempts at reconciliation, the separation of science and art was so complete by the 20th century that C. P. Snow was able to define them accurately as two separate worlds.

A new initiative has now begun to reconstitute a whole and unified world inclusive of art, science, and technology. Science began it, and technology and the arts are now responding. This is no sentimental movement to recapture some nostalgic past. It is motivated by recent discoveries of science which suggest that the core of natural reality may be governed by principles which have hitherto been regarded as spiritual or artistic. Artists, similarly, are finding that accurate scientific knowledge and technological skills are essential to any creation that pretends to express the spirit of modern mankind. Not merely affection, but need, has moved science and art to approach their new alliance and technology to reappraise its relationships with both.

To understand what is happening, it may help to review events of the recent past which have affected the way scientists and artists think about their work and about one another. Many of the cultural and intellectual barriers which have separated art and science since the Renaissance have crumbled lately, and it may be useful to glance at their wreckage before stepping over them.

THE TECHNOLOGICAL CONQUEST OF NATURE

The joint venture of science and technology to conquer nature has turned out to be much more complicated than Francis Bacon supposed. Harnessing the resources and processes of nature has made life more convenient for many people, but each new exploitation of nature must, it turns out, be paid for by a corresponding decline of the world's overall environmental stability. Oil produces power and pollution, nuclear weapons produce military victories and dangerous genetic mutations, supercities and superagriculture produce toxic water and toxic air, efficient whaling eliminates whales. The management of natural systems for human benefit is crawling with ironies as each new achievement paves the way for new disasters. Even medicine, laboring to save lives and to relieve human suffering, has succeeded so well that it bears a heavy responsibility for the overpopulation which could overwhelm us all. Nature has at last made it clear that conquering her is a self-destructive enterprise. Enlightened technology can no longer honestly believe that the conquest of nature is its main purpose. It always was a rather naive idea, more congenial to political orators and chambers of commerce than to serious thinkers.

THE SCIENTIFIC SEARCH FOR TRUTH

Most scientists would probably prefer, anyway, to think of themselves not as conquerors but as seekers after truth. Discovering objective facts through the rigorous application of logical and scientific methods is a noble goal that does not seem to hurt anybody. Unfortunately, the search for truth has increasingly seemed an illusory quest. Objectivity, the pedestal of scientific method, turns out to be shaky ground according to science's own testimony. Werner Heisenberg, for instance, reminded his fellow quantum physicists that "reality varies, depending upon whether we observe it or not" and that "what we observe is not nature in itself but nature exposed to our method of questioning."[2] Illusions resulting from scientific methods of questioning have recently received much attention from historians and philosophers of science like Thomas Kuhn, Stephen Toulmin, and Arthur Koestler. One result has been a loss of some faith among scientists that learning the truth about nature is an achievable goal. Thoughtful scientists now are becoming vaguely discontented as they consider the discouraging possibility that their labors may reveal more about the processes of their own minds than they reveal about the processes of nonhuman nature.

RIGHT AND LEFT BRAINS

The mind itself has come under close scientific scrutiny of late, and those results are disturbing, too. Logic and objectivity are not

[2] Werner Heisenberg, *Physics and Philosophy* (New York, 1958), pp. 52, 58.

the only important mental activities, and they may not be the most dependable, especially when they are divorced from emotive functions. The brain's two hemispheres—the left specialized for linear functions and the right for synthetic functions—must continually interact with one another across the bridge of the corpus callosum whether we like it or not. Scientific and technological attempts to use only analytical mental processes, to the exclusion of subjective and emotional states, have never succeeded well, because they contradict the basic integrating tendencies of the human brain. Artistic movements emphasizing subjectivity to the exclusion of fact and logic are doomed to fail for the same reason. All human mental activities, but especially science and art, require close collaboration between linear and synthetic modes of thinking and free traffic between the left and right hemispheres of the brain.

One-sided people who try to be totally scientific or totally artistic usually create misery for themselves and dangerous inbalances in their creative work. Isaac Newton's neuroses and Van Gogh's depressions are well-known examples, but more often than not such half-brained specializations result in mediocrity rather than greatness. Universities harbor many exclusive hatreds between dedicated scientists and dedicated artists, most of whom are semi-developed people capable of no more than plodding work. Psychologists, I am told, have even named one neurosis "the Cal-Tech syndrome": a psychological disorder common among physicists and mathematicians who have become incapable of bridging the gap between their scientific work and the nonscientific world in which they must live. A corresponding "bohemian syndrome" has long been recognized among artists who indulge in identity crises and emotional excesses in ignorance of the facts of life and nature. In the long run, neither group amounts to much artistically or scientifically, and all lead unhappy lives.

People who are exclusively artistic in their approach to life either make fools of themselves or destroy themselves. They deserve mankind's pity, and they usually get a generous share of it. People who are exclusively logical and scientific, however, have a way of destroying not themselves, but the people and the world around them. Literary artists have long known that the clearest way to portray absolute evil is through a character who is intellectually brilliant and supremely rational. Shakespeare's Iago is such a character, and so are Milton's Satan, Goethe's Mephistopheles, and Dostoyevsky's Ivan Karamazov.

In the past two centuries, such figures have most frequently been pictured as scientists or engineers. From Dr. Frankenstein to the many evil geniuses of modern science fiction, the brilliant scientific-technological mind is a thing to be feared. More recently, the rational intellect lacking emotional and artistic balance has been pictured as perhaps the most dangerous technological creation ever: the logical but soulless computer. Computers evoke in us an ancient and well-founded fear of the terrors of pure intellect severed from any spiritual or artistic influences. They are ultimate images of left-brained linear logic lacking right-brained artistry and synthesis.

SHAKY INSTITUTIONS

Many other barriers which have separated science and the arts have been laboriously erected since the Renaissance by the institutions responsible for organizing and disseminating knowledge. Happily, these too are falling apart. For the past century or so, these institutions have operated on the assumption that specialization is essential to success. Physics departments in universities are supposed to produce physicists who will spend their lives doing only physics. Departments of art or English supposedly grind out artists and writers in the same way, though they are more likely to produce teachers. Schools of engineering are often cut off from both groups. Specializations thus branch and multiply, becoming ever nittier and grittier and ever more isolated from other nits and grits around them.

A century of this experiment has produced enormous piles of knowledge but pitifully little wisdom or understanding. Many of the most able minds of our time are beginning now to realize that a lot of learning is a dangerous thing, especially if it is held in ignorance of complementary forms of knowledge deriving from other kinds and levels of experience. And even the most specialized specialists have been amazed recently to find that, when they have dug to the very bottoms of their mine-shaft disciplines, what they find there is a network of unsuspected tunnels connecting with other kinds of knowledge with which they are unprepared to cope. The spectacular success of the specialization of knowledge in the modern era has led to the discovery that disciplinary specialization is probably a poor way to institutionalize human learning.

KNOWLEDGE AS PERSONAL PROPERTY

Academic fields, as universities have defined them, are pieces of property resembling farmers' fields. Scholars tend their fences, repel outsiders, and cultivate their hybrid intellectual blossoms with personal love and care. Whatever they harvest from their academic fields is regarded as their personal property, marketable in exchange for tenure, status, and identity. The system leads scholars to believe that knowledge is as much a part of their personality as their gender, their childhood, or their personal habits. Like personality, one's knowledge then must be differentiated as much as possible from the knowledge of others in the interests of ego satisfaction. Anyone who can manage to master obscure truths that few others can understand has gained an important survival advantage over the competition. Academics from the sciences, the arts, and technology have long played that game with one another, each claiming possession of special truths which only the initiated can appreciate.

Knowledge is not going to let us use it that way very much longer. While it was immature, it tolerated artificial boundaries and let itself to the service of personal ambitions. Now it is flexing its muscles against the disciplinary restraints imposed upon it by institutions of learning. Without regard for the aspirations of professors or the restraining categories of academic departments and foundations, knowledge is seeking a renewed wholeness and integrity that has been denied it for several centuries. Existing institutional and professional structures can no longer contain it.

REASONS FOR REUNION

The disappearance of many impediments separating science and art is not enough in itself to bring about their reunion. Reunions require positive reasons to occur, not merely the absence of reasons not to. Science and art are eyeing one another these days with a fresh sense of mutual need based upon their own hopes for fulfilling themselves. If they are to find one another again, it will be because each has found itself to be inadequate in isolation and because there are sound reasons to expect that a reunion will strengthen and complete both partners. Within the intricate workings of modern art and science, such needs are now noticeably stirring.

There is another world between art and science called technology. Many scientists and artists refuse to claim it for fear that they might be regarded as ordinary practical people. But from Francis Bacon to NSF grantsmanship, science has always earned support because of its promise to do something that people want done, and that requires technological application of its discoveries. Artists, too, sometimes like to think that they inhabit a sphere of pure mind or spirit where tools and craft are of small importance. But no great sculptor has been known to gnaw and claw his materials into shape, and Bach could never have written a fugue unless someone had invented the keyboard. Technology is the medium through which both science and art deliver their messages. If science and art are to be remarried, technology will build the chapel, light the altar, and maybe even throw the rice (though it is not likely to be that kind of a marriage).

Technology takes from and contributes to both science and art. It also pursues its independent existence, creating forms and meanings of its own. Its sphere includes the sensible (in both senses of the term) and the useful. But there is nothing in technology's history that excludes it from the beautiful as well. As Cyril Stanley Smith has reminded us, "Engineers, if not exactly aesthetes, have always had a rich and valid aesthetic experience in building their structures and devising their machines."[3] Engineers' efforts are often aesthetically pleasing, whether in the graceful curves of freeway ramps or in the mind-bending displays of kinetic and computer art. Technology has an identity of its own and is not merely applied art or science. Its sharp focus is upon useful experience in the tangible world, for it is not a symbol-making activity as art and science are. Useful experience has a special beauty all its own.

Cyril Smith has also made it clear that the separation of art, science, and technology into distinct categories of thought and activity is a recent innovation which runs contrary to most of human history. Science and technology, he demonstrates, arose first in an artistic atmosphere: "Over and over again scientifically im-

[3]Cyril Stanley Smith, "Art, Technology, and Science: Notes on Their Historical Interaction," *Technology and Culture* 11, no. 4 (October 1970): 537.

portant properties of matter and technologically important ways of making and using them have been discovered or developed in an environment which suggests the dominance of aesthetic motivation.''[4] Beauty, more than power, profit, convenience, or even truth, has until recently been the basic goal of science and technology, and perhaps with a bit of luck and care it can become so again.

TECHNOLOGY AND SCIENCE IN ART

Paleolithic engineers applied their knowledge of physics to make stone tools and their knowledge of chemistry to prepare color dyes. Ancient technology made possible the great works of art displayed on cave walls 30,000 years ago when art, science, and technology were one unified activity. Now, as then, artists and engineers need to understand the nature of the materials they work with and the possibilities of form that are inherent in them. That understanding is called science, whoever happens to hold it and however it is acquired.

Visual artists need the factual knowledge that physics and chemistry offer concerning matter, energy, space, and time. They need also the new medical and psychological insights into the nature of sensation and perception. The material world and human perceptual equipment are the stuff of visual art which science has lately illuminated more than ever before. Anyone who hopes to shape matter into art is obliged to face the strange new space that science describes.

Music occupies space, too, but it also organizes time. The sounds and rhythms of music connect the inner and outer spaces through which the world conducts its business. Periodicity and patterns in time are complex systemic structures with rules and limitations which apply equally to linear accelerators, to symphony orchestras, and to pods of singing whales. Music, it appears, is a nearly universal activity of all living creatures, and there is reason to suspect that forms of musical sound are important features of biological survival. Musicians need to learn about music in nature from biologists and about the nature of music from physicists.

The arts of language have always depended upon scientific knowledge. Current science provides the images which constitute literary reality, whether it be in Homer's exact knowledge of human anatomy or in Dante's vision of universal cosmology. Even Walt Whitman, I have heard, has done a creditable job of expressing the second law of thermodynamics in poetry.[5] Literature expresses the human ecology of the world as it is understood at any given time. Science fiction not only describes the impact of technology upon the past and present lives of people, but has also proved to be an accurate and reliable predictor of future cultural changes which will grow from technolgical innovations.[6]

Technology is the common ground where art and science commonly meet. A student of mine once wrote an essay in which he said that Johann Sebastian Bach ''was the father of twenty children and practiced on a spinster in the attic.'' Bach's musical powers, like his sexual abilities, were greater than the resources available to him. Bach may have loved the keyboard on his ''spinster,'' but he was always demanding that it perform beyond its capabilities. He would surely have loved to switch on a synthesizer for himself and play with the greater range of sound that computer technology provides. And, if Michelangelo could have projected a light show onto the ceiling of the Sistine Chapel, he might have saved himself many a backache and created an even more inspirational setting for worship. Artists are always quick to seize upon the expressive possibilities of new tools and ideas, and they generally use them much better than the military, business, or political customers of science and technology. As art uses technology, it becomes a new form of symbolic expression, not merely a means toward greater power or convenience.

ART IN SCIENCE AND TECHNOLOGY

Science has seemed to need art less than art needed science during their centuries of separation. Although scientists are often inclined to read poetry and enjoy music, they seldom try to incorporate such interests into their professional work, feeling perhaps that art represents emotional or spiritual influences inappropriate to the rigorous logic of the laboratory. Only now are scientists and engineers beginning to suspect that laboratories are also workshops of the

[4] Ibid., p. 498.

[5] Thomas Blackburn, ''Science and the Other Gloom,'' *North American Review* 260, no. 2 (Fall 1975): 13-16.

[6] Isaac Asimov, ''The Science Fiction Writer as Prophet,'' address before the American Association for the Advancement of Science, February 1975, New York.

spirit, where the elements and processes of nature resemble music more than mathematics—or rather, where music and mathematics have become indistinguishable.

Some forty years ago, Sir James Jeans observed that "the stream of knowledge is heading towards a non-mechanical reality, the universe begins to look more like a great thought than like a great machine."[7] Sir James's voice sounded lonely then, but it has since acquired a polyphonic chorus of scientists who are busy elaborating the harmonic implications of his statement. Machines, however great, may be disassembled for piecemeal examination; great thoughts must be grasped whole or not at all. Ironically, science's successful attempts to identify the elementary parts of the world machine have revealed particles whose characteristics are those of art and thought.

Contemporary physics has borrowed an old vocabulary from art to describe the new realities found in its bubble chambers. When Murray Gell-Mann needed a name for an elusive new elementary particle with perplexing properties, he borrowed a term from that encyclopedia of 20th-century perplexity, James Joyce's *Finnegans Wake*. With acknowledgements to Joyce, Gell-Mann named the particles "quarks," and received the 1969 Nobel Prize for his quarkwork. ("Quark" is also a German word for stinky soft cheeses of low social status.) This linguistic breakthrough may have been as significant as the breakthrough in quantum physics which the Nobel Prize recognized, for it has licensed scientists to think about matter in rich new metaphoric images. Science is now busily and gleefully freeing itself from the language of machines which has long circumscribed its expression and adopting in its place a language of suggestive nuances.

Quarks, like ice cream cones, come in several *flavors* and *colors*. Sheldon Glashow has noted that one quark flavor is further distinguished because it also possesses a quality called *charm*. Quarks, some of them charmed, make up particles called hadrons, which group themselves into eight quantum numbers according to a process known as "the eightfold way." Students of Buddhist philosophy will recognize that the eightfold way is also the Buddha's description of the essential path toward spiritual enlightenment. Some new particles, like leptons and muons, are said to "feel" the electromagnetic forces which act upon them, and some of them suffer the feeling of "strangeness" because of the anomalously long duration of their lives (a split nanosecond or so). Three quarks are thought to have peculiar charges known as "up," "down," and "sideways." Since quarks come in only three flavors, we are told, they can only make one married pair, leaving one quark "unwed." Glashow argues that "the scheme could be made much tidier if there were a fourth quark flavor, in order to provide a partner for the unwed quark." As one might guess, the search is now on for the bachelor particle needed to complete this image of marital harmony: the charmed quark.[8]

All of these terms and images, of course, have rather precise meanings for the physicists who use them, not to be confused with the ordinary conversational use of such language. It is important, however, that all of these terms convey states of human emotions and that they represent artistic, emotional, or spiritual experiences. They also express fun and fancifulness, both necessary ingredients of an artistic view of life which have been conspicuously absent in much of the science and technology of the past few centuries. Physicists seem to be groping imaginatively for a way to convey their growing sense that the interior of matter resembles the interior of the human spirit. Matter and spirit are coming together happily at the center of things.

Physicists were first to need metaphoric language for the official description of scientific facts, but biological scientists will probably join them soon. Living organisms, too, are threatening to display levels of depth and complexity which will soon surely burst the seams of conventional mechanistic explanations. The DNA and RNA come closer each day to playing roles in life like the role formerly attributed to God. Systems ecology and cybernetics describe life forms that resemble the subtle and delicately balanced forms created by a Homer or a Shakespeare. Intricacies of animal structure and behavior rival the most creative accomplishments of the human mind, as Karl von Frisch

---

[7]Sir James Jeans, *The Mysterious Universe* (London, 1937), p. 122 (quoted in Arthur Koestler, *The Roots of Concidence* [London, 1972], p. 48).

[8]A lucid and delightful account of quarkdom is Sheldon Lee Glashow's article "Quarks with Color and Flavor" (*Scientific American* 233, no. 4 [October 1975], pp. 38-50), from which the information in this paragraph comes.

has demonstrated in his recent provocative study of animal architecture.[9] It is no surprise to discover that biology operates in agreement with aesthetic principles, but no time before ours has come so close to understanding just how it does so.

SPIRITUAL TECHNOLOGIES

Studies in parapsychology over the past century have also confirmed that the world resembles a thought more than a machine. Parapsychology begins to look like a new order of technology, where practical actions and the rearrangement of matter proceed according to hitherto unsuspected dynamic laws. Technology will require fresh concepts and a new language to cope with forces which have always been called "spiritual" or "mystic" and so have been outside the concerns of engineers. Extrasensory perception and psychokinesis are no longer confined to kooky occultism but now promise to become sturdy new planks in the bridge between matter and spirit.

Somehow, energy is translatable into consciousness, and consciousness in turn influences paths of energy in living and nonliving matter. The unknown factors in physics, biology, and parapsychology increasingly resemble one another, giving rise to the suspicion that a mystery common to all three is about to be confronted. Perhaps this explains why the memberships in the world's major associations for research in parapsychology—in Britain, Russia, and the United States—are dominated by eminent scientists and engineers. Arthur Koestler has analyzed this strange new alliance between scientific and psychic research and concludes that "the reapproachment between the conceptual world of parapsychology and that of modern physics is an important step towards the demolition of the greatest superstition of our age—the materialistic clockwork universe of early nineteenth century physics."[10] Koestler adds that it also promises to reveal a new world where mental and emotional life agree in principle and practice with the laws governing physical and mechanical processes.

THE SURVIVAL VALUE OF FORM

Thanks to these fresh new trends in science, technology, and art, it may be possible to give up some of our old illusions about all three. We might as well admit that science is not the ultimate key to unlock the secrets of the universe, as its early prophets predicted. Neither is art going to corner the market on experiences of human emotion, beauty, or spiritual transcendence. And technology is not likely to transform all of material reality to suit human convenience, as some people once thought possible. Science, art, and technology are valuable and complementary methods for understanding the world and for adapting human beings to its given conditions. Together they must now teach us the survival value of form.

In recent centuries, science and art have pursued idealistic concepts of form, as if it were their business to rearrange reality to fit some preconceived notion of how things ought to be. Now both have a new task: to provide knowledge of, and emotional identification with, a deep and genuine external reality. Technology must show how humanity can participate in and contribute to the world's processes, not merely how to escape or manipulate those processes for supposed human benefit. The new forms will be adaptive, not triumphant. Like spiders and termites, humans must learn to weave elegant webs and construct elaborate apartment houses using materials in accordance with nature and mindful of its restraints. To achieve such appropriate forms, art and scientific technology must collaborate, and they must become much more knowledgeable and skilled than they have been before.

A sense of form is required to appreciate the values of sameness and otherness. Males and females recognize both their common humanity and their attractive differences, and that is why love is possible between them. Love of nature also requires precise knowledge of its processes, including those which are alien to human wishes as well as those which gratify us. If nature is to tolerate the otherness of humanity, we will have to act much more in resonance with its established rules of etiquette to prove that we belong among its creatures. Art, science, and technology can also appreciate their differences while encouraging the sameness of their goals: to help mankind feel at home on the earth and to live in knowing acceptance of its conditions.

THE NEW ALLIANCE

If I read correctly the current state of science and art, their long separation is nearing its end. As might be expected, their new alliance will

[9]Karl von Frisch, *Animal Architecture* (New York, 1974).
[10]Koestler, p. 77.

be founded upon quite different principles than those governing their last known union several centuries ago. This will be a thoroughly modern union, much freer and better informed than before.

Technology is more than merely an intermediary between science and art, for only technology can build bridges between the symbols of science and art and the real spaces where people and other creatures must live. The union of science and art will surely be expressed through technology, but technology will also express itself independently. The three together are complementary processes in the elaborate game of human consciousness, the purpose of which is unknown.

The union of science and art will not in itself provide a cure for the woes of the world. It will not solve problems of energy, overpopulation, environmental degradation, or injustice, nor will it restore order and good sense to a world suffering from mental and spiritual breakdown. Augmented by imaginative technologies, however, it could bring together two powerful allies interested in making the world seem more intelligible, more beautiful, and more useful than it now is for most people. And it could unite once more the most creative powers of human consciousness. Then, when we respond to the world's problems, we might do so as whole and well-balanced creatures in whom beauty, knowledge, and the skill to do useful work are united.

6

Technology and sex

From *Love and Will*

ROLLO MAY

■ Rollo May *is one of the country's leading clinical psychologists. Among his many books are* Man's Search For Himself, The Meaning of Anxiety, *and* Love and Will, *from which these selections are taken.*

In the first of the excerpts from Love and Will *included here, Professor May appraises the paradoxes which attend the relation of sex to love in our contemporary technological society.*

One of these paradoxes is that there should be so much anxiety about sexual matters in this post-Freudian, permissive milieu. It is no longer sexual repression which causes these difficulties, but the lack of it, a fear that one cannot "perform" in socially expected ways.

A second paradox is what May sees as the loss of passion in the face of ever more emphasis on sexual technique. Sex has become quantified.

Further, May sees a new puritanism in the wings, a phenomenon whose features include alienation from one's body, the separation of emotion from reason, and the use of the body as a machine.

In "The Revolt Against Sex" May argues against McLuhan's thesis that we are moving into an age in which sex will be diffused into a relation to a newly sensualized environment. He sees instead a time of overfamiliarity with sex, a time in which sex has become "boring" to many.

In the final excerpt, May contrasts technology, with its quantification of experience and its requirements of predictability, regularity, and tight observance of clock time, to the traditional notions of eros, which fight against all these elements. Eros, he concludes, is the vitality of a culture. But technology in its present form undermines eros and, consequently, that vitality.

There are four kinds of love in Western tradition. One is *sex,* or what we call lust, libido. The second is *eros,* the drive of love to procreate or create—the urge, as the Greeks put it, toward higher forms of being and relationship. A third is *philia,* or friendship, brotherly love. The fourth is *agape* or *caritas* as the Latins called it, the love which is devoted to the welfare of the other, the prototype of which is the love of God for man. Every human experience of authentic love is a blending, in varying proportions, of these four.

We begin with sex not only because that is where our society begins but also because that is where every man's biological existence begins as well. Each of us owes his being to the fact that at some moment in history a man and a woman leapt the gap, in T. S. Eliot's words, "between the desire and the spasm." Regardless of how much sex may be banalized in our society, it still remains the power of procreation, the drive which perpetuates the race, the source at once of the human being's most intense pleasure and his most pervasive anxiety. It can, in its daimonic form, hurl the individual into sloughs of despond, and, when allied with eros, it can lift him out of his despondency into orbits of ectasy.

The ancients took sex, or lust, for granted just as they took death for granted. It is only in the contemporary age that we have succeeded, on a fairly broad scale, in singling out sex for our chief concern and have required it to carry the weight of all four forms of love. Regardless of Freud's overextension of sexual phenomena as such—in which he is but the voice of the struggle of thesis and antithesis of modern history—it remains true that sexuality is basic to the ongoing power of the race and surely has the *importance* Freud gave it, if not the *extension*. Trivialize sex in our novels and dramas as we will, or defend ourselves from its power by cynicism and playing it cool as we wish, sexual passion remains ready at any moment to catch us off guard and prove that it is still the *mysterium tremendum*.

But as soon as we look at the relation of sex and love in our time, we find ourselves immediately caught up in a whirlpool of contradictions. Let us, therefore, get our bearings by beginning with a brief phenomenological sketch of the strange paradoxes which surround sex in our society.

SEXUAL WILDERNESS

In Victorian times, when the denial of sexual impulses, feelings, and drives was the mode and one would not talk about sex in polite company, an aura of sanctifying repulsiveness surrounded the whole topic. Males and females dealt with each other as though neither possessed sexual organs. William James, that redoubtable crusader who was far ahead of his time on every other topic, treated sex with the polite aversion characteristic of the turn of the century. In the whole two volumes of his epoch-making *Principles of Psychology,* only one page is devoted to sex, at the end of which he adds, "These details are a little unpleasant to discuss. . . ."[1] But William Blake's warning a century before Victorianism, that "He who desires but acts not, breeds pestilence," was amply demonstrated by the later psychotherapists. Freud, a Victorian who did look at sex, was right in his description of the morass of neurotic symptoms which resulted from cutting off so vital a part of the human body and the self.

Then, in the 1920s, a radical change occurred almost overnight. The belief became a militant dogma in liberal circles that the opposite of repression—namely, sex education, freedom of talking, feeling, and expression—would have healthy effects, and obviously constituted the only stand for the enlightened person. In an amazingly short period following World War I, we shifted from acting as though sex did not exist at all to being obsessed with it. We now placed more emphasis on sex than any society since that of ancient Rome, and some scholars believe we are more preoccupied with sex than any other people in all of history. Today, far from not talking about sex, we might well seem, to a visitor from Mars dropping into Times Square, to have no other topic of communication.

And this is not solely an American obsession. Across the ocean in England, for example, "from bishops to biologist, everyone is in on the act." A perceptive front-page article in *The Times Literary Supplement,* London, goes on to point to the "whole turgid flood of post-Kinsey utilitarianism and post-Chatterley moral uplift. Open any newspaper, any day (Sunday in particular), and the odds are you will find some pundit treating the public to his views on contraception, abortion, adultery, obscene publications, homosexuality between consenting adults or (if all else fails) contemporary moral patterns among our adolescents."[2]

Partly as a result of this radical shift, many therapists today rarely see patients who exhibit repression of sex in the manner of Freud's pre-World War I hysterical patients. In fact, we find in the people who come for help just the opposite: a great deal of talk about sex, a great deal of sexual activity, practically no one complaining of cultural prohibitions over going to bed as often or with as many partners as one wishes. But what our patients do complain of is lack of feeling and passion. "The curious thing about this ferment of discussion is how little anyone seems to be *enjoying* emancipation."[3] So much sex and so little meaning or even fun in it!

Where the Victorian didn't want anyone to know that he or she had sexual feelings, we are ashamed if we do not. Before 1910, if you called a lady "sexy" she would be insulted; nowadays, she prizes the compliment and rewards you by turning her charms in your direction. Our patients often have the problems of frigidity and impotence, but the strange and poignant thing we observe is how desperately

---

1. William James, *Principles of Psychology* (New York: Dover Publications, 1950; originally published by Henry Holt, 1890), **2:**439.

2. *Atlas* (November, 1965), p. 302. Reprinted from *The Times Literary Supplement,* London.
3. *Ibid.*

they struggle not to let anyone find out they don't feel sexually. The Victorian nice man or woman was guilty if he or she did experience sex; now we are guilty if we *don't*.

One paradox, therefore, is that enlightenment has not solved the sexual problems in our culture. To be sure, there are important positive results of the new enlightenment, chiefly in increased freedom for the individual. Most external problems are eased: sexual knowledge can be bought in any bookstore, contraception is available everywhere except in Boston where it is still believed, as the English countess averred on her wedding night, that sex is "too good for the common people." Couples can, without guilt and generally without squeamishness, discuss their sexual relationship and undertake to make it more mutually gratifying and meaningful. Let these gains not be underestimated. External social anxiety and guilt have lessened; dull would be the man who did not rejoice in this.

But *internal* anxiety and guilt have increased. And in some ways these are more morbid, harder to handle, and impose a heavier burden upon the individual than external anxiety and guilt.

The challenge a woman used to face from men was simple and direct—would she or would she not go to bed?—a direct issue of how she stood vis-à-vis cultural mores. But the question men ask now is no longer, "Will she or won't she?" but "Can she or can't she?" The challenge is shifted to the woman's personal adequacy, namely, her own capacity to have the vaunted orgasm—which should resemble a *grand mal* seizure. Though we might agree that the second question places the problem of sexual decision more where it should be, we cannot overlook the fact that the first question is much easier for the person to handle. In my practice, one woman was afraid to go to bed for fear that the man "won't find me very good at making love." Another was afraid because "I don't even know how to do it," assuming that her lover would hold this against her. Another was scared to death of the second marriage for fear that she wouldn't be able to have the orgasm as she had not in her first. Often the woman's hesitation is formulated as, "He won't like me well enough to come back again."

In past decades you could blame society's strict mores and preserve your own self-esteem by telling yourself what you did or didn't do was society's fault and not yours. And this would give you some time in which to decide what you do want to do, or to let yourself grow into a decision. But when the question is simply how you can perform, your own sense of adequacy and self-esteem is called immediately into question, and the whole weight of the encounter is shifted inward to how you can meet the test.

College students, in their fights with college authorities about hours girls are to be permitted in the men's rooms, are curiously blind to the fact that rules are often a boon. Rules give the student time to find himself. He has the leeway to consider a way of behaving without being committed before he is ready, to try on for size, to venture into relationships tentatively—which is part of any growing up. Better to have the lack of commitment direct and open rather than to go into sexual relations under pressure—doing violence to his feelings by having physical commitment without psychological. He may flaunt the rules; but at least they give some structure to be flaunted. My point is true whether he obeys the rule or not. Many contemporary students, understandably anxious because of their new sexual freedom, repress this anxiety ("one should *like* freedom") and then compensate for the additional anxiety the repression gives them by attacking the parietal authorities for not giving them more freedom!

What we did not see in our short-sighted liberalism in sex was that throwing the individual into an unbounded and empty sea of free choice does not in itself give freedom, but is more apt to increase inner conflict. The sexual freedom to which we were devoted fell short of being fully human.

In the arts, we have also been discovering what an illusion it was to believe that mere freedom would solve our problem. Consider, for example, the drama. In an article entitled "Is Sex Kaput?," Howard Taubman, former drama critic of *The New York Times,* summarized what we have all observed in drama after drama: "Engaging in sex was like setting out to shop on a dull afternoon; desire had nothing to do with it and even curiosity was faint."[4] Consider also the novel. In the "revolt against the Victorians," writes Leon Edel, "the ex-

4. Howard Taubman, "Is Sex Kaput?," *New York Times,* sect. 2, January 17, 1965.

tremists have had their day. Thus far they have impoverished the novel rather than enriched it."[5] Edel perceptively brings out the crucial point that in sheer realistic "enlightenment" there has occurred a *dehumanization* of sex in fiction. There are "sexual encounters in Zola," he insists, "which have more truth in them than any D. H. Lawrence described—and also more humanity."[6]

The battle against censorship and for freedom of expression surely was a great battle to win, but has it not become a new straitjacket? The writers, both novelists and dramatists, "would rather hock their typewriters than turn in a manuscript without the obligatory scenes of unsparing anatomical documentation of their characters' sexual behavior. . . ."[7] Our "dogmatic enlightenment" is self-defeating: it ends up destroying the very sexual passion it set out to protect. In the great tide of realistic chronicling, we forgot, on the stage and in the novel and even in psychotherapy, that imagination is the lifeblood of eros, and that realism is neither sexual nor erotic. Indeed, there is nothing *less* sexy than sheer nakedness, as a random hour at any nudist camp will prove. It requires the infusion of the imagination (which I shall later call intentionality) to transmute physiology and anatomy into *interpersonal* experience—into art, into passion, into eros in a million forms which has the power to shake or charm us.

Could it not be that an "enlightenment" which reduces itself to sheer realistic detail is itself an escape from the anxiety involved in the relation of human imagination or erotic passion?

SALVATION THROUGH TECHNIQUE

A second paradox is that *the new emphasis on technique in sex and love-making backfires.* It often occurs to me that there is an inverse relationship between the number of how-to-do-it books perused by a person or rolling off the presses in a society and the amount of sexual passion or even pleasure experienced by the persons involved. Certainly nothing is wrong with technique as such, in playing golf or acting or making love. But the emphasis beyond a certain point on technique in sex makes for a mechanistic attitude toward love-making, and goes along with alienation, feelings of loneliness, and depersonalization.

One aspect of the alienation is that the lover, with his age-old art, tends to be superseded by the computer operator with his modern efficiency. Couples place great emphasis on bookkeeping and timetables in their love-making—a practice confirmed and standardized by Kinsey. If they fall behind schedule they become anxious and feel impelled to go to bed whether they want to or not. My colleague, Dr. John Schimel, observes, "My patients have endured stoically, or without noticing, remarkably destructive treatment at the hands of their spouses, but they have experienced falling behind in the sexual time-table as a loss of love."[8] The man feels he is somehow losing his masculine status if he does not perform up to schedule, and the woman that she has lost her feminine attractiveness if too long a period goes by without the man at least making a pass at her. The phrase "between men," which women use about their affairs, similarly suggests a gap in time like the *entr' acte*. Elaborate accounting- and ledger-book lists—how often this week have we made love? did he (or she) pay the right amount of attention to me during the evening? was the foreplay long enough?—make one wonder how the spontaneity of this most spontaneous act can possibly survive. The computer hovers in the stage wings of the drama of lovemaking the way Freud said one's parents used to.

It is not surprising then, in this preoccupation with techniques, that the questions typically asked about an act of love-making are not, Was there passion of meaning or pleasure in the act? but, How well did I perform?[9] Take, for example, what Cyril Connolly calls "the tyranny of the orgasm," and the preoccupation with achieving a simultaneous orgasm, which is an-

5. Leon Edel, "Sex and the Novel," *New York Times*, sect. 7, pt. I, November 1, 1964.

6. *Ibid.*

7. See Taubman, "Is Sex Kaput?"

8. John L. Schimel, "Ideology and Sexual Practices," *Sexual Behavior and the Law*, ed. Ralph Slovenko (Springfield, Ill.: Charles C Thomas, 1965), pp. 195, 197.

9. Sometimes a woman patient will report to me, in the course of describing how a man tried to seduce her, that he cites as part of his seduction line how efficient a lover he is, and he promises to perform the act eminently satisfactorily for her. (Imagine Mozart's Don Giovanni offering such an argument!) In fairness to elemental human nature, I must add that as far as I can remember, the women reported that this "advance billing" did not add to the seducers' chances of success.

other aspect of the alienation. I confess that when people talk about the "apocalyptic orgasm," I find myself wondering, Why do they have to try so hard? What abyss of self-doubt, what inner void of loneliness, are they trying to cover up by this great concern with grandiose effects?

Even the sexologists, whose attitude is generally the more sex the merrier, are raising their eyebrows these days about the anxious overemphasis on achieving the orgasm and the great importance attached to "satisfying" the partner. A man makes a point of asking the woman if she "made it," or if she is "all right," or uses some other euphemism for an experience for which obviously no euphemism is possible. We men are reminded by Simone de Beauvoir and other women who try to interpret the love act that this is the last thing in the world a woman wants to be asked at that moment. Furthermore, the technical preoccupation robs the woman of exactly what she wants most of all, physically and emotionally, namely the man's spontaneous abandon at the moment of climax. This abandon gives her whatever thrill or ecstasy she and the experience are capable of. When we cut through all the rigmarole about roles and performance, what still remains is how amazingly important the sheer fact of intimacy of relationship is—the meeting, the growing closeness with the excitement of not knowing where it will lead, the assertion of the self, and the giving of the self—in making a sexual encounter memorable. Is it not this intimacy that makes us return to the event in memory again and again when we need to be warmed by whatever hearths life makes available?

It is a strange thing in our society that what goes into building a relationship—the sharing of tastes, fantasies, dreams, hopes for the future, and fears from the past—seems to make people more shy and vulnerable than going to bed with each other. They are more wary of the tenderness that goes with psychological and spiritual nakedness than they are of the physical nakedness in sexual intimacy.

THE NEW PURITANISM

The third paradox is that our highly-vaunted sexual freedom has turned out to be a new form of puritanism. I spell it with a small "p" because I do not wish to confuse this with the original Puritanism. That, as in the passion of Hester and Dimmesdale in Hawthorne's *The Scarlet Letter*, was a very different thing.[10] I refer to puritanism as it came down via our Victorian grandparents and became allied with industrialism and emotional and moral compartmentalization.

I define this puritanism as consisting of three elements. First, *a state of alienation from the body*. Second, *the separation of emotion from*

10. That the actual Puritans in the sixteenth and seventeenth centuries were a different breed from those who represented the deteriorated forms in our century can be seen in a number of sources. Roland H. Bainton in the chapter "Puritanism and the Modern Period," of his book *What Christianity Says about Sex, Love and Marriage* (New York: Reflection Books, Association Press, 1957), writes "The Puritan ideal for the relations of man and wife was summed up in the words, 'a tender respectiveness.'" He quotes Thomas Hooker: "The man whose heart is endeared to the woman he loves, he dreams of her in the night, hath her in his eye and apprehension when he awakes, museth on her as he sits at table, walks with her when he travels and parlies with her in each place he comes." Ronald Mushat Frye, in a thoughtful paper, "The Teachings of Classical Puritanism on Conjugal Love," *Studies from the Renaissance*, II (1955), submits conclusive evidence that Classical Puritanism inculcated a view of sexual life in marriage as the "Crown of all our bliss," "Founded in Reason, Loyal, Just, and Pure" (p. 149). He believes that "the fact remains that the education of England in a more liberal view of married love in the sixteenth and early seventeenth centuries was in large part the work of that party within English Protestantism which is called Puritan" (p. 149). The Puritans were against lust and acting on physical attraction outside of marriage, but they as strongly believed in the sexual side of marriage and believed it the duty of all people to keep this alive all their lives. It was a later confusion which associated them with the asceticism of continence in marriage. Frye states, "In the course of a wide reading of Puritan and other Protestant writers in the sixteenth and early seventeenth centuries, I have found nothing but opposition to this type of ascetic 'perfection'" (p. 152).

One has only to look carefully at the New England churches built by the Puritans and in the Puritan heritage to see the great refinement and dignity of form which surely implies a passionate attitude toward life. They had the dignity of controlled passion, which may have made possible an actual living with passion in contrast to our present pattern of expressing and dispersing all passion. The deterioration of Puritanism into our modern secular attitudes was caused by the confluence of three trends: industrialism, Victorian emotional compartmentalization, and the secularization of all religious attitudes. The first introduced the specific mechanical model; the second introduced the emotional dishonesty which Freud analyzed so well; and the third took away the depth dimensions of religion and made the concerns how one "behaved" in such matters as smoking, drinking, and sex in the superficial forms which we are attacking above. (For a view of the delightful love letters between husband and wife in this period, see the two volume biography of John Adams by Page Smith. See also the writings on the Puritans by Perry Miller.)

reason. And third, *the use of the body as a machine*.

In our new puritanism, bad health is equated with sin.[11] Sin used to mean giving in to one's sexual desires; it now means not having full sexual expression. Our contemporary puritan holds that it is immoral *not* to express your libido. Apparently this is true on both sides of the ocean: "There are few more depressing sights," the London *Times Literary Supplement* writes, "than a progressive intellectual determined to end up in bed with someone from a sense of moral duty. . . . There is no more high-minded puritan in the world than your modern advocate of salvation through properly directed passion. . . ."[12] A woman used to be guilty if she went to bed with a man; now she feels vaguely guilty if after a certain number of dates she still refrains; her sin is "morbid repression," refusing to "give." And the partner, who is always completely enlightened (or at least pretends to be) refuses to allay her guilt by getting overtly angry at her (if she could fight him on the issue, the conflict would be a lot easier for her). But he stands broadmindedly by, ready at the end of every date to undertake a crusade to assist her out of her fallen state. And this, of course, makes her "no" all the more guilt-producing for her.

This all means, of course, that people not only have to learn to perform sexually but have to make sure, at the same time, that they can do so without letting themselves go in passion or unseemly commitment—the latter of which may be interpreted as exerting an unhealthy demand upon the partner. *The Victorian person sought to have love without falling into sex; the modern person seeks to have sex without falling into love*.

I once diverted myself by drawing an impressionistic sketch of the attitude of the contemporary enlightened person toward sex and love. I would like to share this picture of what I call the new sophisticate:

> The new sophisticate is not castrated by society, but like Origen is self-castrated. Sex and the body are for him not something to be and live out, but tools to be cultivated like a T.V. announcer's voice. The new sophisticate expresses his passion by devoting himself passionately to the moral principle of dispersing all passion, loving everybody until love has no power left to scare anyone. He is deathly afraid of his passions unless they are kept under leash, and the theory of total expression is precisely his leash. His dogma of liberty is his repression; and his principle of full libidinal health, full sexual satisfaction, is his denial of eros. The old Puritans repressed sex and were passionate; our new puritan represses passion and is sexual. His purpose is to hold back the body, to try to make nature a slave. The new sophisticate's rigid principle of full freedom is not freedom but a new straitjacket. He does all this because he is afraid of his body and his compassionate roots in nature, afraid of the soil and his procreative power. He is our latter-day Baconian devoted to gaining power *over* nature, gaining knowledge in order to get more power. And you gain power over sexuality (like working the slave until all zest for revolt is squeezed out of him) precisely by the role of full expression. Sex becomes our tool like the caveman's bow and arrows, crowbar, or adz. Sex, the new machine, the *Machina Ultima*.

This new puritanism has crept into contemporary psychiatry and psychology. It is argued in some books on the counseling of married couples that the therapist ought to use only the term "fuck" when discussing sexual intercourse, and to insist the patients use it; for any other word plays into the patients' dissimulation. What is significant here is not the use of the term itself: surely the sheer lust, animal but self-conscious, and bodily abandon which is rightly called fucking is not be be left out of the spectrum of human experience. But the interesting thing is that the use of the once-forbidden word is now made into an *ought*—a duty for the moral reason of honesty. To be sure, it *is* dissimulation to deny the biological side of copulation. But it is also dissimulation to use the term fuck for the sexual experience when what we seek is a relationship of personal intimacy which is more than a release of sexual tension, a personal intimacy which will be remembered tomorrow and many weeks after tomorrow. The former is dissimulation in the service of inhibition; the latter is dissimulation in the service of alienation of the self, a defense of the self against the anxiety of intimate relationship. As the former was the particular problem of Freud's day, the latter is the particular problem of ours.

The new puritanism brings with it a depersonalization of our whole language. Instead of making love, we "have sex"; in contrast to intercourse, we "screw"; instead of going to bed, we "lay" someone or (heaven help the English language as well as ourselves!) we "are laid."

11. This formulation was originally suggested to me by Dr. Ludwig Lefebre.
12. *Atlas* (November, 1965), p. 302.

This alienation has become so much the order of the day that in some psychotherapeutic training schools, young psychiatrists and psychologists are taught that it is "therapeutic" to use solely the four-letter words in sessions; the patient is probably masking some repression if he talks about making love; so it becomes our righteous duty—the new puritanism incarnate!—to let him know he only fucks. Everyone seems so intent on sweeping away the last vestiges of Victorian prudishness that we entirely forget that these different words refer to different kinds of human experience. Probably most people have experienced the different forms of sexual relationship described by the different terms and don't have much difficulty distinguishing among them. I am not making a value judgment among these different experiences; they are all appropriate to their own kinds of relationship. Every woman wants at some time to be "laid"—transported, carried away, "made" to have passion when at first she has none, as in the famous scene between Rhett Butler and Scarlett O'Hara in *Gone with the Wind*. But if being "laid" is all that ever happens in her sexual life, then her experience of personal alienation and rejection of sex are just around the corner. If the therapist does not appreciate these diverse kinds of experience, he will be presiding at the shrinking and truncating of the patient's consciousness, and will be confirming the narrowing of the patient's bodily awareness as well as his or her capacity for relationship. This is the chief criticism of the new puritanism: it grossly limits feelings, it blocks the infinite variety and richness of the act, and it makes for emotional impoverishment.

It is not surprising that the new puritanism develops smoldering hostility among the members of our society. And that hostility, in turn, comes out frequently in references to the sexual act itself. We say "go fuck yourself" or "fuck you" as a term of contempt to show that the other is of no value whatever beyond being used and tossed aside. The biological lust is here in its *reductio ad absurdum*. Indeed, the word fuck is the most common expletive in our contemporary language to express violent hostility. I do not think this is by accident. . . .

THE REVOLT AGAINST SEX

With the confusion of motives in sex that we have noted above—almost every motive being present in the act except the desire to make love—it is no wonder that there is a diminution of feeling and that passion has lessened almost to the vanishing point. This diminution of feeling often takes the form of a kind of anesthesia (now with no need of ointment) in people who can perform the mechanical aspects of the sexual act very well. We are becoming used to the plaint from the couch or patient's chair that "We made love, but I didn't feel anything." Again, the poets tell us the same things as our patients. T. S. Eliot writes in *The Waste Land* that after "lovely woman stoops to folly," and the carbuncular clerk who seduced her at tea leaves,

> She turns and looks a moment in the glass,
> Hardly aware of her departed lover;
> Her brain allows one half-formed thought to pass;
> "Well now that's done: and I'm glad it's over."
> When lovely woman stoops to folly and
> Paces about her room again, alone,
> She smoothes her hair with automatic hand,
> And puts a record on the gramophone.
>
> (III:249-256)

Sex is the "last frontier," David Riesman meaningfully phrases it in *The Lonely Crowd*. Gerald Sykes, in the same vein, remarks, "In a world gone grey with market reports, time studies, tax regulations and path lab analyses, the rebel finds sex to be the one green thing."[13] It is surely true that the zest, adventure, and trying out of one's strength, the discovering of vast and exciting new areas of feeling and experience in one's self and in one's relations to others, and the validation of the self that goes with these are indeed "frontier experiences." They are rightly and normally present in sexuality as part of the psychosocial development of every person. Sex in our society did, in fact, have this power for several decades after the 1920s, when almost every other activity was becoming "other-directed," jaded, emptied of zest and adventure. But for various reasons—one of them being that sex by itself had to carry the weight for the validation of the personality in practically all other realms as well—the frontier freshness, newness, and challenge become more and more lost.

For we are now living in the post-Riesman age, and are experiencing the long-run implications of Riesman's "other-directed" behavior, the radar-reflected way of life. The last frontier has become a teeming Las Vegas and no frontier at all. Young people can no longer get a

[13]Gerald Sykes, *The Cool Millennium* (New York, 1967).

bootlegged feeling of personal identity out of revolting in sexuality since there is nothing there to revolt against. Studies of drug addiction among young people report them as saying that the revolt against parents, the social "kick of feeling their own oats" which they used to get from sex, they now have to get from drugs. One such study indicates that students express a "certain boredom with sex, while drugs are synonymous with excitement, curiosity, forbidden adventure, and society's abounding permissiveness."[14]

It no longer sounds new when we discover that for many young people what used to be called love-making is now experienced as a futile "panting palm to palm," in Aldous Huxley's predictive phrase; that they tell us that it is hard for them to understand what the poets were talking about, and that we should so often hear the disappointed refrain, "We went to bed but it wasn't any good."

Nothing to revolt against, did I say? Well, there is obviously one thing left to revolt against, and that is sex itself. The frontier, the establishing of identity, the validation of the self can be, and not infrequently does become for some people, a revolt against sexuality entirely. I am certainly not advocating this. What I wish to indicate is that the very revolt against sex—this modern Lysistrata in robot's dress—is rumbling at the gates of our cities or, if not rumbling, at least hovering. The sexual revolution comes finally back on itself not with a bang but a whimper.

Thus it is not surprising that, as sex becomes more machinelike, with passion irrelevant and then even pleasure diminishing, the problem has come full circle. And we find, *mirabile dictu,* a progression from an *anesthetic* attitude to an *antiseptic* one. Sexual contact itself then tends to get put on the shelf and to be avoided. This is another and surely least constructive aspect of the new puritanism: it returns, finally, to a new asceticism. This is said graphically in a charming limerick that seems to have sprung up on some sophisticated campus:

The word has come down from the Dean
That with the aid of the teaching machine,
  King Oedipus Rex
  Could have learned about sex
Without ever touching the Queen.

Marshall McLuhan, among others, welcomes this revolt against sex. "Sex as we now think of it may soon be dead," write McLuhan and Leonard. "Sexual concepts, ideals and practices already are being altered almost beyond recognition. . . . The foldout playmate in *Playboy* magazine—she of outsize breast and buttocks, pictured in sharp detail—signals the death throes of a departing age."[15] McLuhan and Leonard then go on to predict that eros will not be lost in the new sexless age but diffused, and that all life will be more erotic than now seems possible.

This last reassurance would be comforting indeed to believe. But as usual, McLuhan's penetrating insights into *present* phenomena are unfortunately placed in a framework of history—"pretribalism" with its so-called lessened distinction between male and female—which has no factual basis at all.[16] And he gives us no evidence whatever for his optimistic prediction that new eros, rather than apathy, will succeed the demise of *vive la difference.* Indeed, there are amazing confusions in this arti-

[14]A survey of students on three college campuses in the New York/New Jersey area conducted by Dr. Sylvia Hertz, chairman of the Essex County Council on Drug Addiction, reported in *The New York Times* on November 26, 1967, that "The use of drugs has become so prominent, that it has relegated sex to second place."

As sex began to lose its power as the arena of proving one's individuality by rebellion and merged with the use of drugs as the new frontier, both then became related to the preoccupation with acts of violence. Efforts crop up anachronistically here and there to use sex as the vehicle for revolt against society. When I was speaking at a college in California, my student chauffeur to the campus told me that there was a society at the college dedicated, as its name indicates, to "Sex Unlimited." I remarked that I hadn't noticed anybody in California trying to limit sex, so what did this society do? He answered that the previous week, the total membership (which turned out to be six or seven students) got undressed at noon and, naked, jumped into the goldfish pool in the center of the campus. The city police then came and hiked them off to jail. My response was that if one wanted to get arrested, that was a good way to do it, but I couldn't see that the experience had a thing in the world to do with sex.

[15]Marshall McLuhan and George G. Leonard, "The Future of Sex," *Look Magazine,* July 25, 1967, p. 58. The article makes a significant point with respect to the polls about sex: "When survey-takers 'prove' that there is no sexual revolution among our young people by showing that the frequency of sexual intercourse has not greatly increased, they are missing the point completely. Indeed, the frequency of intercourse may decrease in the future *because of* a real revolution in attitudes toward, feelings about and uses of sex, especially concerning the roles of the male and female" (p. 57).

[16]Not being an anthropologist, I conferred with Ashley Montague on this point. The judgment was expressed orally to me.

cle arising from McLuhan's and Leonard's worship of the new electric age. In likening Twiggy to an X-ray as against Sophia Loren to a Rubens, they ask, "And what does an X-ray of a woman reveal? Not a realistic picture, but a deep, involving image. Not a specialized female, but a *human being.*"[17] Well! An X-ray actually reveals not a human being at all but a depersonalized, fragmentized segment of bone or tissue which can be read only by a highly specialized technician and from which we could never in a thousand years recognize a human being or any man or woman we know, let alone one we love. Such a "reassuring" view of the future is frightening and depressing in the extreme.

And may I not be permitted to prefer Sophia Loren over Twiggy for an idle erotic daydream without being read out of the New Society?

Our future is taken more seriously by the participants in the discussion on this topic at the Center for the Study of Democratic Institutions at Santa Barbara. Their report, called "The A-Sexual Society," frankly faces the fact that "we are hurtling into, not a bisexual or a multisexual, but an a-sexual society: the boys grow long hair and the girls wear pants. . . . Romance will disappear; in fact, it has almost disappeared now. . . . Given the guaranteed Annual Income and The Pill, will women choose to marry? Why should they?"[18] Mrs. Eleanor Garth, a participant in the discussion and writer of the report, goes on to point out the radical change that may well occur in having and rearing children. "What of the time when the fertilized ovum can be implanted in the womb of a mercenary, and one's progeny selected from a sperm-bank? Will the lady choose to reproduce her husband, if there still are such things? . . . No problems, no jealousy, no love-transference. . . . And what of the children, incubated under glass? . . . Will communal love develop the human qualities that we assume emerge from the present rearing of children? Will women under these conditions lose the survival drive and become as death-oriented as the present generation of American men? . . . I don't raise the question in advocacy," she adds, "I consider some of the possibilities horrifying."[19]

[17] McLuhan and Leonard, p. 58. The words are italicized by McLuhan and Leonard.

[18] Eleanor Garth, "The A-Sexual Society," *Center Diary,* published by the Center for the Study of Democratic Institutions, 15, November-December, 1966, p. 43.

[19] *Ibid.*

Mrs. Garth and her colleagues at the Center recognize that the real issue underlying this revolution is not what one does with sexual organs and sexual functions per se, but what happens to man's humanity. "What disturbs me is the real possibility of the disappearance of our humane, life-giving qualities with the speed of developments in the life sciences, and the fact that no one seems to be discussing the alternative possibilities for good and evil in these developments."[20]

The purpose of our discussion in this book is precisely to raise the questions of the alternative possibilities for good and evil—that is, the destruction or the enhancement of the qualities which constitute man's "humane, life-giving qualities." . . .

EROS SICKENING

The Eros we have been discussing is that of the classical age, when he was still the creative power and the bridge between men and gods. But this "healthy" Eros deteriorated. Plato's understanding of Eros is a middle form of the concept, standing between Hesiod's view of Eros as the powerful and original creator and the later deteriorated form in which Eros becomes a sickly child. These three aspects of Eros are also accurate reflections of psychological archetypes of human experience: each of us at different times has the experience of Eros as creator, as mediator, and as banal playboy. Our age is by no means the first to experience the banalization of love, and to find that without passion, love sickens.

. . . the ancient Greeks . . . put into the quintessential language of myth the insights which spring from the archetypes of the human psyche. Eros, the child of Ares and Aphrodite, "did not grow as other children, but remained a small, rosy, chubby child, with gauzy wings and rougish, dimpled face." After telling us that the alarmed mother was informed, "Love cannot grow without Passion," the myth goes on:

> In vain the goddess strove to catch the concealed meaning of this answer. It was only revealed to her when Anteros, god of passion, was born. When with his brother, Eros grew and flourished, until he became a handsome slender youth; but when separated from him, he invariably resumed his childish form and mischievous habits.[21]

[20] *Ibid.*

[21] Helene A. Guerber, *Myths of Greece and Rome* (London, British Book Centre, 1907), p. 86.

Within these disarmingly naïve sentences, with which the Greeks were wont to clothe their most profound wisdom, lie several points which are crucial for our problems now. One is that Eros is the child of *Ares* as well as Aphrodite. This is to say that love is inseparably connected with aggression.

Another is that the Eros which had been the powerful creator in Hesiod's time, causing the barren earth to spring up with green trees and breathing the spirit of life into man, has now deteriorated into a child, a rosy, chubby, playful creature, sometimes a mere fat infant playing with his bow and arrows. We see him represented as an effete Cupid in so many of the paintings of the seventeenth and eighteenth centuries as well as in ancient times. "In archaic art Eros is represented as a beautiful winged youth and tends to be made younger and younger until by the Hellenistic period he is an infant." In Alexandrine poetry, he degenerates into a mischievous child.[22] There must be something within Eros' own nature to cause this deterioration, for it is present already in the myth which, while later than the Hesiod version, still dates from long before Greek civilization disintegrated.

This brings us to the very heart of what has also gone wrong in our day: eros has lost passion, and has become insipid, childish, banal.

As is so often the case, the myth reveals a critical conflict in the roots of human experience, true for the Greeks and true for us: we engage in a flight from eros, the once powerful, original source of being, to sex, the mischievous plaything. Eros is demoted to the function of a pretty bartender, serving grapes and wine, a stimulator for dalliance whose task is to keep life endlessly sensuous on a bank of soft clouds. He stands not for the creative use of power—sexual, procreative, and other—but for the immediacy of gratification. And, *mirabile dictu,* we discover that the myth proclaims exactly what we have seen happening in our own day: *eros, then, even loses interest in sex.* In one version of the myth, Aphrodite tries to find him to get him up and about his business of spreading love with his bow and arrows. And, teenage loafer that he has become, he is off gambling with Ganymede and cheating at the cards.

Gone is the spirit of the life-giving arrows, gone the creature who could breathe spirit into man and woman, gone the powerful Dionysian festivals, gone the frenzied dancing and the mysteries that moved the initiates more than the vaunted drugs of our mechanical age, gone even the bucolic intoxication. Eros now playboy indeed! Bacchanal with Pepsi-Cola.

Is this what civilization always does—tames Eros to make him fit the needs of the society to perpetuate itself? Changes him from the power that brings to birth new being and ideas and passion, weakens him till he is no longer the creative force that breaks old forms asunder to make new ones? Tames him until he stands for the goal of perpetual ease, dalliance, affluence, and, ultimately, apathy?[23]

In this respect we confront a new and specific problem in our Western world—*the war between eros and technology.* There is no war between *sex* and technology: our technical inventions help sex to be safe, available, and efficient as demonstrated from birth-control pills all the way to the how-to-do-it books. Sex and technology join together to achieve "adjustment"; with the full release of tension over the weekend, you can work better in the button-down world on Monday. Sensual needs and their gratification are not at war with technology, at least in any immediate sense (whether they are in the long run is another question).

But it is not at all clear that technology and *eros* are compatible, or can even live without perpetual warfare. The lover, like the poet, is a menace on the assembly line. Eros breaks existing forms and creates new ones and that, natu-

---

[22] "Eros," *Encyclopaedia Britannica,* vol. VIII (1947), p. 695.

[23] Rollo May, in a review of Vance Packard's *The Sexual Wilderness: The Contemporary Upheaval in Male-Female Relationships* (New York, David McKay Company, 1968), appearing in *The New York Times Book Review,* October 13, 1968: "Packard here cites J. D. Unwin's massive, if almost forgotten, 'Sex and Culture' (1934), a study of 80 uncivilized societies and also a number of historically advanced cultures. Unwin sought to correlate various societies' sexual permissiveness with their energy for civilized advancement. He concluded that the 'amount of cultural ascent of the primitive societies closely paralleled the amount of limitation they placed upon the nonmarital sexual opportunity.' Virtually all the civilized societies Unwin examined—the Babylonians, Athenians, Romans, Anglo-Saxons, and English—began their historical careers in a 'state of absolute monogamy.' The one exception was the Moors, where a specific religious sanction supported polygamy. 'Any human society,' Unwin writes, 'is free to choose either to display great energy or to enjoy sexual freedom; the evidence is that it cannot do both for more than one generation.' Packard points out that this is supported in different ways by other historians and anthropologists, such as Carl C. Zimmerman, Arnold J. Toynbee, Charles Winick and Pitirim A. Sorokin."

rally, is a threat to technology. Technology requires regularity, predictability, and runs by the clock. The untamed eros fights against all concepts and confines of time.

Eros is the impetus in building civilizations. But the civilization then turns on its progenitor and disciplines the erotic impulses. This can still work toward the increase and expansion of consciousness. The erotic impulses can and should have some discipline: the gospel of the free expression of every impulse disperses experience like a river with no banks, its water spilled and wasted as it flows in every direction. The discipline of eros provides *forms* in which we can develop and which protect us from unbearable anxiety. Freud believed that the disciplining of eros was necessary for a culture, and that it was from the repression and sublimation of erotic impulses that the power came out of which civilizations were built. De Rougement, for one of the few times, here agrees with Freud; he does not forget

> that without the sexual discipline which the so-called puritanical tendencies have imposed on us since Europe first existed, there would be nothing more in our civilization than in those nations known as underdeveloped, and no doubt less: there would be neither work, organized effort nor the technology which has created the present world. There would also not be the problem of eroticism! The erotic authors forget this fact quite naively, committed as they are to their poetic or moralizing passion, which too often alienates them from the true nature of the "facts of life," and their complex links with economy, society, and culture.[24]

[24]From Denis de Rougement's *The Myths of Love* (New York, Pantheon Books, 1963), quoted in *Atlas*, November, 1965, p. 306.

But there comes a point (and this is the challenge facing modern technological Western man) when the cult of technique destroys feeling, undermines passion, and blots out individual identity. The technologically efficient lover, defeated in the contradiction which is copulation without eros, is ultimately the impotent one. He has lost the power to be carried away; he knows only too well what he is doing. At this point, technology diminishes consciousness and demolishes eros. Tools are no longer an enlargement of consciousness but a substitute for it and, indeed, tend to repress and truncate it.

Must civilization always tame eros to keep the society from breaking up again? Hesiod lived in the strongly fomenting, archaic sixth century, closer to the sources of culture and the moments of gestation and birth, when the procreative powers were at work, and man *had* to live with chaos and form it into something new. But then, with the growing need for stabilization, the daimonic and tragic elements tended to be buried. Insight into the downfall of civilizations is revealed here. We see effete Athens set up for the more primitive Macedonians, they in turn for the Romans, and the Romans in turn for the Huns. And we for the yellow and black races?

Eros is the center of the vitality of a culture—its heart and soul. And when release of tension takes the place of creative eros, the downfall of the civilization is assured.

The technologies of man-made sex

ROBERT T. FRANCOEUR

■ Robert T. Francoeur *is Professor of Human Sexuality and Embryology at Fairleigh Dickinson University.*

In contrast to the technological innovations which spawned the tool, the machine, and the computer, and which Francoeur calls "parapersonal," or indirect, the technologies of sex are deeply personal, "affecting every man and woman alive today."

The rapid acceleration of advances in science and medicine, the equalization of the sexes, and the explosion of information about matters once considered "private" have combined to generate a sexual revolution.

There are some unique characteristics apparent in the revolution we are witnessing today in human sexual behavior. In other social revolutions triggered by man's new technologies it was difficult, if not impossible, to diagnose the patient's condition while the fever of change still afflicted him as an individual and as a member of society. This is not the situation with our sexual revolution. For various reasons the basic character and trends in the sexual revolution are already evident, even to the casual observer outside the deliberate and scientific disciplines of sociology, family relations, anthropology, and psychology.

One unique characteristic of the sexual revolution is the swiftness of its impact on the average citizen in the Western world and in many developing nations. Previous technological revolutions have created their own cultures and modified human behavioral patterns along with the structure of society, the relations of men and women, and the pattern of the family. The discovery of fire and agriculture, the invention of wheels, gunpowder, steam and internal combustion engines, automobiles, nuclear power, television, and space travel—each of these technologies has triggered revolutions in man's vision and handling of his world. Adaptation to these new technologies naturally reverberates throughout the whole fabric of human culture to modify our patterns of family life and basic human relations. Sometimes these changes were hardly noticed; at other times they were more apparent and drastic. The fragmentation of the sexes and the emergence of sexual roles that came with the advent of agriculture and urbanization over ten thousand years ago modified the relations of men and women almost imperceptibly because the shift to urban life was spread over many generations and centuries. In some areas this shift is just now touching the last remnants of nomadic and tribal cultures. A more recent innovation, the automobile, has had a more devastating impact because it practically destroyed the Victorian American extended family and created the new reality of the so-called nuclear family of mobile post–World War I America. This shift took only four or five decades to sweep a continent.

But agriculture, urbanization, automobiles, television, and industrialization are technologies whose practical results, applications and impact are *parapersonal*. They are out there, so to speak, and do not really impinge directly on each person in a culture or society. They are tools *we use,* extensions of our senses and muscles, but only *indirectly* do we incorporate them into our personalities or allow them to affect our thinking and behavior. They mold the periphery of our lives. Their impact consequently has been at best the creation of a gradual change, at worst the birth of a painfully rapid upheaval.

The technologies of sex . . . are quite different. The new technologies of reproduction, contraception, sexual modification, and genetic designing crash into the very intimate nature of the human, exploding our concepts of male and female. They throw into chaos our deepest images of ourselves as sexual persons, our most treasured and seemingly stable images of male-female relations, marriage, parenthood, and family. *These technologies affect every man and woman alive today*. They are creating a revolution which amounts to *a period of apocalyptic continuity,* probably the first discontinuity of such magnitude to occur since humankind appeared two million years ago.

The reason for this radical discontinuity, I believe, can be traced to more than just the fact that reproductive and genetic technologies now threaten our basic images of man and

□ From Robert T. and Anna K. Francoeur, eds., *The Future of Sexual Relations,* © 1974, pp. 3-11. Reprinted by permission of Prentice-Hall, Inc., Englewood Cliffs, New Jersey.

woman as sexual persons who relate in set structures of marriage and family. The discontinuity is also due to the interplay of three basic factors which may seem discrete and unrelated, but which are in reality closely interwoven.

First is the extraordinarily rapid acceleration in the advance of science and medicine. Ninety percent of all the scientists in human history are living today. An equal portion of important scientific and medical discoveries has occurred in the last century or so.

If our recent discoveries of the contraceptive pill, the intrauterine devices, the minipills, vasectomy and intervas valves, sperm banks, and other contraceptive hardware had occurred a thousand years ago, their advent would have been spread over centuries. Lazaro Spallanzani, one of the earliest experimenters in reproductive technology, successfully inseminated some frogs and dogs artificially in 1776 and 1780. Thirty years passed before the technique was applied to a woman in England, and another 150 years passed before this technique became a fairly common mode of human reproduction. If artificial insemination were to be discovered today it would be known and practiced around the world within five or ten years. This is exactly what happened with the technology of the contraceptive pill. And with ninety percent of the scientists alive and working today, it is small wonder that the pace of scientific and medical research today has entered what the statistician would call an exponential or geometric growth phase. Within a decade or two our technology has plunged us into the midst of *man-made sex:* artificial insemination, frozen sperm and egg banks for humans, embryo transplants, surrogate mothers, genetic selection and modifications, predetermination of fetal sex, conception control, legalized abortion, and transexual operation are already in use. And tomorrow holds the probabilities of artificial wombs, asexual forms of human reproduction, and genetic engineering. Most of these new technologies have developed within this century!

The second social factor or vector is the equalization of the sexes. This socioeconomic revolution is epitomized but hardly encompassed in the phrase *women's liberation*. As Teilhard de Chardin repeatedly stated, this century is witnessing a critical threshold in the hominization of the human race: the emergence of women as human beings and as independent persons socially and economically. Throughout most of human history the brute animal strength and aggression of the human male has denied true humanity to women, constantly defining them as incomplete creatures who depend on the male for their identity. As dependent persons, women have traditionally drawn their identity from their roles as wives and mothers. Now, because of women's accelerating economic independence, this "relative" existence is changing. With women no longer chained to their roles as domestic support systems for the male, our whole society changes. This is why the so-called sexual revolution is a revolution that affects our whole social fabric and culture.

Finally, these two vectors—devastating enough if taken alone or together—combine with our transistorized, near-instantaneous communications network. As McLuhan and others have pointed out, we are fast becoming a global tribe in which communications satellites, transistorized battery-powered radios, and television sets reach the most isolated cultures. Today, if a group of doctors and scientists are trying to solve the problem of childless couples in which the wife's oviducts are blocked, all it takes is an interview with one of these patients on the evening television news and millions are soon made aware of the potential of artificial inovulation or embryo transplantation. *Look* magazine picks it up for a cover story, *Family Circle* and *Woman's Day* peddle the message in the supermarket, television talk shows carry it to night owls and the housewife at her morning coffee break.

In my own files I have several dozen inquiries from women around the country who feel that embryo transplants might provide a satisfactory solution to their own childless condition. In several cases, female neighbors or relatives have volunteered to carry the child. Attitudes are changing, and it could well be that surrogate motherhood will become a socially acceptable option, an alternate way of parenthood in the decades ahead.

One final illustration will conclude our look at the impact of television and the printed word on our changing images of male and female, marriage, and parenthood. The case that I have chosen is not uncommon and will likely become more common in the years ahead. Paul Grossman, a fifty-three-year-old music teacher from Plainfield, New Jersey, his wife of some twenty years, Ruth, and their three daughters went through a crisis in the spring of 1971. At the age of fifty, Paul finally reached the breaking point in coping with an affliction he had suffered with since his earliest memories. Born with a perfectly normal male anatomy and genetic constitution, he unfortunately did not have a matching male psychosexual identity, probably due to a perinatal hormone imbalance that allowed female imprinting of the brain. When the joy of his fiftieth birthday party dissolved, Paul broke down and discussed his condition and decision with his wife and teenage daughters. Within weeks Paul had arranged with Dr. Harry Benjamin's clinic in New York City for a transexual operation. After the operation, *Paula* returned to her teaching position where the principal allowed her to continue provided she wore her customary male garb. But the contract for the coming school year created more serious problems and resulted in her being denied a contract despite her tenured position. The case is now in the courts, and eventually may reach the United States Supreme Court for decision on job discrimination on the basis of sex.

Again, however, the mass media comes into play as a molder of public opinion and an agent for rapidly changing our traditional concepts. Thousands of transexual operations have been performed in the United States since the first one in 1931. Paula's case is unique only in that she took a public stand and remained very visible. Her case has become internationally known as a result of her many appearances on television, newspaper interviews, and public lectures around the country. More recently, Jan Morris, a renowned British journalist, has told the story of her transition from male to female status in the best seller *Conundrum,* thus adding to the public impact of this medical procedure.

TODAY'S TECHNOLOGIES

—The exponential growth of science, technology, and medicine

—The liberation of women as persons

—The near-instantaneous global network of television, communications satellites, and the printed word

These are the complicating factors in our revolution. But what about the actual raw materials, the technological advances in human reproduction that directly modify and revolutionize our traditional images of male and female, parenthood, and the relationships of men and women?

To answer this question let us look first at what is already available, already a part of our world:

Artificial insemination, which allows conception and parenthood without genital intercourse, was first accomplished in 1776. In 1799, an English woman gave birth to the first human conceived with what then became known as "ethereal copulation." When British scientists successfully froze semen without damage in 1949, artificial insemination became a very practical reproductive technology for all animals. With domesticated animals, artificial insemination with frozen semen is the common mode of reproduction for cattle, turkeys, and sheep. On the human level, common estimates suggest that somewhere around one percent of all the children born in the United States today are conceived by artificial insemination.

Frozen human sperm banks had their commercial start in October 1971 when a fourteen-year-old private sperm bank in Minnesota decided to open a public service branch in New York City. In the months that followed, human sperm banks began to appear in many locations around the United States. Eighteen such facilities were functioning by the end of 1972, with four more scheduled to open in early 1973.

Embryo transplantation (artificial inovulation) became a reality for domesticated animals in the early 1950s. The first successful *in vitro* fertilization of a human egg to be followed by transfer to a surrogate mother was done by Landrum B. Shettles at Columbia Presbyterian Hospital in New York City in late 1970. Similar experiments with women have been attempted in England and Australia. As of mid-1973 no human fetus had yet gone full term after transplantation, but this is an almost inevitable outcome of these experiments. It may already be a reality in Dr. Edwards' laboratory at Cambridge, where a two-year silence on embryo transplant experiments has followed some initially promising announcements.

Artificial wombs are under active research in a dozen laboratories around the world in attempts to understand the nature of pregnancy and birth. At the National Institute of Heart and Lung Diseases in Bethesda, Maryland a premature human fetus was maintained in an artificial womb for several days. Experiments with lamb fetuses have been far more successful and promising. The Russians claimed in the late 1960s to have maintained a human fetus for six months before terminating the gestation.

Asexual reproduction, parthenogenesis (virgin birth) and cloning from a single parent made their appearance in 1896 when French scientists produced virgin-born sea urchins, and in 1952 when Briggs and King first transplanted an adult nucleus into enucleated frog eggs. Parthenogenetic rabbits came in 1939 with the work of John Rock, and virgin-born mice in 1970 through the efforts of some Polish scientists.

Predetermination of fetal sex made a major step towards reality when Landrum B. Shettles combined several well-known facts about sperm morphology and physiology into a six-point "recipe" for boys or girls which he claims is eighty-five to ninety percent accurate. His "recipe" appeared in a cover story in *Look* magazine in 1970 and then in a full-length book entitled: *Your Baby's Sex: Now You Can Choose*.

Transexual operations began in 1930 when a Dutch artist underwent a complete plastic surgery transformation from male to female anatomy. In its first three years of operation the Johns Hopkins Gender Identification Clinic received inquiries from over 1500 persons interested in a transexual operation. Doctors at Stanford University estimate that probably two thousand such operations were performed in the United States in the last four years.

Uterine transplants have been partially successful in work with rhesus monkeys, but not yet with humans as of 1974. In Brazil, a woman had a normal pregnancy after becoming the recipient of an *ovary transplant*. One *transplant of a vagina* has been reported in Greece, where a forty-five-year-old widow allowed the transplantation of her vagina to her daughter who was born with none.

Embryo fusions, resulting in what geneticists call allophenic offspring, have been produced by Beatrice Mintz working at the Institute for Cancer Research in Philadelphia. Very young mouse embryos can be taken from their mother's oviducts after normal mating and conception. Even when two or three embryos are produced in two or three mothers of different genetic strains—black, white, mottled, hooded—they can be fused together into a single embryo and transferred to the womb of a surrogate mother, where they develop and are born normally. Such embryos have four or even six genetic parents of *both* sexes.

Minimenstruation has been developed out of abortion technologies so that women can use suction machines to remove the menstrual endometrium and thereby reduce the normal several days of menstruation to a brief fifteen minutes.

Frozen embryos are another reality of reproductive technology. Dr. Wittingham has successfully frozen young mouse embryos in liquid nitrogen for upwards of eight days. When they were thawed, Wittingham was able to transplant them to surrogate mothers, in whose wombs they developed normally. After birth, these same exfrozen mice were indistinguishable from their nonfrozen siblings.

Animal/plant hybrids are in their earliest stages of development, with tissue-culture hybridizing of chloroplasts from African violets and spinach with mouse cells, and the hybridizing of chromosome material from viral and mouse sources with living humans. Several pioneering experiments are now under way to determine whether viral hereditary material can be hybridized with the defective chromosomal material of children with arginemia, a serious type of mental retardation due to a missing genetic unit.

New species and sexes of animals are another product of our reproductive technology that is just beginning. James Danielli reported his success in 1971 in combining the nucleus of one species of amoeba with the cell membrane of a second species and the cytoplasm of a third species. The result was a totally new species of amoeba with hereditary information from all three species. Emil Witschi has created a new kind of female frog. *Xenopus laevis* females normally have *ZW* sex-determining chromosomes and the males *ZZ*. Z and W are used for the sex chromosomes instead of X and Y when the female determines the offspring's sex by producing two kinds of sperm. By experimentally reversing the sex of developing tadpoles and then mating the products, Witschi produced a normal female with *WW* chromosomes.

This brief inventory covers the main areas of our genetic and reproductive technologies as they stand in early 1974.

MALE/FEMALE ECOSYSTEMS

Important as our reproductive and contraceptive technologies are in molding our social structures, values, attitudes, and behavioral patterns, *they do not stand alone*. The technologies are part of a total ecosystem in which we and our children live. Our present images of what it means to be male and female, of marriage and parenthood—our attitudes, values, and expectations—have been molded and modified by countless factors in our environment other than the technological. These I want to touch on, however briefly, because they also are vital in understanding and appreciating the future of human sexuality, marriage patterns, and parenthood.

Our traditional male/female images and patriarchal form of monogamy evolved when mankind enjoyed a relatively short *life expectancy*. A million years ago the average human lived eighteen years. Two thousand years ago, average life expectancy was up to twenty-two years, and in the middle ages, up to thirty-three years. Today, our life expectancies are in the low and middle seventies for Americans. In some developing nations life expectancies have doubled in one or two generations. What does this do to marriage, the family, and our images and expectations for ourselves? What if our life expectancy goes over a hundred years?

Infant mortality rates have dropped rapidly with the advances of modern medicine and the discovery of antibiotics. So also has the size of our families. In colonial America the average family had ten to twenty children, with twenty to thirty not uncommon. Today, the average American family is down to 2.01 children, with many couples having no children at all.

Mobility has affected our images and expectations and values. The average American moves once every five years and fourteen times in his lifetime.

The average work week has dropped within two centuries from six days, twelve to fourteen hours per day, to a common five-day, forty-hour work week. A growing trend is evident towards four-day, ten-hour-a-day weeks, or even three-day, twelve-hour-a-day weeks. *Retirement,* once unknown, is now down to sixty years, pushing into the fifties, and projected within a couple of decades to be down into the forties. Leisure affects us very much in what we value and expect out of life.

In 1940, seventeen percent of all American wives worked outside the home. Today nearly half of all American wives are working.

As the economic basis and function of the monogamous family continue to fade in the posttechnological age, so does the link between marriage and parenthood. One out of every six minors in the United States is in the custody of a single parent. Half the children born in the District of Columbia in 1973 were born out-of-wedlock. The number of single men and women under thirty-five has doubled in ten years to fifty-six percent for men and forty-five percent for women. Single persons, homosexuals and lesbian couples can now legally adopt or raise children in many states.

In the past couple of centuries better nutrition and other factors have lowered the average age of puberty for both men and women from around twenty to the very early teens. Simultaneously, social factors, including the need for more education, have led to the postponement of marriage into the early twenties, five or six years later than it occurred a few centuries ago. The sexually mature single person is a relatively new phenomenon in human society.

Finally, our divorce rate has moved from one divorce for every seven marriages in 1920 to more than 3.8 divorces for every ten marriages in 1972.

When all our present and emerging technologies with their psychological impact are combined with the broader environmental factors just listed, one can easily see why we are in a state of apocalyptic discontinuity which is already producing radical changes in the relationships of men and women.

Hot and cool sex—closed and open marriage

ROBERT T. and ANNA K. FRANCOEUR

■ Anna K. Francoeur, *whose profession is budgetary accounting, has done graduate work in history and has written extensively on the nature of sexual relations. (For a note on Robert T. Francoeur, see introduction to preceding essay.)*

The Francoeurs use Marshall McLuhan's terms "hot" and "cool" to contrast sexual attitudes. "Hot" sex is "machine" sex, depersonalized and male dominated. It is sex in which one's partner is regarded as property. "Cool" sex, on the other hand, is characterized as egalitarian, synergistic, and sensually diffused.

SEXUAL MYTHS

Every culture is nourished by its own piebald collage of attitudes, images, expectations, and values, some quite congenial and others uncomfortably restless in the overall culture. In each culture a unique consciousness is woven, mostly by weavers unconscious of their raw materials, or even of the final social fabric into which they have plaited themselves. Threads rise out of the past, emerging sometimes gently, sometimes with twisted violence, but they inevitably combine their web to catch up their unwitting weavers in the final fabric.

What we propose to do in this essay is trace for you the beginning and the end of our cultural fabric. We want to examine the past, present, and future of one crucial thread in this fabric, the male/female relationship and the structure of marriage.

We Americans have no clear vision or image of where we are going. We hardly have any vision of the past either. In fact, our youth delight in the instantaneous present, scorning the past and shunning tomorrow. We need, then, an image of our past. We have to understand and appreciate the values, attitudes, and expectations of the past in which we are rooted. We need a social myth of the past. But we also need an image of the future, a social myth of values and expectations that will provide inspiration and direction as we and our culture evolve with cataclysmic pace.

At the 1973 Groves Conference on Marriage and the Family, one workshop was devoted to discussion of semantics, labels, and models which seem to be emerging in our culture. Labels and models are almost an essential tool in every culture before we can understand and discuss novelties in human behavior. We have to relate the novel to what we know by experience. But first we must verbalize what we know by experience. We have to label and describe our past attitudes, values, and models in male/female relations before we can project and describe the attitudes, values, and expectations we seem to be working towards in sex and marriage.

In this context, Marshall McLuhan and George B. Leonard suggest a basic model in their . . . picture of "The Future of Sex." What we propose to do here is expand and develop the basic insight of McLuhan and Leonard into two detailed social myths: Hot Sex, which encompasses the sexual attitudes that have characterized our American and European cultures for the past century in particular, and Cool Sex, the futuristic image of where we seem to be headed.

HOT SEX

Some may bristle at the judgmental tone implied in labeling the sexual consciousness of most Americans and Europeans today as Hot Sex. But the label is appropriate as a cultural myth and stereotype provided we exercise caution. The portrait we propose here for Hot Sex is an abstraction, a composite. Do not expect to see all the portrait's details in the behavior of a single person you know well, be that individual a friend, enemy, spouse, or yourself. But each of us exhibits a variety of these Hot-Sex traits simply because we have all been raised in this culture and breathe its attitudes and values.

McLuhan and Leonard [have] sketched the roots of the Hot-Sex mentality, so we need not cover again that important territory. What we need to go into are the details of the Hot-Sex attitudes, values, and expectations.

Hot Sex, like hot media, presents a very clearly defined picture. The anatomical precision of the blown-up Playmate of the Month

□ From Robert T. and Anna K. Francoeur, eds., *The Future of Sexual Relations*, © 1974, pp. 30-41. Reprinted by permission of Prentice-Hall, Inc., Englewood Cliffs, New Jersey.

with no genital or mammary detail left to the imagination is the ideal Hot-Sex female. In the male stereotype, the raw aggressiveness of Loren Hardeman I (whose very name has a Hot-Sex significance) offers an excellent model of the Hot-Sex male in Harold Robbins' novel *The Betsey*. The high definition of Hot-Sex masculinity and femininity is precisely the anonymity of genital interlocking, the belief that sex is nothing except penis and vagina.

A curious circus of obsessions and anxieties appears when sex is segregated from persons and from everyday life, when it is reduced to genitality. The Hot-Sex obsession with genital intercourse shows in our common, spontaneous equation of intimacy with genital intercourse. It turns up in our common, often expressed concern with the size of the penis or breast. Loren Hardeman's French playmate of the moment paid him the ultimate Hot-Sex compliment when "She stared at it in wonder. 'C'est formidable. Un vrai Canon!'"

Hot Sex is the American fascination with what appears to be an unlimited variety of "perfect" sexual techniques, positions, and combinations, all of which must be experienced if one is to be "with it." Hot Sex is the worried adolescent quest for mutual orgasm at all costs, the anxious, frantic search for multiple orgasm with the "perfect" partner, or, more honestly, the perfect organ. Hot-Sex relationships, like those in *The Last Tango in Paris,* are casual in their impersonalism: in the dark, one hole is the same as any other, one rod the same as any other—and often the same even when faced in the light of day. Hot Sex often merely moves the fig leaf to the face.

Hot Sex is curiously haunted by the virginal siren. Attraction and fear combine in the male's desire to be the first to deflower his property. Virginity, like the person you marry or are in love with, is basically a commodity. As a commodity, virginity is undamaged, intact property. As long as no penis has *violated* a girl's vaginal canal, or taken possession of it, she remains a virgin. She may very well have masturbated to mutual orgasm with a dozen males, but as long as no male has taken possession of her "innermost treasure," she remains *technically* a virgin. The wives in Bangladesh were quickly discarded by their husbands after their rape by invading soldiers: spoiled property is worthless. Male virginity, on the other hand, is frowned upon in a Hot-Sex culture—not unexpectedly since Hot Sex is patriarchal and holds to a double standard. The female's role is to provide man with his pleasure and comfort, as well as a domestic support system.

When human sexuality is segmented from life and clearly defined in terms of genitals, it naturally has to be scheduled, arranged, planned, both in time and place. The evolution of the bedroom (McLuhan/Leonard, p. 17), and the plotted affair pursued at night in some hidden motel are evidence of the explosive, volatile, and tenuous place Hot Sex holds in everyday life.

Hot Sex is *Last-Tango* screwing in the most depersonalized sex-object way. In classic style, Loren Hardeman "poised over her, like a giant animal blocking out the light until all she could see was him . . . as he slammed into her with the force of the giant body press she had seen working in his factory on a tour the day before."

In a Hot-Sex culture there is also a fig-leafed obsession with nudity and the naked female figure. Female nudity is accepted or tolerated for the sake of art, or the erotic enjoyment of the male, but male nudity is frowned upon. Even in the privacy of one's home and family circle, not infrequently even with one's spouse, the privacy of one's body is clothed from view.

Hot Sex is entropic, self-destructive, because it lives by *possessing* and conquering sex objects. It is entropic also because its life blood is the vital compulsion to perform: the destructive pressure on the male to screw every chance he gets and the equally destructive pressure on the female to satisfy the male ego with the benediction of a mutual orgasm, or better, of mutual and multiple orgasms.

Genital Hot Sex is an end unto itself: sex *for* fun, *for* ego satisfaction, *for* ego building. In Hot Sex one can escape the unbearable burden of time and aging simply by multiplying experiences. Like the passenger on the train in Marcel Proust's *Remembrance of Things Past,* the Hot-Sex male darts desperately from one window to the next, from one conquest to the next, in hopes of encompassing and proving his manhood in a good scorecard with ever mounting conquests.

The *real* sex in a Hot-Sex value system is the forbidden fruit, the thrill of cheating, the escape from boredom and routine of everyday relations into the romantic wonderland of an affair. Even in that modern version of the socially acceptable infidelity, the swinger and mate-swapper are not up to dealing with real involvement, real intimacy. The swinger allows genital infidelity, provided it does not involve persons. The swinger does not resolve his or her jealousy and instead retains a possessiveness that allows a temporary, safety-valve swing while still viewing any real intimacy as a threat and potential competition. Hot Sex views any human relationship in terms of competition and threat.

In a Hot-Sex culture, marital fidelity is reduced to what I did not do in Dubuque. Marital fidelity is synonymous with genital exclusivity. And since sexuality is implied in any intimacy, every sexually mature single person poses a threat. The unmarried, widowed, divorced, obviously are sex-starved because one cannot possibly be a fulfilled male or female without constant sexual genital experiences. Intimacy of any kind with the unmarried must be resolved (in the hypocrisy of American life) as soon as possible by marriage. Likewise, the extramarital relationship must be resolved either in divorce and remarriage, or in termination of the affair. With the forbidden-fruit romance of the affair weighting the value scale in favor of divorce and remarriage, a Hot-Sex culture can only view the single person as a threat to all married people.

In this same vein, couples exist—*not* sexually mature persons who happen also to be married. Couples go everywhere together. Married couples are safe because they are asexually symbiotic in their togetherness. They go everywhere with their spouses and would never think of leaving their spouse home and going out with another person of the opposite sex.

In a Hot-Sex culture based on propertied relations, it is the male, the husband, who is free and self-defined. Because a wife belongs to her husband, her identity is drawn from her husband. The father gives away his daughter at a wedding; no one gives away the groom. Many men continue to respond emotionally at even the suggestion of a threat to their property rights, as did one television personality in a discussion of open marriage: "I'll strangle the bastard who puts his paws on my property!"

Monogamy, life-long and sexually exclusive, is obviously the sole way of adult life (though the husband is allowed certain freedom and license not open to his wife). Every adult is urged, even compelled, into marriage as soon as possible. The slightest thought of an alternate to traditional monogamy is taboo, for monogamy (with the wife as a domestic support system) is the inviolable monolithic foundation of all our economics, politics, and social structure.

Hot Sex is dualistic and gnostic. It really despises or, at best, tolerates the body. The human body is enshrouded with countless taboos that restrict touching and body contact of all but the most "innocent" and "nonintimate" type. As a result, Hot Sex is sterile, antiseptic, and antisensual. It is cut off from the whole person, isolated from nature and the cosmos. For the earthy, cosmic myths of tribal, nomadic cultures with their cool-sex attitudes toward the body, the hot-sex mentality substitutes only the frail, treacherous lure of the great orgasm hunt, the Parsiphalian quest of the perfect partner with the perfect organ and technique, the *Love-Story* myth, and the belief that despite its segregation from life—and, in fact, because of this segregation—Hot Sex, genital sex, is IT.

In brief, Hot Sex is male-dominated, double-standarded, intercourse-obsessed, property-oriented, and clearly stereotyped in its sexist images and models.

COOL SEX

After centuries of gestating in our increasingly hot-sex culture, a new set of attitudes, values, and expectations is finally breaking through to the surface. This is happening not primarily, as the mass media and common image has it, among the college generation, but rather among those married couples in their thirties and forties, where a certain amount of financial independence has joined with the perspective only experience can bring to allow a critical look at their traditional Hot-Sex attitudes and expectations. The courtly love of medieval times and the Renaissance began the degenitalization of Hot Sex, setting the stage for an emasculation of American Hot Sex centuries later and putting the western world on the path to a new culture, the Leonard/McLuhan Cool Sex.

In this new framework, the relationship of men and women can once again be expressed validly as a peer relationship, between two evolving, maturing, sexual, and unique persons. In this Cool-Sex consciousness, men, and especially women, have to become conscious of themselves as individuals with a real existence outside their socially imposed, stereotyped roles. Men and women must realize that creation, our creation as human beings, is an ongoing process of becoming. In many respects, as Paul Klee suggested, "Becoming is superior to being." Masculinity and femininity cannot be defined clearly in a Cool-Sex culture because they are still in the process of being evolved by every sexual person individually. Maleness and femaleness are not subject to the high definition of a Hot-Sex culture. They are fluid realities, constantly changing and constantly being created, not as rigid archetypes but as scintillating process incarnations, each expression with its own unique value.

The consciousness of Cool Sex requires a degree of self-identity. Men or women must first of all be somewhat secure in their image of themselves as persons, without relying on the blessing of society's stereotypes. In Maslow's language, they must be self-actualizing and free of the need to borrow their identity and direction from society or another individual. The ability to stand alone, with some real degree of psychological and emotional maturity is essential to a Cool-Sex mentality.

Given the low definition of masculinity in Cool Sex, it is no longer possible for a man to judge his maleness and identity in terms of multiple conquests and (male) progeny. Nor can a woman find her identity as a person in the phrases "his wife" and "their mother," or in the number of times she has produced pattering feet.

Cool Sex means considering and accepting for others *and for oneself* the *real* possibility of alternatives to the traditional hot-sex stereotypes of breadwinner, housewife, parent, married *couple,* "good girl" (fair white maiden), and double standards. Since sexual persons in a cool-sex culture are not defined in terms of their roles, men and women are free to explore and express their own personalities with as little role playing as possible and with a minimum of imperatives other than the basic rule of not exploiting others in any way, as objects, sexual or otherwise.

Cool sexuality is expressed in integrated, holistic behavior which accepts the human body wholeheartedly and fully. It is neither disturbed by nudity nor scandalized by "immodesty." Cool sexual consciousness celebrates the body in the tradition of Solomon's *Song of Songs,* and the Woodstock–Watkins Glen generation. It is involved and intimate, simultaneously inclusive and embracing rather than exclusive, possessive, and jealous. It takes into consideration *all* the needs and responsibilities of *all* the persons involved in or affected by a relationship. It diffuses the genital spotlight over the whole body. It tends to integrate a whole range of bodily intimacies, touching, nudity, and sensuality, along with genital intercourse, into the total framework of daily living.

Human relations guided by a Cool-Sex consciousness are synergistic, rather than entropic. Cool Sex means neither the end nor the lack of emotions, intense feelings, concern, or warmth. What it does mean is that relations are not taken in terms of possession or competition.

Marriage, in particular, is modified in a cool-sex culture. We are convinced, despite the predictions of monogamy's impending doom, of the value of long-term one-to-one commitments and relations. But the self-destructive, unreal exclusiveness of the Victorian romantic monogamy has got to change radically. We have already accepted in our culture a major modification in one of the two basic values in marriage. We have found it easier to modify our value of life-long monogamy with no-fault divorce, divorce, and remarriage rather than modify the exclusivity of traditional marriages. But we are beginning to question whether this is the more humane and growth-oriented adaptation. Its ease, when compared with exploring adaptations and modifications in the traditional marital exclusivity, may destroy human relations and hinder personal growth far more than any modification in marital exclusivity. Our view is that long-term pair bonds can evolve as primary relationships within the realistic context of today's world with its increasing mobility, stretching life expectancies, contraceptives, and the socioeconomic liberation of women.

In a Cool-Sex culture, the nuclear family and exclusive couple will, more often than not, after about five years during which they become secure in their own primary relationship, come to accept an openness and flexibility unheard of in a Hot-Sex culture. This openness would accept a variety of intimate relations *on all levels,* for *both* husband and wife, including the *possibility* of genitally expressed relations, within the orbit of the primary relationship and complementing it. This open, flexible type of marriage, with its multilateral pluralism of intimacies, would be far more functional than the rigid couple pattern of past generations, though it also involves certain new psychological and emotional demands and risks which we have explored in detail in *Hot and Cool Sex: Cultures in Conflict* (1974).

The Reverend William Genné, of the National Council of Churches' Family Life Bureau, and Drs. Rustum and Della Roy, authors of *Honest Sex,* several years ago coined the label *comarital* to describe a relationship which parallels, complements, and reinforces the primary relationship of a married couple. This comarital relationship may be a very personal and intimately emotional friendship, even one involving genital expression. More recently we have suggested the label of *satellite relationship* since the word *comarital* gives a bias toward the marital end and does not cover the situation in which a secondary relationship complements a primary nonmarital relationship. The need for, and function of, satellite relationships that complement long-term primary relations are based on the complexities of modern life, the varieties of educational backgrounds, and the inevitable, not necessarily parallel, growth and change in personal expectations. When one's education, expectations, and background are limited, it is fairly easy to find one person who can completely and totally meet your growth potential for life. With short life expectancies, most men and women were forced into serial monogamy with several mates by the early death of a spouse in childbirth, war, plague, or pestilence. Today, as our life span stretches and we become richer in background and expectations, it may well be that one person will be unable to meet all our changing needs over fifty or more years.

The choice for our generation, then, seems to fall between a traditional, sexually exclusive *series* of relationships—modular, monogamous marriages which explode after five or ten years when their Hot-Sex expectations are frustrated and unfulfilled—and, on the other side, some sort of open marriage based on the Cool-Sex attitudes which neither reduce marital fidelity to sexual exclusivity nor view one's partner as property.

The satellite relationship, however, only becomes possible when both partners in a primary relation are secure in their own self-identity and in their primary pair-bonding. Property can be lost, but a personal commitment is not necessarily destroyed by a secondary commitment. The satellite relation, then, is not the explosive affair, but a constructive, complementary relationship open to married and single persons, husbands and wives alike, in a context and social structure in which relations are synergistic, reinforcing, and strengthening rather than competitive and entropic.

This expansion of human relations to integrate new modes and expressions of intimacy and community within the couple marriage seems to recapture a consciousness that apparently found its way from the earliest aborigines into the early biblical tradition only to be lost as urbanization began its slow triumph in the Western Judaeo-Christian tradition. The early Hebrews had no words for sex or sexual intercourse until they borrowed these fragmenting terms and concepts from the more urban thinkers of Persia, Greece, and Rome. The original biblical tradition, and the Hebrew language even today, speaks of the engaging, pleasuring, person-integrating relationship between a man and woman not as "making love" nor as genital intercourse. Rather they used the simple, rich word *yahda* ("knowing"). This is no Victorian euphemism, but rather a clear indication of the holistic approach to human relations existing among the early nomadic Hebrews before patriarchal, sexist distinctions took over.

Few modern western writers have captured this cool-sex consciousness. Most of them cannot even deal with the present tensions of ordinary people caught in the transition from hot to cool. Witness, for instance, the inability of the gifted John Updike, both in *Couples* and *Rabbit Redux,* to deal with extramarital intimacies and relations in anything but a totally negative, threatening Hot-Sex framework.

Only in science fiction or the utopian essay/novels of Robert Rimmer do we catch some glimpse of what Cool-Sex attitudes and values might be. Rimmer's classic *The Harrad Experiment* and his latest novel, *Thursday, My Love,* are prime examples, as is Robert Heinlein's haunting *Stranger in a Strange Land.* Heinlein uses satire, humor, and fantasy to tell the story of Valentine Michael Smith, the son of the first humans to land on Mars. After being raised and educated by Martians when his parents died, Mike returns to this earth with a later expedition only to be shocked by the Hot-Sex mentality of earthlings. The Martian pattern of male-female relations he knew was communal and multisensual, with no sharp cultural distinction between male and female roles. What earthlings call sexual intercourse and reduce to genital coupling, Mike views as "grokking" or "growing closer," a kind of demierotic relating and interpersonal knowing in the original biblical sense.

Similar examples turned up in the cinema in the early 1970s. *The Graduate, The Summer of '42,* and *Carnal Knowledge* contain beautiful expressions of the Hot- and Cool-Sex myths in their main characters.

In books about human sexual behavior, *The Joy of Sex,* which Alexander Comfort edited for Crown Publishers (1973), is outstanding. Its line drawings and soft, detailed illustrations communicate a cool-sex mentality better than any other book of this type we know.

In facing the problems of language and new words to describe our emerging values and attitudes, it is well to recall that in most tribal and aboriginal cultures, there is no word for illegitimacy because all children are considered as young persons in their own right and not the property of any set of parents. Often too, because of the parity of men and women in tribal cultures, there is no word for adultery. Social taboos do limit sexual behavior for the good of the community, but not because the wife or daughter is the property of some male.

Cool Sex, then, is egalitarian, single-standarded, sensually-diffused, and oriented towards intimacy and open, synergistic relations with persons.

HOT AND COOL SEX, CLOSED AND OPEN MARRIAGE

We have discovered some fascinating parallels between our social models of Hot and Cool Sex and the descriptions of Closed and Open Marriage proposed by anthropologists Nena and George O'Neill in *Open Marriage.*

With the help of friends at Sandstone, a unique Cool-Sex experiment in California, we worked out an itemized list of characteristics for Hot and Cool Sex. In a chapter on "Synergy," the O'Neills also itemize the details of Closed and Open Marriages. Rearranging the O'Neills' tables and ours on the basis of 1) definitions, 2) value systems, 3) behavioral structures, and 4) concerns produces a very informative picture:

Definitions:

Hot Sex: high definition; clear sex-role stereotyping; sex equated with genital coupling; segregation of sex in time and place; numerous strong imperatives from society; highly structured with many "games."

Closed Marriage: static framework; rigid role prescriptions; highly calculating; change is threatening.

Cool Sex: low definition; sexuality coextensive with personality; diffused sensuality; spontaneous; light structure and few social games; few social imperatives; little, if any role stereotyping; self-actualization encouraged.

Open Marriage: dynamic framework; flexible in its roles; adaptable to change; spontaneous.

Value systems:

Hot Sex: patriarchal, with aggressive male dominating passive female; double moral standard.

Closed Marriage: unequal status of husband and wife; selfhood subjugated to couplehood; bondage.

Cool Sex: equalitarian; partnership; friendship; single moral standard.

Open Marriage: equality of stature; personal identity; freedom.

Behavioral structures:

Hot Sex: property-oriented; possessively closed; casual and impersonal; physical sex segregated from life, emotions, and responsibility; grossly selective of playmates; screwing sex objects for conquest; genital hedonism.

Closed Marriage: couples locked together; smothering, with limited growth potential; possession shuts out others; closed, self-limiting energy system.

Cool Sex: person-oriented; open; inclusive; involved and intimate; sex integrated in everyday life; finely selective in all relations; sex viewed as communications and "knowing" sexual persons.

Open Marriage: openness between spouses; freedom that incorporates others; growth-oriented, open companionships; privacy for self-growth; individual autonomy in an open, expanding energy system.

Concerns:

Hot Sex: orgasm-obsessed; performance pressures; fidelity means sexual exclusivity; extramarital relations are an escape; nudity taboo; emotions and senses feared; sexuality basically feared; entropic because property can be used up; alcohol and drug-altered states common; personal space a prime concern.

Closed Marriage: limited love; deception and game-playing; conditional, static trust; exclusion of others; inhibitive; degenerative and subtractive relations in a closed, self-limiting energy system.

Cool Sex: engaging, pleasuring communications; sexual relations accepted but truly optional; fidelity means commitment and responsibility; comarital relations, with or without genital expression, growth reinforcing; senses and emotions embraced; nudity, even in groups, optional; synergistic relationships; grokking; drug-altered states not common.

Open Marriage: honesty; truth; open love; open trust;

openness to others; creativeness; expansive and additive relations in an open, expanding energy system.

Within the emerging pluralism of male-female relations in our global city, many patterns of marriage will be increasingly tolerated because some people find them functional. However, we are convinced that the majority of people will find themselves confronted with a choice between two patterns. One of these is the traditional romantic, patriarchal, sexually exclusive marriage which will adapt to the new environmental system of today and tomorrow by modification in practice, if not immediately in theory. Multiple divorce and remarriage, serial polygamy, will be the result, with all the other attitudes, values, and expectations of the Hot-Sex myth retained. The alternative will be some sort of flexible, dynamic Open Marriage with a new dynamic definition of fidelity as "loving concern" and commitment. In this second pattern, "forsaking all others" will be the element of traditional marriage that is modified.

The question is which modification will be the most humane, most functional, most growth-promoting, and for which people. Not all can handle the Cool-Sex values and attitudes, nor will all continue to opt for the Hot-Sex values and attitudes. The trend, however, is definitely towards pluralism in our marital patterns, with the growing popularity of new values of fidelity and commitment, new risks and new challenges.

7

Technology assessment

Technology assessment in a sharp social focus

HENRYK SKOLIMOWSKI

■ Henryk Skolimowski *is a member of the faculty of the Department of Humanities of The College of Engineering, University of Michigan. He has published extensively on the subject of technology assessment.*

In his important essay Professor Skolimowski argues that technology assessment is a branch of a larger discipline, which he calls philosophy of technology. This is the case, he suggests, because a genuine assessment of technology is a "sociomoral," and therefore a philosophical, enterprise. Like philosophy, it should avoid apologetics. It should be critical with respect to its area of inquiry. For these reasons technology assessment cannot be a branch of technology itself.

My concern with Technology Assessment is born out of my larger concern with Philosophy of Technology. What is Philosophy of Technology? It is a systematic reflection on the nature of contemporary technology, its role and function in society and civilization at large. It may be said without exaggeration that technology is the major force shaping the destiny of the present western civilization: thus shaping the destiny of Society, and therefore, to a large degree, shaping the destiny of its individuals. It is quite obvious that technology is not only a collection of tools, but a vital social and cultural force determining our future. It is not an assembly of gadgets, but a part of our world view, indeed an intrinsic part of the western mentality: whenever we westerners think technology, we invariably think "manipulation" and "control". The primary locus of Philosophy of Technology is not a simple accumulation of insights, a merely analytical and dispassionate inquiry, but an attempt to find some answers to the most urgent social, moral and human dilemmas of our times—the dilemmas which have largely been caused by the relentless unfolding of technological progress. In short, Philosophy of Technology is the philosophy for our times.

The relation between Technology Assessment and Philosophy of Technology is quite clear: the former is a subclass of the latter. Genuine Technology Assessment is, and must be, a form of a sociomoral (therefore philosophical) reflection on the large scale unintended consequences of technology at large. As such it is inextricably bound to a larger corpus of philosophy which I call Philosophy of Technology.

Moreover, unless and until Technology Assessment is seen in a broader social and philosophic framework, it is bound to be a one-sided apologia for the prowess of existing technology. Genuine Technology Assessment must be essentially critical, not apologetic, with regard to Technology. The idea of Technology Assessment was born at the time when we realized that in addition to its positive aspects, Technology may and does bring about negative consequences. The recognition of both these aspects of technology is the core of Technology Assessment. Since positive aspects are easily perceptible, it is quite natural that we should be more concerned with negative aspects while doing our assessment.

□ From Technological Forecasting and Social Change, **8**(1976), 421-425. Reprinted by permission of Elsevier North Holland, Inc.

During its first few years of existence, Technology Assessment, being by and large performed by people with stakes in existing technologies, tended to be too apologetic and insufficiently critical of technological projects and processes. This situation is alarming because Technology Assessment is too important a social tool to be left in the hands of technocrats and their predominantly technical criteria.

It suffices to look at various existing "technology assessments" in order to be at once persuaded that present technology assessment is unfortunately, more often than not, an adjunct to technology itself, a set of technical procedures generated by technology and ultimately serving its purposes. Ironically, many a time the same people who had developed a given technology or a given process, later themselves are assessing this technology, usually applying predominantly technical criteria. It will perhaps be more adequate than harsh to say that quite often technology assessments are fraudulant from a social and human point of view, for while paying lipservice to "social aspects" the overall tenor, methodology, and conclusions are technical: a technical exercise performed by technicians.

Technology Assessment is not *another* branch of technology. Technology Assessment is not an insignificant evaluation of some aspects of Technology. Technology Assessment is a social critique of Technology at large. This critique may be vital to the survival of the technological society; or should we say: may be vital for the preservation of society and its evolution towards a post-technological society. Therefore, this critique must not be left in the hands of those who are often themselves responsible for creating powerful, but sometimes lethal tools. We must therefore be aware that Technology Assessment does not degenerate into a servile adjunct of Technology. We must also be aware of another danger, of a much subtler nature. We are all *technicians*. Our attitudes and our mentality are profoundly effected by the ideals of technology, and by the assumptions on which technology and the whole civilization are based. We must be therefore aware that even when we genuinely try to assess Technology from a social or a moral point of view "the process," as one astute critic has put it, "may be totally slanted in favor of the assumptions underlying the technological civilization."

In the first issue of the periodical *Technology Assessment* there is a comprehensive survey of the field by Genevieve J. Knezo—"Technology Assessment: A Bibliographic Review"—and in my opinion it shows a distinctive bias towards the treatment of Technology Assessment as an adjunct of Technology. It is slanted in a subtle and often an explicit way in favor of the assumptions underlying the technological civilization, of which it is supposed to be an assessment. Knezo writes, for instance: "The growing literature on technology assessment takes a variety of forms. Some of it is of an emotional, neo-luddite, and polemic nature designed to arouse and mold mass public opinion to the 'uncontrollable' hazards of technology, such as Muller's *The Children of Frankenstein: A Primer of Modern Technology and Values,* or Schwartz's *Overskill: The Decline of Technology in Modern Civilization,* or Douglas' *The Technological Threat.* Much of it, on the other hand, usefully serves to inform the public, through a responsible press, of the pros and cons of a public issue of national importance" [1]. It would thus appear from Knezo's statement that no literature on Technology Assessment should be of a "polemic nature", and that the literature that does so is not a "responsible press". The pro-technology bias of Knezo is too striking to accept her statements and sentiments as unconditionally valid. More importantly, we should really be aware of the fact that so-called anti-technology literature (and what does that mean in the ultimate analysis?) is often pro-human literature. Whoever stops at the level of its apparently anti-technological bias, and does not see its positive thrust, engages in a superficial analysis. In any case, "pro" or "anti" literature, whatever the subject matter, must not be dismissed off-hand before we examine its claims and arguments.

It is quite clear that Technology Assessment has not yet found its identity. However it is too important a field to be left in the hands of the apologists of Technology. I would therefore like to propose what I call Skolimowski's laws of Technology Assessment as a set of guidelines for a genuine process of assessing Technology at large. Here are the laws:

1. No system can adequately assess itself.
2. The more satisfactory is the assessment from the quantitative point of view, the less valuable it is from the social point of view.
3. All genuine assessment must terminate in value terms.

4. The "real expertise" in Technology Assessment is social and moral not technical.

Regarding No. 1. "No system can adequately assess itself" is equivalent to saying that systems must be assessed by agencies outside the system. The very essence of assessment is just that: it is not something that follows from the system, but something that X-rays the system from outside. Can an X-ray machine X-ray itself? No. In the same sense no system can *adequately assess* itself. There is one exception however. And this exception is the human being. If we think about the human being as a system, then it is the only system capable of adequate self-assessment. (Some people would doubt even that.) Because of his ability to judge himself on the variety of meta-levels simultaneously, because of his self-consciousness, and because of his peculiarity to self-judge himself, the human being may be the only system capable of adequate self-assessment. The history of philosophy is an impressive record of the human being self-assessing himself as a species being, as a social being, as an individual being. Returning to the first law: the criteria for assessing a given technological system cannot be derived from this system and cannot be limited to the system, because then there is an apology for the system, not an assessment of it.

Regarding No. 2. "The more satisfactory is the assessment from the quantitative point of view, the less valuable it is from the social point of view". This law will raise many a hair and objection. I am not saying that it is analytically true, that is to say, that there is a logical relationship between the two. Logically speaking, there is no reason for quantitative perfection to be achieved at the expense of qualitative judgments. But I am suggesting that this is a contingent law, based on actual observations. It *just* happens to be the case that the more successful we are in quantifying, the more successful we are in eliminating qualitative judgments. One immediate reason for this state of affairs is: within the present technological system the process of quantification has almost invariably served the cause of instrumental values, so that quantification came to mean an exclusion of qualitative, or intrinsic values. Unless it is demonstrably *shown* that quantification can serve the cause of intrinsic values, the second law holds.

Quantification is often considered to be the hallmark of "good science". But is this contention justified? Perhaps we have accepted it too readily and uncritically. The undoubted successes of social science via quantitative methods are not as striking as they might appear, for they are concerned with relatively unimportant social phenomena, such as voting behavior, while they are negligible in explaining deep-seated value and social preference factors which alone helps us to understand the actual behavior of societies. In present social science for the sake of easy quantification we often, much too often, trivialize important social phenomena in order to fit them into our quantitative schemata. Given the present state of art, one can suggest that we are unable to quantify intrinsic values without changing their nature in the process of quantification. The onus is, again, on the shoulders of the advocates of quantification.

Regarding No. 3. "All genuine assessment must terminate in value terms". This law is, of course, a consequence of law 2. But it must be clearly stated because very often instrumental values usurp for themselves the place of ultimate values. It is often argued that the maximization of the economic value, or raising of the standard of living for all is the ultimate value which justifies all our pursuits. This is a favoraite strategy for subverting genuine assessments by instrumental or economic ones. Another strategy is to argue that "A strong capacity in applying adequate techniques of technology assessment is necessary in both the executive and the legislative branch". Then methodology takes precedence over values and we gently ride on the high horse of quantitative techniques towards the instrumental paradise. Genuine assessments must be moral, human and social assessments, related to some intrinsic values in which ultimate ends of man's life are expressed, and not merely means leading to these ends. Here again the importance of a larger philosophical basis, whether we call it philosophy of technology or not is irrelevant, is of vital importance.

Regarding No. 4. "The real expertise in Technology Assessment is social and moral not technical". Such a formulation may strike one as a contradiction in terms, for expertise is something technical, while morality and sensitivity are considered the qualities of the mind and the heart clearly lying outside "spheres of expertise". This is a real dilemma which must be met face to face. We must be able to evolve a set of rules and criteria based on moral and social sensitivity which alone will make sense of technology assessment in the long run, and on the scale of the whole society and the entire civilization. Perhaps what is needed is a new social science as a guardian and executor of Technology Assessment. Such a social science does not exist; and it undoubtedly will take a while before we work it out. Such a social science will be directly opposed to Skinner's behaviorism.

The laws here suggested need not be taken too seriously. It might even follow from my discourse that there cannot be simple laws concerning the evaluation of such extraordinarily complex and difficult matters as the interaction of technology with society. Yet, in formulating these laws, as heuristic guidelines, we can provide a focus and a frame of reference for discerning what is really important in our thinking about Technology Assessment.

Technology Assessment is sometimes abbreviated T.A. On the other hand, we have another movement which is called Alternative Technology, and it is abbreviated as A.T. Looking at the symbols we see that one is the reversal of the other. These symbols correspond to their respective realities. For while Technology Assessment has, by and large, been an arm and justification of the *status quo,* Alternative Technology has, by and large, been purused by the opponents of the *status quo.*

Alternative Technology is known under a variety of names: soft technology, low-impact technology, ecologically sound technology. This kind of technology has been pursued by a large variety of heterogeneous groups which, though their ideological biases may be questioned, have seriously taken into account the idea that present big power technology produces too much destruction as it unfolds its "progress". For this reason, these various groups attempted to re-think comprehensively a whole set of relationships concerning technology and environment, and technology vis-a-vis society and the human individual. As a result of this re-thinking, they came to the conclusion that we have to change the *nature* of present technology. Hence the idea of Alternative Technology. The most comprehensive formulation of the field, so far, can be found by the reader in David Dickson's *Alternative Technology* (Fontana Original, London, 1974).

Alternative Technology is not a mere extravaganza of the hippies, confused drop-outs, half-baked revolutionaries (although it is sometimes associated with those). Alternative Technology is a profound critique of the entire existing technology. It is profound because it not only offers a critique of the existing system, but also suggests a thoroughgoing alternative. Alternative Technology is Technology Assessment through and through. The sooner we realize that the more we can learn from this movement, whose individuals often possess rare and sharp insights into the nature of present reality.

The end of the thing is often close to its beginning. So let me finish where I started. Technology Assessment is a branch of Philosophy of Technology. Philosophy of Technology is, in a sense, an extension of philosophy of science. But Philosophy of Technology has been lamentably neglected by philosophers of science. In concentrating on exclusively technical aspects of their disciplines, philosophers of science have lost touch with new epistemological and moral realities of the scientific-technological world. The result is: philosophy of technology has already superseded philosophy of science, but philosophers of science do not know it yet.

Should anyone wish to argue that the views expressed here represent "an opinion of a starry-eyed philosopher, good in theory but impossible in practice", I would wish to point out that these opinions are not at all idiosyncratic. The distinguished historian Lynn White, who is perhaps more accomplished in interpreting technology in cultural contexts than any living historian, seems to be quite in accord with my views. He writes: "Systems analysts are caught in Descartes's dualism between the measurable *res extensa* and the incommensurable *res cogitans,* but they lack his pineal gland to connect what he thought were two sorts of reality. In the long run the entire Cartesian assumption must be abandoned for recognition *that quantity is only one of the qualities and that all decisions, including the quantitative, are inherently* qualitative. [italics, mine: H.S.] That such a statement to some ears has an ominously Aristotelian ring does not automatically refute it.

There is a second present defect in the art of technology assessment: the lack of a sense of depth in time; this may be called the Hudson Institute syndrome. It is understandable not only because most social sciences normally take a flat contemporary view of phenomena, but also because the concrete problems set before systems analysts for solution look toward future action and discourage probing the genesis of things'' [2]. Technology assessment is coming of age. Let us not allow it to become a quantitative caricature. These remarks do not pretend to be any definitive guidelines, but, one should like to hope, the beginning of a debate on an important subject.

REFERENCES

1. Genevieve J. Knezo, Technology assessment: A bibliographical review, *Technology Assessment* **1** (1), 67.
2. Lynn White, Technology assessment from the stance of a medieval historian, *Technol. Forecast. Soc. Change* 6 (4), 357-369 (1974).

From *New Reformation*

PAUL GOODMAN

▪ Paul Goodman *was until his death in 1972 a noted author, lecturer, and psychotherapist. His major works include* Growing Up Absurd, *published in 1960, and* New Reformation, *1979, from which this selection is taken.*

Since technology is properly a branch of moral philosophy, not of science (a claim similar to that made by Skolimowski) Mr. Goodman argues in this selection that it is time to retrench, to slow the pace of technology in order to allow time for assessment. He calls for the raising of the role of the ''technologist'' to a legitimate profession, which would propose, assess, and certainly remind all citizens of technological societies that they have ''fallen willingly under the dominion'' of technology.

Whether or not it draws on new scientific research, technology is a branch of moral philosophy, not of science. It aims at prudent goods for the commonwealth, to provide efficient means for these goods. At present, however, ''scientific technology'' occupies a bastard position, in the universities, in funding, and in the public mind. It is half tied to the theoretical sciences and half treated as mere know-how for political and commercial purposes. It has no principles of its own. To remedy this—so Karl Jaspers in Europe and Robert Hutchins in America have urged—technology must have its proper place on the faculty as a learned profession important in modern society, along with medicine, law, the humanities, and natural philosophy, learning from them and having something to teach them. As a moral philosopher, a technician should be able to criticize the programs given him to implement. As a professional in a community of learned professionals, a technologist must have a different kind of training and develop a different character from what we see at present among technicians and engineers. He should know something of the social sciences, law, the fine arts, and medicine, as well as relevant natural sciences.

Prudence is foresight, caution, utility. Thus it is up to the technologists, not merely to regulatory agencies of the government, to provide for safety and to think about remote effects. This is what Ralph Nader sometimes says and Rachel Carson used to ask. An important aspect of caution is flexibility, to avoid the pyramiding catastrophe that occurs when something goes wrong in interlocking technologies, as in urban power failures. Naturally, to take responsibility often requires standing up to the front office, urban politicians, and the Pentagon, and technologists must organize themselves in order to have power to do it.

Often it is pretty clear that a technology has been oversold, like the cars. Then even though the public, seduced by advertising, wants more, technologists must balk, as any professional does when his client wants what isn't good for him. We are now repeating the same self-defeating congestion with the planes and airports: the more the technology is oversold, the less immediate utility it provides, the greater the costs, and the more damaging the remote effects. As this becomes evident, it is time for technologists to confer with sociologists and economists and ask deeper questions. Is so much travel necessary? Are there ways to diminish it? Instead, the recent history of technology has consisted largely of desperate efforts

□ New York: Random House, Inc., 1970, pp. 7-20.

to remedy situations caused by previous overapplications of technology.

Technologists should certainly have a say about simple waste, for even in an affluent society there are priorities—consider the supersonic transport, which has little to recommend it. But the Moon shot has presented the more usual dilemma of authentic conflicting claims. I myself believe that space exploration is a great human adventure, with immense esthetic and moral benefits, whatever the scientific or utilitarian uses. It must be pursued. Yet the context and auspices have been such that perhaps it would be better if it were not pursued. (This is discussed in Chapter 2.)

These days, perhaps the chief moral criterion of a philosophic technology is modesty, having a sense of the whole and not obtruding more than a particular function warrants. Immodesty is always a danger of free enterprise, but when the same disposition to market is financed by big corporations, technologists rush into production with solutions that swamp the environment. This applies to the packaging and garbage, freeways that bulldoze neighborhoods, high rises that destroy landscape, wiping out species for a passing fashion, strip mining, scrapping an expensive machine rather than making a minor repair, draining a watershed for irrigation because (as in southern California) the cultivable land has been covered by asphalt. Given this disposition, it is not surprising that we defoliate a forest in order to expose a guerilla and spray tear gas from a helicopter on a crowded campus.

Since we are technologically overcommitted, a good general maxim in advanced countries at present is to innovate in order to simplify, but otherwise to innovate as sparingly as possible. Every advanced country is overtechnologized; past a certain point, the quality of life diminishes with new "improvements." Yet no country is rightly technologized, making efficient use of available techniques. There are ingenious devices for unimportant functions, stressful mazes for essential functions, and drastic dislocation when anything goes wrong, which happens with increasing frequency. To add to the complexity, the mass of people tend to become incompetent, and dependent on repairmen. Indeed, unrepairability except by experts has become a desideratum of industrial design.

When I speak of slowing down or cutting back, the issue is not whether research and making working models should be encouraged or not. They should be, in every direction, and given a blank check. The point is to resist the temptation to apply every new device without a second thought. But the big corporate organization of research and development makes prudence and modesty very difficult; it is necessary to get big contracts and rush into production in order to pay the salaries of the big team, and to keep the team from dispersing. Like bureaucracies, technological organizations are finally run to maintain themselves in being, as a team, but they are more dangerous because in capitalist countries they are in a competitive arena and must stir up business.

It used to be the classical socialist objection to capitalism that it curtailed innovation and production in order to make the most out of existing capital. This objection still holds, of course—a serious example is the foot-dragging about producing an electric or steam car which, according to Ford, will take thirty years, though models adequate for urban use are ready for production at present. But by and large, the present menace of free enterprise is proving to be the same as its past glory, its fantastic productivity, its technological explosion. And this is not the classic overproduction that creates a glut on the market: it is overproduction that burdens life and the environment.

I mean the maxim of simplification quite strictly, to simplify the *technical* system. I am unimpressed by the argument that what is technically more complicated is really economically or politically simpler. For example, by complicating the packaging we improve the supermarkets; by throwing away the machine rather than repairing it we give cheaper and faster service all around; or even, by expanding the economy with trivial innovations, we increase employment, allay discontent, save on welfare. Such ideas may be profitable for private companies or political parties, but for society they have created an accelerating rat race. The technical structure of the environment is too important to be a political or economic pawn: the effect on the quality of life is too disastrous. The hidden social costs are not calculated: the auto graveyards, the torn-up streets, the longer miles of commuting, the advertising, the inflation, etc. As I pointed out in *People or Personnel,* a country with a fourth of our per capita income, such as Ireland, is not less well off; in some respects it is much richer, in some respects a little poorer. If possible, it is better to solve political problems by political means. For instance, if teaching machines and audio-visual aids are indeed educative, well and good; but if school boards hope to use them just to save money on teachers, then they are not good at all—nor do they save money.

Of course, the goals of right technology must come to terms with other values of society. I am not a technocrat. But the advantage of raising technology to be a responsible learned profession with its own principles is that it can have a voice in the debate and argue for *its* proper contribution to the community. Consider the important case of modular sizes in building, or prefabrication of a unit bathroom: these conflict with the short-run interests of manufacturers and craft unions, yet to deny them is technically an abomination. The usual recourse is for a government agency to set standards: such agencies accommodate to interests that have a strong voice; and at present technologists have no voice.

The crucial need for technological simplification, however, is not in the advanced countries—which can afford their clutter and probably deserve it—but in underdeveloped countries which must rapidly innovate in order to diminish disease, drudgery, and starvation. They cannot afford to make mistakes. It is now widely conceded that the technological aid we have given to such areas according to our own high style—a style often demanded by the native ruling groups—has done more harm than good. Even when, frequently if not usually, aid has been benevolent, without strings attached—and not military, and not dumping—it has nevertheless disrupted ways of life, fomented tribal wars, accelerated urbanization, decreased the food supply, gone to waste for lack of skills to use it, developed a do-nothing elite.

By contrast, a group of international scientists called Intermediate Technology argue that what is needed is techniques that use only native labor, resources, traditional customs, and teachable know-how, with the simple aim of remedying drudgery, disease, and hunger, so that people can then develop further in their own style. This avoids cultural imperialism. Such intermediate techniques may be quite primitive, on a level unknown among us for a couple of centuries, and yet they may pose extremely subtle problems, requiring exquisite scientific research and political and human understanding, to devise a very simple technology. Here is a reported case (by E. F. Schumacher, which I trust I remember accurately). In Botswana, a very poor country, pasture was overgrazed, but the economy could be salvaged if the land was fenced. There was no local material for fencing, and imported fencing was prohibitively expensive. The solution was to find a formula and technique to make posts out of mud, and a pedagogic method to teach people how to do it.

In *The Two Cultures,* C. P. Snow berated the humanists for their irrelevance when two-thirds of mankind are starving and what is needed is science and technology. The humanities have perhaps been irrelevant; but unless technology is itself more humanistic and philosophical, it too is of no use. There is only one culture.

And, let me make a remark about amenity as a technical criterion. It is discouraging to see the concern about beautifying a highway and banning billboards, and about the cosmetic appearance of cars, when there is no regard for the ugliness of bumper-to-bumper traffic and the suffering of the drivers. Or the concern for preserving a historical landmark while the neighborhood is torn up and the city has no shape. Without moral philosophy, people have nothing but sentiments.

III

The complement to prudent technology is the ecological approach to science. To simplify the technical system and modestly pinpoint our artificial intervention in the environment is to make it possible for the environment to survive in its complexity, evolved for a billion years, whereas the overwhelming instant intervention of tightly interlocked and bulldozing technology has already disrupted many of the delicate sequences and balances. The calculable consequences are already frightening, but of course we don't know enough, and won't in the foreseeable future, to predict the remote effects of much of what we have done.

Cyberneticists come to the same cautious thinking. The use of computers has enabled us to carry out crashingly inept programs on the basis of willful analyses; but we have also become increasingly alert to the fact that things respond, systematically, continually, cumulatively; they cannot simply be manipulated or pushed around. Whether bacteria, weeds, bugs, the technologically unemployed, or unpleasant thoughts, we cannot simply eliminate and forget them; repressed, they return in new forms. A complicated system works most efficiently if its parts readjust themselves decentrally, with a minimum of central intervention or control, except in cases of breakdown. Usually there is an advantage in a central clearing house of information about the gross total situation, but technical decision and execution require more minute local information. The fantastically rehearsed Moon landing hung on a last-second correction on the spot. To make decisions in headquarters means to rely on information from the field that is cumulatively abstract and may be irrelevant, and to execute by chain-of-command is to use standards that cumulatively do not fit the abilities of real individuals in concrete situations. By and large it is better, given a sense of the whole picture, for those in the field to decide what to do and to do it (compare *People or Personnel,* Chapter 3). But with organisms too, this has long been the bias of psychosomatic medicine, the Wisdom of the Body, as Cannon called it. To cite a classic experiment of Ralph Hefferline of Columbia: A subject is wired to suffer an annoying regular buzz, which can be delayed and finally eliminated if he makes a precise but unlikely gesture, say, by twisting his ankle in a certain way; then it is found that he adjusts more quickly if he is *not* told the method and it is left to his spontaneous twitching than if he is told and tries deliberately to help himself—he adjusts better without conscious control, either the experimenter's or his own.

Technological modesty, fittingness, is not negative. It is the ecological wisdom of cooperating with Nature rather than trying to master her. (The personification of "Nature" is linguistic wisdom.) A well-known example is the long-run superiority of partial pest control in farming by using biological rather than chemical deterrents. The living defenders work harder, at the right moment, and with more pinpointed targets. But let me give another example because it is so lovely (I have forgotten the name of my informant): A tribe in Yucatan educates its children to identify and pull up all weeds in the region; then what is left is a garden of useful plants that have chosen to be there and that now thrive.

In the life sciences there are at present two opposite trends in methodology. The rule is still to increase experimental intervention; but there is also a considerable revival of old-fashioned naturalism, mainly watching and thinking, with very modest intervention. Thus, in medicine, there is new diagnostic machinery, new drugs, spectacular surgery; but there is also a new respect for family practice with a psychosomatic background, and a strong push, among young doctors and students, for a social-psychological and sociological approach, aimed at prevention and building up resistance. In psychology, the operant conditioners multiply and refine their machinery to give maximum control of the organism and the environment (I have not heard of any dramatic discoveries, but likely I don't understand); on the other hand, the most interesting psychology in recent years has certainly come from animal naturalists: studies of the pecking order, territoriality, learning to control aggression, language of the bees, overcrowding among rats, communication of dolphins.

On a fair judgment, both contrasting approaches give positive results. The logical scientific problem that arises is, What is there in the nature of things that makes a certain method, or even moral attitude, work well or poorly in a given case? This question is not much studied. Every scientist seems to know what *the* scientific method is.

"In the pure glow of molecular biology," says Barry Commoner, "studying the biology of sewage is a dull and distasteful exercise hardly worth the attention of a modern biologist. [But] the systems which are at risk in the environment are natural and because they are natural, complex. For this reason they are not readily approached by the atomistic methodology which is so characteristic of much of modern biological research. Any new basic knowledge which is expected to elucidate environmental biology, and guide our efforts to cope with the balance of nature, must be relevant to the natural biological systems which are the arena in which these problems exist."

Another contrast of style, extremely relevant at present, is that between Big Science and old-fashioned shoestring science. There is plenty of research, with corresponding technology, that can be done only by Big Science; yet much, and perhaps most, of science will always be shoestring science, for which it is absurd to use the fancy and expensive equipment that has gotten to be the fashion.

Consider urban medicine. The problem, given a shortage of doctors and facilities, is how to improve the level of mass health, the vital statistics, and yet to practice medicine which aims at the maximum possible health for each person. Perhaps the most efficient use of Big Science technology for the general health would be to have compulsory biennial check-ups, as we inspect cars, for early diagnosis and to forestall chronic conditions and their accumulating costs. But up to now, Dr. Michael Halberstam cautions me, mass diagnosis has not paid off as much as hoped. For this an excellent machine would be a total diagnostic bus that would visit the neighborhoods—as we do chest X-rays. It could be designed by Bell Lab, for instance. On the other hand, for actual treatment and especially for convalescence, the evidence seems to be that small personalized hospitals are best. And to revive family practice, maybe the right idea is to offer a doctor a splendid suite in a public housing project. Here, big corporations might best keep out of it.

It is fantastically expensive to provide and run a hospital bed; yet very many of the beds (up to a third?) are occupied by cases, e.g. tonsillectomies, that could better be dealt with at home if conditions are good, or in tiny infirmaries on each street.

Our contemporary practice makes little sense. We have expensive technology stored in specialists' offices and big hospitals which is unavailable for mass use in the neighborhoods; yet every individual, even if he is quite rich, finds it almost impossible to get attention for himself as an individual whole organism in his setting. He is sent from specialist to specialist and exists as a bag of symptoms and a file of test scores.

In automating, there is an analogous dilemma of how to cope with masses of people and get economies of scale without losing the individual at great consequent human and economic cost. A question of immense importance for the immediate future is, Which functions should be automated or organized to use business machines, and which should not? This question also is not getting asked, and the present disposition is that the sky is the limit for extraction, refining, manufacturing, processing, packaging, transportation, clerical work, ticketing, transactions, information retrieval, recruitment, middle management, evaluation, diagnosis, instruction, and even research and invention. Whether the machines can do all these kinds of jobs and more is partly an empirical question, but it also partly depends on what is meant by doing a job. Very often, for example in college admissions, machines are acquired for putative economies (which do not eventuate), but the true reason is that an overgrown and overcentralized organization cannot be administered without them. The technology conceals the essential trouble, perhaps that there is no community of the faculty and that students are treated like things. The function is badly performed, and finally the system breaks down anyway. I doubt that enterprises in which interpersonal relations are very important are suited to much programming.

But worse, what can happen is that the real function of an enterprise is subtly altered to make it suitable for the mechanical system. (For example, ''information retrieval'' is taken as an adequate replacement for critical scholarship.) Incommensurable factors, individual differences, local context, the weighing of evidence, are quietly overlooked, though they may be of the essence. The system, with its subtly transformed purposes, seems to run very smoothly, it is productive, and it is more and more out of line with the nature of things and the real problems. Meantime the system is geared in with other enterprises of society, and its products are taken at face value. Thus, major public policy may depend on welfare or unemployment statistics which, as they are tabulated, are not about anything real. In such a case, the particular system may not break down; the whole society may explode.

I need hardly point out that American society is peculiarly liable to the corruption of inauthenticity. Busily producing phony products, it lives by public relations, abstract ideals, front politics, show-business communications, mandarin credentials. It is preeminently overtechnologized. And computer technologists especially suffer the euphoria of being in a new and rapidly expanding field. It is so astonishing that a robot can do the job at all, or seem to do it, that it is easy to blink at the fact that he is doing it badly or isn't really doing quite the job.

IV

The current political assumption is that scientists and inventors, and even social scientists, are value-neutral, but that their discoveries are ''applied'' by those who make decisions for the nation. Counter to this, I have been insinuating into the reader's mind a kind of Jeffersonian democracy or guild socialism (I am really an anarchist), namely, that scientists and inventors and other workmen are responsible for the uses of the work they do, and they ought to be competent to judge these uses and have a say in deciding them. They usually are competent. To give a poignant example, Ford assembly-line workers, according to Harvey Swados who worked with them, are accurately critical of the glut of cars, but they have no way to vent their dissatisfaction with their useless occupation except to leave nuts and bolts to rattle in the body.

My bias is also pluralistic. Instead of the few national goals of a few decision-makers, I think that there are many goods in many activities of life, and many professions and other interest groups each with its own criteria and goals, that must be taken into account. It is better not to organize too tightly, or there is unnecessary trouble. A society that distributes power widely is superficially conflictful but fundamentally stable.

Research and development ought to be widely decentralized, the national fund for them being distributed through thousands of centers of initiative and decision. This would not be chaotic. We seem to have forgotten that for four hundred years, Western science majestically progressed with no central direction whatever, yet with exquisite international coordination, little duplication, almost nothing getting lost, in constant communication despite slow facilities. The reason was simply that all scientists wanted to get on with the same enterprise of testing the boundaries of knowledge, and they relied on one another.

And it is noteworthy that something similar holds also in invention and innovation, even in recent decades when there has been such a concentration of funding and apparent concentration of opportunity. The majority of big advances have still come from independents, partnerships, and tiny companies (evidence published by the Senate Subcommittee on Antitrust and Monopoly, May 1965). To name a few, jet engines, xerography, automatic transmission, cellophane, air conditioning, quick freeze, antibiotics, and tranquilizers. Big technological teams must have disadvantages that outweigh their advantages—such as lack of single-mindedness, poor communications, awkward scheduling, not to speak of enormous overhead and offices full of idle people or people doing busywork. Naturally, big corporations have taken over the innovations, but the Senate evidence is that 90 percent of the government subsidy has gone for last-stage development for production, which they ought to have paid for out of their own pockets.

In the exploding technology, a remarkable phenomenon has been that enterprising young fellows split off from big firms, form small companies of their own, and succeed mightily. A recent study of such cases along Route 128 shows that the salient characteristic of the independents is that their fathers were independents!

We now have a theory that we have learned to learn, and that we can program technical progress, directed by a central planning board. But this doesn't make it so. The essence of the new still seems to be that nobody has thought of it before, and the ones who get ideas are those in direct contact with the work. *Too precise* a preconception of what is wanted discourages creativity more than it channels it; and bureaucratic memoranda from distant directors don't help. This is especially true when, as at present, so much of the preconception of what is wanted comes from desperate political anxiety in emergencies. Solutions that emerge from such an attitude rarely strike out on new paths, but rather repeat traditional thinking with new gimmicks; they tend to compound the problem. A priceless advantage of widespread decentralization is that it engages more minds, and more mind, instead of a few panicky (or greedy) corporate minds.

A homespun advantage of small groups, according to the Senate testimony, is that coworkers can talk to one another, without schedules, reports, clock-watching, and face-saving.

An important hope in decentralizing science is to develop knowledgeable citizens, and provide not only a bigger pool of scientists and inventors but also a public better able to protect itself and know how to judge the enormous budgets asked for. The safety of the environment is too important to be left to scientists, even ecologists. During the last decades of the nineteenth century and the first decade of the twentieth, the heyday of public faith in the beneficent religion of science and invention, say, from Pasteur and Huxley to Edison and the Wright Brothers, philosophers of science had a vision of a "scientific way of life," one in which people would be objective, respectful of evidence, accurate, free of superstition and taboo, immune to irrational authority, experimental. All would be well, is the impression one gets from Thomas Huxley, if everybody knew the splendid ninth edition of the *Encyclopedia Britannica* with its articles by Darwin and Clerk Maxwell. Veblen put his faith in the modesty and matter-of-factness of Engineers to govern. Louis Sullivan and Frank Lloyd Wright spoke for an austere functionalism and respect for the nature of materials and industrial processes. Patrick Geddes thought that new technology would finally get us out of the horrors of the Industrial Revolution and produce good communities. John Dewey devised a system of education to rear pragmatic and experimental citizens who would be at home in the new technological world, rather than estranged from it. Now fifty years later, we are in the swamp of a scientific and technological environment, and there are more scientists alive, etc., etc. But the mention of the "scientific way of life" seems like black humor.

Many of those who have grown up since 1945 and have never seen any other state of science and technology, assume that rationalism itself is totally evil and dehumanizing. It is probably more significant than we like to think that they go in for astrology and the Book of Changes, as well as inducing psychedelic dreams by technological means. Jacques Ellul, a more philosophic critic than the hippies, tries to show that technology is necessarily overcontrolling, standardizing, and voraciously inclusive, so that there is no place for freedom. But I doubt that any of this is intrinsic to science and technology. The crude history has been, rather, that they have fallen, willingly, under the dominion of money and power. Like Christianity or communism, the scientific way of life has never been tried. And, as in the other two cases, we have gotten the horrors of abusing a good idea, *corruptio optimi pessima*.

Technology assessment the benefits . . . the costs . . . the consequences

JOSEPH COATES

■ Joseph Coates, *formerly Program Manager in the Exploratory Research and Assessment Office of the National Science Foundation in Washington, D.C., currently heads a consulting firm in Washington.*

For Coates, technology assessment takes place inside the realm of technology itself. It involves an inquiry into the secondary and tertiary effects of a technological decision as opposed to its primary or intended effects.

After providing several reasons why we as a society should equip ourselves with such an "early warning system," he sketches some of the important consequences of technology assessment and discusses ways in which assessments can be facilitated.

Technology assessment may be defined as the systematic study of the effects on society that may occur when a technology is introduced, extended, or modified, with special emphasis on the impacts that are unintended, indirect, and delayed.

Ideally, the assessment of a technology should anticipate and evaluate the impacts of a new technology on all sectors of society. So far, however, only a few such full-fledged technology assessments have been conducted. Instead, there is a long history of partial assessments, generally limited to impacts on the economy and, more recently, the environment.

Technology assessment emphasizes the secondary or tertiary effects of new technology rather than the primary (intended) effects, because:

1. In the long run, the unintended and indirect effects may be the most significant;
2. Undesirable secondary consequences often are unnecessary and may be prevented by proper planning;
3. First-order impacts usually are subject to extensive study in the planning stage. In building a bridge, dredging a canal, introducing enzyme detergents, or electrifying a railroad, the first order effects—those intended as the primary goal of the effort—are generally explicitly planned for, and costed out in the initial plan. Technology assessment focuses on the question of what *else* may happen when the technology is introduced.

☐ From *The Futurist*, 5(1971), 225-231. Reprinted by permission of the World Future Society, 4916 St. Elmo Avenue, Washington, D.C. 20014.

WHAT EXACTLY IS TECHNOLOGY?

There is nearly universal agreement that technology includes physical artifacts, such as power plants, telecommunication systems, and airplanes, as well as the physical activities or actions that alter the environment, such as building construction and the siting of oil refineries. But sometimes the definition is expanded to include social tools and techniques. Unquestionably, social technology is a major source of secondary impacts on society. Consider, for example, the far-reaching consequences of such social instruments as:

1. The "pay-as-you-go" (withholding) income tax, proposed by Beardsley Ruml during World War II, and now a generally accepted part of the tax system.
2. The Morrill Land-Grant College Act of 1862.
3. The invention and application of farm subsidies.
4. The county agent system.

Just as a minor alteration in a physical technology may have profound side effects, so a minor alteration in a nation's social technology can have many important consequences undreamed of by the legislators or others who approve the alteration. One curious example of an unintended effect is the impact of the U.S. social security system on the living arrangements of elderly persons. Since social security payments are reduced when two people get married, large numbers of elderly couples live together without benefit of clergy. The retirement communities of Florida are said to rival university communities in the frequency with which this "shacking up" occurs, albeit with a different age group.

Another example of the secondary effects of a social technology may be the alleged flight of whites from U.S. cities. It has been argued that the movement of whites to the suburbs and the concentration of blacks in the core cities resulted in large measure from government housing policies immediately after World War II. Thanks to Veterans Administration and Federal Housing Administration policies, veterans had great economic incentives to buy new homes—but not older houses in the city. This meant that the veterans preferentially chose housing in the suburbs where new homes were being built, abandoning the city to the lower-income blacks who crowded into the older, cheaper houses.

TRENDS MAKE TECHNOLOGY ASSESSMENT NECESSARY

Several trends make it virtually mandatory that society develop new early warning techniques, and better aids to planning and decision-making:

1. The growing complexity of society. Our social institutions have become so interwoven and interdependent that a disorder in a single component may create havoc throughout the society, as in 1965 when a power failure blacked out the northeastern United States, causing a major dislocation in all sectors of life. The interconnectedness of the various elements in our society is also well illustrated by waterways, which are used not only for transportation but for water supplies, recreation, and dumping. A river is often a convenient place to dump industrial wastes, but this practice may kill the fish and even endanger human health.

2. Man's increasing power over nature. For much of recorded history, man has waged a relentless war against nature, but the forces of nature, while challenged, were rarely beatable. One of the rare instances in which man was able to conquer nature—much to his own long-term detriment—was the deforestation of the ancient world. In the time of Homer, the hills of Greece were covered with trees, as were the hills of Sicily and Southern Italy which the Greeks colonized. But the trees were cut down to build ships, cook food, and heat homes, and to make way for crops. The forests disappeared; the domestic animals overgrazed the land; the rich top soil washed away. Today the once lush area is largely barren.

3. Scientific knowledge has improved man's ability to anticipate more reliably the future consequences of his actions. The ability to anticipate makes it possible to explore the future, to anticipate desirable outcomes of human intervention, and to promote those that are good while guarding against those that are bad.

4. The larger scale of human enterprise. Dam building, weather modification, and space exploration are conducted on such an enormous scale that they demand gigantic investments, involve long planning periods, and engender intractable if not irreversible consequences. Thus it becomes essential to take into consideration at the planning stage any incidental or secondary effects these enterprises may have. The cost of correcting an undesirable side effect is often far higher than designing the effect out of the system at the beginning.

5. The throw-away society. As a result of the exponential growth of technological capability, the U.S. economy has moved in the last two generations from a parsimonious, conserving mode to one in which things are manufactured, distributed, utilized and thrown away in stupendous quantities. This is a recent development in human history, and poses new problems. As late as the French revolution most garments bought and sold in Europe were second hand; when finally worn out, they were eagerly snapped up by the rag paper industry.

The throw-away economy fostered by our technical capabilities and prosperity has created huge demands for raw materials on the one hand while creating a stupendous waste disposal problem on the other. Our knowledge of the limitations of natural resources and of the dangers of pollution have stimulated a general interest in managing our technology more wisely.

6. The shift in societal values. One consequence of man's increasing prosperity has been a shift in his values. Where once he was concerned with getting enough food and shelter, he now seeks amenities, such as more beautiful surroundings. Smoke belching from a factory smokestack once symbolized jobs and prosperity, and filled people with gladness; now it means air pollution, and fills people with horror. Well-fed, comfortably housed by technology, and more concerned with leisure and recreation, many people now are outraged by technology's less esthetic side effects.

7. A boost from Congress. Passage of the National Environmental Policy Act of 1969 (public Law 91-190) has enormously stimulated technology assessment. Besides establishing the Council for Environmental Quality, the law has required every U.S. Government agency that is planning a project to file with the Council an assessment of the impact that the project may have on the environment.

In the words of Section 102 of the Act, each agency must

> . . . include in every recommendation or report on proposals for legislation and other major Federal actions significantly affecting the quality of the human environment, a detailed statement by the responsible official on:
>
> (i) the environmental impact of the proposed action.
> (ii) any adverse environmental effects which cannot be avoided should the proposal be implemented.
> (iii) alternatives to the proposed action.
> (iv) the relationship between local short-term use of man's environment and the maintenance and enhancement of long-term productivity, and
> (v) any irreversible and irretrievable commitments of resources which would be involved if the proposed action should be implemented.

Preparation of these environmental impact statements has created a general demand throughout the Federal system for the systematic exploration of the secondary impacts of projects, and a consequent demand for the development of methodologies, techniques, approaches, and protocol. Much of what is occurring as a result of the Environmental Policy Act will benefit and influence assessments in general.

WHAT HAPPENS AFTER A TECHNOLOGY IS ASSESSED?

The assessment of a technological intervention may have a variety of important consequences:

1. The project is modified. Unless an assessment thoroughly condemns a project and demonstrates intrinsic disutilities that far outweigh the advantages, the most likely result is a modification of the project to make it more useful, safer, effective, or aesthetically satisfying. For example, the assessment of the proposed Alaska pipeline has temporarily stopped the project, but it is hard to believe that oil will not eventually be drained from Alaska to meet the needs of the U.S. economy. Thus the long-term consequence of the assessment will probably be a modification of the pipeline (or the substitution of some other transport technology) to make the project less risky to the environment.

2. The technology will come under continuing surveillance. Since assessments explore the unknown and uncertain, they are necessarily conjectural, however much they may be based on solid scientific reasoning. Consequently, one may anticipate that an assessment will often be followed by a program to monitor the consequences of the technological intervention.

3. Research and development are stimulated. An assessment often will stimulate research and development in many ways:

a) Research may be indicated to define more reliably the risks indicated by the assessment. For example, if an assessment suggests that a power plant will raise the temperature of a river, careful research may be needed to determine exactly how much a given temperature rise will affect the biota in that river.

b) The assessment may stimulate research to forestall the undesirable second-order effects that are foreseen. In the case of a power plant, research might be done to determine how to get rid of heat in a safe, and perhaps even profitable manner.

c) An assessment may lead to research aimed at finding alternative methods of meeting the need that would be served by the proposed technology. If an insecticide appears likely to harm livestock or human beings, entomologists might be enlisted to find predators that could bring the insects under control.

d) Research may be needed to find corrective measures to deal with adverse effects. For at least 200 years it has been known that the unchecked growth of population can lead to disaster, and it was also fully predictable that the expansion of world-wide public health measures (assuming there were no reduction in the birth rate) would yield a great population expansion and a substantial pressure against resources. Unfortunately, it is only in the last five years that well-funded research has been undertaken in an effort to treat population growth as a public health problem. Hopefully, more incisive technology assessments will speed up the needed research in similar instances in the future.

4. Controls may be established. Regulations, taxes, prohibitions and other controls may be instituted when an assessment indicates that a particular technology would have undesirable effects. At present, there is a growing literature on the possibly adverse effects of such new technologies as genetic engineering and electrophysiological control of behavior. (See "Physico-chemical Control of the Mind" by Henry Clark in the August 1971 issue of *THE FUTURIST*.) As these assessments become more refined and the realities of the technology become clearer, government may restrict or limit some applications of these techniques.

5. Technology may be encouraged to move into new areas. An assessment may suggest that a proposed technology promises major potential benefits to society, and should therefore be applied in many ways that were not originally envisioned. Transparent adhesive tape was developed originally as a method of repairing books; only later did its myriad uses become clear. An assessment may suggest that the technology be applied experimentally in given areas before it is applied throughout the society. In the socio-technological area, American society is experimenting with various methods of supplementing the income of poor people, in an effort to find a way that will not only meet their immediate needs for food and shelter, but will give them adequate incentive to get out of the poverty cycle.

6. New laws may be required. An assessment may suggest the need for legislation or may clarify the impact of proposed legislation. Some observers concerned about genetic engineering and new communications technology have suggested that a new Bill of Rights may be necessary to protect individual liberties, much as the present Bill of Rights, adopted in the 18th century, brought the law into harmony with such technologies as the printing press and firearms.

7. A technology may be blocked. One of the fears associated with technology assessment is that it may become "technology harassment" or even "technology arrestment." Already the automobile and electric power industries, mine owners, farmers, and cattlemen face demands—often backed by law—that they meet new standards, especially as regards their impact on the environment. The risk of arrestment is perhaps greatest in those new technologies where flamboyantly expressed fears seem to aim at outright bans or stringent curbs.

Hopefully, technology assessment will generate a more even-handed regulation of technology, and thus will forestall the over-reaction and polarized positions that characterize many current controversies.

THREE VIEWS OF TECHNOLOGY ASSESSMENT

Although there is general agreement that technology assessment should be an aid to decision-making, there are three distinct views concerning how it should function.

1. An advocate's tool. One view is that technology assessment is a means of supporting whatever position an advocate chooses to take. Those who hold this position tend to see the courts and regulatory agencies as taking the lead in managing technology. Since the law operates on an adversary basis, technology assessment should be structured to fit that pattern.

2. A neutral analysis. Technology assessment may also be viewed as a neutral application of scientific analysis to future outcomes and alternatives. It should be exhaustive, comprehensive, and balanced, as it explores alternative outcomes and weighs their relative probabilities and effects. Once an assessment has been made, it becomes raw material for advocates of any predisposition to consider and use.

3. A search for desirable choices. The third view goes a step beyond the even-handed analysis of consequences, and maintains that technology assessment should highlight various desirable policy options, although the back-up analysis should be available so that the reasons for supporting a given option should be clear.

I believe that the legitimate role for assessment is in the second or third category. Certainly when an assessment is conducted with public money, it should provide a balanced look at all alternatives, options, and possible outcomes. Some policy alternatives may be highlighted for decision-making purposes, but the analysis and the means of arriving at the options should not be omitted. If an assessment is not even-handed, it becomes merely another weapon in a rancorous dispute. One of the major difficulties with many informal assessments made by public interest groups is that they do not represent conclusions drawn from an assessment of consequences, but reflect rather an attitude of "I've made up my mind; tell me why I'm right." If more balanced assessments come to be the rule, one may anticipate a rise in their value in the formulation of wise social policies.

HOW TECHNOLOGY ASSESSMENTS ARE MADE

There is as yet no general method or technique for conducting a technology assessment, and it is not likely that a single technique will ever dominate the field, partly because any given assessment is likely to be strongly influenced by the technology under investigation. However, every assessment should meet certain criteria:

1. It should consider all aspects of the technology and identify certain areas for intensive examination.
2. It should not mirror the prejudices and preconceptions of the investigators, the sponsoring institution, or the ultimate users. (This is a particularly difficult condition to meet.)
3. The method used in making the assessment should put a premium on imagination, understanding, the ability to anticipate and visualize future consequences, and the ability to step beyond preconceptions and self-interest.

The nature of the technology to be assessed will strongly influence the methodology to be used in the assessment. For example, an assessment of snow enhancement—the use of techniques to increase snowfall and consequently the runoff into rivers—might start by tracing two physical systems: the meteorologic system (the movement of air masses) and the hydrologic system (the movement of waters in lakes, rivers, etc.), and then look to see how these physical systems might impact on vegetation, animal life, human activities, etc.

Special techniques for locating possible points of impact are now used. For instance, the U.S. Geological Survey's Circular 645 presents a *General Procedure for Evaluating Environmental Impacts,* which includes an 88-by-100-square matrix of actions and impacts. An analogous but more general approach has been developed by George Washington University's Program of Policy Studies in Science and Technology, directed by Louis H. Mayo. These methods systematically suggest likely points of impact but do not indicate that an impact will in fact occur or that it will be significant. Supplementary techniques are required to make that determination.

Other methods include the use of expert panels, the Delphi technique, and cross-impact matrices. These three methods depend ultimately on individual judgments. In some assessments, modeling and simulation techniques have proved useful. Economic tools are, of course, very powerful and central to most analyses.

ASSESSMENTS VARY WIDELY IN COST

A full-fledged technology assessment, not involving experimental work or the generation of new data, costs between $150,000 and $250,000, or the equivalent of four-to-six senior professional man-years of effort, according to my estimates. Lower-cost assessments may be conducted on smaller projects with more definable impacts or in preliminary assessments designed to prepare for a larger effort. On the other hand, the costs may run 10 times higher for a complex technology involving the detailed consequence of various technical alternatives, such as the

THE EFFECTS OF TECHNOLOGY

The intended effect of a technology is rarely, if ever, the only impact it has on human life. Unintended, unknown, and delayed consequences may prove even more important in the long run than the direct and intended effects. At times, different technologies can have similar unintended consequences that combine to have a serious impact undreamed of by the users of the technology. The following table suggests how the automobile, television, and refrigerators may have helped to break down community life.

Assuming that these technologies are indeed increasing the isolation of the nuclear family, one may wish for ways to block this effect, or soften its impact. Perhaps new technology may be called into play to help people in a community to become better acquainted with each other and to interact more, thereby lessening their isolation, increasing community spirit, and decreasing the divorce rate.

	First-order consequences	Second-order consequences	Third-order consequences	Fourth-order consequences	Fifth-order consequences	Sixth-order consequences
Automobile	People have a means of traving rapidly, easily, cheaply, privately, door-to-door.	People patronize stores at greater distances from their homes. These are generally bigger stores that have large clienteles.	Residents of a community do not meet so often and therefore do not know each other so well.	Strangers to each other, community members find it difficult to unite to deal with common problems. Individuals find themselves increasingly isolated and alienated from their neighbors.	Isolated from their neighbors, members of a family depend more on each other for satisfaction of most of their psychological needs.	When spouses are unable to meet heavy psychological demands that each makes on the other, frustration occurs. This may lead to divorce.
Improved refrigeration	Food can be kept for longer periods in the home.	People stay home more, because they don't need to go to stores.	Same as above; also, more free time for wife.	Same as above. (Also, additional free time increases demand for recreation and entertainment.)	Same as above.	Same as above.
Television	People have a new source of entertainment and enlightenment in their homes.	People stay home more, rather than going out to local clubs and bars where they would meet their fellows.	Same as above. (Also, people become less dependent on other people for entertainment.)	Same as above.	Same as above.	Same as above.

laying of a pipeline or the establishment of a communications system.

Here are the costs of some recent assessments:

Assessment of a proposal to extend the Kennedy Jetport into Jamaica Bay. (Conducted by the National Academy of Sciences-National Academy of Engineering for the Port of New York Authority).	$300,000 (approx.)
Assessment of a Snow Enhancement Project for the Colorado River Valley. Conducted by the Stanford Research Institute for the National Science Foundation. (Note: Figure does not include a substantial contribution from the Institute itself).	$150,000
A Civil Aviation Research and Development Policy Study. Conducted by George Washington University for the U.S. Department of Transportation and the National Aeronautics and Space Administration.	$220,000

The scope of these projects is dwarfed by the efforts to determine the impacts of the proposed Alaska pipeline, in order to fulfill the requirements of the Environmental Policy Act. This assessment has already cost an estimated $7 to $8 million, not to mention the millions lost due to the delay of construction.

FIELDING AN EFFECTIVE TECHNOLOGY ASSESSMENT TEAM

A technology assessment almost always requires a team effort. The team must include a wide range of talents, representing experience in such fields as engineering, chemistry, physics, biology, etc. Furthermore, the team must have an interdisciplinary capability, that is, an ability to work on problems for which there is no well-defined body of knowledge. Meeting the *inter*disciplinary requirement is far more difficult than meeting the *multi*disciplinary requirement, because it is easier to identify and assemble specialists than it is to identify and assemble people who can work effectively in the areas not covered by specialists.

To obtain an assessment that is comprehensive, balanced, imaginative, and broad in scope, the team leader must be a man of broad intellectual capability who can move comfortably into new areas, work with divergent talents, and make different personalities effectively interrelate with each other.

If the assessment is to be useful to decision-makers, it must be timely, balanced, and independent. The assessment must be available at the time the decision is made. It should consider all the possible impacts, both positive and negative. And it should not be influenced by the biases of the investigators or their sponsors.

The team must take adequate account of legal, economic, political, and social effects. For this reason, most assessment teams should include lawyers and social scientists.

ASSESSMENT AND ORGANIZATIONS

What organizations should sponsor technology assessments, and what organizations should carry them out?

Any Government agency may need assessments. Industry may also find it desirable to fund assessments either to determine its own options or to anticipate public reaction and legal constraints that it may encounter. Both government and industry have the task of establishing an atmosphere of independence for the assessment and of assuring those who are to be influenced by the assessment that it is in fact independent. For this reason, many agencies turn to outside groups in order to achieve the desired degree of independence. The outside organization may be a university, a nonprofit research institute, or a commercial research organization. Farming out the assessment has the additional advantage of enabling the sponsor to see a variety of candidates display their approaches and talent.

Universities have certain advantages over research institutes when it comes to making assessments—and some disadvantages. Generally speaking, a university has a broad base of intellectual talent and a depth of knowledge and expertise, but typically a university will lack the management capability to marshal an effective, integrated team that can deliver a timely, comprehensive, balanced assessment. The university reward system generally favors individual performance (especially publication in scholarly journals) rather than policy-oriented team efforts. In many universities, however, the leadership needed to field effective teams does exist, and in other cases a university may have centers that can provide the organization, leadership, and environment for a focused study.

Research institutes are usually experienced in comprehensive, systematic approaches, and this gives them an important advantage in conducting assessments. Furthermore, they are prac-

II PROPOSED ACTIONS WHICH MAY CAUSE ENVIRONMENTAL IMPACT

INSTRUCTIONS

1. Identify all actions (located across the top of the matrix) that are part of the proposed project.
2. Under each of the proposed actions, place a slash at the intersection with each item on the side of the matrix if an impact is possible.
3. Having completed the matrix, in the upper left-hand corner of each box with a slash, place a number from 1 to 10 which indicates the MAGNITUDE of the possible impact: 10 represents the greatest magnitude of impact and 1, the least (no zeroes). Before each number place + if the impact would be beneficial. In the lower right-hand corner of the box place a number from 1 to 10 which indicates the IMPORTANCE of the possible impact (e.g., regional vs. local) 10 represents the greatest importance and 1, the least (no zeroes).
4. The text which accompanies the matrix should be a discussion of the significant impacts, those columns and rows with large numbers of boxes marked and individual boxes with the larger numbers.

SAMPLE MATRIX

	a	b	c	d	e
a	2/1	/			8/5
b		7/2	8/8	3/1	9/7

A. MODIFICATION OF REGIME

PROPOSED ACTIONS	a. Exotic flora or fauna introduction	b. Biological controls	c. Modification of habitat	d. Alteration of ground cover	e. Alteration of ground water hydrology	f. Alteration of drainage	g. River control and flow modification	h. Canalization	i. Irrigation	j. Weather modification	k. Burning	l. Surface or paving	m. Noise and vibration
ARACTERISTICS I. EARTH a. Mineral resources													
b. Construction material													
c. Soils													
d. Land form													
e. Force fields and background radiation													
f. Unique physical features													
TER a. Surface													
b. Ocean													
c. Underground													

B. LAND TRANSFORMATION AND CONSTRUCTION

PROPOSED ACTIONS	a. Urbanization	b. Industrial sites and buildings	c. Airports	d. Highways and bridges	e. Roads and trails	f. Railroads	g. Cables and lifts	h. Transmission lines, pipelines and corridors	i. Barriers including fencing	j. Channel dredging and straightening	k. Channel revetments	l. Canals	m. Dams and impoundments	n. Piers, seawalls, marinas, and sea terminals	o. Offshore structures	p. Recreational structures	q. Blasting and drilling	r. Cut and fill	s. Tunnels and underground structures
ARACTERISTICS I. EARTH a. Mineral resources																			
b. Construction material																			
c. Soils																			
d. Land form																			
e. Force fields and background radiation																			
f. Unique physical features																			
TER a. Surface																			
b. Ocean																			
c. Underground																			

C. RESOURCE EXTRACTION

PROPOSED ACTIONS	a. Blasting and drilling	b. Surface excavation	c. Subsurface excavation and retorting	d. Well drilling and fluid removal	e. Dredging	f. Clear cutting and other lumbering	g. Commercial fishing and hunting
ARACTERISTICS I. EARTH a. Mineral resources							
b. Construction material							
c. Soils							
d. Land form							
e. Force fields and background radiation							
f. Unique physical features							
TER a. Surface							
b. Ocean							
c. Underground							

D. PROCESSING

PROPOSED ACTIONS	a. Farming	b. Ranching and grazing	c. Feed lots	d. Dairying	e. Energy generation	f. Mineral processing	g. Metallurgical industry	h. Chemical industry	i. Textile industry	j. Automobile and aircraft	k. Oil refining	l. Food	m. Lumbering	n. Pulp and paper	o. Product storage
ARACTERISTICS I. EARTH a. Mineral resources															
b. Construction material															
c. Soils															
d. Land form															
e. Force fields and background radiation															
f. Unique physical features															
TER a. Surface															
b. Ocean															
c. Underground															

E. LAND ALTERATION / F. RESO RENE

PROPOSED ACTIONS	E. a. Erosion control and terracing	E. b. Mine sealing and waste control	E. c. Strip mining rehabilitation	E. d. Landscaping	E. e. Harbor dredging	E. f. Marsh fill and drainage	F. a. Reforestation	F. b. Wildlife stocking and management	F. c. Ground water recharge
ARACTERISTICS I. EARTH a. Mineral resources									
b. Construction material									
c. Soils									
d. Land form									
e. Force fields and background radiation									
f. Unique physical features									
TER a. Surface									
b. Ocean									
c. Underground									

The complexities of an assessment are suggested by an 8,800-square matrix prepared by the U.S. Geological Survey for organizations evaluating a project's impact on the environment. (Only the upper left hand corner of the 28-by-30 inch table is shown here.) The chart allows a person making an assessment to indicate how such actions as reforestation or weed control may affect birds, monuments, climate, camping, and other aspects of the environment.

ticed in the art of delivering usable outputs in timely fashion. Though they lack the relatively superior intellectual capabilities of the university, they may call upon consultants to augment their efforts.

Joint ventures involving both universities and research institutes may offer one of the best approaches to utilizing the systems and management capabilities of one and the intellectual inputs of the other.

Public interest groups also engage in making assessments, though these are generally extremely informal and meagerly funded. Often the public interest group operates largely on donated information and volunteer services. A major goal in the technology assessment field should be to increase the professional standards and even-handedness of assessments by public interest and make them more informed participants in the public policy process.

Government agencies (particularly under the pressure to meet the requirements of the Environmental Protection Act) are generating in-house capabilities for making some assessments, as well as more sophistication in farming these out to external organizations.

Industry conducts some assessments, but most industry assessments are not full-fledged efforts, and really are not much removed from market surveys. However, industry is very close-mouthed about its assessments, so relatively little is known about their quality, methods, and usefulness.

TEACHING TECHNOLOGY ASSESSMENT

A growing number of universities now offer courses in technology assessment. Dean Harvey Brooks of the Harvard School of Engineering, who led a National Academy of Engineering study of technology assessment, is now teaching a course in the subject. So too are physicist Raymond Bowers at Cornell, electrical engineer J. P. Ruina at Massachusetts Institute of Technology, and physicist David Goldman at the University of Maryland. There are also courses at the University of California at Berkeley and York University in Toronto.

UNSETTLED ISSUES IN TECHNOLOGY ASSESSMENT

As the methodology of technology assessment expands, a number of unresolved issues are coming into sharper focus.

1. The importance of independence. Central to any assessment is the independence of the team from the preconceptions and interests, real or imagined, of the sponsor and the participants. The U.S. Atomic Energy Commission has been responsible both for developing nuclear power facilities and for assessing any risks incurred; many observers have felt that it could not under those circumstances make an unbiased assessment.

2. Involvement of professionals. The professional societies of America generally have been slow to provide mechanisms for making their skills available for assessment of technology. They have given only lip service to the demands for professional responsibility in meeting society's needs.

3. Assessment by public interest groups. The thousands of public interest groups in America frequently behave as if they were operating merely as advocates. A major need is to encourage voluntary organizations to become involved in—if not actually to sponsor—professional, balanced, even-handed assessments of the issues with which they are concerned, to base their policy positions on those assessments and to use the assessments when they try to influence the decision process.

4. Benefits as well as dangers need to be revealed. Assessment has grown up under the need to deal with problems posed by new technology. But a full-fledged assessment should reveal positive as well as negative aspects of a technology. We need to encourage an awareness of future benefits as well as future dangers and disutilities that may arise from a new technology.

There is a feeling in some quarters that any risk posed by a new technology should make it ineligible for consideration; the potential benefits are ignored. This unbalanced attitude needs to be corrected.

In the future we shall have increasingly sophisticated technology. To manage it successfully, we will need to know more about the full range of its impacts on our life. We will then be able to choose more wisely the right kind of technology for each job that needs doing. During the years ahead, government and industry will become increasingly aware that assessing technology is just as important as inventing technology. Thus technology assessment will probably become an increasingly important field in the years ahead.

Assessing technology assessment

DAVID M. KIEFER

■ David M. Kiefer *is Senior Editor of* Chemical and Engineering News.

In this assessment of technology assessment, Mr. Kiefer argues that we must find new means of deciding which technologies should be used. Traditional indicators, such as the profit motive and the mechanism of the market, have failed to take into account "side effects, by-products, spillovers and trade-offs" among the different forms of technology that compete to perform a given task. He criticizes the "revolving door" between industry and its regulatory agencies, and calls for "corporate Cassandras, or "long-range debunking groups" within the overall corporate structure.

About eight years ago, in speaking to a gathering of distinguished scientists, President Kennedy commented: "Everytime you scientists make a major invention, we politicians have to invent a new institution to cope with it." That pretty well sums up the idea behind technology assessment.

In theory at least, technology assessment is easy to understand and difficult to fault. It is a reasoned response to the stress that a rapidly changing and expanding technology puts on our complex and increasingly industrialized, urbanized, and densely populated society. It attempts to make the process of coping with technological development more systematic and rational.

Technology assessment can be viewed as a mixture of early warning signals and visions of opportunity. Or as a device for protecting man from his own technological creativity. Or as a formal mechanism for allocating scientific resources, setting technological priorities, and seeking more benign alternatives for technologies already in use. Or as an attempt to control and direct emerging technologies so as to maximize the public benefits while minimizing public risks.

What could appear less controversial?

No wonder, then, that the idea has caught on or that it is winning over more and more people, especially on Capitol Hill and within some of the federal agencies, in the academic world, or from public interest groups, who are anxious to try to put it to practical work—and the sooner the better. They are convinced that if we don't try it, not only will many of our present problems become deeper but that we will be faced with an expanding array of newly emerging problems which will only make those of today pale in comparison.

Just what would be the best way to implement technology assessment is not quite clear, of course. While we are swept up in the onrush of technological change, we don't really know yet how to foresee its effects upon society. Nor, for that matter, do we really know how to evaluate those effects in any meaningful or objective manner. We might all agree that we must somehow learn to inhibit or reject uses of technology that are harmful or detrimental. But we are unable to identify those uses or measure their hazards in many cases before the technology is actually used. We find it difficult to sort out costs and benefits clearly so that we may encourage the good and hold back or modify the bad. Lacking standards of social progress and lacking established, widely accepted national priorities and goals, we tend to drift from crisis to crisis on a sea of change?

WHAT'S NEW ABOUT TECHNOLOGY ASSESSMENT?

There is much about the idea of assessment that may seem old hat, at least when it is interpreted in a broad context. After all, businessmen have long had to assess laboratory projects and the development and commercialization of new products and processes in one way or another. That is what research management, investment analysis, market research, commercial development, and long-range planning are all about.

The Federal Government, too, has been interested in the consequences of new technology for more than a century. Many agencies in Washington now have the assessment of technology as a large part of their mission: the Food and Drug Administration, the Federal Communications Commission, the Federal Aviation

☐ From *The Futurist*, 5(1971), 234-239. Reprinted by permission of the World Future Society, 4916 St. Elmo Avenue, Washington, D.C. 20014.

Authority, the Atomic Energy Commission are but a few of the institutions that politicians have set up to cope with technology.

But until now, with rare exception, whatever assessments have been made have really only been half-assessments (or less), done on a trial-and-error, hit-or-miss, ad hoc basis. In business, at least, they have mostly been limited to such questions as: Is a new development technologically feasible? Will it be economically profitable?

Technology assessment, in the sense now coming into vogue, would focus on more than just the direct or primary effects which traditionally are the ones that have been subjected to intensive study because they are the direct objectives of the technology.

Technology assessment would scuritinize (in addition to the primary effects) the interactions, side effects, by-products, spillovers, and trade-offs among several developing technologies or between a new technology and society at large and the environment. It is the emphasis on adding indirect or second- and higher-order effects and social impacts to the cost-benefit equation that is the novel aspect of technology assessment. These second-order effects, in the long run, may affect society more deeply than the intended primary effects. Yet if they are unwanted, they can be controlled or removed more easily if they are identified early in the development process.

What it gets down to is that the profit motive and the traditional market mechanism that we have long relied upon no longer appear to be sufficient for the task of sorting out what technology should be used—and how. The first-order effects of damming a river, or launching a supersonic transport, or introducing a new detergent additive or pesticide, or building a highway may stand out clearly after the customary cost-benefit analysis. Technology assessment would try to get at what else might happen but be overlooked, whether it be beneficial or harmful.

MANY GROUPS CLAIM TO BE DOING TECHNOLOGY ASSESSMENT

But technology assessment is a concept that is still groping for scope and definition. And since it does have an appealing sound of relevance about it, at a time when to be relevant is to be on the side of the angels, many people seem eager to appropriate the term for their own purposes, or to apply it as a way to add new luster to whatever they may have been already doing in the past.

Thus businessmen, if they have paid the idea any heed at all, may view it as merely a more sophisticated term for market research and long-range planning. Futures researchers may see it as only an extension of technological forecasting or systems analysis. Environmentalists can envision it as a new way to apply leverage against polluters. Social activists think they may have a new tool with which they can chip away at the establishment's power structure. Antitechnology crusaders look on it as a weapon to combat big science. And it is difficult to escape the feeling that some people in government and the academic world think they may have uncovered a new means for bureaucratic empire building or opened up a new channel for research grants. Obviously there is something in technology assessment for everyone.

The term *technology assessment* itself was coined about six years ago by the staff of the Subcommittee on Science, Research, and Development of the U.S. House of Representatives. The subcommittee chairman, Emilio Daddario, then a Congressman from Connecticut, provided much of the initial thrust behind the plans to set up a formal assessment body within the Government.

"Our goal," Daddario stated, "is a capability for policy determination in applied science and technology which will be anticipatory and adaptive rather than reactionary and symptomatic."

Daddario envisioned an assessment mechanism that "would identify all impacts of a program; establish cause and effect relationships, were possible; determine alternative programs to achieve the same goal and point out the impacts; measure and compare sums of good and bad impacts; and present findings from the analysis. In the initial step, one would place the technology within the total social framework and identify all impacts in the natural, social, economic, legal, and political sectors. Direct effects would be separated from derivative effects."

Now this is a very large order indeed. A full-fledged technology assessment would have to seek answers for some very searching questions:

How will an innovation be used, not just today but in the future?

What will be the consequences of those uses, direct and indirect, for good or bad, on any part of society or the environment?

What responses or interactions or cross-impacts can be expected from other areas of science and technology?

How do the tonic effects balance out against the toxic effects?

Are the effects irreversible either in the short run or the long run—or are we painting ourselves into a corner once we have introduced the new technology?

What are our options? Could the benefits of a new development be achieved at less cost or less risk by some alternative?

It is possible, of course, that the answers for many questions of this type are already in existence and remain only to be uncovered by a diligent search of the pertinent scientific and technical literature, reinforced perhaps by a limited amount of original, independent laboratory or field research and testing. This might be the case, especially, when the assessment is restricted to a relatively narrow innovation or to a field that is relatively stable and well-documented. Until assessment really becomes more widely accepted and well grounded in experience, therefore, much of the work may actually be a glorified literature search that would carefully document data currently available and evaluate it as to relevancy, authenticity, acceptability, etc. The intent would not be just to supply a greater amount of information to policy-makers, but rather to improve the quality, pertinence, and completeness of the information on which the policy-makers must base their decisions. Such a painstaking analysis of existing information sources would be adequate for spelling out incipient dangers and drawbacks posed by many new technological developments, particuarly those that represent relatively small advances from preceding technology.

Even a more wide-ranging technology assessment must be geared to a comprehensive data bank of available current information. For one thing, it is necessary to develop a monitoring system able to identify emerging technologies as early as possible and to trigger full-fledged technology assessments. In addition some sort of ongoing monitoring system will be needed to track the effects of developing technologies upon society.

ASSESSMENT CHALLENGES FORECASTERS

But it is nevertheless clear that if technology assessment is to probe the future in order to fullfill many of the broader objectives that its proponents have in mind, then it demands more than just a documentation of existing information dealing with present technology. Assessment must also suggest where emerging technological developments or long-range technological trends are likely to lead.

Hence the forecasting of technology is essential, both to uncover potential problems that otherwise would not be foreseen and to disclose unappreciated opportunities.

But can technological forecasting meet this challenge? It too is a discipline—or better yet, an art form—that is still in its formative years. Its antecedents lie in the books of the science-fiction writers and the speculations of science popularizers during the first half of the 20th century. The development of technological forecasting as a formal endeavor stems largely from the need, following World War II, for organizing and planning huge military and aerospace research programs. Only within the past decade has technological forecasting come into its own.

Some very interesting techniques have been developed to forecast technological change more effectively (Delphi studies, cross-impact analysis, relevance trees, scenario writing, envelope curves, etc.). But by and large these techniques produce results that must still be accepted with a considerable degree of faith, for their reliability remains largely untested. One can cite examples of uncanny predictions made in the past by Jules Verne, H. G. Wells, and others, but the overall record for foretelling technological change is spotty.

In 1936, for example, a well-known American educator and engineer estimated that the ultimate speed of airplanes "might well approach 500 miles per hour." In 1939, an admiral of the U.S. Navy declared that "as far as sinking a ship with a bomb is concerned, you just can't do it." In 1945, Vannevar Bush, then president of the Carnegie Institution, discounted the intercontinental ballistic missile, "In my opinion, such a thing is impossible for many years." Then there was the Astronomer Royal who said, in 1956, "Space travel is utter bilge." (See "Blunders of Negative Forecasting" by Joseph Martino in *THE FUTURIST*, December 1968.)

Who foresaw how the development of the transistor would spur widespread use of computers or influence the economic development of several Far Eastern countries? Who foresaw the impact of DDT on birdlife? Who, at the turn of the century, envisioned that the automobile would become the major source of urban air pollution and that it would dramatically alter the way in which people spend their spare time, where they live, the location of retailing activities, and even the puberty and fertility rites of youth? A quotation from a turn-of-the-century issue of *Scientific American* is pertinent:

> The improvement in city conditions by the general adoption of the motor car can hardly be overestimated. Streets clean, dustless, and odorless, with light rubber-tired vehicles moving swiftly and noiselessly over their smooth expanse, would eliminate a greater part of the nervousness, distraction, and strain of modern metropolitan life.

There's a clear-sighted vision of the future!

MEASURE OF "GROSS NATIONAL HAPPINESS" IS NEEDED

While technological forecasting is an uncertain link in the technology assessment process, an even greater lack is the absence of effective means for evaluating the manner in which technology interacts with society. What is needed is some meaningful measure of social change and some index by which to gage social well-being—a gross national happiness index, as it were. But there is at present no good set of social indexes or indicators by which we can measure the quality of life in a way similar to economic indexes and indicators that measure our economic growth.

Even if we could measure social change, we have no real standards by which we can judge our progress or set our course. Until we have formulated national goals and priorities that can be well defined and widely agreed upon, we will be incapable for the most part of assessing technology in a definitive way and in terms of what society should do about it.

WHO IS TO SET GOALS FOR SOCIETY?

The goal-setting process may be particularly frustrating in a democracy. Who is to do it? Political parties? Bureaucrats? Pressure groups? Some undefined elite?

And can we impose our ambitions, values, and desires on generations still unborn? How would goals formulated in the days of Queen Victoria meet the needs and problems we face today?

On the other hand, if we don't somehow fashion objectives to which we can aspire, are we not condemning our children to a future world that is nothing more than a continuation of the present, with all its inadequacies? In a world that is increasingly crowded and technologically complex, present shortcomings are likely to be magnified severalfold, perhaps to the point where they become irreversible and intolerable. *Do we really have any choice but to make choices, even though we can perceive only dimly—if at all—all the eventual consequences of those choices?*

Technology assessment offers us the hope of making technology our servant rather than our master. But who is to do technology assessment, and how and where will it be done?

Let's take a brief look at what is going on now.

Two bills have been introduced in Congress to set up agencies specifically charged with technology assessment. One of these identical with a proposal introduced by former Congressman Doddario last year, would establish an Office of Technology Assessment as an arm of Congress to conduct technology assessments with assistance from the General Accounting Office, the Congressional Research Service (a branch of the Library of Congress), and the National Science Foundation. This proposal has already been approved by the House Committee on Science and Astronautics. Identical legislation has been introduced in the Senate.

A second bill, introduced by Senators Warren Magnuson (Washington) and Philip Hart (Michigan), would establish an independent Technology Assessment Commission. Just how this commission would function is still not very clear, but it would be an agency separate from all present branches of government. Some people have viewed it as a fourth branch of government, but one with rather nebulous powers.

No strong opposition to technology assessment has surfaced in Congress, but there is no apparent sense of urgency to assure quick passage of pending legislation. Nevertheless, interest on Capitol Hill has sparked activity in other government agencies. The Office of Science and Technology, for example, has sponsored five pilot studies in technology assessment at Mitre Corp., completed this summer and aimed at demonstrating the feasibility of assessment methods. The National Science Foundation, through its RANN program, has funded to the tune of about $2.5 million, about three dozen grants with a technology assessment slant. Among technologies being studied under NSF grants are off-shore oil drilling, nuclear control, solid waste management, and the Big Sky recreational development. The first assessment to be completed deals with the effects of seeding clouds to increase snow fall, and hence runoff, in the upper Colorado River basin.

The National Environmental Policy Act of 1969 also has been spurring federal assessment activities. The law requires that all federal agencies assess the impact of technological programs for which they are responsible on the environment.

Despite all the talk and some evidence of activity, multidisciplinary, full impact technology assessments are hard to find anywhere, either within government or outside. What evidence is available suggests that except for traditional cost-benefit investment and economic studies, technology assessment is still receiving more lip service than implementation. And where they have been undertaken, assessments have centered on pollution control or other environmental matters, land use, and the like. The quality of such assessments varies tremendously. Many have been superficial, based out-of-hand on available engineering data, and apparently intended more to promote a given program than to appraise it in an even-handed way.

It must be stressed that while an evaluation of the role of technology in the degradation of the environment is a very important part of any overall assessment, it is all too easy but wrong to view technology assessment as just another means for controlling pollution. Certainly there is a strong overlap between assessment and environmental protection. But they are by no means identical. Technology assessment demands a more comprehensive approach. If it were to concentrate on environmental problems alone, many by-products of innovation which could be even more hazardous or undesirable would be overlooked.

What it boils down to is that no present organization may be really capable of doing the job. Government agencies, private industry, professional societies, the universities, the non-profit institutions and think tanks—all have inherent limitations—some as a result of the way in which they are organized, others because of limitations of perspective. Wherever you look, there are questions of bias, self-interest conflict of purpose, competitive pressures, limited objectives, inadequate incentives, narrow or tunnel vision, and insufficient power.

Can the Federal Aviation Authority, with its mission of improving aviation, objectively judge the economic and social impact of an SST? Federal regulatory agencies all too often have tended to be captured by the very groups they were set up to control.

Can the assessment of the ecological effects of a new pesticide be made in an impartial manner by manufacturers of agricultural chemicals? Or the allocation of national resources for research and development be left to a group made up solely of scientists and engineers?

People working on a project inherently are biased toward moving that project forward. Even if they view their own interests in a totally neutral, balanced, impartial fashion, they may still fail to see how their component fits into the overall system.

There is also the question of credibility. No matter how even-handed an assessment, it is likely to be suspect if the assessor is presumed to have a self-serving interest in the outcome. Most organizations, whether in private industry or the government, have been set up to promote rather than regulate a given technology.

On the other hand, technology assessment is not a job for dilettantes or do-gooders. It demands expertise and intellectual discipline rather than the superficiality which has been the hallmark of most social interest groups. Where are such experts to come from? It seems unlikely, certainly, that any single body within the Government could muster within itself a staff with talent ranging across the full spectrum of scientific and sociological disciplines that a solidly grounded assessment would seem to require. Because assessment is clearly a broadly interdisciplinary function, it is likely to remain an *ad hoc* function as well. Much of the needed expertise obviously resides within private industry, although how to bring it into focus without raising serious conflict-of-interest questions is still unclear.

Of course, the task of assessment might seem to fall to the universities, with their pool of presumably disinterested scholars. But few, if any, universities today are really organized or have the managerial capabilities to do the job. The interdisciplinary, mission-oriented applied research that is intrinsic to technology assessment is quite foreign to the universities' traditional structure built along relatively rigid disciplinary lines.

There is another question of the "who" type. How many engineers and scientists capable of working on assessments are also interested in such assignments, especially considering how thankless they are likely to be. The task of critical evaluation, with all its negative aspects, is likely to seem less creative and less intellectually stimulating then the laboratory research and process design that most scientists and engineers have been traditionally trained to do.

Ideally, what seems to be needed is an infusion of the spirit of technology assessment into the total fabric of our thinking and way of life.

CORPORATE CASSANDRAS ARE NEEDED

Certainly industry needs a technology stance of its own if only for self-preservation so that it may respond to pressures either from the government or from private interest groups. *Maybe what it needs is a counter commercial development staff or long-range debunking group within the overall corporate structure. Such a group would assess business objectives and priorities not merely in the customary terms of short-range profits and sales growth but also in terms of social responsibility and consequences.* Clearly, this will be no easy function to fill. It will demand the asking of embarrassing questions and throwing up of road-blocks before pet projects. The role of corporate Cassandra will hardly help anyone to win friends in the executive suite. But industry may soon find that it cannot afford, either from an economic or a public relations standpoint, not to have a few Cassandras on the payroll, lest it be clobbered by outside pressures.

Much of the responsibility for making assessments and putting them to use in controlling technological progress must, of course, rest with government. The function of government, after all, is to set ground rules and establish priorities within which private groups may operate.

Businessmen's fears of technology assessment are not unreasonable. By adding new uncertainties to the research and development equation that is already strewn with risks and ambiguities, assessment could well discourage private investment and undercut innovation. By adding new costs and delays in an increasingly competitive world, assessment could well weaken our ability to meet challenges from overseas. It certainly will not be easy to force businessmen to account for all the indirect consequences and spillovers that they long have been accustomed to ignore or pass on to the public at large.

And before we become too enthralled with the idea of assessing the full impact of indirect effects of technology, we would do well to take a close look at the possible first- and second-order effects of the assessment process itself on the fragile and poorly understood process of innovation.

It would be unfortunate, too, if any attempt to assess technology were to be used merely as an excuse for a general assault on science and engineering. Yet many advocates of assessment, especially in the academic world, seem to take the benefits of technology for granted while spotlighting its faults and miscarriages. These adovcates tend to stress, in vivid if not extravagant terms, what is wrong with science, and their manner often alienates many people in the scientific and industrial world.

But if the scientific and industrial community remain unequipped to sort out good uses of technology from bad, all science and engineering may suffer. A growing impatience with the failures of technology could turn the public against science, causing very heavy social costs. Many remedies for past failures will only be found through the introduction of still newer and more sophisticated technology.

It would be a great mistake to put too much stress on the risks and negative aspects of technology assessment. If it is done with fair even-handedness, there is no reason why assessment should not promote the use of unappreciated and unemployed technology so that on balance it will enhance our well-being and reduce the long-term cost of innovation. It should help to stem the waste that results from poorly planned, unproductive, and unfeasible programs of research and development. At the same time, it should spur the development of beneficial technologies that might otherwise be overlooked because they seem to fall too far outside the marketplace economy to warrant exploitation. Such technologies need the advocacy that well-designed technology assessment could generate. Thus technology assessment can be viewed as a rectifying as well as a regulating process—a process that adds a new social and economic dimension to technological planning.

Technological choices must be made. The question is, Will such choices continue to be made in a haphazard, cavalier, slipshod, profit-centered, disorganized manner? Or can our policies and priorities be established in a more rational, deliberate way? Can we learn to identify and weigh the trade-offs in the decisions that we must make, and recognize how a decision made today may irrevocably affect the decisions we may want to make tomorrow?

The advocates of technology assessment are seeking hard-headed, practical methods to do just these things.

Part two

SOME SALIENT VIEWS ON TECHNOLOGY

Part Two contains the views of some prominent thinkers on the subject of technology. Since these views are quite diverse, it is useful to provide the reader with a historical account that places them in the proper perspective. This introduction therefore considers various views on technology, from ancient times to our present day, and interprets them in light of their historical and philosophical context. It presents only one possible way of understanding historical data concerning technology, but one that provides an interesting perspective on the views included in Part Two.

• • •

The diversity of attitudes towards technology has become more prominent lately as the impact of technology on society has intensified and spread to every aspect of our social and private lives (for more on this see Part One). But even in Greek times this diversity was already detectable. Together, Homer, Plato, and Aristotle illustrate it clearly. Homer admired technological excellence and devoted long passages of his verse to revealing the aesthetic dimension of outstanding technological acts. He delighted in describing such acts in great detail, unlocking in the process the secrets of their beauty. The following passage was selected by Samuel Florman as an outstanding example of Homer's ability to transform a simple technological act into a poetic experience.[1] Here Homer describes how the hero Odysseus built a raft with the assistance of Kalypso (sic) the nymph.

> She gave him a great ax that was fitting to his palms and headed
> with bronze, with a double edge each way, and fitted inside it
> a very beautiful handle of olive wood, well hafted;
> then she gave him a well-finished adze, and led the way onward
> to the far end of the island where there were trees, tall grown,
> alder and black poplar and fir that towered to the heaven,
> but all gone dry long ago and dead, so they would float lightly.
> But when she had shown him where the tall trees grew, Kalypso,
> shining among divinities, went back to her own house
> while he turned to cutting his timbers and quickly had his work finished.
> He threw down twenty in all, and trimmed them well with his bronze ax,
> and planed them expertly, and trued them straight to a chalkline.
> Kalypso, the shining goddess, at that time came back, bringing him
> an auger, and he bored through them all and pinned them together
> with dowels, and then with cords he lashed his raft together.
> And as great as is the bottom of a broad cargo-carrying ship,

when a man well skilled in carpentry fashions it, such was
the size of the broad raft made for himself by Odysseus.
Next, setting up the deck boards and fitting them to close uprights
he worked them on, and closed in the ends with sweeping gunwales.
Then he fashioned the mast, with an upper deck fitted to it,
and made in addition a steering oar by which to direct her,
and fenced her in down the whole length with wattles of osier
to keep the water out, and expended much timber upon this.

Florman remarks on Homer's delight in describing the building process down to the minutest detail. Let us look at some of these details. Odysseus' ax was "fitting in his palms" with "a very beautiful handle of olive wood." The trees that Odysseus felled "towered to the heaven" but were nonetheless "all gone dry long ago and dead, so they would float lightly." "He trimmed them well with his bronze ax, and planed them expertly." The beauty that emerges from this description is the beauty of a job well done, with skill and pride. What Odysseus finally builds is not just a "raft," but rather a raft carrying the distinctive mark of Odysseus' excellent workmanship. Such technological achievements Homer never ceased to admire.

Centuries later, another famous Greek, the philosopher Plato, took quite a different position with respect to technology. To understand this position, we must first familiarize ourselves with some of Plato's other views. Plato argued that true knowledge of things in our material world is not possible because this world is mutable—that is, changeable. True knowledge, he said, is possible of forms or ideas. Forms are understood to be general ideas or concepts that are exemplified in our world but are not themselves in it. For example, the form "green" is exemplified in this world by a green apple. The apple may change color and become red, but the form "green" itself as a general idea or concept remains unchanged. Forms are eternal, immutable, and real. The world we live in is not. This view led Plato to conclude in the *Phaedo* and the *Republic* that knowledge or science is possible only through conceptual inquiries, whereas data obtained through the world of everyday life has the mere status of opinion, not true knowledge.[2]

It follows from this position that the technological knowledge of our world does not deserve the status of knowledge at all. It is mere opinion, and as such it is inferior to philosophical knowledge based on contemplating the forms. Furthermore, technological activity is itself an inferior kind of activity, for as Hannah Arendt*[3] explains in her discussion of this Platonic doctrine, "work makes perishable and spoils the excellence of what remained eternal so long as it was the object of mere contemplation." Thus, instead of the earlier delighted admiration of Homer at well-fabricated things, the Platonic conception substituted mild disdain. A fabricated thing is inherently defective and inferior, no matter how well made it is. Its fabrication merely "spoils the excellence" of the external ideas exemplified in that fabricated thing.

Aristotle, Plato's student, rejected his teacher's two-world view. But he did not reject the idea that there are things other than the sensible particulars of our everyday world, namely forms. Unlike Plato, Aristotle thought of these forms as imbedded in the particulars. Reality for him was in this world. Nevertheless, knowledge for Aristotle remained as it was for Plato, knowledge of the universal and the real. But combined with his view of universals and reality, knowledge became tied more closely to the empirical world. As would be expected, this led to a different attitude towards technological knowledge. Since knowledge of this world could in fact be attained, technological knowledge was not discounted by Aristotle as mere opinion. He regarded it as true knowledge, but of a lower kind. "For contemplation is at once the highest form of activity (since the intellect is the highest thing in us, and the objects with which the intellect deals are the highest things that can be known)."[4] Technological activity was regarded as a way of achieving certain human ends. It had no intrinsic value; rather its value was seen as determined by the value of those human ends toward which it was directed.[5] Compared with its status during Homeric times, this was no modest loss for technology, but from a "spoiler of excellence" under the Platonic conception, it did move up to a neutral status in the Aristotelian framework.

• • •

The divisions concerning technology continued throughout the Middle Ages to the present day. Samuel Florman claims that underlying these later divisions was not only Greek influence but also religion. To support this claim he begins by presenting what he takes to be the "antitechnologists'" argument against technology:

> It has become a cliché that technology is an obstacle in the path of those who seek to find an existential sense of themselves. The antitechnologists have preached this message. . . . So have priests, philosophers, and poets. The argument seems self-evident. If people are entranced with trinkets, how can they plumb the depths of their spirit? . . . If they seek comfort, how can they expect to find truth?[6]

This is of course an oversimplification of what Florman calls the antitechnologist position, but it is not altogether wrong. Seyyed Hossein Nasr* uses a variant of this argument to denounce modern technological society. The malaise of modern technological society is seen by Nasr as a result of man having forgotten his true (though not necessarily existential) self—in other words, of his having forgotten who he really is.†

Having characterized the antitechnological position in this manner, Florman attempts to show its relation to religion. He argues that the New Testament is a major souce of this "antimaterialism," although he allows for the fact that the relation of Christianity to technology is not without ambiguity:

> Christianity's assumption that man is intended to dominate the earth conflicts with current ecological concerns. Also, the Christian attitude toward work is somewhat ambivalent. Man is instructed not to labor toward material goals, yet the Benedictines teach that idleness is the enemy of the soul. Often, indeed, the so-called Protestant work ethic is blamed for our worst technological excesses.
>
> In the broader view, however, it can be seen that the effect of our Greek and Christian heritage has been to convince us that materialism is a defect in human nature.[7]

This attitude toward materialism is contrasted by Florman to the one exhibited in the Old Testament, in which, "as in the *Iliad* and *Odyssey,* we find ourselves in an ancient, barren landscape, where man-made objects are the subject of wonder and delight . . ."[8]

Again, though the broadest outlines of Florman's argument are correct, the argument oversimplifies the attitude of Christianity to technology. Indeed, it could be argued that there is no one such attitude, and that the ambiguities he sees are in fact expressions of different attitudes within Christianity. In the next few sections a discussion of the attitudes of religion toward technology will be conducted in light of Florman's criticisms.

• • •

† Authors often use the word "man" to refer to the whole human species. With the advent of feminism (see the last part of this introduction) such usage became unacceptable. However, in this introduction I shall persist in using the old terminology. Several reasons could be given to justify this choice. For example, one could argue that since the quotations that appear in the introduction utilize sexist language, it is awkward to keep shifting to feminist language in the rest of the text. Or one could argue that no adequate substitute for the older usage has been found. I do not find any of these reasons convincing. My terminology stems from my belief that the history of technology has been predominantly the history of male technology, and hence the older usage is appropriate.

It is true that Jesus placed minimal emphasis on material posessions and disdained the gluttonous pursuit of profit. But on the other hand, he did not scorn man's material needs either. When the multitudes who came to listen to him grew hungry, he did not send them away to the villages, as his disciples suggested, to feed themselves.[9] Nor did he preach to them on the virtues of subduing bodily needs and desires. Instead, he fed them.

Again, Jesus healed the sick on several occasions, and on one occasion resurrected the dead.[10] The resurrection of Lazarus came after Jesus was moved by the grief of the dead man's sister and friends. For when he saw their weeping he was troubled. And although the other sister, Martha, had already assured him that she knew that Lazarus would rise again in the resurrection at the last day, Jesus did not wait until then, nor did he counsel the grievers to do so. Compare this with the story of Kisa Gotami in Buddhist literature.[11] The woman lost her only son. In her grief she refused to part with him. Instead, she took him on her hip, and went about from house to house seeking medicine that would bring him back to life. Finally she sought the help of the Buddha. He expressed a willingness to help on the condition that she provide him with tiny grains of mustard from any house in the village which was never visited by death. When she failed to find such a house, she of course realized the inevitability of death and finally parted with her son.

Again, a Buddhist, like Jesus, would not have asked the hungry crowd to leave. But he would have counselled them to endure the pangs of hunger in the pursuit of knowledge and would not have performed a miracle to feed them. This illustrates that while Christianity's teachings are less materialistic than those of Judaism, they are more materialistic than those of Buddhism. Thus it is a mistake to suggest that Christianity is "*anti*materialistic."

Two prominent Christian figures in medieval times—namely, St. Augustine and St. Thomas Aquinas—held epistemological views that are of consequence to this discussion. They both accepted the possibility and desirability of having knowledge of corporeal things. For St. Augustine "action, by which we use temporal things well, differs from contemplation of eternal things; and the latter is reckoned to wisdom, the former to knowledge."[12] The ideal state of affairs for St. Augustine exists when a person attends chiefly to the development of his contemplative knowledge while at the same time directing his reason to the good use of material things, "without which this life does not go on."[13] What is important here is not to allow this balance to be disturbed so that one's attention to material things surpasses his attention to the eternal things. Thus it would seem that technology, insofar as it is directed to ward the good use of material things, is quite desirable, so long as its importance is kept within perspective.

Aquinas' view is even more materialistic, for he saw sense experience as the starting point of knowledge. It is from this starting point that man's knowledge of universals is developed. Aquinas held a teleological view of activities. Activities undertaken in the light of reason must be directed toward goals. They can be judged as good or bad in light of a goal, its attainment, and the means used for its attainment. Thus if the goal of one's technological activity is the realization of the potentialities and gifts God has endowed one with, then it would seem that this activity would be regarded as good, on the condition that the means used for achieving this goal of self-realization are also acceptable. On the other hand, if the goal of a technological activity is not the manifestation and admiration of the beauty of God's creation but rather destruction, it would seem that that technological activity would be seen as bad. This teleological view is, of course, quite Aristotelian, and hardly "antimaterialistic."

In the two selections presented under the subtitle "religious," the authors clearly do not condemn the realm of the technical as evil in itself. Gabriel Marcel* emphasizes that the realm of the technical is not evil in itself. Indeed, he argues that technique is the expression of something good, since it is no more than the application of our gift of reason to reality. But what Marcel does criticize is the concrete relationship that tends to grow between man and technical progress, a relationship that gives rise to covetousness and envy. In particular Marcel draws attention to the specific values of the capitalist society, which is oriented toward the accumulation of material wealth for the sake, not of God, but of the capitalist. This is partly what Nicholas Berdyaev* has in mind when he announces that technique has long ceased to be neutral and that the world is being dehumanized by the monstrous power of technique. His solution is that technique be subordinated to spirit and the spiritual values of life.

This attitude towards technological activity is not as different from the Judaic attitude as Florman would have us believe. In Exodus, we are told of a technological activity that the people of Moses engaged in but that did not meet with his or God's favor.[14] They broke off their golden earrings and brought them to Aaron, who fashioned out of the gold a molten calf to be worshipped by the people of Israel. But Moses was not taken by the artifact. Without pausing to decide whether the artifact was well made, or whether it revealed any skills of workmanship, the angry patriarch took the golden calf, burnt it in the fire, and then ground it to powder. It was a sin to worship other gods. Thus Moses' view of technology is also teleological. The technological activity is good or bad in light of the end it serves and the means it utilizes. This tradition of a teleological view of technological activity permeates the Old Testament. True, Solomon's temple was built by skillful craftsmen from gold, silver, and cedar, as Florman points out.[15] But Solomon could allow himself this indulgence because the temple was the temple of God. It was an expression of Solomon's love and respect for God. But Solomon himself knew that the highest values were not material. That is why when he prayed to God he asked not for riches but for wisdom, for understanding.

Let us go back to the ambiguities that Florman finds in Christianity. He says that the Christian attitude towards work is ambivalent. On the one hand, ''man is instructed not to labor toward material goals.'' On the other hand, the ''Benedictines teach that idleness is the enemy of the soul. Often, indeed, the so-called Protestant work ethic is blamed for our worst technological excesses.''

The Benedictine code enjoins the monk not to be idle. It states, ''Idleness is the great enemy of the soul, therefore the monks should always be occupied, either in manual labor or *in holy reading''* [emphasis added].[16] Given the full text of this part of the code, the ambiguity is lessened if not eliminated. What is required is that the monk refrain from idleness. But the opposite of idleness is not only manual labor, but also religious self-education, holy reading. As argued earlier, a good Christian, and especially a monk, does not direct his attention to the accumulation of material possessions, but he is expected to attend to his basic needs and the needs of his community.And that requires labor. Once this is attended to, a monk can turn away from idleness by reading.

A more complex situation is associated with the Protestant work ethic because it involves the notion of a ''calling,'' or a task set by God. One must work hard in one's calling, for wasted time reduces labor for the glory of God. One important measure of the usefulness of a calling, and consequently its favor in the eyes of God, is private profitableness. This is reminiscent of the story of Solomon, whose piety was rewarded by riches. As Max Weber says, ''For if that God, whose hand the Puritan sees in all the occurrences of life, shows one of His elect a chance of profit, he must do it with a purpose. Hence the faithful Christian must follow the call by taking advantage of the opportunity.''[17] Thus the Puritan must work and increase his profit in the service of God. This means that as his wealth increases, the good Puritan must use that wealth in accordance with piety. For not only is his labour in a calling an outward expression of brotherly love, but the wealth it engenders must be used rationally for the needs of the individual, his family, and his community, not for material luxury, which is condemned as idolatry of the flesh.

Puritanism then does not deny man's material needs. It does not deny him a comfortable life for himself and his family. But it does condemn the excess of material luxury and the gluttony for material posessions. For man is only a trustee of the profits God has blessed him with, and hence is responsible for spending them wisely in the service of God. This of course encourages acts of philanthropy in the community. It also leads, as Weber explains, to capitalism.[18] Neither consequence is antimaterialistic.

So far it has been argued that contrary to Florman's claim, Christianity, though not strongly materialistic, is nevertheless not antimaterialistic. The discussion of the ambiguities Florman found in Christianity bolstered this view. Now it will be argued that such "weak" materialism is not by necessity antitechnological. The proof for this claim is immediately obvious in the case of the Puritan work ethic. If technology increases profit, or serves the needs of the community, then God must have shown us that opportunity for a purpose, and we must follow it up. This position supports technological activity and is hardly antitechnologist.

The proof of this claim for the general case is no less obvious. In Christianity one must attend to his material needs without indulging himself to the point where his attention is distracted from the spiritual realm. But he must attend to his material needs. That is accepted and recognized. However, what are these needs?

The answer to this question varies from one historical epoch to another and from one geographical area to the next. In the age of modern technology, "needs" have proliferated enormously, and in fact they have perhaps surpassed what an impoverished Third World person would regard as luxury. There is now a need for an automobile, for a house with plumbing, for a clock, and so on. But the basic human needs are reasonably stable, and it is those needs that Christianity refers to. Among them are the needs for food, shelter, and physical well-being. These alone already justify an enormous amount of technology: agricultural technology, engineering technology, and medical technology, among others. Some of these technologies are less directly involved in serving these needs than others, but so long as they are involved they can be justified. What cannot be justified under this view is an indulgent technology which is not directed toward the service of humanity. But to say this is not to take an antitechnological position.

To emphasize further the general distinction between "weak" materialism and an antitechnological attitude, let us look at one interesting historical example from outside the Judeo-Christian societies.

Toward the end of the tenth century, a group of Muslim scholars formed a secret brotherhood in Basra, Iraq, called "Ikhwān al-Safā." These scholars carried on and then recorded some of the most serious investigations concerning man, nature, and God. The Ikhwān regarded human beings in this world as prisoners who can only be freed through knowledge. Thus they devoted themselves to research in numerous areas, ranging from astronomy and embryology to alchemy, accounting, and agriculture. Their aim was twofold: to reveal to all men, first, the beauty and harmony of the universe and, second, the necessity of going beyond mere material existence.[19]

Alchemy, a predecessor of chemistry, is often characterized as the enterprise of attempting to change base metals into gold. This description is a vulgarization of the alchemical tradition. The alchemical tradition, which upheld the cosmological doctrine of the unity of Being, regarded alchemical activity as an attempt to better understand God's widsom through unlocking some of the secrets of his creation. The attempt to unlock these secrets was seen to require patience, persistence, and human wisdom. The secrets could not be discovered except after a long period of hard dedicated work and training, which led the individual up the ladder of spiritual maturity. A person interested only in the ultimate material product of alchemy, namely, gold, was expected to lose patience long before he approached an understanding of the basic principles involved. To ensure that this knowledge did not fall into the wrong hands, alchemists always wrote in riddles.

The Ikhwān are an example of an other-worldly tradition which nevertheless concentrated heavily on the material world and its transformation as a way of understanding the higher spiritual reality. Thus this tradition was both weakly materialist and pro-technological at the same time. Some may object to using alchemy as an example of technological activity and prefer to refer to it as witchcraft. But we are only following common usage here. For example, Ortega* describes magic as ''nothing but a kind of technology, albeit a frustrated and illusory one.'' He classifies it as belonging to the stage of primitive technology, as does Theodore Roszak.[20]

We are now left with the most interesting assertion in Florman's list of ambiguities in Christian attitudes toward technology. It involves Christianity's assumption that man is intended to dominate the earth, an assumption which Florman sees as conflicting with current ecological concerns.

This assumption is based on what God tells man in Genesis: ''And God said, let us make man in our image, after our likeness: and let them have dominion over the fish of the sea, and over the fowl of the air, and over the cattle, and over all the earth, and over every creeping thing that creepeth upon the earth.''[21] The assumption belongs to both Judaism and Christianity. It also belongs to Islam, since the Quran contains a similar assertion. It is a most interesting assumption in light of our earlier discussion of modern man's view of the goal of science and technology as one of mastery over nature. The passage in Genesis legitimizes that view with the full force of God's words. Thus to aim at dominating nature is not necessarily an expression of vanity or greed. It is the natural order of things. To criticize it is to dispute divine wisdom. So the difference between Judaism, Christianity, and Islam is not one of whether man ought to dominate nature, but rather only one of assessing the value of such a dominion.

This might seem to contradict certain well-known religious views. For example the Muslim philosopher Avicenna describes the relationship of man to nature and the cosmos as one of love.[22] But this contradiction is only apparent, for love and dominion in these three religious traditions are not regarded as incompatible. This is best illustrated in the biblical relation of a man to his wife, and of a father to his son. The man or father is seen as the master of the household, and yet he is also depicted as loving, as in the examples of Hosea and Abraham.

Even after man proved himself unworthy of God's trust by eating from the tree of knowledge of good and evil, God did not revoke man's privilege of dominion over earth and everything on it. Man's punishment was a lifetime of toil and sorrow, followed by death. But God did have some apprehensions about man whom he created in his own image: ''And the Lord God said, Behold, the man is become as one of us, to know good and evil: and now, lest he put forth his hand, and take also of the tree of life, and eat, and live forever: Therefore the Lord God sent him forth from the garden of Eden . . .''[23]

What emerges from this passage is the picture of man as a creature of God who was created in God's image, and who upon eating the forbidden fruit of the tree of knowledge of good and evil became so increasingly Godlike that God had to banish him from the garden of Eden before man got to the tree of life and became immortal, thus eroding further the differences between him and God, his maker. This can be regarded as an ancient version of the modern problems of autonomous technology. God worried that his own version of the ''Frankenstein monster'' might get out of control. God's solution to the problem was to cause the stored-up energies of the creature to be expended in a struggle for survival in a hostile environment, until ''self-destruct'' mechanisms succeeded in overtaking the exhausted creature. This explains the ''fall'' from Eden, a perfectly friendly environment, and the twin curse of toil and mortality. But at no time did God contemplate the destruction of Adam and Eve, the most sophisticated product of his creation, before they duplicated themselves through reproduction. For God was proud of his abilities as a maker. In Genesis we are told that God "saw everything that he had made, and, behold, *it was* very good."[24] When his technology went out of control in his most sophisticated creation, God's solution was not to destroy the product altogether, but to alter some of the variables involved in the situation in such a way as to contain the problem.

• • •

Even before religion informed man that mortality was his punishment for disobeying God's words, man was already concerned about his mortality, and searching for ways to overcome it. This search took various forms. In pre-Socratic times, attention was focused on producing things, including words and deeds, that can last beyond the life-span of their producers. By producing things that seemed to belong to the ages, like Homer's *Odyssey,* man believed that he could immortalize himself. This attitude toward production related to technological activity in a direct way. Insofar as certain technological activities may have provided the possibility of immortality, they became desirable. Surely, not all kinds of technological activity provided one with such a possibility; the more mundane kinds were therefore left for the slaves to perform.

But as the historical process unfolded, it became obvious that immortalization through fabrication was hardly satisfactory. Not only did works themselves disintegrate sooner or later, but often, when they temporarily escaped the onslaught of time, they ended up immortalizing either themselves or someone other than their producers. More often than not the individual identity and life story of the producers were completely lost.

Christianity succeeded in providing a better solution. Man did not need to immortalize himself by building monuments; instead, through modesty and humility in the service of God, he could gain for himself something much better—eternal life. Thus, technological progress during these times was pursued mainly for reasons other than the desire for immortality. Technological progress was acceptable, so long as it was for a good purpose, but was held secondary or auxiliary to the important goal, namely spiritual progress.

This solution could not be tested by historical events. It was not possible to ascertain whether in fact good men went to heaven. That had to be taken on faith, and men became restless when forced to entrust such a major concern to the unknown. At the same time, accumulated scientific theories in the West continued to present ever stronger reasons to doubt the biblical accounts of creation and of man himself. The Copernican, Newtonian, and Darwinian revolutions—and later, psychoanalysis—all inflicted deep wounds on the human ego, as Freud observed.[25] Man in Western society discovered that he was not the center of the universe as he had thought; that the universe had its own laws, which were not particularly oriented to the service of man; that man's creation was the result of an evolutionary process rather than an act of God. Also, man in Western society discovered that he did not know what was going on even within himself, let alone the rest of the universe. His faith was sorely tested.

The secularization of science had already started. Together with the long history of political conflict between Church and State, the stage was finally set for a secularized society. The otherworldly restraints on technological progress weakened steadily, and men in Western society had to face the emerging new reality.

It was during this turbulent era that Nietzsche wrote that "God is dead."[26] In doing so, he was reflecting the deepening religious crisis, and the growing concern of man with a world in which he suddenly found himself alone. Nietzsche argued that with the renunciation of religion, man lost the right to his claim of supernatural dignity. Such a dignity, therefore, could no longer be regarded as a given fact; instead it had to be regarded as a goal to strive for. But such a goal, Nietzsche argued, was not easy to approach. For the person who succeeds in approaching it must stop being merely human. He must be hard against himself, transcending his limitations in order to become creative. He must reject the Christian "slave-morality," characterized by humility, meekness, and charity, and overcome his individual weaknesses. Nietzsche called such a person the *Übermensch,* but because he did not define in absolute terms the traits of the *Übermensch,* the concept was subject to individual interpretation.

To complicate matters, a central concept of Nietzsche's philosophy, the Will-to-Power was interpreted as a drive for domination characteristic of the *Übermensch*. In fact, Nietzsche saw the Will-to-Power as a basic trait to be found in all men, strong and weak alike. The misinterpreation of this concept as well as that of the *Übermensch* led to erroneous popular interpretations of Nietzsche's philosophy. For example, it is recounted that Nietzsche's sister told Hitler that he was what her brother had in mind when he wrote about the *Übermensch*.[27]

It is indicative of the values of the times that such interpretations were taken seriously. These interpretations were based on two significant claims: first, that with the "death of God" a special breed of men, the strong, inherit the earth and, second, that the relation of these men to others is one of unrestrained domination.

The claim that the relation of men to others was one of unrestrained domination was not altogether new. Already in the *Phenomenology of Mind,* Hegel,* who died shortly before Nietzsche's birth, had described what came to be known as "the master-slave dialectic." In a famour passage, Hegel developed an analysis of what happens when two "consciousnesses" (or individuals) meet. Hegel claimed that each of them attempts to elicit recognition from the other; and to that end a struggle till death is commenced. But then the winner realizes the impossibility of eliciting recognition from a dead man. Thus he abandons the struggle till death and replaces it with the "dialectial cancelling" of the adversary—that is, keeping the adversary alive after the latter has negated his own independence and subjugated his freedom to the victor.

This is only the second stage of Hegel's master-slave dialectic, but one can already notice that this dialectic was conceived outside the traditional framework of Christian values. The dialectic presented here is in fact a power struggle for supremacy. If we attempt to superimpose the notion of Nietzsche on this dialectic, it will be immediately obvious that under the original interpretation of Nietzsche the *Übermensch* is the one who wins his battles for recognition, no matter how ruthless they are.

It must be pointed out that Hegel also speaks of a higher dialectic among men, which rejects domination as a mode of relating to one another. Nevertheless, in *The Philosophy of Right* he described even the relation of men to women as essentially one of domination justified by Hegel's perceived inferiority of women.[28] Thus Hegel seemed to endorse the second stage of his master-slave dialectic.

If we place the original interpretation of Nietzsche's philosophy in the context of our discussion on immortality, the situation may be summed up as follows. With the "death of God" proclaimed, modern man in Western society found himself thrown into a world he did not choose and deprived of a right he thought was his—the right given to him in Genesis, namely that of dominion over nature. Furthermore, his hopes for either immortality or everlasting life were dashed by his assessment of the accumulated historical evidence. So he set out to cope with his anguish and solitude by taking his life into his own hands. This is one meaning of the *Übermensch* myth. The other is that in the absence of diety all restraints on one's power are removed. So when modern man in Western society set out to re-establish his dominion over earth on a new foundation, he did not shrink from turning his undisguised power against men whom he regarded as inferior adversaries.

It is reasonable to expect that with the apparent abandonment of the active pursuit of the goals of immortality, man's attention turned to his life in the material world. For, as far as he was concerned, this was the only life he had, and he had better make the best of it. So, in an age in which the most effective and reliable tool in Western society was technological power, such power became an instrument not only for reducing his toil but also for producing luxuries. Man in Western society was finally able to celebrate material joys without the fear of sin and punishment. After centuries of self-deprivation, he could indulge in earthly pleasures.

But technological production required two things at least besides machines and capital: raw material and workers. This justified the spread of capitalism and colonialism in the world. With the "master-morality" already advocated, it was possible to ignore the restraints of Christian "slave-morality" and pursue the goals of unlimited power both at home and abroad. An examination of the social, economic, and political developments that were taking place at the time will confirm this view of the dynamics of technological development.

• • •

While scientific discoveries in the West were forcing man into an "existential" crisis, characterized by anguish and the search for new values, technological innovations were changing the shape of his daily life. Medieval inventions like the mechanical clock and the printing press were already finding their way to his daily life. So were the blast furnace, the steam engine, the railway, the mine, and the factory. These inventions, according to John Kenneth Galbraith,* together with the discovery of abundant land supplies in America, South Africa, and Australia, led to a significant shift of power in society. While in feudal days land was the most desirable possession as a source of power and prestige in the state, in the age of factories and machines capital became more desirable. For the person with capital could purchase not only factories and machines but also labor and land as well.

With the emergence of capital as the source of power in Western society, the capitalist of course aimed at maximizing his capital, which meant in effect maximizing his profits and savings. One obvious place for saving was in the area of labor. The worker as a commodity (an object with an exchange value in the marketplace) had replaced the serf. On the face of it, this was an improvement, but in fact the worker, unlike the serf, lacked the minimal security of the land and the customary protection of the lord. He was on his own, and the responsibility of the capitalist toward him ended with the day's labor. Furthermore, the capitalist saved by paying the worker as little as possible for the long hours at the factory. Mumford* relates in some detail the working and living conditions of workers and their families during this period —conditions that promoted disease, ignorance, and starvation.

Nevertheless, it is wrong to assume that capitalists were idling in their mansions while workers toiled in the factory. Indeed, capitalists often worked just as hard as the workers in order to accumulate the desired capital. At times, they even starved themselves, although in their case this was done by choice and not out of necessity.

Finally, this struggle for more gain and consequently more power took more advanced forms. With the steady industrialization of Western society, competition increased. But competition drives prices down. Thus the capitalist had to find a way to keep his profits from diminishing. To this end, two trends in the capitalist system appeared: a drive for increased concentration of production and a drive for expansion of foreign markets.

By concentrating production in bigger and bigger enterprises the capitalist substantially improved his competitive abilities and profits while at the same time reducing the risk arising from market fluctuations. For example, by combining several branches of an industry in one enterprise the capitalist was able at times to act as his own supplier, thus reducing his cost of production and increasing his profits. Also, by diversifying, he was in a better position to absorb unexpected losses in some branches of the industry. Such considerations led to the emergence of larger enterprises.

With the development of expensive and sophisticated technology, the complexity of the business enterprise increased and new reasons for the formation of large enterprises appeared. Today it is not possible to manufacture certain technologically sophisticated products without the detailed and varied expertise that only a large organization can supply. To use an example from Galbraith, we can no longer produce an automobile in somebody's machine shop. Yet the original Ford automobile was produced in this fashion.[29] The Dodge brothers' machine shop supplied their automobile's engine and chassis. Since then advances in metallurgy, design, mechanical engineering, and other areas, have all combined to necessitate the presence of an organizational structure which coordinates and finances specialists in these various areas. So, to this extent the size of the enterprise is crucial—despite the fact that such an army of specialists is often overinflated in size, oddly enough, for pecuniary reasons. For example, much of the redesigning activity and metallurgic research is directed not toward producing a superior automobile but rather toward making the present one look obsolete or outmoded. This would of course encourage sales of new automobiles and improve the company's profits. Similarly, a large part of the specialists' energies and the company's funds go toward advertising, which aims at influencing the attitudes of the consumer to the benefit of the company. Nevertheless, Galbraith's example succeeds in illustrating the new reasons for the formation of large organizations.

It is important to note that as a company grows its sheer size becomes an additional source of substantial power. Thus, "Bigger is better." The Chrysler corporation, for example, is having financial difficulties. By common standards, if a business cannot make a profit, it should close its doors. That is what happens to the small business enterprise. The story is different for a large corporation. A corporation as huge as Chrysler cannot close down without having a substantial negative impact on the society's economic scene. The extent to which the folding of a corporation injures society becomes a measure of the power of that corporation in that society. Through sheer size a corporation manages to enmesh itself so deeply into the social and political fabric of a society that the possibility of its collapse becomes dreaded by everyone. The government then steps in to avoid an economic and political crisis. Such a step was taken several years ago in the case of Lockheed, an airplane manufacturer.

But this is not the only way in which the government and the corporation help each other. A project may be too monumental for a private enterprise to undertake without some help or protection from the state. If the state deems that project desirable, it can provide that enterprise with the help it needs. For example, few sophisticated weapons are developed by U.S. companies today without prior commitment on the part of the government to purchase the weapons or to permit other countries to purchase them. In some cases, like NASA, the state takes over altogether, contracting firms for various parts of projects. Thus, while the corporation needs the protection of the state, the state needs the technological "know-how" and capabilities of the corporation, especially in the case of war.

So, in order to maximize his profits and hence his power, the capitalist increases the size of his enterprise. Historically, this has led to the birth of the corporation, which gradually replaced the individual capitalist as the wielder of power in society. The capitalist's power, while still measured by the size of his bank account, is now also being increasingly measured by his "weight," his "pull," and his "contacts" with corporations or their boards and management. But while the pursuit of contacts, and other similar goals, is being combined with the pursuit of profit in today's society, the power principle itself remains supreme. However, it has become an organizational instead of an individual goal; and with the increasing interdependence between corporations and government, society is moving closer to a state-capitalism characterized by the concentration of power in one overall structure, the corporate-governmental complex. Where the means of production are totally owned by the state, as in the U.S.S.R., the overall structure is simply the state.

One serious consequence of the increase in size, production, and efficiency in the modern industrial state is the effect this has on the worker. During the early stages of industrialization, as Mumford points out, capitalism generated conditions of poverty, ignorance, and disease among the workers. But at the advanced stage of industrialization the effect is of a completely different sort. Fromm* argues that today's worker, who is a mere appendage of the machine, feels powerless, lonely, and anxious. He has been reduced to a state of passivity, conformity, and alienation. Thus, though educated, well-fed, and well-paid, the worker remains unhappy. A malaise hovers over society. Industrializaiton, which brought power to the state, renders the majority of the citizens discontented.

• • •

The sharp competition in the local markets forced the capitalist to look elsewhere for marketing his goods. Overseas colonies had already been tapped as a rich source of raw materials. Now they looked additionally attractive as potential markets. This led industrial countries like Britain, France, and Germany to increase their colonial acquisitions substantially. Thus during this period colonialism and imperialism received an unprecendented impetus and set in motion a series of political events that have been unfolding internationally to the present day.

One person whose ideas on colonialism affected the course of history was V. I. Lenin. Lenin argued that while the notion of colonization disturbed leading British politicians at the time, they nevertheless came to accept it out of purely economic and political considerations. As a case in point he mentioned the following remarks made by Cecil Rhodes, the famous British magnate, which were related by Rhodes' close friend, the journalist Stead:

> I was in the East End of London yesterday and attended a meeting of the unemployed. I listened to the wild speeches, which were just a cry for 'bread,' 'bread,' 'bread,' and on my way home I pondered over the scene and I became more than ever convinced of the importance of imperialism . . . . My cherished idea is a solution for the social problem, i.e., in order to save a 40,000,000 inhabitants of the United Kingdom from a bloody civil war, we colonial statesmen must acquire new lands for settling the surplus population, to provide new markets for the goods produced in the factories and mines. The Empire, as I have always said, is a bread and butter question. If you want to avoid civil war, you must become imperialists.[30]

This again underlines the growing relation between the state and the capitalist. That growing relation had been manifested earlier in the military force that the state used abroad to back up its economic as well as political interests. But as the goal of using the colonies as new markets emerged, a shift in colonial policies became necessary. Frantz Fanon explains that "what the factory owners and finance magnates of the mother country expect from their government is not that it should decimate the colonial peoples, but that it should safeguard with the help of economic conventions their own 'legitimate interests.'"[31]

Ultimately, the colonial occupation ended altogether, but the economic interests remained. This may seem surprising given the sharp criticisms of the brutal policies of colonialists all over the world, but the dynamics of technology and industrialization again explain why this is so. In a world divided between the technologically advanced and the "underdeveloped," the relation of economic dependency gains in complexity. Not only did the capitalist need raw materials, markets, and (later) cheap labor, but the "native" himself needed the technological "know-how" possessed by the West, to improve his living conditions. Thus the relationship of dependence ran both ways.

But the native was not pleased by what he was receiving from the Western capitalist, and regarded the exchange as unjust. Although his reasons for taking this point of view were varied, his major complaint was that the Western corporations were not willing to relinquish control of foreign markets and that they went to great lengths to preserve the technological "backwardness" of his country. In "Growth is Not Development," Samir Amin* works out the details of this claim.

Zbigniew Brzezinski,* currently National Security Advisor to President Jimmy Carter, also worries about this situation. He observes that about seventy-five percent of the world's computers operate in the United States, and that approximately eighty percent of all scientific and technical discoveries made during the last few decades originated in the United States. Thus the technological gap, according to Brzezinski, between the United States and other countries, including Europe, is widening. The hopes of bridging it by imitation are doomed. All of this contributes to the development of an ideology of rejection in the developing nations. This, Brzezinski warns, could lead to polarization on a global scale between the developed and underdeveloped countries, culminating possibly in confrontation and chaos. Daniel Bell,* in "The Future World Disorder" echoes Brzezinski's concerns. One must therefore conclude that it is no longer possible to retain absolute control in the world through the use of technology without becoming exposed to monumental risks. The dialectic of history has shown modern man that the quality of life can no longer be maintained by unlimited power, which would only invite global wars and destruction. The principle of power must be limited in order to preserve life itself. Two events in modern history illustrate in shocking ways the truth of this claim.

• • •

To understand these events, we have to examine the views of an infamous political leader, Adolf Hitler, with whom the concept of the Will-to-Power found its crudest interpretation. In his book *Mein Kampf* Hitler argues that the strong must rule the weak and refrain from uniting with him, in order to preserve his superior stature. But Hitler's strong man was not totally self-made. He had to be born an Aryan, for Hitler combined his interpretation of the Will-to-Power with the concept of racial supremacy and argued that the Aryan race was superior to any other.[32] This ideological combination allowed Hitler to mobilize the whole state in the pursuit of power; and the state, as we argued earlier, is necessary when power is heavily dependent on sophisticated technology. Furthermore, this same ideology served him in waging a struggle till death against his adversaries both at home and abroad. The only groups Hitler did not regard as his adversaries (the Aryan Germans and their international allies) were exactly those whom he could not do without in his quest for power. Everyone else was dispensable sooner or later.

The internal adversaries of Hitler's state were the Jews, and against them he unleashed his most diabolical scheme. That scheme, which was later expanded beyond the German borders, was based on technologically advanced standards of "efficiency" and "organization." It brought death to millions of people in a disciplined, impersonal, and technologically modern way. Human dignity and emotions were dismissed as Nazi soldiers went about their business, following orders, gassing humans and cremating bodies. In the process, they exhibited an amazing sense of detachment, which enabled them to reduce human suffering to the category of mere data. Rudolf Hess, a commander of Auschwitz, the infamous concentration camp, discussed at the Nuremberg trials the superiority of the gas he employed to "exterminate" people.

> I visited Treblinka to find out how they carried out their extermination. The camp commandant at Treblinka told me that he had liquidated 80,000 in the course of half a year. He was principally concerned with liquidating all the Jews from the Warsaw ghetto.
>
> He used monoxide gas and I did not think that his methods were very efficient. So when I set up the extermination building at Auschwitz, I used Zyklon B, which was a crystallized prussic acid which we dropped into the death chamber from a small opening. It took from three to fifteen minutes to kill the people in the death chamber, depending upon climatic conditions.[33]

The same attitude carried over to the German business establishment. Various companies competed to supply the most efficient and durable equipment needed to carry out the mass murders.

In the midst of conducting a holocaust at home and a war abroad, the Nazi scientific and technological establishment continued its attempts to develop superior weapons that would realize the Führer's dream of Aryan supremacy. In 1939, in the United States, the father of the nuclear age, Albert Einstein, sent a letter to then President Roosevelt in which he warned of the ongoing nuclear research in Germany.[34] Alerted by Einstein, the United States started the process that eventually produced the atomic bomb. By then, an ally of Germany, Imperial Japan, was at war with the United States, and Germany was already losing the war. On August 6, 1945, the United States dropped over Hiroshima the first atomic bomb ever used in warfare in the history of mankind. It was followed by another over Nagasaki. Both were major cities in Japan. A woman writer, Yōko Ōta described her experience:

> I just could not understand why our surroundings had changed so greatly in one instant . . . I thought it might have been something which had nothing to do with the war, the collapse of the earth which it was said would take place at the end of the world, and which I had read about as a child.[35]

An eyewitness described what he saw:

> The appearance of people was . . . well, they all had skin blackened by burns . . . they had no hair because their hair was burned, and at a glance you couldn't tell whether you were looking at them from in front or in back . . . They held their arms bent [forward] like this [he proceeded to demonstrate their position] . . . and their skin—not only on their hands, but on their faces and bodies too—hung down . . . If there had been only one or two such people . . . perhaps I wouldn't have had such a strong impression. But wherever I walked I met these people . . . Many of them died along the road—I can still picture them in my mind—like walking ghosts. They didn't look like people of this world . . . They had a special way of walking—very slowly . . . I myself was one of them.[36]

The shock caused by these two holocausts was universal. Technology had reached such a stage of development that unless restraints were placed again on its use, the result could be catastrophic for humanity as a whole. Bertrand Russell wrote Einstein: "The thing to emphasize is that war may well mean the extinction of life on this planet."[37] Scientists, more than anyone else, were suffering guilt and taking a second look at their work, but they were not alone in their concern. The whole world joined in. The new weapons had caused the world to shrink. Devastation could now be universal.

It became clear that by removing all constraints on the use of power the whole of humanity was endangered. Man suppressed his hopes for immortality only to find his one remaining refuge, this life, threatened more than ever with destruction. His aim was not to destroy this life; it was to enjoy it. Yet in the totally unshackled pursuit of power he had achieved the exact opposite. Thus the principle of unrestrained power had to be modified. In the absence of God restraints were imposed in the name of humanism, the secular version of religious morality. Yet even with the advent of an uneasy world peace, interest in technology persisted and warnings against its evils continued to be sounded.

Jacques Ellul,* the famous French historian, declared that technique was a Frankenstein monster gone out of control. It may seem that such a position is simply an abdication of man's responsibility, but that was not Ellul's intention nor the intention of other well-known anti-technologists. What they wanted to do was simply to call attention to the most serious as well as the most common kinds of adverse effects technology has on society during peacetime, especially those effects which slip by unnoticed, such as boredom, conformity, and passivity, as well as smog, congestion, and radioactive contamination. None of these writers advocated an all-out return to nature, which would be impossible in light of the fact that human nature itself has become highly dependent on artificial surroundings. It is in this light that the Autonomous Technology Debate (see Part Three) should be viewed.

• • •

The notion of restraining man's power is not readily acceptable, given man's inclination to inherit the earth now that God has been declared dead. In the following sections I shall discuss very briefly three approaches to the problem of limiting man's power. These approaches are provided by (1) theoretical Marxism, which advocates the end of one man's economic power over another; (2) feminism, which advocates the end of man's power over woman; and (3) ecological thought, which advocates a limit on man's domination of nature. The three approaches are interrelated in many ways, and each has been in existence for some time. Nevertheless, the new developments discussed earlier gave them a new impetus and a sense of urgency.

Perhaps the best-known work of Marx is *Das Capital,* in which he delivers a scathing critique of British capitalism in the nineteenth century. In that work, Marx studies the minutest details of human relationships under the factory system and reveals the extent of exploitation and misery to which the worker was subjected. One might conclude that Marx regarded industrialization as an evil to be eliminated, but nothing could be further from the truth. In *The German Ideology,* Marx unequivocally states his position towards technology. He notes that liberation can be achieved only in the real world and by real means, and that "slavery cannot be abolished without the steam-engine and the mule jenny."[38] He then adds that in general people cannot be liberated so long as their basic needs go ummet. Thus Marx in his critique of capitalism was not rejecting technological progress, but only the kind of human relations that had historically become associated with it. The development of advanced machinery was supposed to reduce the toil of the worker, and shorten the working day. But, Marx complained, this had not taken place. Instead, the capitalist used advanced machinery to increase his profits at the expense of the worker, who continued working long hours.[39]

It is this aspect of industrialization that Marx finds objectionable. As a solution, Marx advocates a classless society, in which the basic relation among humans is one of social cooperation and not of domination. He argues that human potential cannot blossom unless man has enough leisure time to exercise his creativity. This cannot happen either when man's basic needs are not met or when he is living under conditions of oppression. Thus Marx calls for the working class to reject the oppression of the capitalist and seize (not destroy) the means of production. Then the rule of the workers will follow as a step toward the abolishment of all rule and the withering away of the state. Although not much is said by Marx about the classless society, one can conclude that it will have two features: First, technology will be present. This will ensure the "liberation" of the individual from necessary work, providing him with leisure time that he will use for developing his potential as an individual. Second, there will be no relation of domination of man by man whatsoever. This is what "classless" means in this context. Clearly, such an approach to the problem of limiting man's power would be, and has been to some extent, attractive to workers in capitalist countries as well as to people of the Third World.

From the very early stages in the development of feminist thought, feminists have criticized the principle of domination in society. That principle was seen as operative not only against slaves, workers, and "natives" but also specifically against women. Recently, many feminists argued (1) that the female was indeed oppressed, even in modern society, and (2) that this oppression could not be removed at this historical juncture by simply liberating the worker. Thus, the principle of domination had to be limited in additional ways.

The first task of recent feminists, then, was to establish that women were indeed oppressed. The oppression of women by men had been taken for centuries as the "natural" order of things but with the rise of recent feminist thought, its naturalness began to be more widely questioned. In her famous book *The Second Sex,* Simone de Beauvoir detailed the historical and multifaceted oppression of women, concluding that even "at the present time, when we women are beginning to take part in the affairs of the world, it is still a world that belongs to men."[40]

Some Marxists argue that the oppression of woman is only part of the oppression of the working class and that as soon as the capitalist's domination of the worker is ended woman will be herself freed, but many feminists nowadays reject this line of thinking. The woman is oppressed as worker, they agree, but there is more to woman's oppression than that. The additional dimension of oppression is tied very closely to the notions of family and patriarchy; thus, it is those institutions that have to be modified in order to put an end to man's oppression of woman. "Patriarchy" is a term used by feminists to describe the male structure of domination over women, including such substructures as the military, industry, technology, and finance.

With the division of the roles of the sexes and the subsequent stereotyping, the technological mode of knowledge, argues Shulamith Firestone,* has come to be identified with the male. Since the technological mode of knowledge has been identified as male, one might expect Firestone and other feminists to be against technology, a male achievement. But in fact, the opposite is true. Firestone argues that recently the technological mode and the "feminine" aesthetic mode have been merging to produce a more humanistic culture. Furthermore, technology today, through contraception and test-tube babies, has supplied the female with the conditions of biological liberation.[41] Thus Firestone is decidedly for technological progress; and so are many other feminists, who argue that patriarchy has kept women out of the area of technology in ways somewhat similar to (though more ancient than) those by which Third World countries have also been excluded.

Immediately after World War II ecological thought was associated primarily with concerns over nuclear testing and its effects on the environment. But its concerns go beyond that to issues of air pollution, stream pollution through the use of chemical fertilizers in nearby fields and the dumping of industrial waste, and many others.

The ecological approach to limiting man's power is based on the belief that "nature knows best" and that man's present intervention in the natural ecosystem, which developed over thousands of years, can only result in breaking the ecocycle and producing negative side effects, unless such intervention is the result of very careful study of the ecosystem. However, the article by Tucker* contests these claims in the case of insecticides.

One may hastily conclude that since ecologists view technology as being essentially an intrusion upon nature, they must be antitechnologists. But again this conclusion is erroneous. A case in point is that of Barry Commoner, who is a leading ecologist. Commoner is not an antitechnologist. Rather, he faults the method of technological and scientific study which reduces the whole to its parts and subparts. In *The Closing Circle* he points out that while a concentrated study of parts and subparts may be fruitful, a correct approach must at some point take a system as a whole into account.[42] Ignoring the whole leads to the omission of vital considerations that could reflect on a proposed technological solution to a problem. Thus what is needed is not the elimination of technology from our society, but a refinement by which the notion of efficiency it relies upon is not that of "narrow" efficiency but that of "global" efficiency involving both society and the environment. The value of a technological innovation should not be measured in terms of individual or corporate effects, but rather in terms of effects on the whole social and ecological system (for more on this topic, see the selection by Fromm).

Furthermore, Commoner argues, the practice of doing something because it can be done must be abandoned. Technological innovations must not be encouraged simply for their own sake. As Archibald Macleish comments:

> After Hiroshima it was obvious that the loyalty of science was not to humanity but to truth—its own truth—and that the law of science was not the law of the good—what humanity thinks of as good, meaning moral, decent, humane—but the law of the possible. What it is possible for science to know science must know. What it is possible for technology to do technology will have done. . . . The frustration—and it is a real and debasing frustration—in which we are mired today will not leave us until we believe in ourselves again, assume again the mastery of our lives, the management of our means.[43]

Thus, to reassume mastery over our own lives, we must restrain technology through social and ecological considerations. Otherwise, we may endanger our very survival.

In facing problems and challenges over the course of several thousand years, man has emerged as a powerful and capable *homo faber*. His problem for the future is to discover how to limit his powers and capabilities in order to enrich further all human life by living in harmony with nature, the universe, and other human beings.

NOTES

1. Samuel Florman, *The Existential Pleasures of Engineering* (New York: St. Martin's Press, Inc., 1976), pp. 106-107.
2. Plato, *The Republic,* trans. Paul Shorey, in *The Collected Dialogues,* ed. Edith Hamilton and Huntington Cairns (New York: Bollingen Foundation, 1964). See for example, Bks. VI and VIII. See also the *Phaedo,* trans. Hugh Tredennick, 72e-76d, in *The Collected Dialogues.*
3. Hannah Arendt, *The Human Condition* (Chicago: The University of Chicago Press, 1969) p. 303.
4. Aristotle, *Nichomachean Ethics,* in *Works,* ed. W. D. Ross, (London: Oxford University Press, 1963), X, 6 (1177a20-21).
5. For an excellent discussion of this aspect of Aristotle's philosophy see Webster F. Hood, "The Aristotelian Versus the Heideggerian Approach to the Problem of "Technology," in *Philosophy and Technology,* ed. Carl Mitcham and Robert Mackey (New York: The Free Press, 1972), pp. 347-63.
6. Florman, p. 101.
7. Florman, p. 103.
8. Florman, p. 110.
9. Matthew 14:15-20.
10. John 11:33-44.
11. E. A. Burtt, ed., *The Teachings of the Compassionate Buddha* (New York: The New American Library, Inc., 1955), pp. 44-46.
12. Saint Augustine, *The Trinity,* trans. A. W. Haddan, revised by W. G. T. Shedd, in *Basic Writings,* ed. Whitney J. Oates (New York: Random House, Inc., 1948), II, Bk XII, Ch. XIV.
13. Saint Augustine, Bk XII, Ch. XIII.
14. Exodus 32:3-4.
15. Florman, p. 111.
16. Quoted by W. T. Jones, *A History of Western Philosophy* (New York: Harcourt, Brace & World, Inc., 1952), p. 397.
17. Max Weber, *The Protestant Ethic and The Spirit of Capitalism,* trans. Talcott Parsons (New York: Charles Scribner's Sons, 1958), p. 162.
18. Weber, pp. 176-77.
19. Seyyed Hossein Nasr, *An Introduction to Islamic Cosmological Doctrines: Conceptions of Nature and Methods Used for its Study by the Ikhwān al-Safa, al-Birūni, and Ibn Sīnā* (Cambridge, Mass.: Belknap Press, 1964), p. 30.
20. Theodore Roszak, *Where the Wasteland Ends: Politics and Transcendence in Postindustrial Society* (Garden City, N.Y.: Anchor Books, 1973), p. 373.

21. Genesis 1:26 (KJ).
22. Nasr, p. 261.
23. Gen. 3:22-23 (KJ).
24. Gen. 1:31 (KJ).
25. Sigmund Freud, *Collected Papers,* authorized trans. under the supervision of Joan Riviere (New York: Basic Books, Inc., Publishers, 1959), IV, pp. 351-55.
26. Friedrich Nietzsche, *The Spake Zarathustra,* trans. T. Common (London: Allen and Unwin, 1967), Part II, XXV.
27. Arthur C. Danto, *Nietzsche as Philosopher* (New York: Macmillan, Inc., 1968), p. 198. This book contains an excellent discussion of the central philosophical concepts of Nietzsche that are presented in this introduction.
28. G. W. F. Hegel, *The Philosophy of Right* (New York: Oxford University Press, 1969), pp. 116, 263-64.
29. John Kenneth Galbraith, *The New Industrial State,* 2nd ed. rev. (New York: The New American Library, Inc., 1972), pp. 30-35.
30. V. I. Lenin, *Essential Works,* ed. Henry M. Christman (New York: Bantam Books, Inc., 1966), p. 229.
31. Frantz Fanon, *The Wretched of the Earth,* trans. Constance Farrington (New York: Grove Press, Inc., 1968), p. 65.
32. Adolf Hitler, *Mein Kampf* (New York: Stackpole Sons, 1939), pp. 278, 282-88.
33. Quoted by Thomas Leckie Jarman, *The Rise and Fall of Nazi Germany* (New York: New York University Press, 1956), p. 968.
34. Otto Nathan and Heinz Norden, eds., *Einstein on Peace* (New York: Simon and Schuster, Inc., 1960), p. 294-96.
35. Quoted by Robert Jay Lifton, *Death in Life: Survivors of Hiroshima* (New York: Random House, Inc., 1967), p. 22-23.
36. Quoted by Lifton, p. 27.
37. Quoted by Nathan and Norden, p. 625.
38. Karl Marx and Frederick Engels, *Collected Works* (New York: International Publishers Co., Inc., 1976), V, p. 38.
39. Karl Marx, *Das Capital,* ed. F. Engels, trans. S. Moore and E. Aveling (New York: International Publishers Co., Inc., 1967), I, p. 371.
40. Simone de Beauvoir, *The Second Sex,* trans. and ed. H. M. Parshley (New York: Bantam Books, Inc., 1970), p. xx.
41. Shulamith Firestone, *The Dialectic of Sex* (New York: Bantam Books, Inc., 1971), p. 197.
42. Barry Commoner, *The Closing Circle: Nature, Man and Technology* (New York: Alfred A. Knopf, Inc., 1971), p. 65.
43. Quoted by Commoner, pp. 180-181.

8

Introductory statements on technology

Toward a philosophy of technology

HANS JONAS

■ Hans Jonas *was born in 1903 in Mönchengladbach, Germany, and moved to the United States in 1955. He is a historian of philosophy and religion, with special interest in the ancient and medieval periods. He is also interested in the areas of technology and ethics, and in the philosophy of organism. He is presently the Alvin Johnson Professor of Philosophy Emeritus, Graduate Faculty, New School for Social Research, New York City. This article is adapted from Jonas' presentation on the occasion of his accepting the second Henry Knowles Beecher award, given by the Institute of Society, Ethics and the Life Sciences for lifetime contributions to ethics and the life sciences. Among his works are* The Phenomenon of Life *and* Philosophical Essays.

In this article Jonas defends the need for a philosophy of technology. Technology, he says, has become "the focal fact of modern life," and he gives striking examples of that fact. Jonas outlines three themes of interest to a philosophy of technology. The first is the formal *dynamics of technology, which "advances by its own 'laws of motion'." This, together with his assertions that "technology dominates our lives" and "technology is destiny," places Jonas on the side of Ellul in the Autonomous Technology Debate (see following article, and also Part Three). The second theme is the* substantive *content of technology—the things technology puts into human use, the powers it confers, and the objectives it makes possible or necessary. The third is the* moral *theme, which pertains to human responsibilities in the face of technological progress. Two salient points appear in the discussion of these themes. The first concerns a trait of modern technology—namely, the ever-continuing process of fitting ends to means. This trait is criticized by Hannah Arendt in "Instrumentality and* Homo Faber*" (see Part Three). The second point concerns the dialectic relation between science and technology.*

Are there philosophical aspects to technology? Of course there are, as there are to all things of importance in human endeavor and destiny. Modern technology touches on almost everything vital to man's existence—material, mental, and spiritual. Indeed, what of man is *not* involved? The way he lives his life and looks at objects, his intercourse with the world and with his peers, his powers and modes of action, kinds of goals, states and changes of society, objectives and forms of politics (including warfare no less than welfare), the sense and quality of life, even man's fate and that of his environment: all these are involved in the technological enterprise as it extends in magnitude and depth. The mere enumeration suggests a staggering host of potentially philosophic themes.

To put it bluntly: if there is a philosophy of science, language, history, and art; if there is social, political, and moral philosophy; philosophy of thought and of action, of reason and passion, of decision and value—all facets of the inclusive philosophy of man—how then could there not be a philosophy of technology, the focal fact of modern life? And at that a philosophy so spacious that it can house portions from all the other branches of philosophy? It is almost a truism, but at the same time so immense a proposition that its challenge staggers the mind. Economy and modesty require that we select, for a beginning, the most obvious from the mul-

□ From *The Hastings Center Report,* 9, No. 1(1979), 34-43.

titude of aspects that invite philosophical attention.

The old but useful distinction of "form" and "matter" allows us to distinguish between these two major themes: (1) the *formal dynamics* of technology as a continuing collective enterprise, which advances by its own "laws of motion"; and (2) the *substantive content* of technology in terms of the things it puts into human use, the powers it confers, the novel objectives it opens up or dictates, and the altered manner of human action by which these objectives are realized.

The first theme considers technology as an abstract whole of movement; the second considers its concrete uses and their impact on our world and our lives. The formal approach will try to grasp the pervasive "process properties" by which modern technology propels itself—through our agency, to be sure—into ever-succeeding and superceding novelty. The material approach will look at the species of novelties themselves, their taxonomy, as it were, and try to make out how the world furnished with them looks. A third, overarching theme is the *moral* side of technology as a burden on human responsibility, especially its long-term effects on the global condition of man and environment. This—my own main preoccupation over the past years—will only be touched upon.

I. THE FORMAL DYNAMICS OF TECHNOLOGY

First some observations about technology's form as an abstract whole of movement. We are concerned with characteristics of *modern* technology and therefore ask first what distinguishes it *formally* from all previous technology. One major distinction is that modern technology is an enterprise and process, whereas earlier technology was a possession and a state. If we roughly describe technology as comprising the use of artificial implements for the business of life, together with their original invention, improvement, and occasional additions, such a tranquil description will do for most of technology through mankind's career (with which it is coeval), but not for modern technology. In the past, generally speaking, a given inventory of tools and procedures used to be fairly constant, tending toward a mutually adjusting, stable equilibrium of ends and means, which—once established—represented for lengthy periods an unchallenged optimum of technical competence.

To be sure, revolutions occurred, but more by accident than by design. The agricultural revolution, the metallurgical revolution that led from the neolithic to the iron age, the rise of cities, and such developments, *happened* rather than were consciously created. Their pace was so slow that only in the time-contraction of historical retrospect do they appear to be "revolutions" (with the misleading connotation that their contemporaries experienced them as such). Even where the change was sudden, as with the introduction first of the chariot, then of armed horsemen into warfare—a violent, if short-lived, revolution indeed—the innovation did not originate from within the military art of the advanced societies that it affected, but was thrust on it from outside by the (much less civilized) peoples of Central Asia. Instead of spreading through the technological universe of their time, other technical breakthroughs, like Phoenician purple-dying, Byzantine "greek fire," Chinese porcelain and silk, and Damascene steel-tempering, remained jealously guarded monopolies of the inventor communities. Still others, like the hydraulic and steam playthings of Alexandrian mechanics, or compass and gunpowder of the Chinese, passed unnoticed in their serious technological potentials.[1]

On the whole (not counting rare upheavals), the great classical civilizations had comparatively early reached a point of technological saturation—the aforementioned "optimum" in equilibrium of means with acknowledged needs and goals—and had little cause later to go beyond it. From there on, convention reigned supreme. From pottery to monumental architecture, from food growing to shipbuilding, from textiles to engines of war, from time measuring to stargazing: tools, techniques, and objectives remained essentially the same over long times; improvements were sporadic and unplanned. Progress therefore—if it occurred at all*—was

*Progress did, in fact, occur even at the heights of classical civilizations. The Roman arch and vault, for example, were distinct engineering advances over the horizontal entablature and flat ceiling of Greek (and Egyptian) architecture, permitting spanning feats and thereby construction objectives not contemplated before (stone bridges, aqueducts, the vast baths and other public halls of Imperial Rome). But materials, tools, and techniques were still the same, the role of human labor and crafts remained unaltered, stonecutting and brickbaking went on as before. An existing technology was enlarged in its scope of performance, but none of its means or even goals made obsolete.

by inconspicuous increments to a universally high level that still excites our admiration and, in historical fact, was more liable to regression than to surpassing. The former at least was the more noted phenomenon, deplored by the epigones with a nostalgic remembrance of a better past (as in the declining Roman world). More important, there was, even in the best and most vigorous times, no proclaimed *idea* of a future of *constant progress* in the arts. Most important, there was never a deliberate method of going about it like "research," the willingness to undergo the risks of trying unorthodox paths, exchanging information widely about the experience, and so on. Least of all was there a "natural science" as a growing body of theory to guide such semitheoretical, prepractical activities, plus their social institutionalization. In routines as well as panoply of instruments, accomplished as they were for the purposes they served, the "arts" seemed as settled as those purposes themselves.*

Traits of modern technology

The exact opposite of this picture holds for modern technology, and this is its first philosophical aspect. Let us begin with some manifest traits.

1. Every new step in whatever direction of whatever technological field tends *not* to approach an equilibrium or saturation point in the process of fitting means to ends (nor is it meant to), but, on the contrary, to give rise, if successful, to further steps in all kinds of direction and with a fluidity of the ends themselves. "Tends to" becomes a compelling "is bound to" with any major or important step (this almost being its criterion); and the innovators themselves expect, beyond the accomplishment, each time, of their immediate task, the constant future repetition of their inventive activity.

2. Every technical innovation is sure to spread quickly through the technological world community, as also do theoretical discoveries in the sciences. The spreading is in terms of knowledge and of practical adoption, the first (and its speed) guaranteed by the universal intercommunication that is itself part of the technological complex, the second enforced by the pressure of competition.

3. The relation of means to ends is not unilinear but circular. Familiar ends of long standing may find better satisfaction by new technologies whose genesis they had inspired. But equally—and increasingly typical—new technologies may suggest, create, even impose new ends, never before conceived, simply by offering their feasibility. (Who had ever wished to have in his living room the Philharmonic orchestra, or open heart surgery, or a helicopter defoliating a Vietnam forest? or to drink his coffee from a disposable plastic cup? or to have artificial insemination, test-tube babies, and host pregnancies? or to see clones of himself and others walking about?) Technology thus adds to the very objectives of human desires, including objectives for technology itself. The last point indicates the dialectics or circularity of the case: once incorporated into the socioeconomic demand diet, ends first gratuitously (perhaps accidentally) generated by technological invention become necessities of life and set technology the task of further perfecting the means of realizing them.

4. Progress, therefore, is not just an ideological gloss on modern technology, and not at all a mere option offered by it, but an inherent drive which acts willy-nilly in the formal automatics of its *modus operandi* as it interacts with society. "Progress" is here not a value term but purely descriptive. We may resent the fact and despise its fruits and yet must go along with it, for—short of a stop by the fiat of total political power, or by a sustained general strike of its clients or some internal collapse of their societies, or by self-destruction through its works (the last, alas, the least unlikely of these)—the juggernaut moves on relentlessly, spawning its always mutated progeny by coping with the challenges and lures of the now. But while not a value term, "progress" here is not a neutral term either, for which we could simply substitute "change." For it is in the nature of the case, or a law of the series, that a later stage is always, in terms of technology itself, *superior* to the preceding stage.* Thus we have here a case of the entropy-defying sort (organic evolution is another), where the internal motion of a system, left to itself and not interfered with,

*One meaning of "classical" is that those civilizations had somehow implicitly "defined" themselves and neither encouraged nor even allowed to pass beyond their innate terms. The—more or less—achieved "equilibrium" was their very pride.

*This only seems to be but is not a value statement, as the reflection on, for example, an ever more destructive atom bomb shows.

leads to ever "higher," not "lower" states of itself. Such at least is the present evidence.* If Napoleon once said, "Politics is destiny," we may well say today, "Technology is destiny."

These points go some way to explicate the initial statement that modern technology, unlike traditional, is an enterprise and not a possession, a process and not a state, a dynamic thrust and not a set of implements and skills. And they already adumbrate certain "laws of motion" for this restless phenomenon. What we have described, let us remember, were formal traits which as yet say little about the contents of the enterprise. We ask two questions of this descriptive picture: *why* is this so, that is, what *causes* the restlessness of modern technology; what is the nature of the thrust? And, what is the philosophical import of the facts so explained?

The nature of restless technology

As we would expect in such a complex phenomenon, the motive forces are many, and some causal hints appeared already in the descriptive account. We have mentioned *pressure of competition*—for profit, but also for power, security, and so forth—as one perpetual mover in the universal appropriation of technical improvements. It is equally operative in their origination, that is, in the process of invention itself, nowadays dependent on constant outside subsidy and even goal-setting: potent interests see to both. War, or the threat of it, has proved an especially powerful agent. The less dramatic, but no less compelling, everyday agents are legion. To keep one's head above the water is their common principle (somewhat paradoxical, in view of an abundance already far surpassing what former ages would have lived with happily ever after). Of pressures other than the competitive ones, we must mention those of population growth and of impending exhaustion of natural resources. Since both phenomena are themselves already by-products of technology (the first by way of medical improvements, the second by the voracity of industry), they offer a good example of the more general truth that to a considerable extent technology itself begets the problems which it is then called upon to overcome by a new forward jump. (The Green Revolution and the development of synthetic substitute materials or of alternate sources of energy come under this heading.) These compulsive pressures for progress, then, would operate even for a technology in a noncompetitive, for example, a socialist setting.

A motive force more autonomous and spontaneous than these almost mechanical pushes with their "sink or swim" imperative would be the pull of the quasi-utopian *vision* of an ever better life, whether vulgarly conceived or nobly, once technology had proved the open-ended capacity for procuring the conditions for it: perceived possibility whetting the appetite ("the American dream," "the revolution of rising expectations"). This less palpable factor is more difficult to appraise, but its playing a role is undeniable. Its deliberate fostering and manipulation by the dream merchants of the industrial-mercantile complex is yet another matter and somewhat taints the spontaneity of the motive, as it also degrades the quality of the dream. It is also moot to what extent the vision itself is *post hoc* rather than *ante hoc,* that is, instilled by the dazzling feats of a technological progress already underway and thus more a response to than a motor of it.

Groping in these obscure regions of motivation, one may as well descend, for an explanation of the dynamism as such, into the Spenglerian mystery of a "Faustian soul" innate in Western culture, that drives it, nonrationally, to infinite novelty and unplumbed possibilities for their own sake; or into the Heideggerian depths of a fateful, metaphysical decision of the will for boundless power over the world of things—a decision equally peculiar to the Western mind: speculative intuitions which do strike a resonance in us, but are beyond proof and disproof.

Surfacing once more, we may also look at the very sober, functional facts of industrialism as such, of production and distribution, output maximization, managerial and labor aspects, which even apart from competitive pressure provide their own incentives for technical progress. Similar observations apply to the requirements of *rule* or control in the vast and populous states of our time, those giant territorial superorganisms which for their very cohesion depend on advanced technology (for example, in information, communication, and transportation, not

*There may conceivably be internal degenerative factors—such as the overloading of finite information-processing capacity—that may bring the (exponential) movement to a halt or even make the system fall apart. We don't know yet.

to speak of weaponry) and thus have a stake in its promotion: the more so, the more centralized they are. This holds for socialist systems no less than for free-market societies. May we conclude from this that even a communist world state, freed from external rivals as well as from internal free-market competition, might still have to push technology ahead for purposes of control on this colossal scale? Marxism, in any case, has its own inbuilt commitment to technological progress beyond necessity. But even disregarding all dynamics of these conjectural kinds, the most monolithic case imaginable would, at any rate, still be exposed to those noncompetitive, natural pressures like population growth and dwindling resources that beset industrialism as such. Thus, it seems, the compulsive element of technological progress may not be bound to its original breeding ground, the capitalist system. Perhaps the odds for an eventual stabilization look somewhat better in a socialist system, provided it is worldwide—and possibly totalitarian in the bargain. As it is, the pluralism we are thankful for ensures the constancy of compulsive advance.

We could go on unravelling the causal skein and would be sure to find many more strands. But none nor all of them, much as they explain, would go to the heart of the matter. For all of them have one premise in common without which they could not operate for long: the premise that there *can* be indefinite progress because there *is* always something new and better to find. The, by no means obvious, givenness of this objective condition is also the pragmatic conviction of the performers in the technological drama; but without its being true, the conviction would help as little as the dream of the alchemists. Unlike theirs, it is backed up by an impressive record of past successes, and for many this is sufficient ground for their belief. (Perhaps holding or not holding it does not even greatly matter.) What makes it more than a sanguine belief, however, is an underlying and well-grounded, theoretical view of the nature of things and of human cognition, according to which they do not set a limit to novelty of discovery and invention, indeed, that they of themselves will at each point offer another opening for the as yet unknown and undone. The corollary conviction, then, is that a technology tailored to a nature and to a knowledge of this indefinite potential ensures its indefinitely continued conversion into the practical powers, each step of it begetting the next, with never a cutoff from internal exhaustion of possibilities.

Only habituation dulls our wonder at this wholly unprecedented belief in virtual "infinity." And by all our present comprehension of reality, the belief is most likely true—at least enough of it to keep the road for innovative technology in the wake of advancing science open for a long time ahead. Unless we understand this ontologic-epistomological premise, we have not understood the inmost agent of technological dynamics, on which the working of all the adventitious causal factors is contingent in the long run.

Let us remember that the virtual infinitude of advance we here seek to explain is in essence different from the always avowed perfectibility of every human accomplishment. Even the undisputed master of his craft always had to admit as possible that he might be surpassed in skill or tools or materials; and no excellence of product ever foreclosed that it might still be bettered, just as today's champion runner must know that his time may one day be beaten. But these are improvements within a given genus, not different in kind from what went before, and they must accrue in diminishing fractions. Clearly, the phenomenon of an exponentially growing *generic* innovation is qualitatively different.

Science as a source of restlessness

The answer lies in the interaction of *science* and *technology* that is the hallmark of modern progress, and thus ultimately in the kind of nature which modern science progressively discloses. For it is here, in the movement of *knowledge,* where relevant novelty first and constantly occurs. This is itself a novelty. To Newtonian physics, nature appeared simple, almost crude, running its show with a few kinds of basic entities and forces by a few universal laws, and the application of those well-known laws to an ever greater variety of composite phenomena promised ever widening knowledge indeed, but no real surprises. Since the mid-nineteenth century, this minimalistic and somehow finished picture of nature has changed with breathtaking acceleration. In a reciprocal interplay with the growing subtlety of exploration (instrumental and conceptual), nature itself stands forth as ever more subtle. The progress of probing makes the object grow richer in modes of operation, not sparer as classical mechanics had expected. And instead of narrowing the margin of the still-undiscovered, science now surprises itself with unlocking dimension after dimension of new depths. The very essence of matter has turned from a blunt, irreducible ultimate to an always reopened challenge for further penetration. No one can say whether this will go on forever, but a suspicion of intrinsic infinity in the very being of things obtrudes itself and therewith an anticipation of unending inquiry of the sort where succeeding steps will not find the same old story again (Descartes' "matter in motion"), but always add new twists to it. If then the art of technology is correlative to the knowledge of nature, technology too acquires from this source that potential of infinity for its innovative advance.

But it is not just that indefinite scientific progress offers the *option* of indefinite technological progress, to be exercised or not as other interests see fit. Rather the cognitive process itself moves by interaction with the technological, and in the most internally vital sense: for its own *theoretical* purpose, science must generate an increasingly sophisticated and physically formidable technology as its tool. What it finds with this help initiates new departures in the practical sphere, and the latter as a whole, that is, technology at work provides with its experiences a large-scale laboratory for science again, a breeding ground for new questions, and so on in an unending cycle. In brief, a mutual feedback operates between science and technology; each requires and propels the other; and as matters now stand, they can only live together or must die together. For the dynamics of technology, with which we are here concerned, this means that (all external promptings apart) an agent of restlessness is implanted in it by its functionally integral bond with science. As long, therefore, as the cognitive impulse lasts, technology is sure to move ahead with it. The cognitive impulse, in its turn, culturally vulnerable in itself, liable to lag or to grow conservative with a treasured canon—that theoretical eros itself no longer lives on the delicate appetite for truth alone, but is spurred on by its hardier offspring, technology, which communicates to it impulsions from the broadest arena of struggling, insistent life. Intellectual curiosity is seconded by interminably self-renewing practical aim.

I am conscious of the conjectural character of some of these thoughts. The revolutions in science over the last fifty years or so are a fact, and so are the revolutionary style they imparted to technology and the reciprocity between the two concurrent streams (nuclear physics is a good example). But whether those scientific revolutions, which hold primacy in the whole syndrome, will be typical for science henceforth—something like a law of motion for its future—or represent only a singular phase in its longer run, is unsure. To the extent, then, that our forecast of incessant novelty for technology was predicated on a guess concerning the future of science, even concerning the nature of things, it is hypothetical, as such extrapolations are bound to be. But even if the recent past did not usher in a state of permanent revolution for science, and the life of theory settles down again to a more sedate pace, the scope for technological innovation will not easily shrink; and what may no longer be a revolution in science, may still revolutionize our lives in its practical impact through technology. "Infinity" being too large a word anyway, let us say that present signs of potential and of incentives point to an indefinite perpetuation and fertility of the technological momentum.

The philosophical implications

It remains to draw philosophical conclusions from our findings, at least to pinpoint aspects of philosophical interest. Some preceding remarks have already been straying into philosophy of science in the technical sense. Of broader issues, two will be ample to provide food for further thought beyond the limitations of this paper. One concerns the status of knowledge in the human scheme, the other the status of technology itself as a human goal, or its tendency to become that from being a means, in a dialectical inversion of the means-end order itself.

Concerning knowledge, it is obvious that the time-honored division of theory and practice has vanished for both sides. The thirst for pure knowledge may persist undiminished, but the involvement of knowing at the heights with doing in the lowlands of life, mediated by technology, has become inextricable; and the aristocratic self-sufficiency of knowing for its own (and the knower's) sake has gone. Nobility has been exchanged for utility. With the possible exception of philosophy, which still can do with paper and pen and tossing thoughts around among peers, all knowledge has become thus tainted, or elevated if you will, whether utility is intended or not. The technological syndrome, in other words, has brought about a thorough *socializing* of the theoretical realm, enlisting it in the service of common need. What used to be the freest of human choices, an extravagance snatched from the pressure of the world—the esoteric life of thought—has become part of the great public play of necessities and a prime necessity in the action of the play.* Remotest abstraction has become enmeshed with nearest concreteness. What this pragmatic functionalization of the once highest indulgence in impractical pursuits portends for the image of man, for the restructuring of a hallowed hierarchy of values, for the idea of "wisdom," and so on, is surely a subject for philosophical pondering.

Concerning technology itself, its actual role in modern life (as distinct from the purely instrumental definition of technology as such) has made the relation of means and ends equivocal all the way up from the daily living to the very vocation of man. There could be no question in former technology that its role was that of humble servant—pride of workmanship and esthetic embellishment of the useful notwithstanding. The Promethean enterprise of modern technology speaks a different language. The word "enterprise" gives the clue, and its unendingness another. We have mentioned that the effect of its innovations is disequilibrating rather than equilibrating with respect to the balance of wants and supply, always breeding its own new wants. This in itself compels the constant attention of the best minds, engaging the full capital of human ingenuity for meeting challenge after challenge and seizing the new chances. It is psychologically natural for that degree of engagement to be invested with the dignity of dominant purpose. Not only does technology dominate our lives in fact, it nourishes also a belief in its being of predominant worth. The sheer grandeur of the enterprise and its seeming infinity inspire enthusiasm and fire ambition. Thus, in addition to spawning new ends (worthy or frivolous) from the mere invention of means, technology as a grand venture tends to establish *itself* as the transcendent end. At least the suggestion is there and casts its spell on the modern mind. At its most modest, it means elevating *homo faber* to the essential aspect of man; at its most extravagant, it means elevating *power* to the position of his dominant and interminable goal. To become ever more masters of the world, to advance from power to power, even if only collectively and perhaps no longer by choice, *can* now be seen to be the chief vocation of mankind. Surely, this again poses philosophical questions that may well lead unto the uncertain grounds of metaphysics or of faith.

I here break off, arbitrarily, the formal account of the technological movement in general, which as yet has told us little of what the enterprise is about. To this subject I now turn, that is, to the new kinds of powers and objectives that technology opens to modern man and the consequently altered quality of human action itself.

II. THE MATERIAL WORKS OF TECHNOLOGY

Technology is a species of power, and we can ask questions about how and on what object any power is exercised. Adopting Aristotle's rule in *de anima* that for understanding a faculty one should begin with its objects, we start from them too—"objects" meaning both the visible *things* technology generates and puts into human use, and the *objectives* they serve. The objects of modern technology are first everything that had always been an object of human artifice and labor: food, clothing, shelter, implements, transportation—all the material necessities and comforts of life. The technological intervention changed at first not the product but its production, in speed, ease, and quantity. However, this is true only of the very first stage of the industrial revolution with which large-scale scientific technology began. For example, the cloth for the steam-driven looms of Lancashire remained the same. Even then, one significant new product was added to the traditional list—the machines themselves, which required an entire new industry with further subsidiary industries to build them. These novel entities, machines—at first capital goods only, not consumer goods—had from the beginning their

*There is a paradoxical side effect to this change of roles. That very science which forfeited its place in the domain of leisure to become a busy toiler in the field of common needs, creates by its toils a growing domain of leisure for the masses, who reap this with the other fruits of technology as an additional (and no less novel) article of forced consumption. Hence leisure, from a privilege of the few, has become a problem for the many to cope with. Science, not idle, provides for the needs of this idleness too: no small part of technology is spent on filling the leisure-time gap which technology itself has made a fact of life.

own impact on man's symbiosis with nature by being consumers themselves. For example: steam-powered water pumps facilitated coal mining, required in turn extra coal for firing their boilers, more coal for the foundries and forges that made those boilers, more for the mining of the requisite iron ore, more for its transportation to the foundries, more—both coal and iron—for the rails and locomotives made in these same foundries, more for the conveyance of the foundries' product to the pitheads and return, and finally more for the distribution of the more abundant coal to the users outside this cycle, among which were increasingly still more machines spawned by the increased availability of coal. Lest it be forgotten over this long chain, we have been speaking of James Watt's modest steam engine for pumping water, out of mine shafts. This syndrome of self-proliferation—by no means a linear chain but an intricate web of reciprocity—has been part of modern technology ever since. To generalize, technology exponentially increases man's drain on nature's resources (of substances and of energy), not only through the multiplication of the final goods for consumption, but also, and perhaps more so, through the production and operation of its own mechanical means. And with these means—machines—it introduced a new category of goods, not for consumption, added to the furniture of our world. That is, among the objects of technology a prominent class is that of technological apparatus itself.

Soon other features also changed the initial picture of a merely mechanized production of familiar commodities. The final products reaching the consumer ceased to be the same, even if still serving the same age-old needs; new needs, or desires, were added by commodities of entirely new kinds which changed the habits of life. Of such commodities, machines themselves became increasingly part of the consumer's daily life to be used directly by himself, as an article not of production but of consumption. My survey can be brief as the facts are familiar.

New kinds of commodities

When I said that the cloth of the mechanized looms of Lancashire remained the same, everyone will have thought of today's synthetic fibre textiles for which the statement surely no longer holds. This is fairly recent, but the general phenomenon starts much earlier, in the synthetic dyes and fertilizers with which the chemical industry—the first to be wholly a fruit of science—began. The original rationale of these technological feats was substitution of artificial for natural materials (for reasons of scarcity or cost), with as nearly as possible the same properties for effective use. But we need only think of plastics to realize that art progressed from substitutes to the creation of really new substances with properties not so found in any natural one, raw or processed, thereby also initiating uses not thought of before and giving rise to new classes of objects to serve them. In chemical (molecular) engineering, man does more than in mechanical (molar) engineering which constructs machinery from natural materials; his intervention is deeper, redesigning the infra-patterns of nature, making substances to specification by arbitrary disposition of molecules. And this, be it noted, is done deductively from the bottom, from the thoroughly analyzed last elements, that is, in a real *via compositiva* after the completed *via resolutiva,* very different from the long-known empirical practice of coaxing substances into new properties, as in metal alloys from the bronze age on. Artificiality or creative engineering with abstract construction invades the heart of matter. This, in molecular biology, points to further, awesome potentialities.

With the sophistication of molecular alchemy we are ahead of our story. Even in straightforward hardware engineering, right in the first blush of the mechanical revolution, the objects of use that came out of the factories did not really remain the same, even where the objectives did. Take the old objective of travel. Railroads and ocean liners are relevantly different from the stage coach and from the sailing ship, not merely in construction and efficiency but in the very feel of the user, making travel a different experience altogether, something one may do for its own sake. Airplanes, finally, leave behind any similarity with former conveyances, except the purpose of getting from here to there, with no experience of what lies in between. And these instrumental objects occupy a prominent, even obtrusive place in our world, far beyond anything wagons and boats ever did. Also they are constantly subject to improvement of design, with obsolescence rather than wear determining their life span.

Or take the oldest, most static of artifacts: human habitation. The multistoried office building of steel, concrete, and glass is a qualitatively different entity from the wood, brick, and stone structures of old. With all that goes into it besides the structures as such—the plumbing and wiring, the elevators, the lighting, heating, and cooling systems—it embodies the end products of a whole spectrum of technologies and far-flung industries, where only at the remote sources human hands still meet with primary materials, no longer recognizable in the final result. The ultimate customer inhabiting the product is ensconced in a shell of thoroughly derivative artifacts (perhaps relieved by a nice piece of driftwood). This transformation into utter artificiality is generally, and increasingly, the effect of technology on the human environment, down to the items of daily use. Only in agriculture has the product so far escaped this transformation by the changed modes of its production. We still eat the meat and rice of our ancestors.*

Then, speaking of the commodities that technology injects into private use, there are machines themselves, those very devices of its own running, originally confined to the economic sphere. This unprecedented novum in the records of individual living started late in the nineteenth century and has since grown to a pervading mass phenomenon in the Western world. The prime example, of course, is the automobile, but we must add to it the whole gamut of household appliances—refrigerators, washers, dryers, vacuum cleaners—by now more common in the lifestyle of the general population than running water or central heating were one hundred years ago. Add lawn mowers and other power tools for home and garden; we are mechanized in our daily chores and recreations (including the toys of our children) with every expectation that new gadgets will continue to arrive.

These paraphernalia are machines in the precise sense that they perform work and consume energy, and their moving parts are of the familiar magnitudes of our perceptual world. But an additional and profoundly different category of technical apparatus was dropped into the lap of the private citizen, not labor-saving and work-performing, partly not even utilitarian, but—with minimal energy input—catering to the senses and the mind: telephone, radio, television, tape recorders, calculators, record players—all the domestic terminals of the electronics industry, the latest arrival on the technological scene. Not only by their insubstantial, mind-addressed output, also by the subvisible, not literally "mechanical" physics of their functioning do these devices differ in kind from all the macroscopic, bodily moving machinery of the classical type. Before inspecting this momentous turn from power engineering, the hallmark of the first industrial revolution, to communication engineering, which almost amounts to a second industrial-technological revolution, we must take a look at its natural base: electricity.

In the march of technology to ever greater artificiality, abstraction, and subtlety, the unlocking of electricity marks a decisive step. Here is a universal force of nature which yet does not naturally appear to man (except in lightning). It is not a datum of uncontrived experience. Its very "appearance" had to wait for science, which contrived the experience for it. Here, then, a technology depended on science for the mere providing of its "object," the entity itself it would deal with—the first case where theory alone, not ordinary experience, wholly preceded practice (repeated later in the case of nuclear energy). And what sort of entity! Heat and steam are familiar objects of sensuous experience, their force bodily displayed in nature; the matter of chemistry is still the concrete, corporeal stuff mankind had always known. But electricity is an abstract object, disembodied, immaterial, unseen; in its usable form, it is entirely an artifact, generated in a subtle transformation from grosser forms of energy (ultimately from heat via motion). Its theory indeed had to be essentially complete before utilization could begin.

Revolutionary as electrical technology was in itself, its purpose was at first the by now conventional one of the industrial revolution in general: to supply motive power for the propulsion of machines. Its advantages lay in the unique versatility of the new force, the ease of its transmission, transformation and distribution—an unsubstantial commodity, no bulk, no weight, instantaneously delivered at the point of consumption. Nothing like it had ever existed before in man's traffic with matter, space, and

---

* Not so, objects my colleague Robert Heilbroner in a letter to me; "I'm sorry to tell you that meat and rice are both *profoundly* influenced by technology. Not even they are left untouched." Correct, but they are at least generically the same (their really profound changes lie far back in the original breeding of domesticated strains from wild ones—as in the case of all cereal plants under cultivation). I am speaking here of an order of transformation in which the results bear no resemblance to the natural materials at their source, nor to any naturally occurring state of them.

time. It made possible the spread of mechanization to every home; this alone was a tremendous boost to the technological tide, at the same time hooking private lives into centralized public networks and thus making them dependent on the functioning of a total system as never before, in fact, for every moment. Remember, you cannot hoard electricity as you can coal and oil, or flour and sugar for that matter.

But something much more unorthodox was to follow. As we all know, the discovery of the universe of electromagnetics caused a revolution in theoretical physics that is still underway. Without it, there would be no relativity theory, no quantum mechanics, no nuclear and subnuclear physics. It also caused a revolution in technology beyond what it contributed, as we noted, to its classical program. The revolution consisted in the passage from electrical to electronic technology which signifies a new level of abstraction in means and ends. It is the difference between power and communication engineering. Its object, the most impalpable of all, is information. Cognitive instruments had been known before—sextant, compass, clock, telescope, microscope, thermometer, all of them for information and not for work. At one time, they were called "philosophical" or "metaphysical" instruments. By the same general criterion, amusing as it may seem, the new electronic information devices, too, could be classed as "philosophical instruments." But those earlier cognitive devices, except the clock, were inert and passive, not generating information actively, as the new instrumentalities do.

Theoretically as well as practically, electronics signifies a genuinely new phase of the scientific-technological revolution. Compared with the sophistication of its theory as well as the delicacy of its apparatus, everything which came before seems crude, almost natural. To appreciate the point, take the man-made satellites now in orbit. In one sense, they are indeed an imitation of celestial mechanics—Newton's laws finally verified by cosmic experiment: astronomy, for millenia the most purely contemplative of the physical sciences, turned into a practical art! Yet, amazing as it is, the astronomic imitation, with all the unleashing of forces and the finesse of techniques that went into it, is the least interesting aspect of those entities. In that respect, they still fall within the terms and feats of classical mechanics (except for the remote-control course corrections).

Their true interest lies in the instruments they carry through the voids of space and in what these do, their measuring, recording, analyzing, computing, their receiving, processing, and transmitting abstract information and even images over cosmic distances. There is nothing in all nature which even remotely foreshadows the kind of things that now ride the heavenly spheres. Man's imitative practical astronomy merely provides the vehicle for something else with which he sovereignly passes beyond all the models and usages of known nature.* That the advent of man portended, in its inner secret of mind and will, a cosmic event was known to religion and philosophy: now it manifests itself as such by fact of things and acts in the visible universe. Electronics indeed creates a range of objects imitating nothing and progressively added to by pure invention.

And no less invented are the ends they serve. Power engineering and chemistry for the most part still answered to the natural needs of man: for food, clothing, shelter, locomotion, and so forth. Communication engineering answers to needs of information and control solely created by the civilization that made this technology possible and, once started, imperative. The novelty of the means continues to engender no less novel ends—both becoming as necessary to the functioning of the civilization that spawned them as they would have been pointless for any former one. The world they help to constitute and which needs computers for its very running is no longer nature supplemented, imitated, improved, transformed, the original habitat made more habitable. In the pervasive mentalization of physical relationships it is a *trans-nature* of human making, but with this inherent paradox: that it threatens the obsolescence of man himself, as increasing automation ousts him from the places of work where he formerly proved his humanhood. And there is a further threat: its strain on nature herself may reach a breaking point.

The last stage of the revolution?

That sentence would make a good dramatic ending. But it is not the end of the story. There

* Note also that in radio technology, the medium of action is nothing material, like wires conducting currents, but the entirely immaterial electromagnetic "field," i.e., space itself. The symbolic picture of "waves" is the last remaining link to the forms of our perceptual world.

may be in the offing another, conceivably the last, stage of the technological revolution, after the mechanical, chemical, electrical, electronic stages we have surveyed, and the nuclear we omitted. All these were based on physics and had to do with what man can put to his use. What about biology? And what about the user himself? Are we, perhaps, on the verge of a technology, based on biological knowledge and wielding an engineering art which, this time, has man himself for its object? This has become a theoretical possibility with the advent of molecular biology and its understanding of genetic programming; and it has been rendered morally possible by the metaphysical neutralizing of man. But the latter, while giving us the license to do as we wish, at the same time denies us the guidance for knowing what to wish. Since the same evolutionary doctrine of which genetics is a cornerstone has deprived us of a valid image of man, the actual techniques, when they are ready, may find us strangely unready for their responsible use. The anti-essentialism of prevailing theory, which knows only of *de facto* outcomes of evolutionary accident and of no valid essences that would give sanction to them, surrenders our being to a freedom without norms. Thus the technological call of the new microbiology is the twofold one of physical feasibility and metaphysical admissibility. Assuming the genetic mechanism to be completely analyzed and its script finally decoded, we can set about rewriting the text. Biologists vary in their estimates of how close we are to the capability; few seem to doubt the right to use it. Judging by the rhetoric of its prophets, the idea of taking our evolution into our own hands is intoxicating even to many scientists.

In any case, the idea of making over man is no longer fantastic, nor interdicted by an inviolable taboo. If and when *that* revolution occurs, if technological power is really going to tinker with the elemental keys on which life will have to play its melody in generations of men to come (perhaps the only such melody in the universe), then a reflection on what is humanly desirable and what should determine the choice—a reflection, in short, on the image of man, becomes an imperative more urgent than any ever inflicted on the understanding of mortal man. Philosophy, it must be confessed, is sadly unprepared for this, its first cosmic task.

III. TOWARD AN ETHICS OF TECHNOLOGY

The last topic has moved naturally from the descriptive and analytic plane, on which the objects of technology are displayed for inspection, onto the evaluative plane where their ethical challenge poses itself for decision. The particular case forced the transition so directly because there the (as yet hypothetical) technological object was man directly. But once removed, man is involved in all the other objects of technology, as these singly and jointly remake the worldly frame of his life, in both the narrower and the wider of its senses: that of the artificial frame of civilization in which social man leads his life proximately, and that of the natural terrestrial environment in which this artifact is embedded and on which it ultimately depends.

Again, because of the magnitude of technological effects on both these vital environments in their totality, both the quality of human life and its very preservation in the future are at stake in the rampage of technology. In short, certainly the "image" of man, and possibly the survival of the species (or of much of it), are in jeopardy. This would summon man's duty to his cause even if the jeopardy were not of his own making. But it is, and, in addition to his ageless obligation to meet the threat of things, he bears for the first time the responsibility of prime agent in the threatening disposition of things. Hence nothing is more natural than the passage from the objects to the ethics of technology, from the things made to the duties of their makers and users.

A similar experience of inevitable passage from analysis of fact to ethical significance, let us remember, befell us toward the end of the first section. As in the case of the matter, so also in the case of the form of the technological dynamics, the image of man appeared at stake. In view of the quasi-automatic compulsion of those dynamics, with their perspective of indefinite progression, every existential and moral question that the objects of technology raise assumes the curiously eschatological quality with which we are becoming familiar from the extrapolating guesses of futurology. But apart from thus raising all challenges of present particular matter to the higher powers of future exponential magnification, the despotic dynamics of the technological movement as such, sweeping its captive movers along in its breathless momentum, poses its own questions to man's axiological conception of himself. Thus, form and matter of technology alike enter into the dimension of ethics.

The questions raised for ethics by the objects of technology are defined by the major areas of their impact and thus fall into such fields of knowledge as ecology (with all its biospheric subdivisions of land, sea, and air), demography economics, biomedical and behavioral sciences (even the psychology of mind pollution by television), and so forth. Not even a sketch of the substantive problems, let alone of ethical policies for dealing with them, can here be attempted. Clearly, for a normative rationale of the latter, ethical theory must plumb the very foundations of value, obligation, and the human good.

The same holds of the different kind of questions raised for ethics by the sheer fact of the formal dynamics of technology. But here, a question of another order is added to the straightforward ethical questions of both kinds, subjecting any resolution of them to a pragmatic proviso of harrowing uncertainty. Given the mastery of the creation over its creators, which yet does not abrogate their responsibility nor silence their vital interest, what are the chances and what are the means of gaining *control* of the process, so that the results of any ethical (or even purely prudential) insights can be translated into effective action? How in short can man's freedom prevail against the determinism he has created for himself? On this most clouded question, whereby hangs not only the effectuality or futility of the ethical search which the facts invite (assuming it to be blessed with *theoretical* success!), but perhaps the future of mankind itself, I will make a few concluding, but—alas—inconclusive, remarks. They are intended to touch on the whole ethical enterprise.

Problematic preconditions of an effective ethics

First, a look at the novel state of determinism. Prima facie, it would seem that the greater and more varied powers bequeathed by technology have expanded the range of choices and hence increased human freedom. For economics, for example, the argument has been made[2] that the uniform compulsion which scarcity and subsistence previously imposed on economic behavior with a virtual denial of alternatives (and hence—conjoined with the universal "maximization" motive of capitalist market competition—gave classical economics at least the appearance of a deterministic "science") has given way to a latitude of indeterminacy. The plenty and powers provided by industrial technology allow a pluralism of choosable alternatives (hence disallow scientific prediction). We are not here concerned with the status of economics as a science. But as to the altered state of things alleged in the argument, I submit that the change means rather that one, relatively homogeneous determinism (thus relatively easy to formalize into a law) has been supplanted by another, more complex, multifarious determinism, namely, that exercised by the human artifact itself upon its creator and user. We, abstractly speaking the possessors of those powers, are concretely subject to their emancipated dynamics and the sheer momentum of our own multitude, the vehicle of those dynamics.

I have spoken elsewhere[3] of the "new realm of necessity" set up, like a second nature, by the feedbacks of our achievements. The almighty we, or Man personified is, alas, an abstraction. *Man* may have become more powerful; *men* very probably the opposite, enmeshed as they are in more dependencies than ever before. What ideal Man now can do is not the same as what real men permit or dictate to be done. And here I am thinking not only of the immanent dynamism, almost automatism, of the impersonal technological complex I have invoked so far, but also of the pathology of its client society. Its compulsions, I fear, are at least as great as were those of unconquered nature. Talk of the blind forces of nature! Are those of the sorcerer's creation less blind? They differ indeed in the serial shape of their causality: the action of nature's forces is cyclical, with periodical recurrence of the same, while that of the technological forces is linear, progressive, cumulative, thus replacing the curse of constant toil with the threat of maturing crisis and possible catastrophe. Apart from this significant vector difference, I seriously wonder whether the tyranny of fate has not become greater, the latitude of spontaneity smaller; and whether man has not actually been weakened in his decision-making capacity by his accretion of collective strength.

However, in speaking, as I have just done, of "his" decision-making capacity, I have been guilty of the same abstraction I had earlier criticized in the use of the term "man." Actually, the subject of the statement was no real or representative individual but Hobbes' "Artificial

Man,'' ''that great Leviathan, called a Common-Wealth,'' or the ''large horse'' to which Socrates likened the city, ''which because of its great size tends to be sluggish and needs stirring by a gadfly.'' Now, the chances of there being such gadflies among the numbers of the commonwealth are today no worse nor better than they have ever been, and in fact they are around and stinging in our field of concern. In that respect, the free spontaneity of personal insight, judgment, and responsible action by speech can be trusted as an ineradicable (if also incalculable) endowment of humanity, and smallness of number is in itself no impediment to shaking public complacency. The problem, however, is not so much complacency or apathy as the counterforces of active, and anything but complacent, interests and the complicity with them of all of us in our daily consumer existence. These interests themselves are factors in the determinism which technology has set up in the space of its sway. The question, then, is that of the possible chances of unselfish insight in the arena of (by nature) selfish *power,* and more particularly: of one long-range, interloping insight against the short-range goals of many incumbent powers. Is there hope that wisdom itself can become power? This renews the thorny old subject of Plato's philosopher-king and—with that inclusion of realism which the utopian Plato did not lack—of the role of myth, not knowledge, in the education of the guardians. Applied to our topic: the *knowledge* of objective dangers and of values endangered, as well as of the technical remedies, is beginning to be there and to be disseminated; but to make it prevail in the marketplace is a matter less of the rational dissemination of truth than of public relations techniques, persuasion, indoctrination, and manipulation, also of unholy alliances, perhaps even conspiracy. The philosopher's descent into the cave may well have to go all the way to ''if you can't lick them, join them.''

That is so not merely because of the active resistance of special interests but because of the optical illusion of the near and the far which condemns the long-range view to impotence against the enticement and threats of the nearby: it is this incurable shortsightedness of animal-human nature more than ill will that makes it difficult to move even those who have no special axe to grind, but still are in countless ways, as we all are, beneficiaries of the untamed system and so have something dear in the present to lose with the inevitable cost of its taming. The taskmaster, I fear, will have to be actual pain beginning to strike, when the far has moved close to the skin and has vulgar optics on its side. Even then, one may resort to palliatives of the hour. In any event, one should try as much as one can to forestall the advent of emergency with its high tax of suffering or, at the least, prepare for it. This is where the scientist can redeem his role in the technological estate.

The incipient knowledge about technological danger trends must be developed, coordinated, systematized, and the full force of computer-aided projection techniques be deployed to determine priorities of action, so as to inform preventive efforts wherever they can be elicited, to minimize the necessary sacrifices, and at the worst to preplan the saving measures which the terror of beginning calamity will eventually make people willing to accept. Even now, hardly a decade after the first stirrings of ''environmental'' consciousness, much of the requisite knowledge, plus the rational persuasion, is available inside and outside academia for any well-meaning powerholder to draw upon. To this, we—the growing band of concerned intellectuals—ought persistently to contribute our bit of competence and passion.

But the real problem is to get the well-meaning into power and have that power as little as possible beholden to the interests which the technological colossus generates on its path. It is the problem of the philosopher-king compounded by the greater magnitude and complexity (also sophistication) of the forces to contend with. Ethically, it becomes a problem of playing the game by its impure rules. For the servant of truth to join in it means to sacrifice some of his time-honored role: he may have to turn apostle or agitator or political operator. This raises moral questions beyond those which technology itself poses, that of sanctioning immoral means for a surpassing end, of giving unto Caesar so as to promote what is not Caesar's. It is the grave question of moral casuistry, or of Dostoevsky's Grand Inquisitor, or of regarding cherished liberties as no longer affordable luxuries (which may well bring the anxious friend of mankind into odious political company)—questions one excusably hesitates to touch but in the further tide of things may not be permitted to evade.

What is, prior to joining the fray, the role of philosophy, that is, of a philosophically grounded ethical knowledge, in all this? The somber note of the last remarks responded to the quasi-apocalyptic prospects of the technological tide, where stark issues of planetary survival loom ahead. There, no philosophical ethics is needed to tell us that disaster must be averted. Mainly, this is the case of the ecological dangers. But there are other, noncatastrophic things afoot in technology where not the existence but the image of man is at stake. They are with us now and will accompany us and be joined by others at every new turn technology may take. Mainly, they are in the biomedical, behavioral, and social fields. They lack the stark simplicity of the survival issue, and there is none of the (at least declaratory) unanimity on them which the spectre of extreme crisis commands. It is here where a philosophical ethics or theory of values has its task. Whether its voice will be listened to in the dispute on policies is not for it to ask; perhaps it cannot even muster an authoritative voice with which to speak—a house divided, as philosophy is. But the philosopher must try for normative knowledge, and if his labors fall predictably short of producing a compelling axiomatics, at least his clarifications can counteract rashness and make people pause for a thoughtful view.

Where not existence but "quality" of life is in question, there is room for honest dissent on goals, time for theory to ponder them, and freedom from the tyranny of the lifeboat situation. Here, philosophy can have its try and its say. Not so on the extremity of the survival issue. The philosopher, to be sure, will also strive for a theoretical grounding of the very proposition that there ought to be men on earth, and that present generations are obligated to the existence of future ones. But such esoteric, ultimate validation of the perpetuity imperative for the species—whether obtainable or not to the satisfaction of reason—is happily not needed for consensus in the face of ultimate threat. Agreement in favor of life is pretheoretical, instinctive, and universal. Averting disaster takes precedence over everything else, including pursuit of the good, and suspends otherwise inviolable prohibitions and rules. All moral standards for individual or group behavior, even demands for individual sacrifice of life, are premised on the continued existence of human life. As I have said elsewhere,[4] "No rules can be devised for the waiving of rules in extremities. As with the famous shipwreck examples of ethical theory, the less said about it, the better."

Never before was there cause for considering the contingency that all mankind may find itself in a lifeboat, but this is exactly what we face when the viability of the planet is at stake. Once the situation becomes desperate, then what there is to do for salvaging it must be done, so that there be life—which "then," after the storm has been weathered, can again be adorned by ethical conduct. The moral inference to be drawn from this lurid eventuality of a moral pause is that we must never allow a lifeboat situation for humanity to arise.[5] One part of the ethics of technology is precisely to guard the space in which any ethics can operate. For the rest, it must grapple with the cross-currents of value in the complexity of life.

A final word on the question of determinism versus freedom which our presentation of the technological syndrome has raised. The best hope of man rests in his most troublesome gift: the spontaneity of human acting which confounds all prediction. As the late Hannah Arendt never tired of stressing: the continuing arrival of newborn individuals in the world assures ever-new beginnings. We should expect to be surprised and to see our predictions come to naught. But those predictions themselves, with their warning voice, can have a vital share in provoking and informing the spontaneity that is going to confound them.

REFERENCES

[1]But as serious an actuality as the Chinese plough "wandered" slowly westward with little traces of its route and finally caused a major, highly beneficial revolution in medieval European agriculture, which almost no one deemed worth recording when it happened (cf. Paul Leser, *Entstehung und Verbreitung des Pfluges,* Münster, 1931; reprint: The International Secretariate for Research on the History of Agricultural Implements, Brede-Lingby, Denmark, 1971).

[2]I here loosely refer to Adolph Lowe, "The Normative Roots of Economic Values," in Sidney Hook, ed., *Human Values and Economic Policy* (New York: New York University Press, 1967) and, more perhaps, to the many discussions I had with Lowe over the years. For my side of the argument, see "Economic Knowledge and the Critique of Goals," in R. L. Heilbroner, ed., *Economic Means and Social Ends* (Englewood Cliffs, N.J.: Prentice-Hall, 1969), reprinted in Hans Jonas, *Philosophical Essays* (Englewood Cliffs, N.J.: Prentice-Hall, 1969), reprinted in Hans Jonas, *Philosophical Essays* (Englewood Cliffs, N.J.: Prentice-Hall, 1974).

[3]"The Practical Uses of Theory," *Social Research* 26 (1959), reprinted in Hans Jonas, *The Phenomenon of Life* (New York, 1966). The reference is to pp. 209-10 in the latter edition.

[4] "Philosophical Reflections on Experimenting with Human Subjects," in Paul A. Freund, ed., *Experimentation with Human Subjects* (New York: George Braziller, 1970), reprinted in Hans Jonas, *Philosophical Essays*. The reference is to pp. 124-25 in the latter edition.

[5] For a comprehensive view of the demands which such a situation or even its approach would make on our social and political values, see Geoffrey Vickers, *Freedom in a Rocking Boat* (London, 1970).

The technological order

JACQUES ELLUL

■ Jacques Ellul *was born in 1912 in Bordeaux. He is a historian and a theological philosopher whose book* The Technological Society, *published in France in 1954, anticipated the contemporary antitechnological movement. Currently he holds the chair of the history of law, in the faculty of law at the University of Bordeaux. He is also a professor of social history at the Bordeaux Institute of Political Studies. Among his other books are* The Ethics of Freedom *and* Propaganda.

In this article Ellul introduces his famous thesis, namely that technique is autonomous, that is, self-determined. To emphasize the power of technique in our society, he argues that it does not influence *all social phenomena. Rather, these are* "situated in" *technique and are* "defined" *through their relation to the technological society. It is clear from Ellul's discussion that "Technique" refers to more than simply machine-related functions; it comprises organizational as well as psycho-sociological techniques. The psycho-sociological techniques "result in the* modification *of men in order to render them happily subordinate to their new environment." It is interesting to evaluate this last assertion in light of Fromm's description of individuals in a technological society as "powerless, lonely and anxious" (see the selection by Fromm later in Part Two).*

Ellul addresses himself to two questions: "Is man able to remain master in a world of means? Can a new civilization appear inclusive of Technique?" His answer to both is "no." In the ensuing discussion he concludes that "there is no possibility of turning back, annulling, or even of arresting technical progress." We are left with an impression of technological determinism. For a discussion of Ellul's position, see the article by Robert Theobald in Part Three.

I. I refer the reader to my book *La Technique* for an account of my general theses on this subject. I shall confine myself here to recapitulating the points which seem to me to be essential to a sociological study of the problem:

1. Technique[1] has become the new and specific *milieu* in which man is required to exist, one which has supplanted the old *milieu*, viz., that of nature.

2. This new technical *milieu* has the following characteristics:

a. It is artificial;
b. It is autonomous with respect to values, ideas, and the state;
c. It is self-determining in a closed circle. Like nature, it is a closed organization which permits it to be self-determinative independently of all human intervention;
d. It grows according to a process which is causal but not directed to ends;
e. It is formed by an accumulation of means which have established primacy over ends;
f. All its parts are mutually implicated to such a degree that it is impossible to separate them or to settle any technical problem in isolation.

3. The development of the individual techniques is an "ambivalent" phenomenon.

4. Since Technique has become the new *milieu*, all social phenomena are situated in it. It is incorrect to say that economics, politics, and the sphere of the cultural are influenced or modified *by* Technique; they are rather situated *in* it, a novel situation modifying all traditional social concepts. Politics, for example, is not modified by Technique as one factor among others which operate upon it; the political world is today *defined* through its relation to the technological society. Traditionally, politics formed a part of a larger social whole; at the present the converse is the case.

5. Technique comprises organizational and psycho-sociological techniques. It is useless to hope that the use of techniques of organization will succeed in compensating for the effects of

☐ Jacques Ellul, "The Technological Order," trans. John Wilkinson. Reprinted from *The Technological Order*, ed. Carl E. Stover (Detroit: Wayne State University Press, 1963), pp. 10-24, by permission of the Wayne State University Press and the author.

techniques in general; or that the use of psycho-sociological techniques will assure mankind ascendancy over the technical phenomenon. In the former case, we will doubtless succeed in averting certain technically induced crises, disorders, and serious social disequilibrations; but this will but confirm the fact that Technique constitutes a closed circle. In the latter case, we will secure human psychic equilibrium in the technological *milieu* by avoiding the psychobiologic pathology resulting from the individual techniques taken singly and thereby attain a certain happiness. But these results will come about through the *adaptation of human beings to the technical milieu*. Psycho-sociological techniques result in the *modification* of men in order to render them happily subordinate to their new environment, and by no means imply any kind of human domination over Technique.

6. The ideas, judgments, beliefs, and myths of the man of today have already been essentially modified by his technical *milieu*. It is no longer possible to reflect that on the one hand, there are techniques which may or may not have an effect on the human being; and, on the other, there is the human being himself who is to attempt to invent means to master his techniques and subordinate them to his own ends by *making a choice* among them. Choices and ends are both based on beliefs, sociological presuppositions, and myths which are a function of the technological society. Modern man's state of mind is completely dominated by technical values, and his goals are represented only by such progress and happiness as is to be achieved through techniques. Modern man in choosing is already incorporated within the technical process and modified in his nature by it. He is no longer in his traditional state of freedom with respect to judgment and choice.

II. To understand the problem posed to us, it is first of all requisite to disembarrass ourselves of certain fake problems.

1. We make too much of the disagreeable features of technical development, for example, urban over-crowding, nervous tension, air pollution, and so forth. I am convinced that all such inconveniences will be done away with by the ongoing evolution of Technique itself, and indeed, that it is only by means of such evolution that this can happen. The inconveniences we emphasize are always dependent on technical solutions, and it is only by means of techniques that they can be solved. This fact leads to the following two considerations:

a. Every solution to some technical inconvenience is able only to reinforce the system of techniques *in their ensemble;*

b. Enmeshed in a process of technical development like our own, the possibilities of human survival are better served by more technique than less, a fact which contributes nothing, however, to the resolution of the basic problem.

2. We hear too often that morals are being threatened by the growth of our techniques. For example, we hear of greater moral decadence in those environments most directly affected technically, say, in working class or urbanized *milieux*. We hear, too, of familial disintegration as a function of techniques. The falseness of this problem consists in contrasting the technological environment with the moral values inculcated by society itself.[2] The presumed opposition between ethical problematics and technological systematics probably at the present is, and certainly in the long run will be, false. The traditional ethical *milieu* and the traditional moral values are admittedly in process of disappearing, and we are witnessing the creation of a *new* technological ethics with its own values. We are witnessing the evolution of a morally consistent system of imperatives and virtues, which tends to replace the traditional system. But man is not necessarily left thereby on a morally inferior level, although a moral relativism is indeed implied—an attitude according to which everything is well, *provided* that the individual obeys some ethic or other. We *could* contest the value of this development *if* we had a clear and adequate concept of what good-in-itself is. But such judgments are impossible on the basis of our general morality. On *that* level, what we are getting is merely a substitution of a new technological morality for a traditional one which Technique has rendered obsolete.

3. We dread the "sterilization" of art through technique. We hear the artist's lack of freedom, calm, and the impossibility of meditation in the technological society. This problem is no more real than the two preceeding. On the contrary, the best artistic production of the present is a result of a close connection between art and Technique. Naturally, new artistic form, expression, and ethic are implied, but this fact does not make art less art than what we traditionally called such. What assuredly is *not* art is a fixation in congealed forms, and a rejection of technical evolution as exemplified, say, in the

neo-classicism of the nineteenth century or in present day "socialist realism." The modern cinema furnishes an artistic response comparable to the Greek theater at its best; and modern music, painting, and poetry express, not a canker, but an authentic esthetic expression of mankind plunged into a new technical *milieu*.

4. One last example of a false problem is our fear that the technological society is completely *eliminating* instinctive human values and powers. It is held that systematization, organization, "rationalized" conditions of labor, overly hygienic living conditions, and the like have a tendency to repress the forces of instinct. For some people the phenomenon of "beatniks," "*blousons noirs*,"[3] and "hooligans" is explained by youth's violent reaction and the protestation of youth's vital force to a society which is overorganized, overordered, overregulated, in short, technicized.[4] But here too, even if the facts are established beyond question, it is very likely that a superior conception of the technological society will result in the integration of these instinctive, creative, and vital forces. Compensatory mechanisms are already coming into play; the increasing appreciation of the aesthetic eroticism of authors like Henry Miller and the rehabilitation of the Marquis de Sade are good examples. The same holds for music like the new jazz forms which are "escapist" and exaltative of instinct; *item*, the latest dances. All these things represent a process of "*défoulement*"[5] which is finding its place in the technological society. In the same way, we are beginning to understand that it is impossible indefinitely to repress or expel religious tendencies and to bring the human race to a perfect rationality. Our fears for our instincts *are* justified to the degree that Technique, instead of provoking conflict, tends rather to *absorb* it, and to *integrate* instinctive and religious forces by giving them a place within its structure, whether it be by an adaptation of Christianity[6] or by the creation of new religious expressions like myths and mystiques which are in full compatibility with the technological society.[7] The Russians have gone farthest in creating a "religion" compatible with Technique by means of their transformation of Communism into a religion.

III. What, then, is the real problem posed to men by the development of the technological society? It comprises two parts: 1. Is man able to remain master[8] in a world of means? 2. Can a new civilization appear inclusive of Technique?

1. The answer to the first question, and the one most often encountered, seem obvious: Man, who exploits the ensemble of means, *is* the master of them. Unfortunately, this manner of viewing matters is purely theoretical and superficial. We must remember the autonomous character of Technique. We must likewise not lose sight of the fact that the human individual himself is to be an ever greater degree the *object* of certain techniques and their procedures. He is the object of pedagogical techniques, psychotechniques, vocational guidance testing, personality and intelligence testing, industrial and group aptitude testing, and so on. In these cases (and in countless others) most men are treated as a collection of objects. But, it might be objected, these techniques are exploited by other men, and the exploiters at least remain masters. In a certain sense this is true; the exploiters *are* masters of the particular techniques they exploit. But, they, too, are subjected to the action of yet other techniques, as, for example, propaganda. Above all, they are spiritually taken over by the technological society; they believe in what they do; they are the most fervent adepts of that society. They themselves have been profoundly technicized. They never in any way affect to despise Technique, which to them is a thing good in itself. They never pretend to assign values to Technique, which to them is in itself an entity working out its own ends. They never claim to subordinate it to any value because for them Technique *is* value.

It may be objected that these individual techniques have as their end the best adaptation of the individual, the best utilization of his abilities, and, in the long run, his happiness. This, in effect, is the objective and the justification of all techniques. (One ought not, of course, to confound man's "happiness" with capacity for mastery with, say, freedom.) If the first of all values is happiness, it is likely that man, thanks to his techniques, will be in a position to attain to a certain state of this good. But happiness does not contain everything it is thought to contain, and *the absolute disparity between happiness and freedom* remains an ever real theme for our reflections. To say that man should remain *subject* rather than *object* in the technological society means two things, viz., that he be capable of giving direction and orientation to Technique, and that, to this end, he be able to master it.

Up to the present he has been able to do neither. As to the first, he is content passively to participate in technical progress, to accept whatever direction it takes automatically, and to admit its autonomous meaning. In the circumstances he can either proclaim this life is an absurdity without meaning or value; *or,* he can predicate a number of indefinitely sophisticated values. But neither attitude accords with the fact of the technical phenomenon any more than it does with the other. Modern declarations of the absurdity of life are not based on modern technological efflorescence, which none (least of all the existentialists) think an absurdity. And the predication of values is a purely theoretical matter, since these values are not equipped with any means for putting them into practice. It is easy to reach agreement on what they are, but it is quite another matter to make them have any effect whatever on the technological society, or to cause them to be accepted in such a way that techniques must evolve in order to realize them. The values spoken of in the technological society are simply there to justify what is; *or,* they are generalities without consequence; *or* technical progress realizes them automatically as a matter of course. Put otherwise, neither of the above alternatives is to be taken seriously.

The second condition *that man be subject rather than object,* i.e., the imperative that he exercise mastery over technical development, is facilely accepted by everyone. But factually it simply does not hold. Even more embarrassing than the question "How?" is the question "Who?" We must ask ourselves realistically and concretely just who is in a position to choose the values which give Technique its justification and to exert mastery over it. If such a person or persons are to be found, it must be in the Western world (inclusive of Russia). They certainly are not to be discovered in the bulk of the world's population which inhabits Africa and Asia, who are, as yet, scarcely contronted by technical problems, and who, in any case, are even less aware of the questions involved than we are.

Is the arbiter we seek to be found among the *philosophers,* those thinking specialists? We well know the small influence these gentry exert upon our society, and how the technicians of every order distrust them and rightly refuse to take their reveries seriously. Even if the philosopher could make his voice heard, he would still have to contrive means of mass education so as to communicate an effective message to the masses.

Can the *technician* himself assume mastery over Technique? The trouble here is that the technician is *always* a specialist and cannot make the slightest claim to have mastered any technique but his own. Those for whom Technique bears its meaning in itself will scarcely discover the values which lend meaning to what they are doing. They will not even look for them. The only thing they can do is to apply their technical specialty and assist in its refinement. They cannot *in principle* dominate the totality of the technical problem or envisage it in its global dimensions. *Ergo,* they are completely incapable of mastering it.

Can the *scientist* do it? There, if anywhere, is the great hope. Does not the scientist dominate our techniques? Is he not an intellectual inclined and fit to put basic questions? Unfortunately, we are obliged to re-examine our hopes here when we look at things as they are. We see quickly enough that the scientist is as specialized as the technician, as incapable of general ideas, and as much out of commission as the philosopher. Think of the scientists who, on one tack or another, have addressed themselves to the technical phenomenon: Einstein, Oppenheimer, Carrel. It is only too clear that the ideas these gentlemen have advanced in the sphere of the philosophic or the spirtual are vague, superficial, and contradictory *in extremis.* They really ought to stick to warnings and proclamations, for as soon as they assay anything else, the other scientists and the technicians rightly refuse to take them seriously, and they even run the risk of losing their reputations as scientists.

Can the *politician* bring it off? In the democracies the politicians are subject to the wishes of their constituents who are primarily concerned with the happiness and well-being which they think Technique assures them. Moreover, the further we get on, the more a conflict shapes up between the politicians and the technicians. We cannot here go into the matter which is just beginning to be the object of serious study.[9] But it would appear that the power of the politician is being (and will continue to be) outclassed by the power of the technician in modern states. Only dictatorships can impose their will on technical evolution. But, on the one hand, human freedom would gain nothing thereby, and, on the other, a dictatorship thirsty for power has no recourse at all but to push toward an excessive development of various techniques at its disposal.

Any of us? An individual can doubtless seek the soundest attitude to dominate the techniques at his disposal. He can inquire after the values to impose on techniques in his use of them, and search out the way to follow in order to remain a man in the fullest sense of the word within a technological society. All this is extremely difficult, but it is far from being useless, since it is apparently the only solution presently possible. But the individual's efforts are powerless to resolve in any way the technical problem in its universality; to accomplish this would mean that *all* men adopt the same values and the same behavior.

2. The second real problem posed by the technological society is whether or not a new civilization can appear which is inclusive of Technique. The elements of this question are as difficult as those of the first. It would obviously be vain to deny all the things that can contribute something useful to a new civilization: security, ease of living, social solidarity, shortening of the work week, social security, and so forth. But a civilization in the strictest sense of the term is not brought into being by all these things.

A threefold contradiction resides between civilization and Technique of which we must be aware if we are to approach the problem correctly:

a. The technical world is the world of material things; it is put together out of material things and with respect to them. When Technique displays any interest in man, it does so by converting him into a material object. The supreme and final authority in the technological society is fact, at once ground and evidence. And when we think on man as he exists in this society it can only be as a being immersed in a universe of objects, machines, and innumerable material things. Technique indeed guarantees him such material happiness as material objects can. But, the technical society is not, and cannot be, a genuinely humanist society since it puts in first place not man but material things. It can only act on man by lessening him and putting him in the way of the quantitative. The radical contradiction referred to exists between technical perfection and human development because such perfection is only to be achieved through quantitative development and necessarily aims exclusively at what is measurable. Human excellence, on the contrary, is of the domain of the qualitative and aims at what is not measurable. Space is lacking here to argue the point that spiritual values cannot evolve as a function of material improvement. The transition from the technically quantitative to the humanly qualitative is an impossible one. In our times, technical growth monopolizes all human forces, passions, intelligences, and virtues in such a way that it is in practice nigh impossible to seek and find anywhere any distinctively human excellence. And if this search is impossible, there cannot be any civilization in the proper sense of the term.

b. Technical growth leads to a growth of power in the sense of technical means incomparably more effective than anything ever before invented, power which has as its object only power, in the widest sense of the word. The possibility of action becomes limitless and absolute. For example, we are confronted for the first time with the possibility of the annihilation of all life on earth, since we have the means to accomplish it. In *every* sphere of action we are faced with just such absolute possibilities. Again, by way of example, governmental techniques, which amalgamate organizational, psychological, and police techniques, tend to lend to government absolute powers. And here I must emphasize a great law which I believe to be essential to the comprehension of the world in which we live, viz., that when power becomes absolute, values disappear. When man is able to accomplish anything at all, there is no value which can be proposed to him; when the means of action are absolute, no goal of action is imaginable. Power eliminates, in proportion to its growth, the boundary between good and evil, between the just and the unjust. We are familiar enough with this phenomenon in totalitarian societies. The distinction between good and evil disappears beginning with the moment that the ground of action (for example the *raison d'état,* or the instinct of the proletariat) claims to have absolute power and thus to incorporate *ipso facto* all value. Thus it is that the growth of technical means tending to absolutism forbids the appearance of values, and condemns to sterility our search for the ethical and the spiritual. Again, where Technique has place, there is the implication of the impossibility of the evolution of civilization.

c. The third and final contradiction is that Technique can never engender freedom. Of course, Technique frees mankind from a whole collection of ancient constraints. It is evident, for example, that it liberates him from the limits imposed on him by time and space; that man, through its agency, is free (or at least tending to become free) from famine, excessive heat and cold, the rhythms of the seasons, and from the gloom of night; that the race is freed from certain social constraints through its commerce with the universe, and from its intellectual limitations through its accumulation of information. But is this what it means really to be free? Other constraints as oppressive and rigorous as the traditional ones are imposed on the human being in today's technological society through the agency of Technique. New limits and technical oppressions have taken the place of the older, natural constraints, and we certainly cannot aver that much has been gained. The problem is deeper—the operation of Technique is the contrary of freedom, an operation of determinism and necessity. Technique is an ensemble of rational and efficient practices; a collection of orders, schemas, and mechanisms. All of this expresses very well a necessary order and a determinate process, but one into which freedom, unorthodoxy, and the sphere of the gratuitous and spontaneous cannot penetrate. All that these last could possibly introduce is discord and disorder. The more technical actions increase in society, the more human autonomy and initiative diminish. The more the human being comes to exist in a world of ever increasing demands (fortified with technical apparatus possessing its own laws to meet these demands), the more he loses any possibility of free choice and individuality in action. This loss is greatly magnified by Technique's character of self-determination, which makes its appearance among us as a kind of fatality and as a species of perpetually exaggerated necessity. But where freedom is excluded in this way, an authentic civilization has little chance. Confronted in this way by the problem, it is clear to us that no solution can exist, in spite of the writings of all the authors who have concerned themselves with it. They all make an unacceptable premise, viz., rejection of Technique and return to a pre-technical society. One may well regret that some value or other of the past, some social or moral form, has disappeared; but, when one attacks the problem of the technical society, one can scarcely make the serious claim to be able to revive the past, a procedure which, in any case, scarcely seems to have been, globally speaking, much of an improvement over the human situation of today. All we know with certainty is that it was different, that the human being confronted other dangers, errors, difficulties, and temptations. Our duty is to occupy ourselves with the dangers, errors, difficulties, and temptations of modern man in the modern world. All regret for the past is vain; every desire to revert to a former social stage is unreal. There is no possibility of turning back, of annulling, or even of arresting technical progress. What is done is done. It is our duty to find our place in our present situation and in no other. Nostalgia has no survival value in the modern world and can only be considered a flight into dreamland.

We shall insist no further on this point. Beyond it, we can divide into two great categories the authors who search for a solution to the problem posed by Technique: The first class is that of those who hold that the problem will solve itself; the second, of those who hold that the problem demands a great effort or even a great modification of the whole man. We shall indicate a number of examples drawn from each class and beg to be excused for choosing to cite principally French authors.

Politicians, scientists and technicians are to be found in the first class. In general, they consider the problem in a very concrete and practical way. Their general notion seems to be that technical progress resolves all difficulties *pari passu* with their appearance, and that it contains within itself the solution to everything. The sufficient condition for them, therefore, is that technical progress be not arrested; everything which plagues us today will disappear tomorrow.

The primary example of these people is furnished by the Marxists, for whom technical progress is the solution to the plight of the proletariat and all its miseries, and to the problem posed by the exploitation of man by man in the capitalistic world. Technical progress, which is for Marx the motive force of history, *necessarily* increases the forces of production, and simultaneously produces a progressive conflict between forward moving factors and stationary social factors like the state, law, ideology, and morality, a conflict occasioning the periodic disappearance of the outmoded factors. Specifically, in the world of the present, conflict necessitates the disappearance of the structures of capitalism, which are so constituted as to be completely unable to absorb the economic results of technical progress, and are hence obliged to vanish. When they do vanish, they of necessity make room for a socialist structure of society corresponding perfectly to the sound and normal utilization of Technique. The Marxist solution to the technical problems is therefore an automatic one since the transition to socialism is *in itself* the solution. Everything is *ex hypothesi* resolved in the socialist society, and humankind finds therein its maturation. Technique, integrated into the socialist society "changes sign": from being destructive it becomes constructive; from being a means of human exploitation it becomes humane; the contradiction between the infrastructures and the suprastructures disappears. In other words, all the admittedly difficult problems raised in the modern world belong to the structure of capitalism and not to that of Technique. On the one hand, it *suffices* that social structures become socialist for social problems to disappear; and on the other, society *must necessarily* become socialist by the very movement of Technique. Technique, therefore, carries in itself the response to all the difficulties it raises.

A second example of this kind of solution is given by a certain number of technicians, for example, Frisch. All difficulties, according to Frisch, will inevitably be resolved by the technical growth which will bring the technicians to power. Technique admittedly raises certain conflicts and problems, but their cause is that the human race remains attached to certain political ideologies and moralities and loyal to certain outmoded and antiquated humanists whose sole visible function is to provoke discord of heart and head, thereby preventing men from adapting themselves and from entering resolutely into the path of technical progress. *Ergo,* men are subject to distortions of life and consciousness which have their origin, *not* in Technique, but in the conflict between Technique and the false values to which men remain attached. These fake values, decrepit sentiments, and outmoded notions must inevitably be eliminated by the invincible progress of Technique. In particular, in the political domain, the majority of crises arise from the fact that men are still wedded to certain antique political forms and ideas, for example, democracy. All problems will be resolved if power is delivered into the hands of the technicians who alone are capable of directing Technique in its entirety and making of it a positive instrument for human service. This is all the more true in that, thanks to the so-called "human techniques" (for example, propaganda) they will be in a position to take account of the human factor in the technical context. The technocrats will be able to use the totality of Technique without destroying the human being, but rather by treating him as he should be treated so as to become simultaneously useful and happy. General power accorded to the technicians become technocrats is the only way out for Frisch, since they are the only ones possessing the necessary competence; and, in any case, they are being carried to power by the current of history, the fact which alone offers a quick enough solution to technical problems. It is impossible to rely on the general improvement of the human species, a process which would take too long and would be too chancy. For the generality of men, it is necessary to take into account that Technique establishes an inevitable discipline, which, on the one hand, they must accept, and, on the other, the technocrats will humanize.

The third and last example (it is possible that there are many more) is furnished by the economists, who, in very different ways, affirm the thesis of the automatic solution. Fourastié is a good example of such economists. For him, the first thing to do is to draw up a balance between that which Technique is able to deliver and that which it may destroy. In his eyes there is no real problem: What Technique can bring to man is incomparably superior to that which it threatens. Moreover, if difficulties *do* exist, they are only temporary ones which will be resolved beneficially, as was the case with the similar difficulties of the last century. Nothing decisive is at stake; man is in no mortal danger. The contrary is the case: Technique produces the foundation, infrastructure, and suprastructure which will enable man really to become man. What we have known up to now can only be called the *prehistory* of a human race so overwhelmed by material cares, famine, and danger, that the truly human never had an opportunity to develop into a civilization worthy of the name. Human intellectual, spiritual, and moral life will, according to Fourastié, never mature except when life is able to start from a complete satisfaction of its material needs, complete security, including security from famine and disease. The growth of Technique, therefore, initiates the genuinely human history of the whole man. This new type of human being will clearly be different from what we have hitherto known; but this fact should occasion no complaint or fear. The new type cannot help being superior to the old in every way, *after* all the traditional (and exclusively material) obstacles to his development have vanished. Thus, progress occurs automatically, and the inevitable role of Technique will be that of guaranteeing such material development as allows the intellectual and spiritual maturation of what has been up to now only potentially present in human nature.

The orientation of the other group of doctrines affirms, on the contrary, that man is dangerously imperiled by technical progress; and that human will, personality, and organization must be set again to rights if society is to be able to guard against the imminent danger. Unfortunately, these doctrines share with their opposites the quality of being too optimistic, in that they affirm that their thesis is even feasible and that man is really capable of the rectifications proposed. I will give three very different examples of this, noting that the attitude in question is generally due to philosophers and theologians.

The orientation of Einstein, and the closely related one of Jules Romains, are well known, viz., that the human being must get technical progress back again into his own hands, admitting that the situation is so complicated and the data so overwhelming that only some kind of "superstate" can possibly accomplish the task. A sort of spiritual power integrated into a world government in possession of indisputable moral authority might be able to master the progression of techniques and to direct human evolution. Einstein's suggestion is the convocation of certain philosopher-scientists, whereas Romains' idea is the establishment of a "Supreme Court of Humanity." Both of these bodies would be organs of meditation, of moral quest, before which temporal powers would be forced to bow. (One thinks, in this connection, of the role of the papacy in medieval Christianity *vis-à-vis* the temporal powers.)

A second example of this kind of orientation is given by Bergson, at the end of his work, *The Two Sources of Morality and Religion*. According to Bergson, initiative can only proceed from humanity, since in Technique there is no *"force des choses."* Technique has conferred disproportionate power on the human being, and a disproportionate extension to his organism. But, "in this disproportionately magnified body, the soul remains what it was, i.e., too small to fill it and too feeble to direct it. Hence the void between the two." Bergson goes on to say that "this enlarged body awaits a supplement of soul, the mechanical demands the mystical," and . . . "that Technique will never render service proportionate to its powers unless humanity, which has bent it earthwards, succeeds by its means in reforming itself and looking heavenwards." This means that humanity has a task to perform, and that man must grow proportionately to his techniques, but that he must *will* it and *force* himself to make the experiment. This experiment is, in Bergson's view, a possibility, and is even favored by that technical growth which allows more material resources to men than ever before. The required "supplement of soul" is therefore of the order of the possible and will suffice for humans to establish mastery over Technique. The same position, it may be added, has in great part been picked up by E. Mounier.

A third example is afforded by a whole group of theologians, most of them Roman Catholic. Man, in his actions in the domain of the technical, is but obeying the vocation assigned him by his Creator. Man, in continuing his work of technical creation, is pursuing the work of his Creator. Thanks to Technique, this man, who was originally created "insufficient," is becoming "adolescent." He is summoned to new responsibilities in this world which do not transcend his powers since they correspond exactly to what God expects of him. Moreover, it is God Himself who through man is the Creator of Technique, which is something not to be taken in itself but in its relation to its Creator. Under such conditions, it is clear that Technique is neither evil nor fraught with evil consequences. On the contrary, it is good and cannot be dangerous to men. It can only become evil to the extent that man turns from God; it is a danger only if its true nature is misapprehended. All the errors and problems visible in today's world result uniquely from the fact that man no longer recognizes his vocation as God's collaborator. If man ceases to adore the "creature" (i.e., Technique) in order to adore the true God; if he turns Technique to God and to His service, the problems must disappear. All of this is considered the more true in that the world transformed by technical activity *must* become the point of departure and the material support of the new creation which is to come at the end of time.

Finally, it is necessary to represent by itself a doctrine which holds at the present a place of some importance in the Western world, i.e., that of Father Teilhard de Chardin, a man who was simultaneously a theologian and a scientist. His doctrine appears as an intermediate between the two tendencies already sketched. For Chardin, evolution in general, since the origin of the universe, has represented a constant progression. First of all, there was a motion toward a diversification of matter and of beings; then, there supervened a motion toward Unity, i.e., a higher Unity. In the biological world, every step forward has been effected when man has passed from a stage of "dispersion" to a stage of "concentration." At the present, technical human progress and the spontaneous movement of life are in agreement and in mutual continuity. They are evolving together toward a higher degree of organization, and this movement manifests the influence of Spirit. Matter, left to itself, is characterized by a necessary and continuous degradation. But on the contrary, we note that progress, advancement, improvement do exist, and, hence, a power contradicting the spontaneous movement of matter, a power of creation and progress exists which is the opposite of matter, i.e., it is Spirit. Spirit has contrived Technique as a means of organizing dispersed matter, in order simultaneously to express progress and to combat the degradation of matter. Technique is producing at the same time a prodigious demographic explosion, i.e., a greater density of human population. By all these means it is bringing forth "communion" among men; and likewise creating from inanimate matter a higher and more organized form of matter which is taking part in the ascension of the cosmos toward God. Granting that it is true that every progression in the physical and biological order is brought about by a condensation of the elements of the preceeding period, what we are witnessing today, according to Chardin, is a condensation, a concentration of the whole human species. Technique, in producing this, possesses a function of unification *inside* humanity, so that humanity becomes able thereby to have access to a sort of unity. Technical progress is therefore synonymous with "socialization," this latter being but the political and economic sign of communion among men, the temporary expression of the "condensation" of the human species into a whole. Technique is the irreversible agent of this condensation; it prepares the new step forward which humanity must make. When men cease to be individual and separate units, and all together form a total and indissoluble communion, then humanity will be a single body. This material concentration is always accompanied by a spiritual, i.e., a maturation of the spirit, the commencement of a new species of life. Thanks to Technique, there is "socialization," the progressive concentration on a planetary scale of disseminated spiritual personalities into a suprapersonal unity. This mutation leads to another Man, spiritual and unique, and means that humanity in its ensemble and in its unity, has attained the supreme goal, i.e., its fusion with that glorious Christ who must appear at the end of time. Thus Chardin holds that in technical progress man is "Christified," and that technical evolution tends inevitably to the "edification" of the cosmic Christ.

It is clear that in Chardin's grandiose perspective, the individual problems, difficulties, and mishaps of Technique are negligible. It is likewise clear how Chardin's doctrine lies midway between the two preceding ones: On the one hand, it affirms a natural and involuntary ascension of man, a process inclusive of biology, history, and the like, evolving as a kind of will of God in which Technique has its proper place; and, on the other, it affirms that the evolution in question implies consciousness, and an intense *involvement* on the part of man who is proceeding to socialization and thus *commiting* himself to this mutation.

We shall not proceed to a critique of these different theories, but content ourselves with noting that all of them appear to repose on a too superficial view of the technical phenomenon; and that they are *practically* inapplicable because they presuppose a certain number of *necessary* conditions which are not given. None of these theories, therefore, can be deemed satisfactory.

NOTES

[1]In his book *La Technique,* Jacques Ellul states he is "in substantial agreement" with H. D. Lasswell's definition of technique: "the ensemble of practices by which one uses available resources in order to achieve certain valued ends." Commenting on Lasswell's definition, Ellul says: "In the examples which Lasswell gives, one discovers that he conceives the terms of his definition in an extremely wide manner. He gives a list of values and the corresponding techniques. For example, he indicates as values riches, power, well-being, affection, and so on, with the techniques of government, production, medicine, the family. This notion of value may seem somewhat novel. The expression is manifestly improper. But this indicates that Lasswell gives to techniques their full scope. Besides, he makes it quite clear that it is necessary to bring into the account not only the ways in which one influences things, but also the ways one influences persons." "Technique" as it is used by Ellul is most nearly equivalent to what we commonly think of as "the technological order" or "the technological society." (Trans.)

[2]Cf. K. Horney.

[3]A kind of French beatnik. (Trans.)

[4]The psychoanalyst Jung has much to say along this line.

[5]An untranslatable French play on words. *Défoulement* is an invented word which presumably expresses the opposite of *refoulement,* i.e., repression.

[6]Teilhard de Chardin represents, in his works, the best example of this.

[7]Examples of such myths are: "Happiness," "Progress," "The Golden Age," etc.

[8]French *sujet.* The usual rendering, "subject," would indicate exactly the contrary of what is meant here, viz., the opposite of "object." The present sense of "subject" is that in virtue of which it governs a grammatical object, for example. (Trans.)

[9]See, for example, the reports of the International Congress for Political Science, October, 1961.

9

Four theories on the origins of technology and civilization

The part played by labour in the transition from ape to man

FREDERICK ENGELS

■ Frederick Engels *(1820-1895) was born in Barmen in the German Rhineland. His father was a textile manufacturer who owned a cotton mill in Manchester, England. Although he is known chiefly for his intellectual association with Marx, Engels was a man of many talents. He was a scholar, linguist, soldier, and businessman. He met Marx in Cologne in 1842 and then in Paris in 1844, when they began their life-long friendship. They worked actively together, and in 1848 they wrote* The Manifesto of the Communist Party. *In 1850 Engels reluctantly joined his father's business in Manchester. This enabled him to extend to Marx some urgently needed financial assistance, which was to continue and increase until Marx's death.*

It was Engels who provided the dialectic foundation for Marxism and deepened its doctrine of historical materialism. Among his well-known works are The Dialectic of Nature *(1925), from which the following selection is taken, and* The Origin of the Family, Private Property and the State *(1884).*

Engels presents here a theory about the origins of technology and civilization. For him, "the decisive step in the transition from ape to man" came when man adopted a more erect posture in walking. This freed his hands, and they became more and more devoted to functions other than the original ones. With the development of the hand, the mastery of nature began. For Engels, this turn of events supplied the basis for the "essential distinction between man and other animals." For the animal "merely uses *external nature," but man "*masters *it." As the hand adapted to new operations, man made tools that enabled him to transform his exclusively vegetable diet into a meat diet. The major effects of that change on the brain, and the resultant technological developments, are discussed in some detail. Engels ventures at one point that "man did not come into existence without a flesh diet." He presumably means that man as a technological being did not emerge historically until the adoption of a meat diet.*

Labour is the source of all wealth, the economists assert. It is this—next to nature, which supplies it with the material that it converts into wealth. But it is also infinitely more than this. It is the primary basic condition for all human existence, and this to such an extent that, in a sense, we have to say that labour created man himself.

Many hundreds of thousands of years ago, during an epoch, not yet definitely determined, of that period of the earth's history which geologists call the Tertiary period, most likely towards the end of it, a specially highly-developed race of anthropoid apes lived somewhere in the tropical zone—probably on a great continent that has now sunk to the bottom of the Indian Ocean. Darwin has given us an approximate description of these ancestors of ours. They were completely covered with hair, they had beards and pointed ears, and they lived in bands in the trees.

Almost certainly as an immediate consequence of their mode of life, for in climbing the hands fulfil quite different functions from the feet, these apes when moving on level ground began to drop the habit of using their

☐ From Frederick Engels, *Dialectics of Nature*, trans. and ed. Clemens Dutt preface and notes by J. B. S. Haldane (New York: International Publishers Co., Inc., 1973), pp. 279-294.

hands and to adopt a more and more erect posture in walking. This was *the decisive step in the transition from ape to man*.

All anthropoid apes of the present day can stand erect and move about on their feet alone, but only in case of need and in a very clumsy way. Their natural gait is in a half-erect posture and includes the use of the hands. The majority rest the knuckles of the fist on the ground and, with legs drawn up, swing the body through their long arms, much as a cripple moves with the aid of crutches. In general, we can to-day still observe among apes all the transition stages from walking on all fours to walking on two legs. But for none of them has the latter method become more than a makeshift.

For erect gait among our hairy ancestors to have become first the rule and in time a necessity presupposes that in the meantime the hands became more and more devoted to other functions. Even among the apes there already prevails a certain separation in the employment of the hands and feet. As already mentioned, in climbing the hands are used differently from the feet. The former serve primarily for collecting and holding food, as already occurs in the use of the fore paws among lower mammals. Many monkeys use their hands to build nests for themselves in the trees or even, like the chimpanzee, to construct roofs between the branches for protection against the weather. With their hands they seize hold of clubs to defend themselves against enemies, or bombard the latter with fruits and stones. In captivity, they carry out with their hands a number of simple operations copied from human beings. But it is just here that one sees how great is the gulf between the undeveloped hand of even the most anthropoid of apes and the human hand that has been highly perfected by the labour of hundreds of thousands of years. The number and general arrangement of the bones and muscles are the same in both; but the hand of the lowest savage can perform hundreds of operations that no monkey's hand can imitate. No simian hand has ever fashioned even the crudest stone knife.

At first, therefore, the operations, for which our ancestors gradually learned to adapt their hands during the many thousands of years of transition from ape to man, could only have been very simple. The lowest savages, even those in whom a regression to a more animal-like condition, with a simultaneous physical degeneration, can be assumed to have occurred, are nevertheless far superior to these transitional beings. Before the first flint could be fashioned into a knife by human hands, a period of time must probably have elapsed in comparison with which the historical period known to us appears insignificant. But the decisive step was taken; *the hand became free* and could henceforth attain ever greater dexterity and skill, and the greater flexibility thus acquired was inherited and increased from generation to generation.

Thus the hand is not only the organ of labour, *it is also the product of labour*. Only by labour, by adaptation to ever new operations, by inheritance of the resulting special development of muscles, ligaments, and, over longer periods of time, bones as well, and by the ever-renewed employment of these inherited improvements in new, more and more complicated operations, has the human hand attained the high degree of perfection that has enabled it to conjure into being the pictures of Raphael, the statues of Thorwaldsen, the music of Paganini.

But the hand did not exist by itself. It was only one member of an entire, highly complex organism. And what benefited the hand benefited also the whole body it served; and this in two ways.

In the first place, the body benefited in consequence of the law of correlation of growth, as Darwin called it. According to this law, particular forms of the individual parts of an organic being are always bound up with certain forms of other parts that apparently have no connection with the first. Thus all animals that have red blood cells without a cell nucleus, and in which the neck is connected to the first vertebra by means of a double articulation (condyles), also without exception possess lacteal glands for suckling their young. Similarly cloven hooves in mammals are regularly associated with the possession of a multiple stomach for rumination. Changes in certain forms involve changes in the form of other parts of the body, although we cannot explain this connection. Perfectly white cats with blue eyes are always, or almost always, deaf. The gradual perfecting of the human hand, and the development that keeps pace with it in the adaptation of the feet for erect gait, has undoubtedly also, by virtue of such correlation, reacted on other parts of the organism. However, this action has as yet been much too little investigated for us to be able to do more here than to state the fact in general terms.

Much more important is the direct, demonstrable reaction of the development of the hand on the rest of the organism. As already said, our simian ancestors were gregarious; it is obviously impossible to seek the derivation of man, the most social of all animals, from non-gregarious immediate ancestors. The mastery over nature, which begins with the development of the hand, with labour, widened man's horizon at every new advance. He was continually discovering new, hitherto unknown, properties of natural objects. On the other hand, the development of labour necessarily helped to bring the members of society closer together by multiplying cases of mutual support, joint activity, and by making clear the advantage of this joint activity to each individual. In short, men in the making arrived at the point where *they had something to say* to one another. The need led to the creation of its organ; the undeveloped larynx of the ape was slowly but surely transformed by means of gradually increased modulation, and the organs of the mouth gradually learned to pronounce one articulate letter after another.

Comparison with animals proves that this explanation of the origin of language from and in the process of labour is the only correct one. The little that even the most highly-developed animals need to communicate to one another can be communicated even without the aid of articulate speech. In a state of nature, no animal feels its inability to speak or to understand human speech. It is quite different when it has been tamed by man. The dog and the horse, by association with man, have developed such a good ear for articulate speech that they easily learn to understand any language within the range of their circle of ideas. Moreover they have acquired the capacity for feelings, such as affection for man, gratitude, etc., which were previously foreign to them. Anyone who has had much to do with such animals will hardly be able to escape the conviction that there are plenty of cases where they *now* feel their inability to speak is a defect, although, unfortunately, it can no longer be remedied owing to their vocal organs being specialised in a definite direction. However, where the organ exists, within certain limits even this inability disappears. The buccal organs of birds are of course radically different from those of man, yet birds are the only animals that can learn to speak; and it is the bird with the most hideous voice, the parrot, that speaks best of all. It need not be objected that the parrot does not understand what it says. It is true that for the sheer pleasure of talking and associating with human beings, the parrot will chatter for hours at a time, continuing to repeat its whole vocabulary. But within the limits of its circle of ideas it can also learn to understand what it is saying. Teach a parrot swear words in such a way that it gets an idea of their significance (one of the great amusements of sailors returning from the tropics); on teasing it one will soon discover that it knows how to use its swear words just as correctly as a Berlin costermonger. Similarly with begging for titbits.

First comes labour, after it, and then side by side with it, articulate speech—these were the two most essential stimuli under the influence of which the brain of the ape gradually changed into that of man, which for all its similarity to the former is far larger and more perfect. Hand in hand with the development of the brain went the development of its most immediate instruments—the sense organs. Just as the gradual development of speech is inevitably accompanied by a corresponding refinement of the organ of hearing, so the development of the brain as a whole is accompanied by a refinement of all the senses. The eagle sees much farther than man, but the human eye sees considerably more in things than does the eye of the eagle. The dog has a far keener sense of smell than man, but it does not distinguish a hundredth part of the odours that for man are definite features of different things. And the sense of touch, which the ape hardly possesses in its crudest initial form, has been developed side by side with the development of the human hand itself, through the medium of labour.

The reaction on labour and speech of the development of the brain and its attendant senses, of the increasing clarity of consciousness, power of abstraction and of judgement, gave an ever-renewed impulse to the further development of both labour and speech. This further development did not reach its conclusion when man finally became distinct from the monkey, but, on the whole, continued to make powerful progress, varying in degree and direction among different peoples and at different times, and here and there even interrupted by a local or temporary regression. This further development has been strongly urged forward, on the one hand, and has been guided along more definite directions on the other hand, owing to a new element which came into play with the appearance of fully-fledged man, viz. *society*.

Hundreds of thousands of years—of no greater significance in the history of the earth than one second in the life of man[1]—certainly elapsed before human society arose out of a band of tree-climbing monkeys. Yet it did finally appear. And what do we find once more as the characteristic difference between the band of monkeys and human society? *Labour*. The ape horde was satisfied to browse over the feeding area determined for it by geographical conditions or the degree of resistance of neighbouring hordes; it undertook migrations and struggles to win new feeding grounds, but it was incapable of extracting from the area which supplied it with food more than the region offered in its natural state, except, perhaps, that the horde unconsciously fertilised the soil with its own excrements. As soon as all possible feeding grounds were occupied, further increase of the monkey population could not occur; the number of animals could at best remain stationary. But all animals waste a great deal of food, and, in addition, destroy in embryo the next generation of the food supply. Unlike the hunter, the wolf does not spare the doe which would provide it with young deer in the next year; the goats in Greece, which graze down the young bushes before they can grow up, have eaten bare all the mountains of the country. This "predatory economy" of animals plays an important part in the gradual transformation of species by forcing them to adapt themselves to other than the usual food, thanks to which their blood acquires a different chemical composition and the whole physical constitution gradually alters, while species that were once established die out. There is no doubt that this predatory economy has powerfully contributed to the gradual evolution of our ancestors into men. In a race of apes that far surpassed all others in intelligence and adaptability, this predatory economy could not help leading to a continual increase in the number of plants used for food and to the devouring of more and more edible parts of these plants. In short, it led to the food becoming more and more varied, hence also the substances entering the body, the chemical premises for the transition to man. But all that was not yet labour in the proper sense of the word. The labour process begins with the making of tools. And what are the most ancient tools that we find—the most ancient judging by the heirlooms of prehistoric man that have been discovered, and by the mode of life of the earliest historical peoples and of the most primitive of contemporary savages? They are hunting and fishing implements, the former at the same time serving as weapons. But hunting and fishing presuppose the transition from an exclusively vegetable diet to the concomitant use of meat, and this is an important step in the transition to man. A *meat diet* contains in an almost ready state the most essential ingredients required by the organism for its metabolism. It shortened the time required, not only for digestion, but also for the other vegetative bodily processes corresponding to those of plant life, and thus gained further time, material, and energy for the active manifestation of animal life in the proper sense of the word. And the further that man in the making became removed from the plant kingdom, the higher he rose also over animals. Just as becoming accustomed to a plant diet side by side with meat has converted wild cats and dogs into the servants of man, so also adaptation to a flesh diet, side by side with a vegetable diet, has considerably contributed to giving bodily strength and independence to man in the making. The most essential effect, however, of a flesh diet was on the brain, which now received a far richer flow of the materials necessary for its nourishment and development, and which therefore could become more rapidly and perfectly developed from generation to generation.[1] With all respect to the vegetarians, it has to be recognised that man did not come into existence without a flesh diet, and if the latter, among all peoples known to us, has led to cannibalism at some time or another (the forefathers of the Berliners, the Weletabians or Wilzians, used to eat their parents as late as the tenth century), that is of no consequence to us to-day.

A meat diet led to two new advances of decisive importance: to the mastery of fire and the taming of animals. The first still further shortened the digestive process, as it provided the mouth with food already as it were semi-digested; the second made meat more copious by opening up a new, more regular source of supply in addition to hunting, and moreover provided, in milk and its products, a new article of food at least as valuable as meat in its composition. Thus, both these advances became directly new means of emancipation for man. It would lead us too far to dwell here in detail on their indirect effects notwithstanding the great importance they have had for the development of man and society.

Just as man learned to consume everything edible, he learned also to live in any climate. He spread over the whole of the habitable world, being the only animal that by its very nature had the power to do so. The other animals that have become accustomed to all climates—domestic animals and vermin—did not become so independently, but only in the wake of man. And the transition from the uniformly hot climate of the original home of man to colder regions, where the year is divided into summer and winter, created new requirements: shelter and clothing as protection against cold and damp, new spheres for labour and hence new forms of activity, which further and further separated man from the animal.

By the co-operation of hands, organs of speech, and brain, not only in each individual, but also in society, human beings became capable of executing more and more complicated operations, and of setting themselves, and achieving, higher and higher aims. With each generation, labour itself became different, more perfect, more diversified. Agriculture was added to hunting and cattle-breeding, then spinning, weaving, metal-working, pottery, and navigation. Along with trade and industry, there appeared finally art and science. From tribes there developed nations and states. Law and politics arose, and with them the fantastic reflection of human things in the human mind: religion. In the face of all these creations, which appeared in the first place to be products of the mind, and which seemed to dominate human society, the more modest productions of the working hand retreated into the background, the more so since the mind that plans the labour process already at a very early stage of development of society (*e.g.* already in the simple family), was able to have the labour that had been planned carried out by other hands than its own. All merit for the swift advance of civilisation was ascribed to the mind, to the development and activity of the brain. Men became accustomed to explain their actions from their thoughts, instead of from their needs—(which in any case are reflected and come to consciousness in the mind)—and so there arose in the course of time that idealistic outlook on the world which, especially since the decline of the ancient world, has dominated men's minds. It still rules them to such a degree that even the most materialistic natural scientists of the Darwinian school are still unable to form any clear idea of the origin of man, because under this ideological influence they do not recognise the part that has been played therein by labour.

Animals, as already indicated, change external nature by their activities just as man does, if not to the same extent, and these changes made by them in their environment, as we have seen, in turn react upon and change their originators. For in nature nothing takes place in isolation. Everything affects every other thing and *vice versa,* and it is usually because this many-sided motion and interaction is forgotten that our natural scientists are prevented from clearly seeing the simplest things. We have seen how goats have prevented the regeneration of forest in Greece; on the island of St. Helena, goats and pigs brought by the first arrivals have succeeded in exterminating almost completely the old vegetation of the island, and so have prepared the soil for the spreading of plants brought later by sailors and colonists. But if animals exert a lasting effect on their environment, it happens unintentionally, and, as far as the animals themselves are concerned, it is an accident. The further men become removed from animals, however, the more their effect on nature assumes the character of a premeditated, planned action directed towards definite ends known in advance. The animal destroys the vegetation of a locality without realising what it is doing. Man destroys it in order to sow field crops on the soil thus released, or to plant trees or vines which he knows will yield many times the amount sown. He transfers useful plants and domestic animals from one country to another and thus changes the flora and fauna of whole continents. More than this. Under artificial cultivation, both plants and animals are so changed by the hand of man that they become unrecognisable. The wild plants from which our grain varieties originated are still being sought in vain. The question of the wild animal from which our dogs are descended, the dogs themselves being so different from one another, or our equally numerous breeds of horses, is still under dispute.

In any case, of course, we have no intention of disputing the ability of animals to act in a planned and premeditated fashion. On the contrary, a planned mode of action exists in embryo wherever protoplasm, living protein, exists and reacts, *i.e.* carries out definite, even if extremely simple, movements as a result of definite external stimuli. Such reaction takes place even where there is as yet no cell at all, far less a nerve cell. The manner in which insectivorous plants capture their prey appears likewise in a certain respect as a planned action, although performed quite unconsciously. In animals the capacity for conscious, planned action develops side by side with the development of the nervous system and among mammals it attains quite a high level. While fox-hunting in England, one can daily observe how unerringly the fox knows how to make use of its excellent knowledge of the locality in order to escape from its pursuers, and how well it knows and turns to account all favourable features of the ground that cause the scent to be interrupted. Among our domestic animals, more highly developed thanks to association with man, every day one can note acts of cunning on exactly the same level as those of children. For, just as the developmental history of the human embryo in the mother's womb is only an abbreviated repetition of the history, extending over millions of years, of the bodily evolution of our animal ancestors, beginning from the worm, so the mental development of the human child is only a still more abbreviated repetition of the intellectual development of these same ancestors, at least of the later ones. But all the planned action of all animals has never resulted in impressing the stamp of their will upon nature. For that, man was required.

In short, the animal merely *uses* external nature, and brings about changes in it simply by his presence; man by his changes makes it serve his ends, *masters* it. This is the final, essential distinction between man and other animals, and once again it is labour that brings about this distinction.

Let us not, however, flatter ourselves overmuch on account of our human conquest over nature. For each such conquest takes its revenge on us. Each of them, it is true, has in the first place the consequences on which we counted, but in the second and third places it has quite different, unforeseen effects which only too often cancel out the first. The people who, in Mesopotamia, Greece, Asia Minor, and elsewhere, destroyed the forests to obtain cultivable land, never dreamed that they were laying the basis for the present devastated condition of these countries, by removing along with the forests the collecting centres and reservoirs of moisture. When, on the southern slopes of the mountains, the Italians of the Alps used up the pine forests so carefully cherished on the northern slopes, they had no inkling that by doing so they were cutting at the roots of the dairy industry in their region; they had still less inkling that they were thereby depriving their mountain springs of water for the greater part of the year, with the effect that these would be able to pour still more furious flood torrents on the plains during the rainy seasons. Those who spread the potato in Europe were not aware that they were at the same time spreading the disease of scrofula. Thus at every step we are reminded that we by no means rule over nature like a conqueror over a foreign people, like someone standing outside nature—but that we, with flesh, blood, and brain, belong to nature, and exist in its midst, and that all our mastery of it consists in the fact that we have the advantage over all other beings of being able to know and correctly apply its laws.

And, in fact, with every day that passes we are learning to understand these laws more correctly, and getting to know both the more immediate and the more remote consequences of our interference with the traditional course of nature. In particular, after the mighty advances of natural science in the present century, we are more and more getting to know, and hence to control, even the more remote natural consequences at least of our more ordinary productive activities. But the more this happens, the more will men not only feel, but also know, their unity with nature, and thus the more impossible will become the senseless and anti-natural idea of a contradiction between mind and matter, man and nature, soul and body, such as arose in Europe after the decline of classic antiquity and which obtained its highest elaboration in Christianity.

But if it has already required the labour of thousands of years for us to learn to some extent to calculate the more remote *natural* consequences of our actions aiming at production, it has been still more difficult in regard to the more remote *social* consequences of these actions. We mentioned the potato and the resulting spread of scrofula. But what is scrofula in comparison with the effect on the living conditions of the masses of the people in whole countries resulting from the workers being reduced to a potato diet, or in comparison with the famine which overtook Ireland in 1847 in consequence of the potato disease, and which put under the earth a million Irishmen, nourished solely or almost exclusively on potatoes, and forced the emigration overseas of two million more? When the Arabs learned to distil alcohol, it never entered their heads that by so doing they were creating one of the chief weapons for the annihilation of the original inhabitants of the still undiscovered American continent. And when afterwards Columbus discovered America, he did not know that by doing so he was giving new life to slavery, which in Europe had long ago been done away with, and laying the basis for the Negro slave traffic. The men who in the seventeenth and eighteenth centuries laboured to create the steam engine had no idea that they were preparing the instrument which more than any other was to revolutionise social conditions throughout the world. Especially in Europe, by concentrating wealth in the hands of a minority, the huge majority being rendered propertyless, this instrument was destined at first to give social and political domination to the bourgeoisie, and then, however, to give rise to a class struggle between bourgeoisie and proletariat, which can end only in the overthrow of the bourgeoisie and the abolition of all class contradictions. But even in this sphere, by long and often cruel experience and by collecting and analysing the historical material, we are gradually learning to get a clear view of the indirect, more remote, social effects of our productive activity, and so the possibility is afforded us of mastering and controlling these effects as well.

To carry out this control requires something more than mere knowledge. It requires a complete revolution in our hitherto existing mode of production, and with it of our whole contemporary social order.

NOTE

[1]A leading authority in this respect, Sir W. Thomson, has calculated that *little more than a hundred million years* could have elapsed since the time when the earth had cooled sufficiently for plants and animals to be able to live on it. [*Note by F. Engels.*]

From *Civilization and Its Discontents*

SIGMUND FREUD

■ Sigmund Freud *(1856-1939) is the founder of psychoanalysis. He was born in Freiberg, Moravia (now part of Czechoslovakia), and moved to Vienna with his parents when he was three years old. There he spent all but the last year of his life. Freud studied medicine, specializing in neurology. To that field he made valuable contributions in 1885 and 1886 before he became specially interested in the psychological aspects of neurology. He went soon after into private practice, devoting his efforts to treating hysterical patients. First he used hypnosis in his treatment; then he discarded it as his psychoanalytic methods developed. In the following years Freud explored the influence of the unconscious on virtually every aspect of human behavior; he became known for emphasizing the sexual origin and content of neurosis. He left many important works. Among these are* The Interpretation of Dreams *(1900) and* Civilization and Its Discontents *(1930), from which this short selection is taken.*

In this passage Freud ventures a most interesting theory about one of the origins of technology and civilization. This theory is almost totally buried in a footnote. He defines "civilization" as "the whole sum of the achievements and the regulations which distinguish our lives from those of our animal ancestors and which serve two purposes–namely to protect men against nature and to adjust their mutual relations." Thus civilization for Freud encompasses the technological as well as the political and social. According to Freud the first acts of civilization are: (1) use of tools, (2) control over fire, and (3) construction of dwelling. The control of fire he judges as an "extraordinary and unexampled achievement" which was gained only when man was finally able to renounce his desires and instincts. Freud's conjecture is that primal man had an infantile desire to put out the fire with a stream of urine. Since the flames were seen as a phallic symbol, this desire amounted to "a kind of sexual act with a male, an enjoyment of sexual potency in a homosexual competition." Once this desire was renounced man was able to use fire instead of putting it out. Thus, this origin of civilization is repressive.

The accumulated repressive aspects of civilization, of which the above is only one, led Freud to devote considerable thought to the contention that "what we call our civilization is largely responsible for our misery, and that we should be much happier if we gave it up and returned to primitive conditions." This contention reappears in various forms in selections in this anthology.

Our enquiry concerning happiness has not so far taught us much that is not already common knowledge. And even if we proceed from it to the problem of why it is so hard for men to be happy, there seems no greater prospect of learning anything new. We have given the answer already by pointing to the three sources from which our suffering comes: the superior power of nature, the feebleness of our own bodies and the inadequacy of the regulations which adjust the mutual relationships of human beings in the family, the state and society. In regard to the first two sources, our judgment cannot hesitate long. It forces us to acknowledge those sources of suffering and to submit to the inevitable. We shall never completely master nature; and our bodily organism, itself a part of that nature, will always remain a transient structure with a limited capacity for adaptation and achievement. This recognition does not have a paralysing effect. On the contrary, it points the direction for our activity. If we cannot remove all suffering, we can remove some, and we can mitigate some: the experience of many thousands of years has convinced us of that. As regards the third source, the social source of suffering, our attitude is a different one. We do not admit it at all; we cannot see why the regulations made by ourselves should not, on the contrary, be a protection and a benefit for every one of us. And yet, when we consider how unsuccessful we have been in precisely this field of prevention of suffering, a suspicion dawns on us that here, too, a piece of unconquerable nature may lie behind—this time a piece of our own psychical constitution.

When we start considering this possibility, we come upon a contention which is so astonishing that we must dwell upon it. This contention holds that what we call our civilization is largely responsible for our misery, and that we should be much happier if we gave it up and returned to primitive conditions. I call this contention astonishing because, in whatever way we may define the concept of civilization, it is a certain fact that all the things with which we seek to protect ourselves against the threats that emanate from the sources of suffering are part of that very civilization. . . .

It is time for us to turn our attention to the nature of this civilization on whose value as a means to happiness doubts have been thrown. We shall not look for a formula in which to express that nature in a few words, until we have learned something by examining it. We shall therefore content ourselves with saying once more that the word "civilization"[1] describes the whole sum of the achievements and the regulations which distinguish our lives from those of our animal ancestors and which serve two purposes—namely to protect men against nature and to adjust their mutual relations.[2] In order to learn more, we will bring together the various features of civilization individually, as they are exhibited in human communities. In doing so, we shall have no hesitation in letting ourselves be guided by linguistic usage or, as it is also called, linguistic feeling, in the conviction that we shall thus be doing justice to inner discernments which still defy expression in abstract terms.

The first stage is easy. We recognize as cultural all activities and resources which are useful to men for making the earth serviceable to them, for protecting them against the violence of the forces of nature, and so on. As regards this side of civilization, there can be scarcely any doubt. If we go back far enough, we find that the first acts of civilization were the use of tools, the gaining of control over fire and the construction of dwellings. Among these, the control over fire stands out as a quite extraordinary and unexampled achievement,[3] while the others opened up paths which man has followed ever since, and the stimulus to which is easily guessed. With every tool man is perfecting his own organs, whether motor or sensory, or is removing the limits to their functioning. Motor power places gigantic forces at his disposal, which, like his muscles, he can employ in any direction; thanks to ships and aircraft neither water nor air can hinder his movements; by means of spectacles he corrects defects in the lens of his own eye; by means of the telescope he sees into the far distance; and by means of the microscope he overcomes the limits of visibility set by the structure of his retina. In the photographic camera he has created an instrument which retains the fleeting visual impressions, just as a gramophone disc retains the equally fleeting auditory ones; both are at bottom materializations of the power he possesses of recollection, his memory. With the help of the telephone he can hear at distances which would be respected as unattainable even in a fairy tale. Writing was in its origin the voice of an absent person; and the dwelling-house was a substitute for the mother's womb, the first lodging, for which in all likelihood man still longs, and in which he was safe and felt at ease.

These things that, by his science and technology, man has brought about on this earth, on which he first appeared as a feeble animal organism and on which each individual of his species must once more make its entry ("oh inch of nature!"[4]) as a helpless suckling—these things do not only sound like a fairy tale, they are an actual fulfilment of every—or of almost every—fairy-tale wish. All these assets he may lay claim to as his cultural acquisition. Long ago he formed an ideal conception of omnipotence and omniscience which he embodied in his gods. To these gods he attributed everything that seemed unattainable to his wishes, or that was forbidden to him. One may say, therefore, that these gods were cultural ideals. To-day he has come very close to the attainment of this ideal, he has almost become a god himself. Only, it is true, in the fashion in which ideals are usually attained according to the general judgment of humanity. Not completely; in some respects not at all, in others only half way. Man has, as it were, become a kind of prosthetic[5] God. When he puts on all his auxiliary organs he is truly magnificent; but those organs have not grown on to him and they still give him much trouble at times. Nevertheless, he is entitled to console himself with the thought that this development will not come to an end precisely with the year 1930 A.D. Future ages will bring with them new and probably unimaginably great advances in this field of civilization and will increase man's likeness to God still more. But in the interests of our investigations, we will not forget that present-day man does not feel happy in his Godlike character.

NOTES

1. "*Kultur*." For the translation of this word see the Editor's Note to *The Future of an Illusion* [Volume XXI of *The Standard Edition of the Complete Psychological Works of Sigmund Freud*, trans. and ed. by James Strachey].
2. See *The Future of an Illusion*.
3. Psycho-analytic material, incomplete as it is and not susceptible to clear interpretation, nevertheless admits of a conjecture—a fantastic-sounding one—about the origin of this human feat. It is as though primal man had the

habit, when he came in contact with fire, of satisfying an infantile desire connected with it, by putting it out with a stream of his urine. The legends that we possess leave no doubt about the originally phallic view taken of tongues of flame as they shoot upwards. Putting out fire by micturating—a theme to which modern giants, Gulliver in Lilliput and Rabelais' Gargantua, still hark back—was therefore a kind of sexual act with a male, an enjoyment of sexual potency in a homosexual competition. The first person to renounce this desire and spare the fire was able to carry it off with him and subdue it to his own use. By damping down the fire of his own sexual excitation, he had tamed the natural force of fire. This great cultural conquest was thus the reward for his renunciation of instinct. Further, it is as though woman had been appointed guardian of the fire which was held captive on the domestic hearth, because her anatomy made it impossible for her to yield to the temptation of this desire. It is remarkable, too, how regularly analytic experience testifies to the connection between ambition, fire and urethral erotism.—[Freud had pointed to the connection between urination and fire as early as in the "Dora" case history (1905*e* [1901]). The connection with ambition came rather later.

4. [In English in the original. This very Shakespearean phrase is not in fact to be found in the canon of Shakespeare. The words "Poore inch of Nature" occur, however, in a novel by George Wilkins, *The Painfull Adventures of Pericles Prince of Tyre,* where they are addressed by Pericles to his infant daughter. This work was first printed in 1608, just after the publication of Shakespeare's play, in which Wilkins has been thought to have had a hand. Freud's unexpected acquaintance with the phrase is explained by its appearance in a discussion of the origins of *Pericles* in Georg Brandes's well-known book on Shakespeare, a copy of the German translation of which had a place in Freud's library. He is known to have greatly admired the Danish critic, and the same book is quoted in his paper on the three caskets (1913*f*).]
5. [A prosthesis is the medical term for an artificial adjunct to the body, to make up for some missing or inadequate part: e.g. false teeth or a false leg.]

From *The Instinct of Workmanship*

THORSTEIN VEBLEN

■ Thorstein Veblen *(1857-1929) was born on a Wisconsin farm. He was a political economist, sociologist, and social theorist; and although he is still well known abroad, he is virtually unknown to college students in the United States today. Veblen developed a highly insightful and comprehensive analysis of American industrial society in the early twentieth century. But his views were not accepted until after his death. He was also interested in developing an inclusive theory of social change. Among his best known books are* The Theory of the Leisure Class *(1899), which contains a severe critique of the existing social order, and* The Instinct of Workmanship *(1914), from which the following selection is excerpted. But he was a scholar, not an activist. His views influenced not only economists and political scientists but also public administrators and policy makers during the Theodore Roosevelt years.*

Technological progress for Veblen is a collective social achievement. True, it is individuals who make innovations, but "every expedient or innovation, great or small, that so is hit upon goes into effect by going into the common stock of technological resources carried by the group." This common stock of technological resources has two origins: habit and instinct. The first permits the accumulation and transference of knowledge; the second colors and shapes that knowledge.

Furthermore, everything else being equal, the growth and nature of technological resources are affected by the natural bent of the society in which they are accumulated. "Any difference of native endowment in this respect between the several races will show itself in the character of their technological achievements as well as in the rate of gain." For this reason "hybrid stocks" seem to be desirable for Veblen. They afford a "wider range of usual variability than the combined extreme limits of the racial types that enter into the composition of the hybrid."

CONTAMINATION OF INSTINCTS IN PRIMITIVE TECHNOLOGY

All instinctive behaviour is subject to development and hence to modification by habit.[1] Such impulsive action as is in no degree intelligent, and so suffers no adaptation through habitual use, is not properly to be called instinctive; it is rather to be classed as tropismatic. In human conduct the effects of habit in this respect are particularly far-reaching. In man the instincts appoint less of a determinate sequence of action, and so leave a more open field for adaptation of behaviour to the circumstances of the case. When instinct enjoins little else than the end of endeavour, leaving the sequence of acts by which this end is to be approached somewhat a matter of open alternatives, the share of reflection, discretion and deliberate adaptation will be correspondingly large. The range and diver-

☐ New York: The Macmillan Co., 1914, pp. 38-40, 103-112.

sity of habituation is also correspondingly enlarged.

In man, too, by the same fact, habit takes on more of a cumulative character, in that the habitual acquirements of the race are handed on from one generation to the next, by tradition, training, education, or whatever general term may best designate that discipline of habituation by which the young acquire what the old have learned. By similar means the like elements of habitual conduct are carried over from one community or one culture to another, leading to further complications. Cumulatively, therefore, habit creates usages, customs, conventions, preconceptions, composite principles of conduct that run back only indirectly to the native predispositions of the race, but that may affect the working-out of any given line of endeavour in much the same way as if these habitual elements were of the nature of a native bias.

Along with this body of derivative standards and canons of conduct, and handed on by the same discipline of habituation, goes a cumulative body of knowledge, made up in part of matter-of-fact acquaintance with phenomena and in greater part of conventional wisdom embodying certain acquired predilections and preconceptions current in the community. Workmanship proceeds on the accumulated knowledge so received and current, and turns it to account in dealing with the material means of life. Whatever passes current in this way as knowledge of facts is turned to account as far as may be, and so it is worked into a customary scheme of ways and means, a system of technology, into which new elements of information or acquaintance with the nature and use of things are incorporated, assimilated as they come.

The scheme of technology so worked out and carried along in the routine of getting a living will be serviceable for current use and have a substantial value for a further advance in technological efficiency somewhat in proportion as the knowledge so embodied in technological practice is effectually of the nature of matter-of-fact. Much of the information derived from experience in industry is likely to be of this matter-of-fact nature; but much of the knowledge made use of for the technological purpose is also of the nature of convention, inference and authentic opinion, arrived at on quite other grounds than workmanlike experience. This alien body of information, or pseudo-information, goes into the grand total of human knowledge quite as freely as any matter of fact, and it is therefore also necessarily taken up and assimilated in that technological equipment of knowledge and proficiency by use of which the work in hand is to be done.

But the experience which yields this useful and pseudo-useful knowledge is got under the impulsion and guidance of one and another of the instincts with which man is endowed, and takes the shape and color given it by the instinctive bias in whose service it is acquired. At the same time, whatever its derivation, the knowledge acquired goes into the aggregate of information drawn on for the ways and means of workmanship. Therefore the habits formed in any line of experience, under the guidance of any given instinctive disposition, will have their effect on the conduct and aims of the workman in all his work and play; so that progress in technological matters is by no means an outcome of the sense of workmanship alone. . . .

THE SAVAGE STATE OF THE INDUSTRIAL ARTS

Technological knowledge is of the nature of a common stock, held and carried forward collectively by the community, which is in this relation to be conceived as a going concern. The state of the industrial arts is a fact of group life, not of individual or private initiative or innovation. It is an affair of the collectivity, not a creative achievement of individuals working self-sufficiently in severalty or in isolation. In the main, the state of the industrial arts is always a heritage out of the past; it is always in process of change, perhaps, but the substantial body of it is knowledge that has come down from earlier generations. New elements of insight and proficiency are continually being added and worked into this common stock by the experience and initiative of the current generation, but such novel elements are always and everywhere slight and inconsequential in comparison with the body of technology that has been carried over from the past.

Each successive move in advance, every new wrinkle of novelty, improvement, invention, adaptation, every further detail of workmanlike innovation, is of course made by individuals and comes out of individual experience and initiative, since the generations of mankind live only in individuals. But each move so made is necessarily made by individuals immersed in the community and exposed to the discipline of group life as it runs in the community, since all life is necessarily group life. The phenomena of human life occur only in this form. It is only as an outcome of this discipline that comes with the routine of group life, and by help of the commonplace knowledge diffused through the community, that any of its members are enabled to make any new move that may in this way be traceable to their individual initiative. Any new technological departure necessarily takes its rise in the workmanlike endeavours of given individuals, but it can do so only by force of their familiarity with the body of knowledge which the group already has in hand. A new departure is always and necessarily an improvement on or alteration in that state of the industrial arts that is already in the keeping of the group at large; and every expedient or innovation, great or small, that so is hit upon goes into effect by going into the common stock of technological resources carried by the group. It can take effect only in this way. Such group solidarity is a necessity of the case, both for the acquirement and use of this immaterial equipment that is spoken of as the state of the industrial arts and for its custody and transmission from generation to generation. . . .

Given the material environment, the rate and character of the technological gains made in any community will depend on the initiative and application of its members, in so far as the growth of institutions has not seriously diverted the genius of the race from its natural bent; it will depend immediately and obviously on individual talent for workmanship—on the workmanlike bent and capacity of the individual members of the community. Therefore any difference of native endowment in this respect between the several races will show itself in the character of their technological achievements as well as in the rate of gain. Races differ among themselves in this matter, both as to the kind and as to the degree of technological proficiency of which they are capable.[2] It is perhaps as needless to insist on this spiritual difference between the various racial stocks as it would be difficult to determine the specific differences that are known to exist, or to exhibit them convincingly in detail. To some such ground much of the distinctive character of different peoples is no doubt to be assigned, though much also may as well be traceable to local peculiarities of environment and of institutional circumstances. Something of the kind, a specific difference in the genius of the people, is by common consent assigned, for instance, in explanation of the pervasive difference in technology and workmanship between the Western culture and the Far East. The like difference in "genius" is still more convincingly shown where different races have long been living near one another under settled cultural conditions.[3]

It should be noted in the same connection that hybrid peoples, such as those of Europe or of Japan, where somewhat widely distinct racial stocks are mingled, should afford a great variety and wide individual variation of native gifts, in workmanship as in other respects. Hybrid stocks, indeed, have a wider range of usual variability than the combined extreme limits of the racial types that enter into the composition of the hybrid. So that a great variety, even aberration and eccentricity, of native gifts is to be looked for in such cases, and this wide range of variation in workmanlike initiative should show itself in the technology of any such peoples. Yet there may still prevail a strikingly determinate difference between any two such hybrid populations, both in the characteristic features of their technology and in their routine workmanship; as is illustrated in the contrast between Japan and the Western nations. These racial differences in point of endowment may be slight in the first instance, but as they work cumulatively their ulterior effect may still be very marked; and they may result in marked differences not only in respect of the character of the technological situation at a given point of time but also in the rate of advance and the direction taken by the technological advance. So in the case of the Far East, as contrasted with the Occidental peoples, the genius of the races engaged has prevailingly taken the direction of proficiency in handicraft, rather than that somewhat crude but efficient recourse to mechanical expedients which chiefly distinguishes the technology of the West.

NOTES

[1]Cf. M. F. Washburn, *The Animal Mind,* ch. x, xi, where the simpler facts of habituation are suggestively presented in conformity with current views of empirical psychology.

[2]On such native differences between the leading races of Europe, cf., e.g., G. V. de Lapouge, *Les Sélections Sociales;* and *l'Aryen;* O. Ammon, *Die Gesellschaftsordnung;* G. Sergi, *Arii e Italici.*

[3]For instance, the Japanese and the Ainu, the Polynesians and the Melanesians, the Cinghalese and the Veddas. On the last named, cf. Seligmann, *The Veddas.*

Man the technician

JOSÉ ORTEGA Y GASSET

■ José Ortega y Gasset *(1883-1955) was a Spanish essayist and philosopher. He was born in Madrid of a patrician family. He taught at the University of Madrid, and was also active as a journalist and a politician. He is known for his analysis of history and modern culture, and especially for his penetrating study of the uniquely modern phenomenon "mass man." His book* The Revolt of the Masses *brought him international recognition. His philosophical views culminated over the years in an existentialist position which regarded the main vocation of the self to be self-realization. He held that the human being is decisively free in his inner self and that man is his history.*

For Ortega man is not an animal with a technological gift. Man is, in his very being, "self-made and autofabricated." Thus fabrication shapes the being of man. The world surrounds man with a web of facilities and difficulties, so that his existence in the world is "an unending struggle to accomodate himself in it." It is through this struggle that man at once defines his being and fabricates it. The origins of technology thus are embedded deep within man's being and his situation in the world. The "formidable and unparalleled character which makes man unique in the universe" is that he is "an entity whose being consists in not yet being," that is, an entity which is defined by its choices of what is not *yet. But these very choices that define man's being are crucial elements in the process of autofabrication. For this reason man is his history.*

EXCURSION TO THE SUBSTRUCTURE OF TECHNOLOGY

The answers which have been given to the question, what is technology, are appallingly superficial; and what is worse, this cannot be blamed on chance. For the same happens to all questions dealing with what is truly human in human beings. There is no way of throwing light upon them until they are tackled in those profound strata from which everything properly human evolves. As long as we continue to speak of the problems that concern man as though we knew what man really is, we shall only succeed in invariably leaving the true issue behind. That is what happens with technology. We must realize into what fundamental depths our argument will lead us. How does it come to pass that there exists in the universe this strange thing called technology, the absolute cosmic fact of man the technician? If we seriously intend to find an answer, we must be ready to plunge into certain unavoidable profundities.

We shall then come upon the fact that an entity in the universe, man, has no other way of existing than by being in another entity, nature or the world. This relation of being one in the other, man in nature, might take on one of three possible aspects. Nature might offer man nothing but facilities for his existence in it. That would mean that the being of man coincides fully with that of nature or, what is the same, that man is a natural being. That is the case of the stone, the plant, and, probably, the animal. If it were that of man, too, he would be without necessities, he would lack nothing, he would not be needy. His desires and their satisfaction would be one and the same. He would wish for nothing that did not exist in the world and, conversely, whatever he wished for would be there of itself, as in the fairy tale of the magic wand. Such an entity could not experience the world as something alien to himself; for the world would offer him no resistance. He would be in the world as though he were in himself.

Or the opposite might happen. The world might offer to man nothing but difficulties, i.e., the being of the world and the being of man might be completely antagonistic. In this case the world would be no abode for man; he could not exist in it, not even for the fraction of a second. There would be no human life and, consequently, no technology.

The third possibility is the one that prevails in reality. Living in the world, man finds that the world surrounds him as an intricate net woven of both facilities and difficulties. Indeed, there are not many things in it which, potentially, are not both. The earth supports him, enabling him to lie down when he is tired and to run when he has to flee. A shipwreck will bring home to him the advantage of the firm earth—a thing grown humble from habitude. But the earth also means distance. Much earth may separate him from the spring when he is thirsty. Or the earth may tower above him as a steep slope

☐ Selection is reprinted from *"History As a System" and Other Essays Toward a Philosophy of History,* by José Ortega y Gasset, with the permission of W. W. Norton & Company, Inc. Trans. by Helene Weyl.

that is hard to climb. This fundamental phenomenon—perhaps the most fundamental of all—that we are surrounded by both facilities and difficulties gives to the reality called human life its peculiar ontological character.

For if man encountered no facilities it would be impossible for him to be in the world, he would not exist, and there would be no problem. Since he finds facilities to rely on, his existence is possible. But this possibility, since he also finds difficulties, is continually challenged, disturbed, imperiled. Hence, man's existence is no passive being in the world; it is an unending struggle to accommodate himself in it. The stone is given its existence; it need not fight for being what it is—a stone in the field. Man has to be himself in spite of unfavorable circumstances; that means he has to make his own existence at every single moment. He is given the abstract possibility of existing, but not the reality. This he has to conquer hour after hour. Man must earn his life, not only economically but metaphysically.

And all this for what reason? Obviously—but this is repeating the same thing in other words—because man's being and nature's being do not fully coincide. Because man's being is made of such strange stuff as to be partly akin to nature and partly not, at once natural and extranatural, a kind of ontological centaur, half immersed in nature, half transcending it. Dante would have likened him to a boat drawn up on the beach with one end of its keel in the water and the other in the sand. What is natural in him is realized by itself; it presents no problem. That is precisely why man does not consider it his true being. His extranatural part, on the other hand, is not there from the outset and of itself; it is but an aspiration, a project of life. And this we feel to be our true being; we call it our personality, our self. Our extra- and antinatural portion, however, must not be interpreted in terms of any of the older spiritual philosophies. I am not interested now in the so-called spirit *(Geist)*, a pretty confused idea laden with speculative wizardry.

If the reader reflects a little upon the meaning of the entity he calls his life, he will find that it is the attempt to carry out a definite program or project of existence. And his self—each man's self—is nothing but this devised program. All we do we do in the service of this program. Thus man begins by being something that has no reality, neither corporeal nor spiritual; he is a project as such, something which is not yet but aspires to be. One may object that there can be no program without somebody having it, without an idea, a mind, a soul, or whatever it is called. I cannot discuss this thoroughly because it would mean embarking on a course of philosophy. But I will say this: although the project of being a great financier has to be conceived of in an idea, "being" the project is different from holding the idea. In fact, I find no difficulty in thinking this idea but I am very far from being this project.

Here we come upon the formidable and unparalleled character which makes man unique in the universe. We are dealing—and let the disquieting strangeness of the case be well noted—with an entity whose being consists not in what it is already, but in what it is not yet, a being that consists in not-yet-being. Everything else in the world is what it is. An entity whose mode of being consists in what it is already, whose potentiality coincides at once with his reality, we call a "thing." Things are given their being ready-made.

In this sense man is not a thing but an aspiration, the aspiration to be this or that. Each epoch, each nation, each individual varies in its own way the general human aspiration.

Now, I hope, all terms of the absolute phenomenon called "my life" will be clearly understood. Existence means, for each of us, the process of realizing, under given conditions, the aspiration we are. We cannot choose the world in which to live. We find ourselves, without our previous consent, embedded in an environment, a here and now. And my environment is made up not only by heaven and earth around me, but by my own body and my own soul. I am not my body; I find myself with it, and with it I must live, be it handsome or ugly, weak or sturdy. Neither am I my soul; I find myself with it and must use it for the purpose of living although it may lack will power or memory and not be of much good. Body and soul are things; but I am a drama, if anything, an unending struggle to be what I have to be. The aspiration or program I am, impresses its peculiar profile on the world about me, and that world reacts to this impress, accepting or resisting it. My aspiration meets with hindrance or with furtherance in my environment.

At this point one remark must be made which would have been misunderstood before. What we call nature, circumstance, or the world is essentially nothing but a conjunction of favorable and adverse conditions encountered by man in the pursuit of this program. The three names are interpretations of ours; what we first come upon is the experience of being hampered or favored in living. We are wont to conceive of nature and world as existing by themselves, independent of man. The concept "thing" likewise refers to something that has a hard and fast being and has it by itself and apart from man. But I repeat, this is the result of an interpretative reaction of our intellect upon what first confronts us. What first confronts us has no being apart from and independent of us; it consists exclusively in presenting facilities and difficulties, that is to say, in what it is in respect to our aspiration. Only in relation to our vital program is something an obstacle or an aid. And according to the aspiration animating us the facilities and difficulties, making up our pure and fundamental environment, will be such or such, greater or smaller.

This explains why to each epoch and even to each individual the world looks different. To the particular profile of our personal project, circumstance answers with another definite profile of facilities and difficulties. The world of the businessman obviously is different from the world of the poet. Where one comes to grief, the other thrives; where one rejoices, the other frets. The two worlds, no doubt, have many elements in common, viz., those which correspond to the generic aspiration of man as a species. But the human species is incomparably less stable and more mutable than any animal species. Men have an intractable way of being enormously unequal in spite of all assurances to the contrary.

LIFE AS AUTOFABRICATION—TECHNOLOGY AND DESIRES

From this point of view human life, the existence of man, appears essentially problematic. To all other entities of the universe existence presents no problem. For existence means actual realization of an essence. It means, for instance, that "being a bull" actually occurs. A bull, if he exists, exists as a bull. For a man, on the contrary, to exist does not mean to exist at once as the man he is, but merely that there exists a possibility of, and an effort towards, accomplishing this. Who of us is all he should be and all he longs to be? In contrast to the rest of creation, man, in existing, has to make his existence. He has to solve the practical problem of transferring into reality the program that is himself. For this reason "my life" is pure task, a thing inexorably to be made. It is not given to me as a present; I have to make it. Life gives me much to do; nay, it is nothing save the "to do" it has in store for me. And this "to do" is not a thing, but action in the most active sense of the word.

In the case of other beings the assumption is that somebody or something, already existing, acts; here we are dealing with an entity that has to act in order to be; its being presupposes action. Man, willy-nilly, is self-made, autofabricated. The word is not unfitting. It emphasizes the fact that in the very root of his essence man finds himself called upon to be an engineer. Life means to him at once and primarily the effort to bring into existence what does not exist offhand, to wit: himself. In short, human life "is" production. By this I mean to say that fundamentally life is not, as has been believed for so many centuries, contemplation, thinking, theory, but action. It is fabrication; and it is thinking, theory, science only because these are needed for its autofabrication, hence secondarily, not primarily. To live . . . that is to find means and ways for realizing the program we are.

The world, the environment, presents itself as *materia prima* and possible machine for this purpose. Since man, in order to exist, has to be in the world and the world does not admit forthwith of the full realization of his being, he sets out to search around for the hidden instrument that may serve his ends. The history of human thinking may be regarded as a long series of observations made to discover what latent possibilities the world offers for the construction of machines. And it is not by chance, as we shall shortly see, that technology properly speaking, technology in the fullness of its maturity, begins around 1600, when man in the course of his theoretical thinking about the world comes to regard it as a machine. Modern technology is linked with the work of Galileo, Descartes, Huygens, i.e., with the mechanical interpretation of the universe. Before that, the corporeal world had been generally believed to be an a-mechanical entity, the ultimate essence of which was constituted by spiritual powers of more or less arbitrary and uncontrollable nature; whereas the world as pure mechanism is the machine of machines.

It is, therefore, a fundamental error to believe that man is an animal endowed with a talent for technology, in other words, that an animal might be transmuted into a man by magically grafting on it the technical gift. The opposite holds: because man has to accomplish a task fundamentally different from that of the animal, an extranatural task, he cannot spend his energies in satisfying his elemental needs, but must stint them in this realm so as to be able to employ them freely in the odd pursuit of realizing his being in the world.

10

Prominent perspectives on technology and civilization

HISTORICAL

The act of invention: causes, contexts, continuities and consequences

LYNN WHITE, Jr.

■ Lynn White *was born in 1907 in San Francisco. A well-known historian of the medieval and Renaissance periods, he is currently Professor Emeritus of History at the University of California at Los Angeles. He is also the president of the Society for the History of Technology. Among his books are* Medieval Technology and Social Change *(1962) and* Medieval Religion and Technology *(1978).*

In the essay reprinted here, White expresses his disapproval of theories about the nature of technological innovation, and about its relation to other activities, that do not take into consideration the concrete historical facts. Furthermore, he observes, even when we do know the facts, they need not add up to an "explanation" of a technological innovation. Therefore, "the best that we can do at present is to work hard to find the facts and then to think cautiously about the facts which have been found." In his attempt to better our understanding of technological innovations—their nature, motivation, conditioning aspects, and effects—White produces interesting historical data. Perhaps the most famous is the data concerning the pennon, whose origins, development, and effects outside the area of technology White makes clear. He also attempts to dispel the belief that once a technological innovation is discovered, it is readily adopted and disseminated. As a case in point, he introduces the Hellenistic discovery of the helix, which did not reach China, despite the Chinese technological sophistication, until modern times. (On this topic see also the second selection by Ortega in this part). White concludes that "a novel technique merely offers opportunity; it does not command." Emmanuel Mesthene later comments on this statement (see Part Three).

The rapidly growing literature on the nature of technological innovation and its relation to other activities is largely rubbish because so few of the relevant concrete facts have thus far been ascertained. It is an inverted pyramid of generalities, the apex of which is very nearly a void. The five plump volumes of *A History of Technology*,[1] edited under the direction of Charles Singer, give the layman a quite false impression of the state of knowledge. They are very useful as a starting point, but they are almost as much a codification of error as of sound information.[2] It is to be feared that the physical weight of these books will be widely interpreted as the weight of authority and that philosophers, sociologists, and others whose personal researches do not lead them into the details of specific technological items may continue to be deceived as to what is known.

Since man is a hypothesizing animal, there is no point in calling for a moratorium on speculation in this area of thought until more firm facts can be accumulated. Indeed, such a moratorium—even if it were possible—would slow down the growth of factual knowledge because hypothesis normally provokes counter-hypoth-

☐ Reprinted from *The Technological Order*, ed. Carl E. Stover (Detroit: Wayne State University Press, 1963), pp. 102-116, by permission of the Wayne State University Press and the author.

eses, and then all factions adduce facts in evidence, often new facts. The best that we can do at present is to work hard to find the facts and then to think cautiously about the facts which have been found.

In view of our ignorance, then, it would seem wise to discuss the problems of the nature, the motivations, the conditioning circumstances, and the effects of the act of invention far less in terms of generality than in terms of specific instances about which something seems to be known.

1. The beginning of wisdom may be to admit that even when we know some facts in the history of technology, these facts are not always fully intelligible, i.e., capable of "explanation," simply because we lack adequate contextual information. The Chumash Indians of the coast of Santa Barbara County built plank boats which were unique in the pre-Columbian New World: their activity was such that the Spanish explorers of California named a Chumash village "La Carpintería."[3] A map will show that this tribe had a particular inducement to venture upon the sea: they were enticed by the largest group of off-shore islands along the Pacific Coast south of Canada. But why did the tribes of South Alaska and British Columbia, of Araucanian Chile, or of the highly accidented eastern coast of the United States never respond to their geography by building plank boats? Geography would seem to be only one element in explanation.

Can a plank-built East Asian boat have drifted on the great arc of currents in the North Pacific to the Santa Barbara region? It is entirely possible; but such boats would have been held together by pegs, whereas the Chumash boats were lashed, like the dhows of the Arabian Sea or like the early Norse ships. Diffusion seems improbable.

Since a group can conceive of nothing which is not first conceived by a person, we are left with the hypothesis of a genius: a Chumash Indian who at some unknown date achieved a break-away from log dugout and reed balsa to the plank boat. But the idea of "genius" is itself an ideological artifact of the age of the Renaisance when painters, sculptors, and architects were trying to raise their social status above that of craftsmen.[4] Does the notion of genius "explain" Chumash plank boats? On the contrary, it would seem to be no more than a traditionally acceptable way of labeling the great Chumash innovation as unintelligible. All we can do is to observe the fact of it and hope that eventually we may grasp the meaning of it.

2. A symbol of the rudimentary nature of our thinking about technology, its development, and its human implications, is the fact that while the *Encyclopaedia Britannica* has an elaborate article on "Alphabet," it contains no discussion of its own organizational presupposition, alphabetization. Alphabetization is the basic invention for the classification and recovery of information: it is fully comparable in significance to the Dewey decimal system and to the new electronic devices for these purposes. Modern big business, big government, big scholarship are inconceivable without alphabetization. One hears that the chief reason why the Chinese Communist regime has decided to Romanize Chinese writing is the inefficiency of trying to classify everything from telephone books to tax registers in terms of 214 radicals of ideographs. Yet we are so blind to the nature of our technical equipment that the world of Western scholars, which uses alphabetization constantly, has produced not even the beginning of a history of it.

Fortunately, Dr. Sterling Dow of Harvard University is now engaged in the task. He tells me that the earliest evidence of alphabetization is found in Greek materials of the third century B.C. In other words, there was a thousand-year gap between the invention of the alphabet as a set of phonetic symbols and the realization that these symbols, and their sequence in individual written words, could be divorced from their phonetic function and used for an entirely different purpose: an arbitrary but very useful convention for storage and retrieval of verbal materials. That we have neglected thus completely the effort to understand so fundamental an invention should give us humility whenever we try to think about the larger aspects of technology.

3. Coinage was one of the most significant and rapidly diffused innovations of Late Antiquity. The dating of it has recently become more conservative than formerly: the earliest extant coins were sealed into the foundation of the temple of Artemis at Ephesus c. 600 B.C., and the invention of coins, i.e., lumps of metal the value of which is officially certified, was presumably made in Lydia not more than a decade earlier.[5]

Here we seem to know something, at least until the next archaeological spades turn up new testimony. But what do we know with any certainty about the impact of coinage? We are compelled to tread the slippery path of *post hoc ergo propter hoc*. There was a great acceleration of commerce in the Aegean, and it is hard to escape the conviction that this movement, which is the economic presupposition of the Periclean Age, was lubricated by the invention of coinage.

If we dare to go this far, we may venture further. Why did the atomic theory of the nature of matter appear so suddenly among the philosophers of the Ionian cities? Their notion that all things are composed of different arrangements of identical atoms of some "element," whether water, fire, ether, or something else, was an intellectual novelty of the first order, yet its sources have not been obvious. The psychological roots of atomism would seem to be found in the saying of Heraclitus of Ephesus that "all things may be reduced to fire, and fire to all things, just as all goods may be turned into gold and gold into all goods."[6] He thought that he was just using a metaphor, but the metaphor had been possible for only a century before he used it.

Here we are faced with a problem of critical method. Apples had been dropping from trees for a considerable period before Newton discovered gravity:[7] we must distinguish cause from occasion. But the appearance of coinage is a phenomenon of a different order from the fall of an apple. The unprecedented element in the general life of sixth-century Ionia, the chief stimulus to the prosperity which provided leisure for the atomistic philosophers, was the invention of coinage: the age of barter was ended. Probably no Ionian was conscious of any connection between this unique new technical instrument and the brainstorms of the local intellectuals. But that a causal relationship did exist can scarcely be doubted, even though it cannot be "proved" but only perceived.

4. Fortunately, however, there are instances of technological devices of which the origins, development, and effects outside the area of technology are quite clear. A case in point is the pennon.[8]

The stirrup is first found in India in the second century B.C. as the big-toe stirrup. For climatic reasons its diffusion to the north was blocked, but it spread wherever India had contact with barefoot aristocracies, from the Philippines and Timor on the east to Ethiopia on the west. The nuclear idea of the stirrup was carried to China on the great Indic culture wave which also spread Buddhism to East Asia, and by the fifth century the shod Chinese were using a foot stirrup.

The stirrup made possible, although it did not require, a new method of fighting with the lance. The unstirrupped rider delivered the blow with the strength of his arm. But stirrups, combined with a saddle equipped with pommel and cantle, welded rider to horse. Now the warrior could lay his lance at rest between his upper arm and body: the blow was delivered not by the arm but by the force of a charging stallion. The stirrup thus substituted horse-power for man-power in battle.

The increase in violence was tremendous. So long as the blow was given by the arm, it was almost impossible to impale one's foe. But in the new style of mounted shock combat, a good hit might put the lance entirely through his body and thus disarm the attacker. This would be dangerous if the victim had friends about. Clearly, a baffle must be provided behind the blade to prevent penetration by the shaft of the lance and thus permit retraction.

Some of the Central Asian peoples attached horse tails behind the blades of lances—this was probably being done by the Bulgars before they invaded Europe. Others nailed a piece of cloth, or pennon, to the shaft behind the blade. When the stirrup reached Western Europe c. 730 A.D., an effort was made to meet the problem by adapting to military purposes the old Roman boar-spear which had a metal crosspiece behind the blade precisely because boars, bears, and leopards had been found to be so ferocious that they would charge up a spear not so equipped.

This was not, however, a satisfactory solution. The new violence of warfare demanded heavier armor. The metal crosspiece of the lance would sometimes get caught in the victim's armor and prevent recovery of the lance. By the early tenth century Europe was using the Central Asian cloth pennon, since even if it got entangled in armor it would rip and enable the victor to retract his weapon.

Until our dimsal age of camouflage, fighting men have always decorated their equipment. The pennons on lances quickly took on color and design. A lance was too long to be taken into a tent conveniently, so a knight usually set it upright outside his tent, and if one were looking for him, one looked first for the flutter of his familiar pennon. Knights riding held their lances erect, and since their increasingly massive armor made recognition difficult, each came to be identified by his pennon. It would seem that is was from the pennon that distinctive "connoissances" were transferred to shield and surcoat. And with the crystallization of the feudal structure, these heraldic devices became hereditary, the symbols of status in European society.

In battle, vassals rallied to the pennon of their liege lord. Since the king was, in theory if not always in practice, the culmination of the feudal hierarchy, his pennon took on a particular aura of emotion: it was the focus of secular loyalty. Gradually a distinction was made between the king's two bodies,[9] his person and his "body politic," the state. But a colored cloth on the shaft of a spear remained the primary symbol of allegiance to either body, and so remains even in politics which have abandoned monarchy. The grimly functional rags first nailed to lance shafts by Asian nomads have had a great destiny. But it is no more remarkable than that of the cross, a hideous implement in the Greco-Roman technology of torture, which was to become the chief symbol of the world's most widespread religion.

In tracing the history of the pennon, and of many other technological items, there is a temptation to convey a sense of inevitability. However, a novel technique merely offers opportunity; it does not command. As has been mentioned, the big-toe stirrup reached Ethiopia. It was still in common use there in the nineteenth century, but at the present time Muslim and European influences have replaced it with the foot stirrup. However, travellers tell me that the Ethiopian gentleman, whose horse is equipped with foot stirrups, rides with only his big toes resting in the stirrups.

5. Indeed, in contemplating the history of technology, and its implications for our understanding of ourselves, one is as frequently astonished by blindness to innovation as by the insights of invention. The Hellenistic discovery of the helix was one of the greatest of technological inspirations. Very quickly it was applied not only to gearing but also to the pumping of water by the so-called Archimedes screw.[10] Somewhat later the holding screw appears in both Roman and Germanic metal work.[11] The helix was taken for granted thenceforth in western technology. Yet Joseph Needham of Cambridge University assures me that, despite the great sophistication of the Chinese in most technical matters, no form of helix was known in East Asia before modern times: it reached India but did not pass the Himalayas. Indeed, I have not been able to locate any such device in the Far East before the early seventeenth century when Archimedes screws, presumably introduced by the Portuguese, were used in Japanese mines.[12]

6. Next to the wheel, the crank is probably the most important single element in machine design, yet until the fifteenth century the history of the crank is a dismal record of inadequate vision of its potentialities.[13] It first appears in China under the Han dynasty, applied to rotary fans for winnowing hulled rice, but its later applications in the Far East were not conspicuous. In the West the crank seems to have developed independently and to have emerged from the hand quern. The earliest querns were fairly heavy, with a handle, or handles, inserted laterally in the upper stone, and the motion was reciprocating. Gradually the stones grew lighter and thinner, so that it was harder to insert the peg-handle horizontally: its angle creeps upward until eventually it stands vertically on top. All the querns found at the Saalburg had horizontal handles, and it is increasingly clear that the vertical peg is post-Roman.

Seated before a quern with a single vertical handle, a person of the twentieth century would give it a continuous rotary motion. It is far from clear that one of the very early Middle Ages would have done so. Crank motion was a kinetic invention more difficult than we can easily conceive. Yet at some point before the time of Louis the Pious the sense of the appropriate motion changed; for out of the rotary quern came a new machine, the rotary grindstone, which (as the Latin term for it, *mola fabri,* shows) is the upper stone of a quern turned on edge and adapted to sharpening. Thus, in Europe at least, crank motion was invented before the crank, and the crank does not appear before the early ninth century. As for the Near East, I find not even the simplest application of the crank until al-Jazarī's book on automata of 1206 A.D.

Once the simple crank was available, its development into the compound crank and connecting rod might have been expected quite quickly. Yet there is no sign of a compound crank until 1335, when the Italian physician of the Queen of France, Guido da Vigevano, in a set of astonishing technological sketches, which Rupert Hall has promised to edit,[14] illustrates three of them.[15] By the fourteenth century Europe was using crankshafts with two simple cranks, one at each end; indeed, this device was known in Cambodia in the thirteenth century. Guido was interested in the problem of self-moving vehicles: paddlewheel boats and fighting towers propelled by windmills or from the inside. For such constricted situations as the inside of a boat or a tower it apparently occurred to him to consolidate the two cranks at the ends of the crankshaft into a compound crank in its middle. It was an inspiration of the first order, yet nothing came of it. Evidently the Queen's physician, despite his technological interests, was socially too far removed from workmen to influence the actual technology of his time. The compound crank's effective appearance was delayed for another three generations. In the 1420's some Flemish carpenter or shipwright invented the bit-and-brace with its compound crank. By c. 1430 a German engineer was applying double compound cranks and connecting rods to machine design: a technological event as significant as the Hellenistic invention of gearing. The idea spread like wildfire, and European applied mechanics was revolutionized.

How can we understand the lateness of the discovery, whether in China or Europe, of even the simple crank, and then the long delay in its wide application and elaboration? Continuous rotary motion is typical of inorganic matter, whereas reciprocating motion is the sole movement found in living things. The crank connects these two kinds of motion; therefore we who are organic find that crank motion does not come easily to us. The great physicist and philosopher Ernst Mach noticed that infants find crank motion hard to learn.[16] Despite the rotary grindstone, even today razors are whetted rather than ground: we find rotary motion a bar to the greatest sensitivity. Perhaps as early as the tenth century the hurdy-gurdy was played with a cranked resined wheel vibrating the strings. But by the thirteenth century the hurdy-gurdy was ceasing to be an instrument for serious music. It yielded to the reciprocating fiddle bow, an introduction of the tenth century which became the foundation of modern European musical development. To use a crank, our tendons and muscles must relate themselves to the motion of galaxies and electrons. From this inhuman adventure our race long recoiled.

7. A sequence originally connected with the crank may serve to illustrate another type of problem in the act of technological innovation: the fact that a simple idea transferred out of its first context may have a vast expansion. The earliest appearance of the crank, as has been mentioned, is found on a Han-dynasty rotary fan to winnow husked rice.[17] The identical apparatus appears in the eighteenth century in the Palatinate,[18] in upper Austria and the Siebenbürgen,[19] and in Sweden.[20] I have not seen the exact channel of this diffusion traced, but it is clearly part of the general Jesuit-inspired *Chinoiserie* of Europe in that age. Similarly, I strongly suspect, but cannot demonstrate, that all subsequent rotary blowers, whether in furnaces, dehydrators, wind tunnels, air conditioning systems, or the simple electric fan, are descended from this Han machine which seems, in China itself, to have produced no progeny.

8. Doubtless when scholarship in the history of technology becomes firmer, another curious device will illustrate the same point. To judge by its wide distribution,[21] the fire piston is an old invention in Malaya. Dr. Thomas Kuhn of the University of California at Berkeley, who has made careful studies of the history of our knowledge of adiabatic heat, assures me that when the fire piston appeared in late eighteenth-century Europe not only for laboratory demonstrations but as a commercial product to light fires, there is no hint in the purely scientific publications that its inspiration was Malayan. But the scientists, curiously, also make no mention of the commercial fire pistons then available. So many Europeans, especially Portuguese and Netherlanders, had been trading, fighting, ruling, and evangelizing in the East Indies for so long a time before the fire piston is found in Europe, that it is hard to believe that the Malayan fire piston was not observed and reported. The realization of its potential in Europe was considerable, culminating in the diesel engine.

9. Why are such nuclear ideas sometimes not exploited in new and wider applications? What sorts of barriers prevent their diffusion? Why, at times, does what appeared to be a successful technological item fall into disuse? The history of the faggoted forging method of producing sword blades[22] may assist our thinking about such questions.

In late Roman times, north of the Alps, Celtic, Slavic, and Germanic metallurgists began to produce swords with laminations produced by welding together bundles of rods of different qualities of iron and steel, hammering the resulting strip thin, folding it over, welding it all together again, and so on. In this way a fairly long blade was produced which had the cutting qualities of steel but the toughness of iron. Although such swords were used at times by barbarian auxiliaries in the Roman army, the Roman legions never adopted them. Yet as soon as the Western Empire crumbled, the short Roman stabbing sword vanished and the laminated slashing blade alone held the field of battle. Can this conservatism in military equipment have been one reason for the failure of the Empire to stop the Germanic invasions? The Germans had adopted the new type of blade with enthusiasm, and by Carolingian times were manufacturing it in quantities in the Rhineland for export to Scandinavia and to Islam where it was much prized. Yet, although such blades were produced marginally as late as the twelfth century, for practical purposes they ceased to be used in Europe in the tenth century. Does the disappearance of such sophisticated swords indicate a decline in medieval metallurgical methods?

We should be cautious in crediting the failure of the Romans to adopt the laminated blade to pure stupidity. The legions seem normally to have fought in very close formation, shield to shield. In such a situation, only a stabbing sword could be effective. The Germans at times used a ''shield wall'' formation, but it was probably a bit more open than the Roman and permitted use of a slashing sword. If the Romans had accepted the new weapon, their entire drill and discipline would have been subject to revision. Unfortunately, we lack studies of the development of Byzantine weapons, sufficiently detailed to let us judge whether, or to what extent, the vigorously surviving Eastern Roman Empire adapted itself to the new military technology.

The famous named swords of Germanic myth, early medieval epic and Wagnerian opera were laminated blades. They were produced by the vast patience and skill of smiths who themselves became legendary. Why did they cease to be made in any number after the tenth century? The answer is found in the rapid increase in the weight of European armor as a result of the consistent Frankish elaboration of the type of mounted shock combat made possible by the stirrup. After the turn of the millenium a sword in Europe had to be very nearly a club with sharp edges: the best of the earlier blades was ineffective against such defenses. The faggoted method of forging blades survived and reached its technical culmination in Japan[23] where, thanks possibly to the fact that archery remained socially appropriate to an aristocrat, mounted shock combat was less emphasized than in Europe and armor remained lighter.

10. Let us now turn to a different problem connected with the act of invention. How do methods develop by the transfer of ideas from one device to another? The origins of the cannon ball and the cannon may prove instructive.[24]

Hellenistic and Roman artillery was activated by the torsion of cords. This was reasonably satisfactory for summer campaigns in the Mediterranean basin, but north of the Alps and in other damper climates the cords tended to lose their resilience. In 1004 A.D. a radically different type of artillery appeared in China with the name *huo p'ao*. It consisted of a large sling-beam pivoted on a frame and actuated by men pulling in unison on ropes attached to the short end of the beam away from the sling. It first appears outside China in a Spanish Christian illumination of the early twelfth century, and from this one might assume diffusion through Islam. But its second appearance is in the northern Crusader army attacking Lisbon in 1147 where a battery of them were operated by shifts of one hundred men for each. It would seem that the Muslim defenders were quite unfamiliar with the new engine of destruction and soon capitulated. This invention, therefore, appears to have reached the West from China not through Islam but directly across Central Asia. Such a path of diffusion is the more credible because by the end of the same century the magnetic needle likewise arrived in the West by the northern route, not as an instrument of navigation but as a means of ascertaining the meridian, and Western Islam got the compass from Italy.[25] When the new artillery arrived in the West it had lost its name. Because of structural analogy, it took on a new name borrowed from a medieval instrument of torture, the ducking stool or *trebuchetum*.

Whatever its merits, the disadvantages of the *huo p'ao* were the amount of man-power required to operate it and the fact that since the gang pulling the ropes would never pull with exactly the same speed and force, missiles could not be aimed with great accuracy. The problem was solved by substituting a huge counterweight at the short end of the sling-beam for the ropes pulled by men. With this device a change in the weight of the caisson of stones or earth, or else a shift of the weight's position in relation to the pivot, would modify the range of the projectile and then keep it uniform, permitting concentration of fire on one spot in the fortifications to be breeched. Between 1187 and 1192 an Arabic treatise written in Syria for Saladin mentions not only Arab, Turkish, and Frankish forms of the primitive trebuchet, but also credits to Iran the invention of the trebuchet with swinging caisson. This ascription, however, must be in error; for from c. 1220 onward oriental sources frequently call this engine *magribī,* i.e., "Western." Morover, while the counterweight artillery has not yet been documented for Europe before 1199, it quickly displaced the older forms of artillery in the West, whereas this new and more effective type of siege machinery became dominant in the Mameluke army only in the second half of the thirteenth century. Thus the trebuchet with counterweights would appear to be a European improvement on the *huo p'ao*. Europe's debt to China was repaid in 1272 when, if we may believe Marco Polo, he and a German technician, helped by a Nestorian Christian, delighted the Great Khan by building trebuchets which speedily reduced a besieged city.

But the very fact that the power of a trebuchet could be so nicely regulated impelled Western military engineers to seek even greater exactitude in artillery attack. They quickly saw that until the weight of projectiles and their friction with the air could be kept uniform, artillery aim would still be variable. As a result, as early as 1244 stones for trebuchets were being cut in the royal arsenals of England calibrated to exact specifications established by an engineer: in other words, the cannon ball before the cannon.

The germinal idea of the cannon is found in the metal tubes from which, at least by the late ninth century, the Byzantines had been shooting Greek fire. It may be that even that early they were also shooting rockets of Greek fire, propelled by the expansion of gases, from bazooka-like metal tubes. When, shortly before 673, the Greek-speaking Syrian refugee engineer Callinicus invented Greek fire, he started the technicians not only of Byzantium but also of Islam, China, and eventually the West in search of ever more combustible mixtures. As chemical methods improved, the saltpeter often used in these compounds became purer, and combustion tended toward explosion. In the thirteenth century one finds, from the Yellow Sea to the Atlantic, incendiary bombs, rockets, firecrackers, and fireballs shot from tubes like Roman candles. The flame and roar of all this has made it marvelously difficult to ascertain just when gunpowder artillery, shooting hard missles from metal tubes, appeared. The first secure evidence is a famous English illumination of 1327 showing a vase-shaped cannon discharging a giant arrow. Moreover, our next certain reference to a gun, a "pot de fer à traire garros de feu" at Rouen in 1338, shows how long it took for technicians to realize that the metal tube, gunpowder, and the calibrated trebuchet missile could be combined. However, iron shot appear at Lucca in 1341; in 1346 in England there were two calibres of lead shot; and balls appear at Toulouse in 1347.

The earliest evidence of cannon in China is extant examples of 1356, 1357, and 1377. It is not necessary to assume the miracle of an almost simultaneous independent Chinese invention of the cannon: enough Europeans were wandering the Yuan realm to have carried it eastward. And it is very strange that the Chinese did not develop the cannon further, or develop hand guns on its analogy. Neither India nor Japan knew cannon until the sixteenth century when they arrived from Europe. As for Islam, despite several claims to the contrary, the first certain use of gunpowder artillery by Muslims comes from Cairo in 1366 and Alexandria in 1376; by 1389 it was common in both Egypt and Syria. Thus there was roughly a forty-year lag in Islam's adoption of the European cannon.

Gunpowder artillery, then, was a complex invention which synthesized and elaborated elements drawn from diverse and sometimes distant sources. Its impact upon Europe was equally complex. Its influences upon other areas of technology such as fortification, metallurgy, and the chemical industries are axiomatic, although they demand much more exact analysis than they have received. The increased expense of war affected tax structures and governmental methods; the new mode of fighting helped to modify social and political relationships. All this has been self-evident for so long a time that perhaps we should begin to ask ourselves whether the obvious is also the true.

For example, it has often been maintained that a large part of the new physics of the seventeenth century sprang from concern with military ballistics. Yet there was continuity between the thought of Galileo or Newton and the fundamental challenge to the Aristotelian theory of impetus which appeared in Franciscus de Marchia's lectures at the University of Paris in the winter of 1319-20,[26] seven years before our first evidence of gunpowder artillery. Moreover, the physicists both of the fourteenth and of the seventeenth centuries were to some extent building upon the criticisms of Aristotle's theory of motion propounded by Philoponus of Alexandria in the age of Justinian, a time when I can detect no new technological stimulus to physical speculation. While most scientists have been aware of current technological problems, and have often talked in terms of them, both science and technology seem to have enjoyed a certain autonomy in their development.

It may well be that continued examination will show that many of the political, economic, and social as well as intellectual developments in Europe which have traditionally been credited to gunpowder artillery were in fact taking place for quite different reasons. But we know of one instance in which the introduction of fire-arms revolutionized an entire society: Japan.[27]

Metallurgical skills were remarkably high in Japan when, in 1543, the Portuguese brought both small arms and cannon to Kyushu. Japanese craftsmen quickly learned from the gunsmiths of European ships how to produce such weapons, and within two or three years were turning them out in great quantity. Military tactics and castle construction were rapidly revised. Nobunaga and his successor, Hideyoshi, seized the new technology of warfare and utilized it to unify all Japan under the shogunate. In Japan, in contrast to Europe, there is no ambiguity about the consequences of the arrival of firearms. But from this fact we must be careful not to argue that the European situation is equally clear if only we would see it so.

11. In examining the origins of gunpowder artillery, we have seen that its roots are multiple, but that all of them (save the European name *trebuchet*) lie in the soil of military technology. It would appear that each area of technology has a certain self-contained quality: borrowings across craft lines are not as frequent as might be expected. Yet they do occur, if exceptionally. A case in point is the fusee.

In the early fifteenth century clock makers tried to develop a portable mechanical timepiece by substituting a spring drive for the weight which powered stationary clocks. But this involved entirely new problems of power control. The weight on a clock exerted equal force at all times, whereas a spring exerts less force in proportion as it uncoils. A new escapement was therefore needed which would exactly compensate for this gradual diminution of power in the drive.

Two solutions were found, the stackfreed and the fusee, the latter being the more satisfactory. Indeed, a leading historian of horology has said of the fusee: "Perhaps no problem in mechanics has ever been solved so simply and so perfectly."[28] The date of its first appearance is much in debate, but we have a diagram of it from 1477.[29] The fusee equalizes the changing force of the mainspring by means of a brake of gut or fine chain which is gradually wound spirally around a conical axle, the force of the brake being dependent upon the leverage of the radius of the cone at any given point and moment. It is a device of great mechanical elegance. Yet the idea did not originate with the clock makers: they borrowed it from the military engineers. In Konrad Keyser's monumental, but still unpublished, treatise on the technology of warfare, *Bellifortis*, completed c. 1405, we find such a conical axle in an apparatus for spanning a heavy crossbow.[30] With very medieval humor, this machine was called "the virgin," presumably because it offered least resistance when the bow was slack and most when it was taut.

• • •

In terms of eleven specific technological acts, or sequences of acts, we have been pondering an abstraction, the act of technological innovation. It is quite possible that there is no such thing to ponder. The analysis of the nature of creativity is one of the chief intellectual commitments of our age. Just as the old unitary concept of "intelligence" is giving way to the notion that the individual's mental capacity consists of a large cluster of various and varying factors mutually affecting each other, so "creativity" may well be a lot of things and not one thing.

Thirteenth century Europe invented the sonnet as a poetic form and the functional button[31] as a means of making civilized life more nearly possible in boreal climes. Since most of us are educated in terms of traditional humanistic presuppositions, we value the sonnet but think that a button is just a button. It is doubtful whether the chilly northerner who invented the button could have invented the sonnet then being produced by his contemporaries in Sicily. It is equally doubtful whether the type of talent required to invent the rhythmic and phonic relationships of the sonnet-pattern is the type of talent needed to perceive the spatial relationships of button and buttonhole. For the button is not obvious until one has seen it, and perhaps not even then. The Chinese never adopted it: they got no further than to adapt the tie-cords of their costumes into elaborate loops to fit over cord-twisted knobs. When the Portuguese brought the button to Japan, the Japanese were delighted with it and took over not only the object itself but also its Portuguese name. Humanistic values, which have been cultivated historically by very specialized groups in quite exceptional circumstances, do not encompass sufficiently the observable human values. The billion or more mothers who, since the thirteenth century, have buttoned their children snugly against winter weather might perceive as much of spirituality in the button as in the sonnet and feel more personal gratitude to the inventor of the former than of the latter. And the historian, concerned not only with art forms but with population, public health, and what S. C. Gilfillan long ago identified as "the coldward course" of culture,[32] must not slight either of these very different manifestations of what would seem to be very different types of creativity.

There is, indeed, no reason to believe that technological creativity is unitary. The unknown Syrian who, in the first century B.C., first blew glass was doing something vastly different from his contemporary who was building the first water-powered mill. For all we know, the kinds of ability required for these two great innovations are as different as those of Picasso and Einstein would seem to be.

The new school of physical anthropologists who maintain that *Homo* is *sapiens* because he is *faber,* that his biological differentiation from the other primates is best understood in relation to tool making, are doubtless exaggerating a provocative thesis. *Homo* is also *ludens, orans,* and much else.[33] But if technology is defined as the systematic modification of the physical environment for human ends, it follows that a more exact understanding of technological innovation is essential to our self-knowledge.

REFERENCES

[1](Oxford, 1954-58).
[2]Cf. the symposium in *Technology and Culture,* I (1960), 229-414.
[3]E. G. Gudde, *California Place Names,* 2nd ed. (Berkeley and Los Angeles, 1960), 52; A. L. Kroeber, "Elements of Culture in Native California," in *The California Indians,* ed. R. F. Heizer and M. A. Whipple (Berkeley and Los Angeles, 1951), 12-13.
[4]E. Zilsel, *Die Entstehung des Geniebegriffes* (Tübingen, 1926).
[5]E. S. G. Robinson, "The Date of the Earliest Coins," *Numismatic Chronicle,* 6th ser., XVI (1956), 4, 8, arbitrarily dates the first coinage c. 64-630 B.C. allowing "the Herodotean interval of a generation" for its diffusion from Lydia to the Ionian cities. But, considering the speed with which coinage appears even in India and China, such an interval is improbable.

D. Kagan, "Pheidon's Aeginetan Coinage," *Transactions and Proceedings of the American Philological Association,* XCI (1960), 121-136, tries to date the first coinage at Aegina before c. 625 B.C. when, he believes, Pheidon died; but the argument is tenuous. The tradition that Pheidon issued a coinage is late, and may well be no more than another example of the Greek tendency to invent culture-heroes. The date of Pheidon's death is uncertain: the belief that he died c. 625 rests solely on the fact that he is not mentioned by Strabo in connection with the war of c. 625-600 B. C.; but if Pheidon, then a very old man, was killed in a revolt of 620 (cf. Kagan's note 21) his participation in this long war would have been so brief and ineffective that Strabo's silence is intelligible.
[6]H. Diels, *Fragmente der Vorsokratiker,* 6th ed. (Berlin, 1951), 171 (B. 90).
[7]The story of the apple is authentic: Newton himself told William Stukeley that when "the notion of gravitation came into his mind [it] was occasion'd by the fall of an apple, as he sat in a contemplative mood"; cf. I. B. Cohen, "Newton in the Light of Recent Scholarship," *Isis,* LI (1960), 490.
[8]The materials on pennons, and other baffles behind the blade of a lance, are found in L. White, jr., *Medieval Technology and Social Change,* (Oxford, 1962), 8, 33, 147, 157.
[9]See the classic work of Ernst Kantorowicz, *The King's Two Bodies,* (Princeton, 1957).
[10]W. Treue, *Kulturgeschichte der Schraube,* (Munich, 1955), 39-43, 57, 109.
[11]F. M. Feldhaus, *Die Technik der Vorzeit, der Geschichtlichen Zeit und der Naturvölker,* (Leipzig, 1914), 984-987.
[12]E. Treptow, "Der älteste Bergbau und seiner Hilfsmittel," *Beiträge zur Geschichte der Technik und Industrie,* VIII (1918), 181, fig. 48; C. N. Bromehead, "Ancient Mining Processes as Illustrated by a Japanese Scroll," *Antiquity,* XVI (1942), 194, 196, 207.
[13]For a detailed history of the crank, cf. White, *op. cit.,* 103-115.
[14]A. R. Hall, "The Military Inventions of Guido da Vigevano," *Actes du VIIIe Congrès International d'Histoire des Sciences,* (Florence, 1958), 966-969.
[15]Bibliothèque Nationale, MS latin 11015, fols. 49^r, 51^v, 52^v. Singer, *op. cit.,* II, figs. 594 and 659, illustrates the first and third of these, but with wrong indications of folio numbers.
[16]H. T. Horwitz, "Uber die Entwicklung der Fahigkeit zum Antreib des Kurbelmechanismus," *Geschichtsblätter fur Technik und Industrie,* XI (1927), 30-31.
[17]White, *op. cit.,* 104 and fig. 4. For what may be a slightly earlier specimen, now in the Seattle Art Museum, see the catalogue of the exhibition *Arts of the Han Dynasty* (New

York, 1961), No. 11, of the Chinese Art Society of America.

[18]I am so informed by Dr. Paul Leser of the Hartford Theological Foundation.

[19]L. Makkai, in *Agrártörténeti Szemle,* I (1957), 42.

[20]P. Leser, "Plow Complex; Culture Change and Cultural Stability," in *Man and Cultures: Selected Papers of the Fifth International Congress of Anthropological and Ethnological Sciences,* ed. A. F. C. Wallace (Philadelphia, 1960), 295.

[21]H. Balfour, "The Fire Piston," in *Anthropological Essays Presented to E. B. Tylor,* (Oxford, 1907), 17-49.

[22]E. Salin, *La Civilisation Mérovingienne,* III (Paris, 1957), 6, 55-115.

[23]C. S. Smith, "A Metallographic Examination of Some Japanese Sword Blades," *Quaderno II del Centro per la Storia della Metallurgia,* (1957), 42-68.

[24]White, *op. cit.,* 96-103, 165.

[25]*Ibid.,* 132.

[26]A. Maier, *Zwei Grundprobleme der scholastischen Naturphilosophie,* 2nd ed. (Rome, 1951), 165, n. 11.

[27]D. M. Brown, "The Impact of Firearms on Japanese Warfare, 1534-98," *Far Eastern Quarterly,* VII (1948), 236-253.

[28]G. Baillie, *Watches,* (London, 1929), 85.

[29]Singer, *op. cit.,* III, fig. 392.

[30]Göttingen University Library, Cod. phil. 63, fol.[V]; cf. F. M. Feldhaus, "Uber den Ursprung vom Federzug und Schnecke," *Deutsche Uhrmacher-Zeitung,* LIV (1930), 720-723.

[31]Some buttons were used in antiquity for ornament, but apparently not for warmth. The first functional buttons are found c. 1235 on the "Adamspforte" of Bamberg Cathedral, and in 1239 on a closely related relief at Bassenheim; cf. E. Panofsky, *Deutsche Plastik des 11. bis 13. Jahrhundert,* (Munich 1924), pl. 74; H. Schnitsler, "Ein unbekanntes Reiterrelief aus dem Kreise des Naumburger Meisters," *Zeitschrift des Deutschen Vereins fur Kunstwissenschaft,* I (1935), 413, fig. 13.

[32]In *The Political Science Quarterly,* XXXV (1920), 393-410.

[33]*Homo ludens* means "man in play," *Homo orans"* means "praying man" [ed. note].

HISTORICAL

The *vita activa* and the modern age

The reversal within the *vita activa* and the vistory of *homo faber*

HANNAH ARENDT

■ Hannah Arendt *(1906-1975) was born in Hannover and immigrated to the United States in 1941. She taught philosophy at several institutions, the last being the New School for Social Research, which she joined in 1967. She remained there until her death. She was an erudite and disciplined thinker who combined intuition with powerful reasoning. She was a critical investigator of political and philosophical developments throughout history. It was in the course of such investigations that she provided her own highly controversial interpretations of these developments. Among her works are* The Origins of Totalitarianism *(1951) and* The Human Condition *(1958), of which the following is an excerpt.*

In the section from which this passage is taken, Arendt discusses the historical shift in man's hierarchy of value. At one time the highest position in that hierarchy was occupied by contemplation. In modern times this is no longer true. Arendt attempts to explain that the significance of this change goes beyond mere replacement of contemplation by fabrication. After all, as Arendt points out "contemplation and fabrication . . . have an inner affinity." That affinity lies in the fact that "contemplation, the beholding of something, was considered to be an inherent element in fabrication as well, inasmuch as the work of the craftsman was guided by the "idea," the model beheld by him before the fabrication process had started as well as after it ended. She points out that the traditional sense of contemplation is rooted in the craftsman's recognition that the models of his fabrication are eternal. Since the craftsman can only imitate and not create, his work can only spoil the excellence of these eternal models. Thus the proper attitude of the craftsman toward the models is to "renounce his capacity for work" and do nothing. He can then behold the models and participate in their eternity. Arendt argues that this kind of contemplation "remains part and parcel of a fabrication process even though it divorced itself from all work."

The significant rupture in the hierarchy occurred when the traditional emphasis on both the product and the model was shifted entirely to the process of production, which was now regarded as the only source of reality. In other words, there was a shift from emphasis on ends to emphasis on means. Contemplation was no longer regarded as a source of truth, and it lost its position in the vita activa *itself. Arendt believes that this shift is the most significant aspect of the change to modern times.*

First among the activities within the *vita activa*[1] to rise to the position formerly occupied by contemplation were the activities of making and fabricating—the prerogatives of *homo faber*. This was natural enough, since it had been an instrument and therefore man in so far as he is a toolmaker that led to the modern revolution. From then on, all scientific progress has been most intimately tied up with the ever more refined development in the manufacture of new tools and instruments. While, for instance, Galileo's experiments with the fall of heavy bodies could have been made at any time in history if men had been inclined to seek truth through experiments, Michelson's experiment with the interferometer at the end of the nineteenth century relied not merely on his "experimental genius" but "required the general advance in technology," and therefore "could not have been made earlier than it was."[2]

□ From *The Human Condition* (Chicago: The University of Chicago Press, 1969), pp. 294-304.

It is not only the paraphernalia of instruments and hence the help man had to enlist from *homo faber* to acquire knowledge that caused these activities to rise from their former humble place in the hierarchy of human capacities. Even more decisive was the element of making and fabricating present in the experiment itself, which produces its own phenomena of observation and therefore depends from the very outset upon man's productive capacities. The use of the experiment for the purpose of knowledge was already the consequence of the conviction that one can know only what he has made himself, for this conviction meant that one might learn about those things man did not make by figuring out and imitating the processes through which they had come into being. The much discussed shift of emphasis in the history of science from the old questions of "what" or "why" something is to the new question of "how" it came into being is a direct consequence of this conviction, and its answer can only be found in the experiment. The experiment repeats the natural process as though man himself were about to make nature's objects, and although in the early stages of the modern age no responsible scientist would have dreamt of the extent to which man actually is capable of "making" nature, he nevertheless from the onset approached it from the standpoint of the One who made it, and this not for practical reasons of technical applicability but exclusively for the "theoretical" reason that certainty in knowledge could not be gained otherwise: "Give me matter and I will build a world from it, that is, give me matter and I will show you how a world developed from it."[3] These words of Kant show in a nutshell the modern blending of making and knowing, whereby it is as though a few centuries of knowing in the mode of making were needed as the apprenticeship to prepare modern man for making what he wanted to know.

Productivity and creativity, which were to become the highest ideals and even the idols of the modern age in its initial stages, are inherent standards of *homo faber,* of man as a builder and fabricator. However, there is another and perhaps even more significant element noticeable in the modern version of these faculties. The shift from the "why" and "what" to the "how" implies that the actual objects of knowledge can no longer be things or eternal motions but must be processes, and that the object of science therefore is no longer nature or the universe but the history, the story of the coming into being, of nature or life or the universe. Long before the modern age developed its unprecedented historical consciousness and the concept of history became dominant in modern philosophy, the natural sciences had developed into historical disciplines, until in the nineteenth century they added to the older disciplines of physics and chemistry, of zoology and botany, the new natural sciences of geology or history of the earth, biology or the history of life, anthropology or the history of human life, and, generally, natural history. In all these instances, development, the key concept of the historical sciences, became the central concept of the physical sciences as well. Nature, because it could be known only in processes which human ingenuity, the ingeniousness of *homo faber,* could repeat and remake in the experiment, became a process,[4] and all particular natural things derived their significance and meaning solely from their functions in the overall process. In the place of the concept of Being we now find the concept of Process. And whereas it is in the nature of Being to appear and thus disclose itself, it is in the nature of Process to remain invisible, to be something whose existence can only be inferred from the presence of certain phenomena. This process was originally the fabrication process which "disappears in the product," and it was based on the experience of *homo faber,* who knew that a production process necessarily precedes the actual existence of every object.

Yet while this insistence on the process of making or the insistence upon considering every thing as the result of a fabrication process is highly characteristic of *homo faber* and his sphere of experience, the exclusive emphasis the modern age placed on it at the expense of all interest in the things, the products themselves, is quite new. It actually transcends the mentality of man as a toolmaker and fabricator, for whom, on the contrary, the production process was a mere means to an end. Here, from the standpoint of *homo faber,* it was as though the means, the production process or development, was more important than the end, the finished product. The reason for this shift of emphasis is obvious: the scientist made only in order to know, not in order to produce things, and the product was a mere by-product, a side effect. Even today all true scientists will agree that the technical applicability of what they are doing is a mere by-product of their endeavor.

The full significance of this reversal of means and ends remained latent as long as the mechanistic world view, the world view of *homo faber* par excellence, was predominant. This view found its most plausible theory in the famous analogy of the relationship between nature and God with the relationship between the watch and the watchmaker. The point in our context is not so much that the eighteenth-century idea of God was obviously formed in the image of *homo faber* as that in this instance the process character of nature was still limited. Although all particular natural things had already been engulfed in the process from which they had come into being, nature as a whole was not yet a process but the more or less stable end product of a divine maker. The image of watch and watchmaker is so strikingly apposite precisely because it contains both the notion of a process character of nature in the image of the movements of the watch and the notion of its still intact object character in the image of the watch itself and its maker.

It is important at this point to remember that

the specifically modern suspicion toward man's truth-receiving capacities, the mistrust of the given, and hence the new confidence in making and introspection which was inspired by the hope that in human consciousness there was a realm where knowing and producing would coincide, did not arise directly from the discovery of the Archimedean point outside the earth in the universe. They were, rather, the necessary consequences of this discovery for the discoverer himself, in so far as he was and remained an earth-bound creature. This close relationship of the modern mentality with philosophical reflection naturally implies that the victory of *homo faber* could not remain restricted to the employment of new methods in the natural sciences, the experiment and the mathematization of scientific inquiry. One of the most plausible consequences to be drawn from Cartesian doubt was to abandon the attempt to understand nature and generally to know about things not produced by man, and to turn instead exclusively to things that owed their existence to man. This kind of argument, in fact, made Vico turn his attention from natural science to history, which he thought to be the only sphere where man could obtain certain knowledge, precisely because he dealt here only with the products of human activity.[5] The modern discovery of history and historical consciousness owed one of its greatest impulses neither to a new enthusiasm for greatness of man, his doings and sufferings, nor to the belief that the meaning of human existence can be found in the story of mankind, but to the despair of human reason, which seemed adequate only when confronted with man-made objects.

Prior to the modern discovery of history but closely connected with it in its impulses are the seventeenth-century attempts to formulate new political philosophies or, rather, to invent the means and instruments with which to "make an artificial animal . . . called a Commonwealth, or State."[6] With Hobbes as with Descartes "the prime mover was doubt,"[7] and the chosen method to establish the "art of man," by which he would make and rule his own world as "God hath made and governs the world" by the art of nature, is also introspection, "to read in himself," since this reading will show him "the similitude of the thoughts and passions of one to the thoughts and passions of another." Here, too, the rules and standards by which to build and judge this most human of human "works of art"[8] do not lie outside of men, are not something men have in common in a worldly reality perceived by the senses or by the mind. They are, rather, inclosed in the inwardness of man, open only to introspection, so that their very validity rests on the assumption that "not . . . the objects of the passions" but the passions themselves are the same in every specimen of the species man-kind. Here again we find the image of the watch, this time applied to the human body and then used for the movements of the passions. The establishment of the Commonwealth, the human creation of "an artificial man," amounts to the building of an "automation [an engine] that moves [itself] by springs and wheels as doth a watch."

In other words, the process which, as we saw, invaded the natural sciences through the experiment, through the attempt to imitate under artificial conditions the process of "making" by which a natural thing came into existence, serves as well or even better as the principle for doing in the realm of human affairs. For here the processes of inner life, found in the passions through introspection, can become the standards and rules for the creation of the "automatic" life of that "artificial man" who is "the great Leviathan." The results yielded by introspection, the only method likely to deliver certain knowledge, are in the nature of movements: only the objects of the senses remain as they are and endure, precede and survive, the act of sensation; only the objects of the passions are permanent and fixed to the extent that they are not devoured by the attainment of some passionate desire; only the objects of thoughts, but never thinking itself, are beyond motion and perishability. Processes, therefore, and not ideas, the models and shapes of the things to be, become the guide for the making and fabricating activities of *homo faber* in the modern age.

NOTES

1. *"Vita activa"* means literally "the active life." With this term Arendt designates three human activities: labor, work, and action [ed. note].
2. Whitehead, *Science and the Modern World,* pp. 116-17.
3. "Gebet mir Materie, ich will eine Welt daraus bauen! das ist, gebet mir Materie, ich will euch zeigen, wie eine Welt daraus entstehen soll" (see Kant's Preface to his *Allgemeine Naiurgeschichte und Theorie des Himmels).*
4. That "nature is a process," that therefore "the ultimate fact for sense-awareness is an event," that natural science deals only with occurrences, happenings, or events, but not with things and that "apart from happenings there is nothing" (see Whitehead, *The Concept of Nature,* pp. 53, 15, 66), belongs among the axioms of modern natural science in all its branches.

5. Vico [*De Nostri Temporis Studiorum Ratione*, ch 4] states explicity why he turned away from natural science. True knowledge of nature is impossible, because not man but God made it; God can know nature with the same certainty man knows geometry: *Geometrica demonstramus quia facimus; si physica demonstrare possemus, faceremus* ("We can prove geometry because we make it; to prove the physical we would have to make it"). This little treatise, written more than fifteen years before the first edition of the *Scienza Nuova* (1725), is interesting in more than one respect. Vico criticizes all existing sciences, but not yet for the sake of his new science of history: what he recommends is the study of moral and political science, which he finds unduly neglected. It must have been much later that the idea occurred to him that history is made by man as nature is made by God. This biographical development, though quite extraordinary in the early eighteenth century, became the rule approximately one hundred years later: each time the modern age had reason to hope for a new political philosophy, it received a philosophy of history instead.
6. Hobbes's Introduction to the *Leviathan*.
7. See Michael Oakeshott's excellent Introduction to the *Leviathan* (Blackwell's Political Texts), p. xiv.
8. *Ibid.*, p. lxiv.

PSYCHOLOGICAL

Where are we now and where are we headed?

ERICH FROMM

■ Erich Fromm *was born in 1900 in Frankfurt, and came to the United States in 1934. He is philosopher, a psychologist, and a writer. As a socialist humanist he concerns himself with the issue of man's isolation, loneliness, and alienation in an industrial society. His works provide a comprehensive attempt to uncover the objective roots of such feelings, and to expound socialist humanist solutions to this general social malaise.*

In this selection Fromm discusses the future technetronic society. It is programmed by two principles. The first is "the maxim that something ought *to be done because it is technically* possible *to do it." The second is "that of maximal efficiency and output." The social consequences of these principles are shown to be varied and to include such significant effects as dehumanization of the individual, increased passivity and conformity, and decreased creativity. This state of affairs leads to a "syndrome of alienation," and to a split between thought and feeling in the individual, causing him to behave like a robot. Fromm concludes that "when the majority of men are like robots, then indeed there will be no problem in building robots who are like men."*

THE PRESENT TECHNOLOGICAL SOCIETY

a. Its principles

The technetronic society may be the system of the future, but it is not yet here; it can develop from what is already here, and it probably will, unless a sufficient number of people see the danger and redirect our course. In order to do so, it is necessary to understand in greater detail the operation of the present technological system and the effect it has on man.

What are the guiding principles of this system as it is today?

It is programed by two principles that direct the efforts and thoughts of everyone working in it: The first principle is the maxim that something *ought* to be done because it is technically *possible* to do it. If it is possible to build nuclear weapons, they must be built even if they might destroy us all. If it is possible to travel to the moon or to the planets, it must be done, even if at the expense of many unfulfilled needs here on earth. This principle means the negation of all values which the humanist tradition has developed. This tradition said that something should be done because it is needed for man, for his growth, joy, and reason, because it is beautiful, good, or true. Once the principle is accepted that something ought to be done because it is technically possible to do it, all other values are dethroned, and technological development becomes the foundation of ethics.[1]

The second principle is that of *maximal efficiency and output*. The requirement of maximal efficiency leads as a consequence to the requirement of minimal individuality. The social machine works more efficiently, so it is believed, if individuals are cut down to purely quantifiable units whose personalities can be expressed on punched cards. These units can be administered more easily by bureaucratic rules because

☐ From pp. 32-46 in *The Revolution of Hope* by Erich Fromm, volume thirty-eight of World Perspective Series, planned and edited by Ruth Nanda Anshen.

they do not make trouble or create friction. In order to reach this result, men must be de-individualized and taught to find their identity in the corporation rather than in themselves.

The question of economic efficiency requires careful thought. The issue of being economically efficient, that is to say, using the smallest possible amount of resources to obtain maximal effect, should be placed in a historical and evolutionary context. The question is obviously more important in a society where real material scarcity is the prime fact of life, and its importance diminishes as the productive powers of a society advance.

A second line of investigation should be a full consideration of the fact that efficiency is only a known element in already existing activities. Since we do not know much about the efficiency or inefficiency of untried approaches, one must be careful in pleading for things as they are on the grounds of efficiency. Furthermore, one must be very careful to think through and specify the area and time period being examined. What may appear efficient by a narrow definition can be highly inefficient if the time and scope of the discussion are broadened. In economics there is increasing awareness of what are called "neighborhood effects"; that is, effects that go beyond the immediate activity and are often neglected in considering benefits and costs. One example would be evaluating the efficiency of a particular industrial project only in terms of the immediate effects on this enterprise—forgetting, for instance, that waste materials deposited in nearby streams and the air represent a costly and a serious inefficiency with regard to the community. We need to clearly develop standards of efficiency that take account of time and society's interest as a whole. Eventually, the human element needs to be taken into account as a basic factor in the system whose efficiency we try to examine.

Dehumanization in the name of efficiency is an all-too-common occurrence; e.g., giant telephone systems employing Brave New World techniques of recording operators' contacts with customers and asking customers to evaluate workers' performance and attitudes, etc.—all aimed at instilling "proper" employee attitude, standardizing service, and increasing efficiency. From the narrow perspective of immediate company purposes, this may yield docile, manageable workers, and thus enhance company efficiency. In terms of the employees, as human beings, the effect is to engender feelings of inadequacy, anxiety, and frustration, which may lead to either indifference or hostility. In broader terms, even efficiency may not be served, since the company and society at large doubtless pay a heavy price for these practices.

Another general practice in organizing work is to constantly remove elements of creativity (involving an element of risk or uncertainty) and group work by dividing and subdividing tasks to the point where no judgment or interpersonal contact remains or is required. Workers and technicians are by no means insensitive to this process. Their frustration is often perceptive and articulate, and comments such as "We are human" and "The work is not fit for human beings" are not uncommon. Again, efficiency in a narrow sense can be demoralizing and costly in individual and social terms.

If we are only concerned with input-output figures, a system may give the impression of efficiency. If we take into account what the given methods do to the human beings in the system, we may discover that they are bored, anxious, depressed, tense, etc. The result would be a twofold one: (1) Their imagination would be hobbled by their psychic pathology, they would be uncreative, their thinking would be routinized and bureaucratic, and hence they would not come up with new ideas and solutions which would contribute to a more productive development of the system; altogether, their energy would be considerably lowered. (2) They would suffer from many physical ills, which are the result of stress and tension; this loss in health is also a loss for the system. Furthermore, if one examines what this tension and anxiety do to them in their relationship to their wives and children, and in their functioning as responsible citizens, it may turn out that for the system as a whole the seemingly efficient method is most inefficient, not only in human terms but also as measured by merely economic criteria.

To sum up: efficiency is desirable in any kind of purposeful activity. But it should be examined in terms of the larger systems, of which the system under study is only a part; it should take account of the human factor within the system. Eventually efficiency as such should not be a *dominant* norm in any kind of enterprise.

The other aspect of the same principle, that of *maximum output,* formulated very simply, maintains that the more we produce of whatever we produce, the better. The success of the economy of the country is measured by its rise of total production. So is the success of a company. Ford may lose several hundred million dollars by the failure of a costly new model, like the Edsel, but this is only a minor mishap as long as the production curve rises. The growth of the economy is visualized in terms of ever-increasing production, and there is no vision of a limit yet where production may be stabilized. The comparison between countries rests upon the same principle. The Soviet Union hopes to surpass the United States by accomplishing a more rapid rise in economic growth.

Not only industrial production is ruled by the principle of continuous and limitless acceleration. The educational system has the same criterion: the more college graduates, the better. The same in sports: every new record is looked upon as progress. Even the attitude toward the weather seems to be determined by the same principle. It is emphasized that this is "the hottest day in the decade," or the coldest, as the case may be, and I suppose some people are comforted for the inconvenience by the proud feeling that they are witnesses to the record temperature. One could go on endlessly giving examples of the concept that constant increase of quantity constitutes the goal of our life; in fact, that it is what is meant by "progress."

Few people raise the question of *quality,* or what all this increase in quantity is good for. This omission is evident in a society which is not centered around man any more, in which one aspect, that of quantity, has choked all others. It is easy to see that the predominance of this principle of "the more the better" leads to an imbalance in the whole system. If all efforts are bent on doing *more,* the quality of living loses all importance, and activities that once were means become ends.[2]

If the overriding economic principle is that we produce more and more, the consumer must be prepared to want—that is, to consume—more and more. Industry does not rely on the consumer's spontaneous desires for more and more commodities. By building in obsolescence it often forces him to buy new things when the old ones could last much longer. By changes in styling of products, dresses, durable goods, and even food, it forces him psychologically to buy more than he might need or want. But industry, in its need for increased production, does not rely on the consumer's needs and wants but to a considerable extent on advertising, which is the most important offensive against the consumer's right to know what he wants. The spending of 16.5 billion dollars on direct advertising in 1966 (in newspapers, magazines, radio, TV) may sound like an irrational and wasteful use of human talents, of paper and print. But it is not irrational in a systhat believes that increasing production and hence consumption is a vital feature of our economic system, without which it would collapse. If we add to the cost of advertising the considerable cost for restyling of durable goods, especially cars, and of packaging, which partly is another form of whetting the consumer's appetite, it is clear that industry is willing to pay a high price for the guarantee of the upward production and sales curve.

The anxiety of industry about what might happen to our economy if our style of life changed is expressed in this brief quote by a leading investment banker:

> Clothing would be purchased for its utility; food would be bought on the basis of economy and nutritional value; automobiles would be stripped to essentials and held by the same owners for the full 10 or 15 years of their useful lives; homes would be built and maintained for their characteristics of shelter, without regard to style or neighborhood. And what would happen to a market dependent upon new models, new styles, new ideas?[3]

b. Its effect on man

What is the effect of this type of organization on man? It reduces man to an appendage of the machine, ruled by its very rhythm and demands. It transforms him into *Homo consumens,* the total consumer, whose only aim is to *have* more and to *use* more. This society produces many useless things, and to the same degree many useless people. Man, as a cog in the production machine, becomes a thing, and ceases to be human. He spends his time doing things in which he is not interested, with people in whom he is not interested, producing things in which he is not interested; and when he is not producing, he is consuming. He is the eternal suckling with the open mouth, "taking in," without effort and without inner activeness, whatever the boredom-preventing (and boredom-producing) industry forces on him—cigarettes, liquor, movies, television, sports, lectures—limited only by what he can afford. But the boredom-preventing industry, that is to say, the gadget-selling industry, the automobile industry, the movie industry, the television industry, and so on, can only succeed in preventing the boredom from being conscious. In fact, they increase the boredom, as a salty drink taken to quench the thirst increases it. However unconscious, boredom remains boredom nevertheless.

The passiveness of man in industrial society today is one of his most characteristic and pathological features. He takes in, he wants to be fed, but he does not move, initiate, he does not digest his food, as it were. He does not reacquire in a productive fashion what he inherited, but he amasses it or consumes it. He suffers from a severe systemic deficiency, not too dissimilar to that which one finds in more extreme forms in depressed people.

Man's passiveness is only one symptom among a total syndrome, which one may call the "syndrome of alienation." Being passive, he does not relate himself to the world actively and is forced to submit to his idols and their demands. Hence, he feels powerless, lonely, and anxious. He has little sense of integrity or self-identity. Conformity seems to be the only way to avoid intolerable anxiety—and even conformity does not always alleviate his anxiety.

No American writer has perceived this dynamism more clearly than Thorstein Veblen. He wrote:

> In all the received formulations of economic theory, whether at the hands of the English economists or those of the continent, the human material with which the inquiry is concerned is conceived in hedonistic terms; that is to say, in terms of a passive and substantially inert and immutably given human nature. . . . The hedonistic conception of man is that of a lightning calculator of pleasures and pains, who oscillates like a homogeneous globule of desire of happiness under the impulse of stimuli that shift him about the area, but leave him intact. He has neither antecedent nor consequent. He is an isolated, definitive human datum, in stable equilibrium except for the buffets of the impinging forces that displace him in one direction or another. Self-imposed in elemental space, he spins symmetrically about his own spiritual axis until the parallelogram of forces bears down upon him, whereupon he follows the line of the resultant. When the force of the impact is spent, he comes to rest, a self contained globule of desire as before. Spiritually, the hedonistic man is not a prime mover. *He is not the seat of a process of living, except in the sense that he is subject to a series of permutations enforced upon him by circumstances external and alien to him.*[4]

Aside from the pathological traits that are rooted in passiveness, there are others which are important for the understanding of today's pathology of normalcy. I am referring to the growing split of cerebral-intellectual function from affective-emotional experience; the split between thought from feeling, mind from the heart, truth from passion.

Logical thought is not rational if it is merely logical[5] and not guided by the concern for life, and by the inquiry into the total process of living in all its concreteness and with all its contradictions. On the other hand, not only thinking but also emotions can be rational. *"Le coeur a ses raisons que la raison ne connaît point,"* as Pascal put it. (The heart has its reasons which reason knows nothing of.) Rationality in emotional life means that the emotions affirm and help the person's psychic structure to maintain a harmonious balance and at the same time to assist its growth. Thus, for instance, irrational love is love which enhances the person's dependency, hence anxiety and hostility. Rational love is a love which relates a person intimately to another, at the same time preserving his independence and integrity.

Reason flows from the blending of rational thought and feeling. If the two functions are torn apart, thinking deteriorates into schizoid intellectual activity, and feeling deteriorates into neurotic life-damaging passions.

The split between thought and affect leads to a sickness, to a low-grade chronic schizophrenia, from which the new man of the technetronic age begins to suffer. In the social sciences it has become fashionable to think about human problems with no reference to the feelings related to these problems. It is assumed that scientific objectivity demands that thoughts and theories concerning man be emptied of all emotional concern with man.

An example of this emotion-free thinking is Herman Kahn's book on thermonuclear warfare. The question is discussed: how many millions of dead Americans are "acceptable" if we use as a criterion the ability to rebuild the economic machine after nuclear war in a reasonably short time so that it is as good as or better than before. Figures for GNP and population increase or decrease are the basic categories in this kind of thinking, while the question of the human results of nuclear war in terms of suffering, pain, brutalization, etc., is left aside.

Kahn's *The Year 2000* is another example of the writing which we may expect in the completely alienated megamachine society. Kahn's concern is that of the figures for production, population increase, and various scenarios for war or peace, as the case may be. He impresses many readers because they mistake the thousands of little data which he combines in every-changing kaleidoscopic pictures for erudition or profundity. They do not notice the basic superficiality in his reasoning and the lack of the human dimension in his description of the future.

When I speak here of low-grade chronic schizophrenia, a brief explanation seems to be needed. Schizophrenia, like any other psychotic state, must be defined not only in psychiatric terms but also in social terms. Schizophrenic experience *beyond* a certain threshold would be considered a sickness in any society, since those suffering from it would be unable to function under any social circumstances (unless the schizophrenic is elevated into the status of a god, shaman, saint, priest, etc.). But there are low-grade chronic forms of psychoses which can be shared by millions of people and which—precisely because they do not go beyond a certain threshold—do not prevent these people from functioning socially. As long as they share their sickness with millions of others, they have the satisfactory feeling of not being alone; in other words, they avoid that sense of complete isolation which is so characteristic of full-fledged psychosis. On the contrary, they look at themselves as normal and at those who have not lost the link between heart and mind as being "crazy." In all low-grade forms of psychoses, the definition of sickness depends on the question as to whether the pathology is shared or not. Just as there is low-grade chronic schizophrenia, so there exist also low-grade chronic paranoia and depression. And there is plenty of evidence that among certain strata of the population, particularly on occasions where a war threatens, the paranoid elements increase but are not felt as pathological as long as they are common.[6]

The tendency to install technical progress as the highest value is linked up not only with our overemphasis on intellect but, most importantly, with a deep emotional attraction to the mechanical, to all that is not alive, to all that is man-made. This attraction to the non-alive, which is in its more extreme form an attraction to death and decay (necrophilia), leads even in its less drastic form to indifference toward life instead of "reverence for life." Those who are attracted to the non-alive are the people who prefer "law and order" to living structure, bureaucratic to spontaneous methods, gadgets to living beings, repetition to originality, neatness to exuberance, hoarding to spending. They want to control life because they are afraid of its uncontrollable spontaneity; they would rather kill it than to expose themselves to it and merge with the world around them. They often gamble with death because they are not rooted in life; their courage is the courage to die and the symbol of their ultimate courage is the Russian roulette.[7] The rate of our automobile accidents and the preparation for thermonuclear war are a testimony to this readiness to gamble with death. And who would not eventually prefer this exciting gamble to the boring unaliveness of the organization man?

One symptom of the attraction of the merely mechanical is the growing popularity, among some scientists and the public, of the idea that it will be possible to construct computers which are no different from man in thinking, feeling, or any other aspect of functioning.[8] The main problem, it seems to me, is not whether such a computer-man can be constructed; it is rather why the idea is becoming so popular in a historical period when nothing seems to be more important than to transform the existing man into a more rational, harmonious, and peace-loving being. One cannot help being suspicious that often the attraction of the computer-man idea is the expression of a flight from life and from humane experience into the mechanical and purely cerebral.

The possibility that we can build robots who are like men belongs, if anywhere, to the future. But the present already shows us men who act like robots. When the majority of men are like robots, then indeed there will be no problem in building robots who are like men. The idea of the manlike computer is a good example of the alternative between the human and the inhuman use of machines. The computer can serve the enhancement of life in many respects. But the idea that it replaces man and life is the manifestation of the pathology of today.

The fascination with the merely mechanical is supplemented by an increasing popularity of conceptions that stress the animal nature of man and the instinctive roots of his emotions or actions. Freud's was such an instinctive psychology; but the importance of his concept of libido is secondary in comparison with his fundamental discovery of the unconscious process in waking life or in sleep. The most popular recent authors who stress instinctual animal heredity, like Konrad Lorenz *(On Aggression)* or Desmond Morris *(The Naked Ape)*, have not offered any new or valuable insights into the specific human problem as Freud has done; they satisfy the wish of many to look at themselves as determined by instincts and thus to camouflage their true and bothersome human problems.[9] The dream of many people seems to be to combine the emotions of a primate with a computerlike brain. If this dream could be fulfilled, the problem of human freedom and of responsibility would seem to disappear. Man's feelings would be determined by his instincts, his reason by the computer; man would not have to give an answer to the questions his existence asks him. Whether one likes the dream or not, its realization is impossible; the naked ape with the computer brain would cease to be human, or rather "he" would not *be*.[10]

Among the technological society's pathogenic effects upon man, two more must be mentioned: the disappearance of *privacy* and of *personal human contact*.

"Privacy" is a complex concept. It was and is a privilege of the middle and upper classes, since its very basis, private space, is costly. This privilege, however, can become a common good with other economic privileges. Aside from this economic factor, it was also based on a hoarding tendency in which *my* private life was *mine* and nobody else's, as was *my* house and any other property. It was also a concomitant of *cant*, of the discrepancy between moral appearances and reality. Yet when all these qualifications are made, privacy still seems to be an important condition for a person's productive development. First of all, because privacy is necessary to collect oneself and to free oneself from the constant "noise" of people's chatter and intrusion, which interferes with one's own mental processes. If all private data are transformed into public data, experiences will tend to become more shallow and more alike. People will be afraid to feel the "wrong thing"; they will become more accessible to psychological manipulation which, through psychological testing, tries to establish norms for "desirable," "normal," "healthy" attitudes. Considering that these tests are applied in order to help the companies and government agencies to find the people with the "best" attitudes, the use of psychological tests, which is by now an almost general condition for getting a good job, constitutes a severe infringement on the citizen's freedom. Unfortunately, a large number of psychologists devote whatever knowledge of man they have to his manipulation in the interests of what the big organization considers efficiency. Thus, psychologists become an important part of the industrial and governmental system while claiming that their activities serve the optimal development of man. This claim is based on the rationalization that what is best for the corporation is best for man. It is important that the managers understand that much of what they get from psychological testing is based on the very limited picture of man which, in fact, management requirements have transmitted to the psychologists, who in turn give it back to management, allegedly as a result of an independent study of man. It hardly needs to be said that the intrusion of privacy may lead to a control of the individual which is more total and could be more devastating than what totalitarian states have demonstrated thus far. Orwell's 1984 will need much assistance from testing, conditioning, and smoothing-out psychologists in order to come true. It is of vital importance to distinguish between a psychology that understands and aims at the well-being of man and a psychology that studies man as an object, with the aim of making him more useful for the technological society.

NOTES

[1] While revising this manuscript I read a paper by Hasan Ozbekhan, "The Triumph of Technology: 'Can' Implies 'Ought.'" This paper, adapted from an invited presentation at MIT and published in mimeographed form by System Development Corporation, Santa Monica, California, was sent to me by the courtesy of Mr. George Weinwurm. As the title indicates, Ozbekhan expresses the same concept as the one I present in the text. His is a brilliant presentation of the problem from the standpoint of an outstanding specialist in the field of management science, and I find it a very encouraging fact that the same idea appears in the work of authors in fields as different as his and mine. I quote a sentence that shows the identity of his concept and the one presented in the text: "Thus, feasibility, which is a strategic concept, becomes elevated into a normative concept, with the result that whatever technological reality indicates we *can* do is taken as implying that we *must* do it" (p. 7).

[2] I find in C. West Churchman's *Challenge to Reason* (New York: McGraw-Hill, 1968) an excellent formulation of the problem:

> "If we explore this idea of a larger and larger model of systems, we may be able to see in what sense completeness represents a challenge to reason. One model that seems to be a good candidate for completeness is called an *allocation* model; it views the world as a system of activities that use resources to "output" usable products.
>
> "The process of reasoning in this model is very simple. One searches for a central quantitative measure of system performance, which has the characteristic: the more of this quantity the better. For example, the more profit a firm makes, the better. The more qualified students a university graduates, the better. The more food

we produce, the better. It will turn out that the particular choice of the measure of system performance is not critical, so long as it is a measure of general concern.

"We take this desirable measure of performance and relate it to the feasible activities of the system. The activities may be the operations of various manufacturing plants, of schools and universities, of farms, and so on. Each significant activity contributes to the desirable quantity in some recognizable way. The contribution, in fact, can often be expressed in a mathematical function that maps the amount of activity onto the amount of the desirable quantity. The more sales of a certain product, the higher the profit of a firm. The more courses we teach, the more graduates we have. The more fertilizer we use, the more food [pp. 156-57]."

[3]Paul Mazur, *The Standards We Raise,* New York, 1953, p. 32.

[4]"Why Is Economics Not an Evolutionary Science?," in *The Place of Science in Modern Civilization and Other Essays* (New York: B. W. Huebsch, 1919), p. 73. (Emphasis added.)

[5]Paranoid thinking is characterized by the fact that it can be completely logical, yet lack any guidance by concern or concrete inquiry into reality; in other words, logic does not exclude madness.

[6]The difference between that which is considered to be sickness and that which is considered to be normal becomes apparent in the following example. If a man declared that in order to free our cities from air pollution, factories, automobiles, airplanes, etc., would have to be destroyed, nobody would doubt that he was insane. But if there is a consensus that in order to protect our life, our freedom, our culture, or that of other nations which we feel obliged to protect, thermonuclear war might be required as a last resort, such opinion appears to be perfectly sane. The difference is not at all in the kind of thinking employed but merely in that the first idea is not shared and hence appears abnormal while the second is shared by millions of people and by powerful governments and hence appears to be normal.

[7]Michael Maccoby has demonstrated the incidence of the life-loving versus the death-loving syndrome in various populations by the application of an "interpretative" questionnaire. Cf. his "Polling Emotional Attitudes in Relation to Political Choices" (to be published).

[8]Dean E. Wooldridge, for instance, in *Mechanical Man* (New York: McGraw-Hill, 1968), writes that it will be possible to manufacture computers synthetically which are "completely undistinguishable from human beings produced in the usual manner" [!] (p. 172). Marvin L. Minsky, a great authority on computers, writes in his book *Computation* (Englewood Cliffs, N.J.: Prentice-Hall, 1967): "There is no reason to suppose machines have any limitations not shared by man" (p. vii).

[9]This criticism of Lorenz refers only to that part of his work in which he deals by analogy with the psychological problems of man, not with his work in the field of animal behavior and instinct theory.

[10]In revising this manuscript I became aware that Lewis Mumford had expressed the same idea in 1954 in *In the Name of Sanity* (New York: Harcourt, Brace & Co.):

"Modern man, therefore, now approaches the last act of his tragedy, and I could not, even if I would, conceal its finality or its horror. We have lived to witness the joining, in intimate partnership, of the automaton and the id, the id rising from the lower depths of the unconscious, and the automaton, the machine-like thinker and the man-like machine, wholly detached from other life-maintaining functions and human reactions, descending from the heights of conscious thought. The first force has proved more brutal, when released from the whole personality, than the most savage of beasts; the other force, so impervious to human emotions, human anxieties, human purposes, so committed to answering only the limited range of questions for which its apparatus was originally loaded, that it lacks the saving intelligence to turn off its own compulsive mechanism, even though it is pushing science as well as civilization to its own doom [p. 198]."

PSYCHOLOGICAL

The dialectic of civilization

HERBERT MARCUSE

■ Herbert Marcuse *(1898-1979) was born in Berlin, Germany, and immigrated to the United States in 1934. He was a philosopher, psychologist, and social critic who devoted most of his life to the fight against fascism. He took a stand against it in Europe, and later wrote a scathing critique on repressive elements in the American industrial society and the Soviet bureaucratic state. In* One Dimensional Man *(1964) he argued that, in the American industrial society, man's nature has been distorted, releasing as a consequence an unusual amount of aggression. He warned against the destruction that can result from such a state of affairs. Among his works are* Reason and Revolution *(1941) and* Eros and Civilization *(1955), from which the following selection is taken.*

In this selection Marcuse creatively expounds the views of Freud. As we already know, Freud viewed civilization as being based primarily on the suppression of sexual instincts. Marcuse observes that for Freud there is no "instinct of workmanship." Work is unpleasant. But "civilization is first of all progress in work." Thus the energy for work must be borrowed from the primary instincts (sexual instincts and destructive instincts). Since civilization is mainly the work of Eros (the life instincts), which derives its strength from the sexual instincts, it follows that the energy used for work is borrowed chiefly from the libido, the reservoir of sexual energy. But by borrowing from the libido, Eros is in turn weakened. This impairs the ability of Eros to effectively "bind" the destructive instincts. As a result "civilization tends towards self-destruction." Marcuse evaluates this Freudian argument, introducing in the process the distinction between "work" and "alienated labor," and his famous concept of "surplus repression."

Civilization is first of all progress in *work*—that is, work for the procurement and augmentation of the necessities of life. This work is normally without satisfaction in itself; to Freud it is unpleasurable, painful. In Freud's metapsychology there is no room for an original "instinct of workmanship," "mastery instinct," etc.[1] The notion of the conservative nature of the instincts under the rule of the pleasure and Nirvana principles strictly precludes such assumptions. When Freud incidentally mentions the "natural human aversion to work,"[2] he only draws the inference from his basic theoretical conception. The instinctual syndrome "unhappiness and work" recurs throughout Freud's writings,[3] and his interpretation of the Prometheus myth is centered on the connection between curbing of sexual passion and civilized work.[4] The basic work in civilization is non-libidinal, is labor; labor is "unpleasantness," and such unpleasantness has to be enforced. "For what motive would induce man to put his sexual energy to other uses if by any disposal of it he could obtain fully satisfying pleasure? He would never let go of this pleasure and would make no further progress."[5] If there is no original "work instinct," then the energy required for (unpleasurable) work must be "withdrawn" from the primary instincts—from the sexual and from the destructive instincts. Since civilization is mainly the work of Eros, it is first of all withdrawal of libido: culture "obtains a great part of the mental energy it needs by subtracting it from sexuality."[6]

But not only the work impulses are thus fed by aim-inhibited sexuality. The specifically "social instincts" (such as the "affectionate relations between parents and children, . . . feelings of friendship, and the emotional ties in marriage") contain impulses which are "held back by internal resistance" from attaining their aims;[7] only by virtue of such renunciation do they become sociable. Each individual contributes his renunciations (first under the impact of external compulsion, then internally), and from "these sources the common stock of the material and ideal wealth of civilization has been accumulated."[8] Although Freud remarks that these social instincts "need not be described as sublimated" (because they have not abandoned their sexual aims but rest content with "certain approximations to satisfaction"), he calls them "closely related" to sublimation.[9] Thus the main sphere of civilization appears as a sphere of *sublimation*. But sublimation involves *desexualization*. Even if and where it draws on a reservoir of "neutral displaceable energy" in the ego and in the id, this neutral energy "proceeds from the narcissistic reservoir

---

☐ From *Eros and Civilization: A Philosophical Inquiry Into Freud* (Boston: The Beacon Press, 1960), pp. 81-88. Reprinted by permission of Penguin Books Ltd, The Beacon Press, and the author.

of libido,'' i.e., it is desexualized Eros.[10] The process of sublimation alters the balance in the instinctual structure. Life is the fusion of Eros and death instinct; in this fusion, Eros has subdued its hostile partner. However:

> After sublimation the erotic component no longer has the power to bind the whole of the destructive elements that were previously combined with it, and these are released in the form of inclinations to aggression and destruction.[11]

Culture demands continuous sublimation; it thereby weakens Eros, the builder of culture. And desexualization, by weakening Eros, unbinds the destructive impulses. Civilization is thus threatened by an instinctual de-fusion, in which the death instinct strives to gain ascendancy over the life instincts. Originating in renunciation and developing under progressive renunciation, civilization tends toward self-destruction.

This argument runs too smooth to be true. A number of objections arise. In the first place, not all work involves desexualization, and not all work is unpleasurable, is renunciation. Secondly, the inhibitions enforced by culture also affect—and perhaps even chiefly affect—the derivatives of the death instinct, aggressiveness and the destruction impulses. In this respect at least, cultural inhibition would accrue to the strength of Eros. Moreover, work in civilization is itself to a great extent *social utilization* of aggressive impulses and is thus work in the service of Eros. An adequate discussion of these problems presupposes that the theory of the instincts is freed from its exclusive orientation on the performance principle, that the image of a non-repressive civilization (which the very achievements of the performance principle suggest) is examined as to its substance. Such an attempt will be made in the last part of this study; here, some tentative clarifications must suffice.

The psychical sources and resources of work, and its relation to sublimation, constitute one of the most neglected areas of psychoanalytic theory. Perhaps nowhere else has psychoanalysis so consistently succumbed to the official ideology of the blessings of ''productivity.''[12] Small wonder then, that in the Neo-Freudian schools, where (as we shall see in the Epilogue) the ideological trends in psychoanalysis triumph over its theory, the tenor of work morality is all-pervasive. The ''orthodox'' discussion is almost in its entirety focused on ''creative'' work, especially art, while work in the realm of necessity—labor—is relegated to the background.

To be sure, there is a mode of work which offers a high degree of libidinal satisfaction, which is pleasurable in its execution. And artistic work, where it is genuine, seems to grow out of a non-repressive instinctual constellation and to envisage non-repressive aims—so much so that the term *sublimation* seems to require considerable modification if applied to this kind of work. But the bulk of the work relations on which civilization rests is of a very different kind. Freud notes that the ''daily work of earning a livelihood affords particular satisfaction when it has been selected by free choice.''[13] However, if ''free choice'' means more than a small selection between pre-established necessities, and if the inclinations and impulses used in work are other than those preshaped by a repressive reality principle, then satisfaction in daily work is only a rare privilege. The work that created and enlarged the material basis of civilization was chiefly labor, alienated labor, painful and miserable—and still is. The performance of such work hardly gratifies *individual* needs and inclinations. It was imposed upon man by brute necessity and brute force; if alienated labor has anything to do with Eros, it must be very indirectly, and with a considerably sublimated and weakened Eros.

But does not the civilized inhibition of *aggressive* impulses in work offset the weakening of Eros? Aggressive as well as libidinal impulses are supposed to be satisfied in work ''by way of sublimation,'' and the culturally beneficial ''sadistic character'' of work has often been emphasized.[14] The development of technics and technological rationality absorbs to a great extent the ''modified'' destructive instincts:

> The instinct of destruction, when tempered and harnessed (as it were, inhibited in its aim) and directed towards objects, is compelled to provide the ego with satisfaction of its needs and with power over nature.[15]

Technics provide the very basis for progress; technological rationality sets the mental and behaviorist pattern for productive performance, and ''power over nature'' has become practically identical with civilization. Is the destructiveness sublimated in these activities sufficiently subdued and diverted to assure the work of Eros? It seems that socially useful destructiveness is less sublimated than socially useful libido. To be sure, the diversion of destructiveness from the ego to the external world secured the growth of civilization. However, extroverted destruction remains destruction: its objects are in most cases actually and violently assailed, deprived of their form, and reconstructed only after partial destruction; units are forcibly divided, and the component parts forcibly rearranged. Nature is literally "violated." Only in certain categories of sublimated aggressiveness (as in surgical practice) does such violation directly strengthen the life of its object. Destructiveness, in extent and intent, seems to be more directly satisfied in civilization than the libido.

However, while the destructive impulses are thus being satisfied, such satisfaction cannot stabilize their energy in the service of Eros. Their destructive force must drive them beyond this servitude and sublimation, for their aim is, not matter, not nature, not any object, but life itself. If they are the derivatives of the death instinct, then they cannot accept as final any "substitutes." Then, through constructive technological destruction, through the constructive violation of nature, the instincts would still operate toward the annihilation of life. The radical hypothesis of *Beyond the Pleasure Principle* would stand: the instincts of self-preservation, self-assertion, and mastery, in so far as they have absorbed this destructiveness, would have the function of assuring the organism's "own path to death." Freud retracted this hypothesis as soon as he had advanced it, but his formulations in *Civilization and Its Discontents* seem to restore its essential content. And the fact that the destruction of life (human and animal) has progressed with the progress of civilization, that cruelty and hatred and the scientific extermination of men have increased in relation to the real possibility of the elimination of oppression—this feature of late industrial civilization would have instinctual roots which perpetuate destructiveness beyond all rationality. The growing mastery of nature then would, with the growing productivity of labor, develop and fulfill the human needs *only as a by-product:* increasing cultural wealth and knowledge would provide the material for progressive destruction and the need for increasing instinctual repression.

This thesis implies the existence of objective criteria for gauging the degree of instinctual repression at a given stage of civilization. However, repression is largely unconscious and automatic, while its degree is measureable only in the light of consciousness. The differential between (phylogenetically necessary) repression and surplus-repression may provide the criteria. Within the total structure of the repressed personality, surplus-repression is that portion which is the result of specific societal conditions sustained in the specific interest of domination. The extent of this surplus-repression provides the standard of measurement: the smaller it is, the less repressive is the stage of civilization. The distinction is equivalent to that between the biological and the historical sources of human suffering. Of the three "sources of human suffering" which Freud enumerates—namely, "the superior force of nature, the disposition to decay of our bodies, and the inadequacy of our methods of regulating human relations in the family, the community and the state"[16]—at least the first and the last are in a strict sense *historical* sources; the superiority of nature and the organization of societal relations have essentially changed in the development of civilization. Consequently, the necessity of repression, and of the suffering derived from it, varies with the maturity of civilization, with the extent of the achieved rational mastery of nature and of society. Objectively, the need for instinctual inhibition and restraint depends on the need for toil and delayed satisfaction. The same and even a reduced scope of instinctual regimentation would constitute a higher degree of repression at a mature stage of civilization, when the need for renunciation and toil is greatly reduced by material and intellectual progress—when civilization could actually afford a considerable release of instinctual energy expended for domination and toil. Scope and intensity of instinctual repression obtain their full significance only in relation to the historically possible extent of freedom.

NOTES

1. Ives Hendrick, "Work and the Pleasure Principle," in *Psychoanalytic Quarterly,* XII (1943), 314.
2. *Civilization and Its Discontents,* p. 34 note.
3. In a letter of April 16, 1896, he speaks of the "moderate misery necessary for intensive work." Ernest Jones, *The Life and Work of Sigmund Freud,* Vol. I (New York: Basic Books, 1953), p. 305.
4. *Civilization and Its Discontents,* pp. 50-51 note; *Collected Papers,* V, 288ff.
5. "The Most Prevalent Form of Degradation in Erotic Life," in *Collected Papers,* IV, 216.
6. *Civilization and Its Discontents,* p. 74.
7. "The Libido Theory," in *Collected Papers,* V, 134.
8. "'Civilized' Sexual Morality . . . ," p. 82.

9. "The Libido Theory," p. 134.
10. *The Ego and the Id* (London: Hogarth Press, 1950), pp. 38, 61-63. See Edward Glover, "Sublimation, Substitution, and Social Anxiety," in *International Journal of Psychoanalysis*, Vol. XII, No. 3 (1931), p. 264.
11. *The Ego and the Id*, p. 80.
12. Ives Hendrick's article cited above is a striking example.
13. *Civilization and Its Discontents*, p. 34 note.
14. See Alfred Winterstein, "Zur Psychologie der Arbeit," in *Imago*, XVIII (1932), 142.
15. *Civilization and Its Discontents*, p. 101.
16. *Civilization and Its Discontents*, p. 43.

SOCIOLOGICAL

The three phases of the machine civilization

LEWIS MUMFORD

■ Lewis Mumford, *born in 1895 in Flushing, New York, is a noted author and social critic whose areas of interest include such topics as cities, technics, and the development of human culture. He has served as an acting editor of* The Sociological Review *(London) and as a co-editor of* The American Caravan. *He has also held teaching posts at various American Universities, and has received numerous honors from prestigious societies in his field. Among his books are* The Myth of the Machine *(vol. I, 1967; vol. II, 1970) and* Technics and Civilization *(1934), from which the following passages are excerpted.*

The author presents here a three-stage theory of the history of modern technology. According to Mumford, the dawn of modern technology "stretches roughly from the year 1000 to 1750," a period that he calls the eotechnic phase. During this early phase, technology was mainly based on two sources of power: wood and water. By the middle of the eighteenth century, the eotechnic phase had faded into the paleotechnic phase, in which the sources of power were mainly coal and iron. While, during the eotechnic phase, technology was evenly distributed over a territory, the new phase ushered in the clustering of industries. Crowded industrial centers appeared, and with them disease, poverty, and other ills.

By the middle of the nineteenth century, many of the major discoveries that ushered in the third phase had already been made. Thus the neotechnic phase, based on electricity and alloys, began to slowly replace the paleotechnic phase. Like the eotechnic phase, this phase did not require the clustering of industries; it promised to restore healthier and happier surroundings to the worker.

1. THE EOTECHNIC PHASE

Technical syncretism

Civilizations are not self-contained organisms. Modern man could not have found his own particular modes of thought or invented his present technical equipment without drawing freely on the cultures that had preceded him or that continued to develop about him.

Each great differentiation in culture seems to be the outcome, in fact, of a process of syncretism. Flinders Petrie, in his discussion of Egyptian civilization, has shown that the admixture which was necessary for its development and fulfillment even had a racial basis; and in the development of Christianity it is plain that the most diverse foreign elements—a Dionysian earth myth, Greek philosophy, Jewish Messianism, Mithraism, Zoroastrianism—all played a part in giving the specific content and even the form to the ultimate collection of myths and offices that became Christianity.

Before this syncretism can take place, the cultures from which the elements are drawn must either be in a state of dissolution, or sufficiently remote in time or space so that single elements can be extracted from the tangled mass of real institutions. Unless this condition existed the elements themselves would not be free, as it were, to move over toward the new pole. Warfare acts as such an agent of dissociation, and in point of time the mechanical renascence of Western Europe was associated with the shock and stir of the Crusades. For what the new civilization picks up is not the complete forms and institutions of a solid culture, but just those fragments that can be transported and transplanted: it uses inventions, patterns, ideas, in the way that the Gothic builders in England used the occasional stones or tiles of the Roman

---

villa in combination with the native flint and in the entirely different forms of a later architecture. If the villa had still been standing and occupied, it could not have been conveniently quarried. It is the death of the original form, or rather, the remaining life in the ruins, that permits the free working over and integration of the elements of other cultures.

One further fact about syncretism must be noted. In the first stages of integration, before a culture has set its own definite mark upon the materials, before invention has crystallized into satisfactory habits and routine, it is free to draw upon the widest sources. The beginning and the end, the first absorption and the final spread and conquest, after the cultural integration has taken place, are over a worldwide realm.

These generalizations apply to the origin of the present-day machine civilization: a creative syncretism of inventions, gathered from the technical debris of other civilizations, made possible the new mechanical body. The waterwheel, in the form of the Noria, had been used by the Egyptians to raise water, and perhaps by the Sumerians for other purposes; certainly in the early part of the Christian era watermills had become fairly common in Rome. The windmill perhaps came from Persia in the eighth century. Paper, the magnetic needle, gunpowder, came from China, the first two by way of the Arabs: algebra came from India through the Arabs, and chemistry and physiology came via the Arabs, too, while geometry and mechanics had their origins in pre-Christian Greece. The steam engine owed its conception to the great inventor and scientist, Hero of Alexandria: it was the translations of his works in the sixteenth century that turned attention to the possibilities of this instrument of power.

In short, most of the important inventions and discoveries that served as the nucleus for further mechanical development, did not arise, as Spengler would have it, out of some mystical inner drive of the Faustian soul: they were wind-blown seeds from other cultures. After the tenth century in Western Europe the ground was, as I have shown, well plowed and harrowed and dragged, ready to receive these seeds; and while the plants themselves were growing, the cultivators of art and science were busy keeping the soil friable. Taking root in medieval culture, in a different climate and soil, these seeds of the machine sported and took on new forms: perhaps, precisely because they had *not* originated in Western Europe and had no natural enemies there, they grew as rapidly and gigantically as the Canada thistle when it made its way onto the South American pampas. But at no point—and this is the important thing to remember—did the machine represent a complete break. So far from being unprepared for in human history, the modern machine age cannot be understood except in terms of a very long and diverse preparation. The notion that a handful of British inventors suddenly made the wheels hum in the eighteenth century is too crude even to dish up as a fairy tale to children.

The technological complex

Looking back over the last thousand years, one can divide the development of the machine and the machine civilization into three successive but *over-lapping and interpenetrating phases:* eotechnic, paleotechnic, neotechnic. The demonstration that industrial civilization was not a single whole, but showed two marked, contrasting phases, was first made by Professor Patrick Geddes and published a generation ago. In defining the paleotechnic and neotechnic phases, he however neglected the important period of preparation, when all the key inventions were either invented or foreshadowed. So, following the archeological parallel he called attention to, I shall call the first period the eotechnic phase: the dawn age of modern technics.

While each of these phases roughly represents a period of human history, it is characterized even more significantly by the fact that it forms a technological complex. Each phase, that is, has its origin in certain definite regions and tends to employ certain special resources and raw materials. Each phase has its specific means of utilizing and generating energy, and its special forms of production. Finally, each phase brings into existence particular types of workers, trains them in particular ways, develops certain aptitudes and discourages others, and draws upon and further develops certain aspects of the social heritage.

Almost any part of a technical complex will point to and symbolize a whole series of relationships within that complex. Take the various types of writing pen. The goose-quill pen, sharpened by the user, is a typical eotechnic product: it indicates the handicraft basis of industry and the close connection with agriculture. Economically it is cheap; technically it is crude, but easily adapted to the style of the user. The steel pen stands equally for the paleotechnic phase: cheap and uniform, if not durable, it is a typical product of the mine, the steel mill and of mass-production. Technically, it is an improvement upon the quill-pen; but to approximate the same adaptability it must be made in half a dozen different standard points and shapes. And finally the fountain pen—though invented as early as the seventeenth century—is a typical neotechnic product. With its barrel of rubber or synthetic resin, with its gold pen, with its automatic action, it points to the finer neotechnic economy: and in its use of the durable iridium tip the fountain pen characteristically lengthens the service of the point and reduces the need for replacement. These respective characteristics are reflected at a hundred points in the typical environment of each phase; for though the various parts of a complex may be invented at various times, the complex itself will not be *in working order* until its major parts are all assembled. Even today the neotechnic complex still awaits a number of inventions necessary to its perfection: in particular an accumulator with six times the voltage and at least the present amperage of the existing types of cell.

Speaking in terms of power and characteristic materials, the eotechnic phase is a water-and-wood complex: the paleotechnic phase is a coal-and-iron complex, and the neotechnic phase is an electricity-and-alloy complex. It was Marx's great contribution as a sociological economist to see and partly to demonstrate that each period of invention and production had its own specific value for civilization, or, as he would have put it, its own historic mission. The machine cannot be divorced from its larger social pattern; for it is this pattern that gives it meaning and purpose. Every period of civilization carries within it the insignificant refuse of past technologies and the important germs of new ones: but the center of growth lies within its own complex.

The dawn-age of our modern technics stretches roughly from the year 1000 to 1750. During this period the dispersed technical advances and suggestions of other civilizations were brought together, and the process of invention and experimental adaptation went on at a slowly accelerating pace. Most of the key inventions necessary to universalize the machine were promoted during this period; there is scarcely an element in the second phase that did not exist as a germ, often as an embryo, frequently as an independent being, in the first phase. This complex reached its climax, technologically speaking, in the seventeenth century, with the foundation of experimental science, laid on a basis of mathematics, fine manipulation, accurate timing, and exact measurement.

The eotechnic phase did not of course come suddenly to an end in the middle of the eighteenth century: just as it reached its climax first of all in Italy in the sixteenth century, in the work of Leonardo and his talented contemporaries, so it came to a delayed fruition in the America of 1850. Two of its finest products, the clipper ship and the Thonet process of making bent-wood furniture, date from the eighteen-thirties. There were parts of the world, like Holland and Denmark, which in many districts slipped directly from an eotechnic into the neotechnic economy, without feeling more than the cold shadow of the paleotechnic cloud.

With respect to human culture as a whole, the eotechnic period, though politically a chequered one, and in its later moments characterized by a deepening degradation of the industrial worker, was one of the most brilliant periods in history. For alongside its great mechanical achievements it built cities, cultivated landscapes, constructed buildings, and painted pictures, which fulfilled, in the realm of human thought and enjoyment, the advances that were being decisively made in the practical life. And if this period failed to establish a just and equitable polity in society at large, there were at least moments in the life of the monastery and the commune that were close to its dream: the afterglow of this life was recorded in More's Utopia and Andreae's Christianopolis.

Noting the underlying unity of eotechnic civilization, through all its superficial changes in costume and creed, one must look upon its successive portions as expressions of a single culture. This point is now being re-enforced by scholars who have come to disbelieve in the notion of the gigantic break supposed to have been made during the Renascence: a contemporary illusion, unduly emphasized by later historians. But one must add a qualification: namely, that with the increasing technical advances of this society there was, for reasons partly independent of the machine itself, a corresponding cultural dissolution and decay. In short, the Renascence was not, socially speaking, the dawn of a new day, but its twilight. The mechanical arts advanced as the humane arts weakened and receded, and it was at the moment when form and civilization had most completely broken up that the tempo of invention became more rapid, and the multiplication of machines and the increase of power took place.

New sources of power

At the bottom of the eotechnic economy stands one important fact: the diminished use of human beings as prime movers, and the separation of the production of energy from its application and immediate control. While the tool still dominated production energy and human skill were united within the craftsman himself: with the separation of these two elements the productive process itself tended toward a greater impersonality, and the machine-tool and the machine developed along with the new engines of power. If power machinery be a criterion, the modern industrial revolution began in the twelfth century and was in full swing by the fifteenth.

The eotechnic period was marked first of all by a steady increase in actual horsepower. This came directly from two pieces of apparatus: first, the introduction of the iron horseshoe, probably in the ninth century, a device that increased the range of the horse, by adapting him to other regions besides the grasslands, and added to his effective pulling power by giving his hoofs a grip. Second: by the tenth century the modern form of harness, in which the pull is met at the shoulder instead of at the neck, was re-invented in Western Europe—it had existed in China as early as 200 B.C.—and by the twelfth century, it had supplanted the inefficient harness the Romans had known. The gain was a considerable one, for the horse was not merely a useful aid in agriculture or a means of transport: he became likewise an improved agent of mechanical production: mills utilizing horsepower directly for grinding corn or for pumping water came into existence all over Europe, sometimes supplementing other forms of non-human power, sometimes serving as the principal source. The increase in the number of horses was made possible, again, by improvements in agriculture and by the opening up of the hitherto sparsely cultivated or primeval forest areas in northern Europe. This created a condition somewhat similar to that which was repeated in America during the pioneering period: the new colonists, with plenty of land at their disposal, were lacking above all in labor power, and were compelled to resort to ingenious labor-saving devices that the better settled regions in the south with their surplus of labor and their easier conditions of living were never forced to invent. This fact perhaps was partly responsible for the high degree of technical initiative that marks the period.

But while horse power ensured the utilization of mechanical methods in regions not otherwise favored by nature, the greatest technical progress came about in regions that had abundant supplies of wind and water. It was along the fast flowing streams, the Rhône and the Danube and the small rapid rivers of Italy, and in the North Sea and Baltic areas, with their strong winds, that this new civilization had its firmest foundations and some of its most splendid cultural expressions. . . .

2. THE PALEOTECHNIC PHASE

England's belated leadership

By the middle of the eighteenth century the fundamental industrial revolution, that which transformed our mode of thinking, our means of production, our manner of living, had been accomplished: the external forces of nature were harnessed and the mills and looms and spindles were working busily through Western Europe. The time had come to consolidate and systematize the great advances that had been made.

At this moment the eotechnic régime was shaken to its foundations. A new movement appeared in industrial society which had been gathering headway almost unnoticed from the fifteenth century on: after 1750 industry passed into a new phase, with a different source of power, different materials, different social objectives. This second revolution multiplied, vulgarized, and spread the methods and goods produced by the first: above all, it was directed toward the quantification of life, and its success could be gauged only in terms of the multiplication table. . . .

The new barbarism

As we have seen, the earlier technical development had not involved a complete breach with the past. On the contrary, it had seized and appropriated and assimilated the technical innovations of other cultures, some very ancient, and the pattern of industry was wrought into the dominant pattern of life itself. Despite all the diligent mining for gold, silver, lead and tin in the sixteenth century, one could not call the civilization itself a mining civilization; and the handicraftsman's world did not change completely when he walked from the workshop to the church, or left the garden behind his house to wander out into the open fields beyond the city's walls.

Paleotechnic industry, on the other hand, arose out of the breakdown of European society and carried the process of disruption to a finish. There was a sharp shift in interest from life values to pecuniary values: the system of interests which only had been latent and which had been restricted in great measure to the merchant and leisure classes now pervaded every walk of life. It was no longer sufficient for industry to provide a lifelihood: it must create an independent fortune: work was no longer a necessary part of living: it became an all-important end. Industry shifted to new regional centers in England: it tended to slip away from the established cities and to escape to decayed boroughs or to rural districts which were outside the field of regulation. Bleak valleys in Yorkshire that supplied water power, dirtier bleaker valleys in other parts of the land which disclosed seams of coal, became the environment of the new industrialism. A landless, traditionless proletariat, which had been steadily gathering since the sixteenth century, was drawn into these new areas and put to work in these new industries: if peasants were not handy, paupers were supplied by willing municipal authorities: if male adults could be dispensed with, women and children were used. These new mill villages and milltowns, barren of even the dead memorials of an older humaner culture, knew no other round and suggested no other outlet, than steady unremitting toil. The operations themselves were repetitive and monotonous; the environment was sordid; the life that was lived in these new centers was empty and barbarous to the last degree. Here the break with the past was complete. People lived and died within sight of the coal pit or the cotton mill in which they spent from fourteen to sixteen hours of their daily life, lived and died without either memory or hope, happy for the crusts that kept them alive or the sleep that brought them the brief uneasy solace of dreams.

Wages, never far above the level of subsistence, were driven down in the new industries by the competition of the machine. So low were they in the early part of the nineteenth century that in the textile trades they even for a while retarded the introduction of the power loom. As if the surplus of workers, ensured by the disfranchisement and pauperization of the agricultural workers, were not enough to re-enforce the Iron Law of Wages, there was an extraordinary rise in the birth-rate. The causes of this initial rise are still obscure; no present theory fully accounts for it. But one of the tangible motives was the fact that unemployed parents were forced to live upon the wages of the young they had begotten. From the chains of poverty and perpetual destitution there was no escape for the new mine worker or factory worker: the servility of the mine, deeply engrained in that occupation, spread to all the accessory employments. It needed both luck and cunning to escape those shackles.

Here was something almost without parallel in the history of civilization: not a lapse into barbarism through the enfeeblement of a higher civilization, but an upthrust into barbarism, aided by the very forces and interests which originally had been directed toward the conquest of the environment and the perfection of human culture. Where and under what conditions did this change take place? And how, when it represented in fact the lowest point in social development Europe had known since the Dark Ages did it come to be looked upon as a humane and beneficial advance? We must answer those questions.

The phase one here defines as paleotechnic reached its highest point, in terms of its own concepts and ends, in England in the middle of the nineteenth century: its cock-crow of triumph was the great industrial exhibition in the new Crystal Palace at Hyde Park in 1851: the first World Exposition, an apparent victory for free trade, free enterprise, free invention, and free access to all the world's markets by the country that boasted already that it was the workshop of the world. From around 1870 onwards the typical interests and preoccupations of the paleotechnic phase have been challenged by later developments in technics itself, and modified by various counterpoises in society. But like the eotechnic phase, it is still with us: indeed, in certain parts of the world, like Japan and China, it even passes for the new, the progressive, the modern, while in Russia an unfortunate residue of paleotechnic concepts and methods has helped misdirect, even partly cripple, the otherwise advanced economy projected by the disciples of Lenin. In the United States the paleotechnic régime did not get under way until the eighteen fifties, almost a century after England; and it reached its highest point at the beginning of the present century, whereas in Germany it dominated the years between 1870 and 1914, and, being carried to perhaps fuller and completer expression, has collapsed with greater rapidity there than in any other part of the world. France, except for its special coal and iron centers, escaped some of the worst defects of the period; while Holland, like Denmark and in part Switzerland, skipped almost directly from an eotechnic into a neotechnic economy, and except in ports like Rotterdam and in the mining districts, vigorously resisted the paleotechnic blight. . . .

Carboniferous capitalism

The great shift in population and industry that took place in the eighteenth century was due to the introduction of coal as a source of mechanical power, to the use of new means of making that power effective—the steam engine—and to new methods of smelting and working up iron. Out of this coal and iron complex, a new civilization developed.

Like so many other elements in the new technical world, the use of coal goes back a considerable distance in history. There is a reference to it in Theophrastus: in 320 B.C. it was used by smiths; while the Chinese not merely used coal for baking porcelain but even employed natural gas for illumination. Coal itself is a unique mineral: apart from the precious metals, it is one of the few unoxidized substances found in nature; at the same time it is one of the most easy to oxidize: weight for weight it is of course much more compact to store and transport than wood.

As early as 1234 the freemen of Newcastle were given a charter to dig for coal, and an ordinance attempting to regulate the coal nuisance in London dates from the fourteenth century. Five hundred years later coal was in general use as a fuel among glassmakers, brewers, distillers, sugar bakers, soap boilers, smiths, dyers, brickmakers, lime burners, founders, and calico printers. But in the meanwhile a more significant use had been found for coal: Dud Dudley at the beginning of the seventeenth century sought to substitute coal for charcoal in the production of iron: this aim was successfully accomplished by a Quaker, Abraham Darby, in 1709. By that invention the high-powered blast furnace became possible; but the method itself did not make its way to Coalbrookdale in Shropshire to Scotland and the North of England until the 1760's. The next development in the making of cast-iron awaited the introduction of a pump which should deliver to the furnace a more effective blast of air: this came with the invention of Watt's steam pump, and the demand for more iron, which followed, in turn increased the demand for coal.

Meanwhile, coal as a fuel for both domestic heating and power was started on a new career. By the end of the eighteenth century coal began to take the place of current sources of energy as an illuminant through Murdock's devices for producing illuminating gas. Wood, wind, water, beeswax, tallow, sperm-oil—all these were displaced steadily by coal and derivatives of coal, albeit an efficient type of burner, that produced by Welsbach, did not appear until electricity was ready to supplant gas for illumination. Coal, which could be mined long in advance of use, and which could be stored up, placed industry almost out of reach of seasonal influences and the caprices of the weather.

In the economy of the earth, the large-scale opening up of coal seams meant that industry was beginning to live for the first time on an accumulation of potential energy, derived from the ferns of the carboniferous period, instead of upon current income. In the abstract, mankind entered into the possession of a capital inheritance more splendid than all the wealth of the Indies; for even at the present rate of use it has been calculated that the present known supplies would last three thousand years. In the concrete, however, the prospects were more limited, and the exploitation of coal carried with it penalties not attached to the extraction of energy from growing plants or from wind and water. As long as the coal seams of England, Wales, the Ruhr, and the Alleghanies were deep and rich the limited terms of this new economy could be overlooked: but as soon as the first easy gains were realized the difficulties of keeping up the process became plain. For mining is a robber industry: the mine owner, as Messrs. Tryon and Eckel point out, is constantly consuming his capital, and as the surface measures are depleted the cost per unit of extracting minerals and ores becomes greater. The mine is the worst possible local base for a permanent civilization: for when the seams are exhausted, the individual mine must be closed down, leaving behind its debris and its deserted sheds and houses. The by-products are a befouled and disorderly environment; the end product is an exhausted one. . . .

The degradation of the worker

Kant's doctrine, that every human being should be treated as an end, not as a means, was formulated precisely at the moment when mechanical industry had begun to treat the worker solely as a means—a means to cheaper mechanical production. Human beings were dealt with in the same spirit of brutality as the landscape: labor was a resource to be exploited, to be mined, to be exhausted, and finally to be discarded. Responsibility for the worker's life and health ended with the cash-payment for the day's labor.

The poor propagated like flies, reached industrial maturity—ten or twelve years of age—promptly, served their term in the new textile mills or the mines, and died inexpensively. During the early paleotechnic period their expectation of life was twenty years less than that of the middle classes. For a number of centuries the degradation of labor had been going on steadily in Europe; at the end of the eighteenth century, thanks to the shrewdness and near-sighted rapacity of the English industrialists, it reached its nadir in England. In other countries, where the paleotechnic system entered later, the same brutality emerged: the English merely set the pace. What were the causes at work?

By the middle of the eighteenth century the handicraft worker had been reduced, in the new industries, into a competitor with the machine. But there was one weak spot in the system: the nature of human beings themselves: for at first they rebelled at the feverish pace, the rigid discipline, the dismal monotony of their tasks. The main difficulty, as Ure pointed out, did not lie so much in the invention of an effective self-acting mechanism as in the "distribution of the different members of the apparatus into one cooperative body, in impelling each organ with its appropriate delicacy and speed, and above all, in training human beings to renounce their desultory habits of work and to identify themselves with the unvarying regularity of the complex automaton." "By the infirmity of human nature," wrote Ure again, "it happens that the more skillful the workman, the more self-willed and intractable he is apt to become, and of course the less fit and component of the mechanical system in which . . . he may do great damage to the whole."

The first requirement for the factory system, then, was the castration of skill. The second was the discipline of starvation. The third was the closing up of alternative occupations by means of land-monopoly and dis-education.

In actual operation, these three requirements were met in reverse order. Poverty and land monopoly kept the workers in the locality that needed them and removed the possibility of their improving their position by migration: while exclusion from craft apprenticeship, together with specialization in subdivided and partitioned mechanical functions, unfitted the machine-worker for the career of pioneer or farmer, even though he might have the opportunity to move into the free lands in the newer parts of the world. Reduced to the function of a cog, the new worker could not operate without being joined to a machine. Since the workers lacked the capitalists' incentives of gain and social opportunity, the only things that kept them bound to the machine were starvation, ignorance, and fear. These three conditions were the foundations of industrial discipline, and they were retained by the directing classes even though the poverty of the worker undermined and periodically ruined the system of mass production which the new factory discipline promoted. Therein lay one of the inherent "contradictions" of the capitalist scheme of production. . . .

3. THE NEOTECHNIC PHASE

The beginnings of neotechnics

The neotechnic phase represents a third definite development in the machine during the last thousand years. It is a true mutation: it differs from the paleotechnic phase almost as white differs from black. But on the other hand, it bears the same relation to the eotechnic phase as the adult form does to the baby.

During the neotechnic phase, the conceptions, the anticipations, the imperious visions of Roger Bacon, Leonardo, Lord Verulam, Porta, Glanvill, and the other philosophers and technicians of that day at last found a local habitation. The first hasty sketches of the fifteenth century were now turned into working drawings: the first guesses were now re-enforced with a technique of verification: the first crude machines were at last carried to perfection in the exquisite mechanical technology of the new age, which gave to motors and turbines properties that had but a century earlier belonged almost exclusively to the clock. The superb animal audacity of Cellini, about to cast his difficult Perseus, or the scarcely less daring work of Michelangelo, constructing the dome of St. Peter's, was replaced by a patient co-operative experimentalism: a whole society was now prepared to do what had heretofore been the burden of solitary individuals.

Now, while the neotechnic phase is a definite physical and social complex, one cannot define it as a period, partly because it has not yet developed its own form and organization, partly because we are still in the midst of it and cannot see its details in their ultimate relationships, and partly because it has not displaced the older régime with anything like the speed and decisiveness that characterized the transformation of the eotechnic order in the late eighteenth century. Emerging from the paleotechnic order, the neotechnic institutions have nevertheless in many cases compromised with it, given way before it, lost their identity by reason of the weight of vested interests that continued to support the obsolete instruments and the anti-social aims of the middle industrial era. *Paleotechnic ideals still largely dominate the industry and the politics of the Western World:* the class struggles and the national struggles are still pushed with relentless vigor. While eotechnic practices linger on as civilizing influences, in gardens and parks and painting and music and the theater, the paleotechnic remains a barbarizing influence. To deny this would be to cling to a fool's paradise. In the seventies Melville framed a question in fumbling verse whose significance has deepened with the intervening years:

> . . . Arts are tools;
> But tools, they say, are to the strong:
> Is Satan weak? Weak is the wrong?
> No blessed augury overrules:
> Your arts advanced in faith's decay:
> You are but drilling the new Hun
> Whose growl even now can some dismay.

To the extent that neotechnic industry has failed to transform the coal-and-iron complex, to the extent that it has failed to secure an adequate foundation for its humaner technology in the community as a whole, to the extent that it has lent its heightened powers to the miner, the financier, the militarist, the possibilities of disruption and chaos have increased.

But the beginnings of the neotechnic phase can nevertheless be approximately fixed. The first definite change, which increased the efficiency of prime movers enormously, multiplying it from three to nine times, was the perfection of the water-turbine by Fourneyron in 1832. . . .

By 1850 a good part of the fundamental scientific discoveries and inventions of the new phase had been made: the electric cell, the storage cell, the dynamo, the motor, the electric lamp, the spectroscope, the doctrine of the conservation of energy. Between 1875 and 1900 the detailed application of these inventions to industrial processes was carried out in the electric power station and the telephone and the radio telegraph. Finally, a series of complementary inventions, the phonograph, the moving picture, the gasoline engine, the steam turbine, the airplane, were all sketched in, if not perfected, by 1900: these in turn effected a radical transformation of the power plant and the factory, and they had further effects in suggesting new principles for the design of cities and for the utilization of the environment as a whole. By 1910 a definite counter-march against paleotechnic methods began in industry itself.

The outlines of the process were blurred by the explosion of the World War and by the sordid disorders and reversions and compensations that followed it. Though the instruments of a neotechnic civilization are now at hand, and though many definite signs of an integration are not lacking, one cannot say confidently that a single region, much less our Western Civilization as a whole, has entirely embraced the neotechnic complex: for the necessary social institutions and the explicit social purposes requisite even for complete technological fulfillment are lacking. The gains in technics are never registered automatically in society: they require equally adroit inventions and adaptations in politics; and the careless habit of attributing to mechanical improvements a direct rôle as instruments of culture and civilization puts a demand upon the machine to which it cannot respond. Lacking a cooperative social intelligence and good-will, our most refined technics promises no more for society's improvement than an electric bulb would promise to a monkey in the midst of a jungle.

True: the industrial world produced during the nineteenth century is either technologically obsolete or socially dead. But unfortunately, its maggoty corpse has produced organisms which in turn may debilitate or possibly kill the new order that should take its place: perhaps leave it a hopeless cripple. One of the first steps, however, toward combating such disastrous results is to realize that even technically the Machine Age does not form a continuous and harmonious unit, that there is a deep gap between the paleotechnic and neotechnic phases, and that the habits of mind and the tactics we have carried over from the old order are obstacles in the way of our developing the new.

The importance of science

The detailed history of the steam engine, the railroad, the textile mill, the iron ship, could be written without more than passing reference to the scientific work of the period. For these devices were made possible largely by the method of empirical practice, by trial and selection: many lives were lost by the explosion of steam-boilers before the safety-valve was generally adopted. And though all these inventions would have been the better for science, they came into existence, for the most part, without its direct aid. It was the practical men in the mines, the factories, the machine shops and the clockmakers' shops and the locksmiths' shops or the curious amateurs with a turn for manipulating materials and imagining new processes, who made them possible. Perhaps the only scientific work that steadily and systematically affected the paleotechnic design was the analysis of the elements of mechanical motion itself.

With the neotechnic phase, two facts of critical importance become plain. First, the scientific method, whose chief advances had been in mathematics and the physical sciences, took possession of other domains of experience: the living organism and human society also became the objects of systematic investigation, and though the work done in these departments was handicapped by the temptation to take over the categories of thought, the modes of investigation, and the special apparatus of quantitative abstraction developed for the isolated physical world, the extension of science here was to have a particularly important effect upon technics. Physiology became for the nineteenth century what mechanics had been for the seventeenth: instead of mechanism forming a pattern for life, living organisms began to form a pattern for mechanism. Whereas the mine dominated the paleotechnic period, it was the vineyard and the farm and the physiological laboratory that directed many of the most fruitful investigations and contributed to some of the most radical inventions and discoveries of the neotechnic phase. . . .

Second only to the more comprehensive attack of the scientific method upon aspects of existence hitherto only feebly touched by it, was the direct application of scientific knowledge to technics and the conduct of life. In the neotechnic phase, the main initiative comes, not from the ingenious inventor, but from the scientist who establishes the general law: the invention is a derivative product. It was Henry who in essentials invented the telegraph, not Morse; it was Faraday who invented the dynamo, not Siemens; it was Oersted who invented the electric motor, not Jacobi; it was Clerk-Maxwell and Hertz who invented the radio telegraph, not Marconi and De Forest. The translation of the scientific knowledge into practical instruments was a mere incident in the process of invention. While distinguished individual inventors like Edison, Baekeland and Sperry remained, the new inventive genius worked on the materials provided by science.

Out of this habit grew a new phenomenon: deliberate and systematic invention. Here was a new material: problem—find a new use for it. Or here was a necessary utility: problem—find the theoretic formula which would permit it to be produced. The ocean cable was finally laid only when Lord Kelvin had contributed the necessary scientific analysis of the problem it presented: the thrust of the propeller shaft on the steamer was finally taken up without clumsy and expensive mechanical devices, only when Michell worked out the behavior of viscous fluids: long distance telephony was made possible only by systematic research by Pupin and others in the Bell Laboratories on the several elements in the problem. Isolated inspiration and empirical fumbling came to count less and less in invention. In a whole series of characteristic neotechnic inventions the thought was father to the wish. And typically, this thought is a collective product. . . .

New sources of energy

The neotechnic phase was marked, to begin with, by the conquest of a new form of energy: electricity. . . .

Unlike coal in long distance transportation, or like steam in local distribution, electricity is much easier to transmit without heavy losses of energy and higher costs. Wires carrying high tension alternating currents can cut across mountains which no road vehicle can pass over; and once an electric power utility is established the rate of deterioration is slow. Moreover, electricity is readily convertible into various forms: the motor, to do mechanical work, the electric lamp, to light, the electric radiator, to heat, the x-ray tube and the ultra-violet light, to penetrate and explore, and the selenium cell, to effect automatic control.

SOCIOLOGICAL

From "Technology, Nature, and Society"

DANIEL BELL

■ Daniel Bell *was born in New York City in 1919. He has been a professor of Sociology at Harvard University since 1969, and was a member of the President's Committee on Technology, Automation, and Economic Progress (1964-1966). He has received many honors and is currently a member of the editorial boards of* Daedalus *and* The American Scholar. *Among his works are* The End of Ideology *(1960) and* The Coming of Post-Industrial Society *(1973).*

Bell takes issue here with Jacques Ellul over the latter's claim that technique is autonomous (see the selection by Ellul earlier in Part Two). Bell claims that "technology, or technique, does not have a life of its own." To support his claim he introduces first a distinction between "technology" on the one hand and "the social 'support system' in which it is imbedded" on the other. Since several kinds of support systems are compatible with technology, Bell concludes that it is false to claim that technology leaves us no choices. He makes a similar distinction between technology and the accounting system that allocates costs. These observations lead Bell to conclude that the problem does not lie with technology, but rather in our ability as a society to make the right choices for that technology. (For related views on this issue, see the Autonomous Technology Debate in Part Three.)

WHAT IS SOCIETY?

The rhetoric of apocalypse haunts our times. Given the recurrence of the Day of Wrath in the Western imagination—when the seven seals are opened and the seven vials pour forth—it may be that great acts of guilt provoke fears of retribution which are projected heavenward as mighty punishments of men. A little more than a decade ago we had the apocalyptic specter (whose reality content was indeed frightening) of a nuclear holocaust, and there was a flood of predictions that a nuclear war was a statistical certainty before the end of the decade. That apocalypse has receded, and other guilts produce other fears. Today it is the ecological crisis, and we find, like the drumroll of Revelation 14 to 16 recording the plagues: *The Doomsday Book, Terracide, Our Plundered Planet, The Chasm Ahead, The Hungry Planet,* and so on.[1]

In the demonology of the time, "the great whore" is technology. It has profaned Mother Nature, it has stripped away the mysteries, it has substituted for the natural environment an artificial environment in which man cannot feel at home.[2] The modern heresy, in the thinking of Jacques Ellul, the French social philosopher whose writing has been the strongest influence in shaping this school of thought, has been to enshrine *la technique* as the ruling principle of society.

☐ In *Technology and the Frontiers of Knowledge,* The Frank Nelson Doubleday Lectures—1972-73 (New York: Doubleday & Company, Inc., 1975), pp. 60-66.

Ellul defines technique as:

> the translation into action of man's concern to master things by means of reason, to account for what is subconscious, make quantitative what is qualitative, make clear and precise the outlines of nature, take hold of chaos and put order into it.

Technique, by its power, takes over the government:

> Theoretically our politicians are at the center of the machinery, but actually they are being progressively eliminated by it. Our statesmen are important satellites of the machine, which, with all its parts and techniques, apparently functions as well without them.

Technique is a new morality which "has placed itself beyond good and evil and has such power and autonomy [that] it in turn has become the judge of what is moral, the creator of a new morality." We have here a new demiurge, an "unnatural" and "blind" logos that in the end enslaves man himself:

> When technique enters into the realm of social life, it collides ceaselessly with the human being. . . . Technique requires predictability and, no less, exactness of prediction. It is necessary, then, that technique prevail over the human being. For technique, this is a matter of life and death. Technique must reduce man to a technical animal, the king of the slaves of technique.[3]

Ellul has painted a reified world in which *la technique* is endowed with anthropomorphic and demonological attributes. (Milton's Satan, someone remarked, is Prometheus with Christian theology.) Many of the criticisms of technology today remind one of Goethe, who rejected Newton's optics on the ground that the microscope and telescope distorted the human scale and confused the mind. The point is well taken, if there is confusion of realms. What the eye can see unaided, and must respond to, is different from the microcosm below and the macrocosm beyond. Necessary distinctions have to be maintained. The difficulty today is that it is the critics of technology who absolutize the dilemmas and have no answers, short of the apocalyptic solutions that sound like the familiar comedy routine "Stop the world, I want to get off."

Against such cosmic anguish one feels almost apologetic for mundane answers. But after the existentialist spasm, there remain the dull and unyielding problems of ordinary, daily life. The point is that technology, or technique, does not have a life of its own. There is no immanent logic of technology, no "imperative" that must be obeyed. Ellul has written: "Technique is a means with a set of rules for the game. . . . There is but one method for its use, one possibility."[4]

But this is patently not so if one distinguishes between technology and the social "support system" in which it is embedded.[5] The automobile and the highway network form a technological system; the way this system is used is a question of social organization. And the relation between the two can vary considerably. We can have a social system that emphasizes the private use of the automobile; money is then spent to provide parking and other facilities necessary to that purpose. On the other hand, arguing that an automobile is a capital expenditure whose "down time" is quite large, and that twenty feet of street space for a single person in one vehicle is a large social waste, we could penalize private auto use and have only a rental and taxi system that would substantially reduce the necessary number of cars. The same technology is compatible with a variety of social organizations, and we choose the one we want to use.

One should also distinguish between technology and the accounting system that allocates costs. Until recently, the social costs generated by different technologies have not been borne by the individuals or firms responsible for them, because the criterion of social accountability was not used. Today that is changing. The technology of the internal-combustion engine is being modified because the government now insists that the pollution it generates be reduced. And the technology is being changed. The energy crisis we face is less a physical shortage than the result of new demands—by consumers, and by socially minded individuals for a different kind of technological use of fuels. If we could burn the high-sulphur fuels used until a few years ago, there would be less of an energy crisis; but there would be more pollution. Here, too, the problem is one of costs and choice.

The source of our predicament is not the "imperatives" of technology but a lack of decision mechanisms for choosing the kinds of technology and social support patterns we want. The venerated teacher of philosophy at City College Morris Raphael Cohen used to pose a question to his students in moral philosophy: If a Moloch God were to offer the human race an invention that would enormously increase each individual's freedom and mobility, but demanded the human sacrifice of thirty-thousand lives (the going price at the time), would you take it? That invention, of course, was the automobile. But we had no mechanisms for assessing its effects and planning for the control of its use. Two hundred years ago, no one "voted" for our present industrial system, as men voted for a polity or a constitution. To this extent, the phrase "the industrial revolution" is deceptive, for there was no single moment when people could decide, as they did politically in 1789 or 1793 or 1917, for or against the new system. And yet today, with our increased awareness of alternates and consequences, we are beginning to make those choices. We can do this by technology assessment, and by social policy which either penalizes or encourages a technological development (e.g., the kind of energy we use) through the mechanism of taxes and subsidy.

A good deal of our intellectual difficulty stems from the way we conceive of society. Émile Durkheim, one of the founding fathers of modern sociology, contributed to this difficulty by saying that society exists *sui generis,* meaning that it could not be reduced to psychological factors. In a crucial sense he was right, but in his formulation he pictured society as an entity, a collective conscience outside the individual, acting as an external constraint on his behavior. And this lent itself to the romantic dualism of the individual versus society.

Society is *sui generis,* a level of complex organizations created by the degree of interdependence and the multiplicity of ties among men.

A traffic jam, as Thomas Schelling has pointed out, is best analyzed not in terms of the individual pathologies of the drivers, but by considering the layout of roads, the pattern of flow into and out of the city, the congestion at particular times because of work scheduling, and so on. Society is not some external artifact, but *a set of social arrangements, created by men,* to regulate normatively the exchange of wants and satisfactions.

The order of society differs from the order of nature. Nature is "out there," without *telos,* and men must discern its binding and constraining laws to refit the world. Society is a moral order, defined by consciousness and purpose, and justified by its ability to satisfy men's needs, material and transcendental. Society is a design that, as men become more and more conscious of its consequences and effects, is subject to reordering and rearrangement in the effort to solve its quandaries. It is a social contract, made not in the past but in the present, in which the constructed rules are obeyed if they seem fair and just.

The problems of modern society arise from its increasing complexity and interdependence—the multiplication of interaction and the spread of syncretism—as old segmentations break down and new arrangements are needed. The resolution of the problem is twofold: to create political and administrative structures that are responsive to the new scales, and to develop a more comprehensive or coherent creed that diverse men can share. The prescription is easy. It is the exegesis, as the listener to Rabbi Hillel finally understood, that is difficult.[6]

NOTES

1. The temper is not restricted to ecologists. Alfred Kazin cites the titles of some recent cultural-social analyses of "our situation": *Reflections on a Sinking Ship, Waiting for the End, The Fire Next Time, The Economy of Death, The Sense of an Ending, On the Edge of History, Thinking About the Unthinkable.*
2. Theodore Roszak, for example, writes: ". . . we must not ignore the fact that there *is* a natural environment—the world of wind and wave, beast and flower, sun and stars—and that preindustrial people lived for millennia in close company with that world, striving to harmonize the things and thoughts of their own making with its non-human forces. Circadian and seasonal rhythms were the first clock people knew, and it was by co-ordinating these fluid organic cycles with their own physiological tempos that they timed their activities. What they ate, they had killed or cultivated with their own hands, staining them with the blood or dirt of their effort. They learned from the flora and fauna of their surroundings, conversed with them, worshiped them, and sacrificed to them. They were convinced that their fate was bound up intimately with these non-human friends and foes, and in their culture they made place for them, honoring their ways."

   What is striking in this evocation of a pagan idyl is the complete neglect of the diseases which wasted most "natural" men, the high infant mortality, the painful, frequent childbirths which debilitated the women, and the recurrent shortages of food and the inadequacies of shelter which made life nasty, brutish, and short.
3. Jacques Ellul, *The Technological Society* (Knopf, New York, 1964), Chapter II, *passim.* What is striking in this unsparing attack on technique is Ellul's omission of any discussion of nature, or how man must live without technique. (The word *nature* does not appear in the index, and there are only a few passing references to the natural world, e.g., p. 79.) As Ellul's translator, John Wilkinson, writes in the Introduction: "In view of the fact that Ellul continually apostrophizes technique as 'unnatural' (except when he calls it the 'new nature'), it might be thought surprising that he has no fixed conception of nature or the natural. The best answer seems to be that he considers 'natural' (in the good sense) *any* environment able to satisfy man's material needs, *if* it leaves him free to use it as means to achieve his individual internally generated ends." Ibid., p. xix.
4. Ibid., p. 97.
5. The distinction is made in the report of the National Academy of Sciences, *Technology: Processes of Assessment and Choice,* published by the Committee on Science and Astronautics, U.S. House of Representatives, July 1969. See p. 16.
6. The traditional story is told that an impatient man once asked Rabbi Hillel to tell him all there was in Judaism while standing on one foot. The Rabbi pondered, and replied: "Do *not* do unto others as you would *not* have them do unto you. All the rest is exegesis."

POLITICAL-ECONOMIC

From "The Problem"

HENRY GEORGE

▪ Henry George *(1839-1897) was born in Philadelphia. He was a well-known American writer, economist, and philosopher. In Dewey's words, "It would require less than the fingers of the two hands to enumerate those who, from Plato down, rank with Henry George among the world's social philosophers." Nevertheless, he is not presently acknowledged as such among philosophers. His main goal as a journalist, writer, and philosopher was to criticize and expose some of the major inequities of his day. His major work is* Progress and Poverty *(1879), from which this selection is taken.*

Henry George is disappointed with technological progress. Speaking of the introduction of machinery to factories, he says "it was natural to expect . . . that labor-saving inventions would lighten the toil and improve the condition of the laborer." But this expectation, along with many others, was dashed. Instead "we find the deepest poverty, the sharpest struggle for existence, and the most of enforced idleness." Technological progress, instead of alleviating human misery, has contributed to it. Thus he finds himself forced to conclude that "social difficulties existing wherever a certain stage of progress has been reached, do not arise from local circumstances, but are, in some way or another, engendered by progress itself." In this statement Henry George is in disagreement with Daniel Bell, among others (see preceding selection in Part Two). He seems to be accepting a form of technological determinism.

The present century has been marked by a prodigious increase in wealth-producing power. The utilization of steam and electricity, the introduction of improved processes and labor-saving machinery, the greater subdivision and grander scale of production, the wonderful facilitation of exchanges, have multiplied enormously the effectiveness of labor.

At the beginning of this marvelous era it was natural to expect, and it was expected, that labor-saving inventions would lighten the toil and improve the condition of the laborer; that the enormous increase in the power of producing wealth would make real poverty a thing of the past. Could a man of the last century—a Franklin or a Priestley—have seen, in a vision of the future, the steamship taking the place of the sailing vessel, the railroad train of the wagon, the reaping machine of the scythe, the threshing machine of the flail; could he have heard the throb of the engines that in obedience to human will, and for the satisfaction of human desire, exert a power greater than that of all the men and all the beasts of burden of the earth combined; could he have seen the forest tree transformed into finished lumber—into doors, sashes, blinds, boxes or barrels, with hardly the touch of a human hand; the great workshops where boots and shoes are turned out by the case with less labor than the old-fashioned cobbler could have put on a sole; the factories where, under the eye of a girl, cotton becomes cloth faster than hundreds of stalwart weavers could have turned it out with their handlooms; could he have seen steam hammers shaping mammoth shafts and mighty anchors, and delicate machinery making tiny watches; the diamond drill cutting through the heart of the rocks, and coal oil sparing the whale; could he have realized the enormous saving of labor resulting from improved facilities of exchange and communication—sheep killed in Australia eaten fresh in England, and the order given by the London banker in the afternoon executed in San Francisco in the morning of the same day; could he have conceived of the hundred thousand improvements which these only suggest, what would he have inferred as to the social condition of mankind?

It would not have seemed like an inference; further than the vision went it would have seemed as though he saw; and his heart would have leaped and his nerves would have thrilled, as one who from a height beholds just ahead of the thirst-stricken caravan the living gleam of rustling woods and the glint of laughing waters. Plainly, in the sight of the imagination, he would have beheld these new forces elevating society from its very foundations, lifting the very poorest above the possibility of want, exempting the very lowest from anxiety for the material needs of life; he would have seen these slaves of the lamp of knowledge taking on themselves the traditional curse, these muscles of iron and sinews of steel making the poorest

☐ From *Progress And Poverty* (New York: The Modern Library, 1938), pp. 3-8.

laborer's life a holiday, in which every high quality and noble impulse could have scope to grow.

And out of these bounteous material conditions he would have seen arising, as necessary sequences, moral conditions realizing the golden age of which mankind have always dreamed. Youth no longer stunted and starved; age no longer harried by avarice; the child at play with the tiger; the man with the muck-rake drinking in the glory of the stars. Foul things fled, fierce things tame; discord turned to harmony! For how could there be greed where all had enough? How could the vice, the crime, the ignorance, the brutality, that spring from poverty and the fear of poverty, exist where poverty had vanished? Who should crouch where all were freemen; who oppress where all were peers?

More or less vague or clear, these have been the hopes, these the dreams born of the improvements which give this wonderful century its preëminence. They have sunk so deeply into the popular mind as radically to change the currents of thought, to recast creeds and displace the most fundamental conceptions. The haunting visions of higher possibilities have not merely gathered splendor and vividness, but their direction has changed—instead of seeing behind the faint tinges of an expiring sunset, all the glory of the daybreak has decked the skies before.

It is true that disappointment has followed disappointment, and that discovery upon discovery, and invention after invention, have neither lessened the toil of those who most need respite, nor brought plenty to the poor. But there have been so many things to which it seemed this failure could be laid, that up to our time the new faith has hardly weakened. We have better appreciated the difficulties to be overcome; but not the less trusted that the tendency of the times was to overcome them.

Now, however, we are coming into collision with facts which there can be no mistaking. From all parts of the civilized world come complaints of industrial depression; of labor condemned to involuntary idleness; of capital massed and wasting; of pecuniary distress among business men; of want and suffering and anxiety among the working classes. All the dull, deadening pain, all the keen, maddening anguish, that to great masses of men are involved in the words "hard times," afflict the world to-day. This state of things, common to communities differing so widely in situation, in political institutions, in fiscal and financial systems, in density of population and in social organization, can hardly be accounted for by local causes. There is distress where large standing armies are maintained, but there is also distress where the standing armies are nominal; there is distress where protective tariffs stupidly and wastefully hamper trade, but there is also distress where trade is nearly free; there is distress where autocratic government yet prevails, but there is also distress where political power is wholly in the hands of the people; in countries where paper is money, and in countries where gold and silver are the only currency. Evidently, beneath all such things as these, we must infer a common cause.

That there is a common cause, and that it is either what we call material progress or something closely connected with material progress, becomes more than an inference when it is noted that the phenomena we class together and speak of as industrial depression are but intensifications of phenomena which always accompany material progress, and which show themselves more clearly and strongly as material progress goes on. Where the conditions to which material progress everywhere tends are most fully realized—that is to say, where population is densest, wealth greatest, and the machinery of production and exchange most highly developed—we find the deepest poverty, the sharpest struggle for existence, and the most of enforced idleness.

It is to the newer countries—that is, to the countries where material progress is yet in its earlier stages—that laborers emigrate in search of higher wages, and capital flows in search of higher interest. It is in the older countries—that is to say, the countries where material progress has reached later stages—that widespread destitution is found in the midst of the greatest abundance. Go into one of the new communities where Anglo-Saxon vigor is just beginning the race of progress; where the machinery of production and exchange is yet rude and inefficient; where the increment of wealth is not yet great enough to enable any class to live in ease and luxury; where the best house is but a cabin of logs or a cloth and paper shanty, and the richest man is forced to daily work—and though you will find an absence of wealth and all its concomitants, you will find no beggars. There is no luxury, but there is no destitution. No one makes an easy living, nor a very good living; but every one *can* make a living, and no one able and willing to work is oppressed by the fear of want.

But just as such a community realizes the conditions which all civilized communities are striving for, and advances in the scale of material progress—just as closer settlement and a more intimate connection with the rest of the world, and greater utilization of labor-saving machinery, make possible greater economies in production and exchange, and wealth in consequence increases, not merely in the aggregate, but in proportion to population—so does poverty take a darker aspect. Some get an infinitely better and easier living, but others find it hard to get a living at all. The "tramp" comes with the locomotive, and almshouses and prisons are as surely the marks of "material progress" as are costly dwellings, rich warehouses, and magnificent churches. Upon streets lighted with gas and patrolled by uniformed policemen, beggars wait for the passer-by, and in the shadow of college, and library, and museum, are gathering the more hideous Huns and fiercer Vandals of whom Macaulay prophesied.

This fact—the great fact that poverty and all its concomitants show themselves in communities just as they develop into the conditions toward which material progress tends—proves that the social difficulties existing wherever a certain stage of progress has been reached, do not arise from local circumstances, but are, in some way or another, engendered by progress itself.

And, unpleasant as it may be to admit it, it is at last becoming evident that the enormous increase in productive power which has marked the present century and is still going on with accelerating ratio, has no tendency to extirpate poverty or to lighten the burdens of those compelled to toil. It simply widens the gulf between Dives and Lazarus, and makes the struggle for existence more intense. The march of invention has clothed mankind with powers of which a century ago the boldest imagination could not have dreamed. But in factories where labor-saving machinery has reached its most wonderful development, little children are at work; wherever the new forces are anything like fully utilized, large classes are maintained by charity or live on the verge of recourse to it; amid the greatest accumulations of wealth, men die of starvation, and puny infants suckle dry breasts; while everywhere the greed of gain, the worship of wealth, shows the force of the fear of want. The promised land flies before us like the mirage. The fruits of the tree of knowledge turn as we grasp them to apples of Sodom that crumble at the touch.

POLITICAL-ECONOMIC

From "Technology, Planning and Organization"

JOHN KENNETH GALBRAITH

■ John Kenneth Galbraith *was born in Ontario, Canada, in 1908. He was educated in Canada, the United States, and England. He taught at Harvard, and in 1959 became the Paul M. Warburg Professor of Economics there. He retired in 1975. Galbraith is the recipient of many awards and honors, and has written several books. Among his books are* The New Industrial State *(1967) and* The Age of Uncertainty *(1977).*

In the essay reprinted in part here, Galbraith traces recent shifts of power in Western societies. During feudal times, Galbraith argues, land was the source of power, for agricultural production then accounted for a large share of all production, and "power to engage in agricultural production rested with land ownership." Furthermore, "to get more land was difficult, and lost land was, as likely as not, irreplaceable."

But with the Industrial Revolution and the discovery of "a munificent supply" of land in the last century, land was dethroned by capital. "The man who owned or supplied the capital now had the strategically important factor of production. Authority over the enterprise, as a result, now passed to him."

Galbraith claims that a third shift of power is taking place in modern society. "Modern economic society can only be understood as an effort, notably successful, to synthesize, by organization, a personality far superior for its purposes to a natural person and with the added advantage of immortality." As a result of this new state of affairs, power has shifted from capital to management, since the latter is now in greater demand than the former.

In the last three decades, evidence has been accumulating of a shift of power from owners to managers within the modern large corporation. The power of the stockholders has seemed increasingly tenuous. A few stockholders assemble in an annual meeting, and a much larger number return proxies, ratifying the decisions of the management including its choices for the Board of Directors to speak for stockholders. So long, at least, as it makes profits—in 1964 none of the largest 100 industrial corporations and only seven of the largest 500 lost money—the position of a management is impregnable. The stockholders are literally powerless. To most economists, as to most lawyers, this whole tendency has seemed of questionable legitimacy. Some, in accordance with the established reaction to seemingly inconvenient truth, have sought to maintain the myth of stockholder power. Others, including all Marxians, have argued that the change is superficial, that capital retains a deeper and more functional control. Some have conceded the change but have deferred judgment as to its significance.[1] Yet others have seen a possibly dangerous usurpation of the legitimate power of capital.[2] No one (of whom I am aware) has questioned the credentials of capital, where power is concerned, or suggested that it might be *durably* in eclipse. If there is power, it was meant to have it.

Yet, over a longer range of time, power over the productive enterprise—and by derivation in the society at large—has shifted radically as between factors of production. The eminence of capital is a relatively recent matter; until about two centuries ago no qualified observer would have doubted that the decisive factor of production was land. The wealth, military power and the sanguinary authority over life and liberty of others that went with land ownership assured its possessor of a position of eminence in his community and of power in the state. These perquisites of land ownership also gave a strong and even controlling direction to history. For the great span of 250 years, until about a hundred years before the discovery of America, it helped inspire the recurrent military campaigns to the East which are called the Crusades. Succor for Byzantium, which was beset by the infidels and redemption of Jerusalem, which had been lost to them, served, without doubt, as a stimulant to religious ardor. But the younger sons of the Frankish nobility badly needed land. Beneath the mantled cross beat hearts soundly attuned to the value of real estate. Baldwin, younger brother of Godfrey of Bouillon, found himself faced on the way to the Holy City with the taxing decision as to whether to continue with the redeeming armies or take up an attractive piece of property at Edessa. He unhesitatingly opted for the latter and, only on the death of his brother, did he leave his fief to become the first King of Jerusalem.

□ From *Values and the Future*, ed. Kurt Baier and Nicholas Rescher (New York: The Free Press, 1969), pp. 355-364.

For four centuries following the discovery of America, appreciation of the strategic role of land gave it an even greater role in history. The Americas were populated—as also the Steppes and the habitable parts of the Antipodes. Once again religion went hand in hand with real property conveyancing, somewhat disguising the role of the latter. Spaniards considered themselves commissioned by God to win the souls of Indians; Puritans believed themselves primarily under obligation to look after their own. For Catholics and Cavaliers the Lord was believed to favor rather large acreages with the opportunity these accorded for custody of (and useful labor by) the aborigines and, as these gave out, of Africans. For Puritans, and Protestants generally, merit lay with the homestead and family farm. But these were details. In the New World, as in the Old, it was assumed that power and responsibility belonged, as right, to men who owned land. Democracy, in its modern meaning, began as a system which gave the suffrage to each and every person who owned land—and to no others.

The economic foundations of this eminence of land, and the incentive to its acquisition, were exceedingly firm. Until comparatively modern times, agricultural production—the provision of food and fiber—accounted for a large share of all production as it still accounts for 70-80 percent of output in countries such as India today. Subject to such rights as law and custom accorded to subordinate tenure, power to engage in agricultural production rested with land ownership. This, *pro tanto,* was power over a very large share of all economic activity.

The other factors of production were not of decisive importance. Agricultural technology was stable and made small use of mechanical power or other capital equipment. Thus a sparse supply of capital was matched, an important but sometimes neglected point, until a couple of hundred years ago by an equally meager opportunity for its use. If implements, work, stock or seed were lost this was not decisive; the modest requirements could be replaced.

The same was true of labor. Its historical tendency had been to keep itself in a condition of comparative abundance. David Ricardo, having regard for experience to that time, could hold in 1817 that "no point is better established than that the supply of labourers will always ultimately be in proportion to the means of supporting them."[3] This was to say that all that might be required would be forthcoming at, or about, the subsistence wage. The labor supply could be easily increased or replaced. But to get more land was difficult, and lost land was, as likely as not, irreplaceable. So land was strategic and not even the philosophers whose ideas ushered in the Industrial Revolution—Smith, and especially Ricardo and Malthus—could envisage a society where this was otherwise.

Then in the last century, in what we all agree to call the advanced countries, land was dethroned. The search for land, set in motion by its strategic role, uncovered a munificent supply. The Americas, Russia, South Africa and Australia were all discovered to have a large, unused and usable supply.

Meanwhile, mechanical inventions and the growth of metallurgical and engineering knowledge were prodigiously expanding opportunities for the employment of capital. From this greater use of capital came greater production and from that production came greater income and savings. It is not clear that in the last century the demand for capital grew more rapidly than the supply. In the new countries, including the United States, capital was generally scarce and the cost was high. In England, however, over most of the century, interest rates were low. But a diminishing proportion of the expanding production was of agricultural products and hence dependent on land. Iron and steel, ships, locomotives, textile machinery, buildings and bridges increasingly dominated the national product. For producing these, command of capital, not land, was what counted. Labor continued to be abundant in most places. Accordingly, the man who owned or supplied the capital now had the strategically important factor of production. Authority over the enterprise, as a result, now passed to him.

So did prestige in the community and political power. At the beginning of the nineteenth century the British Parliament was still dominated by the landed great; by the end of the century its premier figure was the Birmingham industrialist and pioneer screw manufacturer Joseph Chamberlain. At the beginning of the century, the United States government was dominated by the Virginia gentlemen; by the end of the century it was profoundly influenced by—depending on one's point of view—the men of enterprise or the malefactors of great wealth. The Senate was called a rich man's club.

This change, a point of much importance for what I am about to say, did not seem natural. George Washington, Thomas Jefferson, and James Madison seemed appropriate to the positions of public power. Public influence exercised by Jay Gould, Collis P. Huntington, J. P. Morgan, Elbert H. Gary, and Andrew Mellon seemed more suspect. The landowners were credited with capacity for action apart from their own interests and action in their own interest—the defense, for example, of slavery—seemed somehow legitimate. The capitalists were not credited with action apart from interest and their interest seemed less legitimate. This contrasting impression has not yet been exorcised from public attitudes or the elementary history books. We may lay it down as a rule that the older the exercise of any power the more benign it will appear and the more recent its assumption the more dangerous it will seem.

While capital in the last century was not scarce, at least in the great industrial centers, it was not in surplus. But in the present day economy, capital is, under most circumstances, abundant. The central task of modern economic policy, as it is most commonly defined, is to insure that all intended savings, at a high level of output, are offset by investment. This is what we have come to call Keynesian economic policy. Failure to invest all savings means unemployment—an excess of labor. So capital and labor have a conjoined tendency to abundance.

Back of this tendency of savings to surplus is a society which, increasingly, emphasizes not the need for frugality but the need for consumption. Saving, so far from being painful, reflects a failure in efforts by industry and the state to promote adequate consumption. Saving is also the product of a strategy by which the industrial enterprise seeks to insure full control of its sources of capital supply and thus to make its use a matter of internal decision. It is an effort which enjoys great success. Nearly three-quarters of capital investment last year was derived from the internal savings of corporations.

Capital, like land before it, owed its power over the enterprise to the difficulty of replacement or addition at the margin. What happens to that power when supply is not only abundant but excessive, when it is a central aim of social policy to offset savings and promote consumption and when it is a basic and successful purpose of business enterprises to exercise the control over the supply of capital that was once the foundation of its authority?

The plausible answer is that it will lose its power to a more strategic factor—one with greater bargaining power at the margin—if there is one. And there is.

Power has passed to what anyone in search of novelty might be forgiven for characterizing as a new factor of production. This is the structure of organization which combines and includes the technical knowledge, talent and experience that modern industrial technology and planning require. This structure is the creature of the modern industrial system and of its technology and planning. It embraces engineers, scientists, sales and advertising specialists, other technical and specialized talent—as well as the conventional leadership of the industrial enterprise. It is on the effectiveness of this structure, as indeed most business doctrine now implicity agrees, that the success of the business enterprise now depends. It can be created or enlarged only with difficulty. In keeping with past experience, the problem of supply at the margin accords *it* power.

The new recipients of power, it will be clear, are not individuals; the new locus of power is collegial or corporate. This fact encounters almost instinctive resistance. The individual has far more standing in our formal culture than the group. An individual has a presumption of accomplishment; a committee has a presumption of inaction. Individuals have souls; corporations are notably soulless. The entrepreneur—individualistic, restless, equipped with vision, guile, and courage—has been the economists' only hero. The great business organization arouses no similar affection. Admission to the economists' heaven is individually and by families; it is not clear that the top management even of an enterprise with an excellent corporate image can yet enter as a group. To be required, in pursuit of truth, to assert the superiority of the group over the individual for important social tasks is a taxing prospect.

Yet it is a necessary task. Modern economic society can only be understood as an effort, notably successful, to synthesize, by organization, a personality far superior for its purposes to a natural person and with the added advantage of immortality.

The need for such synthetic personality begins *first* with the fact that in modern industry a large number of decisions, and *all* that are important, require information possessed by more than one man. All important decisions draw on the specialized scientific and technical knowledge; on the accumulated information or experience; and on the artistic or intuitive reaction of several or many persons. The final decision will be informed only as it draws on all whose information is relevant. And there is the further important requirement that this information must be properly weighed to assess its relevance and its reliability. There must be, in other words, a mechanism for drawing on the information of numerous individuals and for measuring the importance and testing the reliability of what each has to offer.

The need to draw on the information of numerous individuals derives first from the *technological* requirements of modern industry. These are not always inordinately sophisticated; a man of moderate genius could, quite conceivably, provide himself with the knowledge of the various branches of metallurgy and chemistry, and of engineering, procurement, production management, quality control, labor relations, styling and merchandising which are involved in the development of a modern automobile. But even moderate genius is in unpredictable supply; and to keep abreast of all the relevant branches of science, engineering, and art would be time consuming. The answer, which allows of the use of far more common talent and with greater predictability of result, is to have men who are appropriately qualified or experienced in each limited area of specialized knowledge or art. Their information is then combined for the design and production of the vehicle. It is the common public impression, greatly encouraged by scientists, engineers and industrialists, that modern scientific, engineering and industrial achievements are the work of a new and quite remarkable race of men. This is pure vanity. The real accomplishment is in taking ordinary men, informing them narrowly but deeply and then devising an organization which combines their knowledge with that of other similarly specialized but equally ordinary men for a highly predictable performance.

The *second* factor requiring the combination of specialized talent derives from large-scale employment of capital in combination with sophisticated technology. This makes imperative planning and accompanying control of environment. The market is, in remarkable degree, an intellectually undemanding institution. The Wisconsin farmer need not anticipate his requirements for fertilizers, pesticides or even machine parts; the market stocks and supplies them. The cost is the same for the farmer of intelligence and the neighbor who under medical examination shows daylight in either ear. There need be no sales strategy; the market takes all his milk at the ruling price. Much of the appeal of the market, to economists at least, has been the way it seems to simplify life.

The extensive use of capital, with advanced technology, greatly reduced the power of the market. Planning, with attendent complexity of task, takes its place. Thus the manufacturer of missiles, space vehicles or modern aircraft must foresee and insure his requirements for specialized plant, specialized talent, arcane materials and intricate components. These the market cannot be counted upon to supply. And there is no open market where these products can be sold. Everything depends on the care with which contracts are sought and nurtured, in Washington. The same complexities hold in only lesser degree for the maker of automobiles, processed foods and detergents. This firm too must foresee requirements and manage the markets for its products. All such planning is dealt with only by highly-qualified men—men who can foresee need and insure the supply of production requirements, relate costs to an appropriate price strategy, see that customers are suitably persuaded to buy what is made available and, at yet higher levels of technology and complexity, see that the state is persuaded.

Technology and planning thus require the extensive combination and testing of information. Much of this is accomplished, in practice, by men talking with each other—by meeting in committee. One can do worse than think of a business organization as a complex of committees. Management consists in recruiting and assigning talent to the right committee, in intervening on occasion to force a decision, and in either announcing the decision or carrying it, as a datum, for a yet larger decision by the next committee.

It must not be supposed that this is an inefficient device. A committee allows men to pool information under circumstances that allow also of immediate probing and discussion to assess the relevance and reliability of the information offered. Loose or foolish talk, or simple uncertainty, is revealed as in no other way. There is also no doubt considerable stimulus to mental effort; men who believe themselves deeply engaged in private thought are usually doing nothing at all. Committees are condemned by those who are caught by the *cliché* that individual effort is somehow superior to group effort; by those whose suspicions are aroused by the fact that for many people group effort is more congenial and pleasant; by those who do not see that the process of extracting, and especially of testing, information has necessarily a somewhat undirected quality—briskly conducted meetings invariably decide matters that were decided beforehand elsewhere; and by those who fail to see that highly-paid men, when sitting around a table as a committee, are not necessarily wasting more time, in the aggregate, than each would waste all by himself. Forthright men frequently react to belief in their own superior capacity for decision by abolishing all committees. They then constitute working parties, task forces, operations centers or executive groups in order to avoid the truly disastrous consequences of deciding matters themselves.

This group decision-making extends deeply into the enterprise; it goes far beyond the group commonly designated as the management. Power, in fact, is *not* closely related to position in the hierarchy of the enterprise. We always carry in our minds an implicit organization chart of the business enterprise. At the top is the Board of Directors and the Board Chairman; next comes the President; next comes the Executive Vice-President; thereafter comes the Department or Divisional Heads—those who preside over the Chevrolet division, large generators, the computer division. Power is presumed to pass down from the pinnacle.

This happens only in organizations with a routine task, such, for example, as the peacetime drill of a platoon. Otherwise the power lies with the individuals who possess the knowledge. If their knowledge is particular and strategic their power becomes very great. Enrico Fermi rode a bicycle to work at Los Alamos. Leslie Groves commanded the whole Manhattan Project. It was Fermi and his colleagues, and not General Groves in his grandeur, who made the decisions of importance.

But it should not be imagined that group decision making is confined to nuclear technology and space mechanics. In our day even simple products are made or packaged or marketed by highly sophisticated methods. For these too power passes into organization. For purposes of pedagogy, I have sometimes illustrated these matters by reference to a technically uncomplicated product, which, unaccountably, neither General Electric nor Westinghouse has yet placed on the market. It is a toaster of standard performance except that it etches on the surface of the toast, in darker carbon, one of a selection of standard messages or designs. For the elegant hostess, monograms would be available, or even a coat of arms; for the devout, there would be at breakfast an appropriate devotional message from the works of Norman Vincent Peale; the patriotic, or worried, would have an aphorism urging vigilance from Mr. J. Edgar Hoover; for modern economists, there would be mathematical design; a restaurant version could sell advertising, or urge the peaceful acceptance of the integration of public eating places.

Conceivably this vision could come from the President of General Electric. But the orderly proliferation of such ideas is the established function of much more lowly men who are charged, specifically, with new product development. At an early stage in the development of the toaster, specialists in style, design and, no doubt, philosophy, art and spelling would have to be accorded a responsible role. No one in a position to authorize the product would do so without a judgment on how the problems of design and inscription were to be solved and the cost. An advance finding would be over-ridden only with caution. All action would be contingent on the work of specialists in market testing and analysis who would determine whether and by what means the toaster could be sold and at what cost for various quantities. They would function as part of a team which would also include merchandising, advertising and dealer relations men. No adverse decision by this group would be over-ruled. Nor, given the notoriety that attaches to missed opportunity, would a favorable decision. It will be evident that nearly all power—initiative, development, rejection or approval—is exercised deep down in the company.

So two great trends have converged. In consequence of advanced technology, highly capitalized production and a capacity through planning to command earnings for the use of the firm, capital has become comparatively abundant. And the imperatives of advanced technology and planning have moved the power of decision from the individual to the group and have moved it deeply into the firm.

NOTES

1. Cf. Edward S. Mason, "The Apologetics of Managerialism," *Journal of Business* (University of Chicago) January, 1958. And "Comment" in *A Survey of Contemporary Economics*, pp. 221-222.
2. Cf. Adolf A. Berle, Jr., *Power Without Property* (New York: Harcourt, Brace and Company, 1959), pp. 98 *et seq*.
3. David Ricardo, "On the Principles of Political Economy and Taxation," *The Works and Correspondence of David Ricardo*, ed. by Piero Sraffa (Cambridge, 1951), p. 292.

POLITICAL-ECONOMIC

From "America in the Technetronic Age: New Questions of Our Time"

ZBIGNIEW BRZEZINSKI

▪ Zbigniew Brzezinski *was born in Warsaw, Poland, in 1928. He came to the United States in 1953, and was naturalized in 1958. He taught at Harvard University, and in 1962 became the director of the Research Institute for International Change at Columbia University. He was also a faculty member of the Russian Institute there. In 1973 he became the director of the influential Trilaterial Commission, and in 1977 he became assistant to the President for national security affairs. Brzezinski has received many honors. Among his books are* Between two Ages *(1970) and* Political Power: USA/USSR *(1966).*

In "America in the Technetronic Age" Brzezinski argues that "we are entering a novel metamorphic phase in human history. The world is on the eve of a transformation more dramatic in its historical and human consequences than that wrought either by the French or the Bolshevik revolutions." Underlying this transformation are new developments in technology, particularly in the areas of computers and communications. These developments are shaping—culturally, socially, and economically—a new society which Brzezinski calls the "technetronic society." This society is substantially different from the industrial society, and Brzezinski discusses several of these differences.

Significantly, the United States is beginning to enter this new phase of human history while countries in Europe remain caught in the industrial phase and while Third World countries remain even farther behind. This state of affairs, Brzezinski argues, is reason for deep concern. He sees the gap developing among these nations as a source of international instability, and he suggests various ways for remedying the problem.

Ours is no longer the conventional revolutionary era; we are entering a novel metamorphic phase in human history. The world is on the eve of a transformation more dramatic in its historic and human consequences than that wrought either by the French or the Bolshevik revolutions. Viewed from a long perspective, these famous revolutions merely scratched the surface of the human condition. The changes they precipitated involved alterations in the distribution of power and property within society; they did not affect the essence of individual and social existence. Life—personal and organised—continued much as before, even though some of its external forms (primarily political) were substantially altered. Shocking though it may sound to their acolytes, by the year 2000 it will be accepted that Robespierre and Lenin were mild reformers.

Unlike the revolutions of the past, the developing metamorphosis will have no charismatic leaders with strident doctrines, but its impact will be far more profound. Most of the change that has so far taken place in human history has been gradual—with the great "revolutions" being mere punctuation marks to a slow, eludible process. In contrast, the approaching transformation will come more rapidly and will have deeper consequences for the way and even perhaps for the meaning of human life than anything experienced by the generations that preceded us.

America is already beginning to experience these changes and in the course of so doing it is becoming a "technetronic" society: a so-

☐ *Encounter* [London], Jan. 1968, pp. 16-19, 23-26.

ciety that is shaped culturally, psychologically, socially and economically by the impact of technology and electronics, particularly computers and communications. The industrial process no longer is the principal determinant of social change, altering the mores, the social structure, and the values of society. This change is separating the United States from the rest of the world, prompting a further fragmentation among an increasingly differentiated mankind, and imposing upon Americans a special obligation to ease the pains of the resulting confrontation.

THE TECHNETRONIC SOCIETY

The far-reaching innovations we are about to experience will be the result primarily of the impact of science and technology on man and his society, especially in the developed world. Recent years have seen a proliferation of exciting and challenging literature on the future. Much of it is serious, and not mere science-fiction.[1] Moreover, both in the United States and, to a lesser degree, in Western Europe a number of systematic, scholarly efforts have been designed to project, predict, and possess what the future holds for us. Curiously very little has been heard on this theme from the Communist World, even though Communist doctrinarians are the first to claim their 19th-century ideology holds a special pass-key to the 21st century.

The work in progress indicates that men living in the developed world will undergo during the next several decades a mutation potentially as basic as that experienced through the slow process of evolution from animal to human experience. The difference, however, is that the process will be telescoped in time—and hence the shock effect of the change may be quite profound. Human conduct will become less spontaneous and less mysterious—more predetermined and subject to deliberate "programming." Man will increasingly possess the capacity to determine the sex of his children, to affect through drugs the extent of their intelligence and to modify and control their personalities. The human brain will acquire expanded powers, with computers becoming as routine an extension of man's reasoning as automobiles have been of man's mobility. The human body will be improved and its durability extended: some estimate that during the next century the average lifespan could reach approximately 120 years.

These developments will have major social impact. The prolongation of life will alter our values, our career patterns, and our social relationships. New forms of social control may be needed to limit the indiscriminate exercise by individuals of their new powers. The possibility of extensive chemical mind-control, the danger of loss of individuality inherent in extensive transplantation, and the feasibility of manipulation of the genetic structure will call for a social definition of common criteria of restraint as well as of utilisation. Scientists predict with some confidence that by the end of this century, computers will reason as well as man, and will be able to engage in "creative" thought; wedded to robots or to "laboratory beings," they could act like humans. The makings of a most complex—and perhaps bitter—philosophical and political dialogue about the nature of man are self-evident in these developments.

Other discoveries and refinements will further alter society as we now know it. The information revolution, including extensive information storage, instant retrieval, and eventually push-button visual and sound availability of needed data in almost any private home, will transform the character of institutionalised collective education. The same techniques could serve to impose well-nigh total political surveillance on every citizen, putting into much sharper relief than is the case today the question of privacy. Cybernetics and automation will revolutionise working habits, with leisure becoming the practice and active work the exception—and a privilege reserved for the most talented. The achievement-oriented society might give way to the amusement-focused society, with essentially spectator spectacles (mass sports, TV) providing an opiate for increasingly purposeless masses.

But while for the masses life will grow longer and time will seem to expand, for the activist élite time will become a rare commodity. Indeed, even the élite's sense of time will alter. Already now speed dictates the pace of our lives—instead of the other way around. As the speed of transportation increases, largely by its own technological momentum, man discovers that he has no choice but to avail himself of that acceleration, either to keep up with others or because he thinks he can thus accomplish more. This will be especially true of the élite, for whom an expansion in leisure time does not seem to be in the cards. Thus as speed expands, time contracts—and the pressures on the élite increase.

By the end of this century the citizens of the more developed countries will live predominantly in cities—hence almost surrounded by man-made environment. Confronting nature could be to them what facing the elements was to our forefathers: meeting the unknown and not necessarily liking it. Enjoying a personal standard of living that (in some countries) may reach almost $10,000 per head, eating artificial food, speedily commuting from one corner of the country to work in another, in continual visual contact with their employer, government, or family, consulting their annual calendars to establish on which day it will rain or shine, our descendants will be shaped almost entirely by what they themselves create and control.

But even short of these far-reaching changes, the transformation that is now taking place is already creating a society increasingly unlike its industrial predecessor.[2] In the industrial society, technical knowledge was applied primarily to one specific end: the acceleration and improvement of production techniques. Social consequences were a later by-product of this paramount concern. In the technetronic society, scientific and technological knowledge, in addition to enhancing productive capabilities, quickly spills over to affect directly almost all aspects of life.

This is particularly evident in the case of the impact of communications and computers. Communications create an extraordinarily interwoven society, in continuous visual, audial, and increasingly close contact among almost all its members—electronically interacting, sharing instantly most intense social experiences, prompting far greater personal involvement, with their consciousnesses shaped in a sporadic manner fundamentally different (as McLuhan has noted) from the literate (or pamphleteering) mode of transmitting information, characteristic of the industrial age. The growing capacity for calculating instantly most complex interactions and the increasing availability of bio-chemical means of human control increase the potential scope of self-conscious direction, and thereby also the pressures to direct, to choose, and to change.

The consequence is a society that differs from the industrial one in a variety of economic, political and social aspects. The following examples may be briefly cited to summarise some of the contrasts:

1. In an industrial society, the mode of production shifts from agriculture to industry, with the use of muscle and animals supplanted by machine-operation. In the technetronic society, industrial employment yields to services, with automation and cybernetics replacing individual operation of machines.

2. Problems of employment and unemployment—not to speak of the earlier stage of the urban socialisation of the post-rural labour force—dominate the relationship between employers, labour, and the market in the industrial society; assuring minimum welfare to the new industrial masses is a source of major concern. In the emerging new society, questions relating to skill-obsolescence, security, vacations, leisure, and profit-sharing dominate the relationship; the matter of psychic well-being of millions of relatively secure but potentially aimless lower-middle class blue collar workers becomes a growing problem.

3. Breaking down traditional barriers to education, thus creating the basic point of departure for social advancement, is a major goal of social reformers in the industrial society. Education, available for limited and specific periods of time, is initially concerned with overcoming illiteracy, and subsequently with technical training, largely based on written, sequential reasoning. In the technetronic society, not only has education become universal but advanced training is available to almost all who have the basic talents. Quantity-training is reinforced by far greater emphasis on quality-selection. The basic problem is to discover the most effective techniques for the rational exploitation of social talent. Latest communication and calculating techniques are applied to that end. The educational process, relying much more on visual and audial devices, becomes extended in time, while the flow of new knowledge necessitates more and more frequent refresher studies.

4. In the industrial society social leadership shifts from the traditional rural-aristocratic to an urban "plutocratic" élite. Newly acquired wealth is its foundation, and intense competition the outlet—as well as the stimulus—for its energy. In the post-industrial technetronic society plutocratic pre-eminence comes under a sustained challenge from the political leadership which itself is increasingly permeated by individuals possessing special skills and intellectual talents. Knowledge becomes a tool of power, and the effective mobilisation of talent an important way for acquiring power.

5. The university in an industrial society—rather in contrast to the medieval times—is an aloof ivory-tower, the repository of irrelevant, even if respected wisdom, and, for only a brief time, the watering fountain for budding members of the established social élite. In the technetronic society, the university becomes an intensely involved *think-tank,* the source of much sustained political planning and social innovation.

6. The turmoil inherent in the shift from the rigidly traditional rural to urban existence engenders an inclination to seek total answers to social dilemmas, thus causing ideologies to thrive in the industrial society.[3] In the technetronic society, increasing ability to reduce social conflicts to quantifiable and measurable dimensions reinforces the trend towards a more pragmatic problem-solving approach to social issues.

7. The activisation of hitherto passive masses makes for intense political conflicts in the industrial society over such matters as disenfranchisement and the right to vote. The issue of political participation is a crucial one. In the technetronic age, the question increasingly is one of ensuring real participation in decisions that seem too complex and too far-removed from the average citizen. Political alienation becomes a problem. Similarly, the issue of political equality of the sexes gives way to a struggle for the sexual equality of women. In the industrial society, woman—the operator of machines—ceases to be physically inferior to the male, a consideration of some importance in rural life, and she begins to demand her political rights. In the emerging society, automation discriminates equally against males and females; intellectual talent is computable; the pill encourages sexual equality.

8. The newly enfranchised masses are coordinated in the industrial society through trade unions and political parties, and integrated by relatively simple and somewhat ideological programmes. Moreover, political attitudes are influenced by appeals to nationalist sentiments, communicated through the massive growth of newspapers, relying, naturally, on native tongues. In the technetronic society, the trend would seem to be towards the aggregation of the individual support of millions of uncoordinated citizens, easily within the reach of magnetic and attractive personalities effectively exploiting the latest communication techniques to manipulate emotions and control reason. Reliance on TV—and hence the tendency to replace language with imagery, with the latter unlimited by national confines (and also including coverage for such matters as hunger in India or war scenes)—tends to create a somewhat more cosmopolitan, though highly impressionistic, involvement in global affairs.

9. Economic power in the industrial society tends to be personalised, either in the shape of great *entrepreneurs* like Henry Ford or bureaucratic industrialisers like Kaganovich in Russia, or Minc in Poland. The tendency towards de-personalisation of economic power is stimulated in the next stage by the appearance of a highly complex interdependence between governmental institutions (including the military), scientific establishments, and industrial organisations. As economic power becomes inseparably linked with political power, it becomes more invisible and the sense of individual futility increases.

10. Relaxation and escapism in the industrial society, in its more intense forms, is a carry-over from the rural drinking bout, in which intimate friends and family would join. Bars and saloons—or fraternities—strive to recreate the atmosphere of intimacy. In the technetronic society social life tends to be so atomised, even though communications (especially TV) make for unprecedented immediacy of social experience, that group intimacy cannot be recreated through the artificial stimulation of externally convivial group behaviour. The new interest in drugs seeks to create intimacy through introspection, allegedly by expanding consciousness.

Eventually, these changes and many others, including the ones that affect much more directly the personality and quality of the human being itself, will make the technetronic society as different from the industrial as the industrial became from the agrarian. . . .

THE TRAUMA OF CONFRONTATION

For the world at large, the appearance of the new technetronic society could have the paradoxical effect of creating more distinct worlds on a planet that is continuously shrinking because of the communications revolution.

While the scientific-technological change will inevitably have some spill-over, not only will the gap between the developed and the underdeveloped worlds probably become wider—especially in the more measurable terms of economic indices—but a *new one* may be developing *within* the industrialised and urban world.

The fact is that America, having left the industrial phase, is today entering a distinct historical era: and one different from that of Western Europe and Japan. This is prompting subtle and still indefinable changes in the American psyche, providing the psycho-cultural bases for the more evident political disagreements between the two sides of the Atlantic. To be sure, there are pockets of innovation or retardation on both sides. Sweden shares with America the problems of leisure, psychic well-being, purposelessness; while Mississippi is experiencing the confrontation with the industrial age in a way not unlike some parts of South-Western Europe. But I believe the broad generalisation still holds true: Europe and America are no longer in the same historical era.

What makes America unique in our time is that it is the first society to experience the future. The confrontation with the new—which will soon include much of what I have outlined—is part of the daily American experience. For better or for worse, the rest of the world learns what is in store for it by observing what happens in the U.S.A.: in the latest scientific discoveries in space, in medicine, or the electric toothbrush in the bathroom; in pop art or LSD, air conditioning or air pollution, old-age problems or juvenile delinquency. The evidence is more elusive in such matters as music, style, values, social mores; but there, too the term "Americanisation" obviously defines the source. Today, America is *the* creative society; the others, consciously and unconsciously, are emulative.

American scientific leadership is particularly strong in the so-called "frontier" industries, involving the most advanced fields of science. It has been estimated that approximately 80% of all scientific and technical discoveries made during the last few decades originated in the United States. About 75% of the world's computers operate in the United States; the American lead in lasers is even more marked; examples of American scientific lead are abundant.

There is reason to assume that this leadership will continue. America has four times as many scientists and research workers as the countries of the European Economic Community combined; three-and-a-half times as many as the Soviet Union. The brain-drain is almost entirely one-way. The United States is also spending more on research: seven times as much as the E.E.C. countries, three-and-a-half times as much as the Soviet Union. Given the fact that scientific development is a dynamic process, it is likely that the gap will widen.[4]

On the social level, American innovation is most strikingly seen in the manner in which the new meritocratic élite is taking over American life, utilising the universities, exploiting the latest techniques of communications, harnessing as rapidly as possible the most recent technological devices. Technetronics dominate American life, but so far nobody else's. This is bound to have social and political—and therefore also psychological—consequences, stimulating a psycho-cultural gap in the developed world.

At the same time, the backward regions of the world are becoming more, rather than less, poor in relation to the developed world. It can be roughly estimated that the per capita income of the underdeveloped world is approximately ten times lower than of America and Europe (and twenty-five times of America itself). By the end of the century, the ratio may be about fifteen-to-one (or possibly thirty-to-one in the case of the U.S.), with the backward nations *at best* approaching the present standards of the very poor European nations but in many cases (*e.g.*, India) probably not even attaining that modest level.

The social élites of these regions, however, will quite naturally tend to assimilate and emulate, as much as their means and power permit, the life-styles of the most advanced world, with which they are, and increasingly will be, in close vicarious contact through global television, movies, travel, education, and international magazines. The international gap will thus have a domestic reflection, with the masses, given the availability even in most backward regions of transistorised radios (and soon television), becoming more and more intensely aware of their deprivation.

It is difficult to conceive how in that context democratic institutions (derived largely from Western experience—but typical only of the more stable and wealthy Western nations) will endure in a country like India, or develop elsewhere. The foreseeable future is more likely to see a turn towards personal dictatorships and some unifying doctrines, in the hope that the combination of the two may preserve the minimum stability necessary for social-economic development. The problem, however, is that whereas in the past ideologies of change gravitated from the developed world to the less, in a way stimulating imitation of the developed world (as was the case with Communism), today the differences between the two worlds are so pronounced that it is difficult to conceive a new ideological wave originating from the developed world, where the tradition of utopian thinking is generally declining.

With the widening gap dooming any hope of imitation, the more likely development is an ideology of rejection of the developed world. Racial hatred could provide the necessary emotional force, exploited by xenophobic and romantic leaders. The writings of Frantz Fanon—violent and racist—are a good example. Such ideologies of rejection, combining racialism with nationalism, would further reduce the chances of meaningful regional cooperation, so essential if technology and science are to be effectively applied. They would certainly widen the existing psychological and emotional gaps. Indeed, one might ask at that point: who is the truer repository of that indefinable quality we call human? The technologically dominant and conditioned technetron, increasingly trained to adjust to leisure, or the more "natural" and backward agrarian, more and more dominated by racial passions and continuously exhorted to work harder, even as his goal of the good life becomes more elusive?

The result could be a modern version on a global scale of the old rural-urban dichotomy. In the past, the strains produced by the shift from an essentially agricultural economy to a more urban one contributed much of the impetus for revolutionary violence.[5] Applied on a global scale, this division could give substance to Lin Piao's bold thesis that:

> Taking the entire globe, if North America and Western Europe can be called "the cities of the world," then Asia, Africa, and Latin America constitute "the rural areas of the world." . . . In a sense, the contemporary world revolution also presents a picture of the encirclement of cities by the rural areas.

In any case, even without envisaging such a dichotomic confrontation, it is fair to say that the underdeveloped regions will be facing increasingly grave problems of political stability and social survival. Indeed (to use a capsule formula), in the developed world, the nature of man as man is threatened; in the underdeveloped, society is. The interaction of the two could produce chaos.

To be sure, the most advanced states will possess ever more deadly means of destruction, possibly even capable of nullifying the consequences of the nuclear proliferation that appears increasingly inevitable. Chemical and biological weapons, death rays, neutron bombs, nerve gases, and a host of other devices, possessed in all their sophisticated variety (as seems likely) only by the two super-states, may impose on the world a measure of stability. Nonetheless, it seems unlikely, given the rivalry between the two principal powers, that a fool-proof system against international violence can be established. Some local wars between the weaker, nationalistically more aroused, poorer nations may occasionally erupt—resulting perhaps even in the total nuclear extinction of one or several smaller nations?—before greater international control is imposed in the wake of the universal moral shock thereby generated.

The underlying problem, however, will be to find a way of avoiding somehow the widening of the cultural and psycho-social gap inherent in the growing differentiation of the world. Even with gradual differentiation throughout human history, it was not until the industrial revolution that sharp differences between societies began to appear. Today, some nations still live in conditions not unlike pre-Christian times; many no different than in the medieval age. Yet soon a few will live in ways so new that it is now difficult to imagine their social and individual ramifications. If the developed world takes a leap—as seems inescapably the case—into a reality that is even more different from ours today than ours is from an Indian village, the gap and its accompanying strains will not narrow.

On the contrary, the instantaneous electronic intermeshing of mankind will make for an intense confrontation, straining social and international peace. In the past, differences were "livable" because of time and distance that separated them. Today, these differences are actually widening while technetronics are eliminating the two insulants of time and distance. The resulting trauma could create almost entirely different perspectives on life, with insecurity, envy, and hostility becoming the dominant emotions for increasingly large numbers of people. A three-way split into rural-backward, urban-industrial, and technetronic ways of life can only further divide man, intensify the existing difficulties to global understanding, and give added vitality to latent or existing conflicts.

The pace of American development both widens the split within mankind and contains the seeds for a constructive response. However, neither military power nor material wealth, both of which America possesses in abundance, can be used directly in responding to the onrushing division in man's thinking, norms, and character. Power, at best, can assure only a relatively stable external environment: the tempering or containing of the potential global civil war; wealth can grease points of socio-economic friction, thereby facilitating development. But as man—especially in the most advanced societies—moves increasingly into the phase of controlling and even creating his environment, increasing attention will have to be given to giving man meaningful content—to improving the quality of life for man *as man*.

> Man has never really tried to use science in the realm of his value systems. Ethical thinking is hard to change, but history demonstrates that it does change. . . . Man does, in limited ways, direct his very important and much more rapid psycho-social education. The evolution of such things as automobiles, airplanes, weapons, legal institutions, corporations, universities, and democratic governments are examples of progressive evolution in the course of time. We have, however, never really tried deliberately to create a better society for man *qua* man. . . .[6]

The urgent need to do just that may compel America to redefine its global posture. During the remainder of this century, given the perspective on the future I have outlined here, America is likely to become less concerned with "fighting communism" or creating "a world safe for diversity" than with helping to develop a common response with the rest of mankind to the implications of a truly new era. This will mean making the massive diffusion of scientific-technological knowledge a principal focus of American involvement in world affairs.

To some extent, the U.S. performs that role already—simply by being what it is. The impact of its reality and its global involvement prompts emulation. The emergence of vast international corporations, mostly originating in the United States, makes for easier transfer of skills, management techniques, marketing procedures, and scientific-technological innovations. The appearance of these corporations in the European market has done much to stimulate Europeans to consider more urgently the need to integrate their resources and to accelerate the pace of their own research and development.

Similarly, returning graduates from American universities have prompted an organisational and intellectual revolution in the academic life of their countries. Changes in the academic life of Britain, Germany, Japan, more recently France, and (to even a greater extent) in the less developed countries, can be traced to the influence of U.S. educational institutions. Indeed, the leading technological institute in Turkey conducts its lectures in "American" and is deliberately imitating, not only in approach but in student-professor relationships, U.S. patterns. Given developments in modern communications, is it not only a matter of time before students at Columbia University and, say, the University of Teheran will be watching, *simultaneously,* the same lecturer?

The appearance of a universal intellectual élite, one that shares certain common values and aspirations, will somewhat compensate for the widening differentiation among men and societies. But it will not resolve the problem posed by that differentiation. In many backward nations tension between what is and what can be will be intensified. Moreover, as Kenneth Boulding observed:

> The network of electronic communication is inevitably producing a world super-culture, and the

relations between this super-culture and the more traditional national and regional cultures of the past remains the great question mark of the next fifty years.[7]

That "super-culture," strongly influenced by American life, with its own universal electronic-computer language, will find it difficult to relate itself to "the more traditional and regional cultures," especially if the basic gap continues to widen.

To cope with that gap, a gradual change in diplomatic style and emphasis may have to follow the redefined emphasis of America's involvement in world affairs. Professional diplomacy will have to yield to intellectual leadership. With government negotiating directly—or quickly dispatching the negotiators—there will be less need for ambassadors who are resident diplomats and more for ambassadors who are capable of serving as creative interpreters of the new age, willing to engage in a meaningful dialogue with the host intellectual community and concerned with promoting the widest possible dissemination of available knowledge. Theirs will be the task to stimulate and to develop scientific-technological programmes of co-operation.

International co-operation will be necessary in almost every facet of life: to reform and to develop more modern educational systems, to promote new sources of food supply, to accelerate economic development, to stimulate technological growth, to control climate, to disseminate new medical knowledge. However, because the new élites have a vested interest in their new nation-states and because of the growing xenophobia among the masses in the third world, the nation-state will remain for a long time the primary focus of loyalty, especially for newly liberated and economically backward peoples. To predict loudly its death, and to act often as if it were dead, could prompt (as it did partially in Europe) an adverse reaction from those whom one would wish to influence. Hence, regionalism will have to be promoted with due deference to the symbolic meaning of national sovereignty—and preferably also by encouraging those concerned themselves to advocate regional approaches.

Even more important will be the stimulation, for the first time in history on a global scale, of the much needed dialogue on what it is about man's life that we wish to safeguard or to promote, and on the relevance of existing moral systems to an age that cannot be fitted into the narrow confines of fading doctrines. The search for new directions—going beyond the tangibles of economic development—could be an appropriate subject for a special world congress, devoted to the technetronic and philosophical problems of the coming age. To these issues no one society, however advanced, is in a position to provide an answer.

NOTES

[1]Perhaps the most useful single source is to be found in the Summer 1967 issue of *Daedalus,* devoted entirely to *"Toward the Year 2000: Work in Progress."* The introduction by Professor Daniel Bell, chairman of the American Academy's Commission on the Year 2000 (of which the present writer is also a member) summarises some of the principal literature on the subject.

[2]See Daniel Bell's pioneering "Notes on the Post-Industrial Society," *The Public Interest,* Nos. 6 and 7, 1967.

[3]The American exception to this rule was due to the absence of the feudal tradition, a point well developed by Louis Hartz in his work *The Liberal Tradition in America* (1955).

[4]In the Soviet case, rigid compartmentalisation between secret military research and industrial research has had a particularly sterile effect of inhibiting spill-over from weapons research into industrial application.

[5]See Barrington Moore's documentation of this in his pioneering study *Social Origins of Dictatorship and Democracy* (1967).

[6]Hudson Hoagland, "Biology, Brains, and Insight," *Columbia University Forum,* Summer 1967.

[7]Kenneth Boulding, "Expecting the Unexpected," *Prospective Changes in Society by 1980* (1960).

RELIGIOUS

From "Technical Progress and Sin"

GABRIEL MARCEL

■ Gabriel Marcel *was born in 1889 in Paris. He was raised in a home dominated by his father's agnosticism and his aunt's liberal protestantism. Nevertheless, Marcel was highly interested in the religious dimension of human experience. In 1929 he converted to Catholicism and became subsequently an intellectual leader in French Catholic circles. During World War I he joined the Red Cross, and his experiences during that period left permanent marks on the direction of his thought. For he realized then the inability of abstract philosophy to cope with the tragic character of human existence. Among his chief works are* The Mystery of Being *(1950) and* Man Against Mass Society *(1951).*

Gabriel Marcel is not an anti-technologist as one might infer from the title of the selection that appears here. On the contrary, "technique is rather something good or the expression of something good, since it amounts to nothing more than a specific instance of our general application of our gift of reason to reality." Nevertheless, Marcel is only too well aware of the relation between technological progress and sin. It is the basis of this relationship that he explores in this selection. According to him, things start going wrong when feelings of power and pride, which an inventor justifiably experiences, "lose their just pretext and their authenticity, in the case of the man who benefits from an invention without having made any contribution towards discovering or perfecting it." This leads first to a kind of "idolatry" of technical products, which weakens the sense of the sacred, and then to "autolatry," that is, self-worship.

Marcel evaluates the pros and cons of communicational advances made possible by technological progress. (Compare these views with those given in Part Two under the heading "Media-oriented".) He also discusses the vice of "envy," which for him is another outgrowth of technological progress. All this leads him to conclude that nothing will save the technological man except "an act of faith."

In the opening paragraph of this selection, Marcel is examining the universal human emotion of indignation that is prompted by wartime atrocities.

This almost universal emotion in the face of horror—an emotion, it may be admitted, that has so far had no appreciable effect in preventing horrors from occurring—is the coming to the surface of a deep sense of piety towards life; and that at an epoch where thought at the more conscious and rationalizing level is being led more and more into denying that life has any "sacred" character; and it is in connection with this spontaneous piety, but as outraging it (and more often than not quite independently of any positive religious attachment, of any link with historical revelation), that these acts, which we have been the witnesses or victims of, seem to us to bear the undeniable mark of sin.

Whatever attempts there may have been in the past to justify war, or at least to recognize a certain spiritual value in war, we ought to proclaim as loudly as possible that war with the face it wears to-day is sin itself. But at the same time we cannot fail to recognize that war is becoming more and more an affair of technicians: it presents to-day the double aspect of destroying whole populations without distinction of age or sex, and of tending more and more to be conducted by a small number of individuals, powerfully equipped, who direct operations from the safe depths of their laboratories. The fate of war and that of technical advancement, in our time, whether or not this conjunction is a merely accidental one, seem to be inextricably linked; and it can be asserted even that, at least in our present phase of history, everything that gives a new impetus to technical research at the same time renders war more radically destructive, and bends it more and more inexorably to what, at the breaking point, would be quite simply the suicide of the human race.

In a strange way, this connection between technical progress and sin becomes clearer if we remember on the one hand that to-day only the State is rich enough to finance the gigantic laboratories in which the new physics is being applied and developed; and on the other hand that, in a world given over like our own to rival imperialisms, the State itself, that "Great Leviathan", to use the phrase of Hobbes, is inevitably led to demand that such researches should be directed towards everything that can increase the power of the State in its coming conflict with its rivals. It is in relation to these facts that we are forced to assert that the growing state-

---

☐ From *Man Against Mass Society*, trans. G. S. Fraser (South Bend, Ind.: Regnery/Gateway, Inc., 1952), pp. 60-66. Reprinted by permission of the publisher.

control of scientific and technical research is one of the worst calamities of our time.

When we reflect on it, however, this tragic situation of ours is very far from appearing a *natural* situation. We cannot say that the realm of the technical is evil in itself or that progress at the technical level ought, as such, to be condemned. Even to pretend that this were so would be to relapse into childishness. We can immediately see, even though it is perhaps impossible to discover the logical basis for this opinion, that it would be absurd to hope to solve the present crisis by closing down the factories and the laboratories for good and all. There is every reason to suppose, on the contrary, that such a step would be the starting point of an almost unimaginable regression for the human race.

The truth is that if we want to state the problem of the relationship between technical progress and sin in acceptable terms we must go back to first principles. In the last analysis, what *is* a technique? It is a group of procedures, methodically elaborated, and consequently capable of being taught and reproduced, and when these procedures are put into operation they assure the achievement of some definite concrete purpose. As I have just been saying, the realm of the technical, as thus defined, is not to be considered as evil in itself; if we think of it in itself, as I have already said, a technique is rather something good or the expression of something good, since it amounts to nothing more than a specific instance of our general application of our gift of reason to reality. To condemn technical progress is, therefore, to utter words empty of meaning. But from the point of view of truth, what we must do is not to cling to our abstract definition but rather to ask ourselves about the concrete relationship that tends to grow up between technical processes on the one hand and human beings on the other; and here things become more complicated.

In so far as a technique is something that we can acquire, it may be compared to a possession—like habit, which is at bottom itself already a technique. And we can at once see that if a man can become the slave of his habits, it is equally probable that he can become the prisoner of his techniques. But we have to go deeper. The truth is that a technique, for the man whose task it is *to invent it,* does not present itself simply as a means; for a time at least, it becomes an end in itself, since it has to be discovered, to be brought into being; and it is easy to understand how a mind absorbed in this task of discovery can be drawn away from any thought of the real purpose to which, in principle, this technique ought to be subordinate. To take a simple example, it is clear enough that a technician to whom, for one reason or another, travelling is impossible or forbidden, might nevertheless devote himself to the improvement of design in motor-cars. I should be tempted to say that all technical progress implies a certain moral and intellectual outlay (of attention, ingenuity, perseverance, and so on) which betrays itself by a feeling of power or of pride; in which fact, of course, there is nothing that is not usual and allowable. Such feelings are the natural accompaniment of inventive activity. But they become unnatural, as we have already seen, they lose their just pretext and their authenticity, in the case of the man who benefits from an invention without having made any contribution towards discovering and perfecting it. We can understand this if we think of the state of mind of certain motorists who acquire a kind of passion for their car, spend their time swapping one car for another, and thus become less and less capable of considering the car as what it is, a means for getting about. The lack of curiosity of the passionate motorist is a fact of common experience. But this remark has a much more general application, and is true for instance also of radio enthusiasts. What we are noticing here is the passage from the realm of the technical, properly so called, to that of a kind of idolatry of which technical products become the object or at least the occasion. And if we follow out this line of reflection, we can see that even this kind of idolatry can degenerate into something worse; it can become *autolatry,* worship of oneself, and often does so in those circles where people can get excited only about records, especially speed records. Certainly, there is a great deal here that we ought to go into more deeply; we could ask ourselves how it is that speed has come to be regarded as an end rather than a means, how it has come to be sought out for its own sake—and we ought to contrast such a state of mind with that of the traveller of the old days, and particularly of the pilgrim, for whom the very slowness of progress was linked to a feeling of veneration. The transformation that has taken place in these matters seems to have even metaphysical significance. In a very general way, we might say that the exaltation of speed records goes hand in hand with a weakening, an attenuation, of the sense of the sacred.

But let us consider another much more general and much more important aspect of the same phenomenon. One might say that the notion of technical progress, at least in our own day, implies above all the notion of progress in communications. The perfecting of means of transport has been to all appearances the condition (while at the same time, of course, one of the effects) of the industrialization which has been proceeding with an accelerating rhythm during the past century. But what we must concentrate our powers of reflection on is just this very notion of communication, taken in a quite external sense. That the world should cease to be divided into many little compartments, that the country folk, in particular, should cease to live, in their own little closed regions, an entirely local life, a life with no relation to that of other neighbouring groups, all that seems to me an infinitely happy transformation, and one which by itself would serve to justify the belief in progress. But we must be careful here. Naturally, it is true to say that this general development of communications *can* or *could*—ought to be able to—produce excellent results: that, for instance, where some new good thing has been discovered, the development of communications guarantees a widespread use of this good thing that would not have been imaginable a century ago. Let us think, for instance, of medicines (serums or penicillin) taken by aeroplane to sick people who, without such outside help, would undoubtedly have died. But this good possibility is only one possibility among many; we ought to ask ourselves whether there are not also evil possibilities whose very principle is to be found just in this perfecting of communication, in a quite external sense, of which we have been talking.

Do we not find, both on the world scale and at the level of national existence, that the development of communications entails a growing uniformity imposed upon our customs and habits? In other words, this perfecting of communications is achieved everywhere at the expense of an individuality which is tending to-day more and more to vanish away: and we are thinking here of beliefs, customs, traditions, as well as of local costumes, local craftsmanship, and so on. If we were taking a quite superficial view of human psychology and history, we might be tempted to say that this elimination of the picturesque is the unavoidable price that we pay for a greater good; for this reduction of habits to a general uniformity might, of course, be the beginning of a genuine unification of mankind. But our contemporary experience allows us to say quite definitely that there is nothing in this argument and that the imposing of uniformity, far from setting men on the path towards a kind of concrete assimilation of the universal, seems on the contrary to develop in them narrowly particular loyalties of a more and more aggressive sort, and to set competing groups against each other.

This might seem quite paradoxical, but reflection clears up the difficulty. Is it not obvious that technical and industrial progress have combined to create for men a kind of lowest common denominator of well-being which becomes an inspirer of covetousness and everywhere gives rise to envy? At the bottom, this lowest common denominator is merely wealth, one might say it is merely cash; but in saying that one should add that, by a very disturbing dialectical process, just as money becomes the lowest common denominator of well-being, so money itself tends to lose all substantial or even apparent reality, to become, in short, a fiction. After all, envy is only possible on the basis of what might be called a common drawing-account; it is less conceivable as existing between individuals and between peoples who have each their own traditions and their own separate genius, of which they are rightly proud. To be sure, this originality of each local and national tradition in respect to every other one has been very far, throughout history, from excluding quarrels and wars; up to a certain point, it has even encouraged them. But these quarrels, these wars, however bloody they may have been, did retain a human character; they did not exclude mutual respect, they made real reconciliations possible. There is nothing in them which at all resembles these attempts at collective extermination of which I spoke at the beginning of this chapter. But, besides all this, it would be of the greatest interest to discover by what odd mechanism ideological conflicts, to-day, conflicts sometimes quite without deep significance, have been able to superimpose themselves on elementary—and alimentary—antagonisms whose sole basis can finally be seen as envy.

It can, of course, always be claimed that this common drawing fund for envy, this lowest common denominator, however regrettable its immediate consequences may be, was none the less necessary, and that in the long run the current growth of uniformity will allow men to form a really organic and harmonious single body. It is difficult to make any judgment on such prophecies. But what must be recognized, it seems to me, speaking in all good faith, is that, if we consider things in a purely rational fashion, we can find no serious reason for expecting an *automatically* favourable outcome to the crisis which mankind is going through to-day. One cannot help observing that those ideological conflicts, which I have just been alluding to, tend to-day, so to say, to *make themselves at home* even in small country villages where, in the past, a friendly good-will prevailed and where to-day we can see the reign of mutual fear and suspicion. It is, of course, still possible to say that this is a purely transitional state of affairs; but the truth is that nobody sees how the state of affairs can be bettered in a way that would suit the aspirations of those who love peace and who also love what Victor Hugo called "concord among citizens". In reality, unless we have recourse to an act of faith, perfectly legitimate in itself and from the religious point of view even requisite, but quite foreign to the spirit of the man of mere technique, we should have to say that the malady from which mankind to-day appears to be suffering is perhaps mortal, and that there is nothing, at the purely human level, which insures our race against that risk of collective suicide . . .

RELIGIOUS

From "Man and Machine"

NICHOLAS BERDYAEV

■ Nicholas Berdyaev *(1874-1948) was born near Kiev in Russia. He was a philosopher and a religious thinker who bridged the gap between religious thought in Russia and the West. He was a leading exponent of Christian existentialism. As a youth he was associated with Marxism. But because his views developed along different lines, he was exiled after the revolution, and he went to Paris in 1924. In his philosophy, his major concern has always been man. He strongly advocated a society which gives each individual the ability to be both free and creative. Among his works are* The New Middle Ages *(1924) and* Slavery and Freedom *(1939).*

In the essay "Man and Machine" *(1933) Berdyaev wages a bitter attack on technique: "The world is being dehumanized as well as dechristianized by the monstrous power of technique." This power is seen as "bound up" not only with the machine, but with capitalism and communism, both of which are rejected by the author. According to Berdyaev, the problem with modern technology is that it deals "terrible blows to man's emotional side, to human feelings." Thus it is the heart and not the spirit which is primarily endangered today. But to combat this situation man must intensify his own spirituality. The human spirit "must not be isolated and dependent only upon itself—it must be united to God." This would strengthen it, and consequently enable it to limit the powers of technique, utilizing technique only for the good of humanity. In this Berdyaev seems to be in agreement with Marcel (see preceding selection).*

Wherein consists the menace of the machine to man, the danger now so clearly apparent? I doubt if it threatens spirit and the spiritual life, but the machine and technique deal terrible blows to man's emotional side, to human feeling, which is on the wane in contemporary civilization. Whereas the old culture threatened man's body, which is neglected and often debilitated, mechanical civilization endangers the heart, which can scarcely bear the contact of cold metal and is unable to live in metallic surroundings. The process of the destruction of the heart as the center of emotional life is characteristic of our times. In the works of such outstanding French writers as Proust and Gide the heart as an integral organ of man's emotional life is inexistent, everything has been decomposed into the intellectual element and sensual feelings. Keyserling is right when he speaks of the destruction of the emotional order in modern civilization and longs for its restoration.[1] Technique strikes fiercely at humanism, the humanist conception of the world, the humanist ideal of man and culture. It seems surprising at first to be told that technique is not so dangerous to

---

☐ From *"The Bourgeois Mind" and Other Essays,* trans. Countess Bennigsen, revised by Donald Attwater (New York: Books For Libraries Press, Inc., 1966), pp. 52-64. Reprinted by Books For Libraries Press. Distributed by Arno Press, Inc.

spirit, yet we may in truth say that ours is the age of technique and spirit, not an age of the heart. The religious significance of contemporary technique consists precisely in the fact that it makes everything a spiritual problem and may lead to the spiritualization of life, for it demands an intensification of spirituality.

Technique has long ceased to be neutral, to be indifferent to spirit and its problems, and, after all, can anything really be neutral? Some things may appear so at a casual glance but, while technique is fatal to the heart, it produces a powerful reaction of the spirit. Through technique man becomes a universal creator, for his former arms seem like childish toys in comparison with the weapons it places in his hands now. This is especially apparent in the field of military technique. The destructive power of the weapons of old was very limited and localized; with cannon, muskets, and sabers neither great human masses nor large towns could be destroyed nor could the very existence of civilization be threatened. All this is now feasible. Peaceful scientists will be able to promote cataclysms not only on a historical but on a cosmic scale; a small group possessing the secrets of technical inventions will be able to tyrannize over the whole of mankind; this is quite plausible, and was foreseen by Renan. When man is given power whereby he may rule the world and wipe out a considerable part of its inhabitants and their culture, then everything depends upon man's spiritual and moral standards, on the question: In whose name will he use this power—of what spirit is he?

Thus we see that this problem of technique inevitably becomes a spiritual and ultimately a religious one, and the future of the human race is in the balance. The miracles of technique are always double, and demand an intensification of the spirit infinitely greater than in former cultural ages. Man's spirituality can no more be organically vegetative; we are faced by the demands of a new herosim, internal and external. Our herosim, bound up with warfare in old times, is now no more; it scarcely existed in the last war; technique demands a new kind of heroism, and we are constantly hearing and reading of its manifestations—scientists leaving their laboratories and studies and flying into the stratosphere or diving to the bottom of the ocean. Human heroism is now connected with cosmic spheres. But, primarily, a strong spirit is needed in order to safeguard man from enslavement and destruction through technique, and in a certain sense we may say it is a question of life or death. We are sometimes haunted by a horrible nightmare: a time may come when machinery will have attained so great a perfection that man would have governed the world through it had he not altogether disappeared from the earth; machines will be working independently, without a hitch and with a maximum of efficiency and results; the last men will become like machines, then they will vanish, partly because they will be unnecessary and also because they will be unable to live and breathe any longer in the mechanized atmosphere; factories will be turning out goods at great speed and airplanes will be flying all over the earth; the wireless will be carrying the sound of music and singing and the speech of the men that once lived; nature will be conquered by technique and this new actuality will be a part of cosmic life. But man himself will be no more, organic life will be no more—a terrible utopia! It rests with man's spirit to escape this fate. The exclusive power of technical organization and machine production is tending toward its goal—inexistence within technical perfection. But we cannot admit an autonomous technique with full freedom of action: it *has* to be subordinated to spirit and the spiritual values of life—as everything else has to be. Only upon one condition can the human spirit cope with this tremendous problem: it must not be isolated and dependent only upon itself—it must be united to God. Then only can man preserve the image and likeness of his maker and be himself preserved. There is the divergence between Christian and technical eschatology.

The power of technique in human life results in a very great change in the prevalent type of religiousness, and we must admit that this is all for the good. In a mechanical age the hereditary, customary, formal, socially established sort of religion is weakened; the religiously-minded man feels less tied to traditional forms, his life demands a spiritually intensified Christianity, free from social influences. Religious life tends to become more personal, it is more painfully attained, and this is not individualism, for the universality and mystical unity of religious consciousness are not sociological.

Yet in another respect the domination of technique may be fatal for religious and spiritual life. Technique conquers time and radically alters our relations to it: man becomes capable of mastering time, but technical actuality subordinates him and his inward life to time's accelerating movement. In the crazy speed of contemporary civilization not one single instant is an end in itself and not a single moment can be fixed as being outside time. There is no exit into an instant *(Augenblick)* in the sense Kirkegaard speaks of it: every moment must speedily be replaced by the next, all remaining in the stream of time and therefore ephemeral. Within each moment there seems to be nothing but motion toward the next one: in itself it is void. Such a conquest of time through speed becomes an enslavement to the current of time, which means that in this relation technical activity is destructive of eternity. Man has no time for it, since what is demanded of him is the quickest passage to the succeeding instant. This does not mean that we must see in the past the eternal which is being destroyed by the future: the past does not belong to eternity any more than does the future—both are in time. In the past, as in the future and at all times, an exit into eternity, the self-sufficient complete instant, is always possible. Time obeys the speed machine, but is not mastered and conquered by it, and man is faced by the question: Will he remain capable of experiencing moments of pure contemplation, of eternity, truth, beauty, God? Unquestionably, man has an active vocation in the world and there is truth in action, but he is also a being capable of contemplation in which there is an element determining his ego. The very act of man's contemplation, his relation to God, contains a creative deed. The formulation of this problem more than ever convinces us that all the ills of modern civilization are due to the discrepancy between the organization of man's soul inherited from other ages and the new technical, mechanical actuality from which he cannot escape. The human soul is unable to stand the speed which contemporary civilization demands and which tends to transform man himself into a machine. It is a painful process. Contemporary man endeavors to strengthen his body through sports, thus fighting anthropological regression. We cannot deny the positive value of sport whereby man reverts to the old Hellenic view of the body, yet sport may become a means of destruction; it will create distortion instead of harmony if not subordinated to his integral idea. By its nature technical civilization is impersonal; it demands man's activity, while denying him the right to a personality, and therefore he experiences an immense difficulty in surviving in such a civilization. In every way *person* is in opposition to *machine*, for person is primarily unity in multiplicity and integrity, it is its own end and refuses to be transformed into a part, a means, an instrument. On the other hand, technical civilization and mechanized society demand that man should be that and nothing else: they strive to destroy his unity and integrity or, in other words, deny him his personality. A fight to the death between this civilization and society and the human person is inevitable; it will be man versus machine. Technique is pitiless to all that lives and exists, and therefore concern for the living and existing has to restrict the power of technique over life.

The machine-mind triumphing in a capitalist civilization begins by perverting the hierarchy of values, and the reinstatement of that hierarchy marks the limitation of the power of the machine. This cannot be done by a reversion to the old structure of the soul and the former natural and organic actuality.[2] The character of modern technical civilization and its influence upon man is inacceptable not only to Christian consciousness but also to man's natural dignity. We are faced by the task of saving the very image of man. He has been called to continue creation and his work represents the eighth day: he was called to be king and master of the earth, yet the work he is doing and to which he was called enslaves him and defaces his image. So a new man appears, with a new structure of the soul, a new image. The man of former days believed himself to be the everlasting man; he was mistaken, for though he possessed an eternal principle he was not eternal: the past is not eternity. A new man is due to appear in the world and the problem consists in the question, not of his relation to the old man, but of his relation to everlasting man, to the eternal in him—and this eternal principle is the divine image and likeness whereby he becomes a person. This is not to be understood statically, for the divine image in man, as in a natural being, is manifested and confirmed dynamically—in this consists the endless struggle against the old man in the name of the new man. But the machine age strives to replace the image and likeness of God by the image and likeness of the machine, and this does not mean the creation of a new man but the destruction and disappearance of man, his substitution by another being with another, non-human, existence. Man created the machine, and this may give him a grand feeling of his own dignity and power, but this pride imperceptibly and gradually leads to his humiliation. All through history man has been changing, he has always been old and new, but throughout the ages he was in contact with eternity and remained man. The new man will finally break away from eternity, will definitely fasten on to the new world he has to possess and conquer, and will cease to be human, though at first he will fail to realize the change. We are witnessing man's dehumanization, and the question is: Is he to be or not to be, not the ancient man who has to be outlived, but just simply man? From the very dawn of human consciousness, as manifested in the Bible and in ancient Greece, this problem has never been posited with such depth and acuteness. European humanism believed in the eternal foundations of human nature, and inherited this belief from the Greco-Roman world. Christianity believes man to be God's creation, bearing his image and likeness and redeemed by his divine son. Both these faiths strengthen European man, who believes himself to be universal, but now they have been shaken; the world is being dehumanized as well as dechristianized by the monstrous power of technique.

This power, like that of the machine, is bound up with capitalism; it originated in the very womb of the capitalist order, and the machine was the strongest weapon for its development. Communism has taken over these things wholesale from capitalist civilization and made a veritable religion of the machine: it worships it as a totem. Undoubtedly, since technique has created capitalism it may also help to conquer it and to create a less unjust social order: it may become a mighty arm in the solution of the social problem. But all will depend on the question, which spirit predominates, of which spirit man will be. Materialistic communism subordinates the problem of man, as a being composed of soul and body, to the problem of society; it is not for man to organize society, but for society to organize man. The truth is the other way around—primacy belongs to man; it is he who has to organize society and the world, and its organization is dependent upon his spirit. Here man is taken not as an individual being but as a social being with a social vocation to fulfill, since only then has he an active and creative vocation. In our days it is usual to hear people, victims of the machine, accuse it, making it responsible for their crippling; this only humiliates man and does not correspond to his dignity. It is not machinery, which is merely man's creation and consequently irresponsible, that is to be blamed, and it is unworthy to transfer responsibility from man to a machine. Man alone is to blame for the awful power that threatens him; it is not the machine which has despiritualized him—he did it himself. The problem has to be transferred from the outward to the inward. A limitation of the power of technique and machinery over human life is a mission of the spirit; therefore, man has to intensify his own spirituality. The machine can become, in human hands, a great asset for the conquest of the elements of nature on the sole condition that man himself becomes a free spirit. A wholesale process of dehumanization is going on and mechanicism is only the projection of this dehumanization. We can see this process in the dehumanizing of physical science. It studies invisible light rays and inaudible sounds, and thereby leads man beyond the limits of his familiar world of light and sound; Einstein carries him beyond the world of space. These discoveries have a positive value and witness to the strength of human consciousness. Dehumanization is a spiritual state, the relation of the spirit to man and to the world.

Christianity liberated man from the bonds of the cosmic infinity that enslaved the ancient world, from the power of natural spirits and demons; it set him upright, strengthened him, made him dependent upon God and not upon nature. But in the science which became accessible when man emancipated himself from nature, on the heights of civilization and technique, he discovers the mysteries of cosmic life formerly hidden from him and the action of energies formerly dormant in the depths of nature. This manifests his power, but it also places him in a precarious position in relation to the universe. His aptitude for organization disorganizes himself internally, and a new problem faces Christianity. Its answer to it presupposes a modification of Christian consciousness in the understanding of man's vocation in the world. The center is in the Christian view of man as such, for we can no longer be satisfied by the patristic, scholastic, or humanistic anthropologies. From the point of view of cognition, a philosophical anthropology becomes a central problem: man and machine, man and organism, man and cosmos, are what it has to deal with. In working out his historical destiny, man traverses many different stages, and invariably his fate is a tragic one. At first he was the slave of nature and valiantly fought for his own preservation, independence, and liberty. He created culture, states, national units, classes, only to become enslaved by his own creations. Now he is entering upon a new period and aims at conquering the irrational social forces; he establishes an organized society and a developed technique, but again becomes enslaved, this time by the machine into which society and himself are becoming transformed. In new and ever newer forms this problem of man's liberation, of his conquest of nature and society, is being restated, and it can only be solved by a consciousness which will place him above them, the human soul above all natural and social forces. Everything that liberates man has to be accepted, and that which enslaves him rejected. This truth about man, his dignity and his calling, is embodied in Christianity, though maybe it has been insufficiently manifested in history and often even perverted. The way of man's final liberation and realization of his vocation is the way to the kingdom of God, which is not only that of Heaven but also the realm of the transfigured earth, the transfigured cosmos.

NOTES

1. See his *Meditations Sud-Americaines*. [This is the French translation of Hermann Alexander Keyserling, *Sudamerikanische Meditationen* (Stuttgart: Deutsche Verlags-Anstalt, 1932). English translation by the author and Theresa Duerr, *South American Meditations on Hell and Heaven in the Soul of Man* (New York: Harper, 1932).]
2. The important book of Gina Lombroso, *La rançon du machinisme*, displays too great a faith in the possibility of a return to a pre-mechanical civilization. [Berdyaev refers to the French translation (Paris: Payot, 1931) of *Le tragedie del progresso* (Torino: Bocca, 1930). English translation by C. Taylor, *The Tragedies of Progress* (New York: Dutton, 1931).]

EXISTENTIAL

From "My Fellowman"

JEAN-PAUL SARTRE

■ Jean-Paul Sartre *(1905-1980) was born in Paris. A philosopher and a writer, Sartre has had the dominant influence in the development of Existentialism. In his works he expresses a passionate interest in human beings. Human beings are for him essentially free. They make themselves through their choices in life. His major work on that theme is* Being and Nothingness *(1943). Sartre fought with the French army in 1940 and was captured by the Germans. This experience tempered his claims about the absolute freedom of human beings and moved him close to a Marxist position. His work* Critique of Dialectic Reason *(1960) attempts to show the underlying harmony between Marxism and Existentialism. Nevertheless, he has been an outspoken critic of the French communist party.*

In this rich selection from Being and Nothingness, *Sartre speaks of the human being's existence as inextricably interwoven with technique. "I am the ends which I have chosen and the techniques which realize them." This is of course reminiscent of the view of another existentialist, Ortega y Gassett. Sartre adds that not only is one's being so related to technique, but in fact the external world as it appears to the individual is itself modified through techniques, for "my factual existence . . . involves my apprehension of the world and of myself through certain techniques." The collectivities to which an individual belongs (nationality, religion, and profession, among others) themselves require the use of certain techniques, so that "the only positive way which I have to exist my factual belonging to these collectivities is the use which I constantly make of the techniques which arise from them." But Sartre's final conclusion is that we are not enslaved by technique; in fact, the only possible foundation of technique is one's freedom to choose. Thus Sartre disagrees with those who argue for technological determinism.*

To live in a world haunted by my fellowman is not only to be able to encounter the Other at every turn of the road; it is also to find myself engaged in a world in which instrumental-complexes can have a meaning which my free project has not first given to them. It means also that in the midst of this world *already* provided with meaning I meet with a meaning which is *mine* and which I have not given to myself, which I discover that I "possess already." Thus when we ask what the original and contingent fact of existing in a world in which "there are" also Others can mean for our situation, the problem thus formulated demands that we study successively three layers of reality which come into play so as to constitute my concrete situation: instruments which are *already* meaningful (a station, a railroad sign, a work of art, a mobilization notice), the meaning which I discover as *already mine* (my nationality, my race, my physical appearance), and finally the Other as a center of reference to which these meanings refer.

Everything would be very simple if I belonged to a world whose meanings were revealed simply in the light of my own ends. In this case I would dispose of things as instruments or as instrumental complexes within the limits of my own choice of myself; it is this choice which would make of the mountain an obstacle difficult to overcome or a spot from which to get a good view of the landscape, etc; the problem would not be posed of knowing what meaning this mountain could have in *itself* since I would be the one by whom meanings come to reality in itself. The problem would again be very much simplified if I were a monad without doors or windows and if I merely knew in some way or other that other monads existed or were possible, each of them conferring new meanings on the things which I see. In this case, which is the one to which philosophers have too often limited themselves in their inquiry, it would be sufficient for me to hold other meanings as *possible,* and finally the plurality of meanings corresponding to the plurality of consciousnesses would coincide very simply for me with the possibility always open to me of making *another choice* of myself. But we have seen that this monadic conception conceals a hidden solipsism precisely because it is going to confuse the plurality of meanings which I can attach to the real and the plurality of meaningful systems each one of which refers to a consciousness which I am not. Moreover on the level of concrete experience this monadic description is revealed as inadequate. There exists, in fact, something in "my" world other than a plurality of possible meanings; there exist objective meanings which are given to me

☐ From *Being and Nothingness,* trans. Hazel E. Barnes (New York: The Citadel Press, 1968), pp. 485-501.

as not having been brought to light by me. I, by whom meanings come to things, I find myself engaged in an *already meaningful* world which reflects to me meanings which I have not put into it.

One may recall, for example, the innumerable host of meanings which are independent of *my choice* and which I discover if I live in a city: streets, houses, shops, streetcars and buses, directing signs, warning sounds, music on the radio, *etc*. In solitude, of course, I should discover the brute and unpredictable existence—*this* rock, for example—and I should limit myself, in short, to making *there be* a rock; that is, that there should be *this* existent here and outside of it nothing. Nevertheless I should confer on it its meaning as "to be climbed," "to be avoided," "to be contemplated," *etc*. When there were the street curves, I discover a building, it is not only a brute existent which I reveal in the world; I do not only cause there to be a "this" qualified in this or that way; but the meaning of the object which is revealed then resists me and remains independent of me. I discover that the property is an apartment house, or a group of offices belonging to the Gas Company, or a prison, *etc*. The meaning here is contingent, independent of my choice; it is presented with the same indifference as the reality of the in-itself; it is made a *thing* and is not distinguished from the *quality* of the in-itself.[1] Similarly the coefficient of adversity in things is revealed to me before being experienced by me. Hosts of notices put me on my guard: "Reduce Speed. Dangerous curve," "Slow. School," "Danger," "Narrow Bridge 100 feet ahead," *etc*. But these meanings while deeply imprinted on things and sharing in their indifferent exteriority—at least in appearance—are nonetheless indications for a conduct to be adopted, and they directly concern me. I shall cross the street in the lanes indicated. I shall go into this particular shop to buy this particular instrument, and a page with directions for using it is given to buyers. Later I shall use this instrument, a pen, for example, to fill out this or that printed form under determined conditions.

Am I not going to find in all this strict limits to my freedom? If I do not follow point by point the directions furnished by others, I shall lose my bearings, I shall take the wrong street, I shall miss my train, *etc*. Moreover these notices are most often imperatives: "Enter here," "Go out here." Such is the meaning of the words "Entrance" and "Exit" painted over doorways. I obey. They come to add to the coefficient of adversity which I cause to be born in things, a strictly human coefficient of adversity. Furthermore if I submit to this organization, I depend on it. The benefits which it provides me can cease; come civil disturbance, a war, and it is always the items of prime necessity which become scarce without my having any hand in it. I am dispossessed, arrested in my projects, deprived of what is necessary in order for me to accomplish my ends. In particular we have observed that directions, instructions, orders, prohibitions, billboards are addressed to me in so far as I am just *anybody;* to the extent that I obey them, that I fall into line, I submit to the goals of a human reality which is just anybody and I realize them by just any techniques. I am therefore modified in my own being since I *am* the ends which I have chosen and the techniques which realize them—to any ends whatsoever, to any techniques whatsoever, any human reality whatsoever. At the same time since the world never appears except through the techniques which I use, the world—it also—is modified. This world, seen through the use which I make of the bicycle, the automobile, the train in order that I may traverse the world, reveals to me a countenance strictly correlative with the means which I employ; therefore it is *the countenance which the world offers to everybody*. Evidently it must follow, someone will say, that my freedom escapes me on every side; there is no longer a *situation* as the organization of a meaningful world around the free choice of my spontaneity; there is a *state* which is imposed upon me. It is this problem which we must now examine.

There is no doubt that my belonging to an inhabited world has the value of a *fact*. It refers to the original fact of the Other's presence in the world, a fact which, as we have seen, can not be deduced from the ontological structure of the for-itself.[2] And although this fact only makes our facticity more deep-rooted, it does not evolve from our facticity in so far as the latter expresses the necessity of the contingency of the for-itself. Rather we must say: the for-itself *exists in fact;* that is, its existence can not be identical with a reality engendered in conformity to a law, nor can it be identical with a free choice. And among the factual characteristics of this "facticity"—*i.e.,* among those which can neither be deduced nor proven but which simply "let themselves be seen"—there is one of these which we call the existence-in-the-world-in-the-presence-of-others. Whether this factual characteristic does or does not need to be recovered by my freedom in order to be efficacious in any manner whatsoever is what we shall discuss a little later. Yet the fact remains that on the level of techniques of appropriating the world, the very *fact* of the Other's existence results in the fact of the collective ownership of techniques. Therefore facticity is expressed on this level by the fact of my appearance in a world which is revealed to me only by collective and already constituted techniques which aim at making me apprehend the world in a form whose meaning has been defined outside of me. These techniques are going to determine my belonging to collectivities: to the *human race,* to the national collectivity, to the professional and to the family group.

It is even necessary to underscore this fact further: outside of my being-for-others—of which we shall speak later—the only positive way which I have *to exist my factual belonging* to these collectivities is the use which I constantly make of the techniques which arise from them. Belonging to the *human race* is defined by the use of very elementary and very general techniques: to know how to walk, to know how to take hold, to know how to pass judgment on the surface and the relative size of perceived objects, to know how to speak, to know how in general to distinguish the true from the false, *etc.* But we do not possess these techniques in this abstract and universal form: to know how to speak is not to know how to pronounce and understand words in general; it is to know how to speak a certain language and by it to manifest one's belonging to humanity *on the level of* the national collectivity. Moreover to know how to speak a language is not to have an abstract and pure knowledge of the language as it is defined by academic dictionaries and grammars; it is to make the language one's own across the peculiar changes and emphasis brought in by one's province, profession, and family. Thus it can be said that the *reality* of our belonging to the human is our *nationality* and that the reality of our nationality is our belonging to the family, to the region, to the profession, *etc.* in the sense that the *reality* of speech is language and that the reality of language is dialect, slang, jargon, *etc.* And conversely the *truth* of the dialect is the language, the *truth* of the language is speech. This means that the concrete techniques by which we manifest our belonging to the family and to the locality refer us to more abstract and more general structures which constitute its meaning and essence; these refer to others still more general until we arrive at the universal and perfectly simple essence of any technique whatsoever by which any being whatsoever appropriates the world.

Thus to be French, for example, is only the *truth* of being a Savoyard. But to be a Savoyard is not simply to inhabit the high valleys of Savoy; it is, among a thousand other things, to ski in the winters, to use the ski as a mode of transportation. And precisely, it is to ski according to the French method, not that of Arlberg or of Norway.[3] But since the mountain and the snowy slopes are apprehended only through a technique, this is precisely to discover the *French* meaning of ski slopes. In fact according to whether one will employ the Norwegian method, which is better for gentle slopes, or the French method which is better for steep slopes, the same slope will appear as steeper or more gentle exactly as an upgrade will appear as more or less steep to the bicyclist according to whether he will "put himself into neutral or low gear." Thus the French skier employs a French "gear" to descend the ski fields, and this "gear" reveals to him a particular type of slope wherever he may be. This is to say that the Swiss or Bavarian Alps, the Telemark, or the Jura will always offer to him a meaning, difficulties, an instrumental complex, or a complex of adversity which are purely French. Similarly it would be easy to show that the majority of attempts to define the working class amount to taking as a criterion production, consumption or a certain type of *Weltanschauung* springing out of an inferiority complex (Marx-Halbwachs-de Man); that is, in all cases certain techniques for the elaboration or the appropriation of the world across which there is offered what we shall be able to call the "proletarian countenance" with its violent oppositions, its great uniform and desert masses, its zones of shadow and its shores of light, the simple and urgent ends which illuminate it.

Now it is evident that although my belonging to a particular class or nation does not derive from my facticity as an ontological structure of my for-itself, my factual existence—*i.e.,* my birth and my place—involves my apprehension of the world and of myself through certain techniques. Now these techniques which I have not chosen confer on the world its meanings. It appears that it is no longer I who decide in terms of my ends whether the world appears to me with the simple, well-marked oppositions of the "proletarian" universe or with the innumerable interwoven nuances of the "bourgeois" world. I am not only thrown face to face with the brute existent. I am thrown into a worker's world, a French world, a world of Lorraine or the South, which offers me its meanings without my having done anything to disclose them.

Let us look more closely. We showed earlier that my nationality is only the *truth* of my belonging to a province, to a family, to a professional group. But must we stop there? If the language is only the truth of the dialect, is the dialect absolutely concrete reality? Is the professional jargon as "they" speak it, or the Alsatian dialect as a linguistic and statistical study enables us to determine its laws—is this the primary phenomenon, the one which finds its foundation in pure fact, in original contingency? Linguistic research can be mistaken here; statistics bring to light constants, phonetic or semantic changes of a given type; they allow us to reconstruct the evolution of a phoneme or a morpheme in a given period so that it appears that the *word* or the *syntactical* rule is an individual reality with its meaning and its history. And in fact individuals seem to have little influence over the evolution of language. Social facts such as invasions, great thoroughfares, commercial relations seem to be the essential causes of linguistic changes. But this is because the question is not placed on the true level of the concrete. Also we find only what we are looking for.

For a long time psychologists have observed that the *word* is not the concrete element of speech—not even the word of the dialect or the word of the family with its particular variation; the elementary structure of speech is the *sentence*. It is within the sentence, in fact, that the word can receive a real function as a designation; outside of the sentence the word is just a propositional function—when it is not a pure and simple rubric designed to group absolutely disparate meanings. Only when it appears in discourse, does it assume a "holophrastic" character, as has often been pointed out. This does not mean that the word can be limited by itself to a precise meaning but that it is integrated in a context as a secondary form in a primary form. The word therefore has only a purely *virtual* existence outside of complex and active organizations which integrate it. It can not exist "in" a consciousness or an unconscious *before* the use which is made of it: the sentence is not *made out of words*. But we need not be content with this. Paulhan has shown in *Fleurs de Tarbes* that entire sentences, "commonplaces," do not, any more than words, pre-exist the use which is made of them. They are mere commonplaces if they are looked at from the outside by a reader who recomposes the paragraph by passing from one sentence to the next, but they lose their banal and conventional character if they are placed within the point of view of the author who saw *the thing to be expressed* and who attended to the most pressing things first by producing an act of designation or recreation without slowing down to consider the very elements of this act. If this is true, then neither the words nor the syntax, nor the "ready-made sentences" pre-exist the use which is made of them. Since the verbal unity is the meaningful sentence, the latter is a constructive act which is conceived only by a transcendence which surpasses and nihilates the given toward an end. To understand the word in the light of the sentence is *very exactly* to understand any given whatsoever in terms of the situation and to understand the situation in the light of the original ends.

To understand a sentence spoken by my companion is, in fact, to understand what he "means"—that is, to espouse his movement of transcendence, to throw myself with him toward possibles, toward ends, and to return again to the ensemble of organized means so as to understand them by their function and their end. The spoken language, moreover, is always interpreted in terms of the situation. References to the weather, to time, to place, to the environment, to the situation of the city, of the province, of the country are given before the word.[4] It is enough for me to have read the papers and to have *seen* Pierre's healthy appearance and anxious expression in order for me to understand the "Things aren't so good" with which he greets me this morning. It is not his health which "is not so good" since he has a rosy complexion, nor is it his business nor his household; it is the situation of our city or of our country. I *knew it already*. In asking him, "How goes it?", I was already outlining an interpretation of his reply; I transported myself already to the four corners of the horizon, ready to *return* from there to Pierre in order to understand him. To listen to conversation is to "speak with," not simply because we imitate in order to interpret, but because we originally project ourselves toward the possibles and because we must understand *in terms of the world*.

But if the sentence pre-exists the word, then we are referred to the *speaker* as the concrete foundation of his speech. A word can indeed seem to have a "life" of its own if one comes upon it in sentences of various epochs. This borrowed life resembles that of an object in a film fantasy; for example, a knife which by itself starts slicing a pear. It is effected by the juxtaposition of instantaneities; it is cinematographic and is constituted in universal time. But if words appear to live when one projects a semantic or morphological film, they are not going to constitute whole sentences; they are only the tracks of the passage of sentences as highways are only the tracks of the passage of pilgrims or caravans. The sentence is a project which can be interpreted only in terms of the nihilation of a given (the very one which one wishes to *designate*) in terms of a posited end (its *designation* which itself supposes other ends in relation to which it is only a means). If the given can not determine the sentence any more than the word can, if on the contrary the sentence is necessary to illuminate the given and to make the word understandable, then the sentence is a moment of the free choice of myself, and it is as such that it is understood by my companion. If a language is the reality of speech, if a dialect or jargon is the reality of language, then the reality of the dialect is the *free act* of designation by which I choose myself as *designating*. And this free act can not be an *assembling* of words. To be sure, if it were a pure assembling or words in conformity with technical prescriptions (grammatical laws), we could speak of factual limits imposed on the freedom of the speaker; these limits would be marked by the material and phonetic nature of the words, the vocabulary of the language employed, the personal vocabulary of the speaker (the *n* words which he has at his command), the "spirit of the language," *etc., etc.* But we have just shown that such is not the case. It has been maintained recently that there is a sort of living order of words, of the dynamic laws of speech, an impersonal life of the logos—in short that speech is a Nature and that to some extent man must obey it in order to make use of it as he does with Nature.[5] But this is because people in considering speech frequently will take speech that is *dead* (*i.e.*, already spoken) and infuse into it an impersonal life and force, affinities and repulsions all of which in fact have been borrowed from the personal freedom of the for-itself which spoke. People have made of speech a *language which speaks all by itself*. This is an error which should not be made with regard to speech or any other technique. If we are to make man arise in the midst of techniques which are applied all by themselves, of a language which speaks itself, of a science which constructs itself, of a city which builds itself according to its own laws, if meanings are fixed in in-itself while we preserve a human transcendence, then the role of man will be reduced to that of a pilot employing the determined forces of winds, waves, and tides in order to direct a ship. But gradually each technique in order to be directed toward human ends will require another technique; for example, to direct a boat, it is necessary to speak. Thus we shall perhaps arrive at the technique of techniques—which in turn will be applied by itself—but we shall have lost forever the possibility of meeting the technician.

If on the other hand, it is by speaking that we cause words to exist, we do not thereby suppress the *necessary technical* connections or the connections *in fact* which are articulated inside the sentence. Better yet, *we found* this necessity. But in order for it to appear, in order for words to enter into relations with one another, in order for them to latch on to one another or repulse one another, it is necessary that they be united in a synthesis which does not come from them. Suppress this synthetic unity and the block which is called "speech" disintegrates; each word returns to its solitude and at the same time loses its unity, being parcelled out among various incommunicable meanings. Thus it is within the free project of the sentence that the laws of speech are organized; it is by speaking that I make grammar. Freedom is the only possible foundation of the laws of language.

Furthermore, *for whom* do the laws of language exist? Paulhan has given the essential answer: they are not for the one who speaks, they are for the one who listens. The person who speaks is only the choice of a *meaning* and apprehends the order of the words only in so far as he *makes it*.[6] The only relations which he will grasp within this organized complex are specifically those which he has established. Consequently if we discover that two (or several) words hold between them not *one* but several defined relations and that there results from this a multiplicity of meanings which are arranged in an hierarchy or opposed to each other—all for one and the same sentence—if, in short, we discover the "Devil's share," this can be only under the two following conditions: (1) The words must have been assembled and presented by a meaningful rapprochement; (2) this synthesis must be seen from *outside—i.e.*, by *The Other* and in the course of a hypothetical deciphering of the possible meanings of this rapprochement. In this case, in fact, each word grasped *first* as a square of meaning is bound to another word similarly apprehended. And the rapprochement will be multivocal. The apprehension of the *true* meaning (*i.e.*, the one expressly willed by the speaker) will be able to put other meanings in the shade or subordinate them, but it will not suppress them. Thus speech, which is a free project *for me*, has specific laws *for others*. And these laws themselves can come into play only within an original synthesis.

Thus we can grasp the clear distinction between the event "sentence" and a natural event. The natural fact is produced in conformity to a law which it manifests but which is a purely external rule of production of which the considered fact is only one example. The "sentence" as an event contains within itself the law of its organization, and it is inside the free project of *designating* that legal (*i.e.*, grammatical) relations can arise between the words. In fact, there can be no laws of speaking before one speaks. And each utterance is a free project of designation issuing from the choice of a personal for-itself and destined to be interpreted in terms of the global situation of this for-itself. What is primary is the situation in terms of which I understand the *meaning* of the sentence; this meaning is not in itself to be considered as a given but rather as an end chosen in a free surpassing of means. Such is the only *reality* which the working linguist can encounter. From the standpoint of this reality a regressive analytical work will be able to bring to light certain more general and more simple structures which are like legal schemata. But these schemata which would function as laws of dialect, for example, are in themselves abstract. Far from presiding over the constitution of the sentence and being the mould into which it flows, they exist only in and through this sentence. In this sense the sentence appears as a free invention of its own laws. We find here simply the original characteristic of every situation; it is by its very surpassing of the given as such (the linguistic apparatus) that the free project of the sentence causes the given to appear as *this* given (these laws of word order and dialectal pronunciation). But the free project of the sentence is precisely a scheme to assume *this given;* it is not just any assumption but is aimed at a not yet existing end across existing means on which it confers their exact meaning as a means.

Thus the sentence is the order of words which become *these words* only by means of their very order. This is indeed what linguists and psychologists have perceived, and their embarrassment can be of use to us here as a counter-proof; they believed that they discovered a circle in the formulation of speaking, for in order to speak it is necessary to know one's thought. But how can we know this thought as a reality made explicit and fixed in concepts except precisely by speaking it? Thus speech refers to thought and thought to speech. But we understand now that there is no circle or rather that this circle—from which linguists and psychologists believed they could escape by the invention of pure psychological idols such as the verbal image or an imageless, wordless thought—is not unique with speech; it is the characteristic of the situation in general. It means nothing else but the ekstatic connection of the present, the future, and the past—that is, the free determination of the existent by the not-yet-existing and the determination of the non-yet-existing by the existent. Once we have established this fact, it will be permissible to uncover abstract operational schemata which will stand as the legal truth of the sentence: the dialectal schema—the schema of the national language—the linguistic schema in general. But these schemata far from pre-existing the concrete sentence are in themselves affected with *Unselbständigkeit* and exist always incarnated and sustained in their very incarnation by a freedom.[7]

It must be understood, of course, that speech is here only the example of one social and universal technique. The same would be true for any other technique. It is the blow of the axe which reveals the axe, it is the hammering which reveals the hammer. It will be permissible in a particular run to reveal the French method of skiing and in this method the general skill of skiing as a human possibility. But this human skill is never anything by itself alone; it exists only *potentially;* it is incarnated and manifested by the *actual* and concrete skill of the skier. This enables us to outline tentatively a solution for the relations of the individual to the race. Without the human race, mankind, there is no truth; that is certain. There would remain only an irrational and contingent swarming of individual choices to which no law could be assigned. If some sort of truth exists capable of unifying the individual choices, it is the human race which can furnish this truth for us. But if the race is the truth of the individual, it can not be a *given* in the individual without profound contradiction. As the laws of speech are sustained by and incarnated in the concrete free project of the sentence, so the human race (as an ensemble of peculiar techniques to define the activity of men) far from pre-existing an individual who would manifest it in the way that this particular fall exemplifies the law of falling bodies, is the ensemble of abstract relations sustained by the free individual choice. The for-itself in order to choose itself as a *person* effects the existence of an internal organization which the for-itself surpasses toward itself, and this internal technical organization is in it the national or the human.

Very well, someone will say. But you have dodged the question. For these linguistic organizations or techniques have not been created by the for-itself so that it may find itself; it has got them from others. The rule for the agreement of participles does not exist, I admit, outside of the free rapprochement of concrete participles in view of an end with a particular designation. But when I employ this rule, I have learned it from others; it is because others in their personal projects cause it to be that I make use of it myself. My speech is then subordinated to the speech of others and ultimately to the national speech.

We should not think of denying this fact. For that matter our problem is not to show that the for-itself is the free foundation of its being; the for-itself is free but *in condition,* and it is the relation of this condition to freedom that we are trying to define by making clear the meaning of the situation. What we have just established, in fact, is only a part of reality. We have shown that the existence of meanings which do not emanate from the for-itself can not constitute an external limit of its freedom. As a for-itself one is not man first in order to be oneself subsequently and one does not constitute oneself as oneself in terms of a human essence given *a priori.* Quite the contrary, it is in its effort to choose itself as a personal self that the for-itself sustains in existence certain social and abstract characteristics which make of it *a man* (or a woman); and the necessary connections which accompany the essential elements of man appear only on the foundation of a free choice; in this sense each for-itself is responsible in its being for the existence of a human race. But it is necessary for us again to stress the undeniable fact that the for-itself can choose itself only beyond certain meanings of which it is not the origin. Each for-itself, in fact, is a for-itself only by choosing itself beyond nationality and race just as it speaks only by choosing the designation beyond the syntax and morphemes. This "beyond" is enough to assure its total independence in relation to the structures which it surpasses; but the fact remains that it constitutes itself as *beyond* in relation to *these* particular structures. What does this mean? It means that the for-itself arises in a world which is a world for other for-itselfs. Such is the *given.* And thereby, as we have seen, the meaning of the world is *alien* to the for-itself. This means simply that each man finds himself in the presence of *meanings* which do not come into the world through him. He arises in a world which is given to him as *already looked-at,* furrowed, explored, worked over in all its meanings, and whose very contexture is already defined by these investigations. In the very act by which he unfolds his time, he temporalizes himself in a world whose temporal meaning is already defined by other temporalizations: this is the fact of simultaneity. We are not dealing here with a limit of freedom; rather it is *in this world* that the for-itself must be free; that is, it must choose itself by taking into account these circumstances and not *ad libitum.* But on the other hand, the for-itself—*i.e.,* man—in rising up *does not merely suffer* the Other's existence; he is compelled to make the Other's existence manifest to himself in the form of a choice. For it is by a choice that he will apprehend the Other as The-Other-as-subject or as The-Other-as-object.[8] Inasmuch as the Other is for him the Other-as-a-look, there can be no question of *techniques* or of foreign meanings; the for-itself experiences itself as an object in the Universe beneath the Other's look. But as soon as the for-itself by surpassing the Other toward its ends makes of him a transcendence-transcended, that which was a free surpassing of the given toward ends appears to it as meaningful, given conduct in the world (fixed in in-itself). The Other-as-object becomes an *indicator of ends* and by its own free project, the For-itself throws itself into a world in which conducts-as-objects designate ends. Thus the Other's presence as a transcended transcendence reveals *given* complexes of means to ends. And as the end decides the means and the means the end by its upsurge in the face of the Other-as-object, the For-itself causes ends in the world to be indicated to itself; it comes into a world peopled by ends. But if consequently the techniques and their ends arise in the look of the For-itself, we must necessarily recognize that it is by means of the free assumption of a position by the For-itself confronting the Other that they become *techniques*. The Other by himself alone can not cause these projects to be revealed to the For-itself as techniques; and due to this fact there *exists for the Other* in so far as he transcends himself toward his possibles, *no technique* but a concrete *doing* which is defined in terms of his individual end. The shoe-repairer who puts a new sole on a shoe does not experience himself as "in the process of applying a technique;" he apprehends the situation as demanding this or that action, that particular piece of leather, as requiring a hammer, *etc*. The For-itself as soon as it assumes a position with respect to the Other, causes techniques to arise in the world as *the conduct of the Other as a transcendence-transcended*. It is at this moment and at this moment only that there appear in the world—bourgeois and workers, French and Germans, in short, men.

Thus the For-itself is responsible for the fact that the Other's conduct is revealed in the world as techniques. The for-itself can not cause the world in which it arises to be furrowed by *this or that particular technique* (it can not make itself appear in a world which is "capitalistic" or "governed by a natural economy" or in a "parasitic civilization"), but it causes that which is lived by the Other as a free project to exist *outside* as technique; the for-itself achieves this precisely by making itself the one by whom an outside comes to the Other. Thus it is by choosing itself and by historicizing itself in the world that the For-itself historicizes the world itself and causes it to be *dated* by its techniques. Henceforth, precisely because the techniques appear as objects, the For-itself can choose to appropriate them. By arising in a world in which Pierre and Paul speak in a certain way, stick to the right when driving a bicycle or a car, *etc.*, and by constituting these free patterns of conduct into meaningful objects, the For-itself is responsible for the fact that *there is* a world in which *they* stick to the right, in which *they* speak French, *etc*. It causes the internal laws of the Other's act, which were originally founded and sustained by a freedom engaged in a project, to become now objective rules of the conduct-as-object; and these rules become universally valid for all analogous conduct, while the supporter of the conduct or the agent-as-object becomes simply *anybody*. This historization, which is the effect of the for-itself's free choice, in no way restricts its freedom; quite the contrary, it is *in this world* and no other that its freedom comes into play; it is in connection with its existence in *this* world that it puts itself into question. For to be free is not to choose the historic world in which one arises—which would have no meaning—but to choose oneself in the world whatever this may be.

In this sense it would be absurd to suppose that a certain *state* of techniques is restrictive to human possibilities. Of course a contemporary of Duns Scotus is ignorant of the use of the automobile or the airplane; but he appears as ignorant *to us* and only from our point of view because we privately apprehend him in terms of a world where the automobile and the airplane exist. For him, who has no relation of any kind with these objects and the techniques which refer to them, there exists a kind of absolute, unthinkable, and undecipherable nothingness. Such a nothingness *can in no way limit* the For-itself which is choosing itself; it can not be apprehended as a lack, no matter how we consider it. The For-itself which historicizes itself in the time of Duns Scotus therefore nihilates itself in the heart of a fullness of being—that is, of a world which like ours is *everything which it can be*. It would be absurd to declare that the Albigenses lacked heavy artillery to use in resisting Simon de Montfort; for the Seigneur de Trencavel or the Comte de Toulouse chose themselves such as they were in a world in which artillery had no place; they viewed politics in that world; they made plans for military resistance in that world; they chose themselves as sympathizers with the Cathari *in that world;* and as they were only what they chose to be, they were *absolutely* in a world as absolutely full as that of the Panzer-divisionen or of the R.A.F.

What is true for material techniques applies as well to more subtle techniques. The fact of existing as a petty noble in Languedoc at the time of Raymond VI is not *determining* if it is placed *in the feudal world* in which this lord exists and in which he chooses himself. It appears as privative only if we commit the error of considering this division of *Francia* and of the Midi from the actual point of view of French unity. The feudal world offered to the vassal lord of Raymond VI infinite possibilities of choice; we do not possess more. A question just as absurd is often posited in a kind of utopian dream: what would Descartes have been if he had known of contemporary physics? This is to suppose that Descartes possesses an *a priori* nature more or less limited and altered by the state of science in his time and that we could transport this brute nature to the contemporary period in which it would react to more extensive and more exact knowledge. But this is to forget that Descartes is what he has chosen to be, that he is an absolute choice of himself from the standpoint of a world of various kinds of knowledge and of techniques which this choice both assumes and illuminates. Descartes is an absolute upsurge at an absolute date and is perfectly unthinkable at another date, for he has made his date by making himself. It is he and not another who has determined the exact state of the mathematical knowledge immediately before him, not by an empty inventory which would be made from no point of view and would be related to no axis of coordination, but by establishing the principles of analytical geometry—that is, by inventing precisely the axis of coordinates which would permit us to define the state of this knowledge. Here again it is free invention and the future which enable us to illuminate the present; it is the perfecting of the technique in view of an end which enables us to evaluate the state of the technique.

Thus when the For-itself affirms itself in the face of the Other-as-object, by the same stroke it reveals *techniques*. Consequently it can appropriate them—that is, *interiorize* them. But suddenly there are the following consequences: (1) By employing a technique, the For-itself surpasses the technique toward its own end; it is always beyond the technique which it employs. (2) The technique which was originally a pure, meaningful conduct fixed in some Other-as-object, now, because it is interiorized, loses its character as a technique and is integrated purely and simply in the free surpassing of the given toward ends; it is recovered and sustained by the freedom which founds it exactly as dialect or speech is sustained by the free project of the sentence. Feudalism as a technical relation between man and man does not exist; it is only a pure abstract, sustained and surpassed by the thousands of individual projects of a particular man who is a liege in relation to his lord. By this we by no means intend to arrive at a sort of historical nominalism. We do not mean that feudalism is the sum of the relations of vassals and suzerains. On the contrary, we hold that it is the abstract structure of these relations; every project of a man of this time must be realized as a surpassing toward the concrete of this abstract moment. It is therefore not necessary to generalize in terms of numerous detailed experiences in order to establish the principles of the feudal technique; this technique exists necessarily and completely in each individual conduct, and it can be brought to light in each case. But it is there only to be surpassed. In the same way the For-itself can not be a person—*i.e.*, choose the ends which it is—without being a man or woman, a member of a national collectivity, of a class, of family, *etc*. But these are abstract structures which the For-itself sustains and surpasses by its project. It makes itself French, a man of a southern province, a workman to order to be *itself* at the horizon of these determinations. Similarly the world which is revealed to the For-itself appears as provided with certain meanings correlative with the techniques adopted. It appears as a world-for-the-Frenchman, a world-for-the-worker, *etc.*, with all the characteristics which would be expected. But these characteristics do not possess *Selbständigkeit*. The world which allows itself to be revealed as French, proletarian, *etc.*, is before all else a world which is illuminated by the For-itself's own ends, its own world.

Nevertheless the Other's existence brings a factual limit to my freedom. This is because of the fact that by means of the upsurge of the Other there appear certain determinations which I *am* without having chosen them. Here I am—Jew, or Aryan, handsome or ugly, one-armed, *etc*. All this I am *for the Other* with no hope of apprehending this meaning which I have *outside* and, still more important, with no hope of changing it. Speech alone will inform me of what I am; again this will never be except as the object of an empty intention; any intuition of it is forever denied me. If my race or my physical appearance were only an image in the Other or the Other's opinion of me, we should soon have done with it; but we have seen that we are dealing with objective characteristics which define me in my being-for-others. As soon as a freedom other than mine arises confronting me, I begin to exist in a new dimension of being; and this time it is not a question of my conferring a meaning on brute existents or of accepting responsibility on my own account for the meaning which Others have conferred on certain objects. It is I myself who see a meaning conferred upon me, and I do not have the recourse of accepting the responsibility for this meaning which I have since it can not be given to me except in the form of an empty indication. Thus something of myself—according to this new dimension—exists in the manner of the *given;* at least *for me,* since this being which I am *is suffered,* it *is* without *being existed*. I learn of it and suffer it in and through the relations which I enter into with others, in and through their conduct with regard to me. I encounter this being at the origin of a thousand prohibitions and a thousand resistances which I bump up against at each instant: Because I am a *minor* I shall not have this or that privilege. Because *I am a Jew* I shall be deprived—in certain societies—of certain possibilities, *etc*. Yet I am unable *in any way* to feel myself as a Jew or as a minor or as a Pariah. It is at this point that I can react against these interdictions by declaring that race, for example, is purely and simply a collective fiction, that only individuals exist. Thus here I suddenly encounter the total alienation of my person: I am something which I have not chosen to be. What is going to be the result of this for the situation?

We must recognize that we have just encountered a *real* limit to our freedom—that is, a way of being which is imposed on us without our freedom being its foundation. Still it is necessary to understand this: the limit imposed does not come from the *action* of others. In a preceding chapter we observed that even torture does not dispossess us of our freedom; when we give in, we do so *freely*. In a more general way the encounter with a prohibition in my path ("No Jews allowed here," or "Jewish restaurant. No Aryans allowed," *etc.*) refers us to the case considered earlier (collective techniques), and this prohibition can have meaning only on and through the foundation of my free choice. In fact according to the free possibilities which I choose, I can disobey the prohibition, pay no attention to it, or, on the contrary, confer upon it a coercive value which it can hold only because of the weight which I attach to it. Of course the prohibition fully retains its character as an "emanation from an alien will;" of course it has for its specific structure the fact of *taking me for an object* and thereby manifesting a transcendence which transcends me. Still the fact remains that it is not incarnated in *my* universe, and it loses its peculiar force of compulsion only within the limits of my own choice and according to whether under any circumstances I prefer life to death or whether, on the contrary, I judge that in certain particular cases death is preferable to certain kinds of life, *etc*. The true limit of my freedom lies purely and simply in the very fact that an Other apprehends me as the Other-as-object and in that second corollary fact that my situation ceases for the Other to be a situation and becomes an objective form in which I exist as an objective structure. It is this alienating process of making an object of my situation which is the constant and specific limit of my situation, just as the making an object of my being-for-itself in being-for-others is the limit of my being. And it is precisely these two characteristic limits which represent the boundaries of my freedom.

In short, by the fact of the Other's existence, I exist in a situation which *has an outside* and which due to this very fact has a dimension of alienation which I can in no way remove from the situation any more than I can act directly upon it. This limit to my freedom is, as we see, posited by the Other's pure and simple existence—that is, by the *fact* that my transcendence exists for a transcendence. Thus we grasp a truth of great importance: we saw earlier, keeping ourselves within the compass of existence-for-itself, that only my freedom can limit my freedom; we see now, when we include the Other's existence in our considerations, that my freedom on this new level finds its limits also in the existence of the Other's freedom. Thus on whatever level we place ourselves, the only limits which a freedom can encounter are found in freedom. Just as thought according to Spinoza can be limited only by thought, so freedom can be limited only by freedom. Its limitation as internal finitude stems from the *fact* that it can not not-be freedom—that is, it is condemned to be free; its limitation as external finitude stems from the *fact* that being freedom, it *is* for other freedoms, freedoms which freely apprehend it in the light of their own ends.

NOTES

1. The in-itself is that which is what it is. Ortega, in the first selection by him in Part Two, explains it as that which is given its existence, like the stone or the plant [ed. note].
2. The for-itself is that which is not what it is, and is what it is not. Ortega explains it as that which is given the abstract possibility of existence, but not the reality—like

man, who is a project, something which is not yet but aspires to be [ed. note].

3. This is a simplification: There are influences and interferences in the matter of technique; the Arlberg method has been prevalent with us for a long time. The reader will easily be able to re-establish the facts in their complexity.
4. We are intentionally oversimplifying. There are influences and interferences. But the reader will be able to re-establish the facts in their complexity. (The French text does not indicate whether this footnote belongs with this sentence or with a sentence in the preceding paragraph. The exact position can hardly be important. Tr.)
5. Brice-Parain: *Essai sur le logos platonicien.*
6. I am simplifying: one can also learn one's own thought from one's sentence. But this is because it is possible to a certain extent to adopt with respect to the sentence the point of view of the Other—exactly as in the case of one's own body.
7. "Unselbständigkeit" may be translated here as "incompleteness" or "the lack of independence." (ed. note).
8. We shall see later that the problem is more complex. But these remarks are sufficient for the present.

EXISTENTIAL

The stages of technology

JOSÉ ORTEGA Y GASSET

■ Ortega *presents in this selection a three-stage theory of the evolution of technology. He rejects off-hand the appearance of inventions as a criterion for the demarcation of these stages. Instead, he proposes what amounts to a theory of the evolution of man's consciousness of his relation to technology. The relation ascends from the unconscious level—that is, the level at which man is unaware of his technological abilities—all the way to the fully conscious level, at which he is completely aware of his technological abilities.*

The first level is a primitive stage characterized by "the technology of chance." In this stage man does not conceive of himself as homo faber; *thus "all primitive technology smacks of magic." This stage is succeeded by a transitional one characterized by the "technology of the craftsman." In this stage man is aware of technological abilities as gifts granted to craftsmen who make them full-time jobs. In the highest stage, characterized by the "technology of the technician," man realizes that his technological abilities belong to him as human being, not to nature and not to a select group of craftsmen. He realizes that they are "a source of practically unlimited human activity."*

Ortega nevertheless issues the following warning: "Technology for all its being a practically unlimited capacity will irretrievably empty the lives of those who are resolved to stake everything on their faith in it and it alone."

TECHNOLOGY OF CHANCE

The subject is difficult. It took me some time to decide upon the principle best suited to distinguish periods of technology. I do not hesitate to reject the one readiest to hand, viz., that we should divide the evolution according to the appearance of certain momentous and characteristic inventions. All I have said in this essay aims to correct the current error of regarding such or such a definite invention as the thing which matters in technology. What really matters and what can bring about a fundamental advance is a change in the general character of technology. No single invention is of such caliber as to bear comparison with the tremendous mass of the integral evolution. We have seen that magnificient advances have been achieved only to be lost again, whether they disappeared completely or whether they had to be rediscovered.

Nay more, an invention may be made sometime and somewhere and still fail to take on its true technical significance. Gunpowder and the printing press, unquestionably two discoveries of great pith and moment, were known in China for centuries without being of much use. It is not before the fifteenth century in Europe that gunpowder and the printing press, the former probably in Lombardy, the latter in Germany, became historical powers. With this in view, when shall we say they were invented? No doubt, they grew effective in history only when they appeared incorporated in the general body of late medieval technology, serving the purposes of the program of life operative in that age. Firearms and the

---

☐ Selection is reprinted from *"History As a System" and Other Essays Toward a Philosophy of History,* by José Ortega Y Gasset, with the permission of W.W. Norton & Company, Inc. Translated by Helene Weyl.

printing press are contemporaries of the compass. They all bear the same marks, so characteristic, as we shall shortly see, of that hour between Gothic and Renaissance, the scientific endeavors of which culminated in Copernicus. The reader will observe that, each in its own manner, they establish contact between man and things at a distance from him. They belong to the instruments of the *actio in distans,* which is at the root of modern technology. The cannon brings distant armies into immediate touch with each other. The compass throws a bridge between man and the cardinal points. The printing press brings the solitary writer into the presence of the infinite orbit of possible readers.

The best principle of delimiting periods in technical evolution is, to my judgment, furnished by the relation between man and technology, in other words by the conception which man in the course of history held, not of this or that particular technology but of the technical function as such. In applying this principle we shall see that it not only clarifies the past, but also throws light on the question we have asked before: how could modern technology give birth to such radical changes, and why is the part it plays in human life unparalleled in any previous age?

Taking this principle as our point of departure we come to discern three main periods in the evolution of technology: technology of chance; technology of the craftsman; technology of the technician.

What I call technology of chance, because in it chance is the engineer responsible for the invention, is the primitive technology of pre- and protohistoric man and of the contemporary savage, viz., of the least-advanced groups of mankind—as the Vedas in Ceylon, the Semang in Borneo, the pigmies in New Guinea and Central Africa, the Australian Negroes, etc.

How does primitive man conceive technology? The answer is easy. He is not aware of his technology as such; he is unconscious of the fact that there is among his faculties one which enables him to refashion nature after his desires.

The repertory of technical acts at the command of primitive man is very small and does not form a body of sufficient volume to stand out against, and be distinguished from, that of his natural acts, which is incomparably more important. That is to say, primitive man is very little man and almost all animal. His technical acts are scattered over and merged into the totality of his natural acts and appear to him as part of his natural life. He finds himself with the ability to light a fire as he finds himself with the ability to walk, swim, use his arms . . . His natural acts are a given stock fixed once and for all; and so are his technical. It does not occur to him that technology is a means of virtually unlimited changes and advances.

The simplicity and scantiness of these pristine technical acts account for their being executed indiscriminately by all members of the community, who all light fires, carve bows and arrows, and so forth. The one differentiation noticeable very early is that women perform certain technical functions and men certain others. But that does not help primitive man to recognize technology as an isolated phenomenon. For the repertory of natural acts is also somewhat different in men and women. That the woman should plow the field—it was she who invented agriculture—appears as natural as that she should bear the children.

Nor does technology at this stage reveal its most characteristic aspect, that of invention. Primitive man is unaware that he has the power of invention; his inventions are not the result of a premeditated and deliberate search. He does not look for them; they seem rather to look for him. In the course of his constant and fortuitous manipulation of objects he may suddenly and by mere chance come upon a new useful device. While for fun or out of sheer restlessness he rubs two sticks together a spark springs up, and a vision of new connections between things will dawn upon him. The stick, which hitherto has served as weapon or support, acquires the new aspect of a thing producing fire. Our savage will be awed, feeling that nature has inadvertently loosed one of its secrets before him. Since fire had always seemed a godlike power, arousing religious emotions, the new fact is prone to take on a magic tinge. All primitive technology smacks of magic. In fact, magic, as we shall shortly see, is nothing but a kind of technology, albeit a frustrated and illusory one.

Primitive man does not look upon himself as the inventor of his inventions. Invention appears to him as another dimension of nature, as part of nature's power to furnish him—

nature furnishing man, not man nature—with certain novel devices. He feels no more responsible for the production of his implements than for that of his hands and feet. He does not conceive of himself as *homo faber*. He is therefore very much in the same situation as Mr. Koehler's monkey when it suddenly notices that the stick in his hands may serve an unforeseen purpose. Mr. Koehler calls this the "aha-impression" after the exclamation of surprise a man utters when coming upon a startling new relation between things. It is obviously a case of the biological law of trial and error applied to the mental sphere. The infusoria "try" various movements and eventually find one with favorable effects on them which they consequently adopt as a function.

The inventions of primitive man, being, as we have seen, products of pure chance, will obey the laws of probability. Given the number of possible independent combinations of things, a certain possibility exists of their presenting themselves some day in such an arrangement as to enable man to see preformed in them a future implement.

TECHNOLOGY AS CRAFTSMANSHIP—TECHNOLOGY OF THE TECHNICIAN

We come to the second stage, the technology of the artisan. This is the technology of Greece, of preimperial Rome, and of the Middle Ages. Here are in swift enumeration some of its essential features.

The repertory of technical acts has grown considerably. But—and this is important—a crisis and setback, or even the sudden disappearance of the principal industrial arts, would not yet be a fatal blow to material life in these societies. The life people lead with all these technical comforts and the life they would have to lead without them are not so radically different as to bar, in case of failure or checks, retreat to a primitive or almost primitive existence. The proportion between the technical and the nontechnical is not yet such as to make the former indispensable for the supporting of life. Man is still relying mainly on nature. At least, and that is what matters, so he himself feels. When technical crises arise he does therefore not realize that they will hamper his life, and consequently fails to meet them in time and with sufficient energy.

Having made this reservation we may now state that technical acts have by this time enormously increased both in number and in complexity. It has become necessary for a definite group of people to take them up systematically and make a full-time job of them. These people are the artisans. Their existence is bound to help man become conscious of technology as an independent entity. He sees the craftsman at work—the cobbler, the blacksmith, the mason, the saddler—and therefore comes to think of technology in terms and in the guise of the technician, the artisan. That is to say, he does not yet know that there is technology, but he knows that there are technicians who perform a peculiar set of activities which are not natural and common to all men.

Socrates in his struggle, which is so appallingly modern, with the people of his time began by trying to convince them that technology is not the same as the technician, that it is an abstract entity of its own not to be mixed up with this or that concrete man who possesses it.

At the second stage of technology everybody knows shoemaking to be a skill peculiar to certain men. It can be greater or smaller and suffer slight variations as do natural skills, running for instance, or swimming or, better still, the flying of a bird, the charging of a bull. That means shoemaking is now recognized as exclusively human and not natural, i.e., animal; but it is still looked upon as a gift granted and fixed once and for all. Since it is something exclusively human it is extranatural, but since it is something fixed and limited, a definite fund not admitting of substantial amplification, it partakes of nature; and thus technology belongs to the nature of man. As man finds himself equipped with the unexchangeable system of his bodily movements, so he finds himself equipped with the fixed system of the "arts." For this is the name technology bears in nations and epochs living on the technical level in question; and this is the original meaning of the Greek word *techne*.

The way technology progresses might disclose that it is an independent and, in principle, unlimited function. But, oddly enough, this fact becomes even less apparent in this than in the primitive period. After all, the few primitive inventions, being so fundamental, must have stood out melodramatically against the workaday routine of animal habits. But in craftsmanship there is no room whatever for a sense of invention. The artisan must learn thoroughly in long apprenticeship—it is the time of masters and apprentices—elaborate usages handed down by long tradition. He is governed by the norm that man must bow to tradition as such. His mind is turned towards the past and closed to novel possibilities. He follows the established routine. Even such modifications and improvements as may be brought about in his craft through continuous and therefore imperceptible shifts present themselves not as fundamental novelties, but rather as differences of personal style and skill. And these styles of certain masters again will spread in the forms of schools and thus retain the outward character of tradition.

We must mention another decisive reason why the idea of technology is not at this time separated from the idea of the person who practices it. Invention has as yet produced only tools and not machines. The first machine in the strict sense of the word—and with it I anticipate the third period—was the weaving machine set up by Robert in 1825. It is the first machine because it is the first tool that works by itself, and by itself produces the object. Herewith technology ceases to be what it was before, handiwork, and becomes mechanical production. In the crafts the tool works as a complement of man; man with his natural actions continues to be the principal agent. In the machine the tool comes to the fore, and now it is no longer the machine that serves man but man who waits on the machine. Working by itself, emancipated from man, the machine, at this stage, finally reveals that technology is a function apart and highly independent of natural man, a function which reaches far beyond the bounds set for him. What a man can do with his fixed animal activities we know beforehand; his scope is limited. But what the machine man is capable of inventing may do, is in principle unlimited.

One more feature of craftsmanship remains to be mentioned which helps to conceal the true character of technology. I mean this: technology implies two things. First, the invention of a plan of activity, of a method or procedure—*mechane,* said the Greeks—and, secondly, the execution of this plan. The former is technology strictly speaking, the latter consists merely in handling the raw material. In short, we have the technician and the worker who between them, performing very different functions, discharge the technical job. The craftsman is both technician and worker; and what appears first is a man at work with his hands, and what appears last, if at all, is the technology behind him. The dissociation of the artisan into his two ingredients, the worker and the technician, is one of the principal symptoms of the technology of the third period.

We have anticipated some of the traits of this technology. We have called it the technology of the technician. Man becomes clearly aware that there is a capacity in him which is totally different from the immutable activities of his natural or animal part. He realizes that technology is not a haphazard discovery, as in the primitive period; that it is not a given and limited skill of some people, the artisans, as in the second period; that it is not this or that definite and therefore fixed "art"; but that it is a source of practically unlimited human activity.

This new insight into technology as such puts man in a situation radically new in his whole history and in a way contrary to all he has experienced before. Hitherto he has been conscious mainly of all the things he is unable to do, i.e., of his deficiencies and limitations. But the conception our time holds of technology—let the reader reflect a moment on his own—places us in a really tragicomic situation. Whenever we imagine some utterly extravagant feat, we catch ourselves in a feeling almost of apprehension lest our reckless dream—say a voyage to the stars—should come true. Who knows but that tomorrow morning's paper will spring upon us the news that it has been possible to send a projectile to the moon by imparting to it a speed great enough to overcome the gravitational attraction. That is to say, present-day man is secretly frightened by his own omnipotence. And this may be another reason why he does not know what he is. For finding himself in principle capable of being almost anything makes it all the harder for him to know what he actually is.

In this connection I want to draw attention to a point which does not properly belong here, that technology for all its being a practically unlimited capacity will irretrievably empty the lives of those who are resolved to stake everything on their faith in it and it alone. To be an engineer and nothing but an engineer means to be potentially everything and actually nothing. Just because of its promise of unlimited possibilities technology is an empty form like the most formalistic logic and is unable to determine the content of life. That is why our time, being the most intensely technical, is also the emptiest in all human history.

RELATION BETWEEN MAN AND TECHNOLOGY IN OUR TIME—THE ENGINEER IN ANTIQUITY

This third stage of technical evolution, which is our own, is characterized by the following features:

Technical acts and achievements have increased enormously. Whereas in the Middle Ages—the era of the artisan—technology and the nature of man counter-balanced each other and the conditions of life made it possible to benefit from the human gift of adapting nature to man without denaturalizing man, in our time the technical devices outweigh the natural ones so gravely that material life would be flatly impossible without them. This is no manner of speaking, it is the literal truth. In *The Revolt of the Masses* I drew attention to the most noteworthy fact that the population of Europe between 500 and 1800 A.D., i.e., for thirteen centuries, never exceeded 180 millions; whereas by now, in little over a century, it has reached 500 millions, not counting those who have emigrated to America. In one century it has grown nearly three and a half times its size. If today 500 million people can live well in a space where 180 lived badly before, it is evident that, whatever the minor causes, the immediate cause and most necessary condition is the perfection of technology. Were technology to suffer a setback, millions of people would perish.

Such fecundity of the human animal could occur only after man had succeeded in interposing between himself and nature a zone of exclusively technical provenance, solid and thick enough to form something like a supernature. Present-day man—I refer not to the individual but to the totality of men—has no choice of whether to live in nature or to take advantage of this supernature. He is as irremediably dependent on, and lodged in, the latter as primitive man is in his natural environment. And that entails certain dangers. Since present-day man, as soon as he opens his eyes to life, finds himself surrounded by a superabundance of technical objects and procedures forming an artificial environment of such compactness that primordial nature is hidden behind it, he will tend to believe that all these things are there in the same way as nature itself is there without further effort on his part: that aspirin and automobiles grow on trees like apples. That is to say, he may easily lose sight of technology and of the conditions—the moral conditions, for example—under which it is produced and return to the primitive attitude of taking it for the gift of nature which is simply there. We thus have the curious fact that, at first, the prodigious expansion of technology made it stand out against the sober background of man's natural activities and allowed him to gain full sight of it, whereas by now its fantastic progress threatens to obscure it again.

Another feature helping man to discover the true character of his own technology we found to be the transition from mere tools to machines, i.e., mechanically working apparatus. A modern factory is a self-sufficient establishment waited on occasionally by a few persons of very modest standing. In consequence, the technician and the worker, who were united in the artisan, have been separated and the technician has grown to be the live expression of technology as such—in a word, the engineer.

Today technology stands before our mind's eye for what it is, apart, unmistakable, isolated, and unobscured by elements other than itself. And this enables certain persons, called engineers, to devote their lives to it. In the paleolithic age or in the Middle Ages technology, that is invention, could not have been a profession because man was ignorant of his own inventive power. Today the engineer embraces as one of the most normal and firmly established forms of activity the occupation of inventor. In contrast to the savage, he knows before he begins to invent that he is capable of doing so, which means that he has "technology" before he has "a technology." To this degree and in this concrete sense our previous assertion holds that technologies are nothing but concrete realizations of the general technical functions of man. The engineer need not wait for chances and favorable odds; he is sure to make discoveries. How can he be?

The question obliges us to say a word about the technique of technology. To some people technique and nothing else is technology. They are right in so far as without technique—the intellectual method operative in technical creation—there is no technology. But with technique alone there is none either. As we have seen before, the existence of a capacity is not enough to put that capacity into action.

I should have liked to talk at leisure and in

detail about both present and past techniques of technology. It is perhaps the subject in which I myself am most interested. But it would have been a mistake to let our investigations gravitate entirely around it. Now that this essay is breathing its last I must be content to give the matter brief consideration—brief, yet, I hope, sufficiently clear.

No doubt, technology could not have expanded so gloriously in these last centuries, nor the machine have replaced the tool, nor the artisan have been split up into his components, the worker and the engineer, had not the method of technology undergone a profound transformation.

Our technical methods are radically different from those of all earlier technologies. How can we best explain the diversity? Perhaps through the following question: now would an engineer of the past, supposing he was a real engineer and his invention was not due to chance but deliberately searched for, go about his task? I will give a schematic and therefore exaggerated example which is, however, historical and not fictitious. The Egyptian architect who built the pyramid of Cheops was confronted with the problem of lifting stone blocks to the highest parts of the monument. Starting as he needs must from the desired end, namely to lift the stones, he looked around for devices to achieve this. "This," I have said, meaning he is concerned with the result as a whole. His mind is absorbed by the final aim in its integrity. He will therefore consider as possible means only such procedures as will bring about the total result at once, in one operation that may take more or less time but which is homogeneous in itself. The unbroken unity of the end prompts him to look for a similarly uniform and undifferentiated means. This accounts for the fact that in the early days of technology the instrument through which an aim is achieved tends to resemble the aim itself. Thus in the construction of the pyramid the stones are raised to the top over another pyramid, an earthen pyramid with a wider base and a more gradual slope, which abuts against the first. Since a solution found through this principle of similitude—*similia similibus*—is not likely to be applicable in many cases, the engineer has no general rule and method to lead him from the intended aim to the adequate means. All he can do is to try out empirically such possibilities as offer more or less hope of serving his purpose. Within the circle defined by his special problem he thus falls back into the attitude of the primitive inventor.

HEGELIAN-MARXIST

From *The Communist Manifesto*

KARL MARX and FREDERICK ENGELS

▪ Karl Marx *(1818-1883) was born in Trier in the Rhineland. He was a revolutionary socialist, a social and economic theorist, and the co-founder, with Engels, of modern socialism. Because of his reputation as a revolutionary writer, he spent most of his life unemployed and in exile. In 1849 he was expelled from Paris. He went to London, where he was to remain the rest of his life. In London he was offered the job of correspondent to the* New York Daily Tribune, *and he accepted it. A series of insightful political analyses resulted from that assignment. It was also in London that Marx wrote his most famous work,* Das Capital *(vol. I, 1867; vols II and III, 1885-1894). In it he presents a detailed and mature critique of capitalism. He lays bare the dehumanizing conditions of factory workers in England at that time, as well as the exploitive relation of the owner of a factory to his workers. Marx argues that capitalism is in its very essence exploitive and thus unacceptable. He calls for a new and free society in which workers control the means of production and in which everyone has an equal opportunity for self-realization.*

In the following selection from the famous Manifesto *(1848), Marx and Engels attempt to explain the forces behind the decline of feudalism and the rise of capitalism. Navigational achievements, which led to colonization and the development of new markets, gave commerce and industry a new impetus to expand rapidly. "Thereupon, steam and machinery revolutionized industrial production. The place of manufacture was taken by the giant Modern Industry, the place of the industrial middle class, by industrial millionaires, . . . the modern bourgeois. The old feudal modes of agricultural and industrial organization became a handicap in the face of industrial progress and had to be removed."*

But because of competition "the bourgeoisie cannot exist without constantly revolutionizing the instruments of production, and thereby the relations of production." Thus with the progress of industrial technology, the worker is steadily robbed of his creativity and individuality. He becomes "an appendage of the machine." His conditions of existence in society worsen rather than improve (for with increased competition, the capitalist is led to increase his profits by paying the worker as little as possible). These conditions finally lead the worker to rise against this mode of social organization and introduce a new one (communism). "What the bourgeoisie, therefore, produces, above all, is its own grave-diggers."

We have here another three-stage theory of the history of technological development—a pre-industrial or agrarian stage associated with feudalism, a modern industrial stage associated with capitalism, and a post-industrial stage associated with communism.

The bourgeoisie cannot exist without constantly revolutionizing the instruments of production, and thereby the relations of production, and with them the whole relations of society. Conservation of the old modes of production in unaltered form was, on the contrary, the first condition of existence for all earlier industrial classes. Constant revolutionizing of production, uninterrupted disturbance of all social conditions, everlasting uncertainty and agitation distinguish the bourgeois epoch from all earlier ones. All fixed, fast frozen relations, with their train of ancient and venerable prejudices and opinions, are swept away, all new formed ones become antiquated before they can ossify. All that is solid melts into the air, all that is holy is profaned, and man is at last compelled to face with sober senses, his real conditions of life, and his relations with his kind.

The need of a constantly expanding market for its products chases the bourgeoisie over the whole surface of the globe. It must nestle everywhere, settle everywhere, establish connections everywhere.

The bourgeoisie has through its exploitation of the world-market given a cosmopolitan character to production and consumption in every country. To the great chagrin of reactionists, it has drawn from under the feet of industry the national ground on which it stood. All old-established national industries have been destroyed or are daily being destroyed. They are dislodged by new industries, whose introduction becomes a life and death question for all civilized nations, by industries that no longer work up indigenous raw material, but raw material drawn from the remotest zones; industries whose products are consumed, not only at home, but in every quarter of the globe.

☐ From *The Communist Manifesto,* in *Capital, The Communist Manifesto and Other Writings by Karl Marx,* ed. and trans. Max Eastman (New York: The Modern Library, 1959), pp. 324-334.

In place of the old wants, satisfied by the productions of the country, we find new wants, requiring for their satisfaction the products of distant lands and climes. In place of the old local and national seclusion and self-sufficiency, we have intercourse in every direction, universal interdependence of nations. And as in material, so also in intellectual production. The intellectual creations of individual nations become common property. National onesidedness and narrowmindedness become more and more impossible, and from the numerous national and local literatures there arises a world-literature.

The bourgeoisie, by the rapid improvement of all instruments of production, by the immensely facilitated means of communication, draws all, even the most barbarian nations into civilization. The cheap prices of its commodities are the heavy artillery with which it batters down all Chinese walls, with which it forces the barbarians' intensely obstinate hatred of foreigners to capitulate. It compels all nations, on pain of extinction, to adopt the bourgeois mode of production; it compels them to introduce what it calls civilization into their midst, *i.e.,* to become bourgeois themselves. In a word, it creates a world after its own image.

The bourgeoisie has subjected the country to the rule of the towns. It has created enormous cities, has greatly increased the urban population as compared with the rural, and has thus rescued a considerable part of the population from the idiocy of rural life. Just as it has made the country dependent on the towns, so it has made barbarian and semi-barbarian countries dependent on civilized ones, nations of peasants on nations of bourgeois, the East on the West.

The bourgeoisie keeps more and more doing away with the scattered state of the population, of the means of production, and of property. It has agglomerated population, centralized means of production, and has concentrated property in a few hands. The necessary consequence of this was political centralization. Independent, or but loosely connected provinces, with separate interests, laws, governments, and systems of taxation, became lumped together in one nation, with one government, one code of laws, one national class interest, one frontier and one customs tariff.

The bourgeoisie, during its rule of scarce one hundred years, has created more massive and more colossal productive forces than have all preceding generations together. Subjection of Nature's forces to man, machinery, application of chemistry to industry and agriculture, steam-navigation, railways, electric telegraphs, clearing of whole continents for cultivation, canalization of rivers, whole populations conjured out of the ground—what earlier century had even a presentiment that such productive forces slumbered in the lap of social labor?

We see then: the means of production and of exchange on whose foundation the bourgeoisie built itself up, were generated in feudal society. At a certain stage in the development of these means of production and of exchange, the conditions under which feudal society produced and exchanged, the feudal organization of agriculture and manufacturing industry, in one word, the feudal relations of property became no longer compatible with the already developed productive forces; they became so many fetters. They had to burst asunder; they were burst asunder.

Into their places stepped free competition, accompanied by social and political constitution adapted to it, and by economical and political sway of the bourgeois class.

A similar movement is going on before our eyes. Modern bourgeois society with its relations of production, of exchange and of property, a society that has conjured up such gigantic means of production and of exchange, is like the sorcerer, who is no longer able to control the powers of the nether world whom he has called up by his spells. For many a decade past, the history of industry and commerce is but the history of the revolt of modern productive forces against modern conditions of production, against the property relations that are the conditions for the existence of the bourgeoisie and of its rule. It is enough to mention the commercial crises that by their periodical return put on its trial, each time more threateningly, the existence of the entire bourgeois society. In these crises a great part not only of the existing products, but also of the previously created productive forces, are periodically destroyed. In these crises there breaks out an epidemic that, in all earlier epochs, would have seemed an absurdity—the epidemic of overproduction. Society suddenly finds itself put back into a state of momentary barbarism; it appears as if a famine, a universal war of devastation, had cut off the supply of every means of subsistence; industry and commerce seem to be destroyed; and why? Because there is too much civilization, too much means of subsistence, too much industry, too much commerce. The productive forces at the disposal of society no longer tend to further the development of the conditions of the bourgeois property; on the contrary, they have become too powerful for these conditions by which they are fettered, and as soon as they overcome these fetters they bring disorder into the whole of bourgeois society, endanger the existence of bourgeois property. The conditions of bourgeois society are too narrow to comprise the wealth created by them. And how does the bourgeoisie get over these crises? On the one hand by enforced destruction of a mass of productive forces; on the other, by the conquest of new markets, and by the more thorough exploitation of the old ones. That is to say, by paving the way for more extensive and more destructive crises, and by diminishing the means whereby crises are prevented.

The weapons with which the bourgeoisie felled feudalism to the ground are now turned against the bourgeoisie itself.

But not only has the bourgeoisie forged the weapons that bring death to itself; it has also called into existence the men who are to wield those weapons—the modern working-class—the proletarians.

In proportion as the bourgeoisie, *i.e.*, capital, is developed, in the same proportion is the proletariat, the modern working-class, developed, a class of laborers who live only so long as they find work, and who find work only so long as their labor increases capital. These laborers, who must sell themselves piecemeal, are a commodity, like every other article of commerce, and are consequently exposed to all the vicissitudes of competition, to all the fluctuations of the market.

Owing to the extensive use of machinery and to division of labor, the work of the proletarians has lost all individual character, and, consequently, all charm for the workman. He becomes an appendage of the machine, and it is only the most simple, most monotonous and most easily acquired knack that is required of him. Hence, the cost of production of a workman is restricted almost entirely to the means of subsistence that he requires for his maintenance, and for the propagation of his race. But the price of a commodity, and also of labor, is equal to its cost of production. In proportion, therefore, as the repulsiveness of the work increases the wage decreases. Nay more, in proportion as the use of machinery and division of labor increases, in the same proportion the burden of toil increases, whether by prolongation of the working hours, by increase of the work enacted in a given time, or by increased speed of the machinery, etc.

Modern industry has converted the little workshop of the patriarchal master into the great factory of the industrial capitalist. Masses of laborers, crowded into factories, are organized like soldiers. As privates of the industrial army they are placed under the command of a perfect hierarchy of officers and sergeants. Not only are they the slaves of the bourgeois class and of the bourgeois state, they are daily and hourly enslaved by the machine, by the overlooker, and, above all, by the individual bourgeois manufacturer himself. The more openly this despotism proclaims gain to be its end and aim, the more petty, the more hateful and the more embittering it is.

The less the skill and exertion or strength implied in manual labor, in other words, the more modern industry becomes developed, the more is the labor of men superseded by that of women. Differences of age and sex have no longer any distinctive social validity for the working class. All are instruments of labor, more or less expensive to use, according to their age and sex.

No sooner is the exploitation of the laborer by the manufacturer, so far at an end, that he receives his wages in cash, than he is set upon by the other portions of the bourgeoisie, the landlord, the shopkeeper, the pawnbroker, etc.

The lower strata of the middle class—the small tradespeople, shopkeepers and retired tradesmen generally, the handicraftsmen and peasants—all these sink gradually into the proletariat, partly because their diminutive capital does not suffice for the scale on which Modern Industry is carried on, and is swamped in the competition with the large capitalists, partly because their specialized skill is rendered worthless by new methods of production. Thus the proletariat is recruited from all classes of the population.

The proletariat goes through various stages of development. With its birth begins its struggle with the bourgeoisie. At first the contest is carried on by individual laborers, then by the workpeople of a factory, then by the operatives of one trade, in one locality, against the individual bourgeois who directly exploits them. They direct their attacks not against the bourgeois conditions of production, but against the instruments of production themselves; they destroy imported wares that compete with their labor, they smash to pieces machinery, they set factories ablaze, they seek to restore by force the vanished status of the workman of the Middle Ages.

At this stage the laborers still form an incoherent mass scattered over the whole country, and broken up by their mutual competition. If anywhere they unite to form more compact bodies, this is not yet the consequence of their own active union, but of the union of the bourgeoisie, which class, in order to attain its own political ends, is compelled to set the whole proletariat in motion, and is moreover yet, for a time, able to do so. At this stage, therefore, the proletarians do not fight their enemies, but the enemies of their enemies, the remnants of absolute monarchy, the landowners, the non-industrial bourgeois, the petty bourgeoisie. Thus the whole historical movement is concentrated in the hands of the bourgeoisie, every victory so obtained is a victory for the bourgeoisie.

But with the development of industry the proletariat not only increases in number; it becomes concentrated in greater masses, its strength grows and it feels that strength more. The various interests and conditions of life within the ranks of the proletariat are more and more equalized, in proportion as machinery obliterates all distinctions of labor, and nearly everywhere reduces wages to the same low level. The growing competition among the bourgeois, and the resulting commercial crisis, make the wages of the workers even more fluctuating. The unceasing improvement of machinery, ever more rapidly developing, makes their livelihood more and more precarious; the collisions between individual workmen and individual bourgeois take more and more the character of collisions between two classes. Thereupon the workers begin to form combinations (Trades' Unions) against the bourgeois; they club together in order to keep up the rate of wages; they found permanent associations in order to make provision beforehand for these occasional revolts. Here and there the contest breaks out into riots.

Now and then the workers are victorious, but only for a time. The real fruit of their battle lies not in the immediate result but in the ever-expanding union of workers. This union is helped on by the improved means of communication that are created by modern industry, and that places the workers of different localities in contact with one another. It was just this contact that was needed to centralize the numerous local struggles, all of the same character, into one national struggle between classes. But every class struggle is a political struggle. And that union, to attain which the burghers of the Middle Ages with their miserable highways, required centuries, the modern proletarians, thanks to railways, achieve in a few years.

This organization of the proletarians into a class, and consequently into a political party, is continually being upset again by the competition between the workers themselves. But it ever rises up again, stronger, firmer, mightier. It compels legislative recognition of particular interests of the workers by taking advantage of the divisions among the bourgeoisie itself. Thus the ten hours' bill in England was carried.

Although collisions between the classes of the old society further, in many ways, the course of development of the proletariat. The bourgeoisie finds itself involved in a constant battle. At first with the aristocracy; later on, with those portions of the bourgeoisie itself whose interests have become antagonistic to the progress of industry; at all times, with the bourgeoisie of foreign countries. In all these battles it sees itself compelled to appeal to the proletariat, to ask for its help, and thus, to drag it into the political arena. The bourgeoisie itself, therefore, supplies the proletariat with its own elements of political and general education; in other words, it furnishes the proletariat with weapons for fighting the bourgeoisie.

Further, as we have already seen, entire sections of the ruling classes are, by the advance of industry, precipitated into the proletariat, or are at least threatened in their conditions of existence. These also supply the proletariat with fresh elements of enlightenment and progress.

Finally, in times when the class-struggle nears the decisive hour, the process of dissolution going on within the ruling class—in fact, within the whole range of an old society—assumes such a violent, glaring character that a small section of the ruling class cuts itself adrift and joins the revolutionary class, the class that holds the future in its hands. Just as, therefore, at an earlier period, a section of the nobility went over to the bourgeoisie, so now a portion of the bourgeoisie goes over to the proletariat, and in particular, a portion of the bourgeois ideologists, who have raised themselves to the level of comprehending theoretically the historical movements as a whole.

Of all the classes that stand face to face with the bourgeoisie to-day the proletariat alone is a really revolutionary class. The other classes decay and finally disappear in the face of modern industry; the proletariat is its special and essential product.

The lower middle class, the small manufacturer, the shopkeeper, the artisan, the peasant, all these fight against the bourgeoisie, to save from extinction their existence as fractions of the middle class. They are therefore not revolutionary, but conservative. Nay, more; they are reactionary, for they try to roll back the wheel of history. If by chance they are revolutionary, they are so only in view of their impending transfer into the proletariat; they thus defend not their present, but their future interests; they desert their own standpoint to place themselves at that of the proletariat.

The "dangerous class," the social scum, that passively rotting mass thrown off by the lowest layers of old society, may, here and there, be swept into the movement by a proletarian revolution; its conditions of life, however, prepare it far more for the part of a bribed tool of reactionary intrigue.

In the conditions of the proletariat, those of the old society at large are already virtually swamped. The proletarian is without property; his relation to his wife and children has no longer anything in common with the bourgeois family relations; modern industrial labor, modern subjection to capital, the same in England as in France, in America as in Germany, has stripped him of every trace of national character. Law, morality, religion, are to him so many bourgeois prejudices, behind which lurk in ambush just as many bourgeois interests.

All the preceding classes that got the upper hand sought to fortify their already acquired status by subjecting society at large to their conditions of appropriation. The proletarians cannot become masters of the productive forces of society, except by abolishing their own previous mode of appropriation, and therby also every other previous mode of appropriation. They have nothing of their own to secure and to fortify; their mission is to destroy all previous securities for and insurances of individual property.

All previous historical movements were movements of minorities, or in the interest of minorities. The proletarian movement is the self-conscious, independent movement of the immense majority. The proletariat, the lowest stratum of our present society, cannot stir, cannot raise itself up without the whole superincumbent strata of official society being sprung into the air.

Though not in substance, yet in form, the struggle of the proletariat with the bourgeoisie is at first a national struggle. The proletariat of each country must, of course, first of all settle matters with its own bourgeoisie.

In depicting the most general phases of the development of the proletariat, we traced the more or less veiled civil war, raging within existing society, up to the point where that war breaks out into open revolution, and where the violent overthrow of the bourgeoisie, lays the foundations for the sway of the proletariat.

Hitherto every form of society has been based, as we have already seen, on the antagonism of oppressing and oppressed classes. But in order to oppress a class, certain conditions must be assured to it under which it can, at least, continue its slavish existence. The serf, in the period of serfdom, raised himself to membership in the commune, just as the petty bourgeois, under the yoke of feudal absolutism, managed to develop into a bourgeois. The modern laborer, on the contrary, instead of rising with the progress of industry, sinks deeper and deeper below the conditions of existence of his own class. He becomes a pauper, and pauperism develops more rapidly than population and wealth. And here it becomes evident that the bourgeoisie is unfit any longer to be the ruling class in society, and to impose its conditions of existence upon society as an overriding law. It is unfit to rule, because it is incompetent to assure an existence to its slave within his slavery, because it cannot help letting him sink into such a state that it has to feed him, instead of being fed by him. Society can no longer live under this bourgeoisie; in other words, its existence is no longer compatible with society.

The essential condition for the existence, and for the sway of the bourgeois class, is the formation and augmentation of capital; the condition for capital is wage labor. Wage labor rests exclusively on competition between the laborers. The advance of industry, whose involuntary promoter is the bourgeoisie, replaces the isolation of the laborers, due to competition, by their involuntary combination, due to association. The development of Modern Industry, therefore, cuts from under its feet the very foundation on which the bourgeoisie produces and appropriates products. What the bourgeoisie therefore produces, above all, are its own grave diggers. Its fall and the victory of the proletariat are equally inevitable.

:GELIAN-MARXIST

·om "Man and His Natural Surroundings"

HAILO MARKOVIĆ

■ Mihailo Marković, *who was born in Yugoslavia in 1923, is an internationally renowned Marxist philosopher. As an officer in Tito's partisan army, Markovic took part in his country's liberation movement during World War II. After the war, he completed his education in Yugoslavia and England, and then taught philosophy at the University of Belgrade. Later he became director of the Institute of Philosophy there. He is currently a visiting professor of philosophy at the University of Pennsylvania. Among his books are* From Affluence to Praxis *(1974) and* The Contemporary Marx *(1974), from which this selection is taken.*

Marković provides two criteria for assessing the interaction between man and the environment: (1) the degree of human control over the natural environment and (2) the degree of realization of human needs and potential creative abilities. He argues for a balance between the two in society. Domination by the first criterion would lead to "an extremely intensive and careless exploitation of the environment." On the other hand, domination by the second criterion "condemns man to a poverty of needs, drudgery and degrading routine labour," among other things. Marković argues that a true Marxist society would achieve a desirable balance, and thus would provide the individual with a true opportunity for self-realization.

The experience which comes from the material and spiritual history of mankind offers two essential criteria for the assessment of interaction between man and the environment.

The first is: *the degree of human control over the natural environment*. From this point of view, priority should be given to those forms of human activity which increase power over natural forces, provide for increasing amounts of energy and usable natural material, enable more rapid and versatile mobility in given space, and which more effectively prevent the occurrence of undesirable natural phenomena; finally, priority should also be given to those activities which enable us to forecast as accurately as possible future natural processes.

The second criterion is the degree of realisation of human needs and of potential creative abilities. Viewed from this aspect, interaction with the natural environment should be directed in such a way that it contributes to the self-discovery and self-identification of every individual, development of all the wealth in his latent abilities, full recognition as a creative being, reinforcement of his communication and solidarity with other members of the society with whom he exercises a mutual influence on the environment.

The first criterion is *technical*, it guarantees a so-called *negative* human freedom (a freedom *from* alien, external powers), and it may also be expressed as a principle of effectiveness and technical rationality. The second criterion is *humanistic*, it guarantees a socalled *positive* freedom (freedom *for* self-determination, self-development), and it may also be expressed as a principle of disalienation and self-realisation.

The crossing of both these criteria gives an instructive typology of the basic forms of interaction between man and the environment. For the *n* values in each criterion we receive 2^n types. For example, if we differentiate, on the one hand, technical *ineffectiveness versus effectiveness* and, on the other hand, a *self-alienating versus self-realising* activity, then we obtain four types of interaction: (1) that which is, simultaneously, technically ineffective and self-alienating (in the sense that it leads

☐ From *The Contemporary Marx* (Nottingham, England: Spokesman Books, 1974), pp. 145-152.

to the creation of a product which escapes human control, hinders creative ability, and leads to degradation and the loss of one's potentialities); this activity is characteristic for an expressly primitive, impoverished, and uncultivated society; (2) an activity which is ineffective but provides for the staisfaction of needs and self-realisation on a relatively low level; this activity was characteristic for some primitive societies with considerably developed culture, such as the ancient Greek or *Maya* civilisations; (3) an activity which is effective but self-alienating, leading to abundance but also to dehumanisation, characteristic of modern industrial society; finally, (4) an activity which is both effective and creative, as it provides for just that type of abundance which is necessary for the overall and rich life of all human senses, intellectual powers, and instincts; this activity has, in the course of history, been possible only for the most highly-developed individuals.

Although these two criteria do not, in principle, exclude each other, one has usually dominated the other in history and in social theory.

The technical criterion requires positive scientific knowledge—knowledge of facts, constant relations and tendencies, and of the most probable possibilities of the future states of the system. This knowledge, in its aspiration toward maximal objectivity, is greatest when the natural environment (whose systems are the most simple), is concerned but is significantly decreased when its subject-matter is the social environment. Positive scientific knowledge of man, however, is extremely limited and scanty. This is one of the causes for the tendency, when assessing the interaction of man and the environment, to over-simplify human aspirations and reduce them to one basic, purely quantitative desire: the desire for expansion, constant growth, and unlimited increase of power. The technical criterion requires a methodology which provides the most adequate procedures for problem-solving in the *framework* of certain *given* value assumptions, but does not go into a critical analysis of these assumptions themselves.

Therefore, in every society in which there exists a significant economic and social inequality, it is inevitable that the ruling value assumptions reflect the interests of the ruling power élite. In this way, the ever-increasing degree of control over the environment can mean, in practice, the increasing realisation of the interests of the ruling social groups (maximisation of profit—in the case of the bourgeoisie; increased power and increasingly more successful mainpulation of human beings—in the case of bureaucracy).

Class and group egoism of this kind, armed with powerful modern technology, leads to an extremely intensive and careless exploitation of the environment, to a destruction of the existent harmony between man and the environment, and to pathological phenomena which, beyond certain limits, may become irreversible.

In some intellectual circles we often encounter a one-sided negation of the technical criterion in favour of the humanistic one.

In a highly-developed industrial society which has transcended material poverty but where the degree of pollution of the natural environment has reached alarming proportions, where most of the population suffer from an apathetic, taciturn satiety and from a pathological passiveness which Paul Goodman called a "disease-nothing-can-be-done," a massive resistance is expressed by a part of the "intelligentsia" and youth. This resistance tends to assume a romantic character because it makes no distinction between technology as a means and technicism as a style of life, between technology in itself and technology which is ill-used by the egoistical ruling group. However, in spite of its romantic, utopian character, this anti-technological tendency has brought certain actual problems into focus and has caused the creation of a new counter culture.[1]

A similar, one-sided resistance to technology has also appeared in underdeveloped, poorly urbanised environments with strong mythological traditions which search for alternatives to the industrial civilisation. Gandhi assessed modern civilisation as being "satanic" because it "worships" Mammon and degrades man morally and spiritually, because of its "mad race for the multiplication of needs," the displacement of the human race by machines, mass-production which "does not take into consideration the real needs of the consumer," the unhealthy life in large cities, the concentration of wealth and power in the hands of the minority.

Gandhi "would welcome the natural extinction of railways and hospitals," he "does not see that the Iron Age was an improvement over the Stone Age,'' that the world is better off for having rapid means for going from place to place; he is ''sickened with the insane desire to destroy space and time, increase animalistic appetites, and travel to the ends of the world in order to satisfy them.'' Gandhi does not believe that boats and telegraphs ''are indispensable for the permanent well-being of the human race.'' Gandhi's criticism of civilisation strikes out at many real evils, it expresses a deep understanding of the specific characteristics of the historical situation of India. Its motives are pure and profoundly moral. Gandhi makes no secret of the fact that his point of view is basically conservative: ''While the Machine Age wants to turn people into machines, I want to return the man-transformed-into machine to his original state.'' And this state is ''a simple but noble life, developing thousands of village homes and living in peace with the world.'' ''Leisure is good and justifiable only up to a certain point. God made man to eat bread earned by the sweat of his brow, and I am afraid of the possibility of plucking everything we need, including food, out of a magician's hat.''[2]

The essential limitation of every humanism founded on a low level of control of the natural environment is that it condemns man to a poverty of needs, drudgery and degrading routine labour, helplessness in the face of natural forces and to a crippling of the majority of potentially creative abilities.

Complete balance between the technical and the humane was achieved for the first time in the ideas of Marx. In an underdeveloped and uncultivated environment, humanism is, in the opinion of Marx, transformed into its opposite—''primitive, crude communism. The abolition of private property in such a society leads to general envy, levelling off, the reduction of everything to a common, relatively low level, and to the negation of man's personality in every sphere.'' The abolition of private property under these conditions leads to ''the abstract negation of the entire world of culture and civilisation, regression to an *unnatural* simplicity of the individual who has not yet achieved private property, much less transcended it.''[3]

Marx's alternative to capitalism was a society in which alienation was transcended, in which ''the free development of the individual is a condition for the free development of all,''[4] in which social processes will be subject to the conscious, planned control of freely-associated individuals, in which there will be the well-rounded development of human needs and a rich life. However, ''this demands such a material basis of the society and such a number of material conditions which, themselves, are again the product of a long and painful history of development.''[5]

In other words, Marx realises the necessity of technological development if human self-realisation is ever to be made possible. Spiritual wealth is conditioned by the abolition of material poverty. Man must master the forces of nature in order to develop freely all of his creative powers. For this reason, Marx is aware of the historical significance of industrialisation, private property, and reification (which are necessary consequences of an intense struggle with the natural environment). He understands that there is no other road to universal human emancipation. For this reason Marx does not reject modern technology and industrialisation, but rather only uncovers their limitations and goes beyond them by his vision of the future.

Technology is and remains a *means* for human self-realisation. Present-day socialist thought has, to a great extent, forgotten this. This has led to the ill-fated inversion of goals and means.

Giving the highest priority to such goals as reaching and overtaking, Western technology may lead to the restoration of some essential structures of the consumption civilisation of the West.

A critical analysis of the criterion of self-realisation leads to the following conclusions:

(1) Self-realisation cannot be reduced to the manifestation of *actual* human abilities only. Actual abilities are conditioned by specific social circumstances, tradition, the particular individual history of each person, the opportunities available to the individual for getting to know himself in the period of his formation—to discover his predispositions, affinities, and talents, to experiment freely with them and to develop them fully in that period when latent predispositions are still flexible, still open to development. A society which still struggles with the natural environment by means of the mechanical and alienated labour of millions of physical workers, and by a narrow professional division of labour, systemati-

cally realises only those abilities which are needed for high technical effectiveness and for the maintenance of a stable social order. However, investigations of human activities under exceptionally favourable conditions (élite education, freedom for leisure, material abundance, etc.) show that man also has *potential* abilities which are the product of the entire course of history and which can be realised by every normal individual who is provided with suitable conditions. Such abilities are: progressive cultivation and enrichment of the senses, rational thinking and problem-solving, imagination, communication, creative activities (the ability for innovation, for bringing beauty to all which is otherwise routine and conventional), self-consciousness (the ability for identification with oneself, for a critical study of oneself and one's role in the society, of freely-chosen turning points in one's basic orientations in life). In modern industrial production, in bureaucratic politics, and in a culture highly popularised on the basis of an *avant garde* of the élite and cheap entertainment for the masses, these potential abilities are blocked or crippled for the majority of individuals.

(2) When we say that in human interaction with the environment the quantity of produced goods and instruments is not essential but rather the satisfaction of human needs, then the problem arises: with which needs are we concerned here? There are many artificial needs which seem to be important or even to take priority, but whose satisfaction leaves us with extreme indifference, emptiness, or even fills us with loathing. In the life of the individual, as well as of entire nations and societies, after a number of decades we again discover that the consciousness of needs was crippled or distorted, that we ran after phantoms which, really, were not indispensable, while during that time the genuine needs remained repressed, unrealised, and perhaps definitely extinguished.

Up to this point, science has not been highly engaged in the problem of human needs. With respect to a typology of needs, much less has been achieved than in the typology of minerals, insects, or of nuclear particles. Still worse, as long as science is predominantly descriptive and value-neutral, it lacks the criteria for making an indispensable distinction between authentic and artificial needs. This problem obviously demands the presence of philosophical assumptions which the positivistic orientation of science tends to avoid.

The prevailing theory in all capitalist societies begins with the principle that the market is the basic regulator of production and that the market should indicate the type and quality of goods desired by the consumers. Therefore, the market is supposed to bring a knowledge of the structure of human needs, which individual companies can take account of in their production plans.

Socialist theory recognises the positivist character of such a solution: the needs manifested on the market are the result of a historically-formed condition. Some of these needs are the consequences of upbringing and of manipulation by a powerful propaganda apparatus; on the other hand, some of those needs which have not yet been expressed should be encouraged and realised. However, the proposed solution proved to be entirely unsatisfactory: if planning committees and political forums are to decide which human needs should be satisfied and developed, and which should not, then they obtain a tremendous power which exceeds every other in history. This is so because (a) hitherto every influence on the formation of the structure of human needs and the scale of priorities was, after all, inadequately organised, unsystematic, and allowed some kind of selection; however, here we are concerned with compulsion by a highly-organised, monolithic, authoritarian will; (b) exercising an influence on human needs, tastes, and aspirations by way of the market and of commercial advertising is at least direct and limited in time, while a bureaucratic influence determines the structure and priority of needs for generations to come.

Complete identification and fulfilment of authentic human needs are possible only under the condition of transcending manipulation by both the modern market economy and by bureaucratic authority. Only under conditions of free experimentation, self-government, and training for self-knowledge and self-improvement is it possible to develop a critical consciousness of artificial needs and to have an orientation which leads to the satisfaction of true needs—those which lead to the realisation of specific, fundamental human abilities.

(3) The concept of self-realisation does not imply the prior existence of an *unchanged* and *closed* human nature. Human nature is constituted by contradictory latent predispositions whose actualisation is not necessary but only possible, dependent upon specific

historical conditions, and open to various possibilities of its further development. It is not necessary to be an existentialist to negate rigid determinism. In order to affirm the possibility of harmony between the technical and humanistic criteria, it is not necessary to deny the fact that human nature contains within itself an element of evil.

The fact that man is still flexible and open to various developmental possibilities is compatible with the fact that previous history has left its mark on him in the form of some basic latent predispositions. There is an anthropological *a priori,* there is a human nature which precedes existence but which is a historical and not transcendental category. It is pluralistic and does not determine existence in a rigid way. It leaves more or less room for human freedom.

Also our hope that, in spite of other unfavourable possibilities in future human history, there is an optimal real possibility that in post-industrial society man will use his power over the natural environment to realise his most creative abilities and develop all the wealth of his authentic needs, is entirely compatible with the recognition that in many of our contemporaries man described by Hobbes—selfish, greedy, aggressive—is still present. However, this component of human nature is a historical product and just as specific historical conditions favoured it and brought it into the centre of attention, changed conditions can block, repress, and transform it into analogous, socially acceptable tendencies of human behaviour.

Optimal interaction between man and the environment is that which, while increasing human control over the natural environment, simultaneously creates conditions for the maximal, historically-possible self-realisation of man.

NOTES

1. Theodore Roszak, *The Making of a Counter Culture,* New York 1968.
2. Gandhi, *Borba nenasiljem* (Struggle by Passive Resistance), 'Komunist' 1970, pp. 153-164.
3. Marx, *Economic and Philosophical Manuscripts.* 'Private Property and Communism', *ad* p. xxxix.
4. Marx and Engels, *Komunisticki manifest* (Communist Manifesto), (*Selected Works,* 'Kultura', 1949, p. 35).
5. *Kapital,* Vol. I, 1, D, § 4 (published by 'Kultura', Belgrade 1948, p. 43).

HEGELIAN-MARXIST

Lordship and bondage

G. W. F. HEGEL

■ Georg Wilhelm Friedrich Hegel *(1770-1831) was born in Stuttgart. He was an idealist philosopher who had an impact on almost every well-known philosopher who succeeded him. He attended a theological seminary at the University of Tübingen. After his graduation he worked for a while as resident tutor for an aristocratic family. In 1800 he moved to Jena, where he accepted a teaching post. It was in Jena that he wrote his most famous book,* The Phenomenology of Mind *(1807). In this work he articulated a complex theory of consciousness. According to Hegel, the Mind (or Absolute Spirit) is the ultimate reality. The experience of the diversity and separateness of material entities in the world only reflects one stage of consciousness in the long journey of the self toward complete self-consciousness. When that ultimate stage is reached, the self will experience itself to be at one with Reality, the infinite Mind.*

In the famous passage that appears here, Hegel discusses the master-slave relationship. The basis of the relationship is the fact that a self-consciousness desires the recognition of another self-consciousness in order to become certain of itself as a true being. The most primitive mode of achieving such recognition is for the self-consciousness to wage a struggle till death with the opposing self-consciousness. For "they prove themselves and each other through a life-and-death struggle." But the end result of such an approach is the destruction of the other self-consciousness, and because of its destruction it is no longer able to supply the necessary recognition. Thus this approach fails.

A higher mode of achieving the needed recognition preserves the life of the other self-consciousness but denies it its independence. Thus the master-slave relationship emerges. But this relationship is not completely satisfactory either. The master again becomes unhappy with his victory, for he realizes that he has only achieved the recognition of a dependent *self-consciousness. On the other hand, the slave through his toil discovers his own independence. Work gives him a feeling of power, because through his labor he gives objects their form. Hegel explains: "This activity giving shape and form, is at the same time the individual existence . . . which now in the work it does is externalized and passes into the condition of permanence," thus apprehending itself directly as an independent being. This means that while the master depends on the slave for the latter's recognition and his mediation with the world of objects in order to assure the master of the certainty of his true being, the slave obtains that assurance directly through his labor. The relationship has been reversed. The master is now dependent; the slave is not, having been liberated through his labor.*

We have here, once more, a description of the evolution of self-consciousness. This evolution is again seen as being related to man's activities as homo faber. *To this extent Ortega and Hegel agree. But while for Ortega fabrication (including autofabrication) defines the very being of a person, for Hegel it liberates him and thus enables him to apprehend his true being. In discussing the decision to fight till death (a form of autofabrication) Hegel says: "The individual, who has not staked his life, may, no doubt, be recognized as a Person; but he has not attained the truth of this recognition as an* independent *self-consciousness" [emphasis added].*

Self-consciousness exists in itself and for itself, in that, and by the fact that it exists for another self-consciousness; that is to say, it *is* only by being acknowledged or "recognized". The conception of this its unity in its duplication, of infinitude realizing itself in self-consciousness, has many sides to it and encloses within it elements of varied significance. Thus its moments must on the one hand be strictly kept apart in detailed distinctiveness, and, on the other, in this distinction must, at the same time, also be taken as not distinguished, or must always be accepted and understood in their opposite sense. This double meaning of what is distinguished lies in the nature of self-consciousness:—of its being infinite, or directly the opposite of the determinateness in which it is fixed. The detailed exposition of the notion of this spiritual unity in its duplication will bring before us the process of Recognition.

Self-consciousness has before it another self-consciousness; it has come outside itself. This has a double significance. First it has lost its own self, since it finds itself as an *other* being; secondly, it has thereby sublated[1] that other, for it does not regard the other as essentially real, but sees its own self in the other.

It must cancel this its other. To do so is the sublation of that first double meaning, and is therefore a second double meaning. First, it must set itself to sublate the other independent being, in order thereby to become certain of itself as true being, secondly, it thereupon proceeds to sublate its own self, for this other is itself.

This sublation in a double sense of its otherness in a double sense is at the same time a return in a double sense into its self. For, firstly, through sublation, it gets back itself, because it becomes one with itself again through the cancelling of *its* otherness; but secondly, it likewise gives otherness back again to the other self-consciousness, for it was aware of being in the other, it cancels this its own being in the other and thus lets the other again go free.

This process of self-consciousness in relation to another self-consciousness has in this manner been represented as the action of one alone. But this action on the part of the one has itself the double significance of being at once its own action and the action of that other as well. For the other is likewise independent, shut up within itself, and there is nothing in it which is not there through itself. The first does not have the object before it only in the passive form characteristic primarily of the object of desire, but as an object existing independently for itself, over which therefore it has no power to do anything for its own behoof, if that object does not *per se* do what the first does to it. The process then is absolutely the double process of both self-consciousnesses. Each sees the other do the same as itself; each itself does what it demands on the part of the other, and for that reason does what it does, only so far as the other does the same. Action from one side only would be useless, because what is to happen can only be brought about by means of both.

The action has then a *double entente* not only in the sense that it is an act done to itself as well as to the other, but also in the sense that the act *simpliciter* is the act of the one as well as of the other regardless of their distinction.

In this movement we see the process repeated which came before us as the play of forces; in the present case, however, it is found in consciousness. What in the former had effect only for us [contemplating experience], holds here for the terms themselves. The middle term is self-consciousness which breaks itself up into the extremes; and each extreme is this interchange of its own determinateness, and complete transition into the opposite. While *qua* consciousness, it no doubt comes outside itself, still, in being outside itself, it is at the same time restrained within itself, it exists for itself, and its self-externalization is for consciousness. *Consciousness* finds that it immediately is and is not another consciousness, as also that this other is for itself only when it cancels itself as existing for itself, and has self-existence only in the self-existence of the other. Each is the mediating term to the other, through which each mediates and unites itself with itself; and each is to itself and to the other an immediate self-existing reality, which, at the same time, exists thus for itself only through this mediation. They recognize themselves as mutually recognizing one another.

This pure conception of recognition, of duplication of self-consciousness within its unity, we must now consider in the way its process appears for self-consciousness. It will, in the first place, present the aspect of the disparity of the two, or the break-up of the middle term into the extremes, which, *qua* extremes, are opposed to one another, and of which one is merely recognized, while the other only recognizes.

Self-consciousness is primarily simple existence for self, self-identity by exclusion of every other form itself. It takes its essential nature and absolute object to be Ego; and in this immediacy, in this bare fact of its self-existence, it is individual. That which for it is other stands as unessential object, as object with the impress and character of negation. But the other is also a self-consciousness; an individual makes its appearance in antithesis to an individual. Appearing thus in their immediacy, they are for each other in the manner of ordinary objects. They are independent individual forms, modes of consciousness that have not risen above the bare level of life (for the existent object here has been determined as life). They are, moreover, forms of consciousness which have not yet accomplished for one another the process of absolute abstraction, of uprooting all immediate existence, and of being merely the bare, negative fact of self-identical consciousness; or, in other words, have not yet revealed themselves to each other as existing purely for themselves, i.e., as self-consciousness. Each is indeed certain of its own self, but not of the other, and hence its own certainty of itself is still without truth. For its truth would be merely that its own individual existence for itself would be shown to it to be an independent object, or, which is the same thing, that the object would be exhibited as this pure certainty of itself. By the notion of recognition, however, this is not possible, except in the form that as the other is for it, so it is for the other; each in its self through its own action and again through the action of the other achieves this pure abstraction of existence for self.

The presentation of itself, however, as pure abstraction of self-consciousness consists in showing itself as a pure negation of its objective form, or in showing that it is fettered to no determinate existence, that it is not bound at all by the particularity everywhere characteristic of existence as such, and is *not* tied up with life. The process of bringing all this out involves a twofold action—action on the part of the other and action on the part of itself. In so far as it is the other's action, each aims at the destruction and death of the other. But in this there is implicated also the second kind of action, self-activity; for the former implies that it risks its own life. The relation of both self-consciousnesses is in this way so constituted that they prove themselves and each other through a life-and-death struggle. They must enter into this struggle, for they must bring their certainty of themselves, the certainty of being for themselves, to the level of objective truth, and make this a fact both in the case of the other and in their own case as well. And it is solely by risking life that freedom is obtained; only thus is it tried and proved that the essential nature of self-consciousness is not bare existence, is not the merely immediate form in which it at first makes its appearance, is not its mere absorption in the expanse of life. Rather it is thereby guaranteed that there is nothing present but what might be taken as a vanishing moment—that self-consciousness is merely pure self-existence, being-for-self. The individual, who has not staked his life, may, no doubt, be recognized as a Person; but he has not attained the truth of this recognition as an independent self-consciousness. In the same way each must aim at the death of the other, as it risks its own life thereby; for that other is to it of no more worth than itself; the other's reality is presented to the former as an external other, as outside itself; it must cancel that externality. The other is a purely existent consciousness and entangled in manifold ways; it must view its otherness as pure existence for itself or as absolute negation.

This trial by death, however, cancels both the truth which was to result from it, and therewith the certainty of self altogether. For just as life is the natural "position" of consciousness, independence without absolute negativity, so death is the natural "negation" of consciousness, negation without independence, which thus remains without the requisite significance of actual recognition. Through death, doubtless, there has arisen the certainty that both did stake their life, and held it lightly both in their own case and in the case of the other; but that is not for those who underwent this struggle. They cancel their consciousness which had its place in this alien element of natural existence; in other words, they cancel themselves and are sublated as terms or extremes seeking to have existence on their own account. But along with this there vanishes from the play of change the essential moment, viz. that of breaking up into extremes with opposite characteristics; and the middle term collapses into a lifeless unity which is broken up into lifeless extremes, merely existent and not opposed. And the two do not mutually give and receive one another back from each other through consciousness; they let one another go quite indifferently, like things. Their act is abstract negation, not the negation characteristic of consciousness, which cancels in such a way that it preserves and maintains what is sublated, and thereby survives its being sublated.

In this experience self-consciousness becomes aware that *life* is as essential to it as pure self-consciousness. In immediate self-consciousness the simple ego is absolute object, which, however, is for us or in itself absolute mediation, and has as its essential moment substantial and solid independence. The dissolution of that simple unity is the result of the first experience; through this there is posited a pure self-consciousness, and a consciousness which is not purely for itself, but for another, i.e. as an existent consciousness, consciousness in the form and shape of thinghood. Both moments are essential, since, in the first instance, they are unlike and opposed, and their reflexion into unity has not yet come to light, they stand as two opposed forms or modes of consciousness. The one is independent, and its essential nature is to be for itself; the other is dependent, and its essence is life or existence for another. The former is the Master, or Lord, the latter the Bondsman.

The master is the consciousness that exists *for itself;* but no longer merely the general notion of existence for self. Rather, it is a consciousness existing on its own account which is mediated with itself through an other consciousness, i.e. through an other whose very nature implies that it is bound up with an independent being or with thinghood in general. The master brings himself into relation to both these moments, to a thing as such, the object of desire, and to the consciousness whose essential character is thinghood. And since the master, is *(a) qua* notion of self-consciousness, an immediate relation of self-existence, but *(b)* is now moreover at the same time mediation, or a being-for-self which is for itself only through an other—he [the master] stands in relation *(a)* immediately to both *(b)* mediately to each through the other. The master relates himself to the bondsman mediately through independent existence, for that is precisely what keeps the bondsman in thrall; it is his chain, from which he could not in the struggle get away, and for that reason he proved himself to be dependent, to have his independence in the shape of thinghood. The master, however, is the power controlling this state of existence, for he has shown in the struggle that he holds it to be merely something negative. Since he is the power dominating existence, while this existence again is the power controlling the other [the bondsman], the master holds, *par consequence,* this other in subordination. In the same way the master relates himself to the thing mediately through the bondsman. The bondsman being a self-consciousness in the broad sense, also takes up a negative attitude to things and cancels them; but the thing is, at the same time, independent for him, and, in consequence, he cannot, with all his negating, get so far as to annihilate it outright and be done with it; that is to say, he merely works on it. To the master, on the other hand, by menas of this mediating process, belongs the immediate relation, in the sense of the pure negation of it, in other words he gets the enjoyment. What mere desire did not attain, he now succeeds in attaining, viz. to have done with the thing, and find satisfaction in enjoyment. Desire alone did not get the length of this, because of the independence of the thing. The master, however, who has interposed the bondsman between it and himself, thereby relates himself merely to the dependence of the thing, and enjoys it without qualification and without reserve. The aspect of its independence he leaves to the bondsman, who labours upon it.

In these two moments, the master gets his recognition through an other consciousness, for in them the latter affirms itself as unessential, both by working upon the thing, and, on the other hand, by the fact of being dependent on a determinate existence; in neither case can this other get the mastery over existence, and succeed in absolutely negating it. We have thus here this moment of recognition, viz. that the other consciousness cancels itself as self-existent, and, *ipso facto,* itself does what the first does to it. In the same way we have the other moment, that this action on the part of the second is the action proper of the first; for what is done by the bondsman is properly an action on the part of the master. The latter exists only for himself, that is his essential nature; he is the negative power without qualification, a power to which the thing is naught. And he is thus the absolutely essential act in this situation, while the bondsman is not so, he is an unessential activity. But for recognition proper there is needed the moment that what the master does to the other he should also do to himself, and what the bondsman does to himself, he should do to the other also. On that account a form of recognition has arisen that is one sided and unequal.

In all this, the unessential consciousness is, for the master, the object which embodies the truth of his certainty of himself. But it is evident that this object does not correspond to its notion; for, just where the master has effectively achieved lordship, he really finds that something has come about quite different from an independent consciousness. It is not an independent, but rather a dependent consciousness that he has achieved. He is thus not assured of self-existence as his truth; he finds that his truth is rather the unessential consciousness, and the fortuitous unessential action of that consciousness.

The truth of the independent consciousness is accordingly the consciousness of the bondsman. This doubtless appears in the first instance outside itself, and not as the truth of self-consciousness. But just as lordship showed its essential nature to be the reverse of what it wants to be, so, too, bondage will, when completed, pass into the opposite of what it immediately is: being a consciousness repressed within itself, it will enter into itself, and change round into real and true independence.

We have seen what bondage is only in relation to lordship. But it is a self-consciousness, and we have now to consider what it is, in this regard, in and for itself. In the first instance, the master is taken to be the essential reality for the state of bondage; hence, for it, the truth is the independent consciousness existing for itself, although this truth is not taken yet as inherent in bondage itself. Still, it does in fact contain within itself this truth of pure negativity and self-existence, because it has experienced this reality within it. For this consciousness was not in peril and fear for this element or that, nor for this or that moment of time, it was afraid for its entire being; it felt the fear of death, the sovereign master. It has been in that experience melted to its inmost soul, has trembled throughout its every fibre, and all that was fixed and steadfast has quaked within it. This complete perturbation of its entire substance, this absolute dissolution of all its stability into fluent continuity, is, however, the simple, ultimate nature of self-consciousness, absolute negativity, pure self-referrent existence, which consequently is involved in this type of consciousness. This moment of pure self-existence is moreover a fact for it; for in the master it finds this as its object. Further, this bondsman's consciousness is not only this total dissolution in a general way; in serving and toiling the bondsman actually carries this out. By serving he cancels in every particular aspect his dependence on and attachment to natural existence, and by his work removes this existence away.

The feeling of absolute power, however, realized both in general and in the particular form of service, is only dissolution implicitly; and albeit the fear of the lord is the beginning of wisdom, consciousness is not therein aware of being self-existent. Through work and labour, however, this consciousness of the bondsman comes to itself. In the moment which corresponds to desire in the case of the master's consciousness, the aspect of the non-essential relation to the thing seemed to fall to the lot of the servant, since the thing there retained its independence. Desire has reserved to itself the pure negating of the object and thereby unalloyed feeling of self. This satisfaction, however, just for that reason is itself only a state of evanescence, for it lacks objectivity or subsistence. Labour, on the other hand, is desire restrained and checked, evanescence delayed and postponed; in other words, labour shapes and fashions the thing. The negative relation to the object passes into the *form* of the object, into something that is permanent and remains; because it is just for the labourer that the object has independence. This negative mediating agency, this activity giving shape and form, is at the same time the individual existence, the pure self-existence of that consciousness, which now in the work it does is externalized and passes into the condition of permanence. The consciousness that toils and serves accordingly attains by this means the direct apprehension of that independent being as its self.

But again, shaping or forming the object has not only the positive significance that the bondsman becomes thereby aware of himself as factually and objectively self-existent; this type of consciousness has also a negative import, in contrast with its first moment, the element of fear. For in shaping the thing it only becomes aware of its own proper negativity, its existence on its own account, as an object, through the fact that it cancels the actual form confronting it. But this objective negative element is precisely the alien, external reality, before which it trembled. Now, however, it destroys this extraneous alien negative, affirms and sets itself up as a negative in the element of permanence, and thereby becomes for itself a self-existent being. In the master, the bondsman feels self-existence to be something external, an objective fact; in fear self-existence is present within himself; in fashioning the thing, self-existence comes to be felt explicitly as his own proper being, and he attains the consciousness that he himself exists in its own right and on its own account *(an und für sich)*. By the fact that the form is objectified, it does not become something other than the consciousness moulding the thing through work; for just that form is his pure self-existence, which therein becomes truly realized. Thus precisely in labour where there seemed to be merely some outsider's mind and ideas involved, the bondsman becomes aware, through this re-discovery of himself by himself, of having and being a "mind of his own".

For this reflexion of self into self the two moments, fear and service in general, as also that of formative activity, are necessary: and at the same time both must exist in a universal manner. Without the discipline of service and obedience, fear remains formal and does not spread over the whole known reality of existence. Without the formative activity shaping the thing, fear remains inward and mute, and consciousness does not become objective for itself. Should consciousness shape and form the thing without the initial state of absolute fear, then it has a merely vain and futile "mind of its own"; for its form or negativity is not negativity *per se,* and hence its formative activity cannot furnish the consciousness of itself as essentially real. If it has endured not absolute fear, but merely some slight anxiety, the negative reality has remained external to it, its substance has not been through and through infected thereby. Since the entire content of its natural consciousness has not tottered and shaken, it is still inherently a determinate mode of being; having a "mind of its own" *(der eigene Sinn)* is simply stubbornness *(Eigensinn)*, a type of freedom which does not get beyond the attitude of bondage. As little as the pure form can become its essential nature, so little is that form, considered as extending over particulars, a universal formative activity, an absolute notion; it is rather a piece of cleverness which has mastery within a certain range, but not over the universal power nor over the entire objective reality.

NOTE

1. "Sublation" means "the dialectic cancelling of the other"—that is, preserving the positive aspects of what is being cancelled, while doing away with the negative ones [ed. note].

PRAGMATIC-LIBERAL

From "Renascent Liberalism"

JOHN DEWEY

■ John Dewey *(1859-1952) was born in Burlington, Vermont. He was a philosopher, an educator, and a social critic. As a reformer he believed that through realistic scientific knowledge and creative imagination humanity could improve its condition. For him the role of philosophy in civilization was an active one. While rooted in a specific culture, philosophy must transcend that culture and point out new ideals for a changing society. Thus Dewey saw philosophy as one important means of directing change in society. Among his books are* Democracy and Education *(1916) and* Philosophy and Civilization *(1931).*

In Liberalism and Social Action *(1935) Dewey expresses concern for the future of liberalism and its values in the face of world conditions. The following selection from this work sounds a familiar note. Like Marković, Dewey values self-realization. For "liberalism is committed to an end that is at once enduring and flexible: the liberation of individuals so that realization of their capacities may be the law of their life." The liberal is committed to a social organization that makes such liberation possible. This is the basis for Dewey's concern with the intense and rapid change occurring in society. The change, which has permeated both public and private life, "does have to be directed. It has to be so controlled that it will move to some end in accordance with the principles of life."*

The most rapid change, according to Dewey, has occurred in the area of industrial habits, while political and legal relations, among others, have lagged behind. Science and technology have introduced an age of potential plenty and corporate organization, but much of society's beliefs and values are still rooted in the previous age of scarcity, insecurity, and small business.

The solution to this uneven social development lies in education. But "the educational task cannot be accomplished merely by working upon men's minds, without action that effects actual change in institutions." Nevertheless, violence as a method of change is ruled out. The liberal is committed instead "to the organization of intelligent action as the chief method."

Nothing is blinder than the supposition that we live in a society and world so static that either nothing new will happen or else it will happen because of the use of violence. Social change is here as a fact, a fact having multifarious forms and marked in intensity. Changes that are revolutionary in effect are in process in every phase of life. Transformations in the family, the church, the school, in science and art, in economic and political relations, are occurring so swiftly that imagination is baffled in attempt to lay hold of them. Flux does not have to be created. But it does have to be directed. It has to be so controlled that it will move to some end in accordance with the principles of life, since life itself is development. Liberalism is committed to an end that is at once enduring and flexible: the liberation of individuals so that realization of their capacities may be the law of their life. It is committed to the use of freed intelligence as the method of directing change. In any case civilization is faced with the problem of uniting the changes that are going on into a

---

□ From *Liberalism and Social Action* (New York: Capricorn Books, 1963), pp. 56-69. Reprinted with the permission of the Center for Dewey Studies, Southern Illinois University, Carbondale.

coherent pattern of social organization. The liberal spirit is marked by its own picture of the pattern that is required: a social organization that will make possible effective liberty and opportunity for personal growth in mind and spirit in all individuals. Its present need is recognition that established material security is a prerequisite of the ends which it cherishes, so that, the basis of life being secure, individuals may actively share in the wealth of cultural resources that now exist and may contribute, each in his own way, to their further enrichment.

The fact of change has been so continual and so intense that it overwhelms our minds. We are bewildered by the spectacle of its rapidity, scope and intensity. It is not surprising that men have protected themselves from the impact of such vast change by resorting to what psychoanalysis has taught us to call rationalizations, in other words, protective fantasies. The Victorian idea that change is a part of an evolution that necessarily leads through successive stages to some preordained divine far-off event is one rationalization. The conception of a sudden, complete, almost catastrophic, transformation, to be brought about by the victory of the proletariat over the class now dominant, is a similar rationalization. But men have met the impact of change in the realm of actuality, mostly by drift and by temporary, usually incoherent, improvisations. Liberalism, like every other theory of life, has suffered from the state of confused uncertainty that is the lot of a world suffering from rapid and varied change for which there is no intellectual and moral preparation.

Because of this lack of mental and moral preparation the impact of swiftly moving changes produced, as I have just said, confusion, uncertainty and drift. Change in patterns of belief, desire and purpose has lagged behind the modification of the external conditions under which men associate. Industrial habits have changed most rapidly; there has followed at considerable distance, change in political relations; alterations in legal relations and methods have lagged even more, while changes in the institutions that deal most directly with patterns of thought and belief have taken place to the least extent. This fact defines the primary, though not by any means the ultimate, responsibility of a liberalism that intends to be a vital force. Its work is first of all education, in the broadest sense of that term. Schooling is a part of the work of education, but education in its full meaning includes all the influences that go to form the attitudes and dispositions (of desire as well as of belief), which constitute dominant habits of mind and character.

Let me mention three changes that have taken place in one of the institutions in which immense shifts have occurred, but that are still relatively external—external in the sense that the pattern of intelligent purpose and emotion has not been correspondingly modified. Civilization existed for most of human history in a state of scarcity in the material basis for a humane life. Our ways of thinking, planning and working have been attuned to this fact. Thanks to science and technology we now live in an age of potential plenty. The immediate effect of the emergence of the new possibility was simply to stimulate, to a point of incredible exaggeration, the striving for the material resources, called wealth, opened to men in the new vista. It is a characteristic of all development, physiological and mental, that when a new force and factor appears, it is first pushed to an extreme. Only when its possibilities have been exhausted (at least relatively) does it take its place in the life perspective. The economic-material phase of life, which belongs in the basal ganglia of society, has usurped for more than a century the cortex of the social body. The habits of desire and effort that were bred in the age of scarcity do not readily subordinate themselves and take the place of the matter-of-course routine that becomes appropriate to them when machines and impersonal power have the capacity to liberate man from bondage to the strivings that were once needed to make secure his physical basis. Even now when there is a vision of an age of abundance and when the vision is supported by hard fact, it is material security as an end that appeals to most rather than the way of living which this security makes possible. Men's minds are still pathetically held in the clutch of old habits and haunted by old memories.

For, in the second place, insecurity is the natural child and the foster child, too, of scarcity. Early liberalism emphasized the importance of insecurity as a fundamentally necessary economic motive, holding that without this goad men would not work, abstain or accumulate. Formulation of this conception was new. But the fact that was formulated was nothing new. It was deeply rooted in the habits that were formed in the long struggle against material scarcity. The system that goes by the name of capitalism is a systematic manifestation of desires and purposes built up in an age of ever threatening want and now carried over into a time of ever increasing potential plenty. The conditions that generate insecurity for the many no longer spring from nature. They are found in institutions and arrangements that are within deliberate human control. Surely this change marks one of the greatest revolutions that has taken place in all human history. Because of it, insecurity is not now the motive to work and sacrifice but to despair. It is not an instigation to put forth energy but to an impotency that can be converted from death into endurance only by charity. But the habits of mind and action that modify institutions to make potential abundance an actuality are still so inchoate that most of us discuss labels like individualism, socialism and communism instead of even perceiving the possibility, much less the necessity for realizing what can and should be.

In the third place, the patterns of belief and purpose that still dominate economic institutions were formed when individuals produced with their hands, alone or in small groups. The notion that society in general is served by the unplanned coincidence of the consequences of a vast multitude of efforts put forth by isolated individuals without reference to any social end, was also something new as a formulation. But it also formulated the working principle of an epoch which the advent of new forces of production was to bring to an end. It demands no great power of intelligence to see that under present conditions the isolated individual is well-nigh helpless. Concentration and corporate organization are the rule. But the concentration and corporate organization are still controlled in their operation by ideas that were institutionalized in eons of separate individual effort. The attempts at coöperation for mutual benefit that are put forth are precious as experimental moves. But that society itself should see to it that a coöperative industrial order be instituted, one that is consonant with the realities of production enforced by an era of machinery and power, is so novel an idea to the general mind that its mere suggestion is hailed with abusive epithets—sometimes with imprisonment.

When, then, I say that the first object of a renascent liberalism is education, I mean that its task is to aid in producing the habits of mind and character, the intellectual and moral patterns, that are somewhere near even with the actual movements of events. It is, I repeat, the split between the latter as they have externally occurred and the ways of desiring, thinking, and of putting emotion and purpose into execution that is the basic cause of present confusion in mind and paralysis in action. The educational task cannot be accomplished merely by working upon men's minds, without action that effects actual change in institutions. The idea that dispositions and attitudes can be altered by merely "moral" means conceived of as something that goes on wholly inside of persons is itself one of the old patterns that has to be changed. Thought, desire and purpose exist in a constant give and take of interaction with environing conditions. But resolute thought is the first step in that change of action that will itself carry further the needed change in patterns of mind and character.

In short, liberalism must now become radical, meaning by "radical" perception of the necessity of thorough-going changes in the set-up of institutions and corresponding activity to bring the changes to pass. For the gulf between what the actual situation makes possible and the actual state itself is so great that it cannot be bridged by piecemeal policies undertaken *ad hoc*. The process of producing the changes will be, in any case, a gradual one. But "reforms" that deal now with this abuse and now with that without having a social goal based upon an inclusive plan, differ entirely from effort at re-forming, in its literal sense, the institutional scheme of things. The liberals of more than a century ago were denounced in their time as subversive radicals, and only when the new economic order was established did they become apologists for the *status quo* or else content with social patchwork. If radicalism be defined as perception of need for radical change, then today any liberalism which is not also radicalism is irrelevant and doomed.

But radicalism also means, in the minds of many, both supporters and opponents, dependence upon use of violence as the main method of effecting drastic changes. Here the liberal parts company. For he is committed to the organization of intelligent action as the chief method. Any frank discussion of the issue must recognize the extent to which those who decry the use of any violence are themselves willing to resort to violence and are ready to put their will into operation. Their fundamental objection is to change in the economic institution that now exists, and for its maintenance they resort to the use of the force that is placed in their hands by this very institution. They do not need to advocate the use of force; their only need is to employ it. Force, rather than intelligence, is built into the procedures of the existing social system, regularly as coercion, in times of crisis as overt violence. The legal system, conspicuously in its penal aspect, more subtly in civil practice, rests upon coercion. Wars are the methods recurrently used in settlement of disputes between nations. One school of radicals dwells upon the fact that in the past the transfer of power in one society has either been accomplished by or attended with violence. But what we need to realize is that physical force is used, at least in the form of coercion, in the very set-up of our society. That the competitive system, which was thought of by early liberals as the means by which the latent abilities of individuals were to be evoked and directed into socially useful channels, is now in fact a state of scarcely disguised battle hardly needs to be dwelt upon. That the control of the means of production by the few in legal possession operates as a standing agency of coercion of the many, may need emphasis in statement, but is surely evident to one who is willing to observe and honestly report the existing scene. It is foolish to regard the political state as the only agency now endowed with coercive power. Its exercise of this power is pale in contrast with that exercised by concentrated and organized property interests.

It is not surprising in view of our standing dependence upon the use of coercive force that at every time of crisis coercion breaks out into open violence. In this country, with its tradition of violence fostered by frontier conditions and by the conditions under which immigration went on during the greater part of our history, resort to violence is especially recurrent on the part of those who are in power. In times of imminent change, our verbal and sentimental worship of the Constitution, with its guarantees of civil liberties of expression, publication and assemblage, readily goes overboard. Often the officials of the law are the worst offenders, acting as agents of some power that rules the economic life of a community. What is said about the value of free speech as a safety valve is then forgotten with the utmost of ease: a comment, perhaps, upon the weakness of the defense of freedom of expression that values it simply as a means of blowing-off steam.

It is not pleasant to face the extent to which, as matter of fact, coercive and violent force is relied upon in the present social system as a means of social control. It is much more agreeable to evade the fact. But unless the fact is acknowledged as a fact in its full depth and breadth, the meaning of dependence upon intelligence as the alternative method of social direction will not be grasped. Failure in acknowledgment signifies, among other things, failure to realize that those who propagate the dogma of dependence upon force have the sanction of much that is already entrenched in the existing system. They would but turn the use of it to opposite ends. The assumption that the method of intelligence already rules and that those who urge the use of violence are introducing a new element into the social picture may not be hypocritical but it is unintelligently unaware of what is actually involved in intelligence as an alternative method of social action.

I begin with an example of what is really involved in the issue. Why is it, apart from our tradition of violence, that liberty of expression is tolerated and even lauded when social affairs seem to be going in a quiet fashion, and yet is so readily destroyed whenever matters grow critical? The general answer, of course, is that at bottom social institutions have habituated us to the use of force in some veiled form. But a part of the answer is found in our ingrained habit of regarding intelligence as an individual possession and its exercise as an individual right. It is false that freedom of inquiry and of expression are not modes of action. They are exceedingly potent modes of action. The reactionary grasps this fact, in practice if not in express idea, more quickly than the liberal, who is too much given to holding that this freedom is innocent of consequences, as well as being a merely individual right. The result is that this liberty is tolerated as long as it does not seem to menace in any way the *status quo* of society. When it does, every effort is put forth to identify the established order with the public good. When this identification is established, it follows that any merely individual right must yield to the general welfare. As long as freedom of thought and speech is claimed as a merely individual right, it will give way, as do other merely personal claims, when it is, or is successfully represented to be, in opposition to the general welfare.

I would not in the least disparage the noble fight waged by early liberals in behalf of individual freedom of thought and expression. We owe more to them than it is possible to record in words. No more eloquent words have ever come from any one than those of Justice Brandeis in the case of a legislative act that in fact restrained freedom of political expression. He said: "Those who won our independence believed that the final end of the State was to make men free to develop their faculties, and that in its government the deliberative faculties should prevail over the arbitrary. They valued liberty both as an end and as a means. They believed liberty to be the secret of happiness and courage to be the secret of liberty. They believed that freedom to think as you will and to speak as you think are means indispensable to the discovery and spread of political truth; that without free speech and assembly discussion would be futile; that with them, discussion affords ordinarily adequate protection against the dissemination of noxious doctrines; that the greatest menace to freedom is an inert people; that public discussion is a political duty; and that this should be a fundamental principle of the American government." This is the creed of a fighting liberalism. But the issue I am raising is connected with the fact that these words are found in a dissenting, a minority opinion of the Supreme Court of the United States. The public function of free individual thought and speech is clearly recognized in the words quoted. But the reception of the truth of the words is met by an obstacle: the old habit of defending liberty of thought and expression as something inhering in individuals apart from and even in opposition to social claims.

Liberalism has to assume the responsibility for making it clear that intelligence is a social asset and is clothed with a function as public as is its origin, in the concrete, in social coöperation. It was Comte who, in reaction against the purely individualistic ideas that seemed to him to underlie the French Revolution, said that in mathematics, physics and astronomy there is no right of private conscience. If we remove the statement from the context of actual scientific procedure, it is dangerous because it is false. The individual inquirer has not only the right but the duty to criticize the ideas, theories and "laws" that are current in science. But if we take the statement in the context of scientific method, it indicates that he carries on this criticism in virtue of a socially generated body of knowledge and by means of methods that are not of private origin and possession. He uses a method that retains public validity even when innovations are introduced in its use and application.

Henry George, speaking of ships that ply the ocean with a velocity of five or six hundred miles a day, remarked, "There is nothing whatever to show that the men who today build and navigate and use such ships are one whit superior in any physical or mental quality to their ancestors, whose best vessel was a coracle of wicker and hide. The enormous improvement which these ships show is not an improvement of human nature; it is an improvement of society—it is due to a wider and fuller union of individual efforts in accomplishment of common ends." This single instance, duly pondered, gives a better idea of the nature of intelligence and its social office than would a volume of abstract dissertation. Consider merely two of the factors that enter in and their social consequences. Consider what is involved in the production of steel, from the first use of fire and then the crude smelting of ore, to the processes that now effect the mass production of steel. Consider also the development of the power of guiding ships across trackless wastes from the day when they hugged the shore, steering by visible sun and stars, to the appliances that now enable a sure course to be taken. It would require a heavy tome to describe the advances in science, in mathematics, astronomy, physics, chemistry, that have made these two things possible. The record would be an account of a vast multitude of coöperative efforts, in which one individual uses the results provided for him by a countless number of individuals, and uses them so as to add to the common and public store. A survey of such facts brings home the actual social character of intelligence as it actually develops and makes its way. Survey of the consequences upon the ways of living of individuals and upon the terms on which men associate together, due to the new method of transportation would take us to the wheat farmer of the prairies, the cattle raiser of the plains, the cotton grower of the South; into a multitude of mills and factories, and to the counting-room of banks, and what would be seen in this country would be repeated in every country of the globe.

MEDIA-ORIENTED

A schoolman's guide to Marshall McLuhan

JOHN M. CULKIN

■ John M. Culkin *is the director of the Center for Communications at Fordham University. In this article he discusses the views of Marshall McLuhan, who is often referred to as "the oracle of the electric age." McLuhan is a professor of English and director of the Center for Culture and Technology at the University of Toronto.*

Culkin states that "this article is an attempt to select and order those elements of McLuhanism which are most relevant to the schools and to provide the schoolman with some new ways of thinking about the schools." According to Culkin, "the student, whose psyche is being programmed for tempo, information, and relevance by his electronic environment, is still being processed in classrooms operating on the postulates of another day." McLuhan says "the aim is to develop an awareness about print and the newer technologies of communication so that we can orchestrate them, minimize their mutual frustrations and clashes, and get the best out of each in the educational process." Culkin discusses some major elements of McLuhan's views that are crucial for developing this awareness.

Education, a seven-year-old assures me, is "how kids learn stuff." Few definitions are as satisfying. It includes all that is essential—a who, a what, and a process. It excludes all the people, places, and things which are only sometimes involved in learning. The economy and accuracy of the definition, however, are more useful in locating the problem than in solving it. We know little enough about *kids,* less about *learning,* and considerably more than we would like to know about *stuff.*

In addition, the whole process of formal schooling is now wrapped inside an environment of speeded-up technological change that is constantly influencing kids and learning and stuff. The jet-speed of this technological revolution, especially in the area of communications, has left us with more reactions to it than reflections about it. Meanwhile back at the school, the student, whose psyche is being programmed for tempo, information, and relevance by his electronic environment, is still being processed in classrooms operating on the postulates of another day. The cold war existing between these two worlds is upsetting for both the student and the schools. One thing is certain: It is hardly a time for educators to plan with nostalgia, timidity, or old formulas. Enter Marshall McLuhan.

He enters from the North, from the University of Toronto, where he teaches English and is Director of the Center for Culture and Technology. He enters with the reputation as "the oracle of the electric age" and as "the most provocative and controversial writer of this generation." More importantly for the schools, he enters as a man with fresh eyes, with new ways of looking at old problems. He is a man who gets his ideas first and judges them later. Most of these ideas are summed up in his book, *Understanding Media.* His critics tried him for not delivering these insights in their most lucid and practical form. It isn't always cricket, however, to ask the same man to crush the grapes and serve the wine. Not all of McLu is nu or tru, but then again neither is *all* of anybody else. This article is an attempt to select and order those elements of McLuhanism which are most relevant to the schools and to provide the schoolman with some new ways of thinking about the schools.

McLuhan's promise is modest enough: "All I have to offer is an enterprise of investigation into a world that's quite unusual and quite unlike any previous world and for which no models of perception will serve." This unexplored world happens to be the present. McLuhan feels that very few men look at the present with a present eye, that they tend to miss the present by translating it into the past, seeing it through a rearview mirror. The unnoticed fact of our present is the electronic environment created by the new communications media. It is as pervasive as the air we breathe (and some would add that it is just as polluted), yet its full import eludes the judgments of common-sense or content-oriented perception. The environments set up by different media are not just containers for people; they are processes that shape people. Such influence is deterministic only if ignored. There is no inevitability as long as there is a willingness to contemplate what is happening.

Theorists can keep reality at arm's length for

☐ From *McLuhan: Pro and Con,* ed. Raymond Rosenthal (New York: Funk & Wagnalls, Inc., 1968), pp. 242-256.

long periods of time. Teachers and administrators can't. They are closeted with reality all day long. In many instances they are co-prisoners with electronic-age students in the old pencil-box cell. And it is the best teachers and the best students who are in the most trouble because they are challenging the system constantly. It is the system that has to come under scrutiny. Teachers and students can say, in the words of the Late Late Show, "Baby, this thing is bigger than both of us." It won't be ameliorated by a few dashes of good will or a little more hard work. It is a question of understanding these new kids and these new media and of getting the schools to deal with the new electronic environment. It's not easy. And the defenders of the old may prove to be the ones least able to defend and preserve the values of the old.

For some people, analysis of these newer technologies automatically implies approbation of them. Their world is so full of *shoulds* that it is hard to squeeze in an *is*. McLuhan suggests a more positive line of exploration:

> At the moment it is important that we understand cause and process. The aim is to develop an awareness about print and the newer technologies of communication so that we can orchestrate them, minimize their mutual frustrations and clashes, and get the best out of each in the educational process. The present conflict leads to elimination of the motive to learn and to diminution of interest in all previous achievement: It leads to loss of the sense of relevance. Without an understanding of media grammars, we cannot hope to achieve a contemporary awareness of the world in which we live.

We have been told that it is the property of true genius to disturb all settled ideas. McLuhan is disturbing in both his medium and his message. His ideas challenge the normal way in which people perceive reality. They can create a very deep and personal threat since they touch on everything in a person's experience. They are just as threatening to the Establishment whose way of life is predicated on the postulates he is questioning. The Establishment has no history of organizing parades to greet its disturbers.

His medium is perhaps more disturbing than his message. From his earliest work he has described his enterprise as "explorations in communication." The word he uses most frequently today is "probe." His books demand a high degree of involvement from the reader. They are poetic and intuitive rather than logical and analytic. Structurally, his unit is the sentence. Most of them are topic sentences—which are left undeveloped. The style is oral and breathless and frequently obscure. It's a different kind of medium.

"The medium is the message," announced McLuhan a dozen years ago in a cryptic and uncompromising aphorism whose meaning is still being explored. The title of his latest book, an illustrated popular paperback treatment of his theories, playfully proclaims that *The Medium Is the Massage*—a title calculated to drive typesetters and critics to hashish and beyond. The original dictum can be looked at in four ways, the third of which includes a massage of importance.

The first meaning would be better communicated orally—"the *medium* is the message." The *medium* is the thing you're missing. Everybody's hooked on content; pay attention to form, structure, framework, *medium*. The play's the thing. The medium's the thing. McLuhan makes the truth stand on its head to attract attention. Why the medium is worthy of attention derives from its other three meanings.

Meaning number two stresses the relation of the medium to the content. The form of communication not only alters the content, but each form also has preferences for certain kinds of messages. Content always exists in some form and is, therefore, to some degree governed by the dynamics of that form. If you don't know the medium, you don't know the message. The insight is neatly summed up by Dr. Edmund Carpenter: "English is a mass medium. All languages are mass media. The new mass media—film, radio, TV—are new languages, their grammars as yet unknown. Each codifies reality differently; each conceals a unique metaphysics. Linguists tell us it's possible to say anything in any language if you use enough words or images, but there's rarely time; the natural course is for a culture to exploit its media biases. . . ."

It is always content-in-form that is mediated. In this sense, the medium is co-message. The third meaning for the M-M formula emphasizes the relation of the medium to the individual psyche. The medium alters the perceptual habits of its users. Independent of the content, the medium itself gets through. Preliterate, literate, and postliterate cultures see the world through different-colored glasses. In the process of delivering content the medium also works over the sensorium of the consumer. To get this subtle insight across, McLuhan punned on m*e*ssage and came up with m*a*ssage. The switch is intended to draw attention to the fact that a medium is not something neutral—it does something to people. It takes hold of them, it jostles them, it bumps them around, it massages them. It opens and closes windows in their sensorium. Proof? Look out the window at the TV generation. They are rediscovering texture, movement, color, and sound as they retribalize the race. TV is a real grabber; it really massages those lazy, unused senses.

The fourth meaning underscores the relation of the medium to society. Whitehead said, "The major advances in civilization are processes that all but wreck the societies in which they occur." The media massage the society as well as the individual. The results pass unnoticed for long periods of time because people tend to view the new as just a little bit more of the old. Whitehead again: "The greatest invention of the nineteenth century was the invention of the method of invention. A new method entered into life. In order to understand our epoch, we can neglect all details of change, such as railways, telegraphs, radios, spinning machines, synthetic dyes. We must concentrate on the method in itself: That is the real novelty which has broken up the foundations of the old civilization." Understanding the medium or process involved is the key to control.

The media shape both content and consumer and do so practically undetected. We recall the story of the Russian worker whose wheelbarrow was searched every day as he left the factory grounds. He was, of course, stealing wheelbarrows. When your medium is your message and they're only investigating content, you can get away with a lot of things—like wheelbarrows, for instance. It's not the picture but the frame. Not the contents but the box. The blank page is not neutral; nor is the classroom.

McLuhan's writings abound with aphorisms, insights, for-instances, and irrelevancies which float loosely around recurring themes. They provide the raw materials of a do-it-yourself kit for tidier types who prefer to do their exploring with clearer charts. What follows is one man's McLuhan served up in barbarously brief form. Five postulates, spanning nearly four thousand years, will serve as the fingers in this endeavor to grasp McLuhan:

(1) 1967 B.C.—*All the senses get into the act*. A conveniently symmetrical year for a thesis that is partially cyclic. It gets us back to man before the Phoenician alphabet. We know from our contemporary ancestors in the jungles of New Guinea and the wastes of the Arctic that preliterate man lives in an all-at-once sense world. The reality that bombards him from all directions is picked up with the omnidirectional antennae of sight, hearing, touch, smell, and taste. Films such as *The Hunters* and *Nanook of the North* depict primitive men tracking game with an across-the-board sensitivity that mystifies Western, literate man. We mystify them too. And it is this cross-mystification that makes inter-cultural abrasions so worthwhile.

Most people presume that their way of perceiving the world is *the* way of perceiving the world. If they hang around with people like themselves, their mode of perception may never be challenged. It is at the poles (literally and figuratively) that the violent contrasts illumine our own unarticulated perceptual prejudices. Toward the North Pole, for example, live Eskimos. A typical Eskimo family consists of a father, a mother, two children, and an anthropologist. When the anthropologist goes into the igloo to study Eskimos, he learns a lot about himself. Eskimos see pictures and maps equally well from all angles. They can draw equally well on top of a table or underneath it. They have phenomenal memories. They travel without visual bearings on their white-on-white world and can sketch cartographically accurate maps of shifting shorelines. They have forty or fifty words for what we call "snow." They live in a world without linearity, a world of acoustic space. They are Eskimos. Their natural way of perceiving the world is different from our natural way of perceiving the world.

Each culture develops its own balance of the senses in response to the demands of its environment. The most generalized formulation of the theory would maintain that the individual's modes of cognition and perception are influenced by the culture he is in, the language he speaks, and the media to which he is exposed. Each culture, as it were, provides its constituents with a custom-made set of goggles. The differences in perception are a question of degree. Some cultures are close enough to each other in perceptual patterns so that the differences pass unnoticed. Other cultural groups, such as the Eskimo and the American teen-ager, are far enough away from us to provide esthetic distance.

(2) *Art imitates life*. In *The Silent Language* Edward T. Hall offers the thesis that all art and technology is an extension of some physical

or psychic element of man. Today man has developed extensions for practically everything he used to do with his body: stone axe for hand, wheel for foot, glasses for eyes, radio for voice and ears. Money is a way of storing energy. This externalizing of individual, specialized functions is now, by definition, at its most advanced stage. Through the electronic media of telegraph, telephone, radio, and television, man has now equipped his world with a nervous system similar to the one within his own body. President Kennedy is shot and the world instantaneously reels from the impact of the bullets. Space and time dissolve under electronic conditions. Current concern for the United Nations, the Common Market, ecumenism, reflects this organic thrust toward the new convergence and unity which is "blowing in the wind." Now in the electric age, our extended faculties and senses constitute a single instantaneous and coexistent field of experience. It's all-at-once. It's shared-by-all. McLuhan calls the world "a global village."

(3) *Life imitates art.* We shape our tools and thereafter they shape us. These extensions of our senses begin to interact with our senses. These media become a massage. The new change in the environment creates a new balance among the senses. No sense operates in isolation. The full sensorium seeks fulfillment in almost every sense experience. And since there is a limited quantum of energy available for any sensory experience, the sense-ratio will differ for different media.

The nature of the sensory effect will be determined by the medium used. McLuhan divides the media according to the quality or definition of their physical signal. The content is not relevant in this kind of analysis. The same picture from the same camera can appear as a glossy photograph or as a newspaper wirephoto. The photograph is well-defined, of excellent pictorial quality, hi-fi within its own medium. McLuhan calls this kind of medium "hot." The newspaper photo is grainy, made up of little dots, low definition. McLuhan calls this kind of medium "cool." Film is hot; television is cool. Radio is hot; telephone is cool. The cool medium or person invites participation and involvement. It leaves room for the response of the consumer. A lecture is hot; all the work is done. A seminar is cool; it gets everyone into the game. Whether all the connections are causal may be debated, but it's interesting that the kids of the cool TV generation want to be so involved and so much a part of what's happening.

(4) *We shaped the alphabet and it shaped us.* In keeping with the McLuhan postulate that "the medium is the message," a literate culture should be more than mildly eager to know what books do to people. Everyone is familiar enough with all the enrichment to living mediated through fine books to allow us to pass on to the subtler effects which might be attributed to the print medium, independent of the content involved. Whether one uses the medium to say that *God is dead* or that *God is love* (--- -- ----), the structure of the medium itself remains unchanged. Nine little black marks with no intrinsic meaning of their own are strung along a line with spaces left after the third and fifth marks. It is this stripping away of meaning that allows us to X-ray the form itself.

As an example, while lecturing to a large audience in a modern hotel in Chicago, a distinguished professor is bitten in the leg by a cobra. The whole experience takes three seconds. He is affected through the touch of the reptile, the gasp of the crowd, the swimming sights before his eyes. His memory, imagination, and emotions come into emergency action. A lot of things happen in three seconds. Two weeks later he is fully recovered and wants to write up the experience in a letter to a colleague. To communicate this experience through print means that it must first be broken down into parts and then mediated, eyedropper fashion, one thing at a time, in an abstract, linear, fragmented, sequential way. That is the essential structure of print. And once a culture uses such a medium for a few centuries, it begins to perceive the world in a one-thing-at-a-time, abstract, linear, fragmented, sequential way. And it shapes its organizations and schools according to the same premises. The form of print has become the form of thought. The medium has become the message.

For centuries now, according to McLuhan, the straight line has been the hidden metaphor of literate man. It was unconsciously but inexorably used as the measure of things. It went unnoticed, unquestioned. It was presumed as natural and universal. It is neither. Like everything else, it is good for the things it is good for. To say that it is not everything is not to say that it is nothing. The electronic media have broken the monopoly of print; they have altered our sensory profiles by heightening our awareness of aural, tactile, and kinetic values.

(5) A.D. 1967—*All the senses want to get in-*

to the act. Print repressed most sense-life in favor of the visual. The end of print's monopoly also marks the end of a visual monopoly. As the early warning system of art and popular culture indicates, all the senses want to get into the act. Some of the excesses in the current excursions into aural, oral, tactile, and kinetic experience may in fact be directly responsive to the sensory deprivation of the print culture. Nature abhors a vacuum. No one glories in the sight of kids totally out of control in reaction to the Beatles. Some say, "What are the Beatles doing to these kids?" Others say, "What have we done to these kids?" All the data aren't in on what it means to be a balanced human being.

Kids are what the game is all about. Given an honest game with enough equipment to go around, it is the mental, emotional, and volitional capacity of the student that most determines the outcome. The whole complicated system of formal education is in business to get through to kids, to motivate kids, to help kids learn stuff. Schools are not in business to label kids, to grade them for the job market or to babysit. They are there to communicate with them.

Communication is a funny business. There isn't as much of it going on as most people think. Many feel that it consists in saying things in the presence of others. Not so. It consists not in saying things but in having things heard. Beautiful English speeches delivered to monolingual Arabs are not beautiful speeches. You have to speak the language of the audience—of the *whom* in the "who-says-what-to-whom" communications diagram. Sometimes the language is lexical (Chinese, Japanese, Portuguese), sometimes it is regional or personal (125th Streetese, Holden Caulfieldese, anybodyese). It has little to do with words and much to do with understanding the audience. The word for good communication is "Whomese"—the language of the audience, of the "whom."

All good communicaters use Whomese. The best writers, film-makers, advertising men, lovers, preachers, and teachers all have the knack for thinking about the hopes, fears, and capacity of the other person and of being able to translate their communication into terms which are *relevant* for that person. Whitehead called "inert ideas" the bane of education. Relevance, however, is one of those subjective words. It doesn't pertian to the object in itself but to the object as perceived by someone. The school may decide that history is *important* for the student, but the role of the teacher is to make history *relevant to* the student.

If *what* has to be tailored to the *whom,* the teacher has to be constantly engaged in audience research. It's not a question of keeping up with the latest slang or of selling out to the current mores of the kids. Neither of these tactics helps either learning or kids. But it is a question of knowing what values are strong in their world, of understanding the obstacles to communication, of sensing their style of life. Communication doesn't have to end there, but it can start nowhere else. If they are tuned in to FM and you are broadcasting on AM, there's no communication. Communication forces you to pay a lot of attention to other people.

McLuhan has been paying a great deal of attention to modern kids. Of necessity they live in the present since they have no theories to diffract or reflect what is happening. They are also the first generation to be born into a world in which there was always television. McLuhan finds them a great deal different from their counterparts at the turn of the century when the electric age was just getting up steam.

A lot of things have happened since 1900 and most of them plug into walls. Today's six-year-old has already learned a lot of stuff by the time he shows up for the first day of school. Soon after his umbilical cord was cut he was planted in front of a TV set "to keep him quiet." He liked it enough there to stay for some 3,000 to 4,000 hours before he started the first grade. By the time he graduates from high school he has clocked 15,000 hours of TV time and 10,800 hours of school time. He lives in a world that bombards him from all sides with information from radios, films, telephones, magazines, recordings, and people. He learns more things from the windows of cars, trains, and even planes. Through travel and communications he has experienced the war in Vietnam, the wide world of sports, the civil rights movement, the death of a President, thousands of commercials, a walk in space, a thousand innocuous shows, and, one may hope, plenty of Captain Kangaroo.

This is all merely descriptive, an effort to lay out what *is,* not what should be. Today's student can hardly be described by any of the old educational analogies comparing him to an empty bucket or a blank page. He comes to the information machine called school and he is already brimming over with information. As he grows his standards for relevance are determined more by what he receives outside the school than what he receives inside. A recent Canadian film tells the story of a bright, articulate middle-class teen-ager who leaves school because there's "no reason to stay." He daydreams about Vietnam while his teacher drones on about the four reasons for the spread of Christianity and the five points such information is worth on the exam. Only the need for a diploma was holding him in school; learning wasn't, and he left. He decided the union ticket wasn't worth the gaff. He left. Some call him a dropout. Some call him a pushout.

The kids have one foot on the dock and one foot on the ferry boat. Living in two centuries makes for that kind of tension. The gap between the classroom and the outside world and the gap between the generations is wider than it has ever been. Those tedious people who quote Socrates on the conduct of the young are trying vainly to reassure themselves that this is just the perennial problem of communication between generations. 'Tain't so. "Today's child is growing up absurd, because he lives in two worlds, and neither of them inclines him to grow up." Says McLuhan in *The Medium Is the Massage,* "Growing up—that is our new work, and it is *total*. Mere instruction will not suffice."

Learning is something that people do for themselves. People, places, and things can facilitate or impede learning; they can't make it happen without some cooperation from the learner. The learner these days comes to school with a vast reservoir of vicarious experiences and loosely related facts; he wants to use all his senses in his learning as an active agent in the process of discovery; he knows that all the answers aren't in. The new learner is the result of the new media, says McLuhan. And a new learner calls for a new kind of learning.

Leo Irrera said, "If God had anticipated the eventual structure of the school system, surely he would have shaped man differently." Kids are being tailored to fit the Procrustean forms of schedules, classrooms, memorizing, testing, etc., which are frequently relics from an obsolete approach to learning. It is the total environment that contains the philosophy of education, not the title page in the school catalogue. And it is the total environment that is invincible because it is invisible to most people. They tend to move things around within the old boxes or to build new and cleaner boxes. They should be asking whether or not there should be a box in the first place.

The new learner, who is the product of the all-at-once electronic environment, often feels out of it in a linear, one-thing-at-a-time school environment. The total environment is now the great teacher; the student has competence models against which to measure the effectiveness of his teachers. Nuclear students in linear schools make for some tense times in education. Students with well-developed interests in science, the arts and humanities, or current events need assistance to suit their pace, not that of the state syllabus. The straight-line theory of development and the uniformity of performance it so frequently encourages just don't fit many needs of the new learner. Interestingly, the one thing most of the current educational innovations share is their break with linear or print-oriented patterns: team teaching, nongraded schools, audio-lingual language training, multimedia learning situations, seminars, student research at all levels of education, individualized learning, and the whole shift of responsibility for learning from the teacher to the student. Needless to say, these are not as widespread as they should be, nor were they brought about through any conscious attention to the premises put forward by McLuhan. Like the print-oriented and linear mentality they now modify, these premises were plagiarized from the atmosphere. McLuhan's value is in the power he gives us to predict and control these changes.

There is too much stuff to learn today. McLuhan calls it an age of "information overload." And the information levels outside the classroom are now higher than those in the classroom. Schools used to have a virtual monopoly on information; now they are part-time competitors in the electronic informational surround. And all human knowledge is expanding at computer speed.

Every choice involves a rejection. If we can't do everything, what priorities will govern our educational policies? "The medium is the message" may not be bad for openers. We can no longer teach kids all about a subject; we can teach them what a subject is all about. We have to introduce them to the form, structure, *Gestalt,* grammar, and process of the knowledge involved. What does a math man do when a math man does do math? This approach to the formal element of a discipline can provide a channel of communication between specialists.

Its focus is not on content or detail but on the postulates, ground rules, frames of reference, and premises of each discipline. It stresses the modes of cognition and perception proper to each field. Most failures in communication are based on disagreement about items which are only corollaries of a larger thesis. It happens between disciplines, individuals, media, and cultures.

The arts play a new role in education because they are explorations in perception. Formerly conceived as a curricular luxury item, they now become a dynamic way of tuning up the sensorium and of providing fresh ways of looking at familiar things. When exploration and discovery become the themes, the old lines between art and science begin to fade. We have to guide students to becoming their own data processors to operate through pattern recognition. The media themselves serve as both aids to learning and as proper objects of study in this search for an all-media literacy. Current interest in film criticism will expand to include all art and communication forms.

And since the knowledge explosion has blown out the walls between subjects, there will be a continued move toward interdisciplinary swapping and understanding. Many of the categorical walls between things are artifacts left over from the packaging days of print. The specialist's life will be even lonelier as we move further from the Gutenberg era. The trends are all toward wholeness and convergence.

These things aren't true just because Marshall McLuhan says they are. They work. They explain problems in education that nobody else is laying a glove on. When presented clearly and with all the necessary examples and footnotes added, they have proven to be a liberating force for hundreds of teachers who were living through the tension of this cultural fission without realizing that the causes for the tension lay outside themselves. McLuhan's relevance for education demands the work of teams of simultaneous translators and researchers who can both shape and substantiate the insights which are scattered through his work. McLuhan didn't invent electricity or put kids in front of TV sets; he is merely trying to describe what's happening out there so that it can be dealt with intelligently. When someone warns you of an oncoming truck, it's frightfully impolite to accuse him of driving the thing. McLuhan can help kids to learn stuff better.

MEDIA-ORIENTED

The problem of responsibility in communications

MARGARET MEAD

■ Margaret Mead *(1901-1978) was born in Philadelphia. As an anthropologist she focused her interests on problems of child rearing, personality, and culture. She travelled extensively around the world and lived among various cultures that she later described in her works. Her chief contribution was in introducing the concept of culture into areas such as education, medicine, and public policy. Among her works are* Coming of Age in Samoa *(1928) and* Culture and Commitment *(1970).*

In the essay that is reprinted in part here, Mead notes that as a result of recent technological advances, the whole world has shrunk into one potential communication system. This underlies her growing concern with communication and the increased possibilities for propaganda. She observes that throughout history methods of propaganda have been used to advance various religious and political ends. "But the addition of modern technological methods, by which the ownership of one radio station may decide the fate of a local revolution, . . . has changed the order of magnitude of the whole problem."

This potential power of communication raises the question of how to use it responsibly. In the United States, Mead observes, "any interest, wishing to "sell" its products or messages to the public, is able to use the full battery of available communication techniques." Such a state of affairs is regarded by Mead as highly disconcerting. To make things worse, "most of the value symbols of American tradition are ready to the hand of the manufacturer of the most trivial debased product." The effects of this situation on society are very serious: confusion, apathy, and perhaps even a swing toward authoritarianism. Thus the issue of responsibility and ethics in the use of communication techniques becomes paramount.

The great contemporary concern with communication problems must be laid not only to the enormous advance in technology and the resulting shrinking of the world into one potential communication system, with all the attendant difficulties of communication across cultural boundaries, but also to the increase in social awareness on the one hand, and the disintegration of the institutionalized centers of responsibility on the other. It is true that, through the centuries, expanding movements and nations have used various methods of propaganda[1] to advance their causes, to convert the unconverted, bring in line the recalcitrant, reconcile the conquered to their lot and the conquerors to their conquering role. It is also true that secular and religious hierarchies have consciously used these methods to advance their avowed and unavowed ends. But the addition of modern technological methods, by which the ownership of one radio station may decide the fate of a local revolution, and a single film or a single voice may reach the whole of the listening and watching world, has changed the order of magnitude of the whole problem.

At the same time development of social science is making it possible for communications to change their character. Instead of the inspired voice of a natural leader, whose zestful "We shall defend our Island, whatever the cost may be. We shall fight on the beaches, we shall fight on the landing ground, we shall fight in the fields and the streets, we shall fight in the hills; we shall never surrender . . . " galvanizing people to action, the appeals can be, to a degree, calculated and planned. Instead of the politician's hunch as to how some program is going over, polls and surveys can be used to bring back accurate information to the source of the propaganda and introduce a corrective. Theories of human nature which are no longer the inexplicit emphases of a coherent culture, but instead the partly rationalized, partly culturally limited formulations of psychological research, can be used as the basis of planned campaigns.[2]

The thinking peoples of the world have been made conscious, during the past quarter of a century, of the power of organized and controlled communication, glimpsing that power both from the point of view of the victim or "target" and of the victimizer, he who wields the powerful weapon. Dissection of the methods of the enemy, the conscious cultivation of an immunity against appeals to one's own emotion, desperate attempts to devise

---

□ From "Some Cultural Approaches to Communication Problems" (pp. 17-26), by Margaret Mead, in "*The Communication of Ideas*, ed. Lyman Bryson. Copyright 1948 by Institute for Religious and Social Studies. Reprinted by permission of Harper & Row, Publishers, Inc.

methods appropriate to a democracy, while we envied totalitarian propagandic controls, have all contributed to the growth of this consciousness in the United States.

But consciousness of the potential power of communication has peculiar implications in the United States, in a country where no institution, neither Church nor State, has any monopoly of the organs of communication. The American, during the past twenty-five years, has seen systems of propagandic control develop in other countries, and even when propagandic moves of extreme importance have actually been promoted within the United States, they have usually been phrased as inspired by Berlin or Tokyo, London or Moscow, rather than as the expression of American attitudes.

The local American emphasis has thus been on resisting high-powered communication pressures, and this has been congruent, not only with the Americans' fear of playing the sucker role *vis-à-vis* other nations, more skilled in international necromancy, but also with the great importance of advertising in the United States. Those European peoples which have felt the impact of modern totalitarian communications had as a background for the experience a past in which Church and State traditionally controlled and manipulated the symbols which could move men to feel and to act. The American on the other hand has experienced instead the manipulation of the same sort of symbols, of patriotism, religious belief, and human strivings after perfection and happiness, by individuals and groups who occupied a very different and far less responsible place in the social hierarchy.

In our American system of communications, any interest, wishing to "sell" its products or message to the public, is able to use the full battery of available communication techniques, radio and film, press and poster. It is characteristic of this system that the symbols used to arouse emotion, evoke attention, and produce action, have come into the hands of those who feel no responsibility toward them. In a society like Bali there is simply no possibility that such a symbol as "The Village," also spoken of as "Mr. Village" and as "God Village," could be used by a casual vendor or rabble rouser. The symbols which evoke responses are used by those whose various positions in the society commit them to a responsible use. But in the United States, most of the value symbols of American tradition are ready to the hand of the manufacturer of the most trivial debased product, or the public relations counsel of the most wildcat and subversive organizations.

The American is used to experiencing the whole symbolic system of his society, in a series of fragmented and contradictory contexts. These beget in him a continually heightened threshold to any sort of appeal (with a recurrent nostalgia for a lost innocence in which his tears could flow simply or his heart swell with uncomplicated emotion) and a casual, non-evaluative attitude toward the power wielded through any communication system. As he straightens his tie and decides not to buy the tie which is being recommended over the radio, or in the street car ad, he gets a sense of immunity which makes him overlook the extent to which he is continually absorbing the ad behind the ad, the deutero[3] contexts of the material which he feels he is resisting.

We may examine the types of learning which result from the various uses of symbols in the United States in terms of: Whose symbol is used? What is the order of relationship between the symbol-possessing group and the group which is using the symbol? What is the nature of the product or message for which the symbol is being used? Who benefits by its use?

As examples of various types of symbol usage, let us consider the use of the symbol of Florence Nightingale, devoted ministrant to suffering and dying humanity. In the first position, a maker of white broadcloth might put out an advertisement which said, "In the great tradition of Florence Nightingale, American nurses are to be found ministering to the suffering. And, needing the very best, in order to fulfill their devoted mission, they use *Blank's* broadcloth for their uniforms, because it wears —through sickness and death." The reader of this advertisement learns that Florence Nightingale is a name to conjure with, that she was admired and respected, and that *Blank's* broadcloth are using her to enhance *their* prestige. To this degree the value of Florence Nightingale's name is increased. But at the same time the reader or listener may also add a footnote, "Trying to tie their old broadcloth on to Florence Nightingale's kite," and the sense of a synthetic, temporary quality of all symbol association is strengthened in his mind.

In the second case, the advocates of a dishonest correspondence course in nursing might use the name of Florence Nightingale in a plea to individuals to rise and follow the lamp once carried aloft by the great Nurse, and prepare themselves, in only twenty lessons, money down in advance, to follow in her footsteps. Here, to the extent that the listener realized that the correspondence course was phony, Florence Nightingale's name would also be shrouded with some of the same feeling of the phonyness, bedraggled and depreciated.

In the third case, a nurses' association might decide to put themselves back of a public education program in chest x-rays for tuberculosis control, and develop a poster in which they placed their great symbol, Florence Nightingale, beside an appeal for support for the local anti-tuberculosis committee. The reader and listeners here recognize that Florence Nightingale is a great and valuable symbol, because those to whom she is a value symbol have themselves used her name to advance some newer and younger cause. This last type is of course characteristic of the historical use of symbols in society. Even when groups which represented religious or political subversion from the point of view of those in power have appropriated to themselves the sacred symbols of those against whom they were fighting, such moves have been made seriously and responsibly by those who believed that their subversion and their heresy were neither subversion nor heresy but political justice and religious truth. Symbols which change hands between orthodox and heterodox, between conservative and liberal, do not suffer by the change as long as each group of users acts responsibly. Instead such exchange is an invaluable ingredient of continuity and consistency within a changing society.

But the advertising agency, the public relations counsel, as institutionalized in our culture, has no responsibility of this sort. An advertising agency, whatever the personal sense of conscientious rectitude of its staff, has one set of functions to perform, to sell the product successfully while keeping within the law. With sufficient sophistication, a refusal to spoil the market, either for the same product in the future, or for other products, might be included within its functions. But our society has no higher jurisdiction to which such agencies owe allegiance. The regulations formulated by patriotic societies to protect the flag have to be respected, or you get into trouble. Religious symbols can be used only if you are sure the churches will not get in your hair. Claims must be muted to the sensitivities of the Pure Food and Drug Administration. If you expect to keep the contract a long time, do not overplay a line which may go sour. If you do not want trouble from your other clients, or other agencies, do not take too obvious a crack at other products or organizations or causes. It is upon such disjointed rules of thumb that the day-by-day manipulation of the responsiveness, the moral potential of the American people, depends.

The National Nutrition Program, administered under federal auspices during the war, was one interesting attempt to deal with this contemporary situation. Agreements were worked out by which advertisers were permitted to use the name of the National Nutrition Program, if, and only if, they acceded to certain conditions, the final ethical sanction for which came from the best scientific knowledge of nutritionists. Advertisers were not permitted to misquote, quote in part, or add to, the gist of the Nutrition theme which had been agreed upon, nor could they use it in association with products of no nutritional value. In spite of many small expediencies which clouded the issues, this was a genuine attempt to supply an ethical sanction, rooted in science and administered by government, to a whole mass of communications on the subject of food and its uses. On a very simple level, this program represented one possible direction in which a country like the United States might move to give ethical form to the almost wholly unregulated mass of communications which now serve the interests of such a variety of groups—one way in which control can be vested in those to whom the symbol belongs.

A continuation of the present state of irresponsibility is exceedingly dangerous because it provides a situation within which steps backward rather than steps forward are so likely to occur. One possible response to the confused state of our symbolic system and the dulling of our responsiveness is an artificial simplification, a demand for the return of control to central authorities who will see to it that there is no more of the haphazard and contradictory use of important symbols. If the only choice open to us appears to be this increasing immunization against any appeal, this increasing apathy and callousness, so that photographs of a thousand murdered innocents no longer have any power to move us, the temptation to swing back to authoritarianism may become increasingly great. If, however, we can go on and formulate a system of responsibility appropriate to the age in which we live, a system which takes into account the state of technology, the type of mixed economy, the democratic aspirations, and the present dulled sensibilities of the American people, we may prevent such a reaction and, instead, move forward.

Any theory of the way in which responsibility for communications must be developed must deal with the problem of intent, with the the beliefs that the communicator has about himself, and about his audience, as well as with the particular constitution and situation of that audience. This facet of the problem is particularly important in America, where the average citizen still identifies his position as a minority one, and so always thinks of power as wielded by THEM, and not by himself or a group to which he belongs. All discussions of the locations of responsibility for the communication stream, in any positive or constructive sense, are likely to stumble over this feeling that responsibility means power, and power is always in the hands of someone else. A set of negative controls, such as the rule that a radio station must discuss both sides of a situation, no matter how imperfectly and destructively each side is presented, is more congenial than any set of positive controls. So also were the teachings of propaganda analysis: the American felt safer in learning how not to respond to a false appeal than in permitting any effective development of appeals which would be so good that he would respond to them.

It therefore seems that it is important to arrive at a phrasing of responsibility which will meet this fear of misused power and develop an ethic of communications within a democracy such as ours. Once a climate of opinion expressing such an ethic begins to develop, appropriate institutional forms may be expected to emerge, either slowly or under intensive cultivation.

Such an ethic might take the form of an insistence that the audience be seen as composed of *whole* individuals, not artificial cutouts from crowd scenes, such as are represented on the dust jacket of a book[4] on radio. It might take the form of insisting that the audience be seen as composed of individuals who could not be manipulated but could only be appealed to in terms of their systematic cultural strengths. It might include a taboo on seeing any individual as the puppet of the propagandist, and focusing instead on the purposeful cultivation of directions of change. It would then be regarded as ethical to try to persuade the American people to drink orange juice, as a pleasant and nutritional drink, by establishing a style of breakfast, a visual preference for oranges, and a moral investment in good nutrition, but not by frightening individual mothers into serving orange juice for fear that they would lose their children's love, or their standing in the community.

Probably the closest analogue for the development of such sanctions can be found in medical ethics, legal ethics, etc., in which a group of self-respecting practitioners constitute themselves as a final court of appeal upon their own behavior. To the extent that advertising, public relations, market research, and the various communication media experts come to hold themselves and be held by the public in greater respect, such internally self-corrective systems might be developed.

If the contention is justified that democratic institutions represent a more complex integration of society, in which greater or different possibilities are accorded to each individual, we must expect corresponding differences between the communication ethics of societies representative of different degrees of feudalism and capitalism in different political combinations. The wholly feudal state may be said to have localized responsibility for communications within a hierarchical status system, and avoided the problem of power over individuals or trends by regarding that system as fixed and immutable. The totalitarian system which has lost the sanctions of feudalism and cannot depend upon the character structure of its citizens, develops monopolistic communication systems which seek to establish a direction in the society, but which in the interval are seen as operating on identified individuals, playing upon their most vulnerable points to bring them in line with a dictated policy. Whether it is claimed that the availability of concentration camps influence the propagandist or merely makes the audience members vulnerable, the interrelationship is there.

Political democracies have, to date, by insisting on negative sanctions, maintained systems in which the individual was the target of many sorts of propagandic themes but in which he was protected by the existence of contradictions in the appeals made to him. Such negative sanctions are better than none, but the target of American advertising is not a dignified human figure.[5] The target of political campaigns in the United States is not a dignified human figure. The limitation on the sense of power of the advertising agency copy writer or the campaign manager has merely been the knowledge that they were opponents in the field, free to act just as irresponsibly as he and free to present an equally contradictory and destructive set of counterappeals.

This negative approach is challenged whenever the country goes to war and wishes to mobilize its citizens toward common goals. It is doubly challenged when branches of the United States Army or the United States government are charged with the task of reeducating peoples who have lived under totalitarian regimes. The resistance of the Germans,[6] for example, to the sort of protection of freedom which is implied in the cultivation of a two-party system, challenges American culture to the development of a more positive ethic.

NOTES

[1]Margaret Mead, "Our Educational Emphasis in Primitive Perspective," in *Education and the Cultural Process,* Charles S. Johnson (ed.). Papers presented at Symposium commemorating the 75th Anniversary of the founding of Fisk University, April-May, 1941. Reprinted from the *American Journal of Sociology,* 48 (May, 1943), 5-12.

[2]Ernest Kris, "Some Problems of War Propaganda," *Psychoanalytic Quarterly.* 12, no. 3, 381-99 (for a discussion of the way in which Nazi propaganda methods drew upon LeBon's psychology of the crowd).

[3]For a discussion of the concept of deutero learning see Gregory Bateson, "Social Planning and the Concept of 'Deutero-Learning,'" *Science, Philosophy and Religion, 2nd Symposium,* Conference on Science, Philosophy and Religion, New York, 1942, pp. 81-97.

[4]Paul F. Lazarfeld and Harry Field, *The People Look at Radio* (Chapel Hill: University of North Carolina Press, 1946).

[5]Constantin Fitz Gibbon, "The Man of Fear," *Atlantic Monthly* (January, 1947), 78-81.

[6]Bertram Schaffner, *Father Land, A Study of Authoritarianism in the German Family* (New York: Columbia University Press, 1948).

FEMINIST

The dialectic within cultural history

SHULAMITH FIRESTONE

■ Shulamith Firestone *was born in Ottawa, Canada, toward the end of World War II. She grew up in the Midwest and received a degree in Fine Arts from the Art Institute of Chicago. She was a founder of the Women's Liberation Movement and, later, editor of* Notes, *a journal of radical feminism. Her book* The Dialectic of Sex *attracted worldwide attention. The following selection is from that book.*

Firestone's chief claim here is that in the history of culture, there is an underlying dialectic of sex. Culture is seen to have evolved historically along two modes: (1) The Aesthetic Mode of culture rests on imagination, and the active search for an alternate, ideal reality. Art and poetry are associated with this mode. (2) The Technological Mode rests on experimentation and the scientific method. It seeks to master nature rather than to construct an alternative reality. Since the first mode is regarded as subjective and intuitive, it corresponds with "female" behavior. The second mode is regarded as objective and logical, hence corresponding to "male" behavior. "Thus the aesthetic is the cultural recreation of that half of the psychological spectrum that has been appropriated to the female, whereas the technological response is the cultural magnification of the male half."

According to Firestone we are now living in the age of the Technological Mode, and the contradictions within it are threatening to explode. Knowledge has developed to the point that it has assumed a life of its own: "The machine has its own momentum." (Firestone seems here to be arguing for this autonomy of technology.) This situation will culminate soon in a sexual revolution that obliterates the divisions between the two modes and integrates them into one richer mode which will give rise to an androgynous culture. Then "control and delay of 'id' satisfaction by the 'ego' will be unnecessary." Thus the repressive aspect of civilization described by both Freud and Marcuse (see earlier selections) is seen by Firestone as disappearing in this new culture.

So far we have treated "culture" as synonymous with "arts and letters" or at its broadest, "humanities." This is a common enough confusion. But it is startling in this context. For we discover that, while only indirectly related to art, women have been entirely excluded from an equally important half of culture: science. If at least with the arts we could find enough material about the relationship of women to culture—whether indirectly as influence, stimulus, or subject matter, or even occasionally as direct participants—to fill at least a chapter, we can hardly find a relationship of women to science worthy of discussion. Perhaps in the broadest sense our statement that women are the emotional force behind all (male) culture holds true—but we are stretching the case to include modern science, where the empirical method specifically demands the exclusion of the scientist's personality from his research. Satisfaction of his emotional needs through a woman in his off hours may make him more stable, and thus steadier on the job, but this is farfetched.

But if even the indirect relationship of women to science is debatable, that there is no direct one is certainly not. One would have to search to find even one woman who had contributed in a major way to scientific culture. Moreover, the situation of women in science is not improving. Even with the work of discovery shifted from the great comprehensive minds of the past to small pragmatic university research teams, there are remarkably few women scientists.[1]

This absence of women at all levels of the scientific disciplines is so commonplace as to lead many (otherwise intelligent) people to attribute it to some deficiency (logic?) in women themselves. Or to women's own predilections for the emotional and subjective over the practical and rational. But the question cannot be so easily dismissed. It is true that women in science are in foreign territory—but how has this situation evolved? Why are there disciplines or branches of inquiry that demand only a "male" mind? Why would a woman, to qualify, have to develop an alien psychology? When and why was the female excluded from this type mind? How and why has science come to be defined as, and restricted to, the "objective"?

I submit that not only were the arts and humanities corrupted by the sex duality, but that modern science has been determined by it. And moreover that *culture reflects this polarity in its very organization*. C. P. Snow was the first to note what had been becoming increasingly obvious: a deep fissure of culture—the liberal arts and the sciences had become incomprehensible to each other. Again, though the universal man of the Renaissance is widely lamented, specialization only increases. These are some of the modern symptoms of a long cultural disease based on the sex dualism. Let us examine the history of culture according to this hypothesis—that there is an underlying dialectic of sex.

I. THE TWO MODES OF CULTURAL HISTORY

For our analysis we shall define culture in the following way: *Culture is the attempt by man to realize the conceivable in the possible*. Man's consciousness of himself within his environment distinguishes him from the lower animals, and turns him into the only animal capable of culture. This consciousness, his highest faculty, allows him to project mentally states of being that do not exist at the moment. Able to construct a past and future, he becomes a creature of time—a historian and a prophet. More than this, he can imagine objects and states of being that have never existed and may never exist in the real world—he becomes a maker of art. Thus, for example, though the ancient Greeks did not know how to fly, still they could imagine it. The myth of Icarus was the formulation in fantasy of their conception of the state "flying."

But man was not only able to project the conceivable into fantasy. He also learned to impose it on reality: by accumulating knowledge, learning experience, about that reality and how to handle it, he could shape it to his liking. This accumulation of skills for controlling the environment, technology, is another means to reaching the same end, the realization of the conceivable in the possible. Thus, in our example, if, in the B.C. era, man could fly on the magic carpet of myth or fantasy, by the twentieth century, his technology, the accumulation of his practical skills, had made it possible for him to fly in actuality—he had invented the airplane. Another example: In the Biblical legend, the Jews, an agricultural people stranded for forty years in the desert, were provided by God with Manna, a miraculous substance that could be transformed at will into food of any color, texture, or taste; modern food processing, especially with the "green revolution," will probably soon create a totally artificial food production, perhaps with this chameleon attribute. Again, in ancient legend, man could imagine mixed species, e.g., the centaur or the unicorn, or hybrid births, like the birth on an animal from a human, or a virgin birth; the current biological revolution, with its increasing knowledge of the reproductive process, could now—if only the first crude stages—create these "monstrosities" in reality. Brownies and elves, the Golem of medieval Jewish lore, Mary Shelley's monster in *Frankenstein*, were the imaginative constructions that preceded by several centuries the corresponding technological acumen. Many other fantastical constructions—ghosts, mental telepathy, Methuselah's age—remain to be realized by modern science.

These two different responses, the idealistic and the scientific, do not merely exist simultaneously: there is a dialogue between the two. The imaginative construction precedes the technological, though often it does not develop until the technological know-how is "in the air." For example, the art of science fiction developed, in the main, only a half-century in advance of, and now coexists with, the scientific revolution that is transforming it into a reality—for example (an innocuous one), the moon flight. The phrases "way out," "far out," "spaced," the observation "it's like something out of science fiction" are common language. In the aesthetic response, because it always develops in advance, and is thus the product of another age, the same realization may take on a sensational or unrealistic cast, e.g., Frankenstein's monster, as opposed to, let us say, General Electric's CAM (Cybernetic Anthropomorphic Machines) Handyman. (An artist can never know in advance just how his vision might be articulated in reality.)

Culture then is the sum of, and the dynamic between, the two modes through which the mind attempts to transcend the limitations and contingencies of reality. These two types of cultural responses entail different methods to achieve the same end, the realization of the conceivable in the possible. In the first,[2] the individual denies the limitations of the given reality by escaping from it altogether, to define, create, his own possible. In the provinces of the imagination, objectified in some way—whether through the development of a visual image within some artificial boundary, say four square feet of canvas, through visual images projected through verbal symbols (poetry), with sound ordered into a sequence (music), or with verbal ideas ordered into a progression (theology, philosophy)—he creates an ideal world governed by his own artificially imposed order and harmony, a structure in which he consciously relates each part to the whole, a static (and therefore "timeless") construction. The degree to which he abstracts his creation from reality is unimportant, for even when he most appears to imitate, he has created an illusion governed by its own—perhaps hidden—set of artificial laws. (Degas said that the artist had to lie in order to tell the truth.) This search for the ideal, realized by means of an artificial medium, we shall call the Aesthetic Mode.

In the second type of cultural response the contingencies of reality are overcome, not through the creation of an alternate reality, but through the mastery of reality's own workings: the laws of nature are exposed, then turned against it, to shape it in accordance with man's conception. If there is a poison, man assumes there is an antidote; if there is a disease, he searches for the cure: every fact of nature that is understood can be used to alter it. But to achieve the ideal through such a procedure takes much longer, and is infinitely more painful, especially in the early stages of knowledge. For the vast and intricate machine of nature must be entirely understood—and there are always fresh and unexpected layers of complexity—before it can be thoroughly controlled. Thus before any solution can be found to the deepest contingencies of the human condition, e.g., death, natural processes of growth and decay must be catalogued, smaller laws related to larger ones. This scientific method (also attempted by Marx and Engels in their materialistic approach to history) is the attempt by man to master nature through the complete understanding of its mechanics. The coaxing of reality to conform with man's conceptual ideal, through the application of information extrapolated from itself, we shall call the Technological Mode.

We have defined culture as the sum of, and the dialectic between, the two different modes through which man can resolve the tension created by the flexibility of his mental faculties within the limitations of his given environment. The correspondence of these two different cultural modes with the two sexes respectively is unmistakable. We have noted how those few women directly creating culture have gravitated to disciplines within the Aesthetic Mode. There is good reason for this: the aesthetic response corresponds with "female" behavior. The same terminology can be applied to either: subjective, intuitive, introverted, wishful, dreamy or fantastic, concerned with the subconscious (the *id*), emotional, even temperamental (hysterical). Correspondingly, the technological response is the masculine response: objective, logical, extroverted, realistic, concerned with the conscious mind (the ego), rational, mechanical, pragmatic and down-to-earth, stable. Thus the aesthetic is the cultural recreation of that half of the psychological specturm that has been appropriated to the female, whereas the technological response is the cultural magnification of the male half.

Just as we have assumed the biological division of the sexes for procreation to be the fundamental "natural" duality from which grows all further division into classes, so we now assume the sex division to be the root of this basic cultural division as well. The interplay between these two cultural responses, the "male" Technological Mode and the "female" Aesthetic Mode, recreates at yet another level the dialectic of the sexes—as well as its superstructure, the caste and the economic-class dialectic. And just as the merging of the divided sexual, racial, and economic classes is a precondition for sexual, racial, or economic revolution respectively, so the merging of the aesthetic with the technological culture is the precondition of a cultural revolution. And just as the revolutionary goal of the sexual, racial, and economic revolutions is, rather than a mere leveling of imbalances of class, an elimination of class categories altogether, so the end result of a cultural revolution must be, not merely the integration of the two streams of culture, but the elimination of cultural categories altogether, the elimination of culture itself as we know it. But before we discuss this ultimate cultural revolution or even the state of cultural division in our own time, let us see how this third level of the sex dialectic—the interaction between the Technological and Aesthetic Modes—operated to determine the flow of cultural history.

• • •

At first technological knowledge accumulated slowly. Gradually man learned to control the crudest aspects of his environment—he discovered the tool, control of fire, the wheel, the melting of ore to make weapons and plows, even, eventually, the alphabet—but these discoveries were few and far between, because as yet he had no systematic way of initiating them. Eventually however, he had gathered enough practical knowledge to build whole systems, e.g., medicine or architecture, to create juridical, political, social, and economic institutions. Civilization developed from the primitive hunting horde into an agricultural society, and finally, through progressive stages, into feudalism, capitalism, and the first attempts at socialism.

But in all this time, man's ability to picture an ideal world was far ahead of his ability to create one. The primary cultural forms of ancient civilizations—religion and its offshoots, mythology, legend, primitive art and magic, prophesy and history—were in the Aesthetic Mode: they imposed only an artificial, imaginary order on a universe still mysterious and chaotic. Even primitive scientific theories were only poetic metaphors for what would later be realized empirically. The science and philosophy and mathematics of classical antiquity, forerunners of modern science, by sheer imaginative prowess, operating in a vacuum independently of material laws, anticipated much of what was later proven: Democritus' atoms and Lucretius' "substance" foreshadowed by thousands of years the discoveries of modern science. But they were realized only within the realm of the imaginary Aesthetic Mode.

In the Middle Ages the Judaeo-Christian heritage was assimilated with pagan culture, to produce medieval religious art and the metaphysics of Thomas Aquinas and the Scholastics. Though concurrently Arab science, an outgrowth of the Greek Alexandrian Period (third century B.C. to seventh century A.D.), was amassing considerable information in such areas as geography, astronomy, physiology, mathematics—a tabulation essential to the later empiricism—there was little dialogue. Western science, with its alchemy, its astrology, the "humours" of medieval medicine, was still in a "pseudo-scientific" stage, or, in our definition, still operating according to the Aesthetic Mode. This medieval aesthetic culture, composed of the Classical and Christian legacies, culminated in the Humanism of the Renaissance.

Until the Renaissance, then, culture occurred in the Aesthetic Mode because, prior to that time, Technology had been so primitive, the body of scientific knowledge so far from complete. In terms of the sex dislectic, this long stage of cultural history corresponds with the matriarchal stage of civilization: The Female Principle—dark, mysterious, uncontrollable—reigned, elevated by man himself, still in awe of unfathomable Nature. Men of culture were its high priests of homage: until and through the Renaissance *all* men of culture were practitioners of the ideal aesthetic mode, thus, in a sense, artists. The Renaissance, the pinnacle of cultural humanism, was the golden age of the Aesthetic (female) Mode.

And also the beginning of its end. By the sixteenth century culture was undergoing a change as profound as the shift from matriarchy to patriarchy in terms of the sex dialectic, and corresponding to the decline of feudalism in the class dialectic. This was the first merging of the aesthetic culture with the technological, in the creation of modern (empirical) science.

In the Renaissance, Aristotelian Scholasticism had remained powerful though the first cracks in the dam were already apparent. But it was not until Francis Bacon, who first proposed to use science to "extend more widely the limits of the power and the greatnesses of man," that the marriage of the Modes was consummated. Bacon and Locke transformed philosophy, the attempt to understand life, from abstract speculation detached from the real world (metaphysics, ethics, theology, aesthetics, logic) to an uncovering of the *real* laws of nature, through proof and demonstration (empirical science).

In the empirical method propounded by Francis Bacon, insight and imagination had to be used only at the earliest stage of the inquiry. Tentative hypotheses would be formed by induction from the facts, and then consequences would be deduced logically and tested for consistency among themselves and for agreement with the primary facts and results of *ad hoc* experiments. The hypothesis would become an accepted theory only after all tests had been passed, and would remain, at least until proven wrong, a theory capable of predicting phenomena to a high degree of probability.

The empirical view held that by recording and tabulating all possible observations and experiments in this manner, the Natural Order would emerge automatically. Though at first the question "why" was still asked as often as the question "how," after information began to accumulate, each discovery building upon the last to complete the jigsaw, the speculative, the intuitive, and the imaginative gradually became less valuable. When once the initial foundations had been laid by men of the stature of Kepler, Galileo, and Newton, thinkers still in the inspired "aesthetic" science tradition, hundreds of anonymous technicians could move to fill in the blanks, leading to, in our own time, the dawn of a golden age of science—to the Technological Mode what the Renaissance had been to the Aesthetic Mode.

II. THE TWO CULTURES TODAY

Now, in 1970, we are experiencing a major scientific breakthrough. The new physics, relativity, and the astrophysical theories of contemporary science had already been realized by the first part of this century. Now, in the latter part, we are arriving, with the help of the electron microscope and other new tools, at similar achievements in biology, biochemistry, and all the life sciences. Important discoveries are made yearly by small, scattered work teams all over the United States, and in other countries as well—of the magnitude of DNA in genetics, or of Urey and Miller's work in the early fifties on the origins of life. Full mastery of the reproductive process is in sight, and there has been significant advance in understanding the basic life and death process. The nature of aging and growth, sleep and hibertion, the chemical functioning of the brain and the development of consciousness and memory are all beginning to be understood in their entirety. This acceleration promises to continue for perhaps another century, however long it takes to achieve the goal of Empiricism: total understanding of the laws of nature.

This amazing accumulation of concrete knowledge in only a few hundred years is the product of philosophy's switch from the Aesthetic to the Technological Mode. The combination of "pure" science, science in the Aesthetic Mode, with pure technology, caused greater progress toward the goal of technology—the realization of the conceivable in the actual—than had been made in thousands of years of previous history.

Empiricism itself is only the means, a quicker and more effective technique, for achieving technology's ultimate cultural goal: the building of the ideal in the real world. One of its own basic dictates is that a certain amount of material must be collected and arranged into categories before any decisive comparison, analysis, or discovery can be made. In this light, the centuries of empirical science have been little more than the building of foundations for the breakthroughs of our own time and the future. The amassing of information and understanding of the laws and mechanical processes of nature ("pure research") is but a means to a larger end: total understanding of Nature in order, ultimately, to transcend it.

In this view of the development and goals of cultural history, Engels' final goal, quoted above in the context of political revolution, is again worthy of quotation:

> The whole sphere of the condition of life which environ man, and have hitherto ruled him, now comes under the dominion and control of man, who for the first time becomes the real conscious Lord of Nature.

Empirical science is to culture what the shift to patriarchy was to the sex dialectic, and what the bourgeois period is to the Marxian dialectic—a latter-day stage prior to revolution. Moreover, the three dialectics are integrally related to one another vertically as well as horizontally: The empirical science growing out of the bourgeoisie (the bourgeois period is in itself a stage of the patriarchal period) follows the humanism of the aristocracy (The Female Principle, the matriarchy) and with its development of the empirical method in order to amass real knowledge (development of modern industry in order to amass capital) eventually puts itself out of business. The body of scientific discovery (the new productive modes) must finally outgrow the empirical (capitalistic) mode of using them.

And just as the internal contradictions of capitalism must become increasingly apparent, so must the internal contradictions of empirical science—as in the development of pure knowledge to the point where it assumes a life of its own, e.g., the atomic bomb. As long as man is still engaged only in the means—the charting of the ways of nature, the gathering of "pure" knowledge—to his final realization, mastery of nature, his knowledge, because it is not complete, is dangerous. So dangerous that many scientists are wondering whether they shouldn't put a lid on certain types of research. But this solution is hopelessly inadequate. The machine of empiricism has its own momentum, and is, for such purposes, completely out of control. Could one actually decide what to discover or not discover? That is, by definition, antithetical to the whole empirical process that Bacon set in motion. Many of the most important discoveries have been practically laboratory accidents, with social implications barely realized by the scientists who stumble into them. For example, as recently as five years ago Professor F. C. Steward of Cornell discovered a process called "cloning": by placing a single carrot cell in a rotating nutrient he was able to grow a whole sheet of identical carrot cells, from which he eventually recreated the same carrot. The understanding of a similar process for more developed animal cells, were it to slip out—as did experiments with "mind-expanding" drugs—could have some awesome implications. Or, again, imagine parthenogenesis, virgin birth, as practiced by the greenfly, actually applied to human fertility.

Another internal contradiction in empirical science: the mechanistic, deterministic, "soulless" scientific world-view, which is the result of the means to, rather than the (inherently noble and often forgotten) ultimate purpose of, Empiricism: the actualization of the ideal in reality.

The cost in humanity is particularly high to the scientist himself, who becomes little more than a cultural technician. For, ironically enough, to properly accumulate knowledge of the universe requires a mentality the very opposite of comprehensive and integrated. Though in the long run the efforts of the individual scientist could lead to domination of the environment in the interest of humanity, temporarily the empirical method demands that its practitioners themselves become "objective," mechanistic, overprecise. The public image of the white-coated Dr. Jekyll with no feelings for his subjects, mere guinea pigs, is not entirely false: there is no room for feelings in the scientist's work; he is forced to eliminate or isolate them in what amounts to an occupational hazard. At best he can resolve this problem by separating his professional from his personal self, by compartmentalizing his emotion. Thus, though often well-versed in an academic way about the arts—the frequency of this, at any rate, is higher than of artists who are well-versed in science—the scientist is generally out of touch with his direct emotions and senses, or, at best, he is emotionally divided. His "private" and "public" life are out of whack; and because his personality is not well-integrated, he can be surprisingly conventional ("Dear, I discovered how to clone people at the lab today. Now we can go skiing at Aspen.") He feels no contradiction in living by convention, even in attending church, for he has never integrated the amazing material of modern science with his daily life. Often it takes the misuse of his discovery to alert him to that connection which he has long since lost in his own mind.

The catalogue of scientific vices is familiar: it duplicates, exaggerates, the catalogue of "male" vices in general. This is to be expected: if the Technological Mode develops from the male principle then it follows that its practitioners would develoop the warpings of the male personality in the extreme. But let us leave science for the moment, winding up for the ultimate cultural revolution, to see what meanwhile had been happening to the aesthetic culture proper.

With philosophy in the broadest classical sense—including "pure" science—defecting, aesthetic culture became increasingly narrow and ingrown, reduced to the arts and humanities in the refined sense that we now know them. Art (hereafter referring to the "liberal arts," especially the arts and letters) had always been, in its very definition, a search for the ideal, removed from the real world. But in primitive days it had been the handmaiden of religion, articulating the common dream, objectifying "other" worlds of the common fantasy, e.g., the art of the Egyptian tombs, to explain and excuse this one. Thus even though it was removed from the real world, it served an important social function: it satisfied artifically those wishes of society that couldn't yet be realized in reality. Though it was patronized and supported only by the aristocracy, the cultured elite, it was never as detached from life as it later became; for the society of those times was, for all practical purposes,synonymous with its ruling class, whether priesthood, monarchy, or nobility. The masses were never considered by "society" to be a legitimate part of humanity, they were slaves, nothing more than human animals, drones, or serfs, without whose labor the small cultured elite could not have maintained itself.

The gradual squeezing out of the aristocracy by the new middle class, the bourgeoisie, signalled the erosion of aesthetic culture. We have seen that capitalism intensified the worst attributes of patriarchalism, how, for example, the nuclear family emerged from the large, loose family household of the past, to reinforce the weakening sex class system, oppressing women and children more intimately than ever before. The cultural mode favored by this new, heavily patriarchal bourgeoisie was the "male" Technological Mode—objective, realistic, factual, "commonsense"—rather than the effeminate, otherworldly, "romantic idealist" Aesthetic Mode. The bourgeoisie, searching for the ideal in the real, soon developed the empirical science that we have described. To the extent that they had any remaining use for aesthetic culture, it was only for "realistic" art, as opposed to the "idealistic" art of classical antiquity, or the abstract religious art of primitive or medieval times. For a time they went in for a literature that described reality—best exemplified by the nineteenth-century novel—and a decorative easel art: still lifes, portraits, family scenes, interiors. Public museums and libraries were built alongside the old salons and private galleries. But with its entrenchment as a secure, even primary, class, the bourgeoisie no longer needed to imitate aristocratic cultivation. More important, with the rapid development of their new science and technology, the little practical value they had for art was eclipsed. Take the scientific development of the camera: The bourgeoisie soon had little need for portrait painters; the little that painters or novelists had been able to do for them, the camera could do better.

"Modern" art was a desperate, but finally self-defeating, retaliation *("épater le bourgeois")* for these injuries: the evaporation of its social function, the severance of the social umbilical cord, the dwindling of the old sources of patronage. The modern art tradition, associated primarily with Picasso and Cézanne, and including all the major schools of the twentieth century—cubism, constructivism, futurism, expressionism, surrealism, abstract expressionism, and so on—is not an authentic expression of modernity as much as it is a reaction to the realism of the bourgeoisie. Post-impressionism deliberately renounced all reality-affirming conventions—indeed the process began with impressionism itself, which broke down the illusion into its formal values, swallowing reality whole and spitting it up again as art—to lead eventually to an art-for-art's-sake so pure, a negation of reality so complete as to make it ultimately meaningless, sterile, even absurd. (Cab drivers *are* philistine: they know a put-on when they see one.) The deliberate violating, deforming, fracturing of the image, called "modern" art, was nothing more than a fifty-year idol smashing—eventually leading to our present cultural impasse.

In the twentieth century, its life blood drained, its social function nullified altogether, art is thrown back on whatever wealthy classes remain, those *nouveaux riches*—particularly in America, still suffering from a cultural inferiority complex—who still need to prove they have "arrived" by evidencing a taste for culture. The sequestering of intellectuals in ivory tower universities, where, except for the sciences, they have little effect on the outside world, no matter how brilliant (and they aren't, because they no longer have the necessary feedback); the abstruse—often literally unintelligible—jargon of the social sciences; the cliquish literary quarterlies with their esoteric poetry; the posh 57th Street galleries and museums (it is no accident that they are right next door to Saks Fifth Avenue and Bonwit Teller) staffed and supplied by, for the most part, fawning rich-widows'-hairdresser types; and not least the vulturous critical establishment thriving on the remains of what was once a great and vital culture—all testify to the death of aesthetic humanism.

For the centuries that Science climbed to new heights, Art decayed. Its forced inbreeding transformed it into a secret code. By definition escapist from reality, it now turned in upon itself to such degree that it gnawed away its own vitals. It became diseased—neurotically self-pitying, self-conscious, focused on the past (as opposed to the futurist orientation of the technological culture) and thus frozen into conventions and academies—orthodoxies of which "avant-garde" is only the latest—pining for remembered glories, the Grand Old Days When Beauty Was In Flower; it became pessimistic and nihilistic, increasingly hostile to the society at large, the "philistines." And when the cocky young Science attempted to woo Art from its ivory tower—eventually garret—with false promises of the courting lover ("You can come down now, we're making the world a better place every day"), Art refused more vehemently than ever to deal with him, much less accept his corrupt gifts, retreating ever deeper into her daydreams—neoclassicism, romanticism, expressionism, surrealism, existentialism.

The individual artist or intellectual saw himself as either a member of an invisible elite, a "highbrow," or as a down-and-outer, mingling with whoever was deemed the dregs of his society. In both cases, whether playing Aristocrat or Bohemian, he was on the margins of the society as a whole. The artist had become a freak. His increasing alienation from the world around him—the new world that science had created was, especially in its primitive stages, an incredible horror, only intensifying his need to escape to the ideal world of art—his lack of an audience, led to a mystique of "genius." Like an ascetic Saint Simeon on his pedestal, the Genius in the Garret was expected to create masterpieces in a vacuum. But his artery to the outside world had been severed. His task, increasingly impossible, often forced him into literal madness, or suicide.

Painted into a corner with nowhere else to go, the artist has got to begin to come to terms with the modern world. He is not too good at it: like an invalid shut away too long, he dosen't know anything about the world anymore, neither politics, nor science, nor even how to live or love. Until now, yes, even now, though less and less so, sublimation, that warping of personality, was commendable: it was the only (albeit indirect) way to achieve fulfillment. But the artistic process has—almost—outlived its usefulness. And its price is high.

The first attempts to confront the modern world have been for the most part misguided. The Bauhaus, a famous example, failed at its objective of replacing an irrelevant easel art (only a few optical illusions and designy chairs mark the grave), ending up with a hybrid, neither art nor science, and certainly not the sum of the two. They failed because they didn't understand science on its own terms: to them, seeing in the old aesthetic way, it was simply a rich new subject matter to be digested whole into the traditional aesthetic system. It is as if one were to see a computer as only a beautifully ordered set of lights and sounds, missing completely the function itself. The scientific experiment is not only beautiful, an elegant structure, another piece of an abstract puzzle, something to be used in the next collage—but scientists, too, in their own way, see science as this abstraction divorced from life—it has a real intrinsic meaning of its own, similar to, but not the same as, the "presence," the *"en-soi,"* of modern painting. Many artists have made the mistake of thus trying to annex science, to incorporate it into their own artistic framework, rather than using it to expand that framework.

Is the current state of aesthetic culture all bleak? No, there have been some progressive developments in contemporary art. We have mentioned how the realistic tradition in painting died with the camera. This tradition had developed over centuries to a level of illusionism with the brush—examine a Bouguereau—that was the equal of, better than, the early photography, then considered only another graphic medium, like etching. The beginning of the new art of film and the realistic tradition of painting overlapped, peaked, in artists like Degas, who used a camera in his work. Then realistic art took a new course: Either it became decadent, academic, divorced from any market and meaning, e.g., the nudes that linger on in art classes and second-rate galleries, or it was fractured into the expressionist or surrealist image, posing an alternate internal or fantastical reality. Meanwhile, however, the young art of film, based on a true synthesis of the Aesthetic and Technological Modes (as Empiricism itself had been), carried on the vital realistic tradition. And just as with the marriage of the divided male and female principles, empirical science bore fruit; so did the medium of film. But, unlike other aesthetic media of the past, it broke down the very division between the artificial and the real, between culture and life itself, on which the Aesthetic Mode is based.

Other related developments: the exploration of artificial materials, e.g., plastics; the attempt to confront plastic culture itself (pop art); the breakdown of traditional categories of media (mixed media), and of the distinctions between art and reality itself (happenings, environments). But I find it difficult to unreservedly call these latter developments progressive: as yet they have produced largely puerile and meaningless works. The artist does not yet know what reality is, let alone how to affect it. Paper cups lined up on the street, pieces of paper thrown into an empty lot, no matter how many ponderous reviews they get in *Art News*, are a waste of time. If these clumsy attempts are at all hopeful, it is only insofar as they are signs of the breakdown of "fine" art.

The merging of the Aesthetic with the Technological Mode will gradually suffocate "pure" high art altogether. The first breakdown of categories, the remerging of art with a (technologized) reality, indicate that we are now in the transitional pre-revolutionary period, in which the three separate cultural streams, technology ("applied science"), "pure research," and "pure" modern art, will melt together—along with the rigid sex categories they reflect.

The sex-based polarity of culture still causes many casualities. If even the "pure" scientist, e.g., nuclear physicist (let alone the "applied" scientist, e.g., engineer), suffers from too much "male," becoming authoritarian, conventional, emotionally insensitive, narrowly unable to understand his own work within the scientific—let alone cultural or social—jigsaw, the artist, in terms of the sex division, has embodied all the imbalances and suffering of the female personality: temperamental, insecure, paranoid, defeatist, narrow. And the recent withholding of reinforcements from behind the front (the larger society) has exaggerated all this enormously; his overdeveloped "id" has nothing left to balance it. Where the pure scientist is "schiz," or worse, *ignorant* of emotional reality altogether, the pure artist *rejects* reality because of its lack of perfection, and, in modern centuries, for its ugliness.[3]

And who suffers the most, the blind (scientist) or the lame (artist)? Culturally, we have had only the choice between one sex role or the other: either a social marginality leading to self-consciousness, introversion, defeatism, pessimism, oversensitivity, and lack of touch with reality, or a split "professionalized" personality, emotional ignorance, the narrow views of the specialist.

CONCLUSION: THE CULTURE-ANTICULTURE REVOLUTION

I have tried to show how the history of culture mirrors the sex dichotomy in its very organization and development. Culture develops not only out of the underlying economic dialectic, but also out of the deeper sex dialectic. Thus, there is not only a horizontal dynamic, but a vertical one as well: each of these three strata forms one more story of the dialectics of history based on the biological dualism. At present we have reached the final stages of Patriarchalism, Capitalism (corporate capitalism), and of the Two Cultures at once. We shall soon have a triplicate set of preconditions for revolution, the absence of which is responsible for the failure of revolutions of the past.

The difference between what is almost possible and what exists is generating revolutionary forces.[4] We are nearing—I believe we shall have, perhaps within a century, if the snowball of empirical knowledge doesn't smash first of its own velocity—a cultural revolution, as well as a sexual and economic one. The cultural revolution, like the economic revolution, must be predicated on the elimination of the (sex) dualism at the origins not only of class, but also of cultural division.

What might this cultural revolution look like? Unlike "cultural revolutions" of the past, it would not be merely a quantitative escalation, more and better culture, in the sense that the Renaissance was a high point of the Aesthetic Mode, or that the present technological breakthrough is the accumulation of centuries of practical knowledge about the real world. Great as they were, neither the Aesthetic nor the Technological culture, even at their respective peaks, ever achieved universality—either it was wholistic but divorced from the real world, or it "achieved progress," at the price of cultural-schizophrenia, and the falseness and dryness of "objectivity." What we shall have in the next cultural revolution is the reintegration of the Male (Technological Mode) with the Female (Aesthetic Mode), to create an androgynous culture surpassing the highs of either cultural stream, or even of the sum of their integrations. More than a marriage, rather an abolition of the cultural categories themselves, a mutual cancellation—a matter-antimatter explosion, ending with a poof! culture itself.

We shall not miss it. We shall no longer need it: by then humanity will have mastered nature totally, will have realized in *actuality* its dreams. With the full achievement of the conceivable in the actual, the surrogate of culture will no longer be necessary. The sublimation process, a detour to wish fulfillment, will give way to direct satisfaction in experience, as felt now only by children, or adults on drugs. (Though normal adults "play" to varying degrees, the example that illustrates more immediately to almost everyone the intense level of this future experience, ranking zero on a scale of accomplishment—"nothing to show for it"—but nevertheless somehow always worth everyone's while, is lovemaking.) Control and delay of "id" satisfaction by the "ego" will be unnecessary; the *id* can live free. Enjoyment will spring directly from being and acting itself, the process of experience, rather than from the quality of achievement. When the male Technological Mode can at least produce in actuality what the female Aesthetic Mode had envisioned, we shall have eliminated the need for either.

NOTES

1. I was struck by this at a recent Women's Liberation workshop scheduled by the science department of a top-level eastern university: of the fifty women present, only one or two were engaged in research, let alone high-level research. The others were lab technicians, graduate assistants, high school science teachers, faculty wives, and the like.
2. The idealistic mode, corresponding roughly to the supra-historical, nonmaterialist "metaphysical" mode of thought against which Marx and Engels revolted.
3. One abstract painter I knew, who had experienced the horrors of North African battlefields in World War II—fields of men (buddies) rotting in the sun with rats darting out of their stomachs—spent years moving a pure beige circle around a pure beige square. In this manner, the "modern" artist denies the ugliness of reality (rats in the stomachs of buddies) in favor of artificial harmonies (circles in squares).
4. Revolutionaries, by definition, are still visionaries of the Aesthetic Mode, the idealists of pragmatic politics.

THIRD WORLD

Technology versus civilization

DENIS GOULET

■ Denis Goulet *was born in 1931 in Fall River, Massachusetts. A noted author, he pioneered a new discipline —the ethics of development. Goulet has lived and worked in Africa, the Middle East, Europe, and Latin America. He holds degrees in philosophy, social planning, and political science. Currently, he is a senior fellow at the Overseas Development Council in Washington, D.C. Among his books are* A New Moral Order: Development Ethics and Liberation Theology *(1974) and* The Crucial Choice: A New Concept in the Theory of Development *(1971).*

In The Uncertain Promise *(1977), Goulet evaluates the problems and promises of the transfer of technology to the Third World. He also discusses questions of policy involved in such transfers. The selection included here contains a critique of Western technology, which "now threatens to annihilate the human species, to destroy the planet's capacity to support life, and to eliminate human meanings in life." And yet, he notes, it is often argued that the world cannot escape a global culture based on Western technology. Quoting John White, Goulet scoffs at this argument as "the last and brillant effort of the white northern world to maintain its cultural dominance in perpetuity, against history, by the pretence that there is no alternative."*

But there is an alternative. It is to allow the "underdeveloped" nations to use their own wisdom in developing fresh outlooks on the relation of technology to society and to help bring forth a new, non-elitist, world order. Thus "the very inability of some poor nations to achieve 'development' may prove a blessing in disguise." Nevertheless, Goulet is not an anti-technologist. His critique is designed to point out the dangers of western technology before it is too late. What he hopes for is a world technology with a human, nonimperialist, face.

Normative consensus over how to deal with change is a vital element in every culture. The term *culture,* as here employed, embraces the way of life of all human groups. It includes all the standardized learning and forms of behavior which others in one's group learn to recognize and expect: language and symbols; multiple forms of organization (family, kin, occupational roles, legitimacy and authority structures, etcetera); heritage (religious, esthetic, ethical, natural). A civilization, in turn, is simply one species in the genus culture, namely,

> that kind of culture which includes the use of writing the presence of cities and of wide political organization and the development of occupational specialization.[1]

Central to the notion of all cultures are collective identity, boundaries of inclusion or exclusion of individuals (whether based on criteria of space, lineage, or blood), continuity, and a common historical experience. To all these traits must be added a shared sense of responsibility for the maintenance, dignity, and freedom of the group. Technology poses a unique challenge to culture because its own value dynamics run counter to the limits posed by cultural identity, by spatial or territorial loyalties, or by consensual norms of

□ From *The Uncertain Promise* (New York: IDOC/North America, 1977), pp. 243-251.

thought and symbolization. The progressive unification of the globe has occurred within a Western framework, but Toynbee believes that "the present Western ascendency in the world is certain not to last."[2] British economist John White explains why:

> By all historical parallels, development in the so-called Third World ought to take the form of the rise of new and competing cultures to contend with the old and dying civilization which is co-terminous with the white western world stretching from California to the Urals. The obvious candidates are in Asia, especially in East Asia, where two societies have succeeded in modernizing on the basis of models of social organization which are historically specific and owe little to the international development industry. Yet two new factors cast doubt on the relevance of the Toynbee-esque model of the challenge and response of competing cultures:
>
> (1) technology;
>
> (2) telecommunications.
>
> These factors open the anti-developmental and rather depressing possibility of a single and unchallengeable global culture. Can there ever again be a new civilization?
>
> The assumption that development is a generalisable concept must be seen in this context. It is far more potent than the crude instruments of 'neo-colonialism.' It is the last and brilliant effort of the white northern world to maintain its cultural dominance in perpetuity, against history, by the pretence that there is no alternative.[3]

Is there truly no alternative to standardized technology? Is advanced industrial society incorrigibly one-dimensional? Notwithstanding its enchantment with modern technology, will the Third World be lured by technology into betraying its deeper values as fully as has the West its own? The very impact of Western technology on other civilizations has helped non-Western peoples re-educate themselves. Out of the clash of values has come the clear lesson that no single nation or people can forever be the center of the universe. And though the West has spread the virus of acquisitiveness and the idolatry of material success everywhere, almost nowhere has the West won the hearts of other peoples. Even those who grasp after the West's tools or material rewards do not hold the West's culture in high esteem. A historical parallel is worth citing here. When Napoleon conquered Egypt, the Muslim historian Al-Gabarti displayed no interest in the Frenchman's technology or material wares.

> Al-Gabarti showed a nicer discrimination. French technology hit him in the eye, but he persisted in waiting for a sign. For him, the touchstone of Western civilization, as of his own, was not technology but justice. This Cairene scholar has apprehended the heart of the matter, the issue which the West has still to fight out within itself.[4]

Toynbee views Western technology as a kind of scaffolding around which all societies are building themselves into a unified world. Yet this Western-built scaffolding is not itself durable:

> The most obvious ingredient in it is technology, and man cannot live by technology alone. In the fullness of time, when the ecumenical house of many mansions stands firmly on its own foundations and the temporary Western scaffolding falls away—as I have no doubt that it will—I believe it will become manifest that the foundations are firm at last because they have been carried down to the bedrock of religion.[5]

The Al-Gabartis of today's Third World no longer seek a sign of justice before adopting the "developed" world's technology; they are wise enough to know that this particular sign will not appear. Nevertheless, they intuitively understand that technology can outlive the "civilization" that diffuses it. Frequently, their vision is more lucid than that of Westerners whose complacency over their technological triumphs blinds them both to the injustices they commit in spreading the *imperium* of technology and to the value impasses the West has created for itself.[6]

Technology now threatens to annihilate the human species, to destroy the planet's capacity to support life, and to eliminate human meanings in life. Small wonder, then, that Innuits (Eskimos)—prototypes of a pretechnological people living at a rudimentary cultural level—deem themselves superior to technologically advanced counterparts. Given the sketchiest training, Innuits master tractors and bulldozers better than the Kabloona—the White Men. They quickly learn how to maintain and repair all types of machinery, and no visitor can ever learn as much as they already know about Artic conditions. As Lord Ritchie-Calder reports:

> That is why they call the Eskimo *Innuit,* the Real Man. They know that *Kabloona* cannot exist in Eskimo country without a welter of civilized equipment such as heated houses, radios, aircraft, supply

ships, and so on, while everything an Eskimo family needs to sustain life under the harshest conditions can be carried on a single dog sledge. When *Kabloona* goes traveling by land it is *Innuit* who must show him the way. So, since he can learn White Man's ways quicker than the White Man can learn his, the Eskimo, without arrogance, knows that he is the Real Man.[7]

Like the Innuits, other Third World culture groups may prove able to master Kabloona's technology more quickly than the White Man can learn Innuit's independence or flexibility. Perhaps only societies which for centuries have respected nature can adapt technology in a non-Promethean mode. Can it be that only cultures which cherish community and kin relationships have long-range survival capacities in a world where competition will prove to be not only socially rapacious but dysfunctional to survival as well? "Conciliatory" speeches from First World leaders purvey a "trickle down" imagery: the rich are to get still richer but, in the process, something will be left over for the poor to improve their lot.[8] This view is hardly calculated to induce, in arenas of global development, a "wisdom to match our sciences." On the contrary, it exacerbates the very inequalities which technology breeds and which in turn reinforce technology's own tendency to become a self-validating end.

In international discussions, "developed" countries display a terminological schizophrenia parallel to the one they employ domestically. The French political theorist Raymond Aron contends that

industrial societies proclaim an egalitarian conception of society; yet at the same time they give rise to collective organizations which are increasingly gigantic and to which individuals are progressively more integrated. They spread an egalitarian conception but create hierarchical structures. Thus every industrial society needs an ideology to fill up the gap between what men live and what, according to ideas, they ought to live. We observe an extreme form of this contradiction in Soviet society where, in the name of an ideology of abundance, consumption is curtailed as much as possible in order to increase the power of the collectivity. And the American ideology which allows the reconciliation of hierarchic structure with the egalitarian ideal is the ancient formula: "Every infantryman carries in his knapsack a field marshal's baton."[9]

Dichotomies between rhetoric and reality flow necessarily from technology's character as simultaneous bearer and destroyer of values. Technologies of persuasion and image-making "transform culture into luxury"[10] and atrophy the capacity to innovate. Technical integration so totally absorbs even revolution that "the supreme luxury of the technical society will be to grant the bonus of useless revolt and of an acquiescent smile,"[11] Scott Buchanan sees Ellul's warning as a summons

to recover our truly scientific understandings, our objective knowledge of our ends and the ends of nature, and our individual and common wills. This might give us back our reverence and love of nature as well as our shrewd ingenuities in exploiting it.[12]

Optimism with respect to developed countries seems unfounded, however, for even in times of crisis they seem unable to demystify technology. As a result, many observers place their hopes in the Third World. The Palestinian physicist A. B. Zahlan observes that

these undeveloped human cultural entities may be structures within which fresh and non-Western relationships between science, technology and man appear that may help resolve the numerous diseases of Western society. In other words, it is in the very interest of Western society and the human race to restrain their cultural imperialism and/or to find measures to promote native creativity in Third World countries.[13]

Indeed the very inability of some poor nations to achieve "development" may prove a blessing in disguise, enabling them to avoid that economic "cannibalism" by which nations devour their own prosperity.[14]

Technological idolatry confirms in societies alienating forms of development. This is no argument for rejecting technology, although technological optimists tend to brand any critique of technology as intellectual Ludditism. Criticism, however, is a plea for cultural wisdom to guide technology. And as E. F. Schumacher writes,

wisdom demands a new orientation of science and technology towards the organic, the gentle, the non-violent, the elegant and beautiful. . . . We must look for a revolution in technology to give us inventions and machines which reverse the destructive trends now threatening us all.[15]

Theorists of social change speak of "viable" and "unviable" nations, warning us that many extant cultures may prove unable to assimilate technology without "losing their soul." Ironically, however, today's technologically "advanced" societies may well be the first to fall victim to generalized anomie, to which they have rendered themselves vulnerable by their pursuit of gigantic size, their compulsive voracity to consume, and their impotence in rewarding creativity except in modes which reinforce technology's sway. The collapse of the industrial world would not surprise Toynbee, however; one recurring theme in his *Study of History* is the existence of an inverse relationship between the cultural level of societies and their degree of technological attainments.[16] Given that any human group's psychic energy is limited, if it channels most of it to solve technological problems, little is left for truly civilizational creativity in esthetic and spiritual domains. The price paid for success in science and technology is often regression on more important fronts, a societal analogue of the tragic persona familiar to our age: the brilliant scientist or industrialist who is emotionally a child and politically an idiot. Toynbee writes that

> man's intellectual and technological achievements have been important to him, not in themselves, but only in so far as they have forced him to face, and grapple with, moral issues which otherwise he might have managed to go on shirking. Modern Science has thus raised moral issues of profound importance, but it has not, and could not have, made any contribution towards solving them. The most important questions that Man must answer are questions on which Science has nothing to say.[17]

The "developed" West may be obliged to return to a hierarchy of values like that which characterized China during the "Middle Ages." Harvard's Everett Mendelsohn, an historian of science, thinks that

> had a visitor from Mars dropped down then, roughly any time from the 5th Century B.C. to the 15th Century A.D., Europe would have seemed the least likely place for the technological revolution to occur . . . for technique to be introduced as *the* rationale of human activity. China, I would guess, would have seemed a much likelier place. Its technology was far more developed; it had a more rationalized commerce and was a more sophisticated bureaucracy. The mandarins made their counterparts in the Vatican look like peasants in terms of the use of knowledge, of written language, of symbolism, and in terms of *their understanding of the position of technique in human life*.[18]

Modern China has turned its back on Confucianism, but its revolutionaries subordinate technique to politics and values. China's early experience with Western technology taught it the lesson that uncritical acceptance of technology leads ultimately to competition, waste, and exploitation. Because technology has to be subordinated to other values, all societies, "developed" and "underdeveloped" alike, will need to revitalize their traditions to serve their future.[19]

One conclusion reached in the present study is that technology can be controlled if it is not sought as an absolute. Paradoxically, technology is indispensable in struggles against the miseries of underdevelopment and against the peculiar ills of overdevelopment. Technology can serve these noble purposes, however, only in those societies in which ideology, values, and decisional structures repudiate the tendency of technology to impose its own logic in striving after goals. Toynbee hopes for the advent of wisdom from efforts by the world's higher religions—Buddhism, Hinduism, Islam, Judaism, and Christianity—to come to terms with universalism and secularism. Lewis Mumford prefers to remind us that civilizations of the past

> did not regard sicentific discovery and technological invention as the sole object of human existence; for I have taken life itself to be the primary phenomenon, and creativity, rather than the 'conquest of nature,' as the ultimate criterion of man's biological and cultural success.[20]

Glorifying life and creativity, however, does not guarantee the fullness of their development. Life also comes to an end, and civilizations too, as Paul Valéry poignantly reminds us, are mortal. And technological creativity can be put to destructive purposes. This danger revives ancient philosophical questions as to the meaning of death, of suffering, of tragedy, of ultimate meaning.[21] All known civilizations have answered these questions in religious, albeit not always in transcendental, terms. Consequently, the religious myth of Prometheus illuminates the destiny of civilizations in a post-technological age.

If humankind is a despairing Prometheus plagued by guilt over having stolen from heaven the devine fire called technology, it cannot avoid being enslaved by its own creation. If, on the other hand, humankind accepts technology as a free gift of the gods enabling the construction of a better world and a closer affinity with the divine, it remains possible for human beings *not* to fall into the idolatry trap.[22] It is no accident that it is precisely within allegedly "one-dimensional" societies like the United States that the strongest voices are heard warning against the twin evils of antitechnological idiocy and romantic technological optimism. Myron Bloy, a theologian and author of *The Crisis of Technological Change,* sees technology bringing new freedoms and new capacities for basing an emerging culture on critically defined norms and values. He explains that, during the technological era,

> God is, in effect, kicking us in the pants and telling us that it is time to grow up. We are given the tools needed to shape a new culture and allowed to use them effectively only in the service of a prophetic committment. . . . There is no assurance that society will accept this challenge rather than hide in increasingly frenzied operationalism or increasingly brittle idealisms until we are overwhelmed by chaos, but these are our only two options.[23]

These two options now confront not the United States alone but the entire world. The first choice is prophetic commitment to peace, justice, material sufficiency for all, ecological integrity, and the rebirth of vital cultural diversity.[24] The alternative, inevitable if the first option is declined, is chaos: exploitative development for the few at the expense of the many war-making, technological servitude, ecological pathology, and the reification of all human values.

This study of values conflicts in technology transfer has attempted to peel away the mystifications which veil the true impact of technology on societies nurturing diverse images of development. Technology is revealed herein as a two-edged sword, simultaneously bearer and destroyer of values. Yet technology is not static: it is a dynamic and expansionist social force which provides a "competitive edge" enabling its possessors to conquer economic, political, and cultural power. Consequently, Third World efforts to harness technology to broader developmental goals are paradigmatic of a still greater task: to create a new world order founded not on elitism, privilege, or force but on effective solidarity in the face of human needs. The gestation of a new world order poses two troubling questions for all societies: Can technology be controlled, and will culture survive?

To these two questions the answer is a qualified yes. But several conditions must first be met. Those who aspire to master technology must learn to look critically and constructively at their own cultural wisdom. This searching look at the past is needed if they are to escape the reductionism which impregnates the technological cast of mind. It is to be hoped that out of the confrontation between past values and present technological necessities may emerge new sources of life, creativity, and organic thinking.

New forms of knowledge must be born. French sociologist Edgar Morin pleads for

> a restructuring of the general shape of knowledge . . . a totally new conception of science itself which will challenge and overturn not only established boundaries among disciplines but the very cornerstones of all paradigms and, in a sense, the scientific institution itself.[25]

Only thus can human knowledge adequately explain "the anthropological trinity of species, society, and the individual."[26]

The revitalization of traditions, values, and wisdoms in the light of modern technological challenges and the construction of new modes of understanding must occur at two levels. While particular loyalties and values are revived, more universal attachments to a global order must also gain sway. World-order thinking is essential, writes Indian economist Rajni Kothari, because

> it is no longer possible to bring about successful change of an enduring kind in one area or country, except in very marginal ways, without taking account of the world context. Even revolutions suffer from this limitation. Similarly, no amount of either pleading or moralizing to restrain standards of consumption or curb 'chauvinist' tendencies is likely to go far in the poorer regions unless at the same time a similar onslaught is directed at the citadels of affluence and the centres of political and military dominance.[27]

New planetary bargains must be struck between the rich and poor, the technologically advanced and those less so.[28]

Can a global order promote just development, technological wisdom, ecological health, and reciprocity among all societies? The options are posited by Reimer in these terms:

> Effective curtailment of world population and of energy and other technological uses will require either a world dictatorship, for which history pro-

vides no model, or an ethical social order for which there is even less historic precedent. Failing control by one of these means, the industrial world cannot survive. If the industrial world breaks down, however, only the same alternatives remain as suitable models for a viable new social order. In this case, however, an additional possibility occurs; namely, that no reconstruction but an indefinite period of barbarism might ensue.[29]

The "developed" West has shaped modern technology and aggressively exported it to other societies, most of whom received it avidly. While processes of technology transfer have solved innumerable problems, they have likewise destroyed many of the cultural values societies need to achieve a wisdom to match their sciences. The tragic truth is, as Mumford writes, that

Western man not merely blighted in some degree every culture that he touched, whether "primitive" or advanced, but he also robbed his own descendants of countless gifts of art and craftmanship, as well as precious knowledge passed on only by word of mouth that disappeared with the dying languages of dying peoples.[30]

Many Third World leaders resignedly accept the destruction of their own cultures in order to gain modernity. A general uneasiness has come to prevail, therefore, in all areas where development is discussed. Visions of brave new worlds are no longer euphoric; even erstwhile champions of development have grown fearful of apocalypse. Especially in the rich world, social critics grow weary and pessimistic and come to fear developmental change.[31] All societies, developed and nondeveloped, are being forced to make what French philosopher J.M. Domenach calls a "return to the tragic."[32] No longer do any certitudes exist regarding the course of technology or the future of humankind. Yet this very obscurity is salutary; our age has learned that easy certitudes are mere tranquilizers peddled in the markets of meaning.

Technology is no panacea for the ills of underdevelopment; even at best its promise is uncertain. And no romantic flight from technology can bring salvation from the alienation specific to "developed" societies. For every historical experience of social change is, as Domenach reminds us, true tragedy "thrusting us to the very heart of those relations which any society has of its own self-image, its language, its history and its future."[33]

As all societies struggle to create a world of genuine development, value conflicts will endure. But these conflicts, like technology itself, can prove beneficial. The key lies in the criteria chosen to decide which values will be destroyed and which will be preserved. Technology is indeed a two-edged sword, at once beneficent and destructive. But so is development itself. So is all of human history.

NOTES

1. James Harvey Robinson, "Civilization and Culture," in *Encyclopedia Britannica,* 14th ed., 23 vols. (Chicago: William Benton, 1969), vol. 5, p. 831.
2. Arnold J. Toynbee, *Civilization on Trial* (New York: Oxford University Press, 1948), p. 158.
3. John White, "What Is Development? And for Whom? (Paper presented at Quaker Conference on "Motive Forces in Development," Hammamet, Tunisia, April 1972), p. 41.
4. Toynbee, *Civilization on Trial,* p. 86.
5. Ibid., p. 91.
6. Cf. Toynbee, *Civilization on Trial,* pp. 70, 158.
7. Lord Ritchie-Calder, *After the Seventh Day: The World Man Created* (New York: Simon and Schuster, 1961), p. 19. For a more detailed and scientific portrait of the adaptability of the Innuits, cf. Sixten S. R. Haraldson, *Evaluation of Alaska Native Health Service* (Report of a study trip, December 1972/January 1973, prepared for the World Health Organization, Geneva).
8. See, for example, the address of Henry Kissinger before the St. Louis World Affairs Council, 12 May 1975. (Text available from US Department of State, Bureau of Media Affairs, Washington. D.C.)

   Although widely hailed as being "generous" and attentive to Third World needs, the speech nonetheless betrays the assumption that 25% of the world's population has every right to 75% of its resources and that the United States, in particular, need feel no uneasiness in controlling at least 30% of these resources to satisfy its mere 6% of the world's peoples. No mention is made of the risk that the pursuit by Americans of still further material wealth can endanger the biosphere and inflict physical and genetic damage on future generations.
9. Raymond Aron, *Dix-huit leçons sur la societé industrielle* (Paris: Gallimard, 1962), p. 361. Translation mine.
10. Jacques Ellul, *The Technological Society,* p. 424.
11. Ibid., p. 427.
12. Scott Buchanan, "Technology as a System of Exploitation," in *The Technological Order,* ed. Carl F. Stover (Detroit: Wayne State University Press, 1963), p. 159.
13. A.B. Zahlan, "Cultural Change and Cultural Transfer: A Preliminary Assessment of Present Conditions" (n.p., 6 September 1973), mimeographed, p. 2.
14. The phrase is Lord Ritchie-Calder's, in his *After the Seventh Day,* p. 427.
15. Quoted in Nicholas Wade, "E. F. Schumacher: Cutting Technology Down to Size," *Science,* vol. 189, no. 4198 (18 July 1975), p. 199.
16. Arnold J. Toynbee, *A Study of History,* 10 vols.,

abridgment by D. C. Somervell, in 2 vols. (New York: Dell, 1965), vol. 1, pp. 59, 379, 382.
17. Ibid., vol. 2, p. 116.
18. Everett Mendelsohn, "The Ethical Implications of Western Technology for Third World Communities" (Address delivered at the Massachusetts Institute of Technology Seminar on Technology and Culture, Cambridge, Mass., 19 November 1974). Italics mine. For a revealing portrait of the differences to which Mendelsohn alludes, see *China in the Sixteenth Century: The Journals of Matteo Ricci, 1583-1610* (New York: Random House, 1953). Also Vincent Cronin, *The Wise Man from the West* (Garden City, N.Y.: Doubleday, 1957).
19. Cf. Mirrit Boutros Ghali, *Tradition for the Future* (Oxford: The Alden Press, 1972).
20. Lewis Mumford, *The Myth of the Machine,* 2 vols. (New York: Harcourt Brace Jovanovich, 1970), vol. 2, *The Pentagon of Power,* foreword.
21. On these themes, see Jeanne Hersch, "Comments on 'Industrial Society and the Good Life,'" in *World Technology and Human Destiny,* ed. Raymond Aron (Ann Arbor: University of Michigan Press, 1963), pp. 195-196.
22. See Thomas Merton, "A Note: Two Faces of Prometheus," in *The Behavior of Titans* (New York: New Directions, 1961), pp. 11-23.
23. Myron B. Bloy, Jr., "Technology and Theology," in *Dialogue on Technology,* ed. Robert Theobald (Indianapolis: Bobbs-Merrill, 1967), p. 89.
24. Cf. Arend T. van Leeuwen, *Prophecy in a Technocratic Era* (New York: Scribner's, 1968).
25. Edgar Morin, *Le paradigme perdu: la nature humaine* (Paris: Editions du Seuil, 1970), p. 229. Translation mine.
26. Ibid., p. 199.
27. Rajni Kothari, *Footsteps into the Future* (New York: The Free Press, 1974), p. 9.
28. See John and Magda C. McHale, *Human Requirements, Supply Levels and Outer Bounds: A Framework for Thinking About the Planetary Bargain* (Policy paper of the Aspen [Colo.] Institute for Humanistic Studies, Program in International Affairs, 1975); also, *The Planetary Bargain: Proposals for a New International Economic Order to Meet Human Needs* (Report of an International Workshop, Aspen Institute for Humanistic Studies, 7 July-1 August 1975)
29. Everett Reimer, "Alternative Futures for the World," unpublished book manuscript (n.p., June 1973), p. 2.
30. Mumford, *The Pentagon of Power,* pp. 10-11.
31. Examples are Robert L. Heilbroner, *An Inquiry into the Human Prospect* (New York: Norton, 1974); Barrington Moore, Jr., *Reflections on Causes of Human Misery* (Boston: Beacon, 1972); and Peter Berger, *Pyramids of Sacrifice* (New York: Basic Books, 1974).
32. Jean-Marie Domenach, *Le rétour du tragique* (Paris: Editions du Seuil, 1967).
33. Ibid, p. 13.

THIRD WORLD

Contemporary Western man between the rim and the axis

SEYYED HOSSEIN NASR

■ Seyyed Hossein Nasr *was born in 1933 in Teheran. He was educated at the Massachusetts Institute of Technology and has taught at institutions all over the world, including Harvard and the American University of Beirut. He was until recently director of the Imperial Iranian Academy of Philosophy and chancellor of Aryamehr University in Tehran. Nasr is interested in the traditional and sacred arts, and has published widely in the areas of philosophy, history, and religion. Among his works are* Science and Civilization in Islam *(1968) and* Encounter of Man and Nature *(1968).*

In Islam and the Plight of Modern Man *(1975), Nasr expounds the Islamic intellectual and spiritual heritage as the way out of the morass in which contemporary Western man finds himself. This short selection discusses the problem of* amnesis, *or forgetfulness, in modern man. "Modern man has simply forgotten who he is" and, like Faust, has sold his soul "to gain dominion over the natural environment" and to create "a situation in which the very control of the environment is turning into its strangulation, bringing in its wake not only ecocide but also, ultimately, suicide."*

The basis of this crisis is man's misconception of his own nature. In his rebellion against heaven, man has attempted to understand his nature through a "scientific" study of fragmented human behaviour. But that has given him only external and superficial knowledge of himself. It has not given him knowledge of the essential characteristics of human nature, including its spiritual dimension. Such knowledge can be achieved only through an awareness of interiority, a direct awareness of the self in the light of God.

The confrontation of man's own inventions and manipulations, in the form of technology, with human culture, as well as the violent effect of the application of man's acquired knowledge of nature to the destruction of the natural environment, have in fact reached such proportions that many people in the modern world, especially in the West, are at last beginning to question the validity of the conception of man held in the Occident since the rise of modern civilization. But, despite this recent awareness, in order to discuss such a vast problem in a meaningful and constructive way, one must begin by clearing the ground of the obstacles which usually prevent the profoundest questions involved from being discussed. Modern man has burned his hands in the fire which he himself kindled when he allowed himself to forget who he is. Having sold his soul in the manner of Faust to gain dominion over the natural environment, he has created a situation in which the very control of the environment is turning into its strangulation, bringing in its wake not only ecocide but also, ultimately, suicide.

The danger is now evident enough not to need repetition. Whereas only two decades ago everyone spoke of man's unlimited possibility for development understood in a physical and materialistic sense, today one speaks of "limits to growth"—a phrase well-known in the West today—or even of an imminent cataclasm. But the concepts and factors according to which the crisis is analyzed, the solutions sought after and even the colours with which the image of an impending doom are depicted are usually all in terms of the very elements that have brought the crisis of modern man into being. The world is still seen as devoid of a spiritual horizon, not because there is no such horizon present, but because he who views the contemporary landscape is most often the man who lives at the rim of the wheel of existence and therefore views all things from the periphery. He remains indifferent to the spokes and completely oblivious of the axis or the Centre, which nevertheless remains ever accessible to him through them.

The problem of the devastation brought upon the environment by technology, the ecological crisis and the like, all issue from the malady of *amnesis* or forgetfulness from which modern man suffers. Modern man has simply forgotten who he is. Living on the periphery of his own existence he has been able to gain a qualitatively superficial but quantitatively staggering knowledge of the world. He has projected the externalized and superficial image of himself upon the world.[1] And then, having come to know the world in such externalized terms, he

☐ From *Islam and the Plight of Modern Man* (London: Longman Group Ltd., 1975), pp. 3-7.

has sought to reconstruct an image of himself based upon this external knowledge. There has been a series of "falls" by means of which man has oscillated in a descending scale between an ever more externalized image of himself and of the world surrounding him, moving ever further from the Centre both of himself and of his cosmic environment. The inner history of the so-called development of modern Western man from his historic background as traditional man—who represents at once his ancestor in time and his centre in space—is a gradual alienation from the Centre and the axis through the spokes of the wheel of existence to the rim, where modern man resides. But just as the existence of the rim presupposes spokes which connect it to the axis of the wheel, so does the very fact of human existence imply the presence of the Centre and the axis and hence an inevitable connection of men of all ages with Man in his primordial and eternal reality as he has been, is, and will continue to be, above all outward changes and transformations.[2]

Nowhere is the tendency of modern man to seek the solution of many problems without considering the factors that have caused these problems in the first place more evident than in the field of the humanities in general and the sciences dealing specifically with man, which are supposed to provide an insight into human nature, in particular. Modern man, having rebelled against Heaven, has created a science based not on the light of the Intellect[3]—as we see in the traditional Islamic sciences—but on the powers of human reason to sift the data of the senses. But the success of this science was so great in its own domain that soon all the other sciences began to ape it, leading to the crass positivism of the past century which caused philosophy as perennially understood to become confused with logical analysis, mental acrobatics or even mere information theory, and the classical fields of the humanities to become converted to quantified social sciences which make even the intuitions of literature about the nature of man inaccessible to many students and seekers today. A number of scientists are in fact among those most critical of the pseudo-humanities being taught in many Western universities in an atmosphere of a psychological and mental sense of inferiority *vis-à-vis* the sciences of nature and mathematics, a "humanities" which tries desperately to become "scientific", only to degenerate into a state of superficiality, not to say triviality.[4] The decadence of the humanities in modern times is caused by man's loss of the direct knowledge of himself and also of the Self that he has always had, and by reliance upon an externalized, indirect knowledge of himself which he seeks to gain from the outside, a literally "superficial" knowledge that is drawn from the rim and is devoid of an awareness of interiority, of the axis of the wheel and of the spokes which stand always before man and connect him like a ray of light to the supernal sun.

It is with a consideration of this background that certain questions created by the confrontation between the traditional concept of man and the "scientific" one must be analyzed and answered. The first of these questions that often arise in people's minds is "What is the relation of piecemeal scientific evidence about human behaviour to what has been called traditionally 'human nature'?" In order to answer this question it is essential to remember that the reality of the human state cannot be exhausted by any of its outward projections. A particular human action or behaviour always reflects a state of being, and its study can lead to a certain kind of knowledge of the state of being of the agent provided there is already an awareness of the whole to which the fragment can be related. Fragmented knowledge of human behaviour is related to human nature in the same way that waves are related to the sea. There is certainly a relationship between them that is both causal and substantial. But unless one has had a vision of the sea in its vastness and illimitable horizons—the sea which reflects the Infinite and its inimitable peace and calm—one cannot gain an essential knowledge of it through the study of its waves. Fragmented knowledge can be related to the whole only when there is already an intellectual vision of the whole.

The careful "scientific" study of fragmented human behaviour is incapable of revealing the profounder aspect of human nature precisely because of an *a priori* limitation that so many branches of the modern behaviouristic sciences of man—veritable pseudo-sciences if there ever were any[5]—have placed on the meaning of the human state itself. There has never been as little knowledge of man, of the *anthropos,* in different human cultures as one finds among most modern anthropologists today. Even the medicine men of Africa (not to speak of the Muslim sages) have had a deeper insight into human nature than the modern behaviourists and their flock, because the former have been concerned with the essential and the latter with the accidental. Now, accidents do possess a reality, but they have a meaning only in relation to the substance which supports them ontologically. Otherwise one could collect accidents and external facts indefinitely without ever reaching the substance, or what is essential. The classical error of modern civilization, to mistake the quantitative accumulation of information for qualitative penetration into the inner meaning of things, applies here as elsewhere. The study of fragmented behaviour without a vision of the human nature which is the cause of this behaviour cannot itself lead to a knowledge of human nature. It can go around the rim of the wheel indefinitely without ever entering upon the spoke to approach the proximity of the axis and the Centre. But if the vision is already present, the gaining of knowledge of external human behaviour can always be an occasion for recollection and a return to the cause by means of the external effect.

In Islamic metaphysics, four basic qualities are attributed to Ultimate Reality, based directly on the Quranic verse, "He is the First and the Last, the Outward and the Inward" (LVII; 3). This attribution, besides other levels of meaning, also has a meaning that is directly pertinent to the present argument. God, the Ultimate Reality, is both the Inward *(al-Bāṭin)* and the Outward *(al-Ẓāhir),* the Centre and the Circumference. The religious man sees God as the Inward; the profane man who has become completely oblivious to the world of the Spirit sees only the Outward, but precisely because of his ignorance of the Centre does not realize that even the outward is a manifestation of the Centre or of the Divine. Hence his fragmented knowledge remains incapable of encompassing the whole of the rim or circumference and therefore, by anticipation, the Centre. A segment of the rim remains nothing more than a figure without a point of reference or Centre, but the whole rim cannot but reflect the Centre. Finally the sage sees God as both the Inward and the Outward. He is able to relate the fragmented external knowledge to the Centre and see in the rim a reflection of the Centre. But this he is able to do only because of his *a priori* awareness of the Centre. Before being able to see the external world—be it the physical world about us or the outer crust of the human psyche—as a manifestation of the Inward, one must already have become attached to the Inward through faith and knowledge.[6] Applying this principle, the sage could thus relate fragmented knowledge to the deeper layers of human nature; but for one who has yet to become aware of the Inward dimension within himself and the Universe about him, fragmented knowledge cannot but remain fragmentary, especially if it is based upon observation of the behaviour of a human collectivity most of whose members themselves live only on the outermost layers of human existence and rarely reflect in their behaviour the deeper dimension of their own being.

This last point leads to an observation that complements the discussion of principles already stated. Western man lives for the most part in a world in which he encounters few people who live on the higher planes of consciousness or in the deeper layers of their being. He is therefore, for the most part, aware of only certain types of human behaviour, as can be readily seen in the writings of most Western social scientists, especially when they make studies of such traditions as Islam. Fragmented knowledge of human behaviour, even if based solely on external observation, could aid modern man to become at least indirectly aware of other dimensions of human nature, provided a study is made of the behaviour of traditional man—of the man who lives in a world with a Centre. The behaviour of traditional men of different societies, especially at the highest level of the saints and sages—be they from the Chinese, the Islamic, the North American Indian or any other traditional background—in the face of great trials, before death, in presence of the beauty of virgin nature and sacred art, or in the throes of love both human and divine, can certainly provide indications of aspects of human nature for the modern observer. Such behaviour can reveal a constancy and permanence within human nature that is truly astonishing and can also be instrumental in depicting the grandeur of man, which has been largely forgotten in a world where he has become a prisoner to the pettiness of his own trivial creations and inventions. Seen in this light, a fragmented knowledge of human behaviour can aid in gaining a knowledge of certain aspects of human nature. But in any case a total knowledge of this nature cannot be achieved except through a knowledge of the Centre or axis, which also "contains" the spokes and the rim. A famous saying of the Prophet of Islam states, "He who knows himself knows his Lord". But precisely because "himself" implies the Self which resides at the Centre of man's being, from another point of view this statement can also be reversed. Man can know himself completely only in the light of God, for the relative cannot be known save with respect to the Absolute.

NOTES

1. It must be remembered that, in the West, man first rebelled against Heaven with the humanism of the Renaissance; only later did the modern sciences come into being. The humanistic anthropology of the Renaissance was a necessary background for the scientific revolution of the seventeenth century and the creation of a science which, although in one sense non-human, is in another sense the most anthropomorphic form of knowledge possible, for it makes human reason and the empirical data based upon the human senses the sole criteria for the validity of all knowledge.

   Concerning the gradual disfiguration of the image of man in the West, see G. Durand, 'Défiguration philosophique et figure traditionnelle de l'homme en Occident', *Eranos-Jahrbuch,* XXXVIII, 1971, pp. 45-93.
2. If such a relation did not exist, it would not even be possible for man to identify himself with other periods of human history, much less for the permanent aspects of human nature to manifest themselves even in the modern world as they have in the past and continue to do today.
3. Throughout this book the word 'intellect' is used in its original Latin sense as *intellectus* or the Greek *nous,* which stands above reason and is able to gain knowledge directly and immediately. Reason is only the reflection of the intellect upon the mirror of the human mind.
4. There is little more pathetic in this type of pseudo-humanities than the attempt now being made in some Islamic countries to introduce this decadence into the very bosom of Islamic culture in the name of progress.

   Certain American scholars such as William Arrowsmith and William Thompson have already criticized what could be called the "pollution of the humanities", but the tendency in this field as in the question of the pollution of the environment is mostly to try to remove the ill effects without curing the underlying causes.
5. In modern times, the occult sciences, whose metaphysical principles have been forgotten, have become known as the "pseudo-sciences", while in reality they contain a profound doctrine concerning the nature of man and the cosmos, provided their symbolism is understood. Much of the social and human sciences today on the contrary veil and hide a total ignorance of human nature with a scientific garb and are, in a sense, the reverse of the occult sciences. Hence they deserve much more than the occult sciences the title of "pseudo-science".
6. This theme is thoroughly analyzed by F. Schuon in his *Dimensions of Islam,* trans. by P. Townsend, London, 1970, Chapter 2. Concerning the sage or the Sufi, he writes: "The Sufi lives under the gaze of *al-Awwal* (the First), *al-Ākhir* (the Last), *al-Zāhir* (the Outward) and *al-Bāṭin* (the Inward). He lives concretely in these metaphysical dimensions as ordinary creatures move in space and time, and as he himself moves in so far as he is a mortal creature. He is consciously the point of intersection where the Divine dimensions meet; unequivocally engaged in the universal drama, he suffers no illusions about impossible avenues of escape, and he never situates himself in the fallacious 'extra-territoriality' of the profane, who imagine that they can live outside spiritual Reality, the only reality there is." pp. 36-37.

THIRD WORLD

Man's nature

RABINDRANATH TAGORE

■ Rabindranath Tagore *(1861-1941) was a Bengali poet, philosopher, and social reformer. He came from an affluent and highly talented Calcutta family. Like Mohandas Gandhi, whom he knew, Rabindranath (as he was known to his people) abhorred violence. He was a simple and gentle person who was full of humor and love of life. His most famous work is* Gitanjali *(song offerings), a collection of poems with an introduction by Yeats, who was highly moved by them. It was this book that earned Tagore his Nobel Prize for literature (1913) and thrust him into international prominence.*

In the following excerpt from another work, The Religion of Man *(1931), Tagore expounds the notion of* dharma, *or the "virtue of a thing." Man has a* dharma; *it is his humanity. Furthermore, "civilization is to express Man's* dharma *and not merely his cleverness, power and possessions." Anything short of this description is not civilization.*

Tagore seems to be suspicious of modern gadgetry in particular. Recalling a childhood playmate who felt superior to his peers for possessing a toy bought from an English shop, Tagore observes: "One thing he failed to realize in his excitement . . . that this temptation [the toy] obscured something a great deal more perfect than his toy . . . the dharma *of the child." The playmate's peers constantly used their imagination for creating new games. But his toy removed his need for such an approach. Tagore thus advocates simplicity in the material aspects of life in order to give the imaginative and spiritual faculties of man the ability to blossom and create a higher civilization.*

From the time when Man became truly conscious of his own self he also became conscious of a mysterious spirit of unity which found its manifestation through him in his society. It is a subtle medium of relationship between individuals, which is not for any utilitarian purpose but for its own ultimate truth, not a sum of arithmetic but a value of life. Somehow Man has felt that this comprehensive spirit of unity has a divine character which could claim the sacrifice of all that is individual in him, that in it dwells his highest meaning transcending his limited self, representing his best freedom.

Man's reverential loyalty to this spirit of unity is expressed in his religion; it is symbolized in the names of his deities. That is why, in the beginning, his gods were tribal gods, even gods of the different communities belonging to the same tribe. With the extension of the consciousness of human unity his God became revealed to him as one and universal, proving that the truth of human unity is the truth of Man's God.

In the Sanskrit language, religion goes by the name *dharma,* which in the derivative meaning implies the principle of relationship that holds us firm, and in its technical sense means the virtue of a thing, the essential quality of it; for instance, heat is the essential quality of fire, though in certain of its stages it may be absent.

Religion consists in the endeavour of men to cultivate and express those qualities which are inherent in the nature of Man the Eternal, and to have faith in him. If these qualities were absolutely natural in individuals, religion could have no purpose. We begin our history with all the original promptings of our brute nature which helps us to fulfil those vital needs of ours that are immediate. But deeper within us there is a current of tendencies which runs in many ways in a contrary direction, the life current of universal humanity. Religion has its function in reconciling the contradiction, by subordinating the brute nature to what we consider as the truth of Man. This is helped when our faith in the Eternal Man, whom we call by different names and imagine in different images, is made strong. The contradiction between the two natures in us is so great that men have willingly sacrificed their vital needs and courted death in order to express their *dharma,* which represents the truth of the Supreme Man.

The vision of the Supreme Man is realized by our imagination, but not created by our mind. More real than individual men, he surpasses each of us in his permeating personality which is transcendental. The procession of his ideas, following his great purpose, is ever moving across obstructive facts towards the perfected truth. We, the individuals, having our place in his composition, may or may not be in conscious harmony with his purpose, may even put obstacles in his path bringing down our doom upon ourselves. But we gain our true religion when we consciously co-operate with him, finding our exceeding joy through suffering and sacrifice. For through our own love for him we are made conscious of a great love that radiates from his being, who is Mahātma, the Supreme Spirit.

The great Chinese sage Lao-tze has said: "One who may die, but will not perish, has life everlasting". It means that he lives in the life of the immortal Man. The urging for this life induces men to go through the struggle for a true survival. And it has been said in our scripture: "Through *adharma* (the negation of *dharma*) man prospers, gains what appears desirable, conquers enemies, but he perishes at the root." In this saying it is suggested that there is a life which is truer for men than their physical life which is transient.

Our life gains what is called "value" in those of its aspects which represent eternal humanity in knowledge, in sympathy, in deeds, in character and creative works. And from the beginning of our history we are seeking, often at the cost of everything else, the value for our life and not merely success; in other words, we are trying to realize in ourselves the immortal Man, so that we may die but not perish. This is the meaning of the utterance in the Upanishad: "*Tam vedyam purusham veda, yatha ma vo mrityuh parivyathah*" —"Realize the Person so that thou mayst not suffer from death."

The meaning of these words is highly paradoxical, and cannot be proved by our senses or our reason, and yet its influence is so strong in men that they have cast away all fear and greed, defied all the instincts that cling to the brute nature, for the sake of acknowledging and preserving a life which belongs to the Eternal Person. It is all the more significant because many of them do not believe in its reality, and yet are ready to fling away for it all that they believe to be final and the only positive fact.

We call this ideal reality "spiritual". That word is vague; nevertheless, through the dim light which reaches us across the barriers of physical existence, we seem to have a stronger faith in the spiritual Man than in the physical; and from the dimmest period of his history, Man has a feeling that the apparent facts of existence are not final; that his supreme welfare depends upon his being able to remain in perfect relationship with some great mystery behind the veil, at the threshold of a larger life, which is for giving him a far higher value than a mere continuation of his physical life in the material world.

Our physical body has its comprehensive reality in the physical world, which may be truly called our universal body, without which our individual body would miss its function. Our physical life realizes its growing meaning through a widening freedom in its relationship with the physical world, and this gives it a greater happiness than the mere pleasure of satisfied needs. We become aware of a profound meaning of our own self at the consciousness of some ideal of perfection, some truth beautiful or majestic which gives us an inner sense of completeness, a heightened sense of our own reality. This strengthens man's faith, effective even if indefinite—his faith in an objective ideal of perfection comprehending the human world. His vision of it has been beautiful or distorted, luminous or obscure, according to the stages of development that his consciousness has attained. But whatever may be the name and nature of his religious creed, man's ideal of human perfection has been based upon a bond of unity running through individuals culminating in a supreme Being who represents the eternal in human personality. In his civilization the perfect expression of this idea produces the wealth of truth which is for the revelation of Man and not merely for the success of life. But when this creative ideal which is *dharma* gives place to some overmastering passion in a large body of men civilization bursts out in an explosive flame, like a star that has lighted its own funeral pyre of boisterous brilliancy.

When I was a child I had the freedom to make my own toys out of trifles and create my own games from imagination. In my happiness my playmates had their full share, in fact the complete enjoyment of my games depended upon their taking part in them. One day, in this paradise of our childhood, entered the temptation from the market world of the adult. A toy brought from an English shop was given to one of our companions; it was perfect, it was big and wonderfully life-like. He became proud of the toy and less mindful of the game; he kept that expensive thing carefully away from us, glorying in his exclusive possession of it, feeling himself superior to his playmates whose toys were cheap. I am sure if he could use the modern language of history he would say that he was more civilized than ourselves to the extent of his owning that ridiculously perfect toy.

One thing he failed to realize in his excitement—a fact which at the moment seemed to him insignificant—that this temptation obscured something a great deal more perfect than his toy, the revelation of the perfect child which ever dwells in the heart of man, in other words, the *dharma* of the child. The toy merely expressed his wealth but not himself, not the child's creative spirit, not the child's generous joy in his play, his identification of himself with others who were his compeers in his play world. Civilization is to express Man's *dharma* and not merely his cleverness, power and possession.

Once there was an occasion for me to motor down to Calcutta from a place a hundred miles away. Something wrong with the mechanism made it necessary for us to have a repeated supply of water almost every half-hour. At the first village where we were compelled to stop, we asked the help of a man to find water for us. It proved quite a task for him, but when we offered him his reward, poor though he was, he refused to accept it. In fifteen other villages the same thing happened. In a hot country, where travellers constantly need water and where the water supply grows scanty in summer, the villagers consider it their duty to offer water to those who need it. They could easily make a business out of it, following the inexorable law of demand and supply. But the ideal which they consider to be their *dharma* has become one with their life. They do not claim any personal merit for possessing it.

Lao-tze, speaking about the man who is truly good, says: "He quickens but owns not. He acts but claims not. Merit he accomplishes but dwells not in it. Since he does not dwell in it, it will never leave him." That which is outside ourselves we can sell; but that which is one with our being we cannot sell. This complete assimilation of truth belongs to the paradise of perfection; it lies beyond the purgatory of self-consciousness. To have reached it proves a long process of civilization.

To be able to take a considerable amount of trouble in order to supply water to a passing stranger and yet never to claim merit or reward for it seems absurdly and negligibly simple compared with the capacity to produce an amazing number of things per minute. A millionaire tourist, ready to corner the food market and grow rich by driving the whole world to the brink of starvation, is sure to feel too superior to notice this simple thing while rushing through our villages at sixty miles an hour.

Yes, it is simple, as simple as it is for a gentleman to be a gentleman; but that simplicity is the product of centuries of culture. That simplicity is difficult of imitation. In a few years' time, it might be possible for me to learn how to make holes in thousands of needles simultaneously by turning a wheel, but to be absolutely simple in one's hospitality to one's enemy, or to a stranger, requires generations of training. Simplicity takes no account of its own value, claims no wages, and therefore those who are enamoured of power do not realize that simplicity of spiritual expression is the highest product of civilization.

A process of disintegration can kill this rare fruit of a higher life, as a whole race of birds possessing some rare beauty can be made extinct by the vulgar power of avarice which has civilized weapons. This fact was clearly proved to me when I found that the only place where a price was expected for the water given to us was a suburb at Calcutta, where life was richer, the water supply easier and more abundant and where progress flowed in numerous channels in all directions. It shows that a harmony of character which the people once had was lost—the harmony with the inner self which is greater in its universality than the self that gives prominence to its personal needs. The latter loses its feeling of beauty and generosity in its calculation of profit; for there it represents exclusively itself and not the universal Man.

There is an utterance in the Atharva Veda, wherein appears the question as to who it was that gave Man his music. Birds repeat their single notes, or a very simple combination of them, but Man builds his world of music and establishes ever new rhythmic relationship of notes. These reveal to him a universal mystery of creation which cannot be described. They bring to him the inner rhythm that transmutes facts into truths. They give him pleasure not merely for his sense of hearing, but for his deeper being, which gains satisfaction in the ideal of perfect unity. Somehow man feels that truth finds its body in such perfection; and when he seeks for his own best revelation he seeks a medium which has the harmonious unity, as has music. Our impulse to give expression to Universal Man produces arts and literature. They in their cadence of lines, colours, movements, words, thoughts, express vastly more than what they appear to be on the surface. They open the windows of our mind to the eternal reality of man. They are the superfluity of wealth of which we claim our common inheritance whatever may be the country and time to which we belong; for they are inspired by the universal mind. And not merely in his arts, but in his own behaviour, the individual must for his excellence give emphasis to an ideal which has some value of truth that ideally belongs to all men. In other words, he should create a music of expression in his conduct and surroundings which makes him represent the supreme Personality. And civilization is the creation of the race, its expression of the univeral Man.

When I first visited Japan I had the opportunity of observing where the two parts of the human sphere strongly contrasted; one, on which grew up the ancient continents of social ideal, standards of beauty, codes of personal behaviour; and the other part, the fluid element, the perpetual current that carried wealth to its shores from all parts of the world. In half a century's time Japan has been able to make her own the mighty spirit of progress which suddenly burst upon her one morning in a storm of insult and menace. China also has had her rousing, when her self-respect was being knocked to pieces through series of helpless years, and I am sure she also will master before long the instrument which hurt her to the quick. But the ideals that imparted life and body to Japanese civilization had been nourished in the reverent hopes of countless generations through ages which were not primarily occupied in an incessant hunt for opportunities. They had those large tracts of leisure in them which are necessary for the blossoming of Life's beauty and the ripening of her wisdom.

On the one hand we can look upon the modern factories in Japan with their numerous mechanical organizations and engines of production and destruction of the latest type. On the other hand, against them we may see some fragile vase, some small piece of silk, some architecture of sublime simplicity, some perfect lyric of bodily movement. We may also notice the Japanese expression of courtesy daily extracting from them a considerable amount of time and trouble. All these have come not from any accurate knowledge of things but from an intense consciousness of the value of reality which takes time for its fullness. What Japan reveals in her skilful manipulation of telegraphic wires and railway lines, of machines for manufacturing things and for killing men, is more or less similar to what we see in other countries which have similar opportunity for training. But in her art of living, her pictures, her code of conduct, the various forms of beauty which her religious and social ideals assume Japan expresses her own personality, her *dharma,* which, in order to be of any worth, must be unique and at the same time represent Man of the Everlasting Life.

Lao-tze has said: "Not knowing the eternal causes passions to rise; and that is evil". He has also said: "Let us die, and yet not perish". For we die when we lose our physical life, we perish when we miss our humanity. And humanity is the *dharma* of human beings.

What is evident in this world is the endless procession of moving things; but what is to be realized, is the supreme human Truth by which the human world is permeated.

We must never forget to-day that a mere movement is not valuable in itself, that it may be a sign of a dangerous form of inertia. We must be reminded that a great upheaval of spirit, a universal realization of true dignity of man once caused by Buddha's teachings in India, started a movement for centuries which produced illumination of literature, art, science and numerous efforts of public beneficence. This was a movement whose motive force was not some additional accession of knowledge or power or urging of some overwhelming passion. It was an inspiration for freedom, the freedom which enables us to realize *dharma,* the truth of Eternal Man.

Lao-tze in one of his utterances has said: "Those who have virtue *(dharma)* attend to their obligations; those who have no virtue attend to their claims." Progress which is not related to an inner *dharma,* but to an attraction which is external, seeks to satisfy our endless claims. But civilization, which is an ideal, gives us the abundant power to renounce which is the power that realizes the infinite and inspires creation.

This great Chinese sage has said: "To increase life is called a blessing." For, the increase of life realizes the eternal life and yet does not transcend the limits of life's unity. The mountain pine grows tall and great, its every inch maintains the rhythm of an inner balance, and therefore even in its seeming extravagance it has the reticent grace of self-control. The tree and its productions belong to the same vital system of cadence; the timber, the flowers, leaves and fruits are one with the tree; their exuberance is not a malady of exaggeration, but a blessing.

THIRD WORLD

Growth is not development

SAMIR AMIN

■ Samir Amin *was born in 1931 in Cairo. An economist, he has taught at the universities of Poitiers, Vincennes, and Dakar. In 1970 he became director of the United Nations African Institute for Economic Development and Planning. Among his books are* Le Développement Inégal *(1973) and* L'Afrique de L'Ouest Bloquée *(1971).*

In the essay that appears here, Amin raises the same question considered by Goulet: Is a global Western civilization based on Western technology inevitable? Like Goulet he thinks that it is not, and that the future of cultural progress lies with the Third World. "This is not the kind of view which the master civilizations like to hear, and when it is voiced, they tend to ignore it."

Amin then embarks on a scathing critique of the "Rostowian 'theory' of the stages of economic growth." This theory, according to Amin, confuses economic growth with development. "Growth is essentially a measure of a few relatively easily identifiable units of output; development is a historical process which encompasses . . . the total life of the nation."

The West is accused of fostering—with the aid of the Third World—economic growth rather than development. This practice has led to increased profits for multinational corporations and the privileged classes in Third World countries, but it has done nothing for the masses, whose basic human needs have gone unmet. This state of affairs is the basis for Amin's demand for a new international order.

Every civilization has a centre and a periphery. As the centre matures, it is the periphery that becomes worthy of our attention. That is where the new forms of social organization are born; usually out of the struggle of the dependent and oppressed people against those that dominate them. Ancient Egypt, ancient Rome and ancient China all stood at the core of strong and creative civilizations, but the power that dominates us all today—capitalism—was not among their creations. It sprang up at first almost unnoticed on the periphery of the civilized world, among the descendants of the "barbarians" of Europe and Japan. Here, then, are lessons which the Europe and America-centred philosophers and economists, who talk about the future of the world, and the forms of social organizations it should adopt, might find well worth their study.

The Euro-American school of economists would have us believe that the prosperity of the world depends on the extension of Western institutions to the developing countries which find themselves on the periphery of the capitalist system. They overlook the fact that the evolution of human societies does not progress in orderly fashion through the extension of institutions from one nation to another or across cultural frontiers. It takes place in spurts of creative energy when a formerly dependent society breaks the bonds that fettered it. The centre has the power, but the future lies with the periphery. This is not the kind of view which the master civilizations like to hear, and when it is voiced, they tend to ignore it. More than a century ago, Alexis de Tocqueville warned that power was shifting away from Europe to America and Russia, but it took the bankers of Queen Victoria, of the Third French Republic and of William II of Germany another 50 years to find out how right he had been.

THE ANGLO-SAXON METAPHYSIC

The same blindness affects those Euro-American centred economists whose education seems to have stopped with the absorption of the Anglo-Saxon metaphysic which sees the universe as peopled by "economic men," all equal in their aspirations, all subject to the same immutable laws of economic behaviour, all obediently and happily fulfilling their pre-ordained functions and acting alternately as "producers" and "consumers." It is a remarkably convenient abstraction which permits the "analyst" to dispense with any notions of culture, nationality, social class, or with the power struggle between those that rule and those that are in a position of dependence. It reduces political and economic analysis to the use of a few simple tools, accessible to anyone who can read compound interest tables and knows how to manipulate a slide-rule. A few statistical formulations—particularly gross domestic product, broken down per head of population—become

□ From *Development Forum,* April 1973.

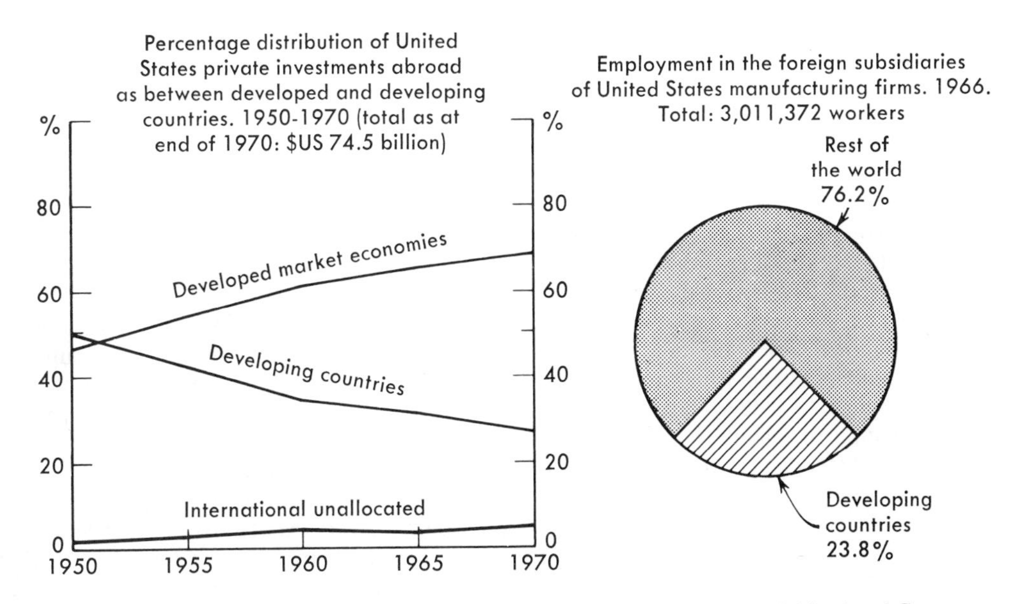

Figures are from an ILO study for the meeting on the Relationship between Multinational Corporations and Social Policy, November 1972. The study remarks that "information on the geographical distribution of foreign investment by countries other than the United States is not readily available, although what information is at hand for the most part parallels the trend shown above."

the ultimate measure of things, the key to the mystery of historical processes. Never mind that it is a measure developed for use in industrial societies and that it makes no sense in the developing, largely rural, often subsistence economies of the Third World; it is an easy statistic to throw around and it avoids the necessity of any further thought and reflection.

The high-priest of this cult, and the man who most perfectly expressed the mechanistic and linear approach to development, is of course Walt W. Rostow with his so-called "theory" of the stages of economic growth.

The Rostowian metaphysic has given rise to an enormous amount of activity in the Euro-American school of economics, in which an endless succession of projections try to establish the future growth rates of far-off lands. Huge wall-maps are marked up in different colour and shadings to distinguish between those countries that will reach the "take-off" point in the year 2,000, those which will make it by 2,050, and those unfortunates who will get there still later. As a means of predicting likely realities of the future the whole thing makes of course no sense at all. Experienced fiction writers could do better.

How could such a gigantic aberration take place? It is basically an unwitting hoax that was perpetrated through a simple piece of sleight-of-hand when the term "economic growth" was first likened to, and then gradually substituted for, the term "development." Today, as the most casual glance at any relevant literature will show, the two are used interchangeably to designate the same thing. Yet they are concepts, which, although related, deal with profoundly different matters. Growth is essentially a measure of a few relatively easily identifiable units of output; development is a historical process which encompasses not only production, but the entire economic and social life of a nation in transition; its health, education, social outlook, the dynamism of its political institutions, in fact the total life of the nation.

The choice of growth over development is no accident. It helps perpetuate an imbalance that has served the capitalist world well. The imbalance concerns the international division of labour under which the industrialized nations of the centre concentrate on high-technology and a high return on manufactured products, while the developing countries of the periphery concentrate on raw materials production and some light manufacturing.

WESTERN ASSISTANCE

The aid policy of the West, focussed as it is on a few major objectives, tries to ensure that this division of labour continues forever. The objectives, stabilization of raw material prices, with buffer stocks to narrow price fluctuations, the promotion of a few "import substitution" industries, the attainment of certain quantitative aid goals, and of course the encouragement of foreign investment, particularly that of the multinational corporations, are all designed to achieve growth. But it is a growth that does not change the pattern, that leaves the division of labour intact and that keeps the economies of the periphery on the same well-trodden tracks.

Growth to maintain the existing international division of labour is not development at all; it is the perpetuation of underdevelopment. The key to the Euro-American concept of a growing economy is its emphasis on exports. The export model, in effect, prescribes that the people of the developing countries shall not work for themselves, but for others. Development, true development, is not directed at exports but at the domestic requirements of the national economy, particularly for the great mass of the people, and the satisfaction of their most urgent needs. This is how development took hold in capitalist countries, such as Europe, North America and Japan. It was also the direction taken by the Soviet Union and China. It is true that China introduced some new and unusual approaches but they were nevertheless aimed at the same overall goals.

Once the development orientation toward the domestic market has been fixed, an element of reality enters the workings of the economy. Real wages become linked to productivity and to technological progress. A fair system of income distribution can get under way. The dynamics of such an economy ensure that production will be channelled into those mass consumption goods most urgently needed by the people. As production and income rise, new productive forces are released and the economy grows in a rational manner. Historically, the first need to be met has been food (the agricultural revolution precedes the industrial one); the next, textiles which have always played a significant part in the development of new and growing economies. Today, with advancing technology, durable consumer goods must be added to the first line advance of a developing country.

The kind of peripheral development which is currently being fostered is a far cry from the model I have just described. It begins with the creation of an export sector, which is heavily encouraged by the industrialized nations. This export sector helps to shape the entire economy, supporting a system of unequal exchange under which the prime task of the economy is to provide cheap labour for export purposes. Wages are as low as local conditions permit, and stand in no relationship to the productivity of the enterprise. As a result, the domestic mass market created by such export activity is narrow and distorted.

With the growth of export activities there is still not a mass market, and no general purchasing power. In their stead, another kind of market develops. This is a luxury market for the privileged classes which derive the full benefits of the export trade. Thus, the principal characteristic of the peripheral economy is a primary concentration on exports and a secondary concentration on luxury goods. The luxury market accomplishes the distortion of the economy which the emphasis on exports has still left undone.

The use of capital, skilled labour, the price structure, the choice of investments, wage rates, etc., if not already set by the export trade, all become subject to demands from the luxury market. The allocation of economic resources becomes topsy-turvy. Any rational person would think that an emerging economy would put first things first, and build a sane and equitable economic structure which meets the needs of the population. In such a scheme the production of luxury goods would come after all other requirements have been met. The peripheral economy reverses the priorities and starts with the luxury industries. The real question then becomes whether it will ever get around to the needs it should have tackled originally.

MARGINAL MASSES

Meanwhile the basic problems of the population go untended. Poverty, unemployment, rising dependence, all shoot up. The broad mass of the population is effectively excluded from the workings of the economy. People become marginal.

Belatedly the international institutions have become concerned. Alarm is voiced about inequities in income distribution and growing unemployment. Their surprise is difficult to understand since these evils are a direct consequence of the economic model that has been pursued. But the warnings do not go uncontradicted. A counter-offensive has been launched by those who benefit from the continued existence of an industrial centre that depends on the support of an underdeveloped periphery. It is claimed that revenues produced by the export-luxury economy are "homothetic"; they in turn lead to productive investment which benefits the masses. The spuriousness of such claims can be easily confirmed by a few visits to African towns in which the slum-dwellers of the "bidonvilles" have risen from one fifth to one half of the total population of the cities within a few years. In a number of Latin American countries, and in Africa as well, the real income of workers and peasants has shrunk, even though the economy has achieved some remarkable records of growth. Growth yes, but growth for whom? For those who benefit from the export-luxury economy, of course.

Their incomes continue to rise and the system they have established feeds on itself to ensure that despite some diversification, industrialization, and so on, the basic direction of the economy will remain the same. Meanwhile the privileged classes not only become "integrated" economically into the world market, but socially become members of the centre. They adopt "European" life-styles and enter the Euro-American culture. Their alienation from their own people becomes complete. The "marginalization" of the masses becomes essential to their survival and further prosperity. Political and cultural forces supplement economic ones to keep the society peripheral.

MYSTERIOUS MULTINATIONALS

The multinational corporations play an important role in this process, and we must be grateful to the United Nations Conference on Trade and Development for having placed the subject on its agenda and for helping to shed some light on it. An aura of mystery surrounds the multinational corporation and it is important that we understand its workings.

One of the claims made by the multinationals is that such corporations transfer technology. They do not. They sell technology and they sell it under highly monopolistic conditions. In a world where everything is bought and sold, technology is a merchandise like any other, although perhaps an unusually complex one. Its constituent parts are machinery and equipment, skilled labour which knows how to use them, knowledge of processes and markets, raw materials and other inputs. The market for technology is a highly controlled one, which gives the multinationals a large element of monopoly in their field. They use it to draw the maximum possible profit from their dealings with the peripheral countries.

The amount spent by the developing world on the acquisition of technology is already considerable—estimated at $1.5 billion a year and rising annually by some 20 per cent. Moreover, the product they buy is by no means always adapted to the conditions under which it will be used. The highly centralized research and development of the multinational companies is not primarily directed at the conditions obtaining on the periphery. When it comes to developing countries, the right questions are not asked. Among them is the important one on how the new technology will affect the broad mass of the people.

The inadequacy of the technology bought makes itself felt in different ways. Growing unemployment is in many cases associated with new technical processes. Moreover, it is by no means the highest kind of technology that is "transferred." The multinationals follow the well-known pattern under which the industries which need low-paid labour are sent abroad while the highly skilled, high return industries—software, electronics, space, atomic and solar energy, to name a few—are kept at home. The experience of countries like Taiwan, Hong Kong and Singapore, which have received "runaway industries," shows that the multinational corporations merely serve to deepen the inequalities in the already existing international division of labour.

The answer clearly is that the developing countries must create their own research and development, not to imitate, but to assimilate and invent the processes that are suitable to their needs. Such an initiative will undoubtedly run into the opposition of the multinationals.

It is regrettable that some of the foremost development economists of our time, notably in Latin America, have not yet agreed on a single view of the multinational corporation. For many years the intellectuals of Latin America have led the way in serious analysis of the problems of underdevelopment. Raúl Prebisch was the first to formulate the concept of the centre and the periphery. F. H. Cardoso, Furtado, Anibal Quijano, Osvaldo Sunkel, Sergio Bagu, Pablo Casanova and others, are associated with the theory of "marginalization" and "dependence." They have enriched social science the world over.

It is difficult to account for the fact that

there should now be different views on the workings of the multinational corporation—and its influence on the elites and their consumption habits—among such trenchant and innovative thinkers. Some hold that its effects are wholly negative and harmful to development, others regard it as having highly beneficial effects and constituting evidence of real development. Is it possible that Euro-American-centred thought processes have invaded even these quarters and have effectively blocked out the view of any civilization other than the extension of consumerism among the rich the world over?

AN INHERENT CONTRADICTION

The arrogant nationalism of the rich should cause us concern; it blinds too many people to the real problems of the Third World. The much-touted integration of the Third World into the world market is doomed to fail through an inherent contradiction. On the one hand it stirs up the desire of the people of the periphery for real development; on the other hand it effectively closes the doors to the growth of national free enterprise systems set up to meet the economic and social needs of the population, and which, historically, are the forerunner of a genuine, universal post-capitalist and socialist society.

As long as this contradiction remains, the countries of the periphery must either create their own groundbreaking forms of social organization or they must perish. This is merely another way of saying that the law of progress through clash of cultures is as valid today as it has ever been. And there can be no doubt that in mankind's search for new creative interrelationships the Third World has much to contribute, if only because its contribution is such an overall precondition of its survival. The rich world has much to lose if it ignores this contribution of the poor and merely goes on congratulating itself on the progress of a civilization which aims at nothing more than a constantly growing consumption of gadgets by an alienated and dependent elite.

''Aid'' is still largely viewed as a palliative designed to give narrow support to the continuation of an unequal international division of labour. One could conceive of a genuine international transfer of resources of the rich to the poor once it is universally accepted that the existing division of labour can be scrapped and a more equitable one take its place. The United Nations could play a role, even a modest one, in bringing about such a change. To do so, the UN would have to become more than a channel of transmission of established ideologies and assume the role of spokesman for the countries of the periphery.

Part three

CONTEMPORARY PERSPECTIVES

The selections in Part Two were for the most part statements about technology from the standpoints of the methods and concerns of particular fields, such as sociology, political science, or religious studies. The selections presented in Part Three will rely upon these "classical" views as a background against which to present some of the more important contemporary debates regarding the nature and function of technology and its relation to human life.

One of the most important and pervasive of these debates is what has been called the "autonomous technology" debate. The point at issue in this exchange can be put in the form of a question: Has technology developed a momentum which expresses certain laws of development that are both inherent in it and inevitable—that is, over which human beings no longer have any control?

Some argue that this is now precisely the human predicament. In *The Technological Society,* which appeared in France in 1954 and in the United States in 1964, Jacques Ellul* expresses the view that technology has become "self-augmenting" and autonomous. "There is," Ellul argues, "an automatic growth (that is, a growth which is not calculated, desired or chosen) of everything which concerns technique. This applies even to men."[1] Ellul suggests that "this is a self-generating process: technique engenders itself. When a new technical form appears it makes possible and conditions a number of others."[2]

This fact of self-augmentation which characterizes technique or technology may, in Ellul's view, be formulated by means of two laws. The first law is the irreversibility of technology. In any given civilization technological progress is irreversible. The instruments developed by technology for human use cannot be un-invented. They are used until they are supplanted by newer or better instruments. The second law is the geometrical, not arithmetical, nature of technological progress. Not only can technological instruments and artifacts not be un-invented, but each spawns a family of related instruments. As Siegfried Giedion reminds us in his masterful book *Mechanization Takes Command,* the invention of the steam engine led to a preoccupation in the 1850's and 1860's with "the harnessing of vapors, steam and gasses to unwonted ends."[3] The infusion of carbon dioxide into bread dough to secure its instantaneous rise, soda water, the Bessemer method of steel production, and even a hot air balloon designed to move a rail car up the side of a mountain—all these were the children of the steam engine.

What then, according to Ellul, is the status of the individual human being in such an autonomous, technologically ordered society?

> The human being is delivered helpless, in respect to life's most important and most trivial affairs, to a power which is in no sense under his control. For there can be no question of a man's controlling the milk he drinks or the bread he eats, any more than of his controlling his government.[4]

For Ellul it is efficiency which offers the key to the success of the program of domination that technique pursues by means of its immutable laws. "Every rejection of a technique judged to be bad entails the application of a new technique, the value of which is estimated from the point of view of efficiency alone."[5] Of Ellul's thesis, the sociologist Robert Merton has written: "The essential point, according to Ellul, is that technique produces all this without plan; no one wills or arranges that it be so."[6]

Ellul tempers some five hundred pages of pessimism in *The Technological Society* with a paragraph of qualification in the foreword to the American edition: "We must look at it [the problem of the autonomy of technology] dialectically, and say that man is indeed determined, but that it is open to him to overcome necessity, and that this *act* is freedom. Freedom is not static, but dynamic; not a vested interest, but a prize continually to be won."[7]

In his widely read book *The Making of a Counter Culture* Theodore Roszak also expresses the view that technology has become autonomous. He writes of a "technocracy" that is "not just a power structure wielding vast material influence" but rather, he suggests, "the expression of a grand cultural imperative, a veritable mystique that is deeply endorsed by the populace."[8] Roszak complains of the size of the component parts of the technocracy:

> In the technocracy, nothing is any longer small or simple or rapidly apparent to the non-technical man. Instead, the scale and intricacy of all human activities—political, economic, cultural—transcends the competence of the amateurish citizen and inexorably demands the attention of specially trained experts.[9]

In addition to this problem of scale, which alienates the "amateur," effectively cutting technology off from its source of appropriate novelty and rendering it undemocratic, other features of the technocracy combine to ensure its dominance of our lives. The technocrats, Roszak argues, have somehow convinced us that "the vital needs of man are (contrary to everything the great souls of history have told us) purely technical in character."[10] Furthermore, even when the technocracy has temporarily failed in one area or another, it maintains a firm guiding hand on the wider situation. Finally, only the experts, who have defined the parameters of our technostructure, can continue to fulfill our needs.

While Ellul argues that technique has become truly autonomous—beyond our control, and even beyond the control of the most powerful groups of technocrats—Roszak intimates that technocrats continue to exercise some measure of control over technology. It is these technocrats, he suggests, who "talk of facts and probabilities and practical solutions. Their politics *is* the technocracy: the relentless quest for efficiency, for even more extensive rational control. Parties and governments may come and go, but the experts stay on forever. Because without them the system does not work."[11]

This difference between Ellul's view of the autonomy of technology and that of Roszak can be brought into sharper focus. For Ellul the only freedom we have is to *act*. This is reminiscent of the discussion of human freedom by Jean-Paul Sartre* in his important work *Being and Nothingness*. Freedom is there described as an act of "choosing oneself."[12] Freedom is not made of dreams and wishes for Sartre, but of the very *act* of choosing.

> Thus we shall not say that a prisoner is always free to go out of prison, which would be absurd, nor that he is free to long for release, which would be an irrelevant truism, but that he is always free to try to escape (or to get himself liberated); that is, that whatever his condition may be, he can project his escape and learn the value of his project by undertaking some action.[13]

There are important similarities between the prison of which Sartre writes and the technological society which is the subject of Ellul's book. In each case the individual is "caught" in a system over which he has little or no control. The individual in the technological society, like the prisoner, wishes or longs to be freed from his bondage, but this wishing or longing does not constitute or bring about emancipation.

While for Sartre and Ellul it is a radical act of will, of choice, that constitutes human freedom in a context in which the human being is bound or incarcerated, such as prison or the technological society, Roszak believes that technological man has freedom of a much more productive and far-reaching variety. The counterculture which had its origins in the 1960's reacted against the technocracy precisely by establishing *communities* of individuals who sought the kind of interaction with one another on a personal level that the technological society sought to deny them. Ellul views autonomous technology as a kind of straitjacket that isolates human beings from one another and that allows only individual acts of courage by way of countering it. But in Roszak's book the tentacles of autonomous technology are not so long and all-embracing. It is possible to free oneself by constructing bridges of personal relations to others in the same predicament. This type of interaction, in Roszak's view, is precisely what the technological society tends to inhibit. For Roszak the genius of the revolt of the counterculture against autonomous technology was that it was not primarily political, but cultural. Political movements are thin and unidimensional forces. They rest on ideological positions which are rationally derived and propagated and which do not have the richness of cultural movements with their multidimensionality, their ability to maximize alternatives, and their tendency to increase participation.

For Ellul, technology is like a vast system of laws. It is like the system of laws we call "nature," but of course it competes with nature and attempts to supplant it at every turn. Although technique is by its very nature artificial, its laws are just as "real"—that is, they exercise a control over us which is as unavoidable as a "law" of nature. For Roszak the situation offers more hope. Technological man, in his view, is like a person who has become addicted to a powerful drug. The "pusher" is in this case the technocrat, who continually devises new ways to "hook" his victim, and at the same time continues to raise the price of his addiction. One can escape the tyranny of the drug, if one chooses to do so, but such an escape is not easy and is often only possible when a community of individuals undertakes that task together, offering one another mutual support.

For Sartre and Ellul salvation from the autonomy of technique is personal. It has no residence in the community. Nor need such personal salvation have any ultimate impact on technology itself. Technology is *beyond* control. For Roszak salvation from technology's autonomy rests firmly on the basis of community endeavor. Such salvation is possible because for him technology is simply *out of control,* and our actions can certainly effect personal and community release from its grip. But what is even more important is that we can change the course of technology by those actions. We can "recall" technology when it has taken a wrong turn. We can put it back on course.

Robert Pirsig* has called our attention to the autonomy of technology in his novel *Zen and the Art of Motorcycle Maintenance*. Technology, for Pirsig, constitutes a double danger. There are those, such as Pirsig's friends John and Sylvia, who ride their motorcycles into the countryside, without benefit of extra spark plugs or wrench, to "get away from technology." The absurdity of this position becomes apparent when the motorcycle fails to start for the trip home.

On the other hand, there are those who conduct seminars in Greek philosophy at the University of Chicago. They have been co-opted by the rationality of technique, quantity, and dialectic. They have failed to recall, and to celebrate, the origins of rationality, and technique, in myth, quality, and rhetoric.

Phaedrus, who functions as Pirsig's alter-ego and as a modern Prometheus, does not bring technical expertise, as did the earlier Prometheus, who brought man fire. He brings instead a personal account of how he has freed himself from the horns of the technological dilemma he describes. There is no recipe for salvation here, just the narrative of one who attempts at once to live inside and outside the technological system. In a manner reminiscent of Herman

Hesse's Steppenwolf, Phaedrus hovers at the edge of technology, an outsider, a critic. At the same time, however, he is a critic who must understand and use technology in his critique.

John Kenneth Galbraith's view of the autonomy of technology is, in important ways, similar to that of Roszak. In his important book *The New Industrial State* Galbraith describes the bondage into which technological man has fallen and suggests means of emancipation. "I am led to the conclusion," he writes,

> that we are becoming servants in thought, as in action, of the machine we have created to serve us. This is in many ways, a comfortable servitude; some will look with wonder, and perhaps even indignation, on anyone who proposes escape. Some people are never content. I am concerned to suggest the general line of emancipation.[14]

Whereas for Ellul the autonomy of technology is modeled on the laws of nature, and for Roszak the model is drug addiction, for Galbraith the autonomy of technology resembles an ignorance which can be overcome by education, or a non-terminal illness which can be cured. Technological man can be healed by exerting pressure on his institutions to reinstantiate a quest after the quality, not just the quantity, of life and by continually questioning the economic establishment, with its ever-growing monopolies and advertising hype.

Another important contribution to the debate concerning the autonomy of technology is that of Karl Marx. Taking his cue from the work of G. W. F. Hegel, Marx believed that there is a kind of necessity attached to the movement of history, and that this necessity certainly includes the movement of technology. It is not machine-like laws, however, which determine the movement of technology. Rather history, and technology, evolve toward the better, the more complex, by means of a series of "jumps," in each of which a new stage emerges out of the brawling forces present in the previous stage.

Is technology autonomous for Marx? It is perhaps more proper to say that, to him, the progress of historical development is autonomous, and that the development of technology is an important part of that process. What is the place of the human being in relation to technology? Human freedom, for Marx, consists in becoming aware of the social processes that shape human life and in enlisting oneself in the historical struggle, which leads to a higher stage of human existence.

A good example of the way in which this necessary movement of history (and technology) works is Marx' discussion of nineteenth-century bourgeois industry, found in Part Two of this book, in the selection from *The Communist Manifesto*.* Marx speaks of the attempts of the reactionary lower middle class to "roll back the wheel of history."[15] But, he concludes, the "development of modern industry . . . cuts from under its feet the very foundation on which the bourgeoisie produces and appropriates his products. What the bourgeoisie, therefore, produces, above all, is its own grave diggers. Its fall and the victory of the proletariat are equally inevitable."[16]

For Marx, then, the movement of history and its subpart technology is determined. Technology, like all else, obeys certain universal, inevitable laws of development.

The contribution of Marshall McLuhan to the autonomous technology debate is also an interesting and important one. He seems to argue for human freedom in the face of technology, but at the same time he is forceful in his argument for a systematic historical development which has in its elements of the determinism that we have seen in Marx. Moreover, one of McLuhan's tenets—perhaps a central one—is that technology is not neutral, that it works us over so completely that we are for the most part unaware of its effects in our lives. It is thus the job of the historian or the sociologist, on the one hand, and of the artist, on the other, to understand technology. The ordinary person is too much a part of technology to understand it or to come to terms with it.

Anthony Quinton[17] has pointed to important similarities between McLuhan's historical account of the development of technology and Marx' history of the economic strife between the classes. For Marx, economic history breaks down into three stages. The first is the stage prior to the rise of capitalism; this stage is in turn divided into three parts: the primitive communism of the tribe, the slave societies of the Orient, and the feudal societies of the medieval West. The second stage of Marx' economic history is capitalism, and the third stage is socialism, which is achieved through the victory of the working classes over their bourgeois oppressors.

The similarity of McLuhan's history of technology to this Marxian schema is uncanny. McLuhan's first stage is the pre-literate era, before the invention of the printing press. He calls this stage the "pre-Gutenberg" era after the inventor of the printing press. This stage is divided into three parts: primitive oral-aural tribal societies, the calligraphy of the Orient, and the manuscript culture of the medieval West. The second stage of technology is the literate, eye-dominated era which extended from the development of printing in 1475 to the early twentieth century. The third stage is the post-literate, electronic, all-information-all-the-time era in which we now live.

Like Marx before him, McLuhan sees certain laws at work within the movement of history. A good example is the "law" by means of which the old technology becomes the content of the new. The first printed books had as their content the old manuscripts, and the content of early television came from radio and the movies. Another example is the "law" by means of which "underdeveloped" societies often "leapfrog" over developed ones when a new form of technology is introduced. The reason for this phenomenon is of course that underdeveloped societies have less to "undo"—that is, they are obliged to make fewer structural changes to allow the development of the new technology.

Certainly not all those who write about contemporary affairs agree with Ellul that technology is beyond control, or even with Roszak that it is out of control. There are many who argue against the thesis that technology has become autonomous and that man has lost his freedom.

Daniel Bell,* for one, criticizes Ellul's determinism as well as his pessimism. In his essay "Technology, Nature, and Society," a portion of which is reprinted in Part Two, Bell argues that "there is no immanent logic of technology, no imperative" that must be obeyed. He holds the view that technology is neutral, and he distinguishes between technologies such as the automobile-highway system and their social support and accounting systems. The automobile-highway system, for example, could engender widely divergent effects depending on the ways in which we choose to deal with it.

Robert Heilbronner,* too, argues against the existence of "immutable laws" as they are propounded by Marx, McLuhan, and Ellul. In his essay "Do Machines Make History?" he argues that although there are "sequences" in technology, many conditioning factors are necessary to such a sequence. Man, according to Heilbronner, is free to choose among significantly different alternatives at any stage of technological development.

Kenneth Boulding* argues that technology is a "superstructure" laid on top of "folk" cultures, from which it draws much of its energy and with which it continues to involve itself in a kind of give and take. Not only is he reluctant to admit that technology has immutable laws (that is, is autonomous) over against human beings, but he warns against the "reificiation" of technology—making technology a "thing." This is of course precisely what Ellul and others have done, ascribing to technique a kind of "thinghood" and then ascribing agency to the thing whose existence they have posited.

Some who are concerned with the state of the environment, such as Barry Commoner, are reluctant to accept Bell's or Heilbronner's claims that human freedom is possible or even enhanced in the context of technology. But they are also reluctant to accept the deterministic picture presented by Ellul, McLuhan, and Marx. They see a special class of technological developments which may ultimately rob man of his freedom by destroying the environment in which he functions, thus obviating many of the choices he now enjoys. Among their concerns are the consequences of the development of nuclear energy and the industrial pollution of our environment.

"My own judgment, based on the evidence now at hand," writes Commoner in 1971, in his book *The Closing Circle,* "is that the present course of environmental degradation, at least in industrialized countries, represents a challenge to essential ecological systems that is so serious that, if continued, it will destroy the capability of the environment to support a reasonably civilized human society."[18]

What kind of freedom, Commoner seems to ask, does a human being who has been poisoned by radioactive waste, or by a nuclear accident, have? Is a human society which has destroyed its own life support system capable of continued free choice?

It is important to see that this position, which subscribes to a limited autonomy of technology, is different from the classical position espoused by Ellul. In this view technology is not now autonomous but may become so in the future unless human beings take quick and effective action. Technology has the capability not only to become a hard and despotic master but also to deprive humanity of life itself.

Implicit in all discussions of the autonomy of technology is the issue that might be called the "technological umbrella." Melvin Kranzberg* and Emmanuel Mesthene,* for example, hold that the progress of technology provides the material conditions for ever wider choices within that system, or what Kranzberg has called "enabling factors." Technology, in this view, functions as a kind of umbrella, protecting human beings, isolating them from the forces which would limit choice, and maximizing human freedom of a particular type.

At the basis of this position are the assumptions that there is no such thing as "radical" or "absolute" human freedom and that all human freedom is freedom within certain contexts determined by random cultural forces, whether they be economic, technological, or religious.

Buttressing this position are those who argue that media—in our time, more specifically popular electronic media—provide a great service by offering "lowest common denominator" images, myths, and metaphors, thus allowing members of a complex society to feel at home in and comfortable with a situation whose complexity would otherwise overwhelm them and render them inactive. Having accepted these conditions, human beings are then free within that system to maximize choice structures and alternatives, but they are also free to sink back into the security offered by the cultural net, to accept the cover of the technological umbrella.

Technology, in this view, has elements of autonomy, but it is an autonomy which threatens only those members of the society whose reactions to technology constitute either extreme acceptance or extreme rejection. The individual who conforms to the national daily average of six and one half hours of television viewing, whose personal attitudes, dreams, and goals are formed by electronic media, obviously represents one of these extremes. For this person, technology has become autonomous, as the entire world of his experience has become autonomous, by default. His interaction with his environment often amounts to no more than choosing between deodorants which come in cans and those which come in jars. It is to this individual that a recent telephone company advertisement appealed when it proudly announced that "The System is the Solution!"

At the other extreme we also find a person for whom technology has become autonomous, but for radically different reasons. This is the individual for whom the technological system is too small and rigid a container for his or her personal experience. Jacob Landau,* for example, has described the difficulties that many working artists have in their attempts to deal with technology. Contemporary literature is filled with characters who have attempted to transcend the claims of technology, without coming to terms with its demands, and who have thereby been destroyed. The protagonist of Robert Pirsig's *Zen and the Art of Motorcycle Maintenance* is one of these. His attempt to transcend the quantitative demands of the technological system becomes a self-destructive search for quality. Pirsig thus shows that technology may become autonomous not only for those who capitulate to it but also for those who fail to take into account its inevitable elements, those who take on the entire system single-handedly.

A whole new generation of writers and artists is beginning to show the way to a humanized, non-autonomous technology, or what E. F. Schumacher* has called "technology with a human face." Avoiding the extremes of total immersion in or self-defeating rejection of technology, they sail that narrow channel of creative involvement with technology, learning to utilize its elements in an ongoing critique of its structure.

• • •

Technology, as it is usually defined, has essentially to do with means, not with the goals or ends for the sake of which those means are employed. Marshall McLuhan's book *Understanding Media* is about the *means* or *media* that aid man in the extension of himself in space and time. And, on a more formal level, Jacques Ellul follows Harold Lasswell in defining technology as "the ensemble of practices by which one uses available resources in order to achieve certain valued ends."[19] He pushes this characterization even farther when he claims that technology is "nothing more than *means* and the *ensemble* of means."[20] But what are these valued ends? How do they originate and what form do they take? What is their relationship to the means by which they are achieved?

These questions are perhaps the most philosophically profound of all those raised in connection with the fact of technology. It is here that man must go beyond the question of his freedom *vis-à-vis* technology, the question of technology's autonomy, to a definition of himself which takes into account his own agency and his place in the world of his experience.

There was a time when the relationship between ends and means appeared to be a simple one. In a world in which it was widely believed that man was created by God to do his bidding, the realm of values, or ends, seemed secure; the central problems were how to identify the best, most efficient pathways to these goals. Samuel Florman, for example, in *The Existential Pleasures of Engineering* writes of the engineers who worked in the late nineteenth and early twentieth centuries, believing that "there was always support to be found in that most venerated of sources—The Bible. For had not God given the earth to man and ordered him to 'subdue it'? Many engineers saw their work as the carrying out of the Christian Mission of subduing the earth for the benefit of man and for the greater glory of God."[21]

For *homo religiosus,* man in a religious world, technology was considered a means to the attainment of values which were external to technology itself. Those who criticized technology from within the religious community criticized only a preoccupation with it at the cost of the original religious commitment. Interestingly, Jacques Ellul's criticism of technique is from this religious, in his case Christian, standpoint. As *homo religiosus* he attacks those whom his translator has called "technolaters"—those who idolize technology at the expense of traditional, transcendent, religious values.

Nicholas Berdayaev* also speaks from within the world of the *homo religiosus*. He decries the machine age, with its attempt to "replace the image and likeness of God by the image and likeness of the machine."[22] "The human soul," he warns, "is unable to stand the speed which contemporary civilization demands, and which tends to transform man himself into a machine."[23]

Gabriel Marcel* also warns us of technological idolatry. "What we are noticing here," he writes, "is the passage from the realm of the technical, properly so called, to that of a kind of idolatry of which technical products become the object or at least the occasion."[24] He suggests that technological idolatry ultimately leads to "autolatry," worship of oneself. This is particularly the case where "speed records" are valued. "In a very general way," he continues, "we might say that the exaltation of speed records goes hand in hand with a weakening, an attenuation, of the sense of the sacred."[25]

To the *homo religiosus,* then, technology becomes dangerous at the moment it ceases to serve the purposes of religious values in order to concern itself with its own values. The move from religion as ground of values to technology as ground of values is the move from *homo religiosus* to *homo faber,* man the maker or fabricator, man the toolmaker.

The concept of man as *homo faber* has been an extremely important one in the post-religious world. This definition of man has been, for example, the presupposition of the anthropological work of the Leakeys in Kenya. If being human can be associated with the use of tools, then an important criterion is available to those, such as the Leakeys, who wish to ascertain the time at which true *Homo* emerged from his relatives among the higher primates.

Samuel Florman's *Existential Pleasures of Engineering* is an encomium to *homo faber.* For Florman, man is essentially the problem solver, and our technological bent "emerges naturally from our earliest cultures, and even from our genetic constitution." So the ends to which technology serves as means are for Florman within technology itself! The primary value, or end, of technology is the solution of perceived problems. "Each new achievement discloses new problems and new possibilities. The allure of these endless vistas "bewitches the engineer of every era."[26] Put more boldly, the "impulse to change the world stirs deep within the engineer."[27]

Florman would of course be horrified to learn that on this issue he is in the same camp with Karl Marx, who also subscribed to a version of the *homo faber* thesis. In his view, man is essentially the tool-making animal, and the material instruments of production have been central in his development as human. "Industry," wrote Marx in his early work, published as *The Economic and Philosophic Manuscripts of 1844,* "is the *actual,* historical relation of nature, and therefore of natural science, to man. If, therefore, industry is conceived as the exoteric revelation of man's *essential powers,* we also gain an understanding of the *human* essence of nature or the *natural* essence of man."[28] "Industry," as Marx uses the term, has to do not only with factories and large-scale production, but with technology in general. Although he does not use the word "technology," his intent is clear. He claims that natural science loses its abstractness through application to human needs. Applied science, fabrication, the improvement of the human condition—all these are elements in what Marx calls man's true "anthropological" nature.

But there are those who are comfortable neither with the more traditional concept of man as *homo religiosus* nor with the more recent concept of man as *homo faber.* They can no longer repair to the divine realm of transcendent values which the *homo religiosus* sets over against technology, but they argue that the concept of man as *homo faber* is an incomplete one, one that allows technology to transform means into ends. "Know-how" is the end of *homo faber,* but one can always ask "Know-how to do what?" (A 1972 *New Yorker* cartoon pictured two men with beards and robes seated on a mountain over which a 747 was flying. "They have the know how," one is saying to the other, "but do they have the know why?") *Homo faber* is hard put to answer the question of why *this* goal and not another. Why nuclear energy and not solar, for example?

Albert Speer, who served as Hitler's armaments minister during World War II, depicts himself in just these terms in his autobiography *Inside The Third Reich.* As a young unemployed architect during the economic depression of the 1930's, Speer was so anxious to have a task, something to *do,* that he never questioned the ends to which his work was being put. He relates, with approval, the contents of an editorial which appeared in a British newspaper during the war, the thrust of which was that Speer, almost uniquely among the members of the Nazi circle of power, was a true technocrat—apolitical, a doer, a manager.

The Dutch anthropologist John Huizinga has argued that the development of human culture has proceeded not on the basis of toolmaking but on the basis of playing, imagining, and pretending. He reminds us that for Aristotle the property most often mentioned as unique to *Homo* was risibility, or the ability to smile and to laugh. In his view, ritual, dancing, and poetry were the early elements of human life which made human cooperation, and hence human culture, possible. Technology, in his view, must be seen as a subspecies of this central play ingredient in human life.

The three essays selected for inclusion under the heading "The *Homo Faber* Debate" represent arguments against the *homo faber* thesis that are both serious and rigorous. The arguments in these essays differ in important ways, but each represents an attempt to go beyond the concept of man as *homo faber*.

Lewis Mumford,* for example, argues that the fabrication of tools was important to man only in the form of a mechanism of support for the development of his true task, the cultivation of "biotechnics." *Homo faber* seeks to bring the organic into line with the mechanical, the artificial, whereas a true biotechnics utilizes the non-organic in the service of the organic. An important supplement to this thesis may be found in Siegfried Giedion's monumental work *Mechanization Takes Command*. Peter Drucker* argues not that the *homo faber* thesis is wrong, but that it does not go far enough: it does not accept the fact that language is as important a tool as the more tangible ones. Drucker's thesis calls to mind George Steiner's encomium to language. "More than fire," he wrote in 1966, "whose power to illumine or to consume, to spread and draw inward, it so strongly resembles, speech is the core of man's mutinous relations to the gods." And Hannah Arendt* argues that the weakness of the *homo faber* concept is that it does not go beyond *means* to *meaning:* meaninglessness on a large scale has emerged in our culture because every end has been transformed to means.

• • •

It is by now clear that the careers of technology and mankind are inextricably intertwined. Mankind is not now, nor has he ever been, in a position to abandon technology completely. But as technology has become more complex, and as new alternatives have been generated by his interaction with it, a new awareness of significantly different technological paths has emerged. The question now facing mankind, especially in Western industrial societies, is not *whether* technology should exist, but *what kind* of technology should exist. What kind of technology is appropriate to his needs and desires?

The question of "appropriate" technology is closely related to, but separate from, the question of the autonomy of technology and the question of means and ends, the *homo faber* issue. If one decides that technology is autonomous in the strong sense in which Ellul thinks it to be, then the question of "appropriate" technology becomes moot. If mankind has lost control, he can no longer choose among technological alternatives. And if *homo* is only *homo faber,* one who creates instruments regardless of the ends to which those instruments are put, then the question of appropriate technology is once again meaningless. In such a case mankind has no means by which to judge appropriateness.

Many argue that technology is not autonomous, that mankind is more than mere *homo faber*. But they see that current large-scale technology is taking mankind onto a spur track which must soon be abandoned if he is to survive. Their arguments take many forms. Martin Heidegger, for example, in a 1966 interview with *Der Spiegel,* addressed what he took to be the central issue of large-scale technology: mankind's alienation from the earth. "Everything," he remarked, "is functioning."

> This is exactly what is so uncanny, that everything is functioning, and that technology tears men loose from the earth and uproots them. I do not know whether you were frightened, but I at any rate was frightened when I saw pictures coming from the moon to the earth. We don't need any atomic bomb. The uprooting of man has already taken place. The only thing we have left is purely technological relationships. This is no longer the earth on which man lives.[29]

What Heidegger was complaining about, and what frightened him, was the scale of current technology. His perception was of a system which dwarfs the human individual and in which he can no longer see his own reflection.

Eric Fromm,* in his book *The Revolution of Hope,* also expresses this fear of the scale of current technology. Man is alienated from his world and from himself, Fromm suggests, because the structures of technology are so large that they are beyond his comprehension, and because the speed-up of the technological environment has overridden man's basic biological and psychological time senses.

Heidegger's and Fromm's analyses of the threatening scale of technology are of course in some ways very abstract ones. A far more concrete complaint has been the growing concern of urbanologists that our great cities have lost their human scale. Many of the "urban renewal" programs of the 1960's, for example, have led to the depopulation of vast areas of our inner cities so that parking lots and garages could be built. The city has become the province of the wheel. The scale of the human foot has been lost.

The problems of large-scale technology perhaps have become most apparent in the programs set up to "export" technology to less developed countries. John Updike parodies this situation in his novel *The Coup,* set in the fictitious African republic of Kush. Hakim Félix Ellelloû, the president of Kush, on one of his tours of the countryside, encounters a man in a white button-down shirt and seersucker suit standing atop a pile of Korn Kurls, Total, Spam, Carnation Instant Milk, and "Kix Trix Chex Pops." This pyramid of junk food, it is explained to Ellelloû by the American aid officer who stands at its apex, is relief for the starving Kushites. Before torching the junk food, and lamentably, the aid officer himself, who refuses to abandon it, Ellelloû explains that the famine in his country has been caused by the export of American technology. Vaccination of livestock has led to larger herds, subsequent overgrazing, and the inevitable depletion of the water sources of the area. There is no drought, he says, simply bad ecology.

Later in the novel, Updike has another American aid officer appear in Kush. Speaking to an official in Ellelloû's government, he says: ". . . you tell the man for me, no problem. Our technical boys can mop up any mess technology creates. All you need is a little developmental input, some dams in the wadis and some intensive replanting with the high-energy pampas grass the guys in the green revolution have come up with."[30] A bit later he continues: "Miracles are everyday business for our boys."[31]

But the green revolution as it has been exported has by no means been a complete success. It was once thought, for example, that the solution to Mexico's food problems was large-scale, capital-intensive farming similar to that which had been successful in the United States. The results of this attempted conversion from labor-intensive to capital-intensive farming, however, have been disastrous. *Campesinos,* displaced from their traditional small-scale farming operations, have drifted toward the cities, especially to Mexico City, creating vast poverty-ridden *barrios* on the fringes. No longer self-sufficient on the land, and without adequate industrial job opportunities, these people have become a major factor in Mexico's rocketing unemployment rate.

One leader in the movement toward a technology which is appropriate in scale and intensity was the late E. F. Schumacher.* In his now famous book *Small is Beautiful* Schumacher argued that "big" technology was centralizing decision-making procedures, thus rendering institutions progressively less democratic. He issued a call for the conscious choice of a different kind of technology, one that would be oriented toward personal growth and development rather than defined by patterns of consumption, one that would overcome the cultural schizophrenia generated by specialism and provide meaningful work designed to benefit the whole person.

Perhaps one of Schumacher's most important concepts is economic in nature: Western industrial society, he argued, has been living off its capital—that is, its non-renewable resources—rather than off the interest of that capital—that is, its renewable resources. This practice, according to Schumacher, makes no more sense for an economic system based on technological development than it does in any other economic system.

In place of the growth economies that are thought by some to be necessary to Western industrial societies, Schumacher argued for an economy stabilized in terms of the production and consumption of material goods but rich in possibilities for personal growth and self-discovery. He called this technological alternative ''intermediate'' technology and claimed that it would reintegrate man into the environment from which he had become estranged by ''full-scale'' or big technology.

It would be a mistake to equate Schumacher's intermediate technology with what has been called ''low technology,'' as opposed to ''high technology.'' Television and videotape systems, for example, are ''high technology,'' as compared, for example, to still photography. Nonetheless, they can play an important part in systems of intermediate technology. Information storage, retrieval, and dissemination are as important to intermediate technology as they are to full-scale technology, but there is an important difference between the ways in which the two technologies handle these functions on a practical level. Rather than using an expensive and complex broadcast system that is exported from an industrialized country to a non-industrialized country, along with specialists to operate the equipment, intermediate technology opts for small-gauge, less complex equipment, such as half or three-quarter inch video systems. Equipment is simple enough that local people can be easily trained in its use, and inexpensive enough so that the system of information dissemination can be in a large number of hands. In this example, then, electronic, ''high'' technology can serve the ends Schumacher details in his account of ''intermediate'' technology (the terminology is confusing because the concept is still in formative stages. Some have suggested that we avoid Schumacher's term ''intermediate'' in favor of ''appropriate'' technology).

Another author who has expressed concern over the scale of technology, especially as that scale mediates between man and his environment, is John Lachs.* He writes of the increasing distance between an act and its effects. We no longer are able to weigh the consequences of our actions because there is a failure of the ''feedback'' mechanism, to use a term from cybernetics, a failure generated by the size and complexity of our technological situation.

Not everyone is impressed with the views of those who endorse ''appropriate'' or ''intermediate'' technology. Samuel Florman,* for example, thinks not only that small is not beautiful, but that it is downright boring and dubious and that it makes little sense. He suggests that there be a struggle for survival among competing forms of technology, a kind of naive technological Darwinism. Furthermore, he questions the importation of political matters into the realm of technology assessment. Florman thinks that technology is simply a matter of efficiency, presumably like economic efficiency, and that decisions about technology can be made outside the political realm.

For Schumacher, as for Amory Lovins, Hazel Henderson, and others who have been instrumental in formulating the tenets of ''appropriate'' technology, political matters are at the heart of the assessment of technological systems. Appropriate technology has among its most important features a commitment to democratic structures in which decision-making powers are dispersed, to a cooperative individualism as opposed to a social Darwinism, and to greater autonomy of local governmental structures which stress global awareness.

• • •

It is appropriate that Part Three contain a discussion of the metaphors of technology. The debates concerning technology's autonomy, the nature of technological man, and the scale of appropriate technology are drenched in metaphors to which it is essential that we attend. The positions of those who represent the various sides of these issues often rest as much on the soundness and originality of metaphors as they do on the structure of arguments.

As D. O. Edge* reminds us, recalling an essay by Douglas Berggren, a metaphor is by definition literally absurd (it is literally absurd, for example, to speak of the brain as a computer or of a person as letting off steam). In addition, a metaphor functions in a context which provides some connection or "link" between its elements: just *any* absurdity is not a metaphor. Metaphors overlap images and create new perceptions by means of radical juxtapositions. Furthermore, metaphors demand tension: absurdity is certainly one form of tension, "stereoscopic" overlay is another. From such tension a number of alternative interpretations of a metaphor may emerge. Metaphors are in this sense negetropic: they create more energy than they consume, and they reverse the usual entropic tendency to disorganization.

Perhaps no analyst of technology has been more attuned to its metaphors than has Marshall McLuhan. His books are replete with descriptions of shifts in metaphor that accompany the transitions between the stages of technology he describes. He invites us, for example, to consider the "horseless carriage" metaphor used in the early days of the automobile, a construction which involves the radical juxtaposition of an assertion and a denial, since carriages up to that time had been horsedrawn. This metaphor, he suggests, is not unlike the one used to describe the process of printing at the beginning of the sixteenth century: "artificial writing" *(artificialiter scribere)*.[32]

McLuhan warns us that new technologies cannibalize old ones for the basic stuff of their metaphors, so that the metaphor is always, as it were, in arrears. Electricity was in the early days of its discovery spoken of in terms of the flow of a river, or the discharge of a gun. Susan Sontag has made this point with respect to the metaphors of photography. Photography took over the language of firearms: one aims, one shoots, the product is a snapshot.

Those who generate and explore the metaphors of an emerging form of technology render a great service to historians, sociologists, and philosophers, and to all analysts of the development of technology. Buckminster Fuller, for example, has mined the metaphor "spaceship earth," and has expressed surprise that even the sophistication of an astronaut yields to talk of being "up here on the moon." He reminds us in this connection that the President of the United States perpetuated this antique spatial metaphor in congratulating the astronauts on "going up to the moon and back down to earth."[33] Fuller captures our imagination when he writes of human beings having within their "cerebral mechanisms the proper atomic radio transceivers to carry on telephatic communication. . . ."

Gene Youngblood, explorer of cinematic and video environments, tells us that "television is earth's superego." He writes that "humanity extends its video Third Eye to the moon and feeds its own image back into its monitors."[34]

And Robert Pirsig turns the motorcycle into a metaphor when he describes the interaction of himself and his friends with their road machines. His book *Zen and the Art of Motorcycle Maintenance* can be viewed as an extended metaphor in which attitudes toward particular technological artifacts take on the dimensions of an attitude toward the whole technological environment.

• • •

The debates that have grown out of attempts of late–twentieth-century human beings to understand and assess their technological environment are of course not limited to the ones presented in Part Three. Other debates concern, for example, the possibility of technological "fixes" or technological solutions to problems once thought strictly social, the role of social engineering in human societies, and the debate between "utopians" and "disutopians."

Nevertheless, the issues raised in the debates presented here are central to the continuing inquiry of human beings into their world of experience. At issue in the autonomous technology debate, for example, is a topic of no less importance than mankind's very freedom. The *homo faber* debate has as its basis fundamental questions of human nature, the characteristics which distinguish mankind from the other animals, and the role of technological activity in characteristics of human life. The "small is beautiful" debate addresses the question of the proper scale of attempts to control the human environment. And the section on technological metaphors deals with the issue of the adequacy of the means of expression available to human beings in the assessment of their technological activities.

These are among the most central and significant questions we can ask about ourselves and our technological milieu. A new world of critical awareness awaits the individual who begins to understand the parameters of these problems.

NOTES

1. Jacques Ellul, *The Technological Society,* trans. John Wilkinson (New York: Alfred A. Knopf, Inc., 1964), p. 87.
2. Ellul, p. 87.
3. Siegfried Giedion, *Mechanization Takes Command* (New York: W. W. Norton & Co., Inc., 1969), p. 185.
4. Ellul, p. 107.
5. Ellul, p. 110.
6. Ellul, p. viii.
7. Ellul, p. xxxiii.
8. Theodore Roszak, *The Making of a Counter Culture* (Garden City, N.Y.: Anchor Books, 1969), p. xiv.
9. Roszak, p. 67.
10. Roszak, p. 11.
11. Roszak, p. 21.
12. Jean-Paul Sartre, *Being and Nothingness,* trans. Hazel Barnes (New York: Philosophical Library, Inc., 1956) p. 479.
13. Sartre, p. 484.
14. John Kenneth Galbraith, *The New Industrial State,* 2nd ed. (New York: The New American Library, Inc., 1972), pp. 26-27.
15. Karl Marx and Friedrich Engels, *The Communist Manifesto,* in *The Marx-Engels Reader,* ed. Robert C. Tucker, 2nd ed. (New York: W. W. Norton & Co., Inc., 1978), p. 482.
16. Marx and Engels, p. 483.
17. Anthony Quinton, "Cut Rate Salvation," *The New York Review of Books,* Nov. 23, 1967, pp. 6-14.
18. Barry Commoner, *The Closing Circle* (New York: Bantam Books, Inc., 1972), p. 215.
19. Ellul, p. 18.
20. Ellul, p. 19.
21. Samuel Florman, *The Existential Pleasures of Engineering* (New York: St. Martin's Press, Inc., 1976), p. 8.
22. Nicholas Berdyaev, *"The Bourgeois Mind" and Other Essays,* trans. Countess Bennigsen, revised by Donald Attwater (New York: Books for Libraries Press, Inc., 1966), p. 212.
23. Berdyaev, p. 211.
24. Gabriel Marcel, *Man Against Society,* trans. G. S. Fraser (South Bend, Ind.: Regnery/Gateway, Inc., 1952) p. 63.
25. Marcel, p. 63.
26. Florman, p. 121.
27. Florman, p. 121.
28. Marx and Engels, *The Economic and Political Manuscripts of 1844,* in Tucker, p. 96.
29. "*Der Spiegel's* Interview with Martin Heidegger on September 23, 1966," trans. Maria P. Alter and John D. Caputo, *Philosophy Today,* 20, 4/4 (1976), p. 227.
30. John Updike, *The Coup* (New York: Alfred A. Knopf, Inc., 1978), pp. 230-231.
31. Updike, p. 231.
32. Marshall McLuhan, *The Gutenberg Galaxy* (New York: The New American Library, Inc., 1969), p. 187.
33. Gene Youngblood, *Expanded Cinema* (New York: E. P. Dutton & Co., Inc., 1970), p. 17.
34. Youngblood, p. 78.

11

The autonomous technology debate

The autonomy of technique

JACQUES ELLUL

■ Jacques Ellul *is Professor of History and Contemporary Sociology at Bordeaux University. His book* The Technological Society, *from which this selection is taken, and his books* Propaganda *and* The Political Illusion *have established him as one of the leading contemporary analysts of technology.*

In this selection he argues that "technique," which he defines, following the American sociologist Harold Lasswell, as the "totality of methods rationally arrived at and having absolute efficiency," has become autonomous. Technique has "become a reality in itself, self-sufficient, with its special laws and its own determinations." The implications of this situation for human freedom are, according to Ellul, disastrous. The individual crumbles before technical efficacy, and human caprice yields to the predictability and ubiquity of the machine. Technique strips man of the sense of mystery, it remakes his life on the assumption that it has been badly made, and it robs him of any sense of the supernatural. For Ellul, the problem is not that man achieves victories in the realm of nature and in his own psychological and social world by means of technique, but rather that technique achieves victory over what is human.

The primary aspect of autonomy is perfectly expressed by Frederick Winslow Taylor, a leading technician. He takes, as his point of departure, the view that the industrial plant is a whole in itself, a "closed organism," an end in itself. Giedion adds: "What is fabricated in this plant and what is the goal of its labor—these are questions outside its design." The complete separation of the goal from the mechanism, the limitation of the problem to the means, and the refusal to interfere in any way with efficiency; all this is clearly expressed by Taylor and lies at the basis of technical autonomy.

Autonomy is the essential condition for the development of technique, as Ernst Kohn-Bramstedt's study of the police clearly indicates. The police must be independent if they are to become efficient. They must form a closed, autonomous organization in order to operate by the most direct and efficient means and not be shackled by subsidiary considerations. And in this autonomy, they must be self-confident in respect to the law. It matters little whether police action is legal, if it is efficient. The rules obeyed by a technical organization are no longer rules of justice or injustice. They are "laws" in a purely technical sense. As far as the police are concerned, the highest stage is reached when the legislature legalizes their independence of the legislature itself and recognizes the primacy of technical laws. This is the opinion of Best, a leading German specialist in police matters.

The autonomy of technique must be examined in different perspectives on the basis of the different spheres in relation to which it has this characteristic. First, technique is autonomous with respect to economics and politics. We have already seen that, at the present, neither economic nor political evolution conditions technical progress. Its progress is likewise independent of the social situation. The converse is actually the case, a point I shall develop at length. Technique elicits and conditions social, political, and economic change. It is the prime mover of all the rest, in spite of any appearance to the contrary and in spite of human pride, which pretends that man's philosophical theories are still determining influences and man's political regimes decisive factors in technical evolution. External necessities no longer determine technique. Technique's own internal necessities are determinative. Technique has become a reality in itself, self-sufficient, with its special laws and its own determinations.

Let us not deceive ourselves on this point. Suppose that the state, for example, intervenes in a technical domain. Either it intervenes for sentimental, theoretical, or intellectual reasons, and the effect of its intervention will be negative or nil; or it intervenes for reasons of political technique, and we have the combined effect of two techniques. There is no other possibility. The historical experience of the last years shows this fully.

To go one step further, technical autonomy is apparent in respect to morality and spiritual values. Technique tolerates no judgment from without and accepts no limitation. It is by virtue of technique rather than science that the great principle has become established: *chacun chez soi*. Morality judges moral problems; as far as technical problems are concerned, it has nothing to say. Only technical criteria are relevant. Technique, in sitting in judgment on itself, is clearly freed from this principal obstacle to human action. (Whether the obstacle is valid is not the question here. For the moment we merely record that it is an obstacle.) Thus, technique theoretically and systematically assures to itself that liberty which it has been able to win practically. Since it has put itself beyond good and evil, it need fear no limitation whatever. It was long claimed that technique was neutral. Today this is no longer a useful distinction. The power and autonomy of technique are so well secured that it, in its turn, has become the judge of what is moral, the creator of a new morality. Thus, it plays the role of creator of a new civilization as well. This morality—internal to technique—is assured of not having to suffer from technique. In any case, in respect to traditional morality, technique affirms itself as an independent power. Man alone is subject, it would seem, to moral judgment. We no longer live in that primitive epoch in which things were good or bad in themselves. Technique in itself is neither, and can therefore do what it will. It is truly autonomous.

However, technique cannot assert its autonomy in respect to physical or biological laws. Instead, it puts them to work; it seeks to dominate them.

Giedion, in his probing study of mechanization and the manufacture of bread, shows that "wherever mechanization encounters a living substance, bacterial or animal, the organic substance determines the laws." For this reason, the mechanization of bakeries was a failure. More subdivisions, intervals, and precautions of various kinds were required in the mechanized bakery than in the non-mechanized bakery. The size of the machines did not save time; it merely gave work to larger numbers of people. Giedion shows how the attempt was made to change the nature of the bread in order to adapt it to mechanical manipulations. In the last resort, the ultimate success of mechanization turned on the transformation of human taste. Whenever technique collides with a natural obstacle, it tends to get around it either by replacing the living organism by a machine, or by modifying the organism so that it no longer presents any specifically organic reaction.

The same phenomenon is evident in yet another area in which technical autonomy asserts itself: the relations between techniques and man. We have already seen, in connection with technical self-augmentation, that technique pursues its own course more and more independently of man. This means that man participates less and less actively in technical creation, which, by the automatic combination of prior elements, becomes a kind of fate. Man is reduced to the level of a catalyst. Better still, he resembles a slug inserted into a slot machine: he starts the operation without participating in it.

But this autonomy with respect to man goes much farther. To the degree that technique must attain its result with mathematical precision, it has for its object the elimination of all human variability and elasticity. It is a commonplace to say that the machine replaces the human being. But it replaces him to a greater degree than has been believed.

Industrial technique will soon succeed in completely replacing the effort of the worker, and it would do so even sooner if capitalism were not an obstacle. The worker, no longer needed to guide or move the machine to action, will be required merely to watch it and to repair it when it breaks down. He will not participate in the work any more than a boxer's manager participates in a prize fight. This is no dream. The automated factory has already been realized for a great number of operations, and it is realizable for a far greater number. Examples multiply from day to day in all areas. Man indicates how this automation and its attendant exclusion of men operates in business offices; for example, in the case of the so-called tabulating machines. The machine itself interprets the data, the elementary bits of information fed into it. It arranges them in texts and distinct numbers. It adds them together and classifies the results in groups and subgroups, and so on. We have here an administrative circuit accomplished by a single, self-controlled machine. It is scarcely necessary to dwell on the astounding growth of automation in the last ten years. The multiple applications of the automatic assembly line, of automatic control of production operations (so-called cybernetics) are well known. Another case in point is the automatic pilot. Until recently the automatic pilot was used only in rectilinear flight; the finer operations were carried out by the living pilot. As early as 1952 the automatic pilot effected the operations of take-off and landing for certain supersonic aircraft. The same kind of feat is performed by automatic direction finders in anit-aircraft defense. Man's role is limited to inspection. This automation results from the development servomechanisms which act as substitutes for human beings in more and more subtle operations by virtue of their "feedback" capacity.

This progressive elimination of man from the circuit must inexorably continue. Is the elimination of man so unavoidably necessary? Certainly! Freeing man from toil is in itself an ideal. Beyond this, every intervention of man, however educated or used to machinery he may be, is a source of error and unpredictability. The combination of man and technique is a happy one only if man has no responsibility. Otherwise, he is ceaselessly tempted to make unpredictable choices and is susceptible to emotional motivations which invalidate the mathematical precision of the machinery. He is also susceptible to fatigue and discouragement. All this disturbs the forward thrust of technique.

Man must have nothing decisive to perform in the course of technical operations; after all, he is the source of error. Political technique is still troubled by certain unpredictable phenomena, in spite of all the precision of the apparatus and the skill of those involved. (But this technique is still in its childhood.) In human reactions, howsoever well calculated they may be, a "coefficient of elasticity" causes imprecision, and imprecision is intolerable to technique. As far as possible, this source of error must be eliminated. Eliminate the individual, and excellent results ensue. Any technical man who is aware of this fact is forced to support the opinions voiced by Robert Jungk, which can be summed up thus: "The individual is a brake on progress." Or: "Considered from the modern technical point of view, man is a useless appendage." For instance, ten per cent of all telephone calls are wrong numbers, due to human error. An excellent use by man of so perfect an apparatus!

Now that statistical operations are carried out by perforated-card machines instead of human beings, they have become exact. Machines no longer perform merely gross operations. They perform a whole complex of subtle ones as well. And before long—what with the electronic brain—they will attain an intellectual power of which man is incapable.

Thus, the "great changing of the guard" is occurring much more extensively than Jacques Duboin envisaged some decades ago. Gaston Bouthoul, a leading sociologist of the phenomena of war, concludes that war breaks out in a social group when there is a "plethora of young men surpassing the indispensable tasks of the economy." When for one reason or another these men are not employed, they become ready for war. It is the multiplication of men who are excluded from working which provokes war. We ought at least to bear this in mind when we boast of the continual decrease in human participation in technical operations.

However, there are spheres in which it is impossible to eliminate human influence. The autonomy of technique then develops in another direction. Technique is not, for example, autonomous in respect to clock time. Machines, like abstract technical laws, are subject to the law of speed, and co-ordination presupposes time adjustment. In his description of the assembly line, Giedion writes: "Extremely precise time tables guide the automatic cooperation of the instruments, which, like the atoms in a planetary system, consist of separate units but gravitate with respect to each other in obedience to their inherent laws." This image shows in a remarkable way how technique became simultaneously independent of man and obedient to the chronometer. Technique obeys its own specific laws, as every machine obeys laws. Each element of the technical complex follows certain laws determined by its relations with the other elements, and these laws are internal to the system and in no way influenced by external factors. It is not a question of causing the human being to disappear, but of making him capitulate, of inducing him to accommodate himself to techniques and not to experience personal feelings and reactions.

No technique is possible when men are free. When technique enters into the realm of social life, it collides ceaselessly with the human being to the degree that the combination of man and technique is unavoidable, and that technical action necessarily results in a determined result. Technique requires predictability and, no less, exactness of prediction. It is necessary, then, that technique prevail over the human being. For technique, this is a matter of life or death. Technique must reduce man to a technical animal, the king of the slaves of technique. Human caprice crumbles before this necessity; there can be no human autonomy in the face of technical autonomy. The individual must be fashioned by techniques, either negatively (by the techniques of understanding man) or positively (by the adaptation of man to the technical framework), in order to wipe out the blots his personal determination introduces into the perfect design of the organization.

But it is requisite that man have certain precise inner characteristics. An extreme example is the atomic worker or the jet pilot. He must be of calm temperament, and even temper, he must be phlegmatic, he must not have too much initiative, and he must be devoid of egotism. The ideal jet pilot is already along in years (perhaps thirty-five) and has a settled direction in life. He flies his jet in the way a good civil servant goes to his office. Human joys and sorrows are fetters on technical aptitude. Jungk cites the case of a test pilot who had to abandon his profession because "his wife behaved in such a way as to lessen his capacity to fly. Every day, when he returned home, he found her shedding tears of joy. Having become in this way accident conscious, he dreaded catastrophe when he had to face a delicate situation." The individual who is a servant of technique must be completely unconscious of himself. Without this quality, his reflexes and his inclinations are not properly adapted to technique.

Moreover, the physiological condition of the individual must answer to technical demands. Jungk gives an impressive picture of the experiments in training and control that jet pilots have to undergo. The pilot is whirled on centrifuges until he "blacks out" (in order to measure his toleration of acceleration). There are catapults, ultrasonic chambers, etc., in which the candidate is forced to undergo unheard-of tortures in order to determine whether he has adequate resistance and whether he is capable of piloting the new machines. That the human organism is, technically speaking, an imperfect one is demonstrated by the experiments. The sufferings the individual endures in these "laboratories" are considered to be due to "biological weaknesses," which must be eliminated. New experiments have pushed even further to determine the reactions of "space pilots" and to prepare these heroes for their roles of tomorrow. This has given birth to new sciences, biometry for example; their one aim is to create the new man, the man adapted to technical functions.

It will be objected that these examples are extreme. This is certainly the case, but to a greater or lesser degree the same problem exists everywhere. And the more technique evolves, the more extreme its character becomes. The object of all the modern "human sciences" (which I will examine later on) is to find answers to these problems.

The enormous effort required to put this technical civilization into motion supposes that all individual effort is directed toward this goal alone and that all social forces are mobilized to attain the mathematically perfect structure of the edifice. ("Mathematically" does not mean "rigidly." The perfect technique is the most adaptable and, consequently, the most plastic one. True technique will know how to maintain the illusion of liberty, choice, and individuality; but these will have been carefully calculated so that they will be integrated into the mathematical reality merely as appearances!) Henceforth it will be wrong for a man to escape this universal effort. It will be inadmissible for any part of the individual not to be integrated in the drive toward technicization; it will be inadmissible that any man even aspire to escape this necessity of the whole society. The individual will no longer be able, materially or spiritually, to disengage himself from society. Materially, he will not be able to release himself because the technical means are so numerous that they invade his whole life and make it impossible for him to escape the collective phenomena. There is no longer an uninhabited place, or any other geographical locale, for the would-be solitary. It is no longer possible to refuse entrance into a community to a highway, a high-tension line, or a dam. It is vain to aspire to live alone when one is obliged to participate in all collective phenomena and to use all the collective's tools, without which it is impossible to earn a bare subsistence. Nothing is gratis any longer in our society; and to live on charity is less and less possible. "Social advantages" are for the workers alone, not for "useless mouths." The solitary is a useless mouth and will have no ration card—up to the day he is transported to a penal colony. (An attempt was made to institute this procedure during the French Revolution, with deportations to Cayenne.)

Spiritually, it will be impossible for the individual to disassociate himself from society. This is due not to the existence of spiritual techniques which have increasing force in our society, but rather to our situation. We are constrained to be "engaged," as the existentialists say, with technique. Positively or negatively, our spiritual attitude is constantly urged, if not determined, by this situation. Only bestiality, because it is unconscious, would seem to escape this situation, and it is itself only a product of the machine.

Every conscious being today is walking the narrow ridge of a decision with regard to technique. He who maintains that he can escape it is either a hypocrite or unconscious. The autonomy of technique forbids the man of today to choose his destiny. Doubtless, someone will ask if it has not always been the case that social conditions, environment, manorial oppression, and the family conditioned man's fate. The answer is, of course, yes. But there is no common denominator between the suppression of ration cards in an authoritarian state and the family pressure of two centuries ago. In the past, when an individual entered into conflict with society, he led a harsh and miserable life that required a vigor which either hardened or broke him. Today the concentration camp and death await him; technique cannot tolerate aberrant activities.

Because of the autonomy of technique, modern man cannot choose his means any more than his ends. In spite of variability and flexibility according to place and circumstance (which are characteristic of technique) there is still only a single employable technique in the given place and time in which an individual is situated. We have already examined the reasons for this.

At this point, we must consider the major consequences of the autonomy of technique. This will bring us to the climax of this analysis.

Technical autonomy explains the "specific weight" with which technique is endowed. It is not a kind of neutral matter, with no direction, quality, or structure. It is a power endowed with its own peculiar force. It refracts in its own specific sense the wills which make use of it and the ends porposed for it. Indeed, independently of the objectives that man pretends to assign to any given technical means, that means always conceals in itself a finality which cannot be evaded. And if there is a competition between this intrinsic finality and an extrinsic end proposed by man, it is always the intrinsic finality which carries the day. If the technique in question is not exactly adapted to a proposed human end, and if an individual pretends that he is adapting the technique to this end, it is generally quickly evident that it is the end which is being modified, not the technique. Of course, this statement must be qualified by what has already been said concerning the endless refinement of techniques and their adaptation. But this adaptation is effected with reference to the techniques concerned and to the conditions of their applicability. It does not depend on external ends. Perrot has demonstrated this in the case of judicial techniques, and Giedion in the case of mechanical techniques. Concerning the over-all problem of the relation between the ends and the means, I take the liberty of referring to my own work, *Présence au monde moderne*.

Once again we are faced with a choice of "all or nothing." If we make use of technique, we must accept the specificity and autonomy of its ends, and the totality of its rules. Our own desires and aspirations can change nothing.

The second consequence of technical autonomy is that it renders technique at once sacrilegious and sacred. (*Sacrilegious* is not used here in the theological but in the sociological sense.) Sociologists have recognized that the world in which man lives is for him not only a material but also a spiritual world; that forces act in it which are unknown and perhaps unknowable; that there are phenomena in it which man interprets as magical; that there are relations and correspondences between things and beings in which material connections are of little consequence. This whole area is mysterious. Mystery (but not in the Catholic sense) is an element of man's life. Jung has shown that it is catastrophic to make superficially clear what is hidden in man's innermost depths. Man must make allowance for a background, a great deep above which lie his reason and his clear consciousness. The mystery of man perhaps creates the mystery of the world he inhabits. Or perhaps this mystery is a reality in itself. There is no way to decide between these two alternatives. But, one way or the other, mystery is a necessity of human life.

Man cannot live without a sense of the secret. The psychoanalysts agree on this point. But the invasion of technique desacralizes the world in which man is called upon to live. For technique nothing is sacred, there is no mystery, no taboo. Autonomy makes this so. Technique does not accept the existence of rules outside itself, or of any norm. Still less will it accept any judgment upon it. As a consequence, no matter where it penetrates, what it does is permitted, lawful, justified.

To a great extent, mystery is desired by man. It is not that he cannot understand, or enter into, or grasp mystery, but that he does not desire to do so. The sacred is what man decides unconsciously to respect. The taboo becomes compelling from a social standpoint, but there is always a factor of adoration and respect which does not derive from compulsion and fear.

Technique worships nothing, respects nothing. It has a single role: to strip off externals, to bring everything to light, and by rational use to transform everything into means. More than science, which limits itself to explaining the "how," technique desacralizes because it demonstrates (by evidence and not by reason, through use and not through books) that mystery does not exist. Science brings to the light of day everything man had believed sacred. Technique takes possession of it and enslaves it. The sacred cannot resist. Science penetrates to the great depths of the sea to photograph the unknown fish of the deep. Technique captures them, hauls them up to see if they are edible—but before they arrive on deck they burst. And why should technique not act thus? It is autonomous and recognizes as barriers only the temporary limits of its action. In its eyes, this terrain, which is for the moment unknown but not mysterious, must be attacked. Far from being restrained by any scruples before the sacred, technique constantly assails it. Everything which is not yet technique becomes so. It is driven onward by itself, by its character of self-augmentation. Technique denies mystery a priori. The mysterious is merely that which has not yet been technicized.

Technique advocates the entire remaking of life and its framework because they have been badly made. Since heredity is full of chance, technique proposes to suppress it so as to engender the kind of men necessary for its ideal of service. The creation of the ideal man will soon be a simple technical operation. It is no longer necessary to rely on the chances of the family or on the personal vigor which is called virtue. Applied biogenetics is an obvious point at which technique desacralizes; but we must not forget psychoanalysis, which holds that dreams, visions, and the psychic life in general are nothing more than objects. Nor must we forget the penetration and exploitation of the earth's secrets. Crash programs, particularly in the United States, are attempting to reconstruct the soil which massive exploitation and the use of chemical fertilizers have impaired. We shall soon discover the functions of chlorophyll and thus entirely transform the conditions of life. Recent investigations in electronic techniques applied to biology have emphasized the importance of DNA and will possibly result in the discovery of the link between the living and the nonliving.

Nothing belongs any longer to the realm of the gods or the supernatural. The individual who lives in the technical milieu knows very well that there is nothing spiritual anywhere. But man cannot live without the sacred. He therefore transfers his sense of the sacred to the very thing which has destroyed its former object: to technique itself. In the world in which we live, technique has become the essential mystery, taking widely diverse forms according to place and race. Those who have preserved some of the notions of magic both admire and fear technique. Radio presents an inexplicable mystery, an obvious and recurrent miracle. It is no less astonishing than the highest manifestations of magic once were, and it is worshipped as an idol would have been worshipped, with the same simplicity and fear.

But custom and the recurrence of the miracle eventually wear out this primitive adoration. It is scarcely found today in European countries; the proletariat, workers and peasants alike, with their motorcycles, radios, and electrical appliances, have an attitude of condescending pride toward the jinn who is their slave. Their ideal is incarnated in certain things which serve them. Yet they retain some feeling of the sacred, in the sense that life is not worth the trouble of living unless a man has these jinns in his home. This attitude goes much further in the case of the conscious segment of the proletariat, among whom technique is seen as a whole and not merely in its occasional aspects. For them, technique is the instrument of liberation for the proletariat. All that is needed is for technique to make a little more headway, and they will be freed proportionately from their chains. Stalin pointed to industrialization as the sole condition for the realization of Communism. Every gain made by technique is a gain for the prletariat. This represents indeed a belief in the sacred. Technique is the god which brings salvation. It is good in its essence. Capitalism is an abomination because on occasion it opposes technique. Technique is the hope of the proletarians; they can have faith in it because its miracles are visible and progressive. A great part of their sense of the mysterious remains attached to it. Karl Marx may have been able to explain rationally how technique would free the proletariat, but the proletariat itself is scarcely equal to a full understanding of this "how." It remains mysterious for them. They retain merely the formula of faith. But their faith addresses itself with enthusiasm to the mysterious agent of their liberation.

The nonintellectual classes of the *bourgeoisie* are perhaps less caught up in this worship of technique. But the technicians of the *Bourgeoisie* are without doubt the ones most powerfully taken with it. For them, technique *is* sacred, since they have no reason to feel a passion for it. Technical men are always disconcerted when one asks them the motives for their faith. No, they do not expect to be liberated; they expect nothing, yet they sacrifice themselves and devote their lives with frenzy to the development of industrial plants and the organization of banks. The happiness of the human race and suchlike nonsense are the commonplaces they allege. But these are no longer of any service even as justifications, and they certainly have nothing at all to do with man's passion for technique.

The technician uses technique perhaps because it is his profession, but he does so with adoration because for him technique is the locus of the sacred. There is neither reason nor explanation in his attitude. The power of technique, mysterious though scientific, which covers the whole earth with its networks of waves, wires, and paper, is to the technician an abstract idol which gives him a reason for living and even for joy. One sign, among many, of the feeling of the sacred that man experiences in the face of technique is the care he takes to treat it with familiarity. Laughter and humor are common human reactions in the presence of the sacred. This is true for primitive peoples; and for the same reason the first atomic bomb was called "Gilda," the giant cyclotron of Los Alamos "Clementine," the atomic piles "water pots," and radioactive contamination "scalding." The technicians of Los Alamos have banned the word *atom* from their vocabulary. These things are significant.

In view of the very different forms of technique, there is no question of a technical religion. But there is associated with it the feeling of the sacred, which expresses itself in different ways. The way differs from man to man, but for all men the feeling of the sacred is expressed in this marvelous instrument of the power instinct which is always joined to mystery and magic. The worker brags about his job because it offers him joyous confirmation of his superiority. The young snob speeds along at 100 m.p.h. in his Porsche. The technician contemplates with satisfaction the gradients of his charts, no matter what their reference is. For these men, technique is in every way sacred: it is the common expression of human power without which they would find themselves poor, alone, naked, and stripped of all pretentions. They would no longer be the heroes, geniuses, or archangels which a motor permits them to be at little expense.

What shall we say of the outburst of frenzy when the Sputnik went into orbit? What of the poems of the Soviets, the metaphysical affirmations of the French, the speculations on the conquest of the universe? What of the identification of this artificial satellite with the sun, or of its invention with the creation of the earth? And, on the other side of the Atlantic, what was the real meaning of the excessive consternation of the Americans? All these bore witness to a marked social attitude with regard to a simple technical fact.

Even people put out of work or ruined by technique, even those who criticize or attack it (without daring to go so far as to turn worshippers against them) have the bad conscience of all iconoclasts. They find neither within nor without themselves a compensating force for the one they call into question. They do not even live in despair, which would be a sign of their freedom. This bad conscience appears to me to be perhaps the most revealing fact about the new sacralization of modern technique.

The characteristics we have examined permit me to assert with confidence that there is no common denominator between the technique of today and that of yesterday. Today we are dealing with an utterly different phenomenon. Those who claim to deduce from man's technical situation in past centuries his situation in this one show that they have grasped nothing of the technical phenomenon. These deductions prove that all their reasonings are without foundation and all their analogies are astigmatic.

The celebrated formula of Alain has been invalidated: "Tools, instruments of necessity, instruments that neither lie nor cheat, tools with which necessity can be subjugated by obeying her, without the help of false laws; tools that make it possible to conquer by obeying." This formula is true of the tool which puts man squarely in contact with a reality that will bear no excuses, in contact with matter to be mastered, and the only way to use it is to obey it. Obedience to the plow and the plane was indeed the only means of dominating earth and wood. But the formula is not true for our techniques. He who serves these techniques enters another realm of necessity. This new necessity is not natural necessity; natural necessity, in fact, no longer exists. It is technique's necessity, which becomes the more constraining the more nature's necessity fades and disappears. It cannot be escaped or mastered. The tool was not false. But technique causes us to penetrate into the innermost realm of falsehood, showing us all the while the noble face of objectivity of result. In this innermost recess, man is no longer able to recognize himself because of the instruments he employs.

The tool enables man to conquer. But, man, dost thou not know there is no more victory which is thy victory? The victory of our days belongs to the tool. The tool alone has the power and carries off the victory. Man bestows on himself the laurel crown, after the example of Napoleon III, who stayed in Paris to plan the strategy of the Crimean War and claimed the bay leaves of the victor.

But this delusion cannot last much longer. The individual obeys and no longer has victory which is his own. He cannot have access even to his apparent triumphs except by becoming himself the object of technique and the offspring of the mating of man and machine. All his accounts are falsified. Alain's definition no longer corresponds to anything in the modern world. In writing this, I have, of course, omitted innumerable facets of our world. There are still artisans, petty tradesmen, butchers, domestics, and small agricultural landowners. But theirs are the faces of yesterday, the more or less hardy survivals of our past. Our world is not made of these static residues of history, and I have attempted to consider only moving forces. In the complexity of the present world, residues do exist, but they have no future and are consequently disappearing.

Only the things which have a future interest us. But how are we to discern them? By making a comparison of three planes of civilization which coexist today: India, Western Europe, and the United States. And by considering the line of historical progression from one to the other—all of this powerfully reinforced by the evolution of the Soviet Union, which is causing history to boil.

In this chapter we have sketched the psychology of the tyrant. Now we must study his biology: the circulatory apparatus, the state; the digestive apparatus, the economy; the cellular tissue, man.

The house that *Homo sapiens* built

ROBERT THEOBALD

▪ Robert Theobald *has written extensively on the subjects of economics and alternative futures. His books include* The Challenge of Abundance, The Guaranteed Income, Futures Conditional, *and* Habit and Habitat.

Jacques Ellul's The Technological Society *was widely read and extensively reviewed when it appeared in English in 1964. Of all its reviews, this is perhaps the best. Theobald offers a careful summary and analysis of Ellul's thesis, and he provides counter-arguments by means of which we may offset Ellul's fundamentally pessimistic position.*

Theobald, for example, argues that man really is *freer when released from degrading toil and that, although one cannot measure freedom in terms of material goods, those goods at least provide a condition for such freedom. Unlike Ellul, Theobald thinks that there is still time to humanize technique.*

The Technological Society is one of the most important books of the second half of the 20th century. In it, Jacques Ellul, Professor of History and Contemporary Sociology at Bordeaux University, convincingly demonstrates that technology, which we continue to conceptualize as the servant of man will overthrow everything that prevents the internal logic of its development, including humanity itself—unless we take the necessary steps to move human society out of the environment that "technique" is creating to meet its own needs.

Ellul devotes his whole volume to an examination of this "technique" which he defines in the following way: ". . . technique is the *totality of methods rationally arrived at and having absolute efficiency* (for a given stage of human development) in *every* field of human activity." Ellul's argument can best be set out using his own words: ". . . in our civilization technique is in no way limited. It has been extended to all spheres and encompasses every activity, including human activities. It has led to a multiplication of means without limit. It has perfected indefinitely the instruments available to man, and put at his disposal an almost limitless variety of intermediaries and auxiliaries. Technique has been extended geographically so that it covers the whole earth." And later: "Within the technical circle nothing else can subsist because technique's proper motion, as Jünger has shown, tends irresistibly toward completeness. To the degree that this completeness is not yet attained, technique is advancing, eliminating every lesser force."

Ellul believes that ". . . there is no common denominator between the technique of today and that of yesterday. Today we are dealing with an utterly different phenomenon. Those who claim to deduce from man's technical situation in past centuries his situation in this one show they have grasped nothing of the technological phenomenon."

For Ellul, the most important consequence of the technical drive is the destruction of all that is human. Examining the progress of technique in production, Ellul states: "In a great corporation, the workers are more than ever enslaved and scarcely in a position to act in a distinctively human way. Even the consumer is frequently imposed upon. The integration of the individual into the technical complex is more complete than ever before."

Ellul also makes the same point at a higher level of analysis:

> It is not in the power of the individual or of the group to decide to follow some method other than the technical. The individual is in a dilemma: either he decides to safeguard his freedom of choice, chooses to use traditional, personal, moral or empirical means, thereby entering into competition with a power against which there is no efficacious defense and before which he must suffer defeat; or he decides to accept technical necessity, in which case he will himself be the victor, but only by submitting irreparably to technical slavery. In effect he has no freedom of choice.

Ellul's thesis is not actually as strong or as complete as he would probably make it today: the original French version, entitled *Le Technique,* was written in 1952 and published in 1954, and it was then impossible to anticipate the speed of development of the computer. Thus, although Ellul deals with the danger of a police state in the technical world, he was not able to examine the effects of a computerized police information network whose present tentative beginnings are now being planned in New York State (*The New York Times,* Sept. 17, 1964) and which could easily be expanded onto a nation-wide basis within the next decade. Nor

☐ From *The Nation,* 199 (1964), 249-252.

did he consider the possibilities of tying the computer into techniques of increasing sales; today, however, a computer program has already been developed which makes it possible to insure that every phone-equipped family within a certain area receives an intermittently repeated tape-recorded message. At present levels of sophistication, the effects of the ongoing computer technology clearly strengthen Ellul's thesis.

Ellul makes the case that technique is presently remaking man and that if this process continues it will eventually sweep aside any obstacles placed in its way including human society—*unless we take steps to end the present subordination of human processes to technological ends*. During a recent conversation with Ellul, he assured me that he had written his book with the precise intention of explicating and illustrating this view, that he was making a hypothetical case to show what would happen *if* mankind allowed itself to be swept along by the technological order. He had previously told me that the *original* version contained a lengthy section discussing ways of releasing men from the technical influence. But this section did not appear in the French edition nor is it included here.

It is deeply unfortunate that only the first part of Ellul's original two-part manuscript is presented in *The Technological Society,* and that the published book therefore reads like a prediction of the inevitable future, rather than a description of one possible line of development and the necessity for its avoidance. The book as presently published not only states that technique may destroy humanity but that it *will* destroy humanity. It is practically impossible to find any remnants of Ellul's privately held view: one of the few remaining in the published text is this: "The technical society must perfect the 'man-machine' system or risk total collapse. Is there no other way out? I am convinced that there is." However, this affirmation of faith is denied repeatedly in the course of the book, for example in the statement: "The most important thing is that man, practically speaking, no longer possesses any means of bringing action to bear upon technique. He is unable to limit it or even to orient it." This must therefore be the position discussed here; I will, however, return to Ellul's private position at the end of this review.

The thesis presented in *The Technological Society* is of course alien to existing technologically oriented modes of thought. Those committed to the implementation of maximum-efficiency systems will be looking for weaknesses in the argument. Unfortunately, these are not too difficult to find. There are occasional exaggerations which are clearly unjustified:

> In the past, when an individual entered into conflict with society, he led a harsh and miserable life that required a vigor which either hardened or broke him. Today the concentration camp and death await him; technique cannot tolerate aberrant activities.

In addition, many readers will feel that Ellul commits the most modern of sins: he nourishes an unrealizable longing for a mythical past. At the beginning of the volume Ellul argues:

> Men now live in conditions that are less than human. Consider the concentration of our great cities, the slums, the lack of space, of air, of time, the gloomy streets, and the sallow lights which confuse night and day. Think of our dehumanized factories, our unsatisfied senses, our working women, our estrangement from nature. Life in such an environment has no meaning.

These are minor flaws within the sweep of Ellul's thesis and do not, in my opinion, invalidate his general point that technique has become the dominant force in the world today, and that it does dehumanize. Ellul essentially argues that mankind cannot abandon technique ("the totality of methods rationally arrived at and having absolute efficiency in every field of human activity") because this would mean the deliberate acceptance of irrationality. "If we disown technique," he says, "we abandon the domain of science and enter into that of hypothesis and theory."

Ellul thus effectively forecloses any argument about the primacy of technique, for it is clear that man cannot abandon rationality. But Ellul leaves suspended the questions he raises: Can rationality and technique justifiably be equated? Will technique, once defined as rationality, necessarily lead to the dire results which he forecasts? The underlying ambiguity of Ellul's position is inherent in the following quotation:

> . . . man . . . does not make a choice of complex and, in some way, human motives. He can decide only in favor of the technique that gives the maximum efficiency. But this is not choice. A machine could effect the same operation.

Such a statement, however, is effectively meaningless. What does Ellul *mean* by efficiency? For the very concept of maximizing efficiency is only meaningful in human affairs if we have some end in view other than the maximization of efficiency itself. Ellul ignores this fact, claiming that: ". . . it is possible to speak of the 'reality' of technique—with its own substance, its own particular mode of being, and a life independent of our power of decision." Such a statement cannot possibly be justified; technique advances because individuals make decisions which advance technique. What Ellul is really describing, therefore, is the crux of our modern dilemma: that our present society is organized in such a way that those making decisions find themselves *forced* to advance technique. Why does this occur? In order to answer this question we must reexamine fundamental Western values and the drives to which they give rise in the context of advancing technology.

In today's world:

(1) Every country must try to be in a position to defend itself against all potential aggressors. As a result, each country must accept and indeed encourage any technical improvement in weaponry. The cybernation revolution, by increasing the efficiency of production, makes it possible to manufacture more weapons without cutting into the production of consumer goods—thus minimizing one of the prime constraints on weaponry production in the past.

(2) Every country must remain competitive so that it is able to export enough to pay for its imports. This requires each nation to accept and indeed encourage any technological development in production. The combination of the cybernation revolution, which has led to unused productive capacity for many goods, plus the research and development revolution, which has led to large-scale innovation, insures that the speed of technological advance required to remain competitive increases constantly.

(3) Means must be found to insure that demand grows as fast as available production. This requires that techniques which will result in increased demand must be accepted and indeed encouraged.

These are the three constraints which together are imposing the technological order upon us. They insure that technique continues to drive forward. But, in my opinion, Ellul totally fails to prove that technique *will* inevitably destroy human society. Societal constraints can always be changed by societies acting on the principle of cooperation, both internally and externally. Indeed, these present constraints not only *can* be changed but *must* be changed, if we are to be able to live in the new era which the continuing development of technique has brought into existence. The nature of this new era is as different from the industrial age as the industrial age was different from the agricultural.

First, there is no longer any real limit to our productive abilities. U Thant, Secretary-General of the United Nations, has expressed this reality in the following terms: "The truth, the central stupendous truth about developed countries today is that they can have, in anything but the shortest run, the kind and scale of resources they decide to have. . . . It is no longer resources that limit decisions. It is the decision that makes the resources. This is the fundamental revolutionary change, perhaps the most revolutionary mankind has ever known." We can provide every individual with the amount of resources required for a decent life by the end of the 20th century.

Second, man need no longer be tied to degrading toil: man's mind is no longer necessary for repetitive physical or mental tasks. Gerard Piel, publisher of *Scientific American,* has put it this way: "The new development of our technology is the replacement of human beings by automatic controls and by the computer that ultimately integrates the functions of the automatic control units *at each point* in the productive process (italics mine). The human muscle began to be disengaged from the productive process at least a hundred years ago. Now the human mind is being disengaged." By the end of the 20th century, we should be able to develop each individual's capacity rather than socialize him in such a way that he will serve the productive system.

Third, modern weaponry provides unlimited destructive power capable of destroying not only civilization but all life. This is now so well known as to need no further discussion.

The constraints which at present insure the future dominance of technique must be eliminated if humanity is to survive. The present requirement that every country must try to be in a position to defend itself against all potential aggressors must be changed and disputes must be settled on the basis of world law. The present requirement that all nations must export enough to pay for their imports must be changed so that those who need additional goods, particularly the poor countries, receive them as a right. The requirement that demand grow as fast as available production must be changed through granting an absolute right to guaranteed income, with the result that production becomes subordinated to the needs and desires of the individual and society. Research has already begun to indicate mankind's possible escape route from the philosophical trap which technique, in its more negative aspects, represents.

We have become aware in recent years that the socio-economic system as a whole is at present favorably organized for the implementation of technique and that, as individuals, we are in a continuing state of anticipation of new technological phenomena: that we are preconditioned to them. It is clear to each of us, as individuals, that we are becoming accustomed to responding to the forces of technique in our daily lives; it is less immediately obvious that these same forces are reorganizing us as a collectivity, that the technical drive acts upon the entire socio-economic system. Individually, we can do little to resist the reforming drive of technique; collectively, we can both resist this drive and create a counter-drive toward a socio-economic system based on human not mechanical principles. But there is the danger, which Ellul discusses at some length, that in order to confront and overcome an impersonal organizing drive we will find ourselves automatically developing a form of organization which is itself impersonal; where individuals act merely as units which respond mechanically, or feed back information as to why they cannot respond, and then are aided in an adjustment process in order to be able to make the technically appropriate response. We must find a means of enabling individuals to express themselves as human beings; we must discover a way of establishing a communication system which will facilitate more than the mere relating of unit to unit in a total system ending toward ever-greater perfection of unit adjustment. It is precisely here that technique can itself be used to combat its own negative aspects, for it is now becoming evident that, through technique, new forms and means of human communication could emerge.

It is perhaps useful to remember that when the potential for new systems of communication first became apparent in the early fifties, it was automatically assumed that such communication would join human beings more closely together. At that time there was little realization of the possibility that the new communication systems which had been developed during the wartime and quasi-wartime periods to enable machine to communicate with machine, would continue to be primarily used to facilitate machine communication and not be adapted for effective human use. It was not recognized that human beings could be separated from one another by intrusive, impersonal communication systems. Curiously, as this unfortunate development proceeded, theoretical discussion of the problems and possibilities of the new communication systems has been increasingly muted.

Ellul's book is unquestionably the most pessimistic description and discussion of machine-to-machine communication systems in print. Recently, Marshal McLuhan, director of the Center for Culture and Technology at the University of Toronto, outlined his appraisal of the possibilities and potential of technique (the Media) in his most recent book *Understanding Media* (see *The Nation,* Oct. 5, 1964). While sharing with Ellul the realistic and fundamental understanding that technique cannot merely be considered as neutral "means," McLuhan argues convincingly that technique calls forth new modes of perception, new ways of thinking, and new reactions in the human mind which can be more valuable for the development of human capabilities than the limited, lineal thinking still current in the industrial age.

In many ways, the argument of McLuhan's book would be acceptable to Ellul as a statement of technique in its positive aspect, and it could, indeed, be held to approximate the position outlined in the second (unpublished) part of *Le Technique*. A reading of both books together is strongly recommended to any interested in the future of mankind.

Technology and human values

MELVIN KRANZBERG

■ Melvin Kranzberg *is Callaway Professor of the History of Technology at Georgia Institute of Technology, and editor of* Technology and Culture.

Professor Kranzberg recounts many of the beneficial aspects of technology, and suggests that there has in fact been enormous progress in the human sphere as a result of it. He argues that technology should be seen as an "enabling factor, not a compulsory mechanism."

Kranzberg's assessment of the future of technology is an optimistic one. His view is that the lot of technological man is better than that of any of his predecessors, not only in terms of health and material goals but in terms of his increasing alternatives and his heightened degree of self-determination.

Just when technology seems on the verge of banishing want and starvation, we find it called into question. The suspicion with which John Ruskin, William Morris, and Herman Melville regarded the machine in the nineteenth century has carried over into the twentieth. Counterbalancing H. G. Wells' belief in the beneficence of technology were the nightmare vision of the future contained in Çapek, Huxley, and Orwell. Most recently, such stalwart upholders of the humane tradition as Jacques Barzun, Lewis Mumford, and F. W. Leavis have exhibited the same ambivalence toward scientific and technological "progress."

Why is it that technology, which ministers to human needs and which has been regarded as a boon to mankind throughout most ages—and is still so regarded in the so-called underdeveloped areas—is viewed with such suspicion? Has technology changed? Has our value system changed?

Man himself is what he is as a result of technological change. Archeology, anthropology, biology, and psychology tell us that man could not have become *homo sapiens* (man the thinker) had he not been at the same time *homo faber* (man the maker). Only by his possession and use of tools can we distinguish between man and "almost man." Man made tools, but tools also made man.

From the very beginning of our species, therefore, man has depended upon his technology. His attempt to master and control his physical environment is enshrined in the legend of Prometheus, who stole fire from the gods to bring it to mankind. Ever since then the march of civilization has been accompanied by technological advance—in terms of materials, from stone to bronze, from bronze to iron, from iron to steel; in terms of energy, from human muscle power to animals, to wind and water, to steam and oil, to rockets and nuclear power; in terms of machines, from hand tools to powered tools, to mass production lines, to computer-controlled factories.

For most of man's history, the term "humanistic" was used to distinguish what was human from what was brutal and coarse, or animal. But now we find that "humanist" is increasingly contrasted with "technology," and "humanistic" is used to distinguish everything else from science and technology. Indeed, the word "humanist" has almost become equated with ignorance of science and technology. The humanities have become the home of anti-rationalists who deny that rational attempts to deal with human material wants have any bearing upon man's pursuit of "the good life."

This change in the meaning of "humanistic," so that it is now equivalent to "spiritual" in contrast to "material," becomes even more confusing when we realize that some non-Western cultures, such as that of India, glorified because of their elevation of spiritual goals over materialistic ends, want to appropriate the technological paraphernalia of Western civilization while repudiating our so-called "humanistic" values. Yet the underdeveloped nations who attempt to borrow our technology are distressed to find that along with it come the humanistic values of our society. In other words, technology is part of the warp and woof of our civilization, and the humanistic cannot be separated from the technological.

The current questioning of the rôle and value of technology is nothing new in human affairs. Technological change has always created tensions within society. Today technological change is outpacing our capacity for social adjustment—a form of William Fielding Ogburn's "cultural lag"—and the pattern of change has become so complex that the "bleeding-heart

□ From *Virginia Quarterly Review,* 40(1964), 578-592.

humanists'' find themselves incapable of understanding it.

A rough parallel can be found by studying the Industrial Revolution in England at the end of the eighteenth and beginning of the nineteenth centuries. Here the machine caused a rapid buildup of tension between owners and workers, a tension which reached its peak when the Luddites naïvely thought that they could combat change by destroying the new machines. Today we are faced with the ''intellectual Luddites.'' They are not so naïve as to believe that social change can be prevented by destroying machines; instead, they attack the philosophical basis for technological change by calling into question its impact on human values.

In the nineteenth century William Morris proposed that Britain go back to cottage industry and an idyllic medieval pastoral life—that had never existed in actuality. John Ruskin also inveighed against the Industrial Revolution. Both believed that the material plenty contributed by machine production carried with it a cultural poverty. Today's ''intellectual Luddites'' are the spiritual descendants of Morris and Ruskin—concerned, articulate, but ineffectual.

It is not only the rate of change that worries critics, however. There is also the fear that contemporary technology represents something qualitatively new and dangerous to human values. Technology was welcomed when machines took the burden off men's backs and eased the strain on human muscles. Now that machines have proceeded to take some burdens off men's minds and to free men from dull and repetitive tasks, there is fear that the machine is making the human being obsolete.

In an ironic sense the machine is making possible the attacks upon it. In the past, technology has been primarily concerned with the human needs of food, clothing, and shelter. It still fulfills those needs, and so successfully that modern Western technology for the first time in history has produced a society which has not only a surplus of goods but a surplus of leisure as well. Living in an economy of abundance made possible by technology, we are now free to question technology in terms of its ability to meet those humanistic needs which go beyond our material desires.

What are these human values which the humanists believe are in such peril from the onslaught of technology? And what has been the actual impact of technology upon them?

Foremost among the humanistic values in Western civilization is our respect for the worth of the individual human being. Our religious concepts, our philosophies, our laws, our schools, and our economic system glorify the individual and are presumably aimed at making it possible for man to realize his full potential as a human being. Placing the individual as the basis of the value system is perhaps unique to Western culture. This elevation of the individual came relatively late—just two or three centuries ago—and even in very recent times, the individual has been subordinated to the social or economic rigidities of a particular nation or ''race.''

The level of technology in any period in history determines the way in which the majority of men earn their living, where they live, and how they spend the major portion of their time. Throughout antiquity—in Egypt, Mesopotamia, Greece, and Rome—manual labor was largely left to slaves, who were considered as property and had few rights as individuals. In the Middle Ages, windmills and waterwheels eased the human physical burden somewhat, but the serfs, who were numerically the largest group in the population, possessed few rights, and those only by virtue of the custom associated with their plot of land, not as individuals. Although the decline of feudalism loosened the bonds of serfdom, not until the political and social revolutions of the seventeenth and eighteenth centuries was the idea accepted that each individual possessed certain inalienable rights and that society would be directed toward the guarantee and maintenance of those rights.

At the same time that the state was being organized in behalf of the individual, great technological changes, lumped together under the term Industrial Revolution, were also taking place. Throughout history the home had been the production center. Farming, weaving, garment making—all were family endeavors centering around hearth and home. Industrialization uprooted this system, moved thousands of families into cities, and made factory labor the source of their livelihood. The Industrial Revolution not only changed the productive mechanisms of society but also transformed the individual's conditions of life.

What else did industrialization do to the individual? As a result of his urbanization, the working man, for the first time, was achieving a degree of political recognition, but the widespread feeling was that the price was the loss of his individuality on the production line. In the factory work was divided into separate units, and this "fracturing of work" took away from the workman much of his pride in craftsmanship. The development of assembly-line techniques has exaggerated this trend, and now automation is separating the worker even further from the end product.

There is no question that this division of labor helped to make possible the enormous production of goods which raised the material conditions of living for the workers engaged in the manufacturing process. Thus the human imperative of satisfying material wants has been in part met by technological progress through the division of labor. But has the humanistic imperative of the dignity and worth of the individual personality in actuality been sacrificed thereby?

At times in the past, and in isolated instances today, the answer would have to be a limited "yes." But in the main, today and in the future the answer is an emphatic "no!" Instead of making man into a robot, as envisioned by Karel Çapek in "R.U.R.," automation is in the process of restoring human dignity by freeing man from the repetitious tasks which dull the human personality. One of the technological imperatives is efficiency; in pursuit of efficiency, industrial engineers and industrial management have been forced to focus upon the individual human being. Modern systems engineering must consider the human parameter in any engineering system. And, despite the dreadful connotation of the term, which makes one think that it is designed to "engineer" human beings, human engineering is actually concerned with the man-machine relationship. Its primary goal is to adapt the machine and the work process to the human mind and body. Thus any threat to the individual arising from technology will tend to be counterbalanced by the technological demand for efficient production which depends upon the proper environment for and the proper use of the individual human being.

Yet it must be recognized that some sacrifice of unbridled individualism is necessary if men are to work together. Technology makes compulsory some form of co-operation and understanding among human beings in order to carry on the productive process. If each worker were to follow his own bent, he might not awaken until noon or he might decide to go fishing instead of going to work; in either case, the production line would cease to function. Technology has therefore necessitated some kind of discipline in behalf of the performance of collective tasks.

However, the threat of technology to the individual lies not only in terms of the social discipline in the actual work process but also in the possible constraints upon human freedom. To some social thinkers, such as Lewis Mumford, certain trends in our contemporary technology point to the development of an "authoritarian technics." In the past this authoritarian technics "raised the ceiling of human achievement in both mass construction and mass destruction." Although marvelously dynamic and productive, authoritarian technics, with its command and utilization of science, is thought to be threatening to displace life, "transferring the attributes of life to the machine and the mechanical collective." In contrast to this is what Mumford calls "democratic technics," a small-scale method of production, "resting mainly on human skill and animal energy but always, even when employing machines, remaining under the active direction of the craftsman or the farmer."

Mr. Mumford may be both right and wrong. There are many ways in which work processes may be organized, even at the most sophisticated levels of technology, and the choice is not necessarily between Mumford's anthitheses. A strong case could be made to the effect that technology, while it may threaten to deprive man of one kind of freedom, actually enlarges human freedom in other dimensions. For millenia men lived under Mr. Mumford's "democratic technics" with its small-scale methods of production—and men were constantly in want. Only in our day with its highly organized, mass-production technology—Mumford's "authoritarian technics"—could men conceive of "freedom from want" as a rational goal and the keystone of international policy, as envisaged in the Atlantic Charter.

If we examine the development of democracy historically, we find that technology, far from destroying democracy, may have actually increased it. Iron has been hailed as the "democratic" metal by some writers, for the replacement of bronze by iron made possible the spread of improved technological devices and a better standard of living for the masses. At a later date we find that the great growth of democracy in the Western world was concomitant with the Industrial Revolution. And there are those Kremlinologists who tell us that an improved standard of living in the Soviet Union, made possible by the application of advanced technology, may help bring about a freer society there.

Technological advance demands democracy in still another way. The institution of slavery, as anti-democratic an institution as one can possibly imagine, has been one of the victims of advancing technology. According to Gerard Piel, "Slavery became immoral when it became technologically obsolete." Despite this oversimplification—which overlooks the possibility of the exploitation of "wage slaves"—the historical fact is that technological developments made outright slavery unprofitable and thereby hastened its end.

Proof of the democratic tendencies of advancing technology is also to be seen in the progress toward integration in the southern United States. There the development of a new industrial South is bringing an end to inequality. Lest we think that this situation is peculiar to our own country, may I also point out that the demands of a modern industrialized society in India have partly breached the caste system.

Besides its political and legal side, democracy has its social aspect. The advance of technology is closing the great gap between rich and poor. In our economy of abundance the poor can enjoy the same entertainment as the rich merely by flicking a switch, can also eat strawberries out of season simply by going to the nearest supermarket, can enjoy clothes which look and wear like those handtailored by a Paris couturier even though mass produced on New York's Seventh Avenue, can step into automobiles which take them to their destination as quickly and surely as the Rolls-Royces of the wealthy. As David Potter has said, "When man gains a satisfactory income, acquires education, dresses himself and his wife in the standard clothes worn by all members of the community, sends his children to school . . . and the system of classes itself, no longer natural, no longer inevitable, begins to seem unjust." Social democracy thus emerges from technological advance.

Another humanistic imperative is the creation and enjoyment of esthetic works—music, literature, art. Ever since the Industrial Revolution, artists, composers, and writers have attacked technology as creating ugliness and stifling artistic creativity. Yet the nineteenth century, when society was coming under the sway of industrial technology, produced a great new form of literary expression, the novel. It was also rich in poetry, in art, and in music. If industrialization is inimical to esthetic elements, how are we to account for the flourishing of the arts in the nineteenth century?

Once we realize that technology is not inherently harmful to the arts, we find two constructive ways in which technology has fulfilled the humanistic demand for cultivation of esthetic works. The first is that an advancing technology has made possible man's cultivation of the arts; the second is that there are esthetic elements within technology itself.

In a society where the subsistence level is marginal, cultivation of the arts is minimal. Only after elementary wants have been satisfied can men have the leisure to produce intellectual and artistic works of quality. In antiquity, the servile work of the masses produced a slight surplus which was appropriated by the ruling groups of society. They utilized this surplus to enjoy the "finer things" of life. Until the Industrial Revolution, literature and the arts were virtually an aristocratic monopoly.

The philosophers of antiquity recognized the need for leisure, made possible only by the surpluses produced by technology, in order to enable man to pursue intellectual and esthetic activity. In his "Politics," Aristotle spoke of leisure as the highest condition of life: "[It] gives pleasure and happiness and enjoyment of life. These are experienced not by the busy man, but by those who have leisure. . . . There are branches of learning and education which we must study merely with the view to leisure spent in intellectual activity, and these are to be valued for their own sake. . . ." Because slaves had no leisure and because their entire time was spent in manual work, Aristotle denied them human status.

The importance of leisure as a basis for cultural activity is recognized today also. Arnold Wesker, the young British playwright, while deploring "this terrible cultural bankruptcy" of modern life, says, "Everything is marvelous about a welfare state but the one thing that is wrong is that you have provided leisure for the community but nothing to fill it in. Today, people have more leisure, more money and better houses. But this is where civilization begins." True, technology has now made it possible for everyone to have leisure for thought, but according to Sebastian de Grazia, our shrinking work week has left us with little net human gain. We have confused Aristotle's true leisure with free time and have filled our free hours with busy work. Instead of using our leisure to attain our highest mental development, as Aristotle would have us do, we spend most of our free time in pursuit of recreation—not to re-create ourselves, but to kill time.

Even though we must make allowances for human frailty, we still can see how the advance of technology has made it possible for more people to engage in esthetic and intellectual activities, transforming these from an aristocratic monopoly to mass participation in the "finer things" of life. Mass culture has been made possible not only by the free time granted by technology but by other technical advances—by long-playing records, by art reproductions available at low cost, by television and movies, the latter two representing new art forms developed through technology. Even in the United States, the attendance at concerts and at museums exceeds the attendance at organized baseball games; more books are being published every year; and the amount of money spent on the theatre and the opera has almost doubled in a decade.

To the humanists, this argument is probably meaningless—largely because it is stated in numerical terms. Technology, they would say, provides only a quantitative rather than a qualitative advance in man's culture. They declaim against the debased esthetic tastes of the masses, and then they turn and attack the technological means by which that taste can be elevated.

My second point in regard to esthetics is the technological imperative which makes for the beauty of man's works in terms of functional design. In the nineteenth century, some poets and artists were much impressed by the scientific advances of the time; they tried to understand and utilize them in their descriptions of the world which they saw about them. Tennyson and Matthew Arnold employed scientific references and metaphors; science became a personal, metaphysical concern. Today we find few poets of the machine.

Our poets have let us down badly in interpreting our times to us. By refusing to consider the impact of science and technology, our men of letters are carrying on their intellectual life in a cultural vacuum. It is no wonder that they can communicate only with one another and complain that they feel alienated from a society whose culture they ignore.

It is shocking to realize that the poets regard the world of technology as a prosaic world. They do not recognize the beauty, the challenges, the human experience—the esthetics, if you will—of technology. Yet they are there, and in large measure. Architecture provides a good example, for men like Frank Lloyd Wright have developed technological means to produce new esthetic delights in the plastic arts. It is indeed ironical that the mathematical precision and symmetry of form which characterize our most advanced technological designs are based upon that mathematics which, along with music, Plato placed at the foundations of humanistic education. It is no wonder that Richard L. Meier claims, "The humanities seem to have cut themselves off from the modern lines of creativity in general cultural activity. Perhaps we are coming to a time when the humanities will be but a minor part of a human endeavor."

If the humanists no longer fulfill their rôle of interpreting nature and society to man, that task will be taken up by the social scientists and psychologists. Instead of reading poetry to acquaint ourselves with the wellsprings of human behavior and man's relations to nature, we will read psychology. Instead of viewing drama and paintings which endeavor to interpret nature and man's relations to society, we will peruse sociological tracts and look at photographs. Indeed, we have already begun to do so, largely because the humanities have abandoned their traditional rôle and left a gap in the fulfillment of this humanistic imperative.

In ethics, too, technological advancement has made possible the fulfillment of some significant goals of society and at the same time has engendered new values.

It is the duty of everyone to aid his neighbor, we have long been told, and the relief of poverty and destitution has ever been a goal of the Christian social ethic. The productivity of today's technology now enables us to translate this into action domestically by social welfare measures and on an international scale through technical assistance programs.

Although technology provides us with the means to fulfill an injunction of Christian ethics, we find ourselves somewhat embarrassed by our abundance of material goods. Our theological preconceptions derived from the past leave us with feelings of guilt about our plenteous economy. The Biblical injunction, that man, through Adam, must earn his living by the sweat of his brow, seems to have been overturned by modern mechanical and electronic technology, and our high level of mass consumption seems to run counter to the Puritan virtues of hard work and thrift.

Actually, the technological developments of the past two centuries have made possible a new ideal of human goals unknown to an earlier society which was theologically preoccupied or pessimistically endowed. The idea of progress, first explicitly stated by Condorcet in the eighteenth century, viewed material well-being as essential to individual liberty and peace. In the course of the nineteenth century, when men could see about them concrete evidence of advances in liberty and material goods, the idea of progress became an accepted part of our Western value system. We have since learned that progress does not automatically guarantee "happiness"; yet we have also learned that an advancing technology is essential to progress in man's material standard of living.

But what progress can there be without peace? And what does technology do for peace? Indeed, in the popular mind, modern technology seems to be identified with rockets, missiles, nuclear weapons, and the very real threat of the destruction of our civilization. Thus technology seems to run counter to the humanist demand that men live together in peace and harmony.

It is true that almost from the very beginnings of human history, warfare has been intimately related to technology. The individual who was stronger in the technology of warfare subjugated his fellows; communities possessing a superior technology could subdue other, less advanced communities; and, as technology made warfare more complex and more costly, it helped to increase the power of the state and the destructiveness of modern armed combat.

Technology has thus produced several paradoxes. Those very advances in military technology which made warfare more horrible arose from the human need for peace and security. While developing more frightening weapons which heightened international suspicions and fears, technology also provided material benefits to ease hunger and destitution and thereby to help do away with those causes of international tension which breed war. And, demanding on the one hand that men live together in peace and stability in order to pursue productive goals, technology at the same time provided men with weapons which made them increasingly dangerous toward their neighbors.

A final paradox derives from the comtemporary rôle of technology in war. Scientific technology has made possible nuclear war and hence the total destruction of mankind. At the same time, the threat of nuclear warfare makes peace essential. Although the elements of dissension among nations still have a lively existence, total warfare may have been avoided during the past fifteen years by the so-called nuclear deterrent. Thus the humanistic imperative of peace may have been served by the demands of military technology.

The Janus-faced aspects of military technology as both enemy and friend of mankind leads to one final question before we can determine whether technological change helps or hinders fulfillment of our human imperatives: is man capable of guiding technology along socially useful channels?

A review of the historical evidence would indicate that technology *per se* can be either good or bad, depending on the use which man makes of it. Technology is an enabling factor, not a compulsory mechanism. As Lynn White says, "Technology opens doors, it does not compel men to enter." In this view, technology is a means, which man is free to employ as he sees fit.

However, a difficulty arises from thinking of technology only as a means. It is not always easy to distinguish between means and ends, and there is always the possibility of means becoming ends in themselves. What is more, the mere availability of technology tempts man to employ it. Man is not compelled to enter Dr. White's open door, but an open door is an invitation. Besides, who determines which doors to open?

The technologist himself may judge a device on the basis of: will it work? Yet the practical decision to apply a technological innovation involves factors which are social, political, and ethical, as well as technical, in character.

The fact is that technology is not sufficient in itself to teach the kind of wisdom that is necessary for "steering the ship of state," choosing our educational objectives, or establishing the principles of "the good life." Even in technical matters C. P. Snow has revealed that the advice given by scientists and technologists to government is liable to the same pitfalls as the advice by anybody in any other field.

Whose is the responsibility for the social and human direction of technology? Instead of attempting to analyze and understand the rôle and meaning of technology, our bleeding-heart humanists retreat to romantic criticism or a self-imposed and incommunicable personal anguish and despair.

To make up for the failure of others to develop the human dimensions of technology, technologists must perforce educate themselves in what Simon Ramo calls "the greater engineering." This involves a more responsive attitude to human and social demands in their ultimate implications, rather than the mere fulfillment of immediate demands. As Dr. Ramo says, "It is this bigger, overall, application of science to serve society that engineering now needs to be concerned with. . . . Now, more than ever, engineering effort must be in proper match to social, industrial, economic, governmental, and psychological needs. So complex is the list of considerations that clearly we are not going to create the best engineers by teaching them the technical facts alone." In other words, some technologists are increasingly aware of the humanistic imperatives and are trying to do something about fulfilling them. Can the same be said of the humanists in understanding the demands of a technological society?

Let us recognize that the problems presented by technology are not simply technical in nature or solution but are the problems of all society and mankind. Technology does not exist in a vacuum; it develops in a social context, as do all other human activities. Let us realize also, that the problems presented by technology must be met by all men. Instead of being a mechanical master which determines man's destiny, or a Frankenstein's monster which threatens to destroy its creator, technology has always been, and still remains, an essential part of man. The question, therefore, is not whether man can master technology, the question is whether man—*homo sapiens* and *homo faber*—can master himself. This is the technological imperative; this is the humanistic imperative.

How technology will shape the future

EMMANUEL G. MESTHENE

■ Emmanuel G. Mesthene *is Professor of Philosophy at Rutgers. He is the author of numerous books and articles on technology, among them* Technological Change, *published in 1970.*

In the article that appears here, first published in 1968, Professor Mesthene is both optimistic and non-deterministic in his analysis of technological development. His optimism leads him to a discussion of the new alternatives, the new possibilities open to individuals participating in a society in which technology is developing. He does not, however, subscribe to the Marxist or Hegelian view that technological developments inevitably *bring about certain specific social changes.*

Those who argue that technological change is inimical to "value," or more specifically, "human value," Mesthene suggests, have mistaken "value" for "stability." This is certainly as unfortunate a mistake as equating change with progress. Thus Mesthene's answer to the question "Is technology autonomous?" is simply that new technological means provide us with new possibilities, but that it is man himself who makes decisions about applications, choosing among alternatives.

There are two ways to approach an understanding of how technology will affect the future. One, which I do not adopt here, is to try to predict the particular technologies of the future along with their particular social effects.[1] The other way is to identify some respects in which technology entails change and to suggest the kinds or patterns of change that, by its nature, it brings about in society. This is what I attempt in the present paper.

1. NEW TECHNOLOGY MEANS CHANGE

It is widely and ritually repeated that a world of technology is a world of change. The statement is true only if the phrase "world of technology" (by which of course we mean our world!) is understood to mean a world characterized by a more or less continuous development of *new* technologies. There is no inherent impetus toward change—no *élan vital*—in tools as such, no matter how many or sophisticated they might be. When new tools emerge, however, there is a strong presumption that there will be changes in nature and in society. I see no such necessity in the technology-culture or technology-society relationship as Karl Marx and his followers have asserted, for example, but I do see, in Hume's words, a rather constant conjunction between technological change and social change as well as a number of good reasons why there should be one.

Technology creates new possibilities

One of the most obvious characteristics of technology is that it brings about changes in physical nature (including changes in the relationships of physical objects and in the patterns of physical processes). Thus, the plow changes the texture of the soil, the wheel speeds up the mobility (change in relative position) of objects, and the smoke-box (or ice-box) inhibits some processes of decay. It would be equally accurate to say that these technologies respectively make possible changes in soil texture, speed of transport, etc.

In these terms, we can define a new technology as one which makes possible a *new* way of inducing a physical change or which creates a wholly new physical possibility that simply did not exist before. A better mousetrap or faster airplane are examples of the former (i.e., new way) and the Salk vaccine or the moon-rocket illustrate the latter (i.e., new possibility). Either kind of new technology will extend the range of what man *can* do, which is what technology is all about.

There is nothing in the nature or fact of a new tool, of course, that requires its use. As Lynn White has observed, "a new device merely opens a door; it does not compel one to enter."[2] But an open door does *invite* one to enter. A house in which a number of new doors have been opened is a different house from what it was before, and the likelihood is high that there will be a consequent change in the behavior of the people who live in it. In the historical if not in the individual dimension, therefore, new technologies make for (entail) change by virtue of creating new possibilities for human action. They do not do so necessarily, but they do so with a very high probability.

Technology alters the mix of choices

A correlative way in which new technology makes for change is by *removing* some options previously available to man. This consequence of technology is derivative and indirect. It is derivative in being dependent on the prior efficacy of technology in creating new possibilities. It is indirect in that the removal of options is not a result of the technology but of the act of choosing a *new* option that the technology has created.[3]

Examples abound. Once select modern plumbing and you render impossible the society of the village pump. Exploitation of industrial technology virtually precludes the choice of a rural or pastoral life. The automobile and airplane buy mobility, but at the expense of stability and constancy of personal relationships.

Since the familiarity of old habits and values counts *ipso facto* in their favor in any competition with alternatives to them, this necessity of sacrificing something old when choosing something new sometimes leads to cries against the tyranny of technique or the cold impersonality of the machine. This reaction overlooks the fact that the same technology that destroys the village green puts the lake in Maine within an hour's distance and that the airplane whose noise disturbs is the same machine that makes the Parthenon available to millions. What those who upbraid technology as such are really complaining about, I think, is the discomfort of real choice that technology imposes.

Opportunity costs are involved in exploiting any opportunity and therefore also the opportunities newly created by technology. Since the latter are new, the costs they involve are unprecedented (i.e., without precedent), too. (They moreover and unfortunately tend to take the form of giving up values so old and habitual that they have come to seem eternal and hallowed.) The presumption, which I have noted, that the new options will be chosen is therefore at the same time a presumption that the choice will be made to pay the new costs. Technology thus begins by a simple adding to the choices available to man and ends by altering the spectrum or mix of his choices. The latter, too, like the former, is change consequent upon technology.

Social change

The first-order effect of technology is thus to multiply and diversify material possibilities and thereby offer new and altered opportunities to man. (As with all opportunities when badly handled, to be sure, new ones created by technology can turn out to have been opportunities to make new mistakes, but they are not the less opportunities for that.)

Since new possibilities and new opportunities generally require new organizations of human effort to realize and exploit them, technology generally has second-order effects that take the form of social change. There have been instances in which changes in technology and in the material culture of a society have not been accompanied by social change, but such cases are rare and exceptional.[4] More generally:

> . . . over the millenia cultures in different environments have changed tremendously, and these changes are basically traceable to new adaptations required by changing technology and productive arrangements. Despite occasional cultural barriers, the useful arts have spread extremely widely, and the instances in which they have not been accepted because of pre-existing cultural patterns are insignificant.[5]

While social change does not *necessarily* follow upon technological change, therefore, it almost always does in fact, thus setting up a strong presumption that it always will. The role of the heavy plow in the organization of rural society and that of the stirrup in the rise of feudalism provide fascinating medieval examples of a nearly direct technology-society relationship.[6] The classic case in our time, of course—which it was Karl Marx's contribution to see so clearly, however badly he clouded his perception by tying it at once to a rigid determinism and to a form of Hegelian absolutism—is the industrial revolution, whose social effects continue to proliferate.

When (as usually) social change does result from the introduction of a new technology, it must, at least in part, be of a sort conducive to exploitation of the new opportunities (possibilities) created by that technology. Otherwise it makes no sense to speak of the social effects of technological change. Social consequences need not be (and surely are not) uniquely and univocally determined by the character of innovation, but they cannot be entirely independent of that character and still be *consequences*. What the advent of nuclear weapons altered was the military organization of the country, not the structure of its communications industry, and the launching of satellites affects international relations much more directly than it does the institutions of organized sport. (A change in international relations may affect international competition in sports, of course, but an "everything is connected with everything else" principle is hardly conducive to meaningful analysis.)

There is a congruence between technology and its effects, in other words, that serves as intellectual ground for all inquiry into the social effects of technology. This congruence has two aspects. I have suggested the first already: the subset of social changes that can result from a given technological innovation is smaller than the set of all possible social changes, although the changes that *do* result are in turn a smaller subset of those that *can* result. That is, in relation to any given innovation, the total spectrum of social consequences can be divided into three classes: (1) impossible (i.e., not a consequence), (2) possible, and (3) actual.

It is the congruence of technology and its social consequences in this sense, I think, that provides the theoretical warrant for the currently fashionable art of "futurology." The more responsible practitioners of this art insist that they do not predict unique future events but rather identify and adumbrate the spectrum of possible (as distinct from impossible) future events or situations (or "futures," as they call them). The effort is warranted by the twin facts that technology constantly alters the mix of possibilities (as previously noted) and that any given technological change promises a spectrum of possible consequences (rather than requiring or being limited to a single consequence).

The second aspect of the congruence between technology and its social consequences is a certain "one-wayness" about the relationship. It is after all the *technology* that creates the new physical possibilities that provide society with new options it can take advantage of by reorganization of its institutions and procedures. The process does not occur the other way around. To be sure, what technologies will be developed at any particular time is dependent on the social organization and system of values that prevail at that particular time. I do not mean to depreciate the *interaction* between technology and society, especially in complex societies where the interaction is recognized and can be guided. But the interaction does not obliterate the identity of the actors. The material initiative remains with technology and the social adaptation to it remains its consequence.

And since the probability is very high (as noted) that innovation will be of social consequence, technology means change in society as well as in nature.

Technology and values

New technology also means a high probability of change in values, for two reasons. First, what we want (our evaluations) and what we cherish (our values) are selections from what it is possible to have. Otherwise morality gives way to fantasy, i.e., to aspiring after the unavailable. When we say that technology makes possible what was not possible before, we say that we can now get (have) what we could not have before. Our old values thus come into question (in the neutral, not the pejorative sense), because (or whenever) we can do something about them by deliberate alteration of their material conditions. Both what we value and the ways in which we decide what we value can change, in other words, because technology can generate new value. This is how technology makes for value change.

Second, a particular technology can portend a particular kind of value change (but not the emergence of a particular value), in exact analogy (see above) to the subset of possible social changes that a new technology may augur (as distinct from both the wider set that includes the impossible and the narrower, actual subset). The reason this is so is that certain attitudes and values are more conducive than others to most effective exploitation of the potentialities of the new tools.

Does that make the argument circular unless appeal is made to value criteria different from the emergent values? Circularity is a danger, but it can be avoided without postulation of two kinds of values. The values existing at any given time will determine the technological choices that a society will make. But these choices will be based on the foreseeable consequences of the new technology. The essence of technology as creative of new possibility, however, means that there is an irreducible element of uncertainty—of unforeseeable consequence—in any innovation.[7] Two evaluations are necessary, therefore: one before and one after the innovation. The first is an evaluation of prospects (of ends-in-view, as John Dewey called them). The second is an evaluation of results (of outcomes actually attained).[8] The uncertainty inherent in technological innovation means there will usually be a difference between the two. To that extent, new technology will lead to value change.

2. PATTERNS OF CHANGE

Our own age is characterized by a deliberate fostering of technological change and, in general, by the growing social role of knowledge. "Every society now lives by innovation and growth; and it is theoretical knowledge that has become the matrix of innovation."[9]

In a modern industrialized society, particularly, there are a number of pressures that conspire toward this result. First, economic pressures argue for the greater efficiency implicit in a new technology. The principal example of this is the continuing process of capital modernization that takes place in industry. Second, there are political pressures that seek the greater absolute effectiveness of a new technology, as in our latest weapons, for example. Third, we turn more and more to the promise of new technology for help in dealing with our social problems. Fourth, there is what I would call the "Everest complex," which spurs action simply because the technology that makes it possible is available. Space vehicles spawn moon programs. There is, finally, an "Apollo syndrome": political and industrial interests engaged in developing a new technology have the vested interest and powerful means needed to urge its adoption and widespread use.

If this social drive to develop evermore new technology is taken in conjunction with the very high probability (argued above) that new technology will result in physical and social (including value) changes, we have the conditions for a world whose defining characteristic is change, i.e., the kind of world I once described as Heraclitean (after the pre-Socratic philosopher Heraclitus, who saw change as the essence of being).[10]

When change becomes that pervasive in the world, it must color the ways in which we understand, organize, and evaluate the world. The sheer fact of change will have an impact on our sensibilities and ideas, our institutions and practices, our politics and values. Most of these have to date developed on the assumption that stability was more characteristic of the world than change; i.e., that change was but a temporary perturbation of stability or a transition to a new (and presumed better or higher) stable state. What happens to them when that fundamental metaphysical assumption is undermined?

The nature of the answer is evoked by anthropologist Evon Vogt in unlikely, twentieth century cold-war terms:

> One of the central tendencies of our United States international policy is to try to maintain, or to restore, homeostatic states or conditions—to keep things in equilibrium. The Communist world, on the other hand, operates with the opposite premise in mind—that social and cultural systems are continually changing. *Since they expect change* [emphasis added], the Communists can look forward to and

take advantage of economic, social, and political upheavals as they occur. They seize the *initiative* while we continually find ourselves in the awkward position of *reacting* to Communist programs and policies.[11]

I have quoted this passage not for the judgment but for the distinction it contains. It shows how farreaching can be the difference between regarding the world as essentially changing or essentially stable. I devote the balance of this paper to suggesting some intellectual, social, and political implications of the *systematic expectation of change*.

Intellectual trends

I have already noted the growing social role of knowledge.[12] Our society values the production and inculcation of knowledge more than ever before, as is evidenced by sharply rising research, development, and education expenditures over the last twenty years. There is an increasing devotion, too, to the systematic use of relevant information in public and private decision-making, as is exemplified by the President's Council of Economic Advisers, by various scientific advisory groups in and out of government, by increasing use of such techniques as systems analysis and program planning and budgeting systems, and by the recent concern with assembling and analyzing a set of "social indicators" to help gauge the social (on the analogy of the economic) health of the nation.

A changing society must put a relatively strong accent on knowledge in order to offset the unfamiliarity and uncertainty that change implies. Traditional ways (beliefs, institutions, procedures, attitudes) may be adequate for dealing with the existent and known. But new technology can be generated and assimilated only if there is technical knowledge about its operation and capabilities, and economic, sociological, and political knowledge about the society into which it will be introduced.

This argues, in turn, for the importance of the social sciences. It is by now reasonably well established that policy-making in many areas can be effective only if it takes account of the findings and potentialities of the natural sciences and of their associated technologies. Starting with economics, we are gradually coming to a similar recognition of the importance of social science to public policy. Research and education in the social sciences are being increasingly supported by public funds, as the natural sciences have been by the military services and the National Science Foundation for the last quarter of a century. Also, both policy-makers and social scientists are seeking new mechanisms of cooperation and exploring the modifications these will require in their respective assumptions and procedures. This trend toward more applied social science is likely to be noticeable in any highly innovative society.

The scientific mores of such a society will also be influenced by the interest in applying technology that defines it. Scientific inquiry (into nature and society) is likely to be motivated by and focused on problems of the society rather than centering mainly around the unsolved puzzles of the scientific disciplines themselves. This does *not* mean, although it *can* if vigilance against political intereference is relaxed, (1) that the resulting science will be any less pure than that proceeding from disinterested curiosity; or (2) that there cannot therefore be any science motivated by curiosity; or (3) that the advancement of scientific knowledge may not be dependent on there always being some. The research into the atomic structure of matter that is undertaken in the interest of developing new materials for supersonic flight is no less basic or pure than the same research undertaken in pursuit of a new and intriguing particle. Even more to the point, social research into voting behavior is not *ipso facto* less basic or pure because it is paid for by an aspiring candidate rather than by a foundation grant.

There is a serious question, in any event, about just how pure is pure in scientific research. One need not subscribe to B. Hessen's out-and-out Marxist postulation of exclusively social and economic origins for Newton's research interests, for example, in order to recognize "the demonstrable fact that the thematics of science in seventeenth century England were in large part determined by the social structure of the time."[13] Nor should we ignore the fashions in science, such as the virtually exclusive emphasis on physics in recent years that was triggered by the military interest in nuclear energy, or the very similar present-day passion for computers and computer science. An innovative society *means* one in which there is a strong interest in bringing the best knowledge of the society to bear on ameliorating its problems and taking advantage of its opportunities. It is not surprising that scientific objectives and choices in that society should be in large measure determined by what those problems and opportunities are.

Another way in which a society of change influences its patterns of inquiry is by putting a premium on the formulation of new questions and in general on the synthetic aspects of knowing. Such a society is by description one that probes at technological and intellectual frontiers, and a frontier, according to the biologist C. H. Waddington, is where "we encounter problems which we cannot yet ask sensible questions."[14] When change is prevalent, in other words, we are frequently in the position of not knowing just what we need to know. A goodly portion of the society's intellectual effort must then be devoted to formulating new research questions (or reformulating old ones in the light of changed circumstances and needs) so that inquiry can remain pertinent.

Three consequences follow. First, there is a need to reexamine the knowledge already available for its meaning in the context of the new questions. This is what I have termed the synthetic aspect of knowing. Second, the need to formulate new questions coupled with the problem (as distinct from discipline) orientation discussed above requires that answers be sought from the intersection of several disciplines. This is the impetus for current emphases on the importance of interdisciplinary or crossdisciplinary inquiry as a supplement to the academic research aimed at expanding knowledge and training scientists. Third, there is a need to institutionalize the function of transferring scientific knowledge to social use. In response, universities are spawning problem- or area-oriented institutes, which surely augur eventual organizational change, and new policy oriented research institutes arise at the borderlines of industry, government, and universities, and at a new no-man's land between public and private.

A fundamental intellectual implication of a world of change is the greater theoretical utility of the concept of progress over that of structure in sociological/cultural analysis. Equilibrium theories of various sorts imply ascription of greater reality to stable socio-cultural patterns than to social change. But as Vogt argues,

> change is basic in social and cultural systems. . . . [E. R.] Leach is fundamentally correct when he states that "every real society is a process in time." Our problem becomes one of describing, conceptualizing, and explaining a set of ongoing processes . . . , [but none of the current approaches is satisfactory] in providing a set of conceptual tools for the description and analysis of the *changing* social and cultural *systems* that we observe.[15]

There is no denial of structure: "Once the processes are understood, the structures manifested at given time-points will emerge with even greater clarity," and Vogt himself goes on to distinguish between short-run "recurrent processes" and long-range and cumulative "directional processes." The former are the repetitive "structural dynamics" of a society. The latter "involve alterations in the structures of social and cultural systems."[16] It is clear that the latter, for Vogt, are more revelatory of the essence of culture and society as changing.[17]

Social trends. Heraclitus's philosophy of universal change was a generalization from his observation of physical nature, as is evident from his appeal to the four elements of ancient physics (fire, earth, water, and air) in support of it.[18] Yet he offered it as a *metaphysical* generalization. Change, flux, is the essential characteristic of all of existence, not of matter only. We should expect to find it central also, therefore, to societies, institutions, values, population patterns, personal careers, etc.

We do. Among the effects of technological change that we are beginning to understand fairly well even now are those (1) on our principal institutions: industry, government, universities; (2) on our production processes and occupational patterns; and (3) on our social and individual environment: our values, educational requirements, group affiliations, physical locations, and personal identities. All of these are in movement (flux). Most are—also and further—*in process;* i.e., there is direction or pattern (structure) to the changes they are undergoing, and the direction is moreover recognizable as a consequence of the growing social role of knowledge induced by proliferation of new technology.

It used to be that industry, government, and universities could operate almost independently of each other. They no longer can, because technical knowledge is increasingly necessary of the successful operation of industry and government, and because universities, as the principal sources and repositories of knowledge must therefore add social service to their traditional roles of research and teaching. This conclusion is supported by (1) the growing importance of research, development, and systematic planning in industry; (2) by the proliferation and growth of knowledge-based industries; (3) by the entry of technical experts into policy-making at all levels of government; (4) by the increasing dependence of effective government (as already noted) on availability of information and analysis of data; (5) by the importance of education and training (and retraining) to successful entry into the society and to maintenance of the economy on an even keel; and (6) by the growth, not only of problem-oriented activities on university campuses, but also of the social role (as consultants, advisory boards, etc.) of university faculties.

The economic affluence that is generated by modern industrial technology accelerates the institutional mixing-up that I have been noting by blurring the heretofore relatively clear distinction between the private and public sectors of the society. As we dispose increasingly of resources not required for production of traditional consumer goods and services, demand is created for such public goods as education, urban improvement, clean air and clean water, and so forth. As a result,

> our national policy assumes that a great deal of our new enterprise is likely to follow from technological developments financed by the government and directed in response to government policy; and many of our most dynamic industries are largely or entirely dependent on doing business with the government through a subordinate relationship that has little resemblance to the traditional market economy.[19] Increasingly it will be recognized that the mature corporation, as it develops, becomes part of the larger administrative complex associated with the state. In time the line between the two will disappear.[20]

This fluidity of institutions and social sectors is not unreminiscent of the apparently more liberal fluidity that Heraclitus immortalized: "You cannot step twice into the same river, for other waters are continually flowing on."[21] But also like the waters of a river, the institutional changes of a technologically active age are not aimless; they have direction, as noted, toward an enhancement of the use of knowledge in society:

> Perhaps it is not too much to say that if the business firm was the key institution of the past hundred years, because of its role in organizing production for the mass creation of products, the university will become the central institution of the next hundred years because of its role as the new [*sic*] source of innovation and knowledge.[22]

The considerable research and public debate of the last few years on the implications of technological change for employment levels and the character or content of work have not been in vain. Positions originally so extreme as to be untenable have been tempered in the process. Few students of the subject believe any longer that the progress of mechanization and automation in industry must lead to an irreversible increase in the level of involuntary unemployment in the society, whether in the forms of unavailability of employment, or of a shortening work week, or of lengthening vacations, or of extension of the period of formal schooling. These developments may occur, either voluntarily or as a result of inadequate education, poor social management, or failure to ameliorate our race problem, but they are not necessary consequences of new industrial technology.

Too much is beginning to be known about what the effects of technology on work and employment in fact are, on the other hand, for them to be adequately accounted for as merely transitional disruptions consequent on industrialization. Explanation in terms of transition (which many academic economists are particularly prone to in dealing with this subject) reveals the "steady-state" or equilibrium assumptions that I earlier cited as antithetical to a thorough-going appreciation of the implications of change. (Considering until how recently economic analysis has tended to treat research and education—i.e., knowledge—as factors exogenous to the traditional production function, in fact, there seems to me to be a clear and urgent need for an economic theory more adequate than previous ones to deal with the implications of technology. J. K. Galbraith's *The New Industrial State* is one of the few works to date intended to be responsive to this need.)

Technological development for most of history has provided substitutes for human muscle power and mechanical skills. Current and projected developments in electronic computers are providing and will continue to provide mechanical substitutes for at least some human mental operations. No technology as yet promises to duplicate human creativity, especially in the artistic sense, if only because we do not yet understand the conditions and functioning of creativity. (This is not to deny that computers, for example, can be useful aids to creative activity.) Nor are there in the offing mechanical equivalents for the emotional dimensions of man. For the foreseeable future, therefore, one may hazard the prediction that distinctively human work will be less and less of the "muscle and elementary mental" kind, and more and more of the "intellectual and artistic creativity and emotions'' kind. (This does *not* mean that only highly inventive or artistic people will be employable in the future. There is much sympathy needed in the world, for example, and the provision of it is neither mechanizable nor requisite of genius. It is illustrative of what I mean by an ''emotional'' service.)

While advancing technology may not displace people by reducing employment in the aggregate, therefore, it unquestionably displaces jobs by rendering them more efficiently performed by machines than by people. What this means for emergent patterns of work and employment has been vividly suggested by Robert Heilbroner:

> What is needed above all is a new expansive group of employments to offer the same absorptive cushion once given by office and service jobs [i.e., before they, too, began to be mechanized]. And if this is the objective, it is not difficult to know where to look to find such employments. We have merely to ask ourselves: what tasks in society are clearly and admittedly undermanned? The answer is provided by every city, in its shortage of adequate housing, its unbeautified and ill-maintained streets and parks; its under-protected citizens, under-educated children, under-cared-for young and old and sick.[23]

A major effect of technological change, in other words, is a new social responsibility to invent and establish mechanisms and procedures of occupational innovation, and that is more than a ''transitional'' problem. It represents a qualitative and permanent alteration in the nature of human society consequent on perception of the ubiquity of change.

This perception and the anticipatory attitude that it implies have some additional consequences, not less important, but as yet less well understood than those for institutional change and occupational patterns. For example, lifetime constancy of trade or profession has been a basis of personal identity and of the sense of individuality. Other bases for this same sense have been identification over time with a particular social group or set of groups as well as with physical or geographical location.

All of these are now subject to Heraclitean flux. The incidence of life-long careers will inevitably lessen, as employing institutions and job contents both change. More than one career per lifetime is likely to be the norm henceforth. Group identities will shift as a result: every occupational change will involve the individual with new professional colleagues, and will often mean a sundering from old friends and cultivation of new ones. Increasing geographical mobility (already so characteristic of advanced industrial society) will not only reinforce these impermanencies, but also shake the sense of identity traditionally associated with ownership and residence upon a piece of land. Even the family may lose some of its influence as bastion of personality, as is already discernible in advanced countries in the decline of the extended family.

I have alluded in the past to the implications for education of a world seen as essentially changing.[24] Education has traditionally had the function of preparing youth to assume full membership in society (1) by imparting a sense for the history and accumulated knowledge of the race; (2) by imbuing the young with a sense of the culture, mores, practices, and values of the group; and (3) by teaching a skill or set of skills necessary to a productive social role. Philosophies of education have accordingly been elaborated on the assumption of stability of values and mores, etc., and on the up-to-now demonstrable principle that one good set of skills well learned could serve a man through a productive lifetime.

I have suggested already that this last principle is undermined by contemporary and foreseeable occupational trends, and the burden of all I have said thus far similarly disputes the assumption of unalterable cultural stabilities. There are significant implications for the enterprise of education, which I shall not examine in detail here. They must include, I think, (1) a decline in the importance of manual skills; (2) a consequent rising emphasis on general techniques of analysis and evaluation of alternatives; (3) training in occupational flexibility; (4) development of management skills; and, in particular, (5) instruction in the potentialities and use of modern intellectual tools.

Above all, perhaps, higher education especially will need to attend more deliberately and systematically than it has in recent decades to developing the reflective, synthetic, speculative, and even the contemplative capacities of men, for understanding may be at a relatively greater premium henceforth than particular knowledge. When we can no longer lean on the world's stabilities, we must be able to rely on new abilities to cope with change and be comfortable with it.

There is an analogous implication for social values and for the human enterprise of valuing. There is concern expressed in many quarters these days about the threat of technology to values. Some writers go so far as to assert an incompatibility between technology and values and to warn that technological progress is tantamount in dehumanization and the destruction of all value.[25]

There is no question, as I noted earlier, that technological change alters the mix of choices available to man and that choices made *ipso facto* preclude other choices that might have been made. Some values are destroyed in this process. It is also the case, no doubt, that some of the choices made are constrained by the very technology that makes them available. In such cases, the loss of value can be tragic, and justly regretted and inveighed against.

I see little but an emotional basis, however, for the fear that technological progress must, of necessity, mean a progressive destruction of value. To the extent that it is more than the usual resistance to change, it has its source, I think, partly in an event in the history of science, but mainly on a fundamental misunderstanding of the nature of value.

A characteristic of modern science which distinguishes it from ancient and medieval science is the absence of the idea of purpose as an explanatory concept. Aristotelian science postulated four "causes" (i.e., principles of explanation) of any process, and the most important (into which the other three eventually collapsed) was the final cause of, or, in the Greek locution, the "for the sake of which" a process occurred. If we now recall that meaning and value are dependent on the having of purpose and the participation in processes calculated to achieve them, we recognize a basic logical compatibility between human values and the ancient concept of science. Science contributed to the understanding necessary to the achievement of human purposes and therefore helped in the generation and preservation of value.

The danger inherent in this approach to science was that the search for truth could be distorted by the politically (or religiously) motivated substitution of purposes alien to the logic of inquiry. This happened increasingly, especially in medieval times, and led to the discredit of purpose as an acceptable explanatory concept. Hence the nearly unanimous modern claim that "science has advanced by getting rid of the idea of purpose, except the abstract purpose of advancing truth and knowledge."[26] The only sense of "cause" thenceforth considered respectable in scientific explanation (at least until David Hume began to raise questions about that one, too) was that of efficient cause, i.e., the "pushing" or motive factor that served to *initiate* a process, without regard to the end of the process that made it intelligible and gave it meaning.

What had seemed to the ancients the natural compatibility of science and value (which they termed the unity of knowledge and virtue) was thus destroyed by the abandonment by science of the concept of end or purpose. The restriction of causation to efficiency made scientific explanation mechanical and wholly independent of meaning, and therefore of considerations of value. There followed the long and dismal history of the science-value dualism, whose contemporary form, I think, is the assertion of an antimony between technology and human values. Nothing in this history, however, justifies the interference too often made from it that the idea of purpose and the ideas of science and technology are antithetical by nature. It may be, indeed, that the increasingly deliberate use we now make of technology for our social objectives will call for a new look at purpose as a factor in the conduct of scientific inquiry.

I believe that the principal source of the view that technology and values are incompatible by nature is the confusion of what is valuable and what is stable. To be sure, the values of a society change more slowly than do the realities of human experience; their viability is inherent in their emergence as values in the first place and in their function as criteria, which means that their adequacy will tend to be judged later rather than earlier. But values do change, as a glance at any history will show. They change more quickly, moreover (as I suggested in an earlier section, above) the more quickly or extensively a society develops and introduces new technology. Since technological change is so prominent a characteristic of our own society, we tend to note inadequacies in our received values more quickly than might have been the case in other times. If that perception be coupled with the conviction (discussed above) that technology and value are inherently inimical to each other, the opinion is reinforced that the advance of technology must mean the decline of value and of the amenities of distinctively human civilization.

While particular values may vary with particular times and particular societies, however, and while they may even at times vary so quickly as to be seen to do so contemporaneously, rather than only in retrospect, the activity of valuing and the social function of values do not change. This is the source of the stability so necessary to human moral experience. It is not to be found, nor should it be sought, exclusively in the familiar values of the past. As the world and society are seen increasingly as processes in constant change under the impact of new technology, value analysis, too, will have to concentrate on its own proper processes: the process of valuation in the individual and the process of value formation and value change in the society. The emphasis will have to shift, in other words, from values to valu*ing*. For it is not particular familiar values as such that are valuable, but the human ability to extract values from experience and to use and cherish them. That *that* value is not threatened by technology; it is only challenged by it.

Political trends

There are a number of respects in which technological change and the intellectual and social changes it brings with it are likely to alter the conditions and patterns of government. I construe government in this connection in the broadest possible sense of the term, i.e., as governance (with a small "g") of a *politeia*. (The value of the Greek term is that it frees us of the narrower and more specialized connotation that usage has attached to the English word "politics" or even the somewhat unclear "polity.") What I seek to encompass by the term, in other words, is the social decision-making function in general, whether exemplified by small or large or public or private groups. (Government with a capital "G" is thus an aspect of the broader concept.) I include in decision-making, moreover, both the values and criteria that govern it and the institutions, mechanisms, procedures, and information by means of which it operates.

Again, the changes that technology purports for government have direction: they enhance the role of government in society and (not surprisingly, given the considerations already adduced) they enhance the role of knowledge in government.

The importance of decision-making will tend to grow relative to other social functions (relative to production, for example, in an affluent society), (1) partly because the frequency with which new possibilities are created in a technologically active age will provide many opportunities for new choices (i.e., new decisions), (2) partly because continuing alteration of the spectrum of available choice alternatives will shorten the useful life of decisions previously made, and (3) partly because the economic affluence consequent on new technology will increase the scope of deliberate public decision-making at the expense of the largely automatic and private charting of society's course by market forces. It is characteristic of our time that the market is increasingly distrusted as a goal-setting mechanism, although there is of course no question of its effectiveness as a signaling and control device.

Some of the ways in which knowledge increasingly enters the fabric of government have been amply noted, both above and in what is by now becoming a fairly voluminous literature on various aspects of the relation of science and public policy. There are other respects in which knowledge (information, technology, science) is bound to have fundamental impact on the structures and processes of decision-making about which, as yet, we know virtually nothing.

The newest information-handling equipment and techniques find their way quickly into the agencies and operations of federal and local government, for example, in the first instance because there are many jobs that they can perform more efficiently than the traditional rows of clerks. But it is notorious that adopting new means in order better to accomplish old ends very often results in the substitution of new ends (inherent in the new means) for old ones.[27] Computers can thus make available information in quantities and forms not previously available to government. This alters traditional relationships between citizens and government (both of whom find it increasingly difficult to "hide" their activities from each other), with consequences for individual privacy, for the modes of government, and for the ethics of and public controls over a new elite of information keepers in the society.

An exciting possibility that is, however, so dimly seen as perhaps to be illusory is that knowledge can widen the area of political consensus. There is no question here of a naive rationalism such as we associate with the eighteenth century Enlightenment. No amount of reason will ever triumph wholly over irrationality, certainly, nor will vested interest fully yield to love of wisdom. Yet there are some political disputes and disagreements, surely, that derive from ignorance of information bearing on an issue or from lack of the means to analyze fully the probable consequences of alternative courses of action. Is it too much to expect that better knowledge may bring about greater political consensus in such cases as these? Is the democratic tenet that an informed public contributes to the commonweal pure political myth? The sociologist S. M. Lipset suggests not:

> [It is] my perhaps overrationalistic belief that a fuller understanding of the various conditions under which a democracy has existed may help men to develop it where it does not now exist.[28]

Robert E. Lane has made the point more generally:

> If we employ the term "ideology" to mean a comprehensive, passionately believed, self-activating view of society, usually organized as a social movement, . . . it makes sense to think of a domain of knowledge distinguishable from a domain of ideology, despite the extent to which they may overlap. Since knowledge and ideology serve somewhat as functional equivalents in orienting a person toward the problems he must face and the policies he must select, the growth of the domain of knowledge causes it to impinge on the domain of ideology.[29]

Lane cites as evidence of his point the 1955 Congress for Cultural Freedom in Milan to which

> scholars and scientists came expecting, indeed inviting, a great confrontation of world views. Under the pressure of economic and social knowledge, a growing body of research, and the codified experience of society, ideological argument tended to give way to technical argument, apparently to the disappointment of some. The debate remained evaluative and partisan, but the domain of ideology was shrunken by the dominance of knowledge.[30]

Daniel Bell, finally, makes a similar point about erstwhile ideological battles in the West:

> In the Western world . . . there is today a rough consensus among intellectuals on political issues: the acceptance of a Welfare State; the desirability of decentralized power; a system of mixed economy and of political pluralism. In that sense . . . the ideological age has ended.[31]

If the technology-values dualism is unwarranted, as I have suggested, it is equally plausible to find no more warrant in a sharp separation between knowledge and political action. Like all dualisms, this one too may have had its origins in the analytic abhorrence of uncertainty. (One is reminded in this connection of the radical dualism that Descartes arrived at as a result of his determination to base his philosophy on the only certain and self-evident principle he could discover.) There certainly is painfully much in political history and political experience to render uncertain a positive correlation between knowledge and political consensus. The correlation is not necessarily absent therefore, and to find it and lead society to act on it may be the greatest challenge yet to political inquiry.

To the extent that technological change expands the spectrum of what man can do, it multiplies the number of choices that society will have to make. These choices will moreover be real ones (rather than results of impersonal market operations) and will therefore have to be made by political means. Since it is unlikely that we will soon be able to predict future opportunities (and their attendant opportunity costs) with any significant degree of reliability, it becomes important to investigate the conditions of a political system (I use the term in the wide sense I assigned to "government" above) with the flexibility and value presuppositions necessary to evaluate alternatives and make choices among them as and when the need arises.

This prescription is analogous, for governance of a changing society to those advanced above for educational policy and for our approach to the problem of social values. In all three cases, the emphasis shifts from allegiance to the known, stable, formulated, and familiar, to a *posture* of expectation of change and readiness to deal with it. It is this kind of shift, occurring across many elements of society, that is the hallmark of a truly Heraclitean age. It is what Vogt seeks to formalize in stressing analysis of process as against the structure of culture and society. The mechanisms, values, attitudes, and procedures called for by a social posture of readiness will be different in kind from those characteristic of a society that sees itself as mature, "arrived," and in stable equilibrium. The most fundamental *political* task of a technological world, in other words, is (as noted earlier) that of systematizing and institutionalizing the social expectation of change.

And that task I think may be a precondition of profiting from our accumulating knowledge of the effects of technological change. To understand those effects is an intellectual problem, but to do something about them and profit from the opportunities that technology offers is a political one. We need above all to gauge the effects of technology on the *politeia* so that we can derive some social value from our knowledge. This, I suppose, is the twentieth century form of the perennial ideal of wedding wisdom and government.

NOTES

1. Herman Kahn and his associates at the Hudson Institute have made a major effort in this direction in *The Year 2000: A Framework for Speculation* (New York: Macmillan, 1967).
2. Lynn White, Jr., *Medieval Technology and Social Change* (New York: Oxford University Press Galaxy Book), p. 28.
3. I deal extensively (albeit in a different context) with the making of new possibilities and the precluding of some choices by making others in E. G. Mesthene, *How Language Makes Us Know* (The Hague: Martinus Nijhoff, 1964), Chapter 3.
4. One such case is described in Evon Z. Vogt, *Modern Homesteaders: The Life of a Twentieth-Century Frontier Community* (Cambridge: The Belknap Press of Harvard University Press, 1955).
5. Julian H. Steward. *Theory of Culture Change: The Methodology of Multilinear Evolution* (Urbana: University of Illinois Press, 1955), p. 37. My citation of Steward is deliberate since he is generally critical of such of his fellow anthropologists as Leslie White and Gordon Childe for adopting strong positions of technological determinism. Yet even Steward says, "White's . . . 'law' that technological development expressed in terms of man's control over energy underlies certain cultural achievements and social changes [has] long been accepted." p. 18.
6. See Lynn White, Jr., pp. 44 f. and 28 ff., respectively, and note especially White's contention (p. 45) that analysis of the influence of the heavy plow has survived all the severe criticisms leveled against it.
7. Robert L. Heilbroner points up this unforeseeable element—he calls it the "indirect effect" of technology—in *The Limits of American Capitalism* (New York: Harper & Row, 1966), p. 97.
8. John Dewey, *Theory of Valuation* [International Encyclopedia of Unified Science, Vol. II, No. 4], (Chicago: University of Chicago Press, 1939). The model of the ends-means continuum that Dewey develops in this work should prove useful in dealing conceptually with the value changes implicit in new technology.
9. Daniel Bell, "Notes on the Post-Industrial Society (I)," *The Public Interest,* No. 6, Winter, 1967, p. 29. See also Robert E. Lane, "The Decline of Politics and Ideology in a Knowledgeable Society," *American Sociological Review,* Vol. 31, No. 5, October, 1966, for evidence (pp. 562-53) and a discussion (passim) of some of the political implications of this development.
10. E. G. Mesthene, "On Understanding Change," *Technology and Culture,* Vol. VI, No. 2, Spring, 1965, pp. 226-27. Donald A. Schon also has recently recalled Heraclitus for a similar descriptive purpose and has stressed how thoroughgoing a revolution of attitudes is implied by recognition of the pervasive character of change; see his *Technology and Change* (New York: Delacorte, 1967), pp. xi ff.
11. Evon Z. Vogt, "On the Concepts of Structure and Process in Cultural Anthropology," *American Anthropologist,* Vol. 62, No. 1 (1960), p. 30n.
12. In addition to Bell and Lane, already cited, see also Lynton W. Caldwell, "Managing the Scientific Super-Culture: The Task of Educational Preparation," *Public Administration Review,* Vol. XXVII, No. 3, June, 1967; Robert L. Heilbroner, Part II, and Alan F. Westin, "Science, Privacy, and Freedom: Issues and Proposals for the 1970's," Part I, *Columbia Law Review,* Vol. 66, No. 6, June, 1966, p. 1010 and passim.
13. The quotation is from Robert K. Merton, *Social Theory and Social Structure* (Glencoe, Ill.: The Free Press, 1949), p. 348. Hessen's analysis is in his "The Social and Economic Roots of Newton's Mechanics," Kniga Ltd. (England). (Science at the Crossroads series, n.d. The paper was read at the Second International Congress of the History of Science and Technology, June 29-July 3, 1931.)
14. Quoted in *Graduate Faculties Newsletter* (Columbia University, March, 1966).
15. E. Z. Vogt, variously from pp. 19, 20.
16. Ibid., pp. 20, 21, 22.
17. The structure-process dualism also has its familiar (and other) philosophical face, of course, which a fuller treatment than this paper allows should not ignore. Such a discussion would recall at least the metaphysical positions that we associate with Aristotle, Hegel, Bergson, and Dewey.
18. E.g., Fragments Nos. 28-34, in Philip Wheelwright, *Heraclitus* (New York: Atheneum, 1964), p. 37. In his commentary on these fragments, Wheelwright makes clear that the element of fire which looms so large in Heraclitus remains a physical actuality for him however much he may also have stressed its symbolic character (ibid., pp. 38-39).
19. Don K. Price, *The Scientific Estate* (Cambridge: The Belknap Press of Harvard University Press, 1965), p. 15.
20. John Kenneth Galbraith, *The New Industrial State* (Boston: Houghton Mifflin, 1967), p. 393.
21. Wheelwright, Fragment 21, p. 29.
22. Daniel Bell, p. 30.
23. Robert L. Heilbroner, "Men and Machines in Perspective," *The Public Interest,* No. 1, Fall, 1965.
24. E. G. Mesthene, "On Understanding Change," pp. 226-27.
25. Examples of such apocalyptic literature are: Jacques Ellul, *The Technological Society* (New York: Knopf, 1964); Donald Michael, *Cybernation: The Silent Conquest* (Santa Barbara, Calif.: Center for the Study of Democratic Institutions, 1962); and Joseph Wood Krutch, "Epitaph for an Age," *The New York Times Magazine,* July 30, 1967.
26. Don K. Price, p. 133.
27. For a more extended discussion, see E. G. Mesthene, "The Impact of Science on Public Policy," *Public Administration Review,* Vol. XXVII, No. 2, June, 1967.
28. Seymour Martin Lipset, *Political Men: The Social Bases of Politics* (New York: Doubleday-Anchor, 1963), p. 455.
29. R. E. Lane, p. 660.
30. Ibid., p. 661.
31. Daniel Bell, *The End of Ideology* (New York: The Free Press, 1960), p. 403 (paperback edition).

Do machines make history?

ROBERT HEILBRONER

▪ Robert Heilbroner *is a member of the faculty of The New School for Social Research. His many books include* The Worldly Philosophers *and* The Limits of American Capitalism.

In this essay Professor Heilbroner, addressing the problem of an autonomous, deterministic technology, discusses two important questions. First, are there "laws of motion" of technology? Second, how does technology affect social relations? In answer to the first question he concludes that there is indeed a "sequence" to technology, but that this sequence is determined by many conditions which are necessary to it. In answer to the second question he argues that technology does influence social structures in the sense that different modes of technology affect the composition and organization of the labor force differently. So Heilbroner ultimately argues for that William James called "soft determinism," or the view that even though technological change does result in change in the social structure, man is nevertheless free to choose among alternative ends.

The hand-mill gives you society with the feudal lord; the steam-mill, society with the industrial capitalist.

Marx, *The Poverty of Philosophy*

That machines make history in some sense—that the level of technology has a direct bearing on the human drama—is of course obvious. That they do not make all of history, however that word be defined, is equally clear. The challenge, then, is to see if one can say something systematic about the matter, to see whether one can order the problem so that it becomes intellectually manageable.

To do so calls at the very beginning for a careful specification of our task. There are a number of important ways in which machines make history that will not concern us here. For example, one can study the impact of technology on the *political* course of history, evidenced most strikingly by the central role played by the technology of war. Or one can study the effect of machines on the *social* attitudes that underlie historical evolution: one thinks of the effect of radio or television on political behavior. Or one can study technology as one of the factors shaping the changeful content of life from one epoch to another: when we speak of "life" in the Middle Ages or today we define an existence much of whose texture and substance is intimately connected with the prevailing technological order.

None of these problems will form the focus of this essay. Instead, I propose to examine the impact of technology on history in another area—an area defined by the famous quotation from Marx that stands beneath our title. The question we are interested in, then, concerns the effect of technology in determining the nature of the *socioeconomic order*. In its simplest terms the question is: did medieval technology bring about feudalism? Is industrial technology the necessary and sufficient condition for capitalism? Or, by extension, will the technology of the computer and the atom constitute the ineluctable cause of a new social order?

Even in this restricted sense, our inquiry promises to be broad and sprawling. Hence, I shall not try to attack it head-on, but to examine it in two stages:

1. If we make the assumption that the hand-mill does "give" us feudalism and the steam-mill capitalism, this places technological change in the position of a prime mover of social history. Can we then explain the "laws of motion" of technology itself? Or to put the question less grandly, can we explain why technology evolves in the sequence it does?

2. Again, taking the Marxian paradigm at face value, exactly what do we mean when we assert that the hand-mill "gives us" society with the feudal lord? Precisely how does the mode of production affect the superstructure of social relationships?

These questions will enable us to test the empirical content—or at least to see if there *is* an empirical content—in the idea of technological determinism. I do not think it will come as a surprise if I announce now that we will find *some* content, and a great deal of missing evidence, in our investigation. What will remain then will be to see if we can place the salvageable elements of the theory in historical perspective—to see, in a word, if we can explain technological determinism historically as well as explain history by technological determinism.

□ Robert Heilbroner, "Do Machines Make History?" *Technology and Culture,* 8(1967), 335-345.

I

We begin with a very difficult question hardly rendered easier by the fact that there exist, to the best of my knowledge, no empirical studies on which to base our speculations. It is the question of whether there is a fixed sequence to technological development and therefore a necessitous path over which technologically developing societies must travel.

I believe there is such a sequence—that the steam-mill follows the hand-mill not by chance but because it is the next "stage" in a technical conquest of nature that follows one and only one grand avenue of advance. To put it differently, I believe that it is impossible to proceed to the age of the steam-mill until one has passed through the age of the hand-mill, and that in turn one cannot move to the age of the hydroelectric plant before one has mastered the steam-mill, nor to the nuclear power age until one has lived through that of electricity.

Before I attempt to justify so sweeping an assertion, let me make a few reservations. To begin with, I am fully conscious that not all societies are interested in developing a technology of production or in channeling to it the same quota of social energy. I am very much aware of the different pressures that different societies exert on the direction in which technology unfolds. Lastly, I am not unmindful of the difference between the discovery of a given machine and its application as a technology—for example, the invention of a steam engine (the aeolipile) by Hero of Alexandria long before its incorporation into a steam-mill. All these problems, to which we will return in our last section, refer however to the way in which technology makes its peace with the social, political, and economic institutions of the society in which it appears. They do not directly affect the contention that there exists a determinate sequence of productive technology for those societies that are interested in originating and applying such a technology.

What evidence do we have for such a view? I would put forward three suggestive pieces of evidence:

1. The simultaneity of invention

The phenomenon of simultaneous discovery is well known.[1] From our view, it argues that the process of discovery takes place along a well-defined frontier of knowledge rather than in grab-bag fashion. Admittedly, the concept of "simultaneity" is impressionistic,[2] but the related phenomenon of technological "clustering" again suggests that technical evolution follows a sequential and determinate rather than random course.[3]

2. The absence of technological leaps

All inventions and innovations, by definition, represent an advance of the art beyond existing base lines. Yet, most advances, particularly in retrospect, appear essentially incremental, evolutionary. If nature makes no sudden leaps, neither, it would appear, does technology. To make my point by exaggeration, we do not find experiments in electricity in the year *1500,* or attempts to extract power from the atom in the year *1700*. On the whole, the development of the technology of production presents a fairly smooth and continuous profile rather than one of jagged peaks and discontinuities.

3. The predictability of technology

There is a long history of technological prediction, some of it ludicrous and some not.[4] What is interesting is that the development of technical progress has always seemed *intrinsically* predictable. This does not mean that we can lay down future timetables of technical discovery, nor does it rule out the possibility of surprises. Yet I venture to state that many scientists would be willing to make *general* predictions as to the nature of technological capability twenty-five or even fifty years ahead. This too

[1]See Robert K. Merton, "Singletons and Multiples in Scientific Discovery: A Chapter in the Sociology of Science," *Proceedings* of the American Philosophical Society, CV (October 1961), 470-86.

[2]See John Jewkes, David Sawers, and Richard Stillerman, *The Sources of Invention* (New York, 1960 [paperback edition]), p. 227, for a skeptical view.

[3]"One can count 21 basically different means of flying, at least eight basic methods of geophysical prospecting; four ways to make uranium explosive; . . . 20 or 30 ways to control birth. . . . If each of these separate inventions were autonomous, i.e., without cause, how could one account for their arriving in these functional groups?" S. C. Gilfillan, "Social Implications of Technological Advance," *Current Sociology,* I (1952), 197. See also Jacob Schmookler, "Economic Sources of Inventive Activity," *Journal of Economic History* (March 1962), pp. 1-20; and Richard Nelson, "The Economics of Invention: A Survey of the Literature," *Journal of Business,* XXXII (April 1959), 101-19.

[4]Jewkes *et al.* (see n. 2) present a catalogue of chastening mistakes (p. 230 f.). On the other hand, for a sober predictive effort, see Francis Bello, "The 1960s: A Forecast of Technology," *Fortune,* LIX (January 1959), 74-78; and Daniel Bell, "The Study of the Future," *Public Interest,* I (Fall 1965), 119-30. Modern attempts at prediction project likely avenues of scientific advance or technological function rather than the feasibility of specific machines.

suggests that technology follows a developmental sequence rather than arriving in a more chancy fashion.

I am aware, needless to say, that these bits of evidence do not constitute anything like a "proof" of my hypothesis. At best they establish the grounds on which a prima facie case of plausibility may be rested. But I should like now to strengthen these grounds by suggesting two deeper-seated reasons why technology *should* display a "structured" history.

The first of these is that a major constraint always operates on the technological capacity of an age, the constraint of its accumulated stock of available knowledge. The application of this knowledge may lag behind its reach; the technology of the hand-mill, for example, was by no means at the frontier of medieval technical knowledge, but technical realization can hardly precede what men generally know (although experiment may incrementally advance both technology and knowledge concurrently). Particularly from the mid-nineteenth century to the present do we sense the loosening constraints on technology stemming from successively yielding barriers of scientific knowledge—loosening constraints that result in the successive arrival of the electrical, chemical, aeronautical, electronic, nuclear, and space stages of technology.[5]

The gradual expansion of knowledge is not, however, the only order-bestowing constraint on the development of technology. A second controlling factor is the material competence of the age, its level of technical expertise. To make a steam engine, for example, requires not only some knowledge of the elastic properties of steam but the ability to cast iron cylinders of considerable dimensions with tolerable accuracy. It is one thing to produce a single steam-machine as an expensive toy, such as the machine depicted by Hero, and another to produce a machine that will produce power economically and effectively. The difficulties experienced by Watt and Boulton in achieving a fit of piston to cylinder illustrate the problems of creating a technology, in contrast with a single machine.

Yet until a metal-working technology was established—indeed, until an embryonic machine-tool industry had taken root—an industrial technology was impossible to create. Furthermore, the competence required to create such a technology does not reside alone in the ability or inability to make a particular machine (one thinks of Babbage's ill-fated calculator as an example of a machine born too soon), but in the ability of many industries to change their products or processes to "fit" a change in one key product or process.

This necessary requirement of technological congruence[6] gives us an additional cause of sequencing. For the ability of many industries to co-operate in producing the equipment needed for a "higher" stage of technology depends not alone on knowledge or sheer skill but on the division of labor and the specialization of industry. And this in turn hinges to a considerable degree on the sheer size of the stock of capital itself. Thus the slow and painful accumulation of capital, from which springs the gradual diversification of industrial function, becomes an independent regulator of the reach of technical capability.

In making this general case for a determinate pattern of technological evolution—at least insofar as that technology is concerned with production—I do not want to claim too much. I am well aware that reasoning about technical sequences is easily faulted as *post hoc ergo propter hoc*. Hence, let me leave this phase of my inquiry by suggesting no more than that the idea of a roughly ordered progression of productive technology seems logical enough to warrant further empirical investigation. To put it as concretely as possible, I do not think it is just by happenstance that the steam-mill follows, and does not precede, the hand-mill, nor is it mere fantasy in our own day when we speak of the coming of the automatic factory. In the future as in the past, the development of the technology of production seems bounded by the constraints of knowledge and capability and thus, in principle at least, open to prediction as a determinable force of the historic process.

II

The second proposition to be investigated is no less difficult than the first. It relates, we will

[5] To be sure, the inquiry now regresses one step and forces us to ask whether there are inherent stages for the expansion of knowledge, at least insofar as it applies to nature. This is a very uncertain question. But having already risked so much, I will hazard the suggestion that the roughly parallel sequential development of scientific understanding in those few cultures that have cultivated it (mainly classical Greece, China, the high Arabian culture, and the West since the Renaissance) makes such a hypothesis possible, provided that one looks to broad outlines and not to inner detail.

[6] The phrase is Richard LaPiere's in *Social Change* (New York, 1965), p. 263 f.

recall, to the explicit statement that a given technology imposes certain social and political characteristics upon the society in which it is found. Is it true that, as Marx wrote in *The German Ideology,* "A certain mode of production, or industrial stage, is always combined with a certain mode of cooperation, or social stage."[7] or as he put it in the sentence immediately preceding our hand-mill, steam-mill paradigm, "In acquiring new productive forces men change their mode of production, and in changing their mode of production they change their way of living—they change all their social relations"?

As before, we must set aside for the moment certain "cultural" aspects of the question. But if we restrict ourselves to the functional relationships directly connected with the process of production itself, I think we can indeed state that the technology of a society imposes a determinate pattern of social relations on that society.

We can, as a matter of fact, distinguish at least two such modes of influence:

1. The composition of the labor force

In order to function, a given technology must be attended by a labor force of a particular kind. Thus, the hand-mill (if we may take this as referring to late medieval technology in general) required a work force composed of skilled or semiskilled craftsmen, who were free to practice their occupations at home or in a small atelier, at times and seasons that varied considerably. By way of contrast, the steam-mill—that is, the technology of the nineteenth century—required a work force composed of semiskilled or unskilled operatives who could work only at the factory site and only at the strict time schedule enforced by turning the machinery on or off. Again, the technology of the electronic age has steadily required a higher proportion of skilled attendants; and the coming technology of automation will still further change the needed mix of skills and the locale of work, and may as well drastically lessen the requirements of labor time itself.

2. The hierarchical organization of work

Different technological apparatuses not only require different labor forces but different orders of supervision and co-ordination. The internal organization of the eighteenth-century handicraft unit, with its typical man-master relationship, presents a social configuration of a wholly different kind from that of the nineteenth-century factory with its men-manager confrontation, and this in turn differs from the internal social structure of the continuous-flow, semi-automated plant of the present. As the intricacy of the production process increases, a much more complex system of internal controls is required to maintain the system in working order.

Does this add up to the proposition that the steam-mill gives us society with the industrial capitalist? Certainly the class characteristics of a particular society are strongly implied in its functional organization. Yet it would seem wise to be very cautious before relating political effects exclusively to functional economic causes. The Soviet Union, for example, proclaims itself to be a socialist society although its technical base resembles that of old-fashioned capitalism. Had Marx written that the steam-mill gives you society with the industrial *manager,* he would have been closer to the truth.

What is less easy to decide is the degree to which the technological infrastructure is responsible for some of the sociological features of society. Is anomie, for instance, a disease of capitalism or of all industrial societies? Is the organization man a creature of monopoly capital or of all bureaucratic industry wherever found? These questions tempt us to look into the problem of the impact of technology on the existential quality of life, an area we have ruled out of bounds for this paper. Suffice it to say that superficial evidence seems to imply that the similar technologies of Russia and America are indeed giving rise to similar social phenomena of this sort.

As with the first portion of our inquiry, it seems advisable to end this section on a note of caution. There is a danger, in discussing the structure of the labor force or the nature of intrafirm organization, of assigning the sole causal efficacy to the visible presence of machinery and of overlooking the invisible influence of other factors at work. Gilfillan, for instance, writes, "engineers have committed such blunders as saying the typewriter brought women to work in offices, and with the typesetting machine made possible the great modern newspaper, forgetting that in Japan there are women office workers and great modern newspapers get-

[7] Karl Marx and Friedrich Engels, *The German Ideology* (London, 1942), p. 18.

ting practically no help from typewriters and typesetting machines."[8] In addition, even where technology seems unquestionably to play the critical role, an independent "social" element unavoidably enters the scene in the *design* of technology, which must take into account such facts as the level of education of the work force or its relative price. In this way the machine will reflect, as much as mould, the social relationships of work.

These caveats urge us to practice what William James called a "soft determinism" with regard to the influence of the machine on social relations. Nevertheless, I would say that our cautions qualify rather than invalidate the thesis that the prevailing level of technology imposes itself powerfully on the structural organization of the productive side of society. A foreknowledge of the shape of the technical core of society fifty years hence may not allow us to describe the political attributes of that society, and may perhaps only hint at its sociological character, but assuredly it presents us with a profile of requirements, both in labor skills and in supervisory needs, that differ considerably from those of today. We cannot say whether the society of the computer will give us the latter-day capitalist or the commissar, but it seems beyond question that it will give us the technician and the bureaucrat.

III

Frequently, during our efforts thus far to demonstrate what is valid and useful in the concept of technological determinism, we have been forced to defer certain aspects of the problem until later. It is time now to turn up the rug and to examine what has been swept under it. Let us try to systematize our qualifications and objections to the basic Marxian paradigm:

1. Technological progress is itself a social activity

A theory of technological determinism must contend with the fact that the very activity of invention and innovation is an attribute of some societies and not of others. The Kalahari bushmen or the tribesmen of New Guinea, for instance, have persisted in a neolithic technology to the present day; the Arabs reached a high degree of technical proficiency in the past and have since suffered a decline; the classical Chinese developed technical expertise in some fields while unaccountably neglecting it in the area of production. What factors serve to encourage or discourage this technical thrust is a problem about which we know extremely little at the present moment.[9]

2. The course of technological advance is responsive to social direction

Whether technology advances in the area of war, the arts, agriculture, or industry depends in part on the rewards, inducements, and incentives offered by society. In this way the direction of technological advance is partially the result of social policy. For example, the system of interchangeable parts, first introduced into France and then independently into England failed to take root in either country for lack of government interest or market stimulus. Its success in America is attributable mainly to government support and to its appeal in a society without guild traditions and with high labor costs.[10] The general *level* of technology may follow an independently determined sequential path, but its areas of application certainly reflect social influences.

3. Technological change must be compatible with existing social conditions

An advance in technology not only must be congruent with the surrounding technology but must also be compatible with the existing economic and other institutions of society. For example, labor-saving machinery will not find ready acceptance in a society where labour is abundant and cheap as a factor of production. Nor would a mass production technique recommend itself to a society that did not have a mass market. Indeed, the presence of slave labor seems generally to inhibit the use of machinery and the presence of expensive labor to accelerate it.[11]

These reflections on the social forces bearing on technical progress tempt us to throw aside the whole notion of technological determinism

[8] Gilfillan (see n. 3), p. 202.

[9] An interesting attempt to find a line of social causation is found in E. Hagen, *The Theory of Social Change* (Homewood, Ill., 1962).

[10] See K. R. Gilbert, "Machine-Tools," in Charles Singer, E. J. Holmyard, A. R. Hall, and Trevor I. Williams (eds.), *A History of Technology* (Oxford, 1958), IV, chap. xiv.

[11] See LaPiere (see n. 6), p. 284; also H. J. Habbakuk, *British and American Technology in the 19th Century* (Cambridge, 1962), *passim*.

as false or misleading.[12] Yet, to relegate technology from an undeserved position of *primum mobile* in history to that of a mediating factor, both acted upon by and acting on the body of society, is not to write off its influence but only to specify its mode of operation with greater precision. Similarly, to admit we understand very little of the cultural factors that give rise to technology does not depreciate its role but focuses our attention on that period of history when technology is clearly a major historic force, namely Western society since 1700.

IV

What is the mediating role played by technology within modern Western society? When we ask this much more modest question, the interaction of society and technology begins to clarify itself for us:

1. The rise of capitalism provided a major stimulus for the development of a technology of production

Not until the emergence of a market system organized around the principle of private property did there also emerge an institution capable of systematically guiding the inventive and innovative abilities of society to the problem of facilitating production. Hence the environment of the eighteenth and nineteenth centuries provided both a novel and an extremely effective encouragement for the development of an *industrial* technology. In addition, the slowly opening political and social framework of late mercantilist society gave rise to social aspirations for which the new technology offered the best chance of realization. It was not only the steam-mill that gave us the industrial capitalist but the rising inventor-manufacturer who gave us the steam-mill.

2. The expansion of technology within the market system took on a new "automatic" aspect

Under the burgeoning market system not alone the initiation of technical improvement but its subsequent adoption and repercussion through the economy was largely governed by market considerations. As a result, both the rise and the proliferation of technology assumed the attributes of an impersonal diffuse "force" bearing on social and economic life. This was all the more pronounced because the political control needed to buffer its disruptive consequences was seriously inhibited by the prevailing laissez-faire ideology.

3. The rise of science gave a new impetus to technology

The period of early capitalism roughly coincided with and provided a congenial setting for the development of an independent source of technological encouragement—the rise of the self-conscious activity of science. The steady expansion of scientific research, dedicated to the exploration of nature's secrets and to their harnessing for social use, provided an increasingly important stimulus for technological advance from the middle of the nineteenth century. Indeed, as the twentieth century has progressed, science has become a major historical force in its own right and is now the indispensable precondition for an effective technology.

• • •

It is for these reasons that technology takes on a special significance in the context of capitalism—or, for that matter, of a socialism based on maximizing production or minimizing costs. For in these societies, both the continuous appearance of technical advance and its diffusion throughout the society assume the attributes of autonomous process, "mysteriously" generated by society and thrust upon its members in a manner as indifferent as it is imperious. This is why, I think, the problem of technological determinism—of how machines make history—comes to us with such insistence despite the ease with which we can disprove its more extreme contentions.

Technological determinism is thus peculiarly a problem of a certain historic epoch—specifically that of high capitalism and low socialism—*in which the forces of technical change have been unleased, but when the agencies for the control or guidance of technology are still rudimentary.*

The point has relevance for the future. The surrender of society to the free play of market forces is now on the wane, but its subservience to the impetus of the scientific ethos is on the rise. The prospect before us is assuredly that of an undiminished and very likely accelerated pace of technical change. From what we can foretell about the direction of this technological

[12] As, for example, in A. Hansen, "The Technological Determination of History," *Quarterly Journal of Economics* (1921), pp. 76-83.

advance and the structural alterations it implies, the pressures in the future will be toward a society marked by a much greater degree of organization and deliberate control. What other political, social, and existential changes the age of the computer will also bring we do not know. What seems certain, however, is that the problem of technological determinism—that is, of the impact of machines on history—will remain germane until there is forged a degree of public control over technology far greater than anything that now exists.

The emerging superculture

KENNETH BOULDING

■ Kenneth Boulding *is Distinguished Professor of Economics in the Institute of Behavioral Sciences, the University of Colorado. Among his many important contributions to the literature of science and technology is* The Meaning of the Twentieth Century.

In the essay that appears here, Professor Boulding argues that far from being "autonomous," technology, or more properly the "superculture"—the technology which transcends folk culture—depends for its system of valuation on traditional, often nontechnological, cultural manifestations. It is his contention that the emerging superculture and the traditional cultures must acquiesce to an interdependence that will assure the longevity of each.

It has been pointed out by B. L. Whorf and a number of writers that one of the problems of those who are trained to think in Indo-European language, is that nouns tend to be substituted for verbs. There seems to be something about the subject-predicate-object structure of the sentence in these languages which inhibits us from talking about activity as such, and which leads us into reification, that is, talking about processes as if they were things. Both the words "technology" and "values" are examples of this peculiar linguistic difficulty. Whether there is any such "thing" as a value it is hard to know in the absence of any secure knowledge about the physical or physiological substructure of the valuation and choice processes of the human nervous system. For all I know, love may be coded into one chemical and hate into another; but up to now at any rate we have not been able to identify these structural forms. What we observe is not values but valuation, that is, an activity which may be inferred from the study of behavior, guided by introspection on the choice process.

Similarly, technology is not a thing. It is also a process, a complex set of ways of doing things with both human and material instruments. Again, perhaps, as a thing it may be represented by some as yet quite unknown structure in somebody's head in terms of knowledge. Up to now at any rate, this carrier cannot be observed directly, and what we observe in technology is people applying means to secure ends.

Among social scientists, economists have probably paid the most attention to the problems involved both in the choice process and in the processes of technology. Oddly enough, the problems are formally rather similar. In his attempt to describe the process of choice, the economist has postulated a utility or welfare function according to which every relevant state of the field or social system is given an ordinal number which indicates an order of preference, first, second, third, and so on. In what is called a strong ordering, each state of the field is given a unique ordinal number; in a weak ordering, different states of the field may have the same ordinal number, in much the way that students may be bracketed in a class list. As the economist sees it, then, the problem of valuation is that of ordering a field of choice and then selecting the first on the order of preference. This is the famous principle of maximizing behavior, as it is called, which is simply a mathematical elaboration of the rather obvious principle that people always do what seems to them best at the time. It has always surprised me, as I have remarked elsewhere, that such a seemingly empty principle should be capable of such enormous mathematical elaboration. It can only be given content, of course, if there are some information processes by which the preference field

□ From *Values and the Future,* ed. Kurt Baier and Nicholas Rescher (New York: Macmillan, Inc., 1969), pp. 336-350.

can be spelled out and the preference function described. Where the field which is to be ordered consists of a set of possible exchanges under a given system of exchange opportunities or prices, certain broad properties of the preference function, at least, can be deduced from the observation of differences in behavior in response to different price systems. This is what is called the "theory of revealed preference." Theoretically, we suppose that we can deduce the preference function of an individual from the differences of his observed behavior under different price structures. In practice, of course, because of the sheer difficulty of observing the behavior of the same individual under different price structures, what we observe is some kind of aggregate behavior of the behavior of different individuals under different price structures, and we deduce from this some kind of aggregate or average preference function. If the preference functions of different individuals are not widely dissimilar, there is some justification for this procedure.

Just as preferences, or the valuation process, is described by economists as a utility function, so technology is defined by a production function. The forms of these two functions, in fact, are highly similar, in fact virtually identical. A production function relates physical inputs of some kind to physical outputs of some kind. In this case, the field consists of all possible or relevant combinations of inputs and the function describes the quantities of outputs which are associated with each combination of inputs. It tells us, for instance, that the quantity x of labor and y of land we will get z of potatoes. In the case of the production function, we can frequently assume not merely ordinal numbering of the product but cardinal numbering. In the case of the utility function, all we know is that of two combinations of inputs, one gives more utility than the other if it is preferred. In the case of the production function, we can usually measure the product directly so that we know not only that one combination of inputs gives more potatoes than another, but we know how much more, and we know, indeed, how great a quantity of potatoes is given in each combination of inputs. Oddly enough for the purposes of price theory, this richness of information about the production function is unnecessary, and all we really need to know to determine the equilibrium price structure is whether any given combination of inputs gives us more, less, or an equal amount of product than another.

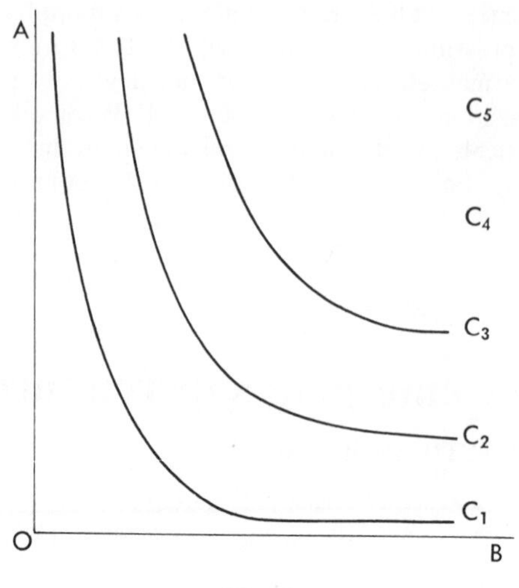

Fig. 1

Figure 1, which is very familiar to economists, illustrates the two concepts. Here we suppose two variables, say inputs in the state of the system, measured along OA and OB. We then postulate a function in the third dimension, of which the curves C_1, C_2, etc. are contours. These are called isoquants in the case of the production function, in which, shall we say, OA measures the quantity of labor, OB the quantity of land, and all combinations on one of the isoquants represents a given quantity of product. In the case of the preference function these are called indifference curves, and represent all combinations of two inputs to which the decision maker is indifferent, that is, which have the same utility. Utility can be thought of as the product of the decision making process, much as potatoes are the product of a production function; and the form of the two functions is likely to be very similar. In the case of the utility function, however, the indifference curves are given merely ordinal numbers so that we know which of any two indifference curves represents the higher utility and is the more preferred. In the case of the production function, the isoquants can be given a cardinal number representing the actual quantity of product.

In elementary economic theory it is generally assumed that both the utility or preference functions on the one hand, and the production functions on the other, are given factors in the social situation. As a first step in the process of analysis, this is quite legitimate, for it is important to deduce the consequences of any given set of preferences and technologies. The moment we try to make the system dynamic, however, it becomes very clear that neither the preferences nor the technologies are given, for they are both derived from a learning process which itself is dependent on the very dynamics of society which we are investigating. This proposition that both values, that is, preference functions, and technologies or production functions are learned is the key to any dynamic theory of society, though unfortunately we know far too little about the learning processes involved. Values and technologies, preference functions and production functions, interact to produce a price system, a system of exchange opportunities, not only in the narrow sense of a system of commodities with which economics usually concerns itself, but in the large sense of the whole social system of terms of trade, that is, the totality of what we give up for what we get. The concept can even be broadened to include what I have elsewhere called the grants economy and the integrative system, by which we give and receive unilateral transfers; for these, too, depend on preference functions and identifications which must be learned.

Because of the greater richness of information which seems to be available at the level of the production functions, the learning process can perhaps be perceived more easily there. Even if we look at a technology as relatively simple as that of peasant agriculture, it is clear that the whole process by which inputs are transformed into outputs is one that must exist in the mind of the producer before any production process will be set in motion. We will not plant seeds unless we have some image in our mind of a process of production by which certain activities of plowing, planting, weeding, fertilizing, and so on will eventually produce a harvest. Men lived on top of rich soil long before they ever thought of devoting it to agriculture, and unless there is an image of a whole productive process in their minds, the sequence of steps required for the process will not be carried out. The same simple principle is true of the engineers and designer; it develops as a set of detailed blueprints and information which organizes an assembly line, and its production requires an enormous amount of communication of specialized knowledge. All human artifacts, indeed all capital, can be regarded as human knowledge imposed on the material world. All processes of production originate in the minds of men and have to be maintained in the minds of men if these processes themselves are to continue. Even the perpetuation of the simplest productive process requires the transmission of an elaborate body of knowledge from one generation to the next, for all human knowledge is lost every generation by the sheer processes of death. Learning is not something which can be done once and for all; it is something which must be repeated to the last detail in every generation, if existing processes are even to continue.

Production functions are not merely transmitted from one generation to the next in an educational process, for what might be described as a net learning process goes on which actually improves them in a developing society. What we call economic development, indeed, is largely a learning process by which improved production processes are learned. Two rather distinct processes are involved here: first, the innovative process by which a new image of a production process is created in the mind of someone which existed in no other mind previously; and second, the educative process by which the image of a production process in one mind is transmitted to another. Both of these are necessary if there is to be development. If there are no innovative processes and education is successful, the knowledge of each generation will be transmitted unimpaired to the next. We should not overlook the possibility of degenerative processes, in which the knowledge of one generation is impaired in transmission to the next, and in which therefore technology declines and productive processes become less productive. The rate of technological development depends almost entirely on the amount of resources which a society devotes to the innovative process and to the educative process. Of these, the innovative process is the most mysterious. We do not really understand the sources of human creativity. Many societies have existed and continue to exist in which all attention is concentrated on the process of transmitting unimpaired the images of one generation into its successor. A good example of this would be the traditional Indian village, or shall we say the Amish society of the United States. The object of the educative process here is to produce children who are exact replicas, in their images, of the parents. We can be pretty sure that innovation will not be carried on unless it is rewarded, and in traditional societies, where the innovator is looked upon with suspicion or even horror as one who violates the ancient dignities and destroys the sacred patterns of the society, innovation is not likely to be successful.

It is a proposition for which there is a good deal of historical evidence that the innovator is likely to be one who is in some sense a "refugee," that is, who is in a degree an alien in the society and yet who has a role and a status that can be accepted. The refugee in the literal sense, that is, one who has been driven out of his previous home and who has sought refuge in another society, is no longer bound by the traditional ways of doing things because his traditional environment is no longer around him. On the other hand, because he is an alien in the society to which he goes, peculiarities of behavior are tolerated in a way that the native does not enjoy. It is not surprising, therefore, that in India and Pakistan it is the refugees who have been vigorous entrepreneurs. It is not surprising also to see the important innovative role that groups like the Jews or the Parsees, the Syrian and Lebanese traders, the Chinese outside China, and other displaced peoples have played in the whole process for world development. There may also be "internal refugees" as well as external ones, the noncomformists like the Quakers and Methodists in England, who played a disproportionate role in the first Industrial Revolution of the eighteenth century, the Samurai, especially the Ronin or masterless Samurai in nineteenth century Japan, the Calvinists in Europe, and so on. These are people who might be described as internal aliens, who are in some sense alienated from the established patterns of the societies in which they live, but who nevertheless have a recognized status as nonconformists and dissenters, which gives them, as it were, license to innovate.

Innovation, of course, is useless unless it is supplemental with a fairly large investment in the educative process. There is no point in having innovations unless they can be imitated. This educative process, of course, is by no means confined to formal education, although as technologies become more complex the role of formal education in transmitting them becomes more important. Even in complex societies like the United States, however, a great deal of the educative process by which innovations are transmitted through the population is quite informal. Transmission takes place by word of mouth or simple observation, through advertising and commercial propaganda, in face to face groups, and so on. The rate of transmission throughout a society of successful innovations depends, of course, to a considerable extent on the value system in the society, and particularly its willingness to innovate. A society which regards all old things as good and new things as bad will be unlikely to innovate in the first place and even if there are innovations, they will take a long time to establish themselves. A society which has the reverse value system, in which new things are regarded as good simply because they are new and old things as bad just because they are old, will put a high value on innovation and innovations will spread rapidly.

The last observation illustrates a principle of the utmost importance, that values and technologies constantly interact on each other in the dynamic processes of society because both are created and transmitted by a common learning process. The learning of values, that is, of preference functions, is less obvious, perhaps, than the learning of technologies; nevertheless, all societies devote a noticeable amount of resources to the process of the transmission of preference functions from one generation to the next, and societies differ enormously in their tolerance of innovation in preference functions. It is clear that there is a certain genetic base for preference functions, but in the case of man, this represents a very small proportion of the total. In the insects and even the birds, the preference functions seem to be generated almost wholly by genetic processes, that is, the genes or genetic structure contains an information code which builds certain preferences into the phenotype which it creates. An oriole has a strong *preference* for building oriole nests, but this preference is not learned from its parents, it is built into the bird by its genes. The same seems to be true of insects, although in the highest insects like the bees there does seem to be a certain process of communication and education. For the most part, however, even the bee does not have to learn how to be a bee—it simply *knows* how to be a bee. As we move towards the mammals and still more towards the primates, the proportion of the value structure which is learned increases. A kitten learns in part how to be a cat from its mother, not from its genes. Monkeys, it would seem, have to learn even such things as sexual behavior from their parents. As we move to man, instincts, that is, the value system which is built in by the genetic system, shrink to a very small proportion of the total. We do seem to have a genetic value system at birth which includes such things as high preference for milk, warmth, and stimulation and low preferences for loud noises, falling, and hunger; but on this slim genetic base we finally achieve preferences for transubstantiation or atheism, surplus value or free private enterprise, oysters, raw fish, olives, alcohol, chastity, and self-immolation. Genetics provides only the vaguest of drives. It seems fair to say today that there are no instincts in man in any detailed sense of the term, and that practically the whole of his value structure is learned from parents, from teachers, from his peers, from the mass media, and from information inputs of all kinds which pour into him both from outside and from within and continue perhaps even in sleep. From the moment of birth, we are the recipients of enormous inputs of information, out of which we gradually build our image of the world in regard to space, time, causality, the future, and values. Our preference functions are not innate; they are a product of our total information input, operating, perhaps, within certain guidelines laid down by our genetic inheritance. The old problem of the relation between heredity and environment has never been solved, it has simply been laid aside because we do not know how to answer it at present. That genetics imposes certain predispositions is very plausible, but it is clear also that in the mass, whatever individual peculiarities may be due to individual differences in genetic structure tend to cancel themselves out, and that the value systems of a culture are transmitted in the processes of that culture by the information inputs which the culture generates. There is nothing, for instance, in the genetic composition of a Japanese American that prevents him from becoming 100 percent American in culture. If he is only 99 percent American, it is because certain physical difference affect a little the way he is treated and the information inputs which he receives. Apart from this slight difference, however, the Nisei will learn to like coffee and eggs and bacon for breakfast as over against the rice, raw egg, and soup of his Japanese cousin, who may be genetically identical.

At this point I believe I can detect a subterranean rumble from the moral philosophers, who are likely to object to what seems to be my identification of values with valuation, and of valuation with preference and choice. Surely, some of them will argue, the choice between good and evil, right and wrong, is qualitatively different from the choice between bacon or sausage for breakfast; and still more, they may argue, concepts of utility or preference functions cannot account for problems of freedom, justice, and still less for mercy, pity, peace, and love. I am not altogether unsympathetic to their indignation against what must seem like economics' imperialism, that is, the attempt on the part of economics to take over not only all the other social sciences but moral philosophy as well. Nevertheless, I am prepared to defend my identification of the problem of values with the problem of preference, though I will gladly concede to the moral philosopher that this can operate on a number of different levels. Thus, while economics tends to assume a preference function, the moral philosopher raises the question of the choice among preference functions themselves. What I think this means in terms of the dynamic process of society is that the learning processes by which we learn our preference functions are not simply arbitrary, nor are they purely relative and culture-bound, but that there are certain selective processes which are in a real sense universal. It seems to be a plain fact of observation that cultural and moral relativism of a pure sort tends to break down when it is pressed too far. It is all very well for the relativist to observe that it is very interesting that some people eat their grandmothers and some do not, but that no one should pass any moral judgment on this interesting difference of behavior. When, however, somebody proposes to eat *him,* there is some tendency for relativism to break down. What we have to recognize here is that in the processes by which we learn our preferences, there are certain information inputs and certain sources of information which are peculiarly salient and effective in the formation of preferences. Preferences are always learned from a reference group, as the social psychologists call it. This may be, in the first instance, the family group with which the individual identifies early, even though the identification, as psychoanalysts have pointed out, is often ambiguous. All societies have produced religious and educational institutions which also are salient in establishing preferences, but these preferences are subject to innovation just as the production function is subject to innovation. The Aztecs had strong preferences for human sacrifice and rather messy religious rituals. Under the joint challenge of the Conquistadores and the Jesuits, Spanish baroque churches celebrating a symbolic sacrifice were substituted in the preference system for the human victims. The relative role of the Conquistadores and the Jesuits in this is almost as difficult to estimate as the role of heredity and environment. We can be pretty sure, however, that one without the other would not have been very effective. In other words, it was a complex mutation-selection process by which old values are challenged by new values, old technologies by new technologies, and indeed these two processes are not very different. The images in the minds of the living, both of preference functions and of production functions, have to be transmitted from generation to generation. In the course of transmission the images are changed, and they both may be changed by innovation or mutations, some of which prosper and spread through the minds of the living, and some which do not.

The problem of the complex interaction between our preferences and the technologies is rendered particularly acute in the present epoch by the fact that we have been going through enormous changes in both technology and values, that is, in our images of what inputs produce what outputs on the one hand and our images also of what states of the system are preferable to others on the other hand. The Great Transition, as I have called it elsewhere, which began with the rise of modern science around the end of the sixteenth century and which continues with ever accelerating change up to today, is a change in the state of man as great, if not greater, as in the change from the neolithic village to urban civilization; and I have described as the change from civilization to post-civilization, civilization being the state of man typified, shall we say, by the Roman Empire or by the poor countries such as Indonesia today. (Certainly if the Roman Empire were around today we would regard it as an extremely poor country and would be giving it aid on a large scale.) This great transition was preceded by a long period of accelerating folk science and folk technology, a period also of slow but continuous change in values and preferences. In the West we can date this preparatory period roughly from the fall of Rome and the rise of the great monastic orders; in China it can be dated from a little earlier, perhaps the beginning of the Han Empire. The so-called Industrial Revolution of the eighteenth century in England and Western Europe represented essentially the culmination of this process of folk technology. It owed very little to science, even though science, as it were, was developing underneath it. The steam engine and the spinning jenny are not in essence very different from the printing press and the clock. The theory of the steam engine, for instance, which is thermodynamics, did not develop until the early nineteenth century, and the steam engine itself clearly owed nothing to it. In the latter half of the nineteenth century, however, the science-based industries began, which could not have developed at all without the previous development of a certain branch of science. Of these, the chemical industry was the first, the electrical industry the second, the biological industry the third, and the nuclear industry the somewhat premature fourth. Today in the developed countries, more than half the economy is producing products and using methods which would have been virtually inconceivable a hundred years ago. Never in the whole course of human history has there been a change as rapid at that which has taken place in the last hundred years.

In this great transition there has been a constant interplay between changing technologies and changing values, both of these being an integral part of the larger process of change in what Teilhard de Chardin calls the "noosphere" or the totality of images of the world in the minds of the living. The interaction between values and technologies is so complex that it is quite impossible to say which precedes the other. It is a hen and egg problem in *n* dimensions. It seems fairly certain, for instance, that there were changes in values, that is, preference systems, which were a necessary prerequisite for the rise of science, in the direction of introducing higher preferences for change, for the authority of nature rather than the authority of sacred books and ancient writers. These changes in values, however, were not unconnected with certain preceding changes in technologies, for instance the rise of the money economy, development of accounting, and the subsequent opportunities for more rational behavior in the light of better information. A strong case can be made out, indeed, that in the origins of science it was the machine that preceded the scientific or mechanical image of the world. The clock, for instance, preceded the Copernican-Newtonian image of the solar system as the great clock; the water pump preceded the discovery of the circulation of the blood, just as the steam engine preceded thermodynamics. The development of the more elaborate folk technology imperceptibly changes the values of the society which used it, and by giving man a little power over the material world perhaps increased his desire for knowledge about it.

On the other hand, there are also changes which occur fairly spontaneously within the system of values and preferences. The Max Weber thesis of the impact of the Reformation and especially of Calvinism on economic activity and technology is well known. Once the authority of the Pope had been challenged and a high value has been placed in the Protestant countries on successful dissent, the legitimation of dissent in general is a fairly easy step. The legitimation of dissent, as we have seen, is an essential element in the innovative process; therefore innovation must be legitimated if it is to be rapid. It is not surprising, therefore, that Luther's initial break with Rome was accompanied by the more radical Anabaptists and followed by Calvin, George Fox, John Wesley, and the more radical religious reformers. It is not surprising either that it was out of this radical nonconformity that the great changes in economic institutions and technology of the seventeenth and eighteenth century largely developed. In this case, the development of a religious value system which stressed the immediacy of personal experience as over against the authority of a pope or even a king, which stressed veracity as a high virtue and which also stressed simplicity of life and thirftiness and the sacredness of the material world, should provide the groundwork for enormous changes in knowledge and technology.

These considerations perhaps throw some light on the puzzling question of why the breakthrough into science and the technology that is related to it and based upon it took place in Europe rather than in China. The great work of Joseph Needham on Chinese technology has opened our minds in the West to the fact that at least up to 1600, the folk technology of China was considerably ahead of that in the West, and that indeed many of the essential developments in the western part of the old world were not only anticipated in China, sometimes by hundreds of years, but in many cases actually derived from China. This seems to be true, for instance, of such things as the stirrup, on which so much of the medieval aristocratic culture was based, which reached Europe from China or Tibet by about the eighth century. Such essential technologies as that of clockwork and printing were likewise discovered in China long before they were discovered in Europe, although here the actual connection is more obscure and the European discoveries may be independent. The exact relation of innovation to imitation remains one of the mysteries of human history.

In spite of the fact, however, that China had been the superior of the West for so long and unquestionably was still the superior of the West, shall we say in 1600—from 1600 on Europe takes an enormous spurt forward under the impact of the rise of modern science, whereas China proceeds in the old slow pace of folk technology. It may be indeed that China was too successful (one very fundamental principle of social science is that nothing fails like success). Perhaps the very disjointed and disintegrated structure of Europe, with its many centers of power, its religious and national divisions, its separation of ecclesiastical from political power, and at the same time its active network of trade and communication (the result partly of its long coastline and waterborne traffic), made that fraction of difference which carried Europe over the watershed into science. China, at any rate did not make this transition. This fact has dominated the history of the last 300 years.

As we compare Europe with China, the subtle and constant interaction between values and technology again becomes apparent. In the fifteenth and sixteenth centuries, for instance, as the Ming voyages indicate, China clearly had the technology to explore the world and to expand its culture, and the fact that it did not do so is almost certainly due to the value systems of its rulers, which favored withdrawal, stability, and staying at home. By contrast, the Spanish and the Portuguese, in what might be described as the last great burst of the Crusades, discovered and colonized America to the West, and Eastward as far as the Philippines and Japan. This was indeed the moment of globalization, the moment in human history at which the earth ceased to be a great plain and became a sphere. All this, indeed, was before the rise of science, but the high values and rewards given to adventure, exploration, and discovery from, say, 1450 to 1600 in the West, unquestionably helped to create the value system which later gave rise to Galileo and his successors. The hens of value produced the eggs of technology; the eggs of technology the hens of value, in an ever-increasing, ever-expanding process of increasing complexity.

By way of conclusion, or perhaps an epilogue, let us take a brief, speculative glance at the implication for future values and perhaps even for future technology of the great transition through which we are now going. Its most obvious and immediate impact is the separation out in the world of two cultural systems, the superculture on the one hand and traditional cultures on the other. It is hardly too much to say that all the major problems of the world today revolve around the tension between these two cultural systems. The superculture is the culture of airports, throughways, skyscrapers, hybrid corn and artificial fertilizers, birth control, and universities. It is worldwide in its scope; in a very real sense all airports are the same airport, all universities the same university. It even has a world language, technical English, and a common ideology, science.

Side by side with the superculture, and interpenetrating it at many points are the various folk cultures, national, religious, ethnic, linguistic, and so on. The tensions between the superculture and traditional cultures are felt at a great many points. We see it, for instance, in the international system, where the superculture has given the traditional cultures of the national states appalling powers of destruction which are threatening the whole future of man. We see it in race relations, where the superculture moves towards uniformity, the absence of discrimination, and differentiation by roles rather than by race or class or other ascribed category. We see it in education, where formal education tends increasingly to become the agent of transmission of the superculture, leaving the transmission of folk culture to the family, the peer group, and more informal organizations. We see it in religion, where the superculture tends towards the secular and traditional culture preserves the sacred.

At a great many points, these tensions between the superculture and traditional cultures produce challenges to traditional values and even disintegration of these values. Family loyalties are replaced by loyalties to larger and more abstract entities; national loyalties are eroded by inconsistencies between the national state and the world order which the supercultural requires; religious loyalties are eroded by new views of man and the universe; political loyalties are eroded by new images of the social system arising out of social sciences.

The picture, however, is not merely one of constant retreat and erosion of traditional values in the fact of the superculture. There is also the transformation and regeneration of traditional values under the impact of the superculture. A strong case can be made, indeed, for the proposition that the superculture itself does not generate the values and preferences which will support it, perhaps because of the very fact that it has to be transmitted through channels of formal education of a more or less authoritarian kind. Traditional culture, on the other hand, is transmitted through the family, the peer group, and intimate relations which are capable of creating much more intense value commitments and stronger preferences than the more abstract and cold-blooded relationships of the superculture. We do not feel towards the airport or the chemistry textbook the degree of emotional involvement that we have with the family, the nation, or the little brown church in the wildwood.

As an integrative system, the superculture is really very weak. Fellow scientists kill each other in national wars almost as enthusiastically as co-religionists. Scientists have not raised money very much to help other scientists, and while they have a certain sense of occupational community, this does not usually go much beyond the rather tenuous bond of the professional association. People die for their countries, even for their faith, but very few people have died for biochemistry. Up to now at any rate, therefore, the ethical values of mankind on the whole have arisen out of the traditional cultures rather than out of the superculture. There is something to be said for the proposition, indeed, that it is only countries which have strong traditional cultures and as a result strong ethical systems which are able to create or adapt to the superculture, and that where the traditional culture is weak, the society will have great difficulty in making adjustments to the superculture. Japan is perhaps one of the best examples of a society in which the traditional culture is very strong and in which it generates principles of ethical judgment and behavior which are friendly to the superculture and which permit it to develop at an enormous rate. In a greater or lesser degree, this is true of all the successfully developing countries. It is by contrast those countries in which the traditional culture cannot adapt itself and produces values which are unfriendly to the superculture that development is most difficult. The contrast between Japan and India in this respect is most instructive. In some cases the traditional culture has proved so incapable of adaptation to the impinging superculture that it has been virtually destroyed. This seems to be the case in China, and to a smaller extent in the other socialist countries.

From the point of view of this paper, communism is a curious phenomenon which represents on the one hand a vehicle for bringing traditional societies into the superculture and which expresses many of the values of the superculture, such as education, equality of status for women, the abolition of castes, and so on. On the other hand, ideologically it represents what is really a prescientific view of society, and its results in a curious fixation of the socialist countries on the attitudes and ideologies of the nineteenth century. Ideologically it is a kind of folk science lying somewhere between an unsophisticated folk image of society on the one hand and empirically based scientific concepts on the other. At certain points, therefore, it may assist, and at other points it may hinder the transition and adaptation of a society to the superculture.

The inability of the superculture to produce adequate values of its own and the adaptability of certain aspects of traditional culture is reflected strongly in the continuing strength of the religious institution in the developed societies. This is nowhere more striking than in the United States, which is at the same time perhaps the furthest advanced towards the superculture and yet is also a society whose history has been characterized by the rise of the numerical strength and power of the churches. What we seem to face in the future, therefore, is a very complex set of mutual adjustments, in which an adapted traditional culture transmitted in the family, the peer group, and the church will create ethical values and preferences which are consistent with the world superculture. If the superculture simply destroys the traditional culture in which it is embedded, it may easily destroy itself. On the other hand, if the traditional culture does not adapt to the superculture, it too may destroy itself. This is a precarious balance, and not all societies may achieve it. The costs of a failure to achieve it, however, are very high, and there is great need, therefore, for widespread self-consciousness about the nature of the problem, and a willingness to put resources into solving it.

12

The "small is beautiful" debate

Technology with a human face

E. F. SCHUMACHER

▪ E. F. Schumacher *was, until his recent death, one of the leading exponents of "intermediate" or "appropriate" technology. The following essay is Chapter 5 from his important book* Small is Beautiful.

Schumacher calls here for a technology that reflects not only the organic world with which it is designed to interact, but the human dimension as well. He argues that technology and economics have too long been in the hands of "experts" who serve only their own narrow interests, especially the desire for growth and a wider domain of influence.

The modern world has been shaped by its metaphysics, which has shaped its education, which in turn has brought forth its science and technology. So, without going back to metaphysics and education, we can say that the modern world has been shaped by technology. It tumbles from crisis to crisis; on all sides there are prophecies of disaster and, indeed, visible signs of breakdown.

If that which has been shaped by technology, and continues to be so shaped, looks sick, it might be wise to have a look at technology itself. If technology is felt to be becoming more and more inhuman, we might do well to consider whether it is possible to have something better—a technology with a human face.

Strange to say, technology, although of course the product of man, tends to develop by its own laws and principles, and these are very different from those of human nature or of living nature in general. Nature always, so to speak, knows where and when to stop. Greater even than the mystery of natural growth is the mystery of the natural cessation of growth. There is measure in all natural things—in their size, speed, or violence. As a result, the system of nature, of which man is a part, tends to be self-balancing, self-adjusting, self-cleansing. Not so with technology, or perhaps I should say: not so with man dominated by technology and specialisation. Technology recognises no self-limiting principle—in terms, for instance, of size, speed, or violence. It therefore does not possess the virtues of being self-balancing, self-adjusting, and self-cleansing. In the subtle system of nature, technology, and in particular the super-technology of the modern world, acts like a foreign body, and there are now numerous signs of rejection.

Suddenly, if not altogether surprisingly, the modern world, shaped by modern technology, finds itself involved in three crises simultaneously. First, human nature revolts against inhuman technological, organisational, and political patterns, which it experiences as suffocating and debilitating; second, the living environment which supports human life aches and groans and gives signs of partial breakdown; and, third, it is clear to anyone fully knowledgeable in the subject matter that the inroads being made into the world's non-renewable resources, particularly those of fossil fuels, are such that serious bottlenecks and virtual exhaustion loom ahead in the quite foreseeable future.

☐ From E. F. Schumacher, *Small Is Beautiful: Economics as if People Mattered* (New York: Harper and Row, London: Blond and Briggs, 1973), pp. 138-151.

Any one of these three crises or illnesses can turn out to be deadly. I do not know which of the three is the most likely to be the direct cause of collapse. What is quite clear is that a way of life that bases itself on materialism, *i.e.* on permanent, limitless expansionism in a finite environment, cannot last long, and that its life expectation is the shorter the more successfully it pursues its expansionist objectives.

If we ask where the tempestuous developments of world industry during the last quarter-century have taken us, the answer is somewhat discouraging. Everywhere the problems seem to be growing faster than the solutions. This seems to apply to the rich countries just as much as to the poor. There is nothing in the experience of the last twenty-five years to suggest that modern technology, as we know it, can really help us to alleviate world poverty, not to mention the problem of unemployment which already reaches levels like thirty per cent in many so-called developing countries, and now threatens to become endemic also in many of the rich countries. In any case, the apparent yet illusory successes of the last twenty-five years cannot be repeated: the threefold crisis of which I have spoken will see to that. So we had better face the question of technology—what does it do and what should it do? Can we develop a technology which really helps us to solve our problems—a technology with a human face?

The primary task of technology, it would seem, is to lighten the burden of work man has to carry in order to stay alive and develop his potential. It is easy enough to see that technology fulfils this purpose when we watch any particular piece of machinery at work—a computer, for instance, can do in seconds what it would take clerks or even mathematicians a very long time, if they can do it at all. It is more difficult to convince oneself of the truth of this simple proposition when one looks at whole societies. When I first began to travel the world, visiting rich and poor countries alike, I was tempted to formulate the first law of economics as follows: "The amount of real leisure a society enjoys tend to be in inverse proportion to the amount of labour-saving machinery it employs." It might be a good idea for the professors of economics to put this proposition into their examination papers and ask their pupils to discuss it. However that may be, the evidence is very strong indeed. If you go from easy-going England to, say, Germany or the United States, you find that people there live under much more strain than here. And if you move to a country like Burma, which is very near to the bottom of the league table of industrial progress, you find that people have an enormous amount of leisure really to enjoy themselves. Of course, as there is so much less labour-saving machinery to help them, they "accomplish" much less than we do; but that is a different point. The fact remains that the burden of living rests much more lightly on their shoulders than on ours.

The question of what technology actually does for us is therefore worthy of investigation. It obviously greatly reduces some kinds of work while it increases other kinds. The type of work which modern technology is most successful in reducing or even eliminating is skillful, productive work of human hands, in touch with real materials of one kind or another. In an advanced industrial society, such work has become exceedingly rare, and to make a decent living by doing such work has become virtually impossible. A great part of the modern neurosis may be due to this very fact; for the human being, defined by Thomas Aquinas as a being with brains and hands, enjoys nothing more than to be creatively, usefully, productively engaged with both his hands and his brains. Today, a person has to be wealthy to be able to enjoy this simple thing, this very great luxury: he has to be able to afford space and good tools; he has to be lucky enough to find a good teacher and plenty of free time to learn and practice. He really has to be rich enough not to need a job; for the number of jobs that would be satisfactory in these respects is very small indeed.

The extent to which modern technology has taken over the work of human hands may be illustrated as follows. We may ask how much of "total social time"—that is to say, the time all of us have together, twenty-four hours a day each—is actually engaged in real production. Rather less than one-half of the total population of this country is, as they say, gainfully occupied, and about one-third of these are actual producers in agriculture, mining, construction, and industry. I do mean *actual producers,* not people who tell other people what to do, or account for the past, or plan for the future, or distribute what other people have produced. In other words, rather less than one-sixth of the total population is engaged in actual production; on average, each of them supports five others beside himself, of which two are gainfully employed on things other than real production and three are not gainfully employed. Now, a fully employed person, allowing for holidays, sickness, and other absence, spends about one-fifth of his total time on his job. It follows that the proportion of "total social time" spent on actual production—in the narrow sense in which I am using the term—is, roughly, one-fifth of one-third of one-half, *i.e.*, 3½ per cent. The other 96½ per cent of "total social time" is spent in other ways, including sleeping, eating, watching television, doing jobs that are not *directly* productive, or just killing time more or less humanely.

Although this bit of figuring work need not be taken to literally, it quite adequately serves to show what technology has enabled us to do: namely, to reduce the amount of time actually spent on production in its most elementary sense to such a tiny percentage of total social time that it pales into insignificance, that it carries no real weight, let alone prestige. When you look at industrial society in this way, you cannot be surprised to find that prestige is carried by those who help fill the other 96½ per cent of total social time, primarily the entertainers but also the executors of Parkinson's Law. In fact, one might put the following proposition to students of sociology: "The prestige carried by people in modern industrial society varies in inverse proportion to their closeness to actual production."

There is a further reason for this. The process of confining productive time to 3½ per cent of total social time has had the inevitable effect of taking all normal human pleasure and satisfaction out of the time spent on this work. Virtually all real production has been turned into an inhuman chore which does not enrich a man but empties him. "From the factory," it has been said, "dead matter goes out improved, whereas men there are corrupted and degraded."

We may say, therefore, that modern technology has deprived man of the kind of work that he enjoys most, creative, useful work with hands and brains, and given him plenty of work of a fragmented kind, most of which he does not enjoy at all. It has multiplied the number of people who are exceedingly busy doing kinds of work which, if it is productive at all, is so only in an indirect or "roundabout" way, and much of which would not be necessary at all if technology were rather less modern. Karl Marx appears to have foreseen much of this when he wrote: "They want production to be limited to useful things, but they forget that the production of too many useful things results in too many useless people", to which we might add: particularly when the processes of production are joyless and boring. All this confirms our suspicion that modern technology, the way it has developed, is developing, and promises further to develop, is showing an increasingly inhuman face, and that we might do well to take stock and reconsider our goals.

Taking stock, we can say that we possess a vast accumulation of new knowledge, splendid scientific techniques to increase it further, and immense experience in its application. All this is truth of a kind. This truthful knowledge, as such, does *not* commit us to a technology of giantism, supersonic speed, violence, and the destruction of human work-enjoyment. The use we have made of our knowledge is only one of its possible uses and, as is now becoming ever more apparent, often an unwise and destructive use.

As I have shown, directly productive time in our society has already been reduced to about 3½ per cent of total social time, and the whole drift of modern technological development is to reduce it further, asymptotically* to zero. Imagine we set outselves a goal in the opposite direction—to increase it sixfold, to about twenty per cent, so that twenty per cent of total social time would be used for actually producing things, employing hands and brains and, naturally, excellent tools. An incredible thought! Even children would be allowed to make themselves useful, even old people. At one-sixth of present-day productivity, we should be producing as much as at present. There would be six times as much time for any piece of work we chose to undertake—enough to make a really good job of it, to enjoy oneself, to produce real quality, even to make things beautiful. Think of the therapeutic value of real work; think of its educational value. No one would then want to raise the school-leaving age or to lower the retirement age, so as to keep people off the labour market. Everybody would be welcome to lend a hand. Everybody would be admitted to what is now the rarest privilege, the opportunity of working usefully, creatively, with his own hands and brains, in his own time, at his own

*Asymptote: A mathematical line continually approaching some curve but never meeting it within a finite distance.

pace—and with excellent tools. Would this mean an enormous extension of working hours? No, people who work in this way do not know the difference between work and leisure. Unless they sleep or eat or occasionally choose to do nothing at all, they are always agreeably, productively engaged. Many of the "on-cost jobs" would simply disappear; I leave it to the reader's imagination to identify them. There would be little need for mindless entertainment or other drugs, and unquestionably much less illness.

Now, it might be said that this is a romantic, a utopian, vision. True enough. What we have today, in modern industrial society, is not romantic and certainly not utopian, as we have it right here. But it is in very deep trouble and holds no promise of survival. We jolly well have to have the courage to dream if we want to survive and give our children a chance of survival. The threefold crisis of which I have spoken will not go away if we simply carry on as before. It will become worse and end in disaster, until or unless we develop a new life-style which is compatible with the real needs of human nature, with the health of living nature around us, and with the resource endowment of the world.

Now, this is indeed a tall order, not because a new life-style to meet these critical requirements and facts is impossible to conceive, but because the present consumer society is like a drug addict who, no matter how miserable he may feel, finds it extremely difficult to get off the hook. The problem children of the world—from this point of view and in spite of many other considerations that could be adduced—are the rich societies and not the poor.

It is almost like a providential blessing that we, the rich countries, have found it in our heart at least to consider the Third World and to try to mitigate its poverty. In spite of the mixture of motives and the persistence of exploitative practices, I think that this fairly recent development in the outlook of the rich is an honourable one. And it could save us; for the poverty of the poor makes it in any case impossible for them successfully to adopt our technology. Of course, they often try to do so, and then have to bear the most dire consequences in terms of mass unemployment, mass migration into cities, rural decay, and intolerable social tensions. They need, in fact, the very thing I am talking about, which we also need: a *different* kind of technology, a technology with a human face, which, instead of making human hands and brains redundant, helps them to become far more productive than they have ever been before.

As Gandhi said, the poor of the world cannot be helped by mass production, only by production by the masses. The system of *mass production,* based on sophisticated, highly capital-intensive, high energy-input dependent, and human labour-saving technology, presupposes that you are already rich, for a great deal of capital investment is needed to establish one single workplace. The system of *production by the masses* mobilises the priceless resources which are possessed by all human beings, their clever brains and skilful hands, *and supports them with first-class tools.* The technology of *mass production* is inherently violent, ecologically damaging, self-defeating in terms of non-renewable resources, and stultifying for the human person. The technology of *production by the masses,* making use of the best of modern knowledge and experience, is conducive to decentralisation, compatible with the laws of ecology, gentle in its use of scarce resources, and designed to serve the human person instead of making him the servant of machines. I have named it *intermediate technology* to signify that it is vastly superior to the primitive technology of bygone ages but at the same time much simpler, cheaper, and freer than the super-technology of the rich. One can also call it self-help technology, or democratic or people's technology—a technology to which everybody can gain admittance and which is not reserved to those already rich and powerful. It will be more fully discussed in later chapters.

Although we are in possession of all requisite knowledge, it still requires a systematic, creative effort to bring this technology into active existence and make it generally visible and available. It is my experience that it is rather more difficult to recapture directness and simplicity than to advance in the direction of ever more sophistication and complexity. Any third-rate engineer or researcher can increase complexity; but it takes a certain flair of real insight to make things simple again. And this insight does not come easily to people who have allowed themselves to become alienated from real, productive work and from the self-balancing system of nature, which never fails to recognise measures and limitation. Any activity which fails to recognise a self-limiting principle is of the devil. In our work with the developing countries we are at least forced to recognise the limitations of poverty, and this work can therefore be a wholesome school for all of us in which, while genuinely trying to help others, we may also gain knowledge and experience how to help ourselves.

I think we can already see the conflict of attitudes which will decide our future. On the one side, I see the people who think they can cope with our threefold crisis by the methods current, only more so; I call them the people of the forward stampede. On the other side, there are people in search of a new life-style, who seek to return to certain basic truths about man and his world; I call them home-comers. Let us admit that the people of the forward stampede, like the devil, have all the best tunes or at least the most popular and familiar tunes. You cannot stand still, they say; standing still means going down; you must go forward; there is nothing wrong with modern technology except that it is as yet incomplete; let us complete it. Dr. Sicco Mansholt, one of the most prominent chiefs of the European Economic Community, may be quoted as a typical representative of this group. "More, further, quicker, richer," he says, "are the watchwords of present-day society." And he thinks we must help people to adapt "for there is no alternative." This is the authentic voice of the forward stampede, which talks in much the same tone as Dostoyevsky's Grand Inquisitor: "Why have you come to hinder us?" They point to the population explosion and to the possibilities of world hunger. Surely, we must take our flight forward and not be fainthearted. If people start protesting and revolting, we shall have to have more police and have them better equipped. If there is trouble with the environment, we shall need more stringent laws against pollution, and faster economic growth to pay for anti-pollution measures. If there are problems about natural resources, we shall turn to synthetics; if there are problems about fossil fuels, we shall move from slow reactors to fast breeders and from fission to fusion. There *are* no insoluble problems. The slogans of the people of the forward stampede burst into the newspaper headlines every day with the message, "a breakthrough a day keeps the crisis at bay."

And what about the other side? This is made up of people who are deeply convinced that technological development has taken a wrong turn and needs to be redirected. The term "home-comer" has, of course, a religious connotation. For it takes a good deal of courage to say "no" to the fashions and fascinations of the age and to question the presuppositions of a civilisation which appears destined to conquer the whole world; the requisite strength can be derived only from deep convictions. If it were derived from nothing more than fear of the future, it would be likely to disappear at the decisive moment. The genuine "home-comer" does not have the best tunes, but he has the most exalted text, nothing less than the Gospels. For him, there could not be a more concise statement of his situation, of *our* situation, than the parable of the prodigal son. Strange to say, the Sermon on the Mount gives pretty precise instructions on how to construct an outlook that could lead to an Economics of Survival.

—How blessed are those who know that they are poor:
the Kingdom of Heaven is theirs.
How blessed are the sorrowful;
they shall find consolation.
—How blessed are those of a gentle spirit;
they shall have the earth for their possession.
—How blessed are those who hunger and thirst to see right prevail;
they shall be satisfied;
—How blessed are the peacemakers;
God shall call them his sons.

It may seem daring to connect these beatitudes with matters of technology and economics. But may it not be that we are in trouble precisely because we have failed for so long to make this connection? It is not difficult to discern what these beatitudes may mean for us today:

—We are poor, not demigods.
—We have plenty to be sorrowful about, and are not emerging into a golden age.
—We need a gentle approach, a non-violent spirit, and small is beautiful.
—We must concern ourselves with justice and see right prevail.
—And all this, only this, can enable us to become peacemakers.

The home-comers base themselves upon a different picture of man from that which motivates the people of the forward stampede. It would be very superficial to say that the latter believe in "growth" while the former do not. In a sense, everybody believes in growth, and rightly so, because growth is an essential feature of life. The whole point, however, is to give to the idea of growth a qualitative determination; for there are always many things that ought to be growing and many things that ought to be diminishing.

Equally, it would be very superficial to say that the home-comers do not believe in progress, which also can be said to be an essential feature of all life. The whole point is to determine what constitutes progress. And the home-comers believe that the direction which modern technology has taken and is continuing to pursue—towards ever-greater size, ever-higher speeds, and ever-increased violence, in defiance of all laws of natural harmony—is the opposite of progress. Hence the call for taking stock and finding a new orientation. The stock-taking indicates that we are destroying our very basis of existence, and the reorientation is based on remembering what human life is really about.

In one way or another everybody will have to take sides in this great conflict. To "leave it to the experts" means to side with the people of the forward stampede. It is widely accepted that politics is too important a matter to be left to experts. Today, the main content of politics is economics, and the main content of economics is technology. If politics cannot be left to the experts, neither can economics and technology.

The case for hope rests on the fact that ordinary people are often able to take a wider view, and a more "humanistic" view, than is normally being taken by experts. The power of ordinary people, who today tend to feel utterly powerless, does not lie in starting new lines of action, but in placing their sympathy and support with minority groups which have already started. I shall give two examples, relevant to the subject here under discussion. One relates to agriculture, still the greatest single activity of man on earth, and the other relates to industrial technology.

Modern agriculture relies on applying to soil, plants, and animals ever-increasing quantities of chemical products, the long-term effect of which on soil fertility and health is subject to very grave doubts. People who raise such doubts are generally confronted with the assertion that the choice lies between "poison or hunger". There are highly successful farmers in many countries who obtain excellent yields without resort to such chemicals and without raising any doubts about long-term soil fertility and health. For the last twenty-five years, a private, voluntary organisation, the Soil Association, has been engaged in exploring the vital relationships between soil, plant, animal, and man; has undertaken and assisted relevant research; and has attempted to keep the public informed about developments in these fields. Neither the successful farmers nor the Soil Association have been able to attract official support or recognition. They have generally been dismissed as "the muck and mystery people", because they are obviously outside the mainstream of modern technological progress. Their methods bear the mark of non-violence and humility towards the infinitely subtle system of natural harmony, and this stands in opposite to the life-style of the modern world. But if we now realise that the modern life-style is putting us into mortal danger, we may find it in our hearts to support and even join these pioneers rather than to ignore or ridicule them.

On the industrial side, there is the Intermediate Technology Development Group. It is engaged in the systematic study on how to help people to help themselves. While its work is primarily concerned with giving technical assistance to the Third World, the results of its research are attracting increasing attention also from those who are concerned about the future of the rich societies. For they show that an intermediate technology, a technology with a human face, is in fact possible; that it is viable; and that it reintegrates the human being, with his skilful hands and creative brain, into the productive process. It serves *production by the masses* instead of *mass production*. Like the Soil Association, it is a private, voluntary organisation depending on public support.

I have no doubt that it is possible to give a new direction to technological development, a direction that shall lead it back to the real needs of man, and that also means: *to the actual size of man*. Man is small, and, therefore, small is beautiful. To go for giantism is to go for self-destruction. And what is the cost of a reorientation? We might remind ourselves that to calculate the cost of survival is perverse. No doubt, a price has to be paid for anything worth while: to redirect technology so that it serves man instead of destroying him requires primarily an effort of the imagination and an abandonment of fear.

Machines that shield us

JOHN LACHS

■ John Lachs *is Professor of Philosophy at Vanderbilt University. His numerous publications include works on American philosophy and contributions to the* Wall Street Journal.

Tools and machines shield us, suggests Professor Lachs, in ways which are detrimental to our freedom and our sense of ourselves. Far from tools being merely neutral instruments with which we do *things, such technological artifacts do things to us. They amplify our ability to act on the world, and they cushion the impact of the world on us. Yet this feature of our technological environment is not generally acknowledged or taken into account. We merely assume the continued existence of these mediating elements, and we proceed as if technology will be able to extricate us from every difficult situation, including those for which technology itself is responsible.*

Some believe that it is never right to use another human as a means. Mediation that is thoroughly institutionalized hides much of this use and a great deal of exploitation. Yet the more steadfast among those of strong moral persuasion may never feel right about mediation unless it is limited to the use of nonrational and preferably even inanimate beings.

Such restriction, of course, entails the loss of civilization. It cannot even be conceived apart from artifically abstract circumstances of the sort under which Robinson Crusoe is imagined to exist. The great classical ideal of perfection, of which God was supposed to be the prime or only instance, is that of solitary independence, of having neither desires nor needs one cannot satisfy alone. Those who think we ought to get along without mediation by our fellows—without society—are tacitly committed to this ideal: they think that each man is somehow a god.

To be sure, mediation in the widest sense does not require that the intermediary between the agent and his act be a person. The most rudimentary forms of mediation, of which there are two, are in fact not human. The first involves the interposition of tools, in a broadly inclusive sense of the term, between oneself and one's action. This is the subject of the current essay. In the second, our own actions serve as means or medium to our ends.

By calling these forms of mediation rudimentary I do not mean to imply that somehow they precede other forms or are more basic. My interest is not in some presumed sequence of development which, even if it ever occurred, is by now lost in the impenetrable haze that hides the history of early man. The likelihood is that every form of mediation is present in each society, though in each perhaps to a different extent. There is, of course, a growth in the complexity of societies, due partly at least to an increase in their size. But to arrange types of mediation in some easy-to-group sequence is not to reproduce their history, just as it is not to tell the story of a square to place it between a triangle and a pentagon.

My purpose is to see how many apparently diverse activities display the same pattern. I do this by attempting to bring disparate phenomena to the unity of a single idea. This must be done not only with caution but with a measure of rational skepticism. For in fitting facts to concepts there is always a bit of slippage and distortion. And, most important of all, there are alternative conceptualizations. An activity may appear as an autonomous fact on one view; on another, it may seem a truncated or extended form of other facts. In such matters a great deal depends on one's point of view. Depending on where we start, for example, the use of human beings to accomplish our ends may be viewed as a consequence or an extension of our use of tools, or our use of tools may be considered a rudimentary form of human mediation. The question of which is the right starting point or correct perspective is misleading. The task is not to find the right concept but to gain a useful one. And usefulness is at least in part a matter of our purposes and goals.

Those who think of mind or consciousness as the essential self may view the body as a sullen tool. I want to start from a less controversial position. Picture a man driving nails with a hammer. There is no doubt that each act of pounding is uniquely his. On the face of it, therefore, he is not at a distance from his deeds. He feels his aim and agency and hence has little difficulty in appropriat-

□ This essay appears for the first time in this volume.

ing every act. The easy exercise of his faculties, the quick succession of exertion and achievement may make his work a spontaneous delight.

It may seem, therefore, that tools do not hide the actions we perform by means of them, whereas human intermediaries normally do. And there appears to be an easy explanation for this: tools are not self-motivating agents. Wherever we find them in operation the intentions and activities of the persons using them are never far behind. In brief, we do not think that tools do anything *for us;* rather, we think that *we do* things with tools.

This view of the matter is, I think, false. It clearly has a point if we remain on the level of simple tools, such as the hammer, directly handled by a single person. But a huge digital computer is also a tool, and it can readily hide the nature of the acts it performs on my behalf. And tools can loom even larger than that and seem more possessed of independent life. The plant manager or vice-president in charge of production has a momentous presence interposed between himself and his acts if his tool is of the magnitude of an oil refinery, in which workers may be viewed as animate appendages of the machine.

Psychic distance of some measure is bound to arise in all these cases. But it is there even in the instance of the simple, directly manipulated tool. Imagine now that our friendly carpenter misses a nail and flattens his own finger. In a flash he discovers what it is like to be on the receiving end of his activity. To the reflective man this is not only a painful but a shocking realization. He simply did not know what force he was exerting in swinging his hammer. He may never have felt the effects of such force on a small, sensitive part of his body. He can suddenly see the hammer in his hand as a nightstick or a tool of medieval torture. The nails he hits might be human heads or breasts, or fingernails broken in their pink and trembling beds. In a rare flash of insight he might see how not suffering with the victim leads to cruelty. For though some atrocity is born of ill design, vastly more results of careless ignorance.

How can use of a simple tool in plain view hide important aspects of our action? The key notion we must accept is that being in plain view is simply not enough. Each action has consequences in all sense modalities. To have anywhere near an adequate firsthand knowledge of it, we must experience it through several of our senses. Eyes, then, are not sufficient to convey full-bodied knowledge of an act, simply because no single sense can be.

But there is more here than the general inadequacy of any one sense. For tools tend to take the place of our bodily organs in dealing with the world. They make it possible for us to act without coming into direct physical contact with what we act on, to affect objects and people even at a considerable remove. Though it takes the body to operate them, our instruments nonetheless shield us from exposure to the butt of action. There is a momentous difference between feeling a small rat brush against one's leg and shooting a lion from a jeep at fifty yards.

All the experience of civilized man points in the direction of acting with dispatch and without involvement. We want to accomplish our ends not only with minimal physical effort but also with minimal bodily exposure. The clearest indication of this is the length to which we go to dress and protect the body, as if it were a weak or fragile object. Requirements of warmth, decency, and convenience do not suffice to account for the use we make of boots or rubber gloves and layer on layer of antiseptic clothing.

We work our tools in plain view and yet they hide the full reality of what we do. The reason is that the more we work with tools, the more we find that we have only a view. Bodily involvement culminating in physical touch is the most potent source of our reality sense. Yet this is precisely what tools interdict. With the immediacy of physical encounter minimized and the immense emotional apparatus of touch detached, we can resort only to sight for information. And sight is a cool and impersonal medium: often it yields facts without feeling. The robust, multifaceted sense of reality fostered by direct physical engagement fades away. We become uninvolved observers of our deeds.

This suggests a neglected avenue of exploration in our search for the source of the dominance of sight in our lives. The use of tools, machines, and man-made instruments, though involving physical acts, nonetheless reduces the exposure and hence the sensitivity of the body. This is what necessitates ever greater reliance on our eyes, while it progressively impoverishes our picture of what we cause.

Tools mediating our actions, then, have two general sorts of effects. They amplify our action on the world, increasing human might and versatility beyond anything a rational observer of our unreason could have predicted. But they also serve as a protective wall that cushions the impact of the world on us, dulls our senses, and destroys direct experience. I have tried to show how these effects are present even in the personal use in plain view of the simplest tools. In an industrial society that places heavy reliance on man-made devices, the consequences become overwhelming.

The craftsman working with tools he understands can come to view them as extensions of his body. He can not only identify them as his, he can identify *with* them. Long experience with them teaches him their power and limits; occasional slips, which turn the tool on his body, render this knowledge vivid and personal. As a result, though his knowledge is not all it could be, the craftsman can at least appropriate his tool-mediated acts. He can view them as his and bear the consequences.

This becomes largely impossible in our industrial world. The devices we use are too complex and too numerous for any one person to understand them. The system of machines on which we rely for continued life seems to be out of scale with our senses. Even those who operate the tools are largely in the dark as to their mode of action. The rest of us sink into comfortable ignorance of the conditions of our life. Such ignorance begins with no thought of these conditions and ends with the thought that there are none.

Our innocence of the operation of our world of tools is so widespread as to appear to be in no need of emphasis. Yet the magnitude of our ignorance of this ignorance demands that we not pass it by without a comment. We are like blissful and brainless fish swimming in an ocean of machinery. Our food and shelter are manufactured, our very air is polluted and purified by our tools. Our bodies are built or crushed, our minds shaped by inanimate devices. We think that nothing but the possession and use of machines can give us respite or satisfaction. We fear that nothing can be done without artificial aid: for lack of microphones the human voice is stilled. We created and we operate the momentous machinery that undergirds civilized life, but it is not true that in any clear sense we are still in charge.

It is not, of course, that the instruments have acquired a mind of their own and are taking us to where their hearts desire. That is never the nature of true slavery. Our bondage is due to our desires. What makes us slaves to our machines is our ingenuous reliance on these means and our abject readiness to suit our ends to feasibility. Slavery is a comfortable state: if it were not pleasant, it would never be. It pays by providing what we want or need. What keeps us bound is not the force of threats but burning want, warm comfort, and the habit of supporting a habit.

Mediating machines make for us a comfortable world of secondhand sensations, a cocoon which lets us hear but distant rumbles in our dome of sleep.

I have tried to show that when we place tools as intermediaries between ourselves and the world, our knowledge even of our direct actions is naturally curtailed. As the tools become more complex and more numerous, we find firsthand experience in general retreat. At first only tactile contact is reduced. But when we turn to actions mediated by vastly complex machines, such as the car, or to those performed on our behalf by means of tools that others operate, such as the building of the car we shall later use, we encounter a general diminution, in each person's life, of knowledge by direct experience.

We can now note two interesting extensions of this effect. The first is that the specialization of tools and facilities goes hand in hand with the specialization of persons. By appropriating sphere after sphere of activity as their unique domain, specialists render our lives ever more constricted and compartmentalized. This closes off whole regions of human experience to most of us. Birth and death, the great terminal events of human life, occur in special places hidden from our ken. Healing and destruction, creativity in the arts, and even simple thinking are normally beyond our reach: each is accomplished by trained and certified experts.

The second point to note is that as direct exposure to the world is curtailed in each of us, certain experiences tend altogether to drop out of the life of the human race. I do not have in mind here only the blatantly evident or trivial cases, such as the experience of trying to make fire by striking stones. Though tools mediate our actions and reduce direct experience, there is a direct experience of operating with tools. As in the course of time certain sorts of tools fall into disuse, the skills required to operate them, the unique and sometimes beautiful experiences derived from working with them all slowly disappear. No one hunts for lunch by tossing rocks at wild boar or a passing cow. There is some loss in this, no doubt. But the view that the gain does not outweigh it is confined to well-fed romantics who have never thrown a stone.

But these are not the only experiences the development of tools eliminates. There are many acts no one does and no one does for us: in the name of efficiency or convenience we have surrendered them to machines. There is hardly a cow milked by hand in the civilized world, and no one participates in the communal hunt. The whole enterprise of providing food has, in fact, been so mechanized that one has to search far and wide for a man who has even a vestigial sense of the struggle with nature for sustenance. In the not too distant future we may face the day when none will remember how to sow or reap or what it is like to kill a sow to eat. Most of us know nothing about these experiences even now. Farms become factories and each meat-processing plant a total system, a huge black box with cows marching in at one end while infinite links of sausage—"untouched by human hands"—hang at the other. This leaves ever more of us in ever more blissful ignorance of what it takes to keep us fed and fat.

One can imagine a child of the future standing in wonder, if wonder will survive, in front of one of these great engines of destruction and life, asking about the mysterious transformation of cow into canned meat. What could one say so he might understand? For this is not just a matter of pulleys and gears and how parts of the machine interact. For him to understand we would have to explain what it is like to kill a dumb but harmless sentient beast in cold blood and for food. No explanation would ever suffice, of course: words cannot capture what the eyes and hands reveal. If the child could see the dim image of terror in those eyes, if he could sense with his whole body the death convulsions of the animal and then nonetheless sit to feed on it, he might spontaneously sense a sacramental element in eating or a deep unity with nature and her ways. Without these experiences the activity of eating remains meaningless and neutral: humans cease to be hungry animals and become empty jugs or nameless tanks to be filled with gas.

There are at least two other fundamental human experiences that are endangered by the spread of machines that mediate our contact with the world. The first is the experience of using the entirety of our bodies in the service of a task. The dog involves its body in the simplest acts: in joy all of its parts shriek happiness, in stalking even its hair attends. By contrast, we compartmentalize our acts: the hand doodles, the foot taps, the chin itches when we listen. The one act for which the total body was essential, that of carrying and balancing heavy weights, was first simplified by tools and then transferred to machines altogether. Now few of our normal tasks work up a sweat; in the civilized world more sweat is due to worry than to work. Perhaps the only activities left that involve the whole body are sex, for those at least who do not view it as a localized sensation, and sport. One is led to wonder if their popularity reflects our need for an honest sweat.

The second threat comes in an area dear to conservationists. As our machines spread, wilderness disappears. Yet there is something terribly important about encountering nature in the raw. The demand not to render species of animals extinct so that we may see them in their natural habitat is not frivolous. We are of nature and she lives in us. We claim not to know what ails us in the city, but is there no hint in finding peace in the movement of a tree or in the passion of a squirrel in the park? Without continued contact with nature, without respect for the rhythm of her acts, we must forever feel without a home. We may not choose to live in her bosom, just as the child may choose to travel from its home. But perhaps we have to learn, as the child does, that to be too far away or to be barred from return by the mother's death is to lead a hopeless, homeless life.

The unaided human body is fragile and relatively weak. The power our tools confer by amplifying our actions and the immunity they promise by protecting us have made us think that this is not really so. We act as if we believed that our strength and resources, even our lives, are infinite. We seem to think there is nothing that cannot be done, that no natural process is immune to suspension or reversal by our might. All arrogance is ultimately arrogance of will. In its final form our arrogance shows itself as the belief that we can accomplish anything, if only we set our minds to it. Taken at face value this claim is simply false. As we grow up each learns with pain the limits of his talents and his time. There are similar limits for mankind, though as a race we seem to lack maturity to face them.

The illusion of our omnipotence leads us to disregard some simple and central facts about the world. The more we forget these facts, the more are attitudes and actions reflect a false appraisal of our state and prospects. Before long, then, we can find our lives out of phase with the movements of the world: our expectations become unreasonable and hence we feel never satisfied. The disregard of these facts leads to grave errors that plague us every day. What makes the errors dangerous is that they destroy even the possibility of happiness. What makes them remarkable is that if stated, they are easily exposed as fallacies: there is no intellect infirm enough to be fooled by them if they are plainly set in view. Yet we can live by such fallacies without knowing the error; they structure our lives and ruin them and in our innocence we cannot find the culprit.

Let me call attention to two of these unspoken fallacies. Both grow directly out of our use of tools and man-made devices to mediate our acts. The first is our belief or hope that human actions are not attended by long-range and irreversible results. We might call this the fallacy of avoidable consequences. Since we have acquired a measure of control over nature, it is easy to suppose that we can always contravene her efficacy and reverse her habitual course. Our technology and much of our social life support the silent conviction that the natural consequences of our acts are optional and may at will be rescinded.

Pills and operations appear, to the casual observer, to go a long way toward undoing the ravages of nature. We routinely convert the damage wrought by catastrophes or culpable acts into the calculable medium of cash and feel that when the insurance pays off the injury is erased. Our sense of the unstructured openness of the future makes it difficult for us to admit being bound by what we have been or done: this is at least a part of the reason for the instability of marriage in the modern world. For we think of human contacts and contracts not only as voluntary in their inception but also as permanently terminable and tentative.

There is a young woman of my acquaintance, known for her eclectic taste in men, who has come to symbolize this error in my mind. She had had two abortions before she reached twenty. She was surprised, even indignant, when she become pregnant for the third time. She viewed her condition not as a natural result of her activities, but as a special calamity in need of explanation beyond what common sense or simple biology could provide. The same surprise and indignation were registered by the student who walked into my office to report that tests done on him had revealed permanent brain damage as a result of the protracted use of drugs. "What do I do now?" he asked in horror but still convinced that there was something he could do to make his problem simply disappear.

We do not seem to be able to accept that sometimes there is nothing we can do. Our acts have natural and inevitable effects. These are not always as spectacular and immediate as a spurt of blood from a severed toe; some of the most striking effects may at first grow in the dark, some of the most lasting ones may be delayed. But ineluctably the consequences return to haunt us. The moment comes when it is too late to clean our streams or to conserve oil. There is nothing we can do to change a mongoloid. And it is altogether impossible to avoid deterioration and death for all of us who were called into this world. The attempt to bring men back from the dead by freezing their bodies until we find a cure for their sickness or the secret of eternal life, is perhaps the most pathetic manifestation of this fallacy.

The fact is that even the pills we take and the devices we use to avoid the unwanted consequences of our actions have unwanted and unavoidable effects. The thalidomide disaster is but one instance of what I have in mind: from aspirin through birth control pills to morphine, the chemicals in our pharmacopoeia are the source of much incidental harm. In counteracting the consequences of our actions we incur new consequences to counteract.

The second fallacy is connected with the first, though the two are not identical. The ease with which we shape the world with tools can make us think that success has no cost. Here, once again, the practices of our society contribute to a great extent to fostering this agreeable illusion. For much of the economy of our country rests on consumption that outstrips all reasonable need. We are caused to buy beyond our means by a massive campaign of advertisers and retailers offering credit to make us think that the best of everything can be ours without effort and at once. When there is cost, it seems trivial by comparison with the benefit conferred: a dollar or two a day for a new car, a deodorant for success in business and romance, a subscription for lasting happiness. Everything seems easy: we can have or be whatever we want without sacrifice or loss or misery.

We might dub this the illusion of costless benefits or the fallacy of free delight. No animal fighting a hostile world alone would ever commit this blunder. For it—or him or her, since we are also animals—adversity will seem a law of nature and life will be a line of obstacles. In the tradition of the barter it may then view pain as the price we have to pay for our pleasures and happiness as the return on suffering. At any rate, each new predicament will remind it that nothing worth having is either free or cheap, that the world always exacts a pound of flesh for its ounce of steak.

It must have been in response to such troubling thoughts that heaven, our boldest ideal, was conceived. There, at least, we can transcend the fatal balancing of loss and gain. To be sure, those fortunate enough to pass the pearly gates were supposed already to have paid the price, cheap as it would seem in retrospect, of eternal bliss. But, and this is the important thing, once entered on this blessed life the soul presumably reaps costless benefits. Exempt from desire and disease, freed of pain, need and loss, it can enjoy each rapture without the dark thought of a day of reckoning.

It is a heaven of this sort on earth we dream when our thoughts are ruled by the fallacy of free delight. Mediation is so widespread in our world that the true cost of our acts and lives is hidden from our ken. Like the bugs that skate on the surface of uncharted waters or like denizens of a decadent Court, we live deprived of the knowledge of what supports us. And should there be cost, we think, our technical know-how will quickly erase the loss. Perhaps by next year we will achieve prevention of the loss and need no remedy. Hidden costs combined with the propaganda that there are really none, the reduction of effort by the use of machines, and our sense of omnipotence add up to the creed that our pleasures are limitless and free.

The final stage in this trend has not yet arrived. It has been foreshadowed in technological utopias; reality suggests fiction to the receptive mind and then copies it to make it all come true. The logic behind such utopias is Orwellian: if reason demands that we get the most for the least, is it not supremely rational to get everything for nothing or at least to feel that the world is ours yet we never paid? The strategy for accomplishing this has two parts. The first is the large-scale use of machines in production to minimize hard work and greatly to increase available consumer goods. The second is the widespread application of psychological techniques. By their means men can be trained to dislike such costly goods as freedom and to love to bear the expense of all the rest. A cost we gladly pay will not appear to be a cost at all; in this way, we can actually come to believe that we lead a free and happy life.

This is what in his intuitive and incoherent way the Savage of *Brave New World* realizes. He senses that that whole society is built on the fallacy of free delight, that there everyone is engineered to disregard costs or to suppose that there are none. People around him think they feel happy, though clearly none of them really is. The ultimate horror in his eyes is the shallowness of perpetual pleasure when measured against the surrender of free action and self-development, which is its cost. In this light, the Savage's otherwise puzzling or pathetic demand for the right to feel unhappy is suddenly revealed as the insistence on individual choice, which requires that costs be manifest. He thinks it ignominious to be duped into believing that all is well and free when in reality we pay for comfort with disfigured lives.

Motivating people to cooperate of their own accord, manipulating them from the inside has always been the dream of tyrannies. Perhaps the only way to accomplish this is by creating the illusion that the government or the social structure it proposes will yield some great and costless benefits. Our technology and social practices antecedently incline us to believe that gratuitous gains are at least possible; in addition, we now have the psychological technology to cause the conviction that a given regime actually provides such benefits. The ingredients, therefore, of a truly extensive and successful tyranny are already at hand.

In such a state, if it were ever to occur, citizens would love their government. Mediation would, of course, not be eliminated, nor would the psychic distance that flows from it be overcome. All questions of agency and responsibility would be submerged in institutionalized ignorance, and ignorance would be guised in a flood of pleasant feeling. The result could well be a generation of men who without being happy would feel as though they were and never really know the difference.

But is there a difference? Perhaps nothing matters beyond how we feel, if our feelings could but be sustained. Synthetic happiness caused by machine, pill, propaganda, or electrodes in the brain is preferable to none. Since it is the only type that can be guaranteed, perhaps it is better than any other. In the end, is liberty more than our sense that we are free? Is self-determination anything beyond the thought that we shape our destiny?

All of this sound strangely plausible. Yet we find that the mere sense of satisfaction does not satisfy.

Small is dubious

SAMUEL C. FLORMAN

▪ Samuel Florman *is best known for his polemic against the critics of large-scale Western technology. His publications include* The Existential Pleasures of Engineering *and a number of articles in various periodicals.*

In "Small is Dubious" Florman attacks E. F. Schumacher and Amory Lovins as being shortsighted in their rejection of technological pluralism. Large technology and small technology can, and in Western industrial societies should, *exist side by side. But Lorman is even more concerned with these writers' importation of moral and political values into the technological arena; he argues that decisions about the future of technological development should instead be made on the basis of cost-benefit analyses. Florman argues for a kind of technological Darwinism in which only the fittest technologies survive.*

Last April, while reading the papers the morning after the President's energy address to the nation, I was struck by a statement attributed to Mr. Carter's pollster and adviser, Patrick Caddell: "The idea that big is bad and that there is something good to smallness is something that the country has come to accept much more today than it did 10 years ago. This has been one of the biggest changes in America over the past decade."

Since the nation had just been exhorted to embark on the most herculean technological, economic, and political enterprises, this reference to smallness seemed to me to be singularly inapt. Waste is to be deplored, of course, and inefficiency. But bigness? I had not realized that the small-is-beautiful philosophy had reached the White House.

A few days after the Carter speech, I had an opportunity to attend a lecture by E. F. Schumacher, the author of *Small Is Beautiful,* the book that, since its publication in 1973, has become the Koran of the antitechnology movement. I listened, bemused, as Dr. Schumacher depicted a United States in which each community would bake its own bread and develop its own resources, a nation of self-reliant craftsmen where interstate transport would practically disappear. The energy crisis could be solved, Schumacher maintained, only by replacing our sprawling network of industrial metropolises with numerous small-scale production centers. Schumacher's audience listened, entranced. It was clear that the energy crisis was giving new life to an idea which otherwise might have died a natural death.

On my way home, I found myself thinking about a telephone call I had received a few weeks earlier from a consultant to the power industry. He was concerned about an article entitled "Energy Strategy: The Road Not Taken?" by Amory B. Lovins, a British physicist, which had appeared in the October 1976 issue of *Foreign Affairs.* The article, which argued the small-is-beautiful position forcefully, had been extensively quoted in the international press, entered into the Congressional Record, discussed in *Business Week,* and been the subject of the most reprint requests ever received by *Foreign Affairs,* surpassing even the famous George Kennan "Mr. X" piece.

Opposition has not been slow to rally. The man who called me put together a collection of rebuttal essays prepared by people prominent in the fields of energy, academe, industry, and labor. This imposing pamphlet has been circulated in large quantities wherever its sponsor fears the Lovins article might have made an impression. It appears that the metaphysical struggle between small and big—reminiscent of the argument over the number of angels that can dance on the head of a pin—has become a real issue.

The small-is-beautiful believers, as exemplified by the Lovins article, commence their campaign with a critique of our existing energy technology, especially our nationwide grid of electrical power. The deficiencies of this system are obvious enough. Electricity is created in huge central plants by boiling water to run generators. Whether the heat that boils the water is furnished by oil, coal, gas, nuclear energy, or even by solar energy, a great deal of energy is wasted in the process, and even more is lost in transmission over long lines. By the time the electricity arrives in our home or factory and is put to use, about two-thirds of the original energy has been dissipated. In addition, the existence of what Lovins calls "the infrastructure" of the power industry itself—tens of thousands of workers occupying enormous office complexes—costs the system more energy, and costs the consumer more money.

The proposed solution, which on first hearing sounds fairly sensible, is the creation of small, efficient energy-creating installations in the buildings where the energy is used, or at most at the medium scale of urban neighborhoods and rural villages. Direct solar plants are the preferred system, although Lovins also mentions small mass-produced diesel generators, wind-driven generators, and several other technologies still in the developmental stage.

Yet, despite the advantages of this system, the new "soft" technologies, to use Lovins's term, would entail the manufacture, transport, and installation of millions of new mechanisms. This cannot but be a monumental undertaking requiring enormous outlays of capital and energy. Then these mechanisms will have to be maintained. We all resent the electric and phone companies, but, when service is interrupted, a crew of competent men arrives on the scene to set things right. Lovins assures us that the solar collectors or windmills in our homes will be serviced by our friendly, independent neighborhood mechanic, a prospect which must chill the blood of anyone who has ever had to have a car repaired or tried to get a plumber in an emergency. As for Americans becoming self-reliant craftsmen, as Schumacher assures us we can, this idea sounds fine in a symposium on the human condition, but it overlooks the enormous practical and psychological difficulties that stand in its way. The recently attempted urban homesteading program, for example, was based on this very appealing concept. Abandoned houses were to be turned over to deserving families at no cost, just as land was made available to homesteaders in the last century. The program failed because most poor families simply were not capable of fixing up the houses.

Another hope of the small-is-beautiful advocates is that great savings can be realized by eliminating the administrations, or "middlemen" of the utility companies. But in the real world it appears that the middleman does perform a useful function. How else can we explain the failure of the cooperative buying movement, which is based on the idea that people can band together to eliminate distribution costs? The shortcomings of large organizations are universally recognized, and "bureaucratic" has long been a synonym for "inefficient." But, like it or not, large organizations with apparently superfluous administrative layers seem to work better than small ones. Chain stores are still in business, while mom-and-pop stores continue to fail. Local power companies, especially, are a vanishing breed. Decisions made in the marketplace do not tell us everything, but they do tell us a lot more than the fantasies of futurist economists.

This is not to say that the situation cannot change. If a handy gadget becomes available that will heat my house economically using wind, water, sunlight, or moonlight, I will rush out to buy it. On the other hand, if the technological breakthroughs come in the power-plant field—perhaps nuclear fusion or direct conversion of sunlight to electricity—then I will be pleased to continue my contractual arrangements with the electric company.

Such an open-minded approach has no appeal to Lovins. Quoting Robert Frost on two roads diverging in a wood, he asserts that we must select one way or the other, since we cannot travel both. The analogy is absurd, since we are a pluralistic society of more than 200 million people, not a solitary poet, and it has been our habit to take every road in sight. Will it be wasteful to build power plants that may soon be obsolete? I think not. If a plant is used for an interim period while other technologies are developed, it will have served its purpose. If it is never used at all, it will still have been a useful component of a contingency plan. When billions of dollars are spent each year on constantly obsolescing weapons which we hope we will never have to use, it does not seem extravagant to ask for some contingency planning for our life-support systems.

Our resources are limited, of course, and we want to allocate them sensibly. At this time it is not clear whether the most promising technologies are "hard" or "soft" or, as is most likely, some combination of both. The "soft" technologies are not being ignored. The Administration's energy program contains incentives for solar heat installations by individual homeowners. Research and development funds are being granted to a multitude of experimental projects. At the same time, we are working on improvements to our conventional systems. What else could a responsible society do? We must assume that the technologies which prevail will be those which prove to be most cost-effective and least hazardous. Improper political pressures may be a factor, but these have a way of cancelling each other out. A new product attracts sophisticated investors, and before long there is a new lobbyist's office in Washington. The struggle for markets and profits creates a jungle in which the fittest technologies are likely to survive.

Technological efficiency, however, is not a standard by which the small-is-beautiful advocates are willing to abide. Lovins makes this clear when he states that even if nuclear power were clean, safe, and economic, "it would still be unattractive because of the political implications of the kind of energy economy it would lock us into." As for making electricity from huge solar collectors in the desert, or from solar energy collected by satellites in outer space—these also will not do, "for they are ingenious high-technology ways to supply energy in a form and at a scale inappropriate to most end-use needs." Finally, he admits straight out that the most important questions of energy strategy "are not mainly technical or economic but rather social and ethical."

So the technological issue is found to be a diversion, not at all the heart of the matter. The *political* consequences of bigness, it would appear, are what we have to fear. A centralized energy system, Lovins tells us, is "less compatible with social diversity and personal freedom of choice" than the small, more pluralistic, approach he favors.

But diversity and freedom, at least in the United States, are protected and encouraged by strong institutions. Exploitation thrives in small towns and in small businesses. Big government and big labor unions, for all their faults, are the means by which we achieve the freedoms we hold so dear.

When big organizations challenge our well-being, as indeed they do—monopolistic corporations, corrupt labor unions, et al.—our protection comes, not from petty insurrections, but from that biggest of all organizations, the federal government. And when big government itself is at fault, the remedy can only be shake-ups and more sensible procedures, not elimination of that bureaucracy which is a crucial element of our democracy. Does it not seem absurd, and quite late in the day, to speak of losing our political freedom through the growth of federally supervised utility companies, when we long ago agreed to give up our individual militias, and entrust the national defense to a national army? The small-is-beautiful philosophy makes just as little sense politically as it does technologically.

The next argument that Schumacher and Lovins present is the social one. Even if large organizations "work" technically and politically, it is claimed, they do not work socially. The subtitle to *Small Is Beautiful* is "Economics As If People Mattered." Only in small social groups, apparently, is it possible for people to "matter." Schumacher and Lovins would not appear to have read such books as *Winesburg, Ohio, Spoon River Anthology,* and *Main Street,* with their picture of the American small town as a petty, cramped, and spiteful community. Cities and small towns will always have their defenders, but the constantly discussed question about whether it is "better" to live in the city, the country, or the suburbs is a matter of taste which cannot be settled by self-appointed intellectual mandarins.

Perhaps what lies at the heart of the new worship of smallness is an increasing revulsion against the ugliness of much of industrial America. Dams, highways, and electric transmission lines, once the symbol of a somewhat naive commercial boosterism, are now depicted as vulgar. But this association of bigness with lack of taste is not warranted. The colossal works of man are no more inherently vulgar than the small works are inherently petty. We prize robustness in life as well as delicacy. Rousseau, coming upon a Roman aqueduct, had this to say:

> The echo of my footsteps under the immense arches made me think I could hear the strong voices of the men who had built it. I felt lost like an in-

sect in the immensity of the work. I felt, along with the sense of my own littleness, something nevertheless which seemed to elevate my soul; I said to myself with a sigh: "Oh! that I had been born a Roman!"

Economic and social arguments aside, Schumacher and Lovins maintain that their philosophy is founded on a base of moral conviction, of thrift, simplicity, and humility. We have sinned by being wasteful, ostentatious, and arrogant. Thus smallness becomes a symbol of virtue.

For a moment, as at every step along the way, we are inclined to agree. The message has an appeal. The problems of our age—the environmental crisis, the energy crisis, the depletion of our natural resources—are, we suspect, caused by our profligacy. Improvidence, it would appear, has become the cardinal sin.

But even the most useful moral precepts—such as patriotism—often have a dark underside. In the present instance, the thrift being preached lends itself to a smallness of spirit. (The day after President Carter's first energy message I heard the radio commentator Paul Harvey question the "waste" of gasoline for busing school children.) The humility proposed evokes those Oriental attitudes which counsel the masses to accept their wretched lot. Such fatalistic beliefs may be useful in adding a measure of serenity to our private lives, but they are insidious elements to inject into debates on public policy.

Much of the debate over big versus small recalls the Lilliputians going to war over the question of whether eggs should be opened at the big or little end. *Smallness,* after all, is a word that is neutral—technologically, politically, socially, aesthetically, and, of course, morally. Its use as a symbol of goodness would be one more entertaining example of human folly were it not for the disturbing consequences of the arguments advanced in its cause.

13

The *homo faber* debate

From *The Myth of the Machine: Technics and Human Development*

LEWIS MUMFORD

Lewis Mumford has been introduced in Part One of this book.

In this selection from The Myth of the Machine, *Mr. Mumford argues that the traditionally accepted picture of man as uniquely the maker of tools,* homo faber, *is both inaccurate and unfortunate. It is a "petrified" notion of man. Since other animals use primitive tools, we must look elsewhere for man's uniqueness. It is at the level of* homo sapiens *that we encounter that uniqueness.*

The last century, we all realize, has witnessed a radical transformation in the entire human environment, largely as a result of the impact of the mathematical and physical sciences upon technology. This shft from an empirical, tradition-bound technics to an experimental mode has opened up such new realms as those of nuclear energy, supersonic transportation, cybernetic intelligence and instantaneous distant communication. Never since the Pyramid Age have such vast physical changes been consummated in so short a time. All these changes have, in turn, produced alterations in the human personality, while still more radical transformations, if this process continue unabated and uncorrected, loom ahead.

In terms of the currently accepted picture of the relation of man to technics, our age is passing from the primeval state of man, marked by his invention of tools and weapons for the purpose of achieving mastery over the forces of nature, to a radically different condition, in which he will have not only conquered nature, but detached himself as far as possible from the organic habitat.

With this new "megatechnics" the dominant minority will create a uniform, all-enveloping, super-planetary structure, designed for automatic operation. Instead of functioning actively as an autonomous personality, man will become a passive, purposeless, machine-conditioned animal whose proper functions, as technicians now interpret man's role, will either be fed into the machine or strictly limited and controlled for the benefit of de-personalized, collective organizations.

My purpose in this book is to question both the assumptions and the predictions upon which our commitment to the present forms of technical and scientific progress, treated as if ends in themselves, has been based. I shall bring forward evidence that casts doubts upon the current theories of man's basic nature which over-rate the part that tools once played—and machines now play—in human development. I shall suggest that not only was Karl Marx in error in giving the material instruments of production the central place and directive function in human development, but that even the seemingly benign interpretation of Teihard de Chardin reads back into the whole story of man the narrow technological rationalism of our own age, and projects into the future a final state in which all the possibilities of human development would come to an end. At that "omega-point'' nothing would be left of man's autonomous original nature, except organized intelligence: a universal and omnipotent layer of abstract mind, loveless and lifeless.

Now, we cannot understand the role that technics has played in human development without a deeper insight into the historic nature of man. Yet that insight has been blurred during the last century because it has been conditioned by a social environment in which a mass of new mechanical inventions had suddenly proliferated, sweeping away ancient processes and institutions, and altering the traditional conception of both human limitations and technical possibilities.

Our predecessors mistakenly coupled their particular mode of mechanical progress with an unjustifiable sense of increasing moral superiority. But our own contemporaries, who have reason to reject this smug Victorian belief in the inevitable improvement of all other human institutions through command of the machine, nevertheless concentrate, with manic fervor, upon the continued expansion of science and technology, as if they alone magically would provide the only means of human salvation. Since our present over-commitment to technics is in part due to a radical misinterpretation of the whole course of human development, the first step toward recovering our balance is to bring under review the main stages of man's emergence from its primal beginnings onward.

Just because man's need for tools is so obvious, we must guard ourselves against overstressing the role of stone tools hundreds of thousands of years before they became functionally differentiated and efficient. In treating tool-making as central to early man's survival, biologists and anthropologists for long underplayed, or neglected, a mass of activities in which many other species were for long more knowledgeable than man. Despite the contrary evidence put forward by R. U. Sayce, Daryll Forde, and André Leroi-Gourhan, there is still a tendency to identify tools and machines with technology: to substitute the part for the whole.

Even in describing only the material components of technics, this practice overlooks the equally vital role of containers: first hearths, pits, traps, cordage; later, baskets, bins, byres, houses, to say nothing of still later collective containers like reservoirs, canals, cities. These static components play an important part in every technology, not least in our own day, with its high-tension transformers, its giant chemical retorts, its atomic reactors.

In any adequate definition of technics, it should be plain that many insects, birds, and mammals had made far more radical innovations in the fabrication of containers, with their intricate nests and bowers, their geometric beehives, their urbanoid anthills and termitaries, their beaver lodges, than man's ancestors had achieved in the making of tools until the emergence of *Homo sapiens*. In short, if technical proficiency alone were sufficient to identify and foster intelligence, man was for long a laggard, compared with many other species. The consequences of this perception should be plain: namely, that there was nothing uniquely human in tool-making until it was modified by linguistic symbols, esthetic designs, and socially transmitted knowledge. At that point, the human brain, not just the hand, was what made a profound difference; and that brain could not possibly have been just a handmade product, since it was already well developed in four-footed creatures like rats, which have no free-fingered hands.

More than a century ago Thomas Carlyle described man as a ''tool-using animal,'' as if this were the one trait that elevated him above the rest of brute creation. This overweighting of tools, weapons, physical apparatus, and machines has obscured the actual path of human development. The definition of man as a tool-using animal, even when corrected to read ''tool-making,'' would have seemed strange to Plato, who attributed man's emergence from a primitive state as much to Marsyas and Orpheus, the makers of music, as to fire-stealing Prometheus, or to Hephaestus, the blacksmith-god, the sole manual worker in the Olympic pantheon.

Yet the description of man as essentially a tool-making animal has become so firmly embedded that the mere finding of the fragments of little primate skulls in the neighborhood of chipped pebbles, as with the Australopithecines of Africa, was deemed sufficient by their finder, Dr. L. S. B. Leakey, to identify the creature as in the direct line of human ascent, despite marked physical divergences from both apes and later men. Since Leakey's sub-hominids had a brain capacity about a third of *Homo sapiens*—less indeed than some apes—the ability to chip and use crude stone tools plainly neither called for nor by itself generated man's rich cerebral equipment.

If the Australopithecines lacked the beginning of other human characteristics, their possession of tools would only prove that at least one other species outside the true genus *Homo* boasted this trait, just as parrots and magpies share the distinctly human achievement of speech, and the bower bird that for colorful decorative embellishment. No single trait, not even tool-making, is sufficient to identify man. What is specially and uniquely human is man's capacity to combine a wide variety of animal propensities into an emergent cultural entity: a human personality.

If the exact functional equivalence of tool-making with utensil-making had been appreciated by earlier investigators, it would have been plain that there was nothing notable about man's hand-made stone artifacts until far along in his development. Even a distant relative of man, the gorilla, puts together a nest of leaves for comfort in sleeping, and will throw a bridge of great fern stalks across a shallow stream, presumably to keep from wetting or scraping his feet. Five-year-old children, who can talk and read and reason, show little aptitude in using tools and still less in making them: so if tool-making were what counted, they could not yet be identified as human.

In early man we have reason to suspect the same kind of facility and the same ineptitude. When we seek for proof of man's genuine superiority to his fellow creatures, we should do well to look for a different kind of evidence than his poor stone tools alone; or rather, we should ask ourselves what activities preoccupied him during those countless years when with the same materials and the same muscular movements he later used so skillfully he might have fashioned better tools.

The answer to this question I shall spell out in detail in the first few chapters; but I shall briefly anticipate the conclusion by saying that there was nothing specifically human in primitive technics, apart from the use and preservation of fire, until man had reconstituted his own physical organs by employing them for functions and purposes quite different from those they had originally served. Probably the first major displacement was the transformation of the quadruped's fore-limbs from specialized organs of locomotion to all-purpose tools for climbing, grasping, striking, tearing, pounding, digging, holding. Early man's hands and pebble tools played a significant part in his development, mainly because, as Du Brul has pointed out, they facilitated the preparatory functions of picking, carrying, and macerating food, and *thus liberated the mouth for speech*.

If man was indeed a tool-maker, he possessed at the beginning one primary, all-purpose tool, more important than any later assemblage: his own mind-activated body, every part of it, including those members that made clubs, hand-axes or wooden spears. To compensate for his extremely primitive working gear, early man had a much more important asset that extended his whole technical horizon: he had a far richer biological equipment than any other animal, a body not specialized for any single activity, and a brain capable of scanning a wider environment and holding all the different parts of his experience together. Precisely because of his extraordinary plasticity and sensitivity, he was able to use a larger portion of both his external environment and his internal, psychosomatic resources.

Through man's overdeveloped and incessantly active brain, he had more mental energy to tap than he needed for survival at a purely animal level; and he was accordingly under the necessity of canalizing that energy, not just into food-getting and sexual reproduction, but into modes of living that would convert this energy more directly and constructively into appropriate cultural—that is, symbolic—forms. Only by creating cultural outlets could he tap and control and fully utilize his own nature.

Cultural "work" by necessity took precedence over manual work. These new activities involved far more than the discipline of hand, muscle, and eye in making and using tools, greatly though they aided man: they likewise demanded a control over all man's natural functions, including his organs of excretion, his upsurging emotions, his promiscuous sexual activities, his tormenting and tempting dreams.

With man's persistent exploration of his own organic capabilities, nose, eyes, ears, tongue, lips, and sexual organs were given new roles to play. Even the hand was no mere horny specialized work-tool: it stroked a lover's body, held a baby close to the breast, made significant gestures, or expressed in shared ritual and ordered dance some otherwise inexpressible sentiment about life or death, a remembered past, or an anxious future. Tool-technics, in fact, is but a fragment of biotechnics: man's total equipment for life.

This gift of free neural energy already

showed itself in man's primate ancestors. Dr. Alison Jolly has recently shown that brain growth in lemurs derived from their athletic playfulness, their mutual grooming, and their enhanced sociability, rather than from tool-using or food-getting habits; while man's exploratory curiosity, his imitativeness, and his manipulativeness, with no thought of ulterior reward, were already visible in his simian relatives. In American usage, "monkey-shines" and "monkeying" are popular identifications of that playfulness and non-utilitarian handling of objects. I shall show that there is even reason to ask whether the standardized patterns observable in early tool-making are not in part derivable from the strictly repetitive motions of ritual, song, and dance, forms that have long existed in a state of perfection among primitive peoples, usually in far more finished style than their tools.

Only a little while ago the Dutch historian, J. Huizinga, in "Homo Ludens" brought forth a mass of evidence to suggest that play, rather than work, was the formative element in human culture: that man's most serious activity belonged to the realm of make-believe. On this showing, ritual and mimesis, sports and games and dramas, released man from his insistent animal attachments; and nothing could demonstrate this better, I would add, than those primitive ceremonies in which he played at being another kind of animal. Long before he had achieved the power to transform the natural environment, man had created a miniature environment, the symbolic field of play, in which every function of life might be refashioned in a strictly human style, as in a game.

So startling was the thesis of "Homo Ludens" that his shocked translator deliberately altered Huizinga's express statement, that all culture was a form of play, into the more obvious conventional notion that play is an element in culture. But the notion that man is neither *Homo sapiens* nor *Homo ludens,* but above all *Homo faber,* man the maker, had taken such firm possession of present-day Western thinkers that even Henri Bergson held it. So certain were nineteenth-century archeologists about the primacy of stone tools and weapons in the "struggle for existence" that when the first paleolithic cave paintings were discovered in Spain in 1879, they were denounced, out of hand, as an outrageous hoax, by "competent authorities" on the ground that Ice Age hunters could not have had the leisure or the mind to produce the elegant art of Altamira.

But mind was exactly what *Homo sapiens* possessed in a singular degree: mind based on the fullest use of all his bodily organs, not just his hands. In this revision of obsolete technological stereotypes, I would go even further: for I submit that at every stage man's inventions and transformations were less for the purpose of increasing the food supply or controlling nature than for utilizing his own immense organic resources and expressing his latent potentialities, in order to fulfill more adequately his superorganic demands and aspirations.

When not curbed by hostile environmental pressures, man's elaboration of symbolic culture answered a more imperative need than that for control over the environment—and, one must infer, largely predated it and for long outpaced it. Among sociologists, Leslie White deserves credit for giving due weight to this fact by his emphasis on "minding" and "symboling," though he has but recovered for the present generation the original insights of the father of anthropology, Edward Tylor.

On this reading, the evolution of language—a culmination of man's more elementary forms of expressing and transmitting meaning—was incomparably more important to further human development than the chipping of a mountain of hand-axes. Besides the relatively simple coordinations required for tool-using, the delicate interplay of the many organs needed for the creation of articulate speech was a far more striking advance. This effort must have occupied a greater part of early man's time, energy, and mental activity, since the ultimate collective product, spoken language, was infinitely more complex and sophisticated at the dawn of civilization than the Egyptian or Mesopotamian kit of tools.

To consider man, then, as primarily a tool-using animal, is to overlook the main chapters of human history. Opposed to this petrified notion, I shall develop the view that man is pre-eminently a mind-making, self-mastering, and self-designing animal; and the primary locus of all his activities lies first in his own organism, and in the social organization through which it finds fuller expression. Until man had made something of himself he could make little of the world around him.

In this process of self-discovery and self-transformation, tools, in the narrow sense, served well as subsidiary instruments, but not as the main operative agent in man's development; for technics has never till our own age dissociated itself from the larger cultural whole in which man, as man, has always functioned. The classic Greek term "*tekhne*" characteristically makes no distinction between industrial production and "fine" or symbolic art; and for the greater part of human history these aspects were inseparable, one side respecting the objective conditions and functions, the other responding to subjective needs.

At its point of origin, technics was related to the whole nature of man, and that nature played a part in every aspect of industry: thus technics, at the beginning, was broadly life-centered, not work-centered, or power-centered. As in any other ecological complex, varied human interests and purposes, different organic needs, restrained the overgrowth of any single component. Though language was man's most potent symbolic expression, it flowed, I shall attempt to show, from the same common source that finally produced the machine: the primeval repetitive order of ritual, a mode of order man was forced to develop, in self-protection, so as to control the tremendous overcharge of psychal energy that his large brain placed at his disposal.

So far from disparaging the role of technics, however, I shall rather demonstrate that once this basic internal organization was established, technics supported and enlarged the capacities for human expression. The discipline of tool-making and tool-using served as a timely correction, on this hypothesis, to the inordinate powers of invention that spoken language gave to man—powers that otherwise unduly inflated the ego and tempted man to substitute magical verbal formulae for efficacious work.

On this interpretation, the specific achievement, which set man apart from even his nearest anthropoid relatives, was the shaping of a new self, visibly different in appearance, in behavior, and in plan of life from his primitive animal forebears. As this differentiation widened and the number of definitely human "identification marks" increased, man speeded the process of his own evolution, achieving through culture in a relatively short span of years changes that other species accomplished laboriously through organic processes, whose results, in contrast to man's cultural modes, could not be easily corrected, improved, or effaced.

Henceforth the main business of man was his own self-transformation, group by group, region by region, culture by culture. This self-transformation not merely rescued man from permanent fixation in his original animal condition, but freed his best-developed organ, his brain, for other tasks than those of ensuring physical survival. The dominant human trait, central to all other traits, is this capacity for conscious, purposeful self-identification, self-transformation, and ultimately for self-understanding.

Every manifestation of human culture, from ritual and speech to costume and social organization, is directed ultimately to the remodelling of the human organism and the expression of the human personality. If it is only now that we belatedly recognize this distinctive feature, it is perhaps because there are widespread indications in contemporary art and politics and technics that man may be on the point of losing it—becoming not a lower animal, but a shapeless, amoeboid nonentity.

In recasting the stereotyped representations of human development, I have fortunately been able to draw upon a growing body of biological and anthropological evidence, which has not until now been correlated or fully interpreted. Yet I am aware, of course, that despite this substantial support the large themes I am about to develop, and even more their speculative subsidiary hypotheses, may well meet with justifiable skepticism; for they have still to undergo competent critical scrutiny. Need I say that so far from starting with a desire to dispute the prevailing orthodox views, I at first respectfully accepted them, since I knew no others? It was only because I could find no clue to modern man's overwhelming commitment to his technology, even at the expense of his health, his physical safety, his mental balance, and his possible future development, that I was driven to reexamine the nature of man and the whole course of technological change.

In addition to discovering the aboriginal field of man's inventiveness, not in his making of external tools, but primarily in the re-fashioning of his own bodily organs, I have undertaken to follow another freshly blazed trail: to examine the broad streak of irrationality that runs all through human history, counter to man's sensible, functionally rational animal inheritance. As compared even with other anthropoids, one might refer without irony to man's superior irrationality. Certainly human development exhibits a chronic disposition to error, mischief, disordered fantasy, hallucination, "original sin," and even socially organized and sanctified misbehavior, such as the practice of human sacrifice and legalized torture. In escaping organic fixations, man forfeited the innate humility and mental stability of less adventurous species. Yet some of his most erratic departures have opened up valuable areas that purely organic evolution, over billions of years, had never explored.

The mischances that followed man's quitting mere animalhood were many, but the rewards were great. Man's proneness to mix his fantasies and projections, his desires and designs, his abstractions and his ideologies, with the commonplaces of daily experience were, we can now see, an important source of his immense creativity. There is no clean dividing line between the irrational and the super-rational; and the handling of these ambivalent gifts has always been a major human problem. One of the reasons that the current utilitarian interpretations of technics and science have been so shallow is that they ignore the fact that this aspect of human culture has been as open to both transcendental aspirations and demonic compulsions as any other part of man's existence—and has never been so open and so vulnerable as today.

The irrational factors that have sometimes constructively prompted, yet too often distorted, man's further development became plain at the moment when the formative elements in paleolithic and neolithic cultures united in the great cultural implosion that took place around the Fourth Millennium B.C.: what is usually called "the rise of civilization." The remarkable fact about this transformation technically is that it was the result, not of mechanical inventions, but of a radically new type of social organization: a product of myth, magic, religion, and the nascent science of astronomy. This implosion of sacred political powers and technological facilities cannot be accounted for by any inventory of the tools, the simple machines, and the technical processes then available. Neither the wheeled wagon, the plow, the potter's wheel, nor the military chariot could of themselves have accomplished the mighty transformations that took place in the great valleys of Egypt, Mesopotamia, and India, and eventually passed, in ripples and waves, to other parts of the planet.

The study of the Pyramid Age I made in preparation for writing "The City in History" unexpectedly revealed that a close parallel existed between the first authoritarian civilizations in the Near East and our own, though most of our contemporaries still regard modern technics, not only as the highest point in man's intellectual development, but as an entirely new phenomenon. On the contrary, I found that what economists lately termed the Machine Age or the Power Age, had its origin, not in the so-called Industrial Revolution of the eighteenth century, but at the very outset in the organization of an archetypal machine composed of human parts.

Two things must be noted about this new mechanism, because they identify it throughout its historic course down to the present. The first is that the organizers of the machine derived their power and authority from a heavenly source. Cosmic order was the basis of this new human order. The exactitude in measurement, the abstract mechanical system, the compulsive regularity of this "megamachine," as I shall call it, sprang directly from astronomical observations and scientific calculations. This inflexible, predictable order, incorporated later in the calendar, was transferred to the regimentation of the human components. As against earlier forms of ritualized order, this mechanized order was external to man. By a combination of divine command and ruthless military coercion, a large population was made to endure grinding poverty and forced labor at mind-dulling repetitive tasks in order to insure "Life, Prosperity, and Health" for the divine or semi-divine ruler and his entourage.

The second point is that the grave social defects of the human machine were partly offset by its superb achievements in flood control and grain production, which laid the ground for an enlarged achievement in every area of human culture: in monumental art, in codified law, in systematically pursued and permanently recorded thought, in the augmentation of all the potentialities of the mind by the assemblage of a varied population, with diverse regional and vocational backgrounds in urban ceremonial centers. Such order, such collective security and abundance, such stimulating cultural mixtures were first achieved in Mesopotamia and Egypt, and later in India, China, Persia, and in the Andean and Mayan cultures: and they were never surpassed until the megamachine was reconstituted in a new form in our own time. Unfortunately these cultural advances were largely offset by equally great social regressions.

Conceptually the instruments of mechanization five thousand years ago were already detached from other human functions and purposes than the constant increase of order, power, predictability, and, above all, control. With this proto-scientific ideology went a corresponding regimentation and degradation of once-autonomous human activities: "mass culture" and "mass control" made their first appearance. With mordant symbolism, the ultimate products of the megamachine in Egypt were colossal tombs, inhabited by mummified corpses; while later in Assyria, as repeatedly in every other expanding empire, the chief testimony to its technical efficiency was a waste of destroyed villages and cities, and poisoned soils: the prototype of similar "civilized" atrocities today. As for the great Egyptian pyramids, what are they but the precise static equivalents of our own space rockets? Both devices for securing, at an extravagant cost, a passage to Heaven for the favored few.

These colossal miscarriages of a dehumanized power-centered culture monotonously soil the pages of history from the rape of Sumer to the blasting of Warsaw and Rotterdam, Tokyo and Hiroshima. Sooner or later, this analysis suggests, we must have the courage to ask ourselves: Is this association of inordinate power and productivity with equally inordinate violence and destruction a purely accidental one?

In the working out of this parallel and in the tracing of the archetypal machine through later Western history, I found that many obscure irrational manifestations in our own highly mechanized and supposedly rational culture became strangely clarified. For in both cases, immense gains in valuable knowledge and usable productivity were cancelled out by equally great increases in ostentatious waste, paranoid hostility, insensate destructiveness, hideous random extermination.

This survey will bring the reader to the threshold of the modern world: the sixteenth century in Western Europe. Though some of the implications of such a study cannot be fully worked out until the events of the last four centuries are re-examined and re-appraised, much that is necessary for understanding—and eventually redirecting—the course of contemporary technics will be already apparent, to a sufficiently perceptive mind, from the earliest chapters on. This widened interpretation of the past is a necessary move toward escaping the dire insufficiencies of current one-generation knowledge. If we do not take the time to review the past we shall not have sufficient insight to understand the present or command the future: for the past never leaves us, and the future is already here.

Work and tools

PETER F. DRUCKER

▪ Peter F. Drucker *has taught at New York University, Bennington, and Claremont Graduate School. His numerous books include* The End of Economic Man, Landmarks of Tomorrow, *and* The Age of Discontinuity.

In the article that appears here, Drucker argues against the homo faber *thesis, but along somewhat different lines from those advanced by Mumford. The uniqueness of being human is not just that tools are used, but that work is done; and by "work" Drucker includes such activities as linguistic behavior and the development of social structures. Man is not, then, uniquely* homo faber, *but* homo faber qui laborat, *man the toolmaker who works.*

I

Man, alone of all animals, is capable of purposeful, nonorganic evolution; he makes tools. This observation by Alfred Russell Wallace, co-discoverer with Darwin of the theory of evolution, may seem obvious if not trite. But it is a profound insight. And though made some seventy or eighty years ago, its implications have yet to be thought through by biologists and technologists.

One such implication is that from a biologist's (or a historian's) point of view, the technologist's identification of tool with material artifact is quite arbitrary. Language, too, is a tool, and so are all abstract concepts. This does not mean that the technologist's definition should be discarded. All human disciplines rest after all on similarly arbitrary distinctions. But it does mean that technologists ought to be conscious of the artificiality of their definition and careful lest it become a barrier rather than a help to knowledge and understanding.

This is particularly relevant for the history of technology, I believe. According to the technologist's definition of "tool," the abacus and the geometer's compass are normally considered technology, but the multiplication table or a table of logarithms are not. Yet this arbitrary division makes all but impossible the understanding of so important a subject as the development of the technology of mathematics. Similarly the technologist's elimination of the fine arts from his field of vision blinds the historian of technology to an understanding of the relationship between scientific knowledge and technology. (See, for instance, volumes III and IV of Singer's monumental *History of Technology*.) For scientific thought and knowledge were married to the fine arts, at least in the West, long before they even got on speaking terms with the mechanical crafts: in the mathematical number theories of the designers of the Gothic cathedral,[1] in the geometric optics of Renaissance painting, or in the acoustics of the great Baroque organs. And Lynn White, Jr., has shown in several recent articles that to understand the history and development of the mechanical devices of the Middle Ages we must understand something so nonmechanical and nonmaterial as the new concept of the dignity and sanctity of labor which St. Benedict first introduced.

Even within the technologist's definition of technology as dealing with mechanical artifacts alone, Wallace's insight has major relevance. The subject matter of technology according to the preface to *A History of Technology* is "how things are done or made"; and most students of technology to my knowledge agree with this. But the Wallace insight leads to a different definition: the subject matter of technology would be "how man does or makes." As to the meaning and end of technology, the same source, again presenting the general view, defines them as "mastery of his (man's) natural environment." Oh no, the Wallace insight would say (and in rather shocked tones): the purpose is to overcome man's own natural, i.e., animal, limitations. Technology enables man, a land-bound biped, without gills, fins or wings, to be at home in the water or in the air. It enables an animal with very poor body insulation, that is, a subtropical animal, to live in all climate zones. It enables one of the weakest and slowest of the primates to add to his own strength that of elephant or ox, and to his own speed that of the horse. It enables him to push his life span from his "natural" twenty years or so to threescore years and ten; it even enables him to forget that natural death is death from predators, disease, starvation, or accident, and to call death from natural

☐ Peter F. Drucker, "Work and Tools," *Technology and Culture,* 1(1959), 28-37.

causes that which has never been observed in wild animals: death from organic decay in old age.[2]

These developments of man have, of course, had impact on his natural environment—though I suspect that until recent days the impact has been very slight indeed. But this impact on nature outside of man is incidental. What really matters is that all these developments alter man's biological capacity—and not through the random genetic mutation of biological evolution but through the purposeful nonorganic development we call "technology."

What I have called here the "Wallace insight," that is, the approach from human biology, thus leads to the conclusion that technology is not about things: tools, processes, and products. It is about work: the specifically human activity by means of which man pushes back the limitations of the iron biological law which condemns all other animals to devote all their time and energy to keeping themselves alive for the next day, if not for the next hour. The same conclusion would be reached, by the way, from any approach, for instance, from that of the anthropologist's "culture," that does not mistake technology for a phenomenon of the physical universe. We might define technology as human action on physical objects or as a set of physical objects characterized by serving human purposes. Either way the realm and subject matter of the study of technology would be human work.

II

For the historian of technology this line of thought might be more than a quibble over definitions. For it leads to the conclusion that the study of the development and history of technology, even in its very narrowest definition as the study of one particular mechanical artifact (either tool or product) or a particular process, would be productive only within an understanding of work and in the context of the history and development of work.

Not only must the available tools and techniques strongly influence what work can and will be done, but how it will be done. Work, its structure, organization, and concepts must in turn powerfully affect tools and techniques and their development. The influence, one would deduce, should be so great as to make it difficult to understand the development of the tool or of the technique unless its relationship to work was known and understood. Whatever evidence we have strongly supports this deduction.

Systematic attempts to study and to improve work only began some seventy-five years ago with Frederick W. Taylor. Until then work had always been taken for granted by everyone—as it is still, apparently, taken for granted by most students of technology. "Scientific management," as Taylor's efforts were called misleadingly ("scientific work study" would have been a better term and would have avoided a great deal of confusion), was not concerned with technology. Indeed, it took tools and techniques largely as given and tried to enable the individual worker to manipulate them more economically, more systematically, and more effectively. And yet this approach resulted almost immediately in major changes and development in tools, processes, and products. The assembly line with its conveyors was an important tool change. An even greater change was the change in process that underlay the switch from building to assembling a product. Today we are beginning to see yet another powerful consequence of Taylor's work on individual operations: the change from organizing production around the doing of things to things, to organizing production around the flow of things and information, the change we call "automation."

A similar, direct impact on tools and techniques is likely to result from another and even more recent approach to the study and improvement of work: the approach called variously "human engineering," "industrial psychology," or "industrial physiology." Scientific management and its descendants study work as operation; human engineering and its allied disciplines are concerned with the relationship between technology and human anatomy, human perception, human nervous system, and human emotion. Fatigue studies were the earliest and most widely known examples; studies of sensory perception and reaction, for instance of airplane pilots, are among the presently most active areas of investigation, as are studies of learning. We have barely scratched the surface here; yet we know already that these studies are leading us to major changes in the theory and design of instruments of measurement and control, and into the re-design of traditional skills, traditional tools, and traditional processes.

But of course we worked on work, if only through trial and error, long before we systematized the job. The best example of scientific management is after all not to be found in our century: it is the alphabet. The assembly line as a concept of work was understood by those unknown geniuses who, at the very beginning of historical time, replaced the aristocratic artist of warfare (portrayed in his last moments of glory by Homer) by the army soldier with his uniform equipment, his few repetitive operations, and his regimented drill. The best example of human engineering is still the long handle that changed the sickle into the scythe, thus belatedly adjusting reaping to the evolutionary change that had much earlier changed man from crouching quadruped into upright biped. Everyone of these developments in work had immediate and powerful impact on tools, process, and product, that is, on the artifacts of technology.

The aspect of work that has probably had the greatest impact on technology is the one we know the least about: the organization of work.

Work, as far back as we have any record of man, has always been both individual and social. The most thoroughly collectivist society history knows, that of Inca Peru, did not succeed in completely collectivizing work; technology, in particular, the making of tools, pottery, textiles, cult objects, remained the work of individuals. It was personally specialized rather than biologically or socially specialized, as is work in a beehive or in an ant heap. The most thoroughgoing individualist society, the perfect market model of classical economics, presupposed a tremendous amount of collective organization in respect to law, money and credit, transportation, and so on. But precisely because individual effort and collective effort must always be calibrated with one another, the organization of work is not determined. To a very considerable extent there are genuine alternatives here, genuine choices. The organization of work, in other words, is in itself one of the major means of that purposeful and non-organic evolution which is specifically human; it is in itself an important tool of man.

Only within the very last decades have we begun to look at the organization of work.[3] But we have already learned that the task, the tools, and the social organization of work are not totally independent but mutually influence and affect one another. We know, for instance, that the almost pre-industrial technology of the New York women's dress industry is the result not of technological, economic, or market conditions but of the social organization of work which is traditional in that industry. The opposite has been proven, too: When we introduce certain tools into locomotive shops, for instance, the traditional organization of work, the organization of the crafts, becomes untenable; and the very skills that made men productive under the old technology now become a major obstacle to their being able to produce at all. A good case can be made for the hypothesis that modern farm implements have made the Russian collective farm socially obsolete as an organization of work, have made it yesterday's socialist solution of farm organization rather than today's, let alone tomorrow's.

This interrelationship between organization of work, tasks, and tools must always have existed. One might even speculate that the explanation for the mysterious time gap between the early introduction of the potter's wheel and the so very late introduction of the spinning wheel lies in the social organization of spinning work as a group task performed, as the Homeric epics describe it, by the mistress working with her daughters and maids. The spinning wheel with its demand for individual concentration on the machinery and its speed is hardly conducive to free social intercourse; even on a narrowly economic basis, the governmental, disciplinary, and educational yields of the spinning bee may well have appeared more valuable than faster and cleaner yarn.

If we know far too little about work and its organization scientifically, we know nothing about it historically. It is not lack of records that explains this, at least not for historical times. Great writers—Hesiod, Aristophanes, Virgil, for instance—have left detailed descriptions. For the early empires and then again for the last seven centuries, beginning with the high Middle Ages, we have an abundance of pictorial material: pottery and relief paintings, woodcuts, etchings, prints. What is lacking is attention and objective study.

III

The political historian or the art historian, still dominated by the prejudices of Hellenism, usually dismisses work as beneath his notice; the historian of technology is "thing-focused." As a result we not only still repeat as fact traditions regarding the organization of work in the past which both our available sources and our knowledge of the organization of work would stamp as old wives' tales, but we also deny ourselves a fuller understanding of the already existing and already collected information regarding the history and use of tools.

One example of this is the lack of attention given to materials-moving and materials-handling equipment. We know that moving things—rather than fabricating things—is the central effort in production. But we have paid little attention to the development of materials-moving and materials-handling equipment.

The Gothic cathedral is another example. H. G. Thomson in *A History of Technology* (II, 384) states, for instance, flatly, "there was no exact medieval equivalent of the specialized architect" in the Middle Ages; there was only "a master mason." But we have overwhelming evidence to the contrary (summarized, for instance, in Simson[4]); the specialized, scientifically trained architect actually dominated. He was sharply distinguished from the master mason by training and social position. Far from being anonymous, as we still commonly assert, he was a famous man, sometimes with an international practice ranging from Scotland to Poland to Sicily. Indeed, he took great pains to make sure that he would be known and remembered, not only in written records but above all by having himself portrayed in the churches he designed in his full regalia as a scientific geometer and designer—something even the best known of today's architects would hesitate to do. Similarly we still repeat early German Romanticism in the belief that the Gothic cathedral was the work of individual craftsmen. But the structural fabric of the cathedral was based on strict uniformity of parts. The men worked to moulds which were collectively held and administered as the property of the guild. Only roofing, ornaments, doors, statuary, windows, and so on, were individual artists' work. Considering both the extreme scarcity of skilled people and the heavy dependence on local, unskilled labor from the countryside, to which all our sources attest, there must also have been a sharp division between the skilled men who made parts and the unskilled who assembled them under the direction of a foreman or a gang boss. There must thus have been a fairly advanced materials-handling technology which is indeed depicted in our sources but neglected by the historians of technology with their uncritical Romanticist bias. And while the moulds to which the craftsman worked are generally mentioned, no one, to my knowledge, has yet investigated so remarkable a tool, and one that so completely contradicts all we otherwise believe we know about medieval work and technology.

I do not mean to suggest that we drop the historical study of tools, processes, and products. We quite obviously need to know much more. I am saying first that the history of work is in itself a big, rewarding, and challenging area which students of technology should be particularly well equipped to tackle. I am saying also that we need work on work if the history of technology is truly to be history and not just the engineer's antiquarianism.

IV

One final question must be asked: Without study and understanding of work, how can we hope to arrive at an understanding of technology?

Singer's great *History of Technology* abandons the attempt to give a comprehensive treatment of its subject with 1850; at that time, the editors tell us, technology became so complex as to defy description, let alone understanding. But it is precisely then that technology began to be a central force and to have major impact both on man's culture and on man's natural environment. To say that we cannot encompass modern technology is very much like saying that medicine stops when the embryo issues from the womb. We need a theory that enables us to organize the variety and complexity of modern tools around some basic, unifying concept.

To a layman who is neither professional historian nor professional technologist, it would moreover appear that even the old technology, the technology before the great explosion of the last hundred years, makes no real sense and cannot be understood, can hardly even be described, without such basic concepts. Every writer on technology acknowledges the extraordinary number, variety, and complexity of factors that play a part in technology and are in turn influenced by it: economy and legal system, political institutions and social values, philosophical abstractions, religious beliefs, and scientific knowledge. No one can know all these, let alone handle them all in their constantly shifting relationship. Yet all of them are part of technology in one way or another, at one time or another.

The typical reaction to such a situation has

of course always been to proclaim one of these factors as *the* determinant—the economy, for instance, or the religious beliefs. We know that this can only lead to complete failure to understand. These factors profoundly influence but do not determine each other; at most they may set limits to each other or create a range of opportunities for each other. Nor can we understand technology in terms of the anthropologist's concept of culture as a stable, complete, and finite balance of these factors. Such a culture may exist among small, primitive, decaying tribes, living in isolation. But this is precisely the reason why they are small, primitive, and decaying. Any viable culture is characterized by capacity for internal self-generated change in the energy level and direction of any one of these factors and in their interrelationships.

Technology, in other words, must be considered as a system,[5] that is, a collection of interrelated and intercommunicating units and activities.

We know that we can study and understand such a system only if we have a unifying focus where the interaction of *all* the forces and factors within the system registers some discernible effect, and where in turn the complexities of the system can be resolved in one theoretical model. Tools, processes, products, are clearly incapable of providing such focus for the understanding of the complex system we call "technology." It is just possible, however, that work might provide the focus, might provide the integration of all these interdependent, yet autonomous variables, might provide one unifying concept which will enable us to understand technology both in itself and in its role, its impact on and relationships with values and institutions, knowledge and beliefs, individual and society.

Such understanding would be of vital importance today. The great, perhaps the central, event of our times is the disappearance of all non-Western societies and cultures under the inundation of Western technology. Yet we have no way of analyzing this process, of predicting what it will do to man, his institutions and values, let alone of controlling it, that is, of specifying with any degree of assurance what needs to be done to make this momentous change productive or at least bearable. We desperately need a real understanding, and a real theory, a real model of technology.

History has never been satisfied to be a mere inventory of what is dead and gone—that indeed is antiquarianism. True history always aims at helping us understand ourselves, at helping us make what shall be. Just as we look to the historian of government for a better understanding of government, and to the historian of art for a better understanding of art, so we are entitled to look to the historian of technology for a better understanding of technology. But how can he give us such an understanding unless he himself has some concept of technology and not merely a collection of individual tools and artifacts? And can he develop such a concept unless work rather than things becomes the focus of his study of technology and of its history?

REFERENCES

[1]S. B. Hamilton only expresses the prevailing view of technologists when he says (in Singer's *A History of Technology,* IV, 469) in respect to the architects of the Gothic cathedral and their patrons that there is "nothing to suggest that either party was driven or pursued by any theory as to what would be beautiful." Yet we have overwhelming and easily accessible evidence to the contrary; both architect and patron were not just "driven," they were actually obsessed by rigorously mathematical theories of structure and beauty. See, for instance, Sedlmayer, *Die Entstehung der Kathedrale* (Zurich, 1950); Von Simson, *The Gothic Cathedral* (New York, 1956); and especially the direct testimony of one of the greatest of the cathedral designers, Abbot Suger of St. Denis, in *Abbot Suger and the Abbey Church of St. Denis,* ed. Erwin Panofsky (Princeton, 1946).

[2]See on this P. B. Medawar, the British biologist, in "Old Age and Natural Death" in his *The Uniqueness of the Individual* (New York, 1957).

[3]Among the studies ought to be mentioned the work of the late Elton Mayo, first in Australia and then at Harvard, especially his two slim books: *The Human Problems of an Industrial Civilization* (Boston, 1933) and *The Social Problems of an Industrial Civilization* (Boston, 1945); the studies of the French sociologist Georges Friedmann, especially his *Industrial Society* (Glencoe, Illinois, 1955); the work carried on at Yale by Charles Walker and his group, especially the book by him and Robert H. Guest: *The Man on the Assembly Line* (Cambridge, Mass., 1952). I understand that studies of the organization of work are also being carried out at the Polish Academy of Science but I have not been able to obtain any of the results.

[4]Von Simson, *op. cit.,* pp. 30 ff.

[5]The word is here used as in Kenneth Boulding's "General Systems Theory—The Skeleton of Science," *Management Science,* II, no. 3 (April 1956), 197, and in the publications of the Society for General Systems Research.

From *The Human Condition*

HANNAH ARENDT

■ Hannah Arendt *was for many years University Professor of Political Philosophy at The New School for Social Research. Her widely read books include* Eichmann in Jerusalem, The Origins of Totalitarianism, *and* The Human Condition.

In the following somewhat difficult but insightful and finely wrought passage from The Human Condition, *Professor Arendt argues that if man were simply* homo faber, *there would be only instrumentability, or means, in his world. But man involves himself with more than just the "in order to" of utility; he is also richly concerned with the "for the sake of"—with ends or meaningfulness. Through brilliant interpretations of Protagoras, Plato, Kant, and Marx, Arendt details the movement from* animal laborans *to* homo faber *and beyond.*

INSTRUMENTALITY AND *HOMO FABER*

The implements and tools of *homo faber,* from which the most fundamental experience of instrumentality arises, determine all work and fabrication. Here it is indeed true that the end justifies the means; it does more, it produces and organizes them. The end justifies the violence done to nature to win the material, as the wood justifies killing the tree and the table justifies destroying the wood. Because of the end product, tools are designed and implements invented, and the same end product organizes the work process itself, decides about the needed specialists, the measure of co-operation, the number of assistants, etc. During the work process, everything is judged in terms of suitability and usefulness for the desired end, and for nothing else.

The same standards of means and end apply to the product itself. Though it is an end with respect to the means by which it was produced and is the end of the fabrication process, it never becomes, so to speak, an end in itself, at least not as long as it remains an object for use. The chair which is the end of carpentering can show its usefulness only by again becoming a means, either as a thing whose durability permits its use as a means for comfortable living or as a means of exchange. The trouble with the utility standard inherent in the very activity of fabrication is that the relationship between means and end on which it relies is very much like a chain whose every end can serve again as a means in some other context. In other words, in a strictly utilitarian world, all ends are bound to be of short duration and to be transformed into means for some further ends.[1]

This perplexity, inherent in all consistent utilitarianism, the philosophy of *homo faber* par excellence, can be diagnosed theoretically as an innate incapacity to understand the distinction between utility and meaningfulness, which we express linguistically by distinguishing between "in order to" and "for the sake of." Thus the ideal of usefulness permeating a society of craftsmen—like the ideal of comfort in a society of laborers or the ideal of acquisition ruling commercial societies—is actually no longer a matter of utility but of meaning. It is "for the sake of" usefulness in general that *homo faber* judges and does everything in terms of "in order to." The ideal of usefulness itself, like the ideals of other societies, can no longer be conceived as something needed in order to have something else; it simply defies questioning about its own use. Obviously there is no answer to the question which Lessing once put to the utilitarian philosophers of his time: "And what is the use of use?" The perplexity of utilitarianism is that it gets caught in the unending chain of means and ends without ever arriving at some principle which could justify the category of means and end, that is, of utility itself. The "in order to" has become the content of the "for the sake of"; in other words, utility established as meaning generates meaninglessness.

Within the category of means and end, and among the experiences of instrumentality which rules over the whole world of use objects and utility, there is not way to end the chain of

1. About the endlessness of the means-end chain (the "*Zweckprogressus in infinitum*") and its inherent destruction of meaning, compare Nietzsche, Aph. 666 in *Wille zur Macht*.

means and ends and prevent all ends from eventually being used again as means, except to declare that one thing or another is "an end in itself." In the world of *homo faber,* where everything must be of some use, that is, must lend itself as an instrument to achieve something else, meaning itself can appear only as an end, as an "end in itself" which actually is either a tautology applying to all ends or a contradiction in terms. For an end, once it is attained, ceases to be an end and loses its capacity to guide and justify the choice of means, to organize and produce them. It has now become an object among objects, that is, it has been added to the huge arsenal of the given from which *homo faber* selects freely his means to pursue his ends. Meaning, on the contrary, must be permanent and lose nothing of its character, whether it is achieved or, rather, found by man or fails man and is missed by him. *Homo faber,* in so far as he is nothing but a fabricator and thinks in no terms but those of means and ends which arise directly out of his work activity, is just as incapable of understanding meaning as the *animal laborans* is incapable of understanding instrumentality. And just as the implements and tools *homo faber* uses to erect the world become for the *animal laborans* the world itself, thus the meaningfulness of this world, which actually is beyond the reach of *homo faber,* becomes for him the paradoxical "end in itself."

The only way out of the dilemma of meaninglessness in all strictly utilitarian philosophy is to turn away from the objective world of use things and fall back upon the subjectivity of use itself. Only in a strictly anthropocentric world, where the user, that is, man himself, becomes the ultimate end which puts a stop to the unending chain of ends and means, can utility as such acquire the dignity of meaningfulness. Yet the tragedy is that in the moment *homo faber* seems to have found fulfilment in terms of his own activity, he begins to degrade the world of things, the end and end product of his own mind and hands; if man the user is the highest end, "the measure of all things," then not only nature, treated by *homo faber* as the almost "worthless material" upon which to work, but the "valuable" things themselves have become mere means, losing thereby their own intrinsic "value."

The anthropocentric utilitarianism of *homo faber* has found its greatest expression in the Kantian formula that no man must ever becomes a means to an end, that every human being is an end in himself. Although we find earlier (for instance, in Locke's insistence that no man can be permitted to possess another man's body or use his bodily strength) an awareness of the fateful consequences which an unhampered and unguided thinking in terms of means and ends must invariably entail in the political realm, it is only in Kant that the philosophy of the earlier stages of the modern age frees itself entirely of the common sense platitudes which we always find where *homo faber* rules the standards of society. The reason is, of course, that Kant did not mean to formulate or conceptualize the tenets of the utilitarianism of his time, but on the contrary wanted first of all to relegate the means-end category to its proper place and prevent its use in the field of political action. His formula, however, can no more deny its origin in utilitarian thinking than his other famous and also inherently paradoxical interpretation of man's attitude toward the only objects that are not "for use," namely works of art, in which he said we take "pleasure without any interest."[2] For the same operation which establishes man as the "supreme end" permits him "if he can [to] subject the whole of nature to it,"[3] that is, to degrade nature and the world into mere means, robbing both of their independent dignity. Not even Kant could solve the perplexity or enlighten the blindness of *homo faber* with respect to the problem of meaning without turning to the paradoxical "end in itself," and this perplexity lies in the fact that while only fabrication with its instrumentality is capable of building a world, this same world becomes as worthless as the employed material, a mere means for further ends, if the standards which governed its coming into being are permitted to rule it after its establishment.

Man, in so far as he is *homo faber,* instrumentalizes, and his instrumentalization implies a degradation of all things into means, their loss of intrinsic and independent value, so that eventually not only the objects of fabrication but also "the earth in general and all forces of nature," which clearly came into being without the help of man and have an existence in-

2. Kant's term is "ein Wohlgefallen ohne alles Interesse" (*Kritik der Urteilskraft* [Cassirer ed.], V, 272).

3. *Ibid.,* p. 515.

dependent of the human world, lose their "value because [they] do not present the reification which comes from work."[4] It was for no other reason than this attitude of *homo faber* to the world that the Greeks in their classical period declared the whole field of the arts and crafts, where men work with instruments and do something not for its own sake but in order to produce something else, to be *banausic,* a term perhaps best translated by "philistine," implying vulgarity of thinking and acting in terms of expediency. The vehemence of this contempt will never cease to startle us if we realize that the great masters of Greek sculpture and architecture were by no means excepted from the verdict.

The issue at stake is, of course, not instrumentality, the use of means to achieve an end, as such, but rather the generalization of the fabrication experience in which usefulness and utility are established as the ultimate standards for life and the world of men. This generalization is inherent in the activity of *homo faber* because the experience of means and end, as it is present in fabrication, does not disappear with the finished product but is extended to its ultimate destination, which is to serve as a use object. The instrumentalization of the whole world and the earth, this limitless devaluation of everything given, this process of growing meaninglessness where every end is transformed into a means and which can be stopped only by making man himself the lord and master of all things, does not directly arise out of the fabrication process; for from the viewpoint of fabrication the finished product is as much an end in itself, as independent durable entity with an existence of its own, as man is an end in himself in Kant's political philosophy. Only in so far as fabrication chiefly fabricates use objects does the finished product again become a means, and only in so far as the life process takes hold of things and uses them for its purposes does the productive and limited instrumentality of fabrication change into the limitless instrumentalization of everything that exists.

It is quite obvious that the Greeks dreaded this devaluation of world and nature with its inherent anthropocentrism—the "absurd" opinion that man is the highest being and that everything else is subject to the exigencies of human life (Aristotle)—no less than they despised the sheer vulgarity of all consistent utilitarianism. To what extent they were aware of the consequences of seeing in *homo faber* the highest human possibility is perhaps best illustrated by Plato's famous argument against Protagoras and his apparently self-evident statement that "man is the measure of all use things *(chrēmata),* of the existence of those that are, and of the nonexistence of those that are not."[5] (Protagoras evidently did not say: "Man is the measure of all things," as tradition and the standard translations have made him say.) The point of the matter is that Plato saw immediately that if one makes man the measure of all things for use, it is man the user and instrumentalizer, and not man the speaker and doer or man the thinker, to whom the world is being related. And since it is in the nature of man the user and instrumentalizer to look upon everything as means to an end—upon every tree as potential wood—this must eventually mean that man becomes the measure not only of things whose existence depends upon him but of literally everything there is.

In this Platonic interpretation, Protagoras in fact sounds like the earliest forerunner of Kant, for if man is the measure of all things, then man is the only thing outside the means-end relationship, the only end in himself who can use everything else as a means. Plato knew quite well that the possibilities of producing use objects and of treating all things of nature as potential use objects are as limitless as the wants and talents of human beings. If one permits the standards of *homo faber* to rule the finished world as they must necessarily rule the coming into being of this world, then *homo faber* will eventually help himself to everything and consider everything that is as a mere means for himself. He will judge every thing

4. "Der Wasserfall, wie die Erde überhaupt, wie alle Naturkraft hat keinen Wert, weil er keine in ihm vergegenständlichte Arbeit darstellt" (*Das Kapital,* III [*Marx-Engels Gesamtausgabe,* Abt. II, Zürich, 1933], 698).

5. *Theaetetus* 152, and *Cratylus* 385E. In these instances, as well as in other ancient quotations of the famous saying, Protagoras is always quoted as follows: *pānton chrēmatōn mentron estin anthrōpos* (see Diels, *Fragmente der Vorsokratiker* [4th ed.; 1922], frag. B1). The word *chrēmata* by no means signifies "all things," but specifically things used or needed or possessed by men. The supposed Protagorean saying, "Man is the measure of all things," would be rendered in Greek rather as *anthrōpos mentron pantōn,* corresponding for instance to Heraclitus' *polemos pater pantōn* ("strife is the father of all things").

as though it belonged to the class of *chrēmata,* of use objects, so that, to follow Plato's own example, the wind will no longer be understood in its own right as a natural force but will be considered exclusively in accordance with human needs for warmth or refreshment—which, of course, means that the wind as something objectively given has been eliminated from human experience. It is because of these consequences that Plato, who at the end of his life recalls once more in the *Laws* the saying of Protagoras, replies with an almost paradoxical formula: not man—who because of his wants and talents wishes to use everything and therefore ends by depriving all things of their intrinsic worth—but "the god is the measure [even] of mere use objects."[6]

THE EXCHANGE MARKET

Marx—in one of many asides which testify to his eminent historical sense—once remarked that Benjamin Franklin's definition of man as a toolmaker is as characteristic of "Yankeedom," that is, of the modern age, as the definition of man as a political animal was for antiquity.[7] The truth of this remark lies in the fact that the modern age was as intent on excluding political man, that is, man who acts and speaks, from its public realm as antiquity was on excluding *homo faber*. In both instances the exclusion was not a matter of course, as was the exclusion of laborers and the propertyless classes until their emancipation in the nineteenth century. The modern age was of course perfectly aware that the political realm was not always and need not necessarily be a mere function of "society," destined to protect the productive, social side of human nature through governmental administration; but it regarded everything beyond the enforcement of law and order as "idle talk" and "vain-glory." The human capacity on which it based its claim of the natural innate productivity of society was the unquestionable productivity of *homo faber*. Conversely, antiquity knew full well types of human communities in which not the citizen of the *polis* and not the *res publica* as such established and determined the content of the public realm, but where the public life of the ordinary man was restricted to "working for the people" at large, that is, to being a *dēmiourgos,* a worker for the people as distinguished from an *oiketēs,* a household laborer and therefore a slave.[8] The hallmark of these non-political communities was that their public place, the *agora,* was not a meeting place of citizens, but a market place where craftsmen could show and exchange their products. In Greece, moreover, it was the ever-frustrated ambition of all tyrants to discourage the citizens from worrying about public affairs, from idling their time away in unproductive *agoreuein* and *politeuesthai,* and to transform the *agora* into an assemblage of shops like the bazaars of oriental despotism. What characterized these market places, and later characterized the medieval cities' trade and craft districts, was that the display of goods for sale was accompanied by a display of their production. "Conspicuous production" (if we may vary Veblen's term) is, in fact, no less a trait of a society of producers than "conspicuous consumption" is a characteristic of a laborers' society.

Unlike the *animal laborans,* whose social life is worldless and herdlike and who therefore is incapable of building or inhabiting a public, worldly realm, *homo faber* is fully capable of having a public realm of his own, even though it may not be a political realm, properly speaking. His public realm is the exchange market, where he can show the products of his hand and receive the esteem which is due him. This inclination to showmanship is closely connected with and probably no less deeply rooted than the "propensity to truck, barter and exchange one thing for another," which, according to Adam Smith, distinguishes man

6. *Laws* 716D quotes the saying of Protagoras textually, except that for the word "man" *(anthrōpos),* "the god" *(ho theos)* appears.

7. *Capital* (Modern Library ed.), p. 358, n. 3.

8. Early medieval history, and particularly the history of the craft guilds, offers a good illustration of the inherent truth in the ancient understanding of laborers as household inmates, as against craftsmen, who were considered workers for the people at large. For the "appearance [of the guilds] marks the second stage in the history of industry, the transition from the family system to the artisan or guild system. In the former there was no class of artisans properly so called . . . because all the needs of a family or other domestic groups . . . were satisfied by the labours of the members of the group itself" (W. J. Ashley, *An Introduction to English Economic History and Theory* [1931], p. 76).

In medieval German, the word *Störer* is an exact equivalent to the Greek word *dēmiourgos.* "Der griechische *dēmiourgos* heisst 'Störer', er geht beim Volk arbeiten, ergeht auf die Stör." *Stör* means *dēmos* ("people"). (See Jost Trier, "Arbeit und Gemeinschaft," *Studium Generale,* Vol. III, No. 11 [November, 1950].)

from animal.[9] The point is that *homo faber,* the builder of the world and the producer of things, can find his proper relationship to other people only by exchanging his products with theirs, because these products themselves are always produced in isolation. The privacy which the early modern age demanded as the supreme right of each member of society was actually the guaranty of isolation, without which no work can be produced. Not the onlookers and spectators on the medieval market places, where the craftsman in his isolation was exposed to the light of the public, but only the rise of the social realm, where the others are not content with beholding, judging, and admiring but wish to be admitted to the company of the craftsman and to participate as equals in the work process, threatened the "splendid isolation" of the worker and eventually undermined the very notions of competence and excellence. This isolation from others is the necessary life condition for every mastership which consists in being alone with the "idea," the mental image of the thing to be. This mastership, unlike political forms of domination, is primarily a mastery of things and material and not of people. The latter, in fact, is quite secondary to the activity of craftsmanship, and the words "worker" and "master"—*ouvrier* and *maître*—were originally used synonymously.[10]

The only company that grows out of workmanship directly is in the need of the master for assistants or in his wish to educate others in his craft. But the distinction between his skill and the unskilled help is temporary, like the distinction between adults and children. There can be hardly anything more alien or even more destructive to workmanship than teamwork, which actually is only a variety of the division of labor and presupposes the "breakdown of operations into their simple constituent motions."[11] The team, the multiheaded subject of all production carried out according to the principle of division of labor, possesses the same togetherness as the parts which form the whole, and each attempt of isolation on the part of the members of the team would be fatal to the production itself. But it is not only this togetherness which the master and workman lacks while actively engaged in production; the specifically political forms of being together with others, acting in concert and speaking with each other, are completely outside the range of his productivity. Only when he stops working and his product is finished can he abandon his isolation.

Historically, the last public realm, the last meeting place which is at least connected with the activity of *homo faber,* is the exchange market on which his products are displayed. The commercial society, characteristic of the earlier stages of the modern age or the beginnings of manufacturing capitalism, sprang from this "conspicuous production" with its concomitant hunger for universal possibilities of truck and barter, and its end came with the rise of labor and the labor society which replaced conspicuous production and its pride with "conspicuous consumption" and its concomitant vanity.

The people who met on the exchange market, to be sure, were no longer the fabricators themselves, and they did not meet as persons but as owners of commodities and exchange values, as Marx abundantly pointed out. In a society where exchange of products has become the chief public activity, even the laborers, because they are confronted with "money or commodity owners," become proprietors, "owners of their labor power." It is only at this point that Marx's famous self-alienation, the degradation of men into commodities, sets in, and this degradation is characteristic of labor's situation in a manufacturing society which judges men not as persons but as producers, according to the quality of their products. A laboring society, on the contrary, judges men according to the functions they perform in the labor process; while labor power in the eyes of *homo*

9. He adds rather emphatically: "Nobody ever saw a dog make a fair and deliberate exchange of one bone for another with another dog" (*Wealth of Nations* [Everyman's ed.], I, 12).

10. E. Levasseur, *Histoire des classes ouvrières et de l'industrie en France avant 1789* (1900): "Le mots maître et ouvrier étaient encore pris comme synonymes au 14e siècle" (p. 564, n. 2), whereas, "au 15e siècle . . .la maîtrise est devenue un titre auquel il n'est premis à tous d'aspirer" (p. 572). Originally, "le mot ouvrier s'appliquait d'ordinaire à quiconque ouvrait, faisait ouvrage, maître ou valet" (p. 309). In the workshops themselves and outside them in social life, there was no great distinction between the master or the owner of the shop and the workers (p. 313). (See also Pierre Brizon, *Histoire du travail et des travailleurs* [4th ed.; 1926], pp. 39 ff.)

11. Charles R. Walker and Robert H. Guest, *The Man on the Assembly Line* (1952), p. 10. Adam Smith's famous description of this principle in pin-making (*op. cit.,* I, 4 ff.) shows clearly how machine work was preceded by the division of labor and derives its principle from it.

faber is only the means to produce the necessarily higher end, that is, either a use object or an object for exchange, laboring society bestows upon labor power the same higher value it reserves for the machine. In other words, this society is only seemingly more "humane," although it is true that under its conditions the price of human labor rises to such an extent that it may seem to be more valued and more valuable than any given material or matter; in fact, it only foreshadows something even more "valuable," namely, the smoother functioning of the machine whose tremendous power of processing first standardizes and then devaluates all things into consumer goods.

Commercial society, or capitalism in its earlier stages when it was still possessed by a fiercely competitive and acquisitive spirit, is still ruled by the standards of *homo faber*. When *homo faber* comes out of his isolation, he appears as a merchant and trader and establishes the exchange market in this capacity. This market must exist prior to the rise of a manufacturing class, which then produces exclusively for the market, that is, produces exchange objects rather than use things. In this process from isolated craftsmanship to manufacturing for the exchange market, the finished end product changes its quality somewhat but not altogether. Durability, which alone determines if a thing can exist as a thing and endure in the world as a distinct entity, remains the supreme criterion, although it no longer makes a thing fit for use but rather fit to "be stored up beforehand" for future exchange.[12]

This is the change in quality reflected in the current distinction between use and exchange value, whereby the latter is related to the former as the merchant and trader is related to the fabricator and manufacturer. In so far as *homo faber* fabricates use objects, he not only produces them in the privacy of isolation but also for the privacy of usage, from which they emerge and appear in the public realm when they become commodities in the exchange market. It has frequently been remarked and unfortunately as frequently been forgotten that value, being "an idea of proportion between the possession of one thing and the possession of another in the conception of man,"[13] "always means value in exchange."[14] For it is only in the exchange market, where everything can be exchanged fore something else, that all things, whether they are products of labor or work, consumer goods or use objects, necessary for the life of the body or the convenience of living or the life of the mind, become "values." This value consists solely in the esteem of the public realm where the things appear as commodities, and it is neither labor, nor work, nor capital, nor profit, nor material, which bestows such value upon an object, but only and exclusively the public realm where it appears to be esteemed, demanded, or neglected. Value is the quality a thing can never possess in privacy but acquires automatically the moment it appears in public. This "marketable value," as Locke very clearly pointed out, has nothing to do with "the intrinsick natural worth of anything"[15] which is an objective quality of the thing itself, "outside the will of the individual purchaser or seller; something attached to the thing itself, existing whether he liked it or not, and that he ought to recognize."[16] This intrinsic worth of a thing can be changed only through the change of the thing itself—thus one ruins the worth of a table by depriving it of one of its legs—whereas "the marketable value" of a commodity is altered by "the alteration of some proportion which

12. Adam Smith, *op. cit.*, II, 241.

13. This definition was given by the Italian economist Abbey Galiani. I quote from Hannah R. Sewall, *The Theory of Value before Adam Smith* (1901) ("Publications of the American Economic Association," 3d Ser., Vol. II, No. 3), p. 92.

14. Alfred Marshall, *Principles of Economics* (1920), I, 8.

15. "Considerations upon the Lowering of Interest and Raising the Value of Money," *Collected Works* (1801), II, 21.

16. W. J. Ashley (*op. cit.*, p. 140) remarks that "the fundamental difference between the medieval and modern point of view . . . is that, with us, value is something entirely subjective; it is what each individual cares to give for a thing. With Aquinas it was something objective." This is true only to an extent, for "the first thing upon which the medieval teachers insist is that value is not determined by the intrinsic excellence of the thing itself, because, if it were, a fly would be more valuable than a pearl as being intrinsically more excellent" (George O'Brien, *An Essay on Medieval Economic Teaching* [1920], p. 109). The discrepancy is resolved if one introduces Locke's distinction between "worth" and "value," calling the former *valor naturalis* and the latter *pretium* and also *valor*. This distinction exists, of course, in all but the most primitive societies, but in the modern age the former disappears more and more in favor of the latter. (For medieval teaching, see also Slater, "Value in Theology and Political Economy," *Irish Ecclesiastical Record* [September, 1901].)

that commodity bears to something else."[17]

Values, in other words, in distinction from things or deeds or ideas, are never the products of a specific human activity, but come into being whenever any such products are drawn into the ever-changing relativity of exchange between the members of society. Nobody, as Marx rightly insisted, seen "in his isolation produces values," and nobody, he could have added, in his isolation cares about them; things or ideas or moral ideals "become values only in their social relationship."[18]

The confusion in classical economics,[19] and the worse confusion arising from the use of the term "value" in philosophy, were originally caused by the fact that the older word "worth," which we still find in Locke, was supplanted by the seemingly more scientific term, "use value." Marx, too, accepted this terminology and, in line with his repugnance to the public realm, saw quite consistently in the change from use value to exchange value the original sin of capitalism. But against these sins of a commercial society, where indeed the exchange market is the most important public place and where therefore every thing becomes an exchangeable value, a commodity, Marx did not summon up the "intrinsick" objective worth of the thing in itself. In its stead he put the function things have in the consuming life process of men which knows neither objective and intrinsic worth nor subjective and socially determined value. In the socialist equal distribution of all goods to all who labor, every tangible thing dissolves into a mere function in the regeneration process of life and labor power.

However, this verbal confusion tells only one part of the story. The reason for Marx's stubborn retention of the term "use value," as well as for the numerous futile attempts to find some objective source—such as labor, or land, or profit—for the birth of values, was that nobody found it easy to accept the simple fact that no "absolute value" exists in the exchange market, which is the proper sphere for values, and that to look for it resembled nothing so much as the attempt to square the circle. The much deplored devaluation of all things, that is, the loss of all intrinsic worth, begins with their transformation into values or commodities, for from this moment on they exist only in relation to some other thing which can be acquired in their stead. Universal relativity, that a thing exists only in relation to other things, and loss of intrinsic worth, that nothing any longer possesses an "objective" value independent of the ever-changing estimations of supply and demand, are inherent in the very concept of value itself.[20] The reason why this development, which seems inevitable in a commercial society, became a deep source of uneasiness and eventually constituted the chief problem of the new science of economics was not even relativity as such, but rather the fact that *homo faber,* whose whole activity is determined by the constant use of yardsticks, measurements, rules, and standards, could not bear the loss of "absolute" standards or yardsticks. For money, which obviously serves as the common denominator for the variety of things so that they can be exchanged for each other, by no means possesses the independent and objective existence, transcending all uses and surviving all manipulation, that the yardstick or any other measurement possesses with regard to the things it is supposed to measure and to the men who handle them.

It is this loss of standards and universal rules, without which no world could ever be erected by man, that Plato already perceived in the Protagorean proposal to establish man, the fabricator of things, and the use he makes of them, as their supreme measure. This shows how closely the relativity of the exchange market

17. Locke, *Second Treatise of Civil Government,* sec. 22.

18. *Das Kapital,* III, 689 (*Marx-Engels Gesamtausgabe,* Part II [Zürich, 1933]).

19. The clearest illustration of the confusion is Ricardo's theory of value, especially his desperate belief in an absolute value. (The interpretations in Gunnar Myrdal, *The Political Element in the Development of Economic Theory* [1953], pp. 66 ff., and Walter A. Weisskopf, *The Psychology of Economics* [1955], ch. 3, are excellent.)

20. The truth of Ashley's remark, which we quoted above (n. 34), lies in the fact that the Middle Ages did not know the exchange market, properly speaking. To the medieval teachers the value of a thing was either determined by its worth or by the objective needs of men—as for instance in Buridan: *valor rerum aestimatur secundum humanam indigentiam*—and the "just price" was normally the result of the common estimate, except that "on account of the varied and corrupt desires of man, it becomes expedient that the medium should be fixed according to the judgment of some wise men" (Gerson *De contractibus* i. 9, quoted from O'Brien, *op. cit.,* pp. 104 ff.). In the absence of an exchange market, it was inconceivable that the value of one thing should consist solely in its relationship or proportion to another thing. The question, therefore, is not so much whether value is objective or subjective, but whether it can be absolute or indicates only the relationship between things.

is connected with the instrumentality arising out of the world of the craftsman and the experience of fabrication. The former, indeed, develops without break and consistently from the latter. Plato's reply, however—not man, a "god is the measure of all things"—would be an empty, moralizing gesture if it were really true, as the modern age assumed, that instrumentality under the disguise of usefulness rules the realm of the finished world as exclusively as it rules the activity through which the world and all things it contains came into being.

THE PERMANENCE OF THE WORLD AND THE WORK OF ART

Among the things that give the human artifice the stability without which it could never be a reliable home for men are a number of objects which are strictly without any utility whatsoever and which, moreover, because they are unique, are not exchangeable and therefore defy equalization through a common denominator such as money; if they enter the exchange market, they can only be arbitrarily priced. Moreover, the proper intercourse with a work of art is certainly not "using" it; on the contrary, it must be removed carefully from the whole context of ordinary use objects to attain its proper place in the world. By the same token, it must be removed from the exigencies and wants of daily life, with which it has less contact than any other thing. Whether this uselessness of art objects has always pertained or whether art formerly served the so-called religious needs of men as ordinary use objects serve more ordinary needs does not enter the argument. Even if the historical origin of art were of an exclusively religious or mythological character, the fact is that art has survived gloriously its severance from religion, magic, and myth.

Because of their outstanding permanence, works of art are the most intensely worldly of all tangible things; their durability is almost untouched by the corroding effect of natural processes, since they are not subject to the use of living creatures, a use which, indeed, far from actualizing their own inherent purpose—as the purpose of a chair is actualized when it is sat upon—can only destroy them. Thus, their durability is of a higher order than that which all things need in order to exist at all; it can attain permanence throughout the ages. In this permanence, the very stability of the human artifice, which, being inhabited and used by mortals, can never be absolute, achieves a representation of its own. Nowhere else does the sheer durability of the world of things appear in such purity and clarity, nowhere else therefore does this thing-world reveal itself so spectacularly as the non-mortal home for mortal beings. It is as though worldly stability had become transparent in the permanence of art, so that a premonition of immortality, not the immortality of the soul or of life but of something immortal achieved by mortal hands, has become tangibly present, to shine and to be seen, to sound and to be heard, to speak and to be read.

The immediate source of the art work is the human capacity for thought, as man's "propensity to truck and barter" is the source of exchange objects, and as his ability to use is the source of use things. These are capacities of man and not mere attributes of the human animal like feelings, wants, and needs, to which they are related and which often constitute their content. Such human properties are as unrelated to the world which man creates as his home on earth as the corresponding properties of other animal species, and if they were to constitute a man-made environment for the human animal, this would be a non-world, the product of emanation rather than of creation. Thought is related to feeling and transforms its mute and inarticulate despondency, as exchange transforms the naked greed of desire and usage transforms the desperate longing of needs—until they all are fit to enter the world and to be transformed into things, to become reified. In each instance, a human capacity which by its very nature is world-open and communicative transcends and releases into the world a passionate intensity from its imprisonment within the self.

In the case of art works, reification is more than mere transformation; it is transfiguration, a veritable metamorphosis is which it is as though the course of nature which wills that all fire burn to ashes is reverted and even dust can burst into flames.[21] Works of art are thought

21. The text refers to a poem by Rilke on art, which under the title "Magic," describes this transfiguration. It reads as follows: "Aus unbeschreiblicher Verwandlung stammen/solche Gebilde—: Fühl! und blaub!/Wir leidens oft: zu Asche werden Flammen,/doch, in der Kunst: zur Flamme wird der Staub./Hier ist Magie. In das Bereich des Zaubers/scheint das gemeine Wort hinaufgestuft . . ./ und ist doch wirklich wie der Ruf des Taubers,/der nach der unsichtbaren Taube ruft" (in *Aus Taschen-Büchern und Merk-Blättern* [1950]).

things, but this does not prevent their being things. The thought process by itself no more produces and fabricates tangible things, such as books, paintings, sculptures, or compositions, than usage by itself produces and fabricates houses and furniture. The reification which occurs in writing something down, painting an image, modeling a figure, or composing a melody is of course related to the thought which preceded it, but what actually makes the thought a reality and fabricates things of thought is the same workmanship which, through the primordial instrument of human hands, builds the other durable things of the human artifice.

We mentioned before that this reification and materialization, without which no thought can become a tangible thing, is always paid for, and that the price is life itself: it is always the "dead letter" in which the "living spirit" must survive, a deadness from which it can be rescued only when the dead letter comes again into contact with a life willing to resurrect it, although this resurrection of the dead shares with all living things that it, too, will die again. This deadness, however, though somehow present in all art and indicating, as it were, the distance between thought's original home in the heart or head of man and its eventual destination in the world, varies in the different arts. In music and poetry, the least "materialistic" of the arts because their "material" consists of sounds and words, reification and the workmanship it demands are kept to a minimum. The young poet and the musical child prodigy can attain a perfection without much training and experience—a phenomenon hardly matched in painting, sculpture, or architecture.

Poetry, whose material is language, is perhaps the most human and least worldly of the arts, the one in which the end product remains closest to the thought that inspired it. The durability of a poem is produced through condensation, so that it is as though language spoken in utmost density and concentration were poetic in itself. Here, remembrance, *Mnemosynē,* the mother of the muses, is directly transformed into memory, and the poet's means to achieve the transformation is rhythm, through which the poem becomes fixed in the recollection almost by itself. It is this closeness to living recollection that enable the poem to remain, to retain its durability, outside the printed or the written page, and though the "quality" of a poem may be subject to a variety of standards, its "memorability" will inevitably determine its durability, that is, its chance to be permanently fixed in the recollection of humanity. Of all things of thought, poetry is closest to thought, and a poem is less a thing than any other work of art; yet even a poem, no matter how long it existed as a living spoken word in the recollection of the bard and those who listened to him, will eventually be "made," that is, written down and transformed into a tangible thing among things, because remembrance and the gift of recollection, from which all desire for imperishability springs, need tangible things to remind them, lest they perish themselves.[22]

Thought and cognition are not the same. Thought, the source of art works, is manifest without transformation or transfiguration in all great philosophy, whereas the chief manifestation of the cognitive processes, by which we acquire and store up knowledge, is the sciences. Cognition always pursues a definite aim, which can be set by practical considerations as well as by "idle curiosity"; but once this aim is reached, the cognitive process has come to an end. Thought, on the contrary, has neither an end nor an aim outside itself, and it does not even produce results; not only the utilitarian philosophy of *homo faber* but also the men of action and the lovers of results in the sciences have never tired of pointing out how entirely "useless" thought is—as useless, indeed, as the works of art it inspires. And not even to these useless products can thought lay claim, for they as well as the great philosophic systems can hardly be called the results of pure thinking, strictly speaking, since it is precisely the thought process which the artist or writing philosopher must interrupt and transform for the materializing reification of his work. The activ-

22. The idiomatic "make a poem" or *faire des vers* for the activity of the poet already relates to this reification. The same is true for the German *dichten,* which probably comes from the Latin *dictare:* "das ausgesonnene geistig Geschaffene niederschreiben oder zum Niederschreiben vorsagen" (Grimm's *Wörterbuch*); the same would be true if the word were derived, as is now suggested by the *Etymologisches Wörterbuch* (1951) of Kluge/Götze, from *tichen,* an old word for *schaffen,* which is perhaps related to the Latin *fingere.* In this case, the poetic activity which produces the poem before it is written down is also understood as "making." Thus Democritus praised the divine genius of Homer, who "framed a cosmos out of all kinds of words"—*epeōn kosmon etektēnato pantoiōn* (Diels, *op. cit.,* B21). The same emphasis on the craftsmanship of poets is present in the Greek idiom for the art of poetry: *tektōnes hymnōn.*

ity of thinking is as relentless and repetitive as life itself, and the question whether thought has any meaning at all constitutes the same unanswerable riddle as the question for the meaning of life; its processes permeate the whole of human existence so intimately that its beginning and end coincide with the beginning and end of human life itself. Thought, therefore, although it inspires the highest worldly productivity of *homo faber,* is by no means his prerogative; it begins to assert itself as his source of inspiration only where he overreaches himself, as it were, and begins to produce useless things, objects which are unrelated to material or intellectual wants, to man's physical needs no less than to his thirst for knowledge. Cognition, on the other hand, belongs to all, and not only to intellectual or artistic work processes; like fabrication itself, it is a process with a beginning and end, whose usefulness can be tested, and which, if it produces no results, has failed, like a carpenter's workmanship has failed when he fabricates a two-legged table. The cognitive processes in the sciences are basically not different from the function of cognition in fabrication; scientific results produced through cognition are added to the human artifice like all other things.

Both thought and cognition, furthermore, must be distinguished from the power of logical reasoning which is manifest in such operations as deductions from axiomatic or self-evident statements, subsumption of particular occurrences under general rules, or the techniques of spinning out consistent chains of conclusions. In these human faculties we are actually confronted with a sort of brain power which in more than one respect resembles nothing so much as the labor power the human animal develops in its metabolism with nature. The mental processes which feed on brain power we usually call intelligence, and this intelligence can indeed be measured by intelligence tests as bodily strength can be measured by other devices. Their laws, the laws of logic, can be discovered like other laws of nature because they are ultimately rooted in the structure of the human brain, and they possess, for the normally healthy individual, the same force of compulsion as the driving necessity which regulates the other functions of our bodies. It is in the structure of the human brain to be compelled to admit that two and two equal four. If it were true that man is an *animal rationale* in the sense in which the modern age understood the term, namely, an animal species which differs from other animals in that it is endowed with superior brain power, then the newly invented electronic machines, which, sometimes to the dismay and sometimes to the confusion of their inventors, are so spectacularly more "intelligent" than human beings, would indeed be *homunculi*. As it is, they are, like all machines, mere substitutes and artificial improvers of human labor power, following the time-honored device of all division of labor to break down every operation into its simplest constituent motions, substituting, for instance, repeated addition for multiplication. The superior power of the machine is manifest in its speed, which is far greater than that of human brain power; because of this superior speed, the machine can dispense with multiplication, which is the pre-electronic technical device to speed up addition. All that the giant computers prove is that the modern age was wrong to believe with Hobbes that rationality, in the sense of "reckoning with consequences," is the highest and most human of man's capacities, and that the life and labor philosophers, Marx or Bergson or Nietzsche, were right to see in this type of intelligence, which they mistook for reason, a mere function of the life process itself, or, as Hume put it, a mere "slave of the passions." Obviously, this brain power and the compelling logical processes it generates are not capable of erecting a world, are as worldless as the compulsory processes of life, labor, and consumption.

One of the striking discrepancies in classical economics is that the same theorists who prided themselves on the consistency of their utilitarian outlook frequently took a very dim view of sheer utility. As a rule, they were well aware that the specific productivity of work lies less in its usefulness than in its capacity for producing durability. By this discrepancy, they tacitly admit the lack of realism in their own utilitarian philosophy. For although the duration of ordinary things is but a feeble reflection of the permanence of which the most worldly of all things, works of art, are capable, something of this quality—which to Plato was divine because it approaches immortality—is inherent in every thing as a thing, and it is precisely this quality or the lack of it that shines forth in its shape and makes it beautiful or ugly. To be sure, an ordinary use object is not and should not be intended to be beautiful; yet whatever has a shape at all and is seen cannot help being either beautiful, ugly, or something in-between. Everything that is, must appear, and nothing can appear without a shape of its own; hence there is in fact no thing that does not in some way transcend its functional use, and its transcendence, its beauty or ugliness, is identical with appearing publicly and being seen. By the same token, namely, in its sheer worldly existence, every thing also transcends the sphere of pure instrumentality once it is completed. The standard by which a thing's excellence is judged is never mere usefulness, as though an ugly table will fulfil the same function as a handsome one, but its adequacy or inadequacy to what it should *look* like, and this is, in Platonic language, nothing but its adequacy or inadequacy to the *eidos* or *idea,* the mental image, or rather the image seen by the inner eye, that preceded its coming into the world and survives its potential destruction. In other words, even use objects are judged not only according to the subjective needs of men but by the objective standards of the world where they will find their place, to last, to be seen, and to be used.

The man-made world of things, the human artifice erected by *homo faber,* becomes a home for mortal men, whose stability will endure and outlast the ever-changing movements of their lives and actions, only insomuch as it transcends both the sheer functionalism of things produced for consumption and the sheer utility of objects produced for use. Life in its non-biological sense, the span of time each man has between birth and death, manifests itself in action and speech, both of which share with life its essential futility. The "doing of great deeds and the speaking of great words" will leave no trace, no product that might endure after the moment of action and the spoken word has passed. If the *animal laborans* needs the help of *homo faber* to ease his labor and remove his pain, and if mortals need his help to erect a home on earth, acting and speaking men need the help of *homo faber* in his highest capacity, that is, the help of the artist, of poets and historiographers, of monument-builders or writers, because without them the only product of their activity, the story they enact and tell, would not survive at all. In order to be what the world is always meant to be, a home for men during their life on earth, the human artifice must be a place fit for action and speech, for activities not only entirely useless for the necessities of life but of an entirely different nature from the manifold activities of fabrication by which the world itself and all things in it are produced. We need not choose here between Plato and Protagoras, or decide whether man or a god should be the measure of all things; what is certain is that the measure can be neither the driving necessity of biological life and labor nor the utilitarian instrumentalism of fabrication and usage.

14

Technology and metaphor

Technological metaphor

D. O. EDGE

▪ D. O. Edge *is director of the Sciences Studies Unit of the University of Edinburgh. Besides his contributions to a philosophy of technology, he has written, with Michael J. Mulkay,* Astronomy Transformed, *an account of the emergence of radio astronomy in Britain.*

In the following essay Professor Edge is concerned with the nature of metaphor in general and more specifically with the ways in which metaphors that attend innovations in technology provide us with new models and patterns of experience. Along the way he discusses the nature of "central cultural metaphor," and he concludes with some remarks concerning the alleged "dehumanizing" influence of science and technology.

". . . Everything's television you can't touch things easily at close quarters . . ."

George Seferis *(1970, p. 68)*

". . . not knowing how it is proper to feel is even more disconcerting than not knowing how it is proper to think."

Lynn White *(1968, p. 171)*

INTRODUCTION

Of the rapidly expanding literature on problems on the interface of science and society, perhaps only one thing can be said with any confidence—namely, that the use of the word "interface" in this discussion is a metaphorical usage. What is more, it is what I propose to call a "technological metaphor", since the currency of the term, if I am not mistaken, stems from its sophisticated technical application in computer technology. That is where the primary reference or current literal focus on the term "interface" is to be found.[1] Its usefulness in this primary domain leads to its metaphorical transfer to other domains, so that the nature of our understanding of the relationship between "science" and "society" comes to partake of features of the known relationship between "hardware" and "software". An uncertain and obscure area is construed in terms of one both familiar and apparently similar. Inappropriate aspects of this transfer may remain unnoticed, in the first flush of clarification that the new metaphor appears to bring.

Discussions of the impact of science and technology seldom refer to this process. It is, in essence, that described by Donald Schon in his important book *Displacement of concepts* (1963) and my remarks in this paper can be thought of as footnotes to Schon's thesis. I want to advance the proposition that one major result of technological innovation is to provide us with new models and patterns of experience that can serve as basic metaphors for thought about men and society; and that the (usually quite unconscious) infiltration of these metaphors into common parlance can have fundamental effects on modes of thought and feeling ("consciousness") within a culture.

I will proceed, first, to some general remarks on, and a brief description of the so-called "tensional theory" of, metaphor; second, to a consideration of some hidden effects of technological metaphor; third, to a suggestion about a central cultural metaphor (the metaphor of knowledge), and the way in which technology can influence it; and, finally, to some remarks on

☐ From *Meaning and Control,* ed. D. O. Edge and J. N. Wolfe (London: Tavistock Publications, Ltd., 1973), pp. 31-59.

the alleged "dehumanizing" influence of science and technology. Any discussion that addresses itself to these issues can hardly avoid being speculative and conjectural, and I do not claim that my own essay will be an exception.

METAPHOR

Chambers Twentieth Century Dictionary offers this definition: "Metaphor: a figure of speech by which a thing is spoken of as being that which it only resembles, as when a ferocious man is called a tiger."

The tone of this definition accords neatly with the common belief that a metaphor is a mere figure of speech. The rejection of this easy belief is central to my position. I take the "figure of speech" to be just the verbal aspect of a process of cognition in which objects are perceived, construed, and acted towards, as if they shared in the *being* of things that the dictionary says they "only resemble". It could be objected that what I wish to discuss is a matter of "models" rather than metaphors. That I prefer to talk of "metaphor" is partly a matter of the linguistic intimacy of metaphorical language (as opposed to the rather abstract and theoretical implications of talk of models), and partly because the use of the term "model" here might itself be taken to be metaphorical.

Rejection of the "mereness" of metaphor is also to be found among those philosophers of science who, in recent years, have turned their attention to the role of metaphor in the development of scientific thought (see Hesse 1966).[2] In summarizing their characterization of metaphor, I will draw on the paper of Douglas Berggren (1962-3), for a specific reason that will emerge later. The main points are as follows:

(a) A metaphor must, by definition, be held to be *literally absurd*. A poet need not propound that "man is a tiger" if this is already a commonplace literal belief; he would not be communicating anything. In like manner, it is absurd to propose that the universe is, literally, a clock (or a billiard-table); that the brain is a computer; that metaphor is a slippery slope, or a double-edged sword; that the Medium is the Message; that poverty (or exploitation) is violence; or that Jesus Christ is the Son of God (or a Sacred Mushroom). Given our usual "literal" meanings of the words involved, all these statements are absurd, although it is clearly *not* necessarily absurd to propound them. What (if anything) they can convey must be teased out with greater care.

(b) Not *every* absurdity yields a metaphor, of course; one needs also "some context . . . which can provide an intelligible connection between the two referents . . . some principle of assimilation which can meaningfully link" them (Berggren 1962-3, p. 240). When these conditions are satisfied, metaphor can lead to "the outreach and extension of meaning through comparison" (Wheelwright 1962, p. 72). But metaphor can also have a more creative role: it can lead to "the creation of new meaning by juxtaposition and synthesis" (Wheelwright ibid.). Metaphor, when acting as a "lens or filter" (Berggren 1962-3, p. 242) can actually *create* similarity. To use a metaphor is to overlap two images; as the metaphor takes its hold, men appear more tigerish—and tigers more human. Our sense of both is subtly altered, by a sort of elision.

(c) These creative (and cognitive) roles of metaphor demand two kinds of "tension". The first, (a) above, is absurdity; the second, (b) above, is characterized by Berggren in terms of "stereoscopic vision" and "non-reducibility". It is not that we switch from seeing man as man to seeing man as a tiger, note a few points of similarity, and then switch back again to our original frame of reference. The essence of a live metaphor is that it sets a puzzle, and suggests all kinds of *possible* relations ("unspoken connections", to use John Wisdom's phrase); the two strangely interacting images are "seen" *together,* overlaid, in much the same way as the brain copes simultaneously with the two images from our eyes. And the tensions arise from the fact that the possibilities inherent in (and thus created by) a live metaphor cannot (again by definition) be reduced to literal statements. A metaphor is viable because of some perceived or felt similarity between two equated objects, but if these similarities could be exhaustively catalogued, the metaphor would be "dead". We do not, after our stereoscopic vision, sum up by saying: "Ah yes, I have now learned that a man is like a tiger in these respects

(A, B, C . . .), and unlike a tiger in those respects (X, Y, Z . . .), and that is all there is to be said.'' With a live metaphor, there is *always* the possibility of ''more to be said''.

(d) ''Myth'' arises when these tensions are not maintained, and when a metaphor, a literal absurdity, is held to be *literally true,* the absurdity remaining unacknowledged. Berggren, however, denies that it is possible to assign any final meaning to the word ''literal''; assertions held to be ''literally true'' may appear ''metaphorical'' in the future (and vice versa).

Two points in Berggren's approach must be noted. First, his technical definition of ''myth'' is not an attempt to characterize all the relevant features of that term, as it is found in, for instance, the work of social anthropologists. The narrative content of myths, their emphasis on origins, their structural form, and so on, are not considered; rather, Berggren is concerned to draw attention to what might be called the ''degree of freedom'' afforded within these linguistic constructions. Myths traditionally describe implausible events and beings, and there is a distinction to be drawn between those people who accept these imaginative scenarios for living with the detachment (and hence freedom) of a metaphorical interpretation, and those who do not. And this leads to the second, and quite general, point. Any definition of ''metaphor'' relies on a shared presupposition of what constitutes the ''literal'', and no *definition* of the ''literal'' can be offered. The situation is that some assertions, we feel, can be made ''with a straight face'' (e.g. ''clouds are water droplets and ice crystals''), while others, when challenged, require more elaborate justification (e.g. ''clouds are a blanket''). The important feature of an analysis based on the literal/metaphorical dimension is its sensitivity to the cognitive implications of this tension. Its logic is dialectical.

With this discussion in the back of our minds, we can proceed to consider the metaphorical impact of technology.

TECHNOLOGICAL METAPHOR

Each technological innovation offers a new kind of human experience. At first, it is entirely strange, and difficult to grasp; but we quickly find in it sufficient familiar features to act as points of reference, and we can then explore it, savour it, come to terms with it, and assimilate it into the pattern of our everyday life. We ''learn to live with it''. Once it is established in this way, it can be the basis of a metaphorical transfer: we then see previously *familiar* things in terms of this novelty. We have acquired a new perceptual tool. But, in using this tool, we cannot guarantee analytic dispassion. The metaphor carries with it, not only structural features that can be assessed as analytically appropriate or inappropriate, but collections of associated emotions, attitudes and dispositions, hopes and fears, and these usually remain unassessed.

It is, moreover, in the nature of technology that its innovations tend to be directed to areas of human life that appear to those responsible to be most in need, and to offer the most promise, of *rationalization* (i.e. of simplification, ordering, more efficient use of power, reduction of disturbing ambiguity, and so on), or of *economic exploitation*. Effective technological innovation, therefore, is likely to affect the experience of individuals at some central focus of their activity and values, at both work and play; inasmuch as it relieves a felt need, and demonstrates the extent of the human capacity for improvement, the experience will be vivid and satisfying. But innovation may not only satisfy an already felt need; it can, by the potentiality it arouses, create a need as it satisfies it, or it may also help to create some other ''need'', impose some other pattern, and arouse some potential conflict, which it cannot relieve, and which is entirely unexpected. These are dramatic possibilities. In the past the natural world has tended to be the dramatic intruder into human affairs; modern technology has an increasing capacity to assume this role, and to play it more often, more insistently, and (inasmuch as its power is altogether tidier) more tellingly. These are precisely the conditions necessary for the kind of new experience that leads to metaphorical transfer—and, what is more, to the transfer of confused associations. In this way, technology can express previously held values and ideals and, by its concrete shape and form, alter and develop (and distort) them.

The advent of the steam-engine, and the development of railways, offers a striking example.[3] We nowadays, quite naturally, say of someone that he is ''letting off steam''. This metaphorical description of human behaviour was quite impossible before the spread of steam-engines (the OED quotes 1818 as the first dated use), but now our familiarity with railway engines and their eccentricities readily allows us to conceive of human anger in these terms. We talk, too, of certain social phenomena as "safety valves" (dated 1797), with the implication that they "reduce the pressure" of society to a "safe level"; or we say, more directly (and reviving a much older turn of phrase), that someone is "working under pressure", with the implication that he is liable, therefore, to "explode". In these instances, we propound the metaphor "the human skull is a steam-boiler": and the metaphor brings with it a heightened sense of danger. For, a century or so ago, exploding boilers were common occurrences; they regularly punctuated the proceedings of coroners' courts and newspaper offices, and the general concern (witnessed by contemporary accounts) led to legislative action.[4] By the use of this technological metaphor the danger of "repressing basic human energies" can be made more vivid, and human and social theorists who analyze civilization in terms of the (potentially dangerous) repression or restraint of primal urges can more easily gain an attentive audience. Could this, perhaps, account, at least in part, for the popularity of some Marxist and Freudian notions?[5] How soon after dieselization will this metaphor die? Perhaps the metaphor is too deeply rooted, and its affective message is already transferring to another technology. The skull/boiler equation, with its attendant social (and educational) implications, may be kept alive by as innocent a device as the hissing pressure-cookers in our kitchens. Or the implication of danger may take an indirect path: since technological explosions (created in man's brain, by the processes of reason) are now in the megaton range, we may have great difficulty in freeing ourselves from the notion that disciplined mental activity and its restraints are inherently dangerous. It is surely significant that a psychedelic experience is commonly referred to as "blowing one's mind". And a recent article in the *Guardian* (22 January 1970), entitled "Their Bomb Disposal School" carried this heading: "'Some children are like Bombs,' said the headmaster at one school meeting, 'they must be given opportunity to explode.' 'And the staff are like a bomb disposal squad,' added Larry."

Some fears are appropriate; but before they can be assessed, their origins and expression must be exposed. We are correct in sensing danger in an overheated steam-boiler, but we should be properly critical of any move to transfer this justified wariness on to our attitude towards a lively child. Metaphor affects this kind of transfer uncritically. We are not aware of the transfer, until it is specifically pointed out. After all, boiler explosions are now relatively rare events; pressure-cookers are accepted, in practice, as a safe addition to the household; and atomic power stations demonstrate the routine control of nuclear forces. "Letting off steam" is itself ambiguous—it signals the inherent danger of the inner pressure, but also demonstrates that this danger is being reduced. Why do we still adopt attitudes more appropriate to the "danger" than to its "control"?

One recalls the uneasy truce that Lovejoy has called "the ethic of the middle link". As Leo Marx (1964, p. 100) describes it, it offers men "an unsatisfactory but nevertheless unavoidable compromise between their animal nature and their rational ideals". This "middle state" theory reconciles two extreme views on the nature of man; it "enjoyed its greatest diffusion and acceptance during the eighteenth century". We can see in the railway steam-engine, with its continual struggle between untamed primal energy and purposeful discipline (as symbolized in the rails), a powerful—and valid—metaphor for a tension that is an essential feature of human nature. What technology has provided us with then, in this relatively brief cultural interlude, is merely a fresh way of sharpening our apprehension of both poles of this essential tension. Alternatively, you might argue that this tension, if it really exists, is not all that important. The ideals of the Enlightenment may have been delusory, and the problem they posed of no consequence. In focusing our attention on this tension, and in acting as the vehicle by which an eighteenth-century dilemma has been vividly transported into the twentieth century, technology may have led us all off the rails.

Further possible examples of technological metaphor can easily be found. One that I think worth mentioning briefly in the company of social scientists concerns the use of the term "structure". Even if sociologists use this term without affective overtones, it is clear that many radical students do not. They often feel estranged from a society that they perceive as rigid and unaccommodating, even hostile. "The structures of society", they constantly reiterate, cannot be altered, and must be destroyed. Where is the current literal focus of this metaphor of "structure"?[6] My guess would be that its common literal association lies with the solid and (apparently) permanent buildings in our cities. The steel frame of every new office block reaffirms that "structures" are unyielding, and designed to last; further, that each new one demands the demolition of old structures. Flexible modifications are less and less the norm. These associations are now, I suggest, wedded to the metaphor. Society is perceived as "structured" (which is probably a realistic and useful insight) and these "structures" are perceived as rigid (which is more debatable). But suppose that new building technology provides us with (literally) more flexible accommodation? If (like the nomadic Hebrews) we all lived in tents, would we (like the Hebrew prophets) tend to view social "structures" as more amenable to change?

In an extended and detailed section, Schon (1963, pp. 111-76) discusses the origin and influence of the metaphors that underlie our traditional theories of the mind, of deciding and of understanding. Some of these come from technology (scales, the use of tools, mechanical devices), some from social inventions (courts, group processes), some from basic human skills (e.g. understanding as vision), and some from metaphysical interpretations of scientific theory (atomism and dynamism). The crucial importance of these reflections is that Schon is here analysing the sources of our thinking about thinking—and about other rational activities, such as planning. It is much more difficult to rid ourselves of the inappropriate aspects of metaphors in this domain, than it is of metaphors employed (as in scientific theory) on the external world. The way we conceive of our thinking *is* the way we think. (Just as the way we conceive of other people is related to the way we conceive ourselves—and is, at least in some important senses, the way we *are*.)[7]

Schon quotes Veblen (1923, pp. 66-7) on the impact of machinery in industry:

> The discipline of the machine process enforces a standardization of conduct and of knowledge in terms of quantitative precision, and inculcates a habit of apprehending and explaining facts in terms of material cause and effect. It involves a valuation of facts, things, relations, and even personal capacity, in terms of force. Its metaphysics is materialism and its point of view is causal sequence.

Schon comments (1963, p. 154):

> Anyone who has lived in industry, and heard men treated regularly in terms of "drive", "forcefulness", "sparking plugs", "brakes", "shifting of gears", "self-starters", "cogs in the machine", and the like, cannot help but admit some of the truth of this.

and he also (pp. 152-3) points to the source of a new metaphor of the mind:

> Although in the realm of physics the Newtonian machine has suffered reverses, it is still dominant in theories of mind. Never since the age of naïve metaphysical mechanism have we been so close to literal belief in the theory of mind as machine. We have seen the birth of the giant computers. From the beginnings these machines were understood in terms of mental functions—perceptions, reasoning, learning, and memory; but it is becoming more and more common to think of the mind and mental processes in terms of computer behaviour. "Information storage", "information retrieval", "input-output", "circuit", "read-out", "digital and analogue", and similar computer-based terms, have gained in currency as ways of talking about thinking, learning, and reasoning. A new science, cybernetics, has grown up to explore the metaphor of mind as computer. The metaphor has been evocative and fruitful. It becomes dangerous only as it begins to go underground.

The phrase "it begins to go underground" is, of course, itself a metaphor—and possibly a dangerous one. In using it Schon is attempting to characterize a shift which, as I understand it, bears a close relationship to Berggren's idea of the lapse from metaphor into myth; there are also echoes of Whitehead's "misplaced concreteness", and McLuhan's "numb stance of the technological idiot", even of "alienation". In all these evocations, there is a sense of a loss of essential tension in our perception of the world, a lapse into easy rationalization.[8] Is this, perhaps, what Leavis (1970) fears?

A recent letter in the *New Scientist* (25 December 1969, p. 658) discusses the need "for some new ethical code to deal with the rapid changes wrought by the growth of science and increasing industrialization". The author continues:

> Might I suggest that the basis for such a code is already taking shape? A new "child" is being born in the very "womb" of science. I refer to the concepts which have been developing within the framework of systems theory and cybernetics . . . Briefly, social and essentially human manifestations can be discussed meaningfully and interpreted, using known and objective things like control, feedback, memory and learning. The human aspects of management, i.e. participation, group dynamics, avoidance of conflict, etc., need no longer be obscured by traditional

ethics. K. Deutsch has proposed definitions for terms such as "freedom", "justice" and "conscience".

The basis of a new morality is there. The difficulty lies in being able to accept it. It is an act of faith, and of Christlike humility, to view ourselves as integral parts of the world in which we live, just complex self-organizing survival machines, equipped with memory and learning units.

Here, surely, is a plea that a useful metaphor be adopted as a myth. Is it mere obscurantism to see this "going underground" of a modern technological metaphor as dangerous?[9]

Philosophers of science who treat metaphor seriously discuss its role in cognition and explanation; they do nto consider its wider, affective impact. Even among those who treat of the metaphorical diffusion of scientific and technical terms into the general culture, such as Schon and Brooks (1965), the emphasis remains on the more detached, intellectual, and analytical consequences. These authors appear to be centrally concerned with *the way we perceive and analyze the world,* and the changes that science and technology have made in our commonsense notions of social, human, and cosmological causation, rather than in *the attitudes we adopt to the world,* and corresponding changes in this domain.

Brooks, for instance, traces some common metaphorical applications (and misapplications) of terms such as "scientific", "evolution", "relativity", "uncertainty", "feedback", and so on. He treats of "the cycle in moral attitudes" as another "possible example" of an unstable feedback system, with a long inherent time-lag, and hence a tendency to "hunt". His approach can perhaps best be conveyed by quoting his concluding sentences:

In the case of the concepts of feedback and information, the ideas appear to have an essentially quantitative and operational significance for social and cultural dynamics, although their application is still in its infancy. The most frequent case is that in which a scientific concept has served as a metaphor for the description of social and political behaviour. This has occurred, for example, in the case of the concepts of relativity, uncertainty, and energy. In other cases, such as evolution and psychoanalysis, the concept has entered even more deeply into our cultural attitudes. (Brooks 1965.)

and in case that last word should suggest a less consciously analytical approach, on referring back we find:

Is it too much to suggest a parallel . . . between the changing scientific interpretations of biological evolution and changing attitudes towards cooperative action in human societies?

which puts the matter firmly within the rational, the intellectual, and the instrumental.

My sense of the incompleteness of this approach may now be clear. I will sharpen it by way of two further examples.

Our kitchen used to harbour a large cooking-range; it now has a compact gas-boiler, beside which there is a spacious alcove, somewhat smaller than a sentry-box. My two young sons often play in this alcove, which they refer to as "the lift". They press imaginary buttons, bend their knees, and shout out half-remembered phrases, such as "Sixth floor: lingerie!" while imaginary doors open and close. What are they doing? It seems to me quite inadequate to say that they are drawing attention to a similarity of shape between our kitchen alcove and the lift in a department store. Rather, their activity maps out one experience (the occasional, and rather frightening, journey in a lift) in terms of another, which is altogether more familiar and secure. What they are extending, I would tentatively suggest, is not so much their cognitive structure, their "model of the world", as the scope of their assured social action. They are learning to accommodate fears, so that they may enlarge the territory that they feel able to explore with confidence. They are using "an intelligible connection between the two referents" in order to mould and adjust their emotional responses; the bleak alcove acquires adventurous overtones, and the crowded lift loses some of its terror. Familiarity with technology, as a metaphorical link, seems to be extending their *emotional* universe.[10]

Or consider Kepler's famous remark that "the universe is not similar to a divine organism, but rather is similar to a clock". This new, scientific, metaphor was the vehicle for a great cognitive advance.[11] It led inexorably to Newton, and to the basis for the whole, powerful structure of modern physics. *But clocks are things you can tinker with.* You buy and sell clocks; wind and adjust them; redesign them; replace them; even, without too great a pang of guilt, smash them. If the world comes to be perceived as genuinely "clocklike", with man in the external role of "clockminder", these other attitudinal changes are likely to follow. The diffusion of clocks throughout our culture, and the changes in our perception that result as science establishes the cognitive worth of the metaphor, will tend to amplify those aspects within our cultural tradition that speak of "man's dominion over nature"; they will also tend to diminish those more tentative, wondering, and reverential attitudes towards nature nowadays commonly classified as "religious", and which stress man's "unity with nature". The world is no longer "divine". Our "common-sense" attitudes, and our whole pattern of behaviour, have changed. It is surely not too great an injustice to the historical data to suggest that this, in essence, seems to have been the sequence of events in modern Western culture.[12]

Nowadays scientific understanding of man as part of a whole ecological system (still, it must be confessed, viewed more as a "mechanism" than as an "organism") suggests, as Brooks (1965) would argue, that we should "change our attitudes". This implication is slowly becoming understood, as an intellectual proposition. But exploitatory attitudes, and the social institutions that have emerged as appropriate to those attitudes, now have strong emotional roots. Mere cognitive insights are insufficient, by themselves, to reverse the emotional tide.

Ultimately, humankind demands some harmony between cognition and affection, between how we think and how we feel. As A. C. Ewing has remarked, "Emotion . . . requires some objective belief, true or false, about the real to support it for long . . .". My argument is that the influence of technological metaphors injected into our complex culture must be assessed on both dimensions, since there is no necessary harmony between them. Our technological innovations and our social institutions are often at war in our consciousness, and there is no easy victory for either side. An apparent conjunction of thought and feeling in one dominant metaphor may be achieved at the expense of a loss of vital tension. We then reach too glib a resolution of the pain of living, and slip into a secular heresy. The power of modern technology can make this heresy doubly "dangerous".

METAPHORS OF KNOWLEDGE

Oscar Handlin (1965), discussing peasant communities, writes:

> Tradition governed both the ways of doing and the ways of knowing. It set the patterns by which the artisan guided his tools, the husbandman his plow. It also gave satisfying responses on the occasions when they wondered why; for it supplied a comprehensive explanation for the affairs of the visible and invisible world. The ways of doing were not identical with the ways of knowing, but they were associated through common reference points in the traditions of the community.

I want to suggest one means of association between "knowing" and "doing" that I take to be valid still in modern society (and probably in all conceivable societies), and to sketch out some of its implications. My suggestion is contained in the following passage by Lynn White (1968, p. vii):[13]

> Those of us who spend our lives scrutinizing the past in order to grasp the present are suspicious of all-inclusive interpretations of human destiny. It is our habit to study any large problem the way we look at a statue: by walking around it. Sometimes we come close to inspect a detail; then we stand back for more perspective. From various stances it looks very different, but the contradictions are not necessarily incongruous, since each view is partial. I have tried to see our new culture from several angles of vision which are deliberately chosen because they are appropriate to some aspect of the subject.

The basic metaphor that White uses here involves a person and a material object. The person is the "knower", the statue the "known". The object is small enough to be seen in the round; it is possible to touch it—ideally, to pick it up. But note that White does not refer to touching the statue: in his use of the metaphor, *all* the information comes to him by *sight*. His "knowing" involves "perspectives", and "angles of vision". The past can only be seen, not touched. And yet he refers to the process as "scrutinizing the past in order to *grasp* the present". The aim is a more intimate, tactile "knowledge" of the present.

My suggestion is this: when we talk of "knowing" a thing, our literal root metaphor is of a man investigating a strange material object. His immediate reaction is to peer at it, and then to pick it up (feel it, weigh it, etc.). This procedure usually tells him all he needs to "know". He will not expect to gain very much useful information by listening to such objects, or smelling them, or tasting them (although, of course, in some cases this information may be crucial). Sight and touch, linked by the act of handling an object, define the basic metaphor of "knowledge". We will, I suggest, attempt to underpin with this metaphor all claims to knowledge, and all the changing conceptions of "understanding", all new techniques of effective action. We will strive to relate these innovations to this literal "touchstone"; it brings together the notions of "knowing how", "knowing of", and "knowing that". Further, I want to suggest that analysis of the interplay of sight and touch within this metaphor is a powerful method of characterizing shifts of consciousness within a culture.

Two objections are immediately raised at this point. One is that our primary access to knowledge is surely through language; we read and listen, and "come to know". But I insist that these channels are *known* to be *secondary;* all our epistemological and ontological arguments are attempts to relate this intellectual activity to the "real world", in which people see stones, and pick them up. We yearn for this kind of direct contact, this intuitive apprehension of what is in fact the case. "Yes, I *see* . . ." we say, as a problem is explained to us; or we demand of someone, not a mere "understanding" of things, but a "proper *grasp* of the phenomena"; we even talk of "grasping a point", as if arguments were tangible objects. Verbal descriptions are no substitute for actually *being there*—so that you, too, can see it and touch it.

The other objection is that we do use metaphors drawn from other senses to express access to knowledge. "I smell a rat" indicates the first apprehension that what appears to be the case is, in fact, otherwise. I grant this, but would point out that when we have properly tracked down the rat, we cite evidence more substantial than its smell to prove our success. In any case, hearing, taste, and smell *do* contribute, even if only peripherally, to our knowledge of a material object, so that a proportion of metaphors based on these senses would be expected. The question is whether or not they predominate.[14]

Sight and touch, then, define dimensions of the social concept of the Man of Knowledge. At one extreme we have the Observer, who deals in abstract understanding; his "knowledge" is based on the metaphor of sight; his lack of contact with the things he "knows" guarantees dispassion and accuracy; he knows *that* something is the case, and that the "that" is there to be "known". Moving away from this end of the scale, we find the man of action, the Actor, the man with a "grasp" of worldly affairs, who demonstrates his knowledge (or "know-how") by his "manipulation" of his environment; his knowledge develops more dynamically; he *moulds* things.

To put the matter this way is to raise sharply the interaction between knowing and doing, for Observers sometimes act, and Actors would be wise, from time to time, to observe. The two conceptions interact—*just as sight and touch interact within an individual*. We have to reconcile different groups of associations. Sight offers much more, and much more exact, information: it operates at a distance, and so is detached, free from involvement; it gives us perspectives, leads us to construct models; it can distinguish more clearly genuine repetition in events; it shows us the principles of structural form, and so leads us on naturally to the conception of ideal forms, of abstractions and generalizations. Touch, on the other hand, is far less precise: it offers us some information of "feel" and texture, weight, hardness and softness, but very little on shape or form (hence the games scouts play, with twenty common objects in a pillowcase); touch does not discriminate very readily; at the same time, it demands contact and involvement. Perhaps surprisingly (and despite the old saw that "seeing is believing") touch is often appealed to as the basic test of *reality,*[15] from Doubting Thomas thrusting his hand into Christ's side to Dr. Johnson idly kicking a stone—whereas, as sceptical philosophers never fail to remind us, sight is theory-laden, and notoriously prone to delusion.

These, and other, muddled and conflicting associations puzzle our brains daily. We resolve the conflicts, individually, as best we can. In society, however, the conflict can be disguised if a metaphor based primarily on one sense comes to dominate our social conception of "knowledge". If we are mutually agreed, the overt tension can be removed. This appears to be the case in Western society, where conceptions of rationality, understanding, and knowledge are dominated by the metaphor of *sight*. This dominance is not created by print, or by any other modern technology: Aristotle, in the first paragraph of his *Metaphysics,* makes the dominance of sight quite explicit, and Plato, obsessed by the essentially visual notion of formal "likenesses", gives the impression, as Professor H. H. Price has put it, "that the world is a tidier place than it is". Schon (1963, pp. 170-6) analyses the dominance of sight in some detail. His conclusion is that: "Common-sense theories of understanding are still based on common-sense theories of seeing."

It would appear that our culture has resolved the conflict of associations by assigning to sight the primary function of gathering information and knowledge. Within such a biased culture, what happens to an individual's touch sense? The touch metaphor has no useful currency within the domain of rationality (apart from the peripheral function of a test of reality or existence); and so our "feelings" delineate a distinct realm, the domain of our "emotions". We come to conceive of "thinking" and "feeling" as essentially unrelated. The general distaste for, and distrust of, "emotions" can be taken as a direct social indicator of the dominance of sight as the metaphorical basis for "reason" and "knowledge". The conflict remains within the individual.

In what way can technology alter this balance? I take it as axiomatic that we can conceive of no technological metaphor that can actually *replace* the central metaphor of knowing. It is human beings, not their artefacts, that "know". If we come to "handle" and "view" all objects by technological remote control, we will (at least at first) see such gadgets as interpositions, "getting in the way of the real thing". However dominant such metaphors as the "computer personality" may become, "sight" and "touch" will remain as they are now—our description of our "primary input channels". (After all, if we come to accept computers literally as "people", we will then presumably talk literally of their "senses".) So the way in which technology affects this fundamental metaphor cannot be the direct ways, which I outlined earlier. We cannot expect a purely technological metaphor to "take over" our basic conception of knowledge and "go underground". However, metaphors deriving from human activity, from "ways of doing", can win this kind of tacit acceptance, and thereby shift the sight/touch balance. Variations in society in the style of effective, knowledge-based, action can modulate the basic scale. We have to consider the whole system of Man-plus-Technological-Devices-dealing-with-Environment, and the predominant sensory associations of this system.

Consider, by way of example, two technological paradigms: the drawing-board and the potter's wheel. The first locates effective social action in an abstract process—the production of blueprints, long-range economic plans, etc. The implications of this system, in the terms I have been discussing, are visual. Within this mode the artist is conceived (as Aristotle appeared to conceive of him) as someone who starts out with an ideal conception clearly in mind, and simply makes it—rather than as one who wrestles with intractable material in order to *evolve* a form through the mutual interplay of conception and realization. The contrast is essentially that between the Platonic Demiurge and the Hebrew God of Creation. The latter's paradigm is the potter's wheel. Its implications are tactile. Here "planning" becomes "moulding"; "reason" (which has a more limited function) is conceived in terms of the intimate feedback and control exercised by a potter's fingers and arms. If we take the emerging modern technological system to be Man-plus-Electric-Communications, then this system is characterized by a quicker and more efficient feedback and control, and a greater "involvement"—therefore its associations make it closer to the paradigm of the potter's wheel—that is to say, to the touch metaphor. This style is more like handling, manipulation. The technology revives the latent conflict that our culture has papered over.

But wait. Need we go to all this trouble? "Quicker and more efficient feedback and control" is the demand when things are perceived to be *changing*. Perhaps the sight metaphor is associated with stability (or, to be precise, with a society that *conceives itself to be stable*), while the touch metaphor becomes important in times of social change (however caused).

We have now, I think, reached the central point. In an individual, sight tends to predominate because of the rapidity with which it can check a range of data; we reassure ourselves with a quick glance around a room and rarely have to explore further. Similarly, within a culture, sight can exclude other, more ambiguous, metaphors; habitual categories harden and simplify. The world appears a tidy place, requiring the merest glance to re-establish confidence—until, that is, things change. When social change crosses a threshold, these neat categories no longer "fit", the habitual pattern collapses, and confusion results.

Now, when sight fails us, we move forward to grasp and touch. In just the same way, within a culture, the touch metaphor has to be revived. As is the way in these matters, the many habitual patterns that continue are forgotten in the face of the few new ones that have to be acquired. When someone says "we are all groping in the dark", we "know how he feels".

The change in society (or, perhaps more exactly, the change in the tempo of society's change) may have a technological "trigger". But the form of the shift I have been characterizing has no necessary connection with the form of the technology. *Every* overall change from apparent stability is a change from sight towards touch within the metaphor of knowledge.[16]

Berggren's essay (1962-3) itself offers a very neat illustration of my thesis. He discusses the role of metaphor in science and poetry. In science he takes the metaphors to be "structural" in content, and the insights they carry are developed by experimental techniques. In this section, he operates entirely within a visual metaphor for knowledge: "stereoscopic vision" and "structure".

However, he has more difficulty in stating what cognitive insights poetic metaphor conveys. He takes its content to be "textural", and he talks of poetry as conveying "the textural feel-of-things", of "construing the world with feeling". That he is here forced beyond "stereoscopic vision" into use of the touch metaphor is, I think, significant; poetry is dealing with those associations that come *with* the language, but are not expressed *in* it. It points to features of the objective world that are not assimilable to the visual mode. But so dominant is the visual metaphor in our social conception of knowledge that Berggren, despite his use of the touch metaphor, concludes: "In a sense . . . it is not the machine, diagram, portrait or specimen, but stereoscopic vision itself which must become the most fundamental root metaphor of metaphysics" (1962-3, p. 471).

Berggren appears to be wrestling with a profound philosophical problem in terms of the conflicting associations and properties of our senses. But that is perhaps what many "profound philosophical problems" *are*. One cannot escape the suggestion that our (scientific) models of the cognitive process itself could usefully be employed as the "fundamental root metaphors of metaphysics".

He is also, it seems to me, propounding a metaphysics appropriate to *change:* he is trying to characterize a process, to define the nature of the puzzle.[17] He has pointed out that our visual perception works by overlapping images, "stereoscopic vision". Some images we take to be so similar as to be identical; others are clearly too dissimilar to be related; metaphor occupies the mid-ground. If, like Plato, we are "similarity-oriented", that mid-ground will shrink; the world will seem tidier than it is, and the imaginative world of metaphor will seem no more than a pleasant decoration. But when our easy identities are blurred, only that imaginative world can redeem our perception. As Berggren argues: ". . . truly creative and non-mythic thought, whether in the arts, the sciences, religion or metaphysics, must be invariably and irreducibly metaphorical" (1962-3, p. 472).

How is this metaphorical mid-ground to be defended and maintained? The problem involves a critical awareness. I suggest that it is kept alive, in society, by a sufficient imprecision in our norms and values to allow change (and the consequent variety in response, belief, and attitude) without perceived deviance and restrictive sanctions; in intellectual life, by the deployment of ambiguous symbols and a greater respect for metaphor and paradox; in religious life and thought, by a shift towards ritualism, with the attendant richness of mythical and apocalyptic images.[18] All this involves a willingness to admit, and in certain respects to accept, what is nowadays often defined as the "irrational"; this willingness is the necessary preliminary to containing and moulding the potentially dangerous forces that that term signifies.

"Imprecision", "ambiguity", even (in a visual culture) "irrationality", are associations of the touch sense. Perhaps the fundamental root metaphor of a metaphysics of change should be stereoscopic vision *plus touch*. An abstract, rootless mysticism would be as inadequate as a blinkered pragmatism; sight and touch, Observer and Actor, must remain in tension.

DEHUMANIZATION

It is often implied that technological metaphor is essentially "dehumanizing". I cannot accept this; I have been describing what I take to be an essentially human function. Initially it operated with "natural" objects and symbols, now it is becoming infiltrated with images derived from "artificial" constructs. I can see no good reason why this transition should necessarily corrupt the function. Lewis Mumford is among the most ardent prophets of corruption; he sees the variety of nature as both inexhaustible and irreplaceable:

> Man's life would be profoundly different if mammals and plants had not evolved together, if trees and grasses had not taken possession of the surface of the earth, if flowering plants and plumed birds, tumbling clouds and vivid sunsets, towering mountains, boundless oceans, starry skies had not captivated his imagination and awakened his mind. Neither the moon or a rocket capsule bears the slightest resemblance to the environment in which man actually thought and throve. Would man have ever dreamed of flight in a world destitute of flying creatures? . . . Long before any richness of culture had been

achieved, nature had provided man with its own master model of inexhaustible creativity, whereby randomness gave way to organization, and organization gradually embodied purpose and significance. This creativity is its own reason for existence and its own reward . . . in their immensely varied totality, they have helped to create man . . . if our descendants reduce this planet to such a denatured state as the bulldozer, the chemical exterminators, and the nuclear bombs and reactors are already doing, then man himself will become equally denatured, that is to say, *dehumanized*. (Mumford 1967, pp. 36-7.)

On the other hand, Harvey Cox, the Harvard theologian, has recently written:[19]

. . . modern man experiences new technologies in ways similar to those in which previous generations sometimes experienced natural forces. The experience (real or imagined) of the jet plane, the transistor radio, the computer, the contraceptive pill, the H-bomb, etc. . . . is an experience with a degree of affective or cathectic intensity, one calling for an adjustment in personal orientation and symbolic appreciation . . . It might even be that these technological artefacts are displacing, or at least complementing, religious objects (such as Christ or Mary) and natural objects (lightning and sun) as the symbolic foci of our elemental affective orientations.

Mumford's writings can be seen as a timely warning against the despoliation of our environment, and a reminder that the transition from natural to technological metaphor is difficult, and not to be carelessly accelerated, for a loss in richness and variety may involve a real loss of freedom; but they offer no argument against facilitating that transition. It is, in any case, already happening.

I have, perhaps, implied in this essay that I equate "dehumanization" with the lapse from metaphor into myth (in Berggren's sense of that term). This would be a half-truth. Many of the most important advances in human thought and achievement have arisen because men have become obsessed with the literal interpretation of a metaphor. Within science itself the drive towards literal elaboration of metaphorical models is of the essence. The Church would be a poor thing without its heretics. To sense literal truth within what would conventionally be held to be the literally absurd has always been the hallmark of "creativity". And yet, when this is acknowledged, the role of "essential tension" in defining our "humanity" must remain.[20]

But perhaps the *definition* of "humanity" will always elude us; all we can hope for is some agreement as to what constitutes the *in*-human. Bruno Bettelheim (1959) has described a classic case of a boy (Joey) who organized his life around the belief that he was a machine, and became a "mechanical boy"; he "plugged himself in" and "switched himself on" before speaking, lapsing into an autistic silence when his "power supply" was removed. His behaviour forced others to act towards him so as to reinforce his pretence. Bettelheim comments:

A human body that functions as if it were a machine and a machine that duplicates human functions are equally fascinating and frightening . . . In any age, when the individual has escaped into a delusional world, he has usually fashioned it from bits and pieces of the world at hand. Joey, in his time and world, chose the machine and froze himself in its image.

Bettelheim's description would be generally agreed to be of a boy suffering some "loss of humanity", but the loss has less to do with the specific model Joey chose than the fact that he "froze himself in its image", lapsing into "myth". A boy less dominated by the machine image, more detached and playful in his adoption of his "delusional world", would be termed "imaginative". But Joey appears to us to have lost his freedom. And what is true of an individual is even more obvious in societies. We all feel a sense of terror at the sight of whole groups acting uncritically under the spell of a myth. Hitler's Germany evokes images of lemmings, of Gadarene swine, of robots, of the less-than-human. Individual mythic obsessions we can tolerate, or even genuinely value and admire, but when an individual comes to see one thing clearly, it is often at the expense of blindness in other matters; this is a danger in a whole community. Once again humanity is seen to be related to variety, and the maintenance of inner tension.

The whole of this discussion can now focus on one final point. Science and technology may provide ready-made symbols for metaphorical transfer. Without railway engines, we could not "let off steam"; we might, of course, find some other metaphor to fulfill (initially) the same function, but it would be difficult to maintain (provided the metaphor achieves any depth at all) that subsequent concepts of human energy would have run the same course. But, this speculation apart, the fact remains that the *reason why* we came to use this particular metaphor is not simply that it is available. The situation also "lends itself" to this expression. The explanation is to be found elsewhere, much as Bettelheim's explanation of Joey's behaviour emphasizes his early childhood experiences, and his relationship with his parents. Joey adopts this "mechanical" strategy as a defence against his emotions, which he has not learned to express socially. The role of technology is incidental—Joey could have chosen other defences.

If science and technology "lend themselves" as imaginative human funk-holes in this way more readily than do other sources of symbolic forms, then we might with some justification "blame" them for our "dehumanization". But this seems to me unproved. If the Man/Technology/Environment system is developing so as to make us talk about a loss of human capacity, the cause (and the cure) may be located elsewhere in the system. The proper study of mankind (and of both humanization and dehumanization) is man himself. And the proper study of the problems I have been attempting to outline in this essay is, I judge, the psychology of perception and the sociology of knowledge: inasmuch as technological metaphor has influenced the development of science itself, it is the historical sociology of scientific knowledge. (Why, for instance, did the clock metaphor not take hold in Chinese society, as it did in the West?) If science can help us elucidate the processes I have been describing in this paper, spell out the myth of metaphor, and help to restore appropriate tensions to our perceptions and our thought, then it will be performing its proper study, and conserving our humanity; else we fall, in Blake's phrase, into "single vision, and Newton's sleep".

NOTES

1. It is, of course, also employed technically elsewhere in science—notably in physical chemistry—but without the same exposure as in computer technology.
2. This book contains a useful bibliography. Miss Hesse notes wistfully: "It is still unfortunately necessary to argue that metaphor is more than a decorative literary device, and that it has cognitive implications whose nature is a proper subject of philosophic discussion."
3. Notable among recent studies are those of Marx (1964), Sussman (1968), and West (1967). Sussman, in particular, quotes vivid instances of the initial disorientation involved in the railway experience, and of the use of railway and steam-engine metaphor in imaginative writing—as in Rudyard Kipling's poem, *MacAndrew's Hymn*.
4. See Burke (1966 & 1968) and Sinclair (1966). For a comment on the German situation, see Burke 1966.
5. An interesting sidelight arises from the (initially surprising) fact that it was not until the last years of the nineteenth century that British scientists, arguing for government support for research, used the idea that the advance of science would "improve the efficiency of the industrial and economic system". The popular rhetorical impact of this idea required (a) the wide diffusion of the notion of aspects of the body politic as closed (mechanical) systems and (b) the elaboration of the notion of "efficiency" into a useful analytical tool; this was achieved by thermodynamicists, prodded by engineers into investigating what went on inside the cylinders of steam-engines.
6. That sociologists acquired the term from biology, via the metaphor of society as an organism, shed its more fluid evolutionary implications, and emphasized the more conservative ideas of homeostasis and equilibrium, is not relevant to this question; I am not concerned with the origin of the term, or its technical use, but rather its primary reference in the lay mind.
7. As David Bohm (1970, p. 168) has recently expressed it, "As we perceive and talk, so we will think and act and, therefore, so we will be."
8. In this the phenomenon is formally equivalent to the notion of "heresy" within the orthodox Christian tradition. Each heresy is an attempt to resolve the tension within a paradoxical doctrine. The traditional doctrine, for instance, that Christ was *both* wholly human *and* wholly divine lapses into, on the one hand, the heresy of Arianism, and, on the other, into Docetism. (Nestorianism and Monophysitism occupy intermediate positions, with partial resolution of the orthodox tension.) Among other polar heresies that might be cited are Pelagianism and Calvinism; Pentecostalism and Legalism; Modernism and Veterism; Legalism and Antinomianism; Essentialism and Existentialism; Quietism and Activism; and Pantheism and Materialism. Orthodoxy walks a desperate tightrope between all these fearsome traps.
9. Connoisseurs of the use of cybernetic and electronic metaphors should also note Gross (1969); another related article is that of Smith (1970) who suggests that "phase transitions and cooperative phenomena in solid state science have analogies in management". See also Mayr (1970).
10. I do not think it altogether outrageous to suggest further that this "mapping" has overtones of social class, for department stores encapsulate class aspirations; and when this association is related to vertical mobility, my children may be acquiring attitudes that will be hard to shake off. Perhaps the experience will help them to understand sociology. Here, technology may be giving a new metaphorical life to certain aspects of "The Great Chain of Being" (as may that notorious technological metaphor, "escalation").
11. This advance has been comprehensively documented and analysed by Dijksterhuis (1961). Aspects have also been usefully discussed by Price (1964), Laudan (1966), and Cohen (1966). A concise account of the historical sequence has been given by Whitrow (1970).

    Although these authors stress different implications of the metaphor, they give ample evidence of its imaginative hold, strikingly exemplified in Descartes's remark that in developing his mechanical account of nature:

    > . . . the example of certain things made by human art was of no little assistance to me; for I recognize no difference between these machines and natural bodies. . . .

    Price stresses the *pre-existence* of a mechanical philosophy of nature, which itself gave rise to the desire to construct such devices as astronomical clocks and human and animal automata—the eventual success of these devices strengthening and developing the older tradition. If clocks had a decisive role to play, he suggests, it may be because "coming to grips with the basic astronomical phenomena probably required a level of sophistication considerably higher than was necessary for a reasonably basic appreciation of the movement of living things". Of Aquinas's remarks that animals could justly be regarded as machines, Price comments:

> Surely, such a near-Cartesian concept could only become possible and convincing when the art of automation-making had reached the point where it was felt that all orderly movement could be reproduced, in principle at least, by a sufficiently complex machine.

This suggests that the metaphor did not so much *introduce* new cognitive insights, as, by increasing human confidence in the possibility, *amplify and strengthen* a predisposition towards "mechanical" explanation.

Dijksterhuis sees the historical transition as passing rapidly through the rigidly mechanical phase characterized by the clock metaphor (with its strictly unscientific, teleological implications) towards the antithetic atomist model, acquiring on the way the "positive analogy" of mechanism in the form of mathematically expressed abstract laws, and a sense of overall order and pattern. He suggests that "mechanics" is a misleading term that might, more appropriately, be "kinetics".

However, whatever the precise role and subsequent fate of the metaphor, it is difficult to avoid the conclusion that a significant innovation in human thought emerged when the experience of a successfully universe-like clock suggested the possibility of a clock-like universe. The clock most often cited is that of Giovanni de Dondi completed in 1364 at Padua; this was primarily a complex mechanical representation of the heavens, and only incidentally a timepiece. (A similar, but less complex, clock was installed at Strasbourg in about 1350.) Whitrow suggests that Nicole Oresme's experience of this clock led him, in his treatise "De commensurabilitate . . .", to emphasize the harmony of the universe with the words:

> For if anyone should make a mechanical clock, would he not move all the wheels as harmoniously as possible?

Whitrow comments:

> This is the earliest instance known of the mechanical simulation of the universe by clockwork suggesting the reciprocal idea that the universe itself is a clock-like machine.

Oresme's treatise is undated (even by Thorndike), but must presumably be between 1364 and 1382.

12. Relevant discussion of these attitudinal changes can be found in Black (1970), White (1962, 1968) and in the Epilogue of Cipolla (1967). See also Hooykaas (1972).
13. Since completing this paper my attention has been drawn to the fact that a similar argument to that advanced in this section is contained in a paper by Walter Ong (1969).
14. Other cultures, of course, may rest their notions on different sensual metaphors. I am told that certain East African tribes conceptualize "understanding" in terms of "hearing". This may simply be because an acute and discriminating sense of hearing is (or has recently been) crucial to their survival in a hostile environment, so that the most important skill is to "hear clearly". I confess to being puzzled by the central Hebrew notion of the "Word". This combines the notion of fundamental knowledge (i.e. knowledge of God) with the technique of effective social action, via the metaphor of the inspired leader, who acts via word of command. In its mystical usage, the Word seems to come close to being an "object", which can be heard and touched, but not seen. Metaphors such as hearing (e.g. the "still small voice") or wrestling (Jacob) are used to characterize the relationship between man and God. These notions are clearly non-visual; if we allow that hearing and touching are, in their ambiguity and impression, very similar, then my suggested fundamental metaphor could, I think, be extended into Hebrew culture.
15. This primacy of touch as a reality-test is very likely to be culturally determined, that is to say, other societies may adopt other priorities. It could be argued that the same is true of the differing associations and properties of sight and touch. We may have to learn, for instance, to assign precision to sight, and to find our touch sense more obscure. After all, blind people can acquire precision of thought via Braille, and blind scientists of distinction are not unknown. The ideas and expectations we have of our senses may have their roots in the precise form in which our technology *allows them to operate*. However, my inclination is to believe that these sensual differences arise from physiological differences, in the structure of brain and sense organs, which are culturally invariant, and common to all members of the species.
16. It has not escaped my attention that my argument here is related to the analysis of Marshall McLuhan, especially as contained in his book *Understanding Media* (1964). Where McLuhan analyses the shifts in emotional balance in a culture by attributing to each technological innovation the property of (apparently literally) "extending one of man's senses", I take the process to consist of changes in associations "carried" by displaced metaphors; in particular, I take the shift that McLuhan characterizes in terms of a transition from a sight-dominated culture to one dominated by touch and hearing (and which he has to "explain" by the arbitrary postulate that TV extends our faculty of *touch,* rather than the more obvious extension of sight) as a shift in the metaphor of knowledge. For McLuhan's one process of "sense extension" I substitute the process of metaphorical transfer. This can be matched to the phenomena with less of a sense of strain. Within it the precise effect of each device can be seen as depending on the extent of its transfer and dominance, and on the subtle and complex group of associations it brings with it. Since these associations are always, to some extent, concealed before the event, and the potential dominance of a new metaphor is difficult to assess, my approach loses in prophetic power and prognostic pretentions—not to mention dramatic effect. What it gains is in its representation and characterization of what is actually happening. For I would contend that many of the effects that McLuhan fumbles in his attempts to characterize fall more naturally into place when the role of metaphor is recognized.

   In particular, I suggest that McLuhan, in propounding that the shift he senses from sight towards touch is determined by the particular form of each new technology, is mistaken. He is simply drawing attention to an underlying feature of the (possibly illusory) idea of "overall social change" *as such*—to a necessary correlate of a situation in which people say "things are changing". That is all that needs to be said.

Or perhaps not quite all. Western society of a century or so ago may have thought itself to be essentially stable, in possession of the simpler generalizations that underlie natural processes. The world looked tidy, bar a few dusty spots that could easily be swept up. That we are now more intimately aware of the complexity of things, and so distrustful of mere tidiness, may be bound up with communications technology (as it is also probably bound up with two World Wars and the Depression). We live, perhaps, in a wider context. Our society is "information-rich". Relaxation in social constraints may be necessary before we can respond to this evidence of complexity, but the evidence must be available. Perhaps, after all, the electronic mass media can be seen as having a role in the switch to a "touch-culture". But if so, ironically, it is for the opposite reason to the one that McLuhan puts forward. He sees the TV image as *deficient* in information (it is an incomplete "mosaic", which "involves" us in its completion), whereas I see it as offering a *redundancy* of information. Our response to its evidence of complexity elicits the touch metaphor and this brings with it the association of involvement and contact. The "inclusive, mythic mode of thought" appropriate to McLuhan's "global village" does not arise because TV "extends our sense of touch", but (if you wish to make TV the cause) because it "overloads our sense of sight".

17. I have referred to this elsewhere (Edge 1970) as "Ximenes-metaphysics", by analogy with the famous and sophisticated crossword published weekly in the *Observer* in which the most difficult thing is always to work out what kind of a puzzle this week's puzzle is, before one can get started at all.
18. See the interesting thesis advanced by Mary Douglas (1970).
19. Quoted in Mesthene (1970).
20. As White (1968, p. 41) notes, "the traditional Christian world view is bifocal".

REFERENCES

Berggren, D. 1962-3. The use and abuse of metaphor, I and II. *Review of Metaphysics* **16,** 237-58, 450-72.

Bettelheim, B. 1959. Joey: a mechanical boy. *Scientific American,* March 1959, 2-9.

Black, J. 1970. *The dominion of man.* Edinburgh: Edinburgh University Press.

Bohm, D. 1970. In *Impact of Science on Society* **20** (2).

Brooks, H. 1965. Scientific concepts and cultural change. *Daedalus* **94** (1), 66-83. Reprinted in G. Holton (ed.), *Science and culture.* Boston: Houghton Mifflin.

Burke, J. G. 1966. Bursting boilers and federal power. *Technology and Culture* **7** (1), 1-23.

Burke, J. G. 1968. Review of Sinclair (1966). *Technology and Culture* **9** (2), 230-2.

Cipolla, C. 1967. *Clocks and culture, 1300-1700.* London: Collins.

Cohen, J. 1966. *Human robots in myth and science.* London: George Allen & Unwin.

Douglas, M. 1970. *Natural symbols.* London: Barrie & Rockcliff.

Dijksterhuis, E. J. 1961. *The mechanization of the world picture.* Oxford: Clarendon Press.

Edge, D. O. 1970. Science and religion: an analysis of issues. In G. Walters (ed.), *Science and religion: the reopening dialogue,* 15-35. Bath: Bath University Press.

Gross, W. A. 1969. A unique explanation of how R & D people interact with associates. *Research Management* **12** (1), 57-71.

Handlin, O. 1965. Science and technology in popular culture. *Daedalus* **94** (1), 156-70. Reprinted in G. Holton (ed.), *Science and culture.* Boston: Houghton Mifflin.

Hesse, M. B. 1966. *Models and analogies in science.* Notre Dame, Indiana: University of Notre Dame Press.

Hooykaas, R. 1972. *Religion and the rise of modern science.* Edinburgh: Scottish Academic Press.

Laudan, L. 1966. The clock metaphor and probabilism. *Annals of Science* **22** (2), 73-104.

Leavis, F. R. 1970. 'Literarism' versus scientism: the misconception and the menace. *Times Literary Supplement,* 23 April 1970, 441-4.

Marx, L. 1964. *The machine in the garden: technology and the pastoral ideal in America.* New York: Oxford University Press.

Mayr, O. 1970. The origins of feedback control. *Scientific American,* October 1970, 110-18.

McLuhan, M. 1964. *Understanding media.* London: Routledge & Kegan Paul; New York: McGraw-Hill.

Mesthene, E. G. 1970. The Harvard programme on technology and society: the Fifth Annual Report. As quoted in *Technology and Society* **5** (4), 146.

Mumford, L. 1967. *The myth of the machine.* London: Secker & Warburg.

Ong, W. J. 1969. World as view and world as event. *American Anthropologist* **71,** 634-47.

Price, D. J. De S. 1964. Automata and the origins of mechanism. *Technology and Culture* **5,** 9-23.

Schon, D. A. 1963. *Displacement of concepts.* London: Tavistock Publications. Since reprinted under the title *Invention and the evolution of ideas.* London: Tavistock Social Science Paperbacks, 1967.

Seferis, G. 1970. Letter to Rex Warner. *Encounter,* February 1970, 68.

Sinclair, B. 1966. *Early research at the Franklin Institute: the investigation into the causes of steam boiler explosions: 1830-1837.* Philadelphia: Franklin Institute.

Smith, W. V. 1970. Research management. *Science* **167,** 957-9.

Sussman, H. L. 1968. *Victorians and the machine.* Cambridge, Mass.: Harvard University Press.

Veblen, T. 1923. *The theory of the business enterprise.* New York: Charles Scribner's Sons.

West, T. R. 1967. *Flesh of steel.* Nashville: Vanderbilt University Press.

Wheelwright, P. 1962. *Metaphor and reality.* Bloomington: Indiana University Press.

White, L. 1962. *Medieval technology and social change.* Oxford: Clarendon Press.

White, L. 1968. *Machina ex Deo.* Cambridge, Mass.: MIT Press.

Whitrow, G. J. 1970. The role of time in the evolution of the scientific world-view. Unpublished Robert Schlapp Lecture, delivered at the University of Edinburgh, 22 May 1970.

From *Zen and the Art of Motorcycle Maintenance*

ROBERT PIRSIG

■ Robert Pirsig *taught rhetoric at Montana State University and at the University of Illinois, Chicago. His book* Zen and The Art of Motorcycle Maintenance, *of which this selection comprises the first two chapters, has been widely read and praised.*

Using the motorcycle as a metaphor for technology itself, Mr. Pirsig explores the mediated world which technology affords those who wish to ignore its profound effects on their lives. His friends John and Sylvia, for example, use the artifacts of technology but feel manipulated by it because of their own fear and indifference.

1

I can see by my watch, without taking my hand from the left grip of the cycle, that it is eight-thirty in the morning. The wind, even at sixty miles an hour, is warm and humid. When it's this hot and muggy at eight-thirty, I'm wondering what it's going to be like in the afternoon.

In the wind are pungent odors from the marshes by the road. We are in an area of the Central Plains filled with thousands of duck hunting sloughs, heading northwest from Minneapolis toward the Dakotas. This highway is an old concrete two-laner that hasn't had much traffic since a four-laner went in parallel to it several years ago. When we pass a marsh the air suddenly becomes cooler. Then, when we are past, it suddenly warms up again.

I'm happy to be riding back into this country. It is a kind of nowhere, famous for nothing at all and has an appeal because of just that. Tensions disappear along old roads like this. We bump along the beat-up concrete between the cattails and stretches of meadow and then more cattails and marsh grass. Here and there is a stretch of open water and if you look closely tails. And turtles. . . . There's a red-winged blackbird.

I whack Chris's knee and point to it.

"What!" he hollers.

"Blackbird!"

He says something I don't hear. "What?" I holler back.

He grabs the back of my helmet and hollers up, "I've seen *lots* of those, Dad!"

"Oh!" I holler back. Then I nod. At age eleven you don't get very impressed with red-winged blackbirds.

You have to get older for that. For me this is all mixed with memories that he doesn't have. Cold mornings long ago when the marsh grass had turned brown and cattails were waving in the northwest wind. The pungent smell then was from muck stirred up by hip boots while we were getting in position for the sun to come up and the duck season to open. Or winters when the sloughs were frozen over and dead and I could walk across the ice and snow between the dead cattails and see nothing but grey skies and dead things and cold. The blackbirds were gone then. But now in July they're back and everything is at its alivest and every foot of these sloughs is humming and cricking and buzzing and chirping, a whole community of millions of living things living out their lives in a kind of benign continuum.

You see things vacationing on a motorcycle in a way that is completely different from any other. In a car you're always in a compartment, and because you're used to it you don't realize that through that car window everything you see is just more TV. You're a passive observer and it is all moving by you boringly in a frame.

On a cycle the frame is gone. You're completely in contact with it all. You're *in* the scene, not just watching it anymore, and the sense of presence is overwhelming. That concrete whizzing by five inches below your foot is the real thing, the same stuff you walk on, it's right there, so blurred you can't focus on it, yet you can put your foot down and touch it anytime, and the whole thing, the whole experience, is never removed from immediate consciousness.

Chris and I are traveling to Montana with some friends riding up ahead, and maybe headed farther than that. Plans are deliberately indefinite, more to travel than to arrive anywhere. We are just vacationing. Secondary roads are preferred. Paved county roads are the best, state highways are next. Freeways are the worst. We want to make good time, but for us now this is measured with emphasis on "good" rather than "time" and when you make that shift in emphasis the whole approach changes. Twisting hilly roads are long in terms of seconds but are much more enjoyable on a cycle where you bank into turns and don't get swung from side to side in any compartment. Roads with little traffic are more enjoyable, as well as safer. Roads free of drive-ins and billboards are better, roads where groves and meadows and orchards and lawns come almost to the shoulder, where kids wave to you when you ride by, where people look from their porches to see who it is, where when you stop to ask directions or information the answer tends to be longer than you want rather than short, where people ask where you're from and how long you've been riding.

It was some years ago that my wife and I and our friends first began to catch on to these roads. We took them once in a while for variety or for a shortcut to another main highway, and each time the scenery was grand and we left the road with a feeling of relaxation and enjoyment. We did this time after time before realizing what should have been obvious: these roads are truly different from the main ones. The whole pace of life and personality of the people who live along them are different. They're not going anywhere. They're not too busy to be courteous. The hereness and nowness of things is something they know all about. It's the others, the ones who moved to the cities years ago and their lost offspring, who have all but forgotten it. The discovery was a real find.

I've wondered why it took us so long to catch on. We saw it and yet we didn't see it. Or rather we were trained *not* to see it. Conned, perhaps, into thinking that the real action was metropolitan and all this was just boring hinterland. It was a puzzling thing. The truth knocks on the door and you say, "Go away, I'm looking for the truth," and so it goes away. Puzzling.

But once we caught on, of course, nothing could keep us off these roads, weekends, evenings, vacations. We have become real secondary-road motorcycle buffs and found there are things you learn as you go.

We have learned how to spot the good ones on a map, for example. If the line wiggles, that's good. That means hills. If it appears to be the main route from a town to a city, that's bad. The best ones always connect nowhere with nowhere and have an alternate that gets you there quicker. If you are going northeast from a large town you never go straight out of town for any long distance. You go out and then start jogging north, then east, then north again, and soon you are on a secondary route that only the local people use.

The main skill is to keep from getting lost. Since the roads are used only by local people who know them by sight nobody complains if the junctions aren't posted. And often they aren't. When they are it's usually a small sign hiding unobtrusively in the weeds and that's all. County-road-sign makers seldom tell you twice. If you miss that sign in the weeds that's *your* problem, not theirs. Moreover, you discover that the highway maps are often inaccurate about county roads. And from time to time you find your "county road" takes you onto a two-rutter and then a single rutter and then into a pasture and stops, or else it takes you into some farmer's backyard.

So we navigate mostly by dead reckoning, and deduction from what clues we find. I keep a compass in one pocket for overcast days when the sun doesn't show directions and have the map mounted in a special carrier on top of the gas tank where I can keep track of miles from the last junction and know what to look for. With those tools and a lack of pressure to "get somewhere" it works out fine and we just about have America all to ourselves.

On Labor Day and Memorial Day weekends we travel for miles on these roads without seeing another vehicle, then cross a federal highway and look at cars strung bumper to bumper to the horizon. Scowling faces inside. Kids crying in the back seat. I keep wishing there were some way to tell them something but they scowl and appear to be in a hurry, and there isn't. . . .

I have seen these marshes a thousand times, yet each time they're new. It's wrong to call them benign. You could just as well call them cruel and senseless, they are all of those things, but the *reality* of them overwhelms halfway conceptions. There! A huge flock of red-winged blackbirds ascends from nests in the cattails, startled by our sound. I swat Chris's knee a second time . . . then I remember he has seen them before.

"What?" he hollers again.

"Nothing."

"Well, *what?*"

"Just checking to see if you're still there," I holler, and nothing more is said.

Unless you're fond of hollering you don't

make great conversations on a running cycle. Instead you spend your time being aware of things and meditating on them. On sights and sounds, on the mood of the weather and things remembered, on the machine and the countryside you're in, thinking about things at great leisure and length without being hurried and without feeling you're losing time.

What I would like to do is use the time that is coming now to talk about some things that have come to mind. We're in such a hurry most of the time we never get much chance to talk. The result is a kind of endless day-to-day shallowness, a monotony that leaves a person wondering years later where all the time went and sorry that it's all gone. Now that we do have some time, and know it, I would like to use the time to talk in some depth about things that seem important.

What is in mind is a sort of Chautauqua—that's the only name I can think of for it—like the traveling tent-show Chautauquas that used to move across America, *this* America, the one that we are now in, an old-time series of popular talks intended to edify and entertain, improve the mind and bring culture and enlightenment to the ears and thoughts of the hearer. The Chautauquas were pushed aside by faster-paced radio, movies and TV, and it seems to me the change was not entirely an improvement. Perhaps because of these changes the stream of national consciousness moves faster now, and is broader, but it seems to run less deep. The old channels cannot contain it and in its search for new ones there seems to be growing havoc and destruction along its banks. In this Chautauqua I would like not to cut any new channels of consciousness but simply dig deeper into old ones that have become silted in with the debris of thoughts grown stale and platitudes too often repeated. "What's new?" is an interesting and broadening eternal question, but one which, if pursued exclusively, results only in an endless parade of trivia and fashion, the silt of tomorrow. I would like, instead, to be concerned with the question "What is best?," a question which cuts deeply rather than broadly, a question whose answers tend to move the silt downstream. There are eras of human history in which the channels of thought have been too deeply cut and no change was possible, and nothing new ever happened, and "best" was a matter of dogma, but that is not the situation now. Now the stream of our common consciousness seems to be obliterating its own banks, losing its central direction and purpose, flooding the lowlands, disconnecting and isolating the highlands and to no particular purpose other than the wasteful fulfillment of its own internal momentum. Some channel deepening seems called for.

Up ahead the other riders, John Sutherland and his wife, Sylvia, have pulled into a roadside picnic area. It's time to stretch. As I pull my machine beside them Sylvia is taking her helmet off and shaking her hair loose, while John puts his BMW up on the stand. Nothing is said. We have been on so many trips together we know from a glance how one another feels. Right now we are just quiet and looking around.

The picnic benches are abandoned at this hour of the morning. We have the whole place to ourselves. John goes across the grass to a cast-iron pump and starts pumping water to drink. Chris wanders down through some trees beyond a grassy knoll to a small stream. I am just staring around.

After a while Sylvia sits down on the wooden picnic bench and straightens out her legs, lifting one at a time slowly without looking up. Long silences mean gloom for her, and I comment on it. She looks up and then looks down again.

"It was all those people in the cars coming the other way," she says. "The first one looked so sad. And then the next one looked exactly the same way, and then the next one and the next one, they were all the same."

"They were just commuting to work."

She perceives well but there was nothing unnatural about it. "Well, you know, *work,*" I repeat. "Monday morning. Half asleep. Who goes to work Monday morning with a grin?"

"It's just that they looked so *lost,*" she says. "Like they were all dead. Like a funeral procession." Then she puts both feet down and leaves them there.

I see what she is saying, but logically it doesn't go anywhere. You work to live and that's what they are doing. "I was watching swamps," I say.

After a while she looks up and says, "What did you see?"

"There was a whole flock of red-winged blackbirds. They rose up suddenly when we went by."

"Oh."

"I was happy to see them again. They tie things together, thoughts and such. You know?"

She thinks for a while and then, with the trees behind her a deep green, she smiles. She understands a peculiar language which has nothing to do with what you are saying. A daughter.

"Yes," she says. "They're beautiful."

"Watch for them," I say.

"All right."

John appears and checks the gear on the cycle. He adjusts some of the ropes and then opens the saddlebag and starts rummaging through. He sets some things on the ground. "If you ever need any rope, don't hesitate," he says. "God, I think I've got about *five* times what I need here."

"Not yet," I answer.

"Matches?" he says, still rummaging. "Sunburn lotion, combs, shoelaces . . . *shoelaces?* What do we need shoelaces for?"

"Let's not start *that,*" Sylvia says. They look at each other deadpan and then both look over at me.

"Shoelaces can break anytime," I say solemnly. They smile, but not at each other.

Chris soon appears and it is time to go. While he gets ready and climbs on, they pull out and Sylvia waves. We are on the highway again, and I watch them gain distance up ahead.

The Chautauqua that is in mind for this trip was inspired by these two many months ago and perhaps, although I don't know, is related to a certain undercurrent of disharmony between them.

Disharmony I suppose is common enough in any marriage, but in their case it seems more tragic. To me, anyway.

It's not a personality clash between them; it's something else, for which neither is to blame, but for which neither has any solution, and for which I'm not sure I have any solution either, just ideas.

The ideas began with what seemed to be a minor difference of opinion between John and me on a matter of small importance: how much one should maintain one's own motorcycle. It seems natural and normal to me to make use of the small tool kits and instruction booklets supplied with each machine, and keep it tuned and adjusted myself. John demurs. He prefers to let a competent mechanic take care of these things so that they are done right. Neither viewpoint is unusual, and this minor difference would never have become magnified if we didn't spend so much time riding together and sitting in country roadhouses drinking beer and talking about whatever comes to mind. What comes to mind, usually, is whatever we've been thinking about in the half hour or forty-five minutes since we last talked to each other. When it's roads or weather or people or old memories or what's in the newspapers, the conversation just naturally builds pleasantly. But whenever the performance of the machine has been on my mind and gets into the conversation, the building stops. The conversation no longer moves forward. There is a silence and a break in the continuity. It is as though two old friends, a Catholic and Protestant, were sitting drinking beer, enjoying life, and the subject of birth control somehow came up. Big freeze-out.

And, of course, when you discover something like that it's like discovering a tooth with a missing filling. You can never leave it alone. You have to probe it, work around it, push on it, think about it, not because it's enjoyable but because it's on your mind and it won't get off your mind. And the more I probe and push on this subject of cycle maintenance the more irritated he gets, and of course that makes me want to probe and push all the more. Not deliberately to irritate him but because the irritation seems symptomatic of something deeper, something under the surface that isn't immediately apparent.

When you're talking birth control, what blocks it and freezes it out is that it's not a matter of more or fewer babies being argued. That's just on the surface. What's underneath is a conflict of faith, of faith in empirical social planning versus faith in the authority of God as revealed by the teachings of the Catholic Church. You can prove the practicality of planned parenthood till you get tired of listening to yourself and it's going to go nowhere because your antagonist isn't buying the assumption that anything socially practical is good per se. Goodness for him has other sources which he values as much as or more than social practicality.

So it is with John. I could preach the practical value and worth of motorcycle maintenance till I'm hoarse and it would make not a dent in him. After two sentences on the subject his goes go completely glassy and he changes the conversation or just looks away. He doesn't want to hear about it.

Sylvia is completely with him on this one. In fact she is even more emphatic. "It's just a whole other thing," she says, when in a thoughtful mood. "Like garbage," she says, when not. They want *not* to understand it. Not to *hear* about it. And the more I try to fathom what makes me enjoy mechanical work and them hate it so, the more elusive it becomes. The ultimate cause of this originally minor difference of opinion appears to run way, way deep.

Inability on their part is ruled out immediately. They are both plenty bright enough. Either one of them could learn to tune a motorcycle in an hour and a half if they put their minds and energy to it, and the saving in money and worry and delay would repay them over and over again for their effort. And they *know* that. Or maybe they don't. I don't know. I never confront them with the question. It's better to just get along.

But I remember once, outside a bar in Savage, Minnesota, on a really scorching day when I just about let loose. We'd been in the bar for about an hour and we came out and the machines were so hot you could hardly get on them. I'm started and ready to go and there's John pumping away on the kick starter. I smell gas like we're next to a refinery and tell him so, thinking this is enough to let him know his engine's flooded.

"Yeah, I smell it too," he says and keeps on pumping. And he pumps and pumps and jumps and pumps and *I* don't know what more to say. Finally, he's really winded and sweat's running down all over his face and he can't pump anymore, and so I suggest taking out the plugs to dry them off and air out the cylinders while we go back for another beer.

Oh my God no! He doesn't want to get into all that stuff.

"All what stuff?"

"Oh, getting out the tools and all that stuff. There's no reason why it shouldn't start. It's a brand-new machine and I'm following the instructions perfectly. See, it's right on full choke like they say."

"Full *choke!*"

"That's what the instructions say."

"That's for when it's *cold!*"

"Well, we've been in there for a half an hour at least," he says.

It kind of shakes me up. "This is a hot day, John," I say. "And they take longer than that to cool off even on a freezing day."

He scratches his head. "Well, why don't they tell you that in the instructions?" He opens the choke and on the second kick it starts. "I guess that was it," he says cheerfully.

And the very next day we were out near the same area and it happened again. This time I was determined not to say a word, and when my wife urged me to go over and help him I shook my head. I told her that until he had a real felt need he was just going to resent help, so we went over and sat in the shade and waited.

I noticed he was being superpolite to Sylvia while he pumped away, meaning he was furious, and she was looking over with a kind of "Ye gods!" look. If he had asked any single question I would have been over in a second to diagnose it, but he wouldn't. It must have been fifteen minutes before he got it started.

Later we were drinking beer again over at Lake Minnetonka and everybody was talking around the table, but he was silent and I could see he was really tied up in knots inside. After all that time. Probably to get them untied he finally said, "You know . . . when it doesn't start like that it just . . . really turns me into a *monster* inside. I just get paranoic about it." This seemed to loosen him up, and he added, "They just had this *one* motorcycle, see? This *lemon*. And they didn't know what to do with it, whether to send it back to the factory or sell it for scrap or what . . . and then at the last moment they saw *me* coming. With eighteen hundred bucks in my pocket. And they knew their problems were over."

In a kind of singsong voice I repeated the plea for tuning and he tried hard to listen. He really tries hard sometimes. But then the block came again and he was off to the bar for another round for all of us and the subject was closed.

He is not stubborn, not narrow-minded, not lazy, not stupid. There was just no easy explanation. So it was left up in the air, a kind of mystery that one gives up on because there is no sense in just going round and round and round looking for an answer that's not there.

It occurred to me that maybe I was the odd one on the subject, but that was disposed of too. Most touring cyclists know how to keep their machines tuned. Car owners usually won't touch the engine, but every town of any size at all has a garage with expensive lifts, special tools and diagnostic equipment that the average owner can't afford. And a car engine is more complex and inaccessible than a cycle engine so there's more sense to this. But for John's cycle, a BMW R60, I'll bet there's not a mechanic between here and Salt Lake City. If his points or plugs burn out, he's done for. I *know* he doesn't have a set of spare points with him. He doesn't know what points are. If it quits on him in western South Dakota or Montana I don't know what he's going to do. Sell it to the Indians maybe. Right now I know what he's doing. He's carefully avoiding giving any thought whatsoever to the subject. The BMW is famous for not giving mechanical problems on the road and that's what he's counting on.

I might have thought this was just a peculiar attitude of theirs about motorcycles but discovered later that it extended to other things. . . . Waiting for them to get going one morning in their kitchen I noticed the sink faucet was dripping and remembered that it was dripping the last time I was there before and that in fact it had been dripping as long as I could remember. I commented on it and John said he had tried to fix it with a new faucet washer but it hadn't worked. That was all he said. The presumption left was that that was the end of the matter. If you try to fix a faucet and your fixing doesn't work then it's just your lot to live with a dripping faucet.

This made me wonder to myself if it got on their nerves, this drip-drip-drip, week in, week out, year in, year out, but I could not notice any irritation or concern about it on their part, and so concluded they just aren't bothered by things like dripping faucets. Some people aren't.

What it was that changed this conclusion, I don't remember . . . some intuition, some insight one day, perhaps it was a subtle change in Sylvia's mood whenever the dripping was particularly loud and she was trying to talk. She has a very soft voice. And one day when she was trying to talk above the dripping and the kids came in and interrupted her she lost her temper at them. It seemed that her anger at the kids would not have been nearly as great if the faucet hadn't also been dripping when she was trying to talk. It was the combined dripping and loud kids that blew her up. What struck me hard then was that she was *not* blaming the faucet, and that she was *deliberately* not blaming the faucet. She wasn't ignoring that faucet at all! She was *suppressing* anger at that faucet and that goddamned dripping faucet was just about *killing* her! But she could not admit the importance of this for some reason.

Why suppress anger at a dripping faucet? I wondered.

Then that patched in with the motorcycle maintenance and one of those light bulbs went on over my head and I thought, Ahhhhhhhh!

It's not the motorcycle maintenance, not the faucet. It's all of technology they can't take. And then all sorts of things started tumbling into place and I knew that was it. Sylvia's irritation at a friend who thought computer programming was "creative." All their drawings and paintings and photographs without a technological thing in them. Of course she's not going to get mad at that faucet, I thought. You always suppress momentary anger at something you deeply and permanently hate. Of course John signs off every time the subject of cycle repair comes up, even when it is obvious he is suffering for it. That's technology. And sure, of course, obviously. It's so simple when you see it. To get away from technology out into the country in the fresh air and sunshine is why they are on the motorcycle in the first place. For me to bring it back to them just at the point and place where they think they have finally escaped it just frosts both of them, tremendously. That's why the conversation always breaks and freezes when the subject comes up.

Other things fit in too. They talk once in a while in as few pained words as possible about "it" or "it all" as in the sentence, "There is just no escape from it." And if I asked, "From what?" the answer might be "The whole thing," or "The whole organized bit," or even "The system." Sylvia once said defensively, "Well, *you* know how to *cope* with it," which puffed me up so much at the time I was embarrassed to ask what "it" was and so remained somewhat puzzled. I thought it was something more mysterious than technology. But now I see that the "it" was mainly, if not entirely, technology. But, that doesn't sound right either. The "it" is a kind of force that gives rise to technology, something undefined, but inhuman, mechanical, lifeless, a blind monster, a death force. Something hideous they are running from but know they can never escape. I'm putting it way too heavily here but in a less emphatic and less defined way this is what it is. Somewhere there are people who understand it and run it but those are technologists, and they speak an inhuman language when describing what they do. It's all parts and relationships of unheard-of things that never make any sense no matter how often you hear about them. And their things, their monster keeps eating up land and polluting their air and lakes, and there is no way to strike back at it, and hardly any way to escape it.

That attitude is not hard to come to. You go through a heavy industrial area of a large city and there it all is, the technology. In front of it are high barbed-wire fences, locked gates, signs saying NO TRESPASSING, and beyond, through sooty air, you see ugly strange shapes of metal and brick whose purpose is unknown, and whose masters you will never see. What it's for you don't know, and why it's there, there's no one to tell, and so all you can feel is alienated, estranged, as though you didn't belong there. Who owns and understands this doesn't want you around. All this technology has somehow made you a stranger in your own land. Its very shape and appearance and mysteriousness say, "Get out." You know there's an explanation for all this somewhere and what it's doing undoubtedly serves mankind in some indirect way but that isn't what you see. What you see is the NO TRESPASSING, KEEP OUT signs and not anything serving people but little people, like ants, serving these strange, incomprehensible shapes. And you think, even if I were a part of this, even if I were not a stranger, I would be just another ant serving the shapes. So the final feeling is hostile, and I think that's ultimately what's involved with this otherwise unexplainable attitude of John and Sylvia. Anything to do with valves and shafts and wrenches is a part of *that* dehumanized world, and they would rather not think about it. They don't want to get into it.

If this is so, they are not alone. There is no question that they have been following their natural feelings in this and not trying to imitate anyone. But many others are also following their natural feelings and not trying to imitate anyone and the natural feelings of very many people are similar on this matter; so that when you look at them collectively, as journalists do, you get the illusion of a mass movement, an antitechnological mass movement, an entire political antitechnological left emerging, looming up from apparently nowhere, saying "Stop the technology. Have it somewhere else. Don't have it here." It is still restrained by a thin web of logic that points out that without the factories there are no jobs or standard of living. But there are human forces stronger than logic. There always have been, and if they become strong can break.

Clichés and stereotypes such as "beatnik" or "hippie" have been invented for the antitechnologists, the antisystem people, and will continue to be. But one does not convert individuals into mass people with the simple coining of a mass term. John and Sylvia are not mass people and neither are most of the others going their way. It is against being a mass person that they seem to be revolting. And they feel that technology has got a lot to do with the forces that are trying to turn them into mass people and they don't like it. So far it's still mostly a passive resistance, flights into the rural areas when they are possible and things like that, but it doesn't always have to be this passive.

I disagree with them about cycle maintenance, but not because I am out of sympathy with their feelings about technology. I just think that their flight from and hatred of technology is self-defeating. The Buddha, the Godhead, resides quite as comfortably in the circuits of a digital computer or the gears of a cycle transmission as he does at the top of a mountain or in the petals of a flower. To think otherwise is to demean the Buddha—which is to demean oneself. That is what I want to talk about in this Chautauqua.

We're out of the marshes now, but the air is still so humid you can look straight up directly at the yellow circle of the sun as if there were smoke or smog in the sky. But we're in the green countryside now. The farmhouses are clean and white and fresh. And there's no smoke or smog.

2

The road winds on and on . . . we stop for rests and lunch, exchange small talk, and settle down to the long ride. The beginning fatigue of afternoon balances the excitement of the first day and we move steadily, not fast, not slow.

We have picked up a southwest side wind, and the cycle cants into the gusts, seemingly by itself, to counter their effect. Lately there's been a sense of something peculiar about this road, apprehension about something, as if we were being watched or followed. But there is not a car anywhere ahead, and in the mirror are only John and Sylvia way behind.

We are not in the Dakotas yet, but the broad fields show we are getting nearer. Some of them are blue with flax blossoms moving in long waves like the surface of the ocean. The sweep of the hills is greater than before and they now dominate everything else, except the sky, which seems wider. Farmhouses in the distance are so small we can hardly see them. The land is beginning to open up.

There is no one place or sharp line where the Central Plains end and the Great Plains begin. It's a gradual change like this that catches you unawares, as if you were sailing out from a choppy coastal harbor, noticed that the waves had taken on a deep swell, and turned back to see that you were out of sight of land. There are fewer trees here and suddenly I am aware they are no longer native. They have been brought here and planted around houses and between fields in rows to break up the wind. But where they haven't been planted there is no underbrush, no second-growth saplings—only grass, sometimes with wildflowers and weeds, but mostly grass. This is grassland now. We are on the prairie.

I have a feeling none of us fully understands what four days on this prairie in July will be like. Memories of car trips across them are always of flatness and great emptiness as far as you can see, extreme monotony and boredom as you drive for hour after hour, getting nowhere, wondering how long this is going to last without a turn in the road, without a change in the land going on and on to the horizon.

John was worried Sylvia would not be up to the discomfort of this and planned to have her fly to Billings, Montana, but Sylvia and I both talked him out of it. I argued that physical discomfort is important only when the mood is wrong. Then you fasten on to whatever thing is uncomfortable and call that the cause. But if the mood is right, then physical discomfort doesn't mean much. And when thinking about Sylvia's moods and feelings, I couldn't see her complaining.

Also, to arrive in the Rocky Mountains by plane would be to see them in one kind of context, as pretty scenery. But to arrive after days of hard travel across the prairies would be to see them in another way, as a goal, a promised land. If John and I and Chris arrived with this feeling and Sylvia arrived seeing them as "nice" and "pretty," there would be more disharmony among us than we would get from the heat and monotony of the Dakotas. Anyway, I like to talk to her and I'm thinking of myself too.

In my mind, when I look at these fields, I say to her, "See? . . . See?" and I think she does. I hope later she will see and feel a thing about these prairies I have given up talking to others about; a thing that exists here because everything else does not and can be noticed because other things are absent. She seems so depressed sometimes by the monotomy and boredom of her city life, I thought maybe in this endless grass and wind she would see a thing that sometimes comes when monotony and boredom are accepted. It's here, but I have no names for it.

Now on the horizon I see something else I don't think the others see. Far off to the southwest—you can see it only from the top of this hill—the sky has a dark edge. Storm coming. That may be what has been bothering me. Deliberately shutting it out of mind, but knowing all along that with this humidity and wind it was more than likely. It's too bad on the first day, but as I said before, on a cycle you're *in* the scene, not just watching it, and storms are definitely part of it.

If it's just thunderheads or broken line squalls you can try to ride around them, but this one isn't. That long dark streak without any preceding cirrus clouds is a cold front. Cold fronts are violent and when they are from the southwest, they are the most violent. Often they contain tornadoes. When they come it's best to just hole up and let them pass over. They don't last long and the cool air behind them makes good riding.

Warm fronts are the worst. They can last for days. I remember Chris and I were on a trip to Canada a few years ago, got about 130 miles and were caught in a warm front of which we had plenty of warning but which we didn't understand. The whole experience was kind of dumb and sad.

We were on a little six-and-one-half-horsepower cycle, way overloaded with luggage and way underloaded with common sense. The machine could do only about forty-five miles per hour wide open against a moderate head wind. It was no touring bike. We reached a large lake in the North Woods the first night and tented amid rainstorms that lasted all night long. I forgot to dig a trench around the tent and at about two in the morning a stream of water came in and soaked both sleeping bags. The next morning we were soggy and depressed and hadn't had much sleep, but I thought that if we just got riding the rain would let up after a while. No such luck. By ten o'clock the sky was so dark all the cars had their headlights on. And then it really came down.

We were wearing the ponchos which had served as a tent the night before. Now they spread out like sails and slowed our speed to thirty miles an hour wide open. The water on the road became two inches deep. Lightning bolts came crashing down all around us. I remember a woman's face looking astonished at us from the window of a passing car, wondering what on earth we were doing on a motorcycle in this weather. I'm sure I couldn't have told her.

The cycle slowed down to twenty-five, then twenty. Then it started hissing, coughing and popping and sputtering until, barely moving at five or six miles an hour, we found an old run-down filling station by some cutover timberland and pulled in.

At the time, like John, I hadn't bothered to learn much about motorcycle maintenance. I remember holding my poncho over my head to keep the rain from the tank and rocking the cycle between my legs. Gas seemed to be sloshing around inside. I looked at the plugs, and looked at the points, and looked at the carburetor, and pumped the kick starter until I was exhausted.

We went into the filling station, which was also a combination beer joint and restaurant, and had a meal of burned-up steak. Then I went back out and tried it again. Chris kept asking questions that started to anger me because he didn't see how serious it was. Finally I saw it was no use, gave it up, and my anger at him disappeared. I explained to him as carefully as I could that it was all over. We weren't going anywhere by cycle on this vacation. Chris suggested things to do like check the gas, which I had done, and find a mechanic. But there weren't any mechanics. Just cutover pine trees and brush and rain.

I sat in the grass with him at the shoulder of the road, defeated, staring into the trees and underbrush. I answered all of Chris's questions patiently and in time they became fewer and fewer. And then Chris finally understood that our cycle trip was really over and began to cry. He was eight then, I think.

We hitchhiked back to our own city and rented a trailer and put it on our car and came up and got the cycle, and hauled it back to our own city and then started out all over again by car. But it wasn't the same. And we didn't really enjoy ourselves much.

Two weeks after the vacation was over, one evening after work, I removed the carburetor to see what was wrong but still couldn't find anything. To clean off the grease before replacing it, I turned the stopcock on the tank for a little gas. Nothing came out. The tank was out of gas. I couldn't believe it. I can still hardly believe it.

I have kicked myself mentally a hundred times for that stupidity and don't think I'll ever really, finally get over it. Evidently what I saw sloshing around was gas in the reserve tank which I had never turned on. I didn't check it carefully because I assumed the rain had caused the engine failure. I didn't understand then how foolish quick assumptions like that are. Now we are on a twenty-eight-horse machine and I take the maintenance of it very seriously.

All of a sudden John passes me, his palm down, signaling a stop. We slow down and look for a place to pull off on the gravelly shoulder. The edge of the concrete is sharp and the gravel is loose and I'm not a bit fond of this maneuver.

Chris asks, "What are we stopping for?"

"I think we missed our turn back there," John says.

I look back and see nothing. "I didn't see any sign," I say.

John shakes his head. "Big as a barn door."

"Really?"

He and Sylvia both nod.

He leans over, studies my map and points to where the turn was and then to a freeway overpass beyond it. "We've already crossed this freeway," he says. I see he is right. Embarrassing. "Go back or go ahead?" I ask.

He thinks about it. "Well, I guess there's really no reason to go back. All right. Let's just go ahead. We'll get there one way or another."

And now tagging along behind them I think, Why should I do a thing like that? I hardly noticed the freeway. And just now I forgot to tell them about the storm. Things are getting a little unsettling.

The storm cloud bank is larger now but it is not moving in as fast as I thought it would. That's not so good. When they come in fast they leave fast. When they come in slow like this you can get stuck for quite a time.

I remove a glove with my teeth, reach down and feel the aluminum side cover of the engine. The temperature is fine. Too warm to leave my hand there, not so hot I get a burn. Nothing wrong there.

On an air-cooled engine like this, extreme overheating can cause a "seizure." This machine has had one . . . in fact, three of them. I check it from time to time the same way I would check a patient who has had a heart attack, even though it seems cured.

In a seizure, the pistons expand from too much heat, become too big for the walls of the cylinders, seize them, melt to them sometimes, and lock the engine and rear wheel and start the whole cycle into a skid. The first time this one seized, my head was pitched over the front wheel and my passenger was almost on top of me. At about thirty it freed up again and started to run but I pulled off the road and stopped to see what was wrong. All my passenger could think to say was "What did you do *that* for?"

I shrugged and was as puzzled as he was, and stood there with the cars whizzing by, just staring. The engine was so hot the air around it shimmered and we could feel the heat radiate. When I put a wet finger on it, it sizzled like a hot iron and we rode home, slowly, with a new sound, a slap that meant the pistons no longer fit and an overhaul was needed.

I took this machine into a shop because I thought it wasn't important enough to justify getting into myself, having to learn all the complicated details and maybe having to order parts and special tools and all that time-dragging stuff when I could get someone else to do it in less time—sort of John's attitude.

The shop was a different scene from the ones I remembered. The mechanics, who had once all seemed like ancient veterans, now looked like children. A radio was going full blast and they were clowning around and talking and seemed not to notice me. When one of them finally came over he barely listened to the piston slap before saying, "Oh yeah. Tappets."

Tappets? I should have known then what was coming.

Two weeks later I paid their bill for 140 dollars, rode the cycle carefully at varying low speeds to wear it in and then after one thousand miles opened it up. At about seventy-five it seized again and freed at thirty, the same as before. When I brought it back they accused me of not breaking it in properly, but after much argument agreed to look into it. They overhauled it again and this time took it out themselves for a high-speed road test.

It seized on *them* this time.

After the third overhaul two months later they replaced the cylinders, put in oversize main carburetor jets, retarded the timing to make it run as coolly as possible and told me, "Don't run it fast."

It was covered with grease and did not start. I found the plugs were diconnected, connected them and started it, and now there really *was* a tappet noise. They hadn't adjusted them. I pointed this out and the kid came with an open-end adjustable wrench, set wrong, and swiftly rounded both of the sheet-aluminum tappet covers, ruining both of them.

I hope we've got some more of those in stock," he said.

I nodded.

He brought out a hammer and cold chisel and started to pound them loose. The chisel punched through the aluminum cover and I could see he was pounding the chisel right into the engine head. On the next blow he missed the chisel completely and struck the head with the hammer, breaking off a portion of two of the cooling fins.

"Just stop," I said politely, feeling this was a bad dream. "Just give me some new covers and I'll take it the way it is."

I got out of there as fast as possible, noisy tappets, shot tappet covers, greasy machine, down the road, and then felt a bad vibration at speeds over twenty. At the curb I discovered two of the four engine-mounting bolts were missing and a nut was missing from the third. The whole engine was hanging on by only one bolt. The overhead-cam chain-tensioner bolt was also missing, meaning it would have been hopeless to try to adjust the tappets anyway. Nightmare.

The thought of John putting his BMW into the hands of one of those people is something I have never brought up with him. Maybe I should.

I found the cause of the seizures a few weeks later, waiting to happen again. It was a little twenty-five-cent pin in the internal oil-delivery system that had been sheared and was preventing oil from reaching the head at high speeds.

The question *why* comes back again and again and has become a major reason for wanting to deliver this Chautauqua. Why did they butcher it so? These were not people running away from technology, like John and Sylvia. These were the technologists themselves. They sat down to do a job and they performed it like chimpanzees. Nothing personal in it. There was no obvious reason for it. And I tried to think back into that shop, that nightmare place, to try to remember anything that could have been the cause.

The radio was a clue. You can't really think hard about what you're doing and listen to the radio at the same time. Maybe they didn't see their job as having anything to do with hard thought, just wrench twiddling. If you can twiddle wrenches while listening to the radio that's more enjoyable.

Their speed was another clue. They were really slopping things around in a hurry and not looking where they slopped them. More money that way—if you don't stop to think that it usually takes longer or comes out worse.

But the biggest clue seemed to be their expressions. They were hard to explain. Good-natured, friendly, easygoing—and uninvolved. They were like spectators. You had the feeling they had just wandered in there themselves and somebody had handed them a wrench. There was no identification with the job. No saying, "I am a mechanic." At 5 P.M. or whenever their eight hours were in, you knew they would cut it off and not have another thought about their work. They were already trying not to have any thoughts about their work *on* the job. In their own way they were achieving the same thing John and Sylvia were, living with technology without really having anything to do with it. Or rather, they had something to do with it, but their own selves were outside of it, detached, removed. They were involved in it but not in such a way as to care.

Not only did these mechanics not find that sheared pin, but it was clearly a mechanic who had sheared it in the first place, by assembling the side cover plate improperly. I remembered the previous owner had said a mechanic had told him the plate was hard to get on. That was why. The shop manual had warned about this, but like the others he was probably in too much of a hurry or he didn't care.

While at work I was thinking about this same lack of care in the digital computer manuals I was editing. Writing and editing technical manuals is what I do for a living the other eleven months of the year and I knew they were full of errors, ambiguities, omissions and information so completely screwed up you had to read them six times to make any sense out of them. But what struck me for the first time was the agreement of these manuals with the spectator attitude I had seen in the shop. These were spectator manuals. It was built into the format of them. Implicit in every line is the idea that "Here is the machine, isolated in time and in space from everything else in the universe. It has no relationship to you, you have no relationship to it, other than to turn certain switches, maintain voltage levels, check for error conditions . . ." and so on. That's it. The mechanics in their attitude toward the machine were really taking no different attitude from the manual's toward the machine, or from the attitude I had when I brought it in there. We were all spectators. And it occurred to me there *is* no manual that deals with the *real* business of motorcycle maintenance, the most important aspect of all. Caring about what you are doing is considered either unimportant or taken for granted.

On this trip I think we should notice it, explore it a little, to see if in that strange separation of what man is from what man does we may have some clues as to what the hell has gone wrong in this twentieth century. I don't want to hurry it. That itself is a poisonous twentieth-century attitude. When you want to hurry something, that means you no longer care about it and want to get on to other things. I just want to get at it slowly, but carefully and thoroughly, with the same attitude I remember was present just before I found that sheared pin. It was that attitude that found it, nothing else.

I suddenly notice the land here has flattened into a Euclidian plane. Not a hill, not a bump anywhere. This means we have entered the Red River Valley. We will soon be into the Dakotas.

Part four

TECHNOLOGY AND THE PROFESSIONS

Part Four focuses on the relation between technology and the professions. Four major professions will be considered: agriculture, medicine, business, and engineering. It first must be emphasized that the relation of technology to these professions is an ancient one. Nothing could be further from the truth than the commonly held belief that the introduction of technology to these and other professions is relatively recent. Technology has been a part of our lives for a very long time. Early man who rubbed two sticks together to produce fire was already engaging in technological activity. So was the hunter with his club, his spear, and later, his bow and arrow. To emphasize the long-standing influence of technology on the professions, some prominent achievements of ancient times will be briefly discussed. But the major portion of this introduction will examine the current connections between technology and the four professions mentioned.

• • •

One of the most outstanding technological achievements of ancient times was the construction of the Great Pyramid of Giza in Egypt. This pyramid is the tomb of King Khufu or Cheops, and was built for him around 2680 B.C. It contained secret passages and chambers, and originally rose to a height of 482 feet. The pyramid, considered one of the Seven Wonders of the Ancient World, covers thirteen acres and was built of approximately 2,300,000 huge limestone blocks, each weighing about two and a half tons. These extraordinarily heavy pieces were brought to the site of the pyramid and then lifted to their proper positions without the benefit of modern equipment. Today, as the Egyptian government prepares to restore the pyramid, which has suffered slow erosion during the past five thousand years, it will again employ the old methods using muscle and rope, rather than modern technology, which so far has not devised a crane that can cope with the shape of the pyramid.

The building of the Great Pyramid of Giza is fascinating because, among other things, it required sophisticated knowledge of two kinds of techniques. The first kind is best referred to as engineering "know-how," which includes the designing of the pyramid and its interior passages and chambers, as well as the devising of ways to execute the design. That the pyramid was designed and the plans executed with a precision that even modern engineers would be proud of is obvious from the detailed description offered by James Henry Breasted in his book *A History of Egypt*. According to Breasted, "some of the masonry finish is so fine that blocks weighing tons are set together with seams of considerable length, showing a joint of one ten-thousandth of an inch . . ."[1]

The second kind of technique that stands out in the building of the Great Pyramid is best described as organizational. The working force was arranged according to a certain hierarchy and a chain of command. This, combined with discipline and the assignment of specific tasks to each level of the hierarchy, created the first "megamachine" in history. Lewis Mumford finds this fact extremely significant because he sees in it the true beginnings of the Machine Age. The machine that was being introduced for the first time was a human machine. All its parts were human, and most were interchangeable. It was capable of generating thousands of horsepower, thus making the building of the pyramid possible without the help of sophisticated machinery. These impressive facts lead Mumford to argue that "this new kind of machine was far more complex than the contemporary potter's wheel or bow drill, and it remained the most advanced type of machine until the invention of the mechanical clock in the fourteenth century."[2]

The importance of this organizational achievement can hardly be overemphasized. Suffice it to say that for the modern economist John Kenneth Galbraith "modern economic society can only be understood as an effort, wholly successful, to synthesize by organization a group personality far superior *for its* purpose to a natural person . . ."[3] That is exactly what characterized the society which built the Great Pyramid of Giza. No one worker could even have begun to move a two-and-a-half–ton block of limestone, but a group armed with technology (as simple as it may have been) could, and in fact did. To emphasize further the parallel between the Egyptian and modern Western society, let us look at this additional passage from Galbraith:

> The real accomplishment of modern science and technology consists in taking ordinary men, informing them narrowly and deeply and then, through appropriate organization, arranging to have their knowledge combined with that of other specialized but equally ordinary men. This dispenses with the need for genius. The resulting performance, though less inspiring, is far more predictable. No individual genius arranged the flights to the moon. It was the work of organization—bureaucracy.[4]

Similarly, the pyramid was not the work of an individual genius. King, engineers, priests, overseers, and workers all contributed to the successful completion of the project. It was a colossal project designed and executed by an organization of ordinary men. Thus long before the age of modern science and technology, engineering and organizational techniques functioned to produce an impressive artifact.

Another interesting technological feat from ancient times is the construction of the Hanging Gardens of Babylon, which are also considered one of the Seven Wonders of the Ancient World. Although they have been destroyed, various writers have described them in detail.

The Hanging Gardens were built by Nebuchadnezzar, king of Babylon (reigned 605-562 B.C.). It is believed that they were built for his wife, a Median princess whose home was in the mountains of Persia. Babylon was flat and hot, and the princess had longed for the cool, green, and fresh surroundings of her native land, so Nebuchadnezzar set out to duplicate that environment as closely as possible. To that end he built an architectural wonder which crowned the roof of the imperial palace with masses of greenery layered in terraces to form hanging gardens.

But the Hanging Gardens were not Babylon's only claim to fame. Earlier, another of its kings, Hammurabi (reigned 1948-1905 B.C.), drew up the world's first code of law, which is of special interest to us because it contains the following laws relating to medical practice:

> If the doctor shall treat a gentleman and shall open an abscess with a bronze knife and shall preserve the eye of the patient, he shall receive ten shekels of silver.
>
> If the doctor shall open an abscess with a bronze knife and shall kill the patient or shall destroy the sight of the eye, his hands shall be cut off.[5]

These passages make clear the existence of medical technology in Babylon as early as the 1900's B.C.

Ancient Egypt also possessed a degree of medical technology. The Ebers Papyrus, written about 1500 B.C. and found in the tomb of Thebes in 1862 by Professor Ebers, contains approximately 900 prescriptions. More recently, the Edwin Smith Papyrus was translated. Its author was a surgeon who noted the role of the brain in controlling the lower limbs, and the role of the heart as the driving power of the human body. Surgical stitching is mentioned in this papyrus for the first time in medical literature.

What has been said of the ancient relation of technology to engineering and medicine can also be said of the relation of technology to agriculture, business, and economics. The wealth of Egypt, especially around the forty-third century B.C., came mainly from grain. Originally the Egyptians used the hoe for cultivating their fields. Then the plow, which utilizes animal power, was invented. This drastically increased the size of the cultivated areas, and was as significant an event in its time as the introduction of modern machinery to agriculture in the nineteenth and twentieth centuries A.D. With the subsequent increase in wealth, a more advanced tax system developed. Since no currency existed, taxes, loans, and business debts were made at first in grain. Later, copper rings were used as money. Those who did not pay their taxes were visited by official "tax collectors."

In the Tigris-Euphrates Valley, we find a similar development as early as the Kingdom of Sumer and Akkad, which preceded the rule of Nebuchadnezzar. The wealth of the area came mainly from barley and wheat. The inhabitants developed dikes, irrigation trenches, and other agricultural techniques to improve their harvest. They were also familiar with copper tools and utensils. During Babylonian times silver pieces came to be exchanged as money, and an enormous body of bookkeeping records was left behind in the form of hardened clay tablets.

• • •

In ancient times the professions were closely tied to religious and magical beliefs, and this state of affairs persisted throughout the Middle Ages into the Renaissance and even later. As mentioned in the introduction to Part Two, professional activities took a secondary place to those of tending to the affairs of the soul. The consequences of such a situation were mixed. Earlier in this book it was argued that religious beliefs and their attendant values succeeded in keeping men's desires for power, possessions, and control within reasonable limits. It may be pointed out now that these beliefs also resulted in many events being regarded as the effects of supernatural causes; investigation of their real causes was thus inhibited. Diseases were often seen as a sign of divine punishment or as an act of possession by the devil. The story of ergot is an excellent example of this mode of reasoning. Leo Vining* informs us that in 857 A.D., according to German chroniclers of that period, the population around Duisberg was ravaged by "a great plague of swollen blisters that consumed the people by a loathsome rot so that their limbs were loosened and fell off before death." This horrible disease, it was much later discovered, was caused by grain contamination with ergot fungus. But people in the Middle Ages refered to it as the "holy fire." When the disease happened to cease at about the same time as a father kneeled before the bones of Saint Anthony, begging that his sick son be spared, the disease was renamed "Saint Anthony's fire." Such a mode of reasoning of course did not encourage the understanding of nature and its workings.

There were other religious beliefs that made scientific progress difficult. Since the body was regarded as the temple of the soul, it had derivative sanctity. Thus, it was not permissible to mutilate it in any way, even after death. Dissecting a corpse for the sake of medical knowledge was not acceptable, and many physicians throughout Europe had to rely on body snatchers to provide them with corpses for their anatomical research.

The story of Galileo, who was forced by the church to retract his pronouncements concerning the heliocentric theory of planetary motion, is widely known. The popular belief was that the earth was immobile and located at the center of the universe and that the other heavenly bodies rotated around it. This view supported the then common literal interpretation of the scriptures, which declared man the center of the universe. Galileo's theory, therefore, was difficult for laymen to believe, but more importantly, it was heretical, and he thus was forced to renounce it.

These and many other incidents, including the burning of witches, frustrated the rapid development of modern science and technology. Finally, however, with the accumulation of data, science and technology were secularized, ushering in the present era (for a more detailed discussion of this historical development, see the introduction to Part Two). In our present era of modern secularized science and technology, innovations are numerous, and the progress of human knowledge is astounding. However, secularized science and technology are encountering problems of their own. The rest of this introduction is devoted to a discussion of some of these problems.

A prefatory comment about this discussion is in order. With the modern alliance of science and technology, it has become possible to do things that man never dreamt of before. New devices are being invented daily. New sources of power are being tapped. Electric appliances, nuclear reactors, alloys, plastics, tractors, computers, kidney machines, brain scanners—all these and much more contribute in different ways to the hectic pace of development in the modern professions. To cope with this hectic pace as well as respond to modern demands for efficiency and profitability, it has become desirable to organize activities within these professions along corporate lines. This fact will be discussed in some detail later; suffice it to observe now that this new technique of organization combines professional goals with those of a successful business venture. As a result, a serious discussion of the professions in this introduction must address itself to this new technique of organization insofar as it has affected these professions and their traditional goals, practices, and values.

• • •

As inventions multiply and new equipment proliferates in the various professions, an individual or a company must keep up or else fall behind through the use of outmoded techniques. Since with each invention new standards of efficiency and sophistication are set, relying on older techniques means less efficient operation. In farming, for example, to fail to keep up with new developments often means less produce and profit per acre than possible. In medicine, new equipment often makes the difference between life and death. In engineering, new techniques form a crucial base for further innovations. In business, the failure to keep up can make economic survival impossible. The constant updating of the professions becomes a need rather than a luxury.

But the cost of constant updating is quite high. Often a piece of equipment becomes obsolete within a year or two of its purchase, and the cost of replacement is continuously rising. For example, a tractor priced at $16,000 in 1974 costs almost twice as much today even though the new version, which may have a better door latch, is not more powerful. In 1896 an x-ray machine cost $50. Today, the more sophisticated and specialized CAT scanner combines x-ray equipment with a computer and a television cathode-ray tube. This diagnostic machine provides the physician with a cross-sectional view of the body and thus allows him to better detect various disorders. It is a great improvement over the old simple x-ray machine, but its current cost is about $700,000.

The effect of these spiraling costs on the professions is to make greater capital outlays necessary. The small farmer and the small businessman can no longer make it in today's world. The engineer must join a corporation, and the physician, whose own office equipment is becoming highly expensive, must count on the backup system of the hospitals in cases requiring more sophisticated equipment.

Increasing costs have thus translated themselves into size. "Bigger is better." A bigger corporation can better absorb a $700,000 bill for one piece of equipment than a single businessman. As a result all professions must organize along corporate lines in order to keep pace.

Even farming, which has traditionally been a small family enterprise, is now also moving in the direction of corporations. This trend has been prompted not only by the cost of machinery but also by the efficiency that the use of machinery demands. A tractor, for example, can work a bigger farm just as easily as a smaller one; and the more acres it works, the smaller the real cost of that tractor to the farmer. In a *Time* article on farming in the United States, Patrick Benedict, president and sole stockholder of Benedict Farms, Inc., argues that "if you're standing still you're really falling behind."[6] To ensure that he does not fall behind, he has continued for years to reinvest his profits and borrow additional money to acquire more land. He is also a firm believer in vertical integration—that is, owning operations related to farm products. He organized 1,600 farmers to raise $20 million, and borrowed another $47 million to buy out a sugar company that used to buy and process his and other farmers' sugar beets. He also teamed with nine other farmers to buy a $1.5 million grain elevator and incorporated it as Northern Grain Company. These two investments have allowed Mr. Benedict to move his farm products, whether for factory processing or sale, at the optimal time, giving him a better chance for increasing his profits.

High interest rates and the rising cost of land, machinery, fertilizers, fuel, and insecticides cut deeply into the farmer's profit. Consider, for example, the cost of fuel, which has increased rapidly for the farmer for two reasons: (1) the price of fuel has been escalating rapidly on the world market, and (2) the industrialization of agriculture has resulted in an increase in the amount of fuel needed to run a farm. Carol and John Steinhart* in "The Energy We Eat" provide us with some idea of how much more fuel is required by industrialized food production. They observe that while in primitive cultures, 5 to 50 calories of food were obtained in return for each calorie invested, industrialized food production requires the investment of 5 to 10 calories of fuel in order to obtain one calorie of food. This situation is regarded by the Steinharts as quite significant—especially if energy costs continue to increase, a highly likely prospect. They warn that it could result in famine in many areas of the world unless we return to less energy-intensive methods of food production.

Some distressing symptoms of the problems that lie ahead are already apparent. In February 1979, 3500 farmers in tractors and mobile homes converged on Washington to demand increased crop subsidies to offset the escalating costs of land, machinery, and fuel. In the summer of that same year, the farmers watched their fuel supplies dwindle as motorists competed for the limited quantities available.

These problems make the farmer's life difficult. For all his hard work Patrick Benedict averages a return of only three and a half percent on his $3.5 million investment. A bank certificate of deposit can return him around nine percent, without any effort. Such data discourages many young farmers and makes them go into other types of work. Even if they were able somehow to secure the needed capital to start a farm, the work is still hard and the returns small. The price of the corn in cornflakes rises from about $1.80 a bushel paid to the farmer to about $37 a bushel paid by the consumer to the grocer. Given the fact that farming is the single largest industry in the United States, according to U.S.D.A. statistics, current developments raise very serious issues concerning the national socioeconomic structure. Joe Flanagan, a farmer, complains in *Progressive Farmer:* "There is something wrong with the system when you trim and trim and trim your expenses, make a good crop, sell it for an above-average price, and still lose money."[7] Or make less profit than a bank's certificate of deposit can produce!

Frances Moore Lappé and Joseph Collins, who wrote *Food First: Beyond the Myth of Scarcity,* blame much of the malaise on agribusiness organizations. They claim that the same increasing concentration of control over land and marketing that directly causes hunger in underdeveloped countries is going on right here in the United States.[8] Let us pause momentarily to take a look at what is going on in the Third World. In *How the Other Half Dies,* Susan George points out that an average of eighty percent of the people in Asia live in the countryside, as do ninety-five percent of the people in some parts of Africa; most of these people are dependent on the land for their living.[9] Yet, she adds, "paradoxically, it is the very people who are living on the land who are not eating enough!"[10] She suggests that part of the explanation for this paradox is the fact that land ownership is a virtual monopoly of the wealthy and politically influential. The capital for farming this group's vast land holdings often comes from agribusiness corporations rooted in the United States. These corporations are interested in cash crops for export, not crops that would help alleviate hunger in the producing countries.[11] Thus while these countries produce, they do not feed their own hungry people. This seems to be a specific example of the kind of complaint raised by Denis Goulet* and Samir Amin.* When exported to other countries, technology (agricultural or otherwise) is used by corporations to serve their own interests, not the interests of the host country.

With this in mind we return to Lappé and Collins' claim that the same concentration of control over land and marketing is taking place in the United States. They produce the following astounding statistics:

> Five and one-half percent of agricultural corporations in the United States operates more than half of all land in farms.
> Almost 90% of vegetable production in the United States is controlled—either directly or through contracts—by major processing corporations.
> Less than 0.2% of all U.S. food manufacturers controls about 50% of the entire industry's assets.[12]

It is these same corporations, we are told, that are spreading their operations in the Third World.

Although these corporations do not necessarily pursue the same detrimental cash crop policy at home as abroad, their thorough vertical integration nevertheless does affect the situation at home. They can easily afford growing corn at about $1.80 a bushel, simply because they are the ones who make it into cornflakes and the ones who then sell it at the grocer's for about $37 a bushel. This is the great advantage of vertical integration. Patrick Benedict, a believer in vertical integration, bought a sugar factory with other farmers to help him market his sugar beet produce. But his grain, as far as we have been informed, remains essentially outside the process of vertical integration (despite his owning a grain elevator), as it does for most family farmers.

For the purchase of only two items—the sugar factory and the grain elevator—Mr. Benedict had to summon the financial powers of 1,609 farmers beside himself and borrow an additional $47 million to produce the necessary grand total of $68.5 million. Compare this figure with the staggering financial power of a single food corporation such as General Foods. Its total identifiable assets (as of March 1979) were about $2.5 billion. About one third of its assets are outside the United States, belonging to major subsidiaries in Europe, Canada, Mexico, Bermuda, Brazil, Venezuela, and the Phillipines. These assets include:

Land	$ 46,862,000
Buildings	424,758,000
Machinery and equipment	934,127,000

Advertising costs for General Foods in one year were $372,770,000.[13] Clearly, Mr. Benedict is a David fighting a Goliath.

The responsibility for this state of affairs must not be shifted automatically to technology. Indeed, as the next part of this discussion shows, the problem has different roots. For while it is true that technology breeds bigger and bigger organizations, it is not clear at all that the ethically and socially significant features of such organizations are necessitated by technology.

The major goal of a present-day corporation is to optimize its own profit. This goal determines the internal values of the organization and its subsequent behavior. For example, it is this goal which results in defining efficiency in terms of internal considerations alone without regard to effects on society (see Erich Fromm* on this subject). But this goal is not necessitated by technology. It is determined by considerations extraneous to technology or even to the imperatives of big organization. These considerations derive mainly from the prevalent ideology of society and its attendant values. In today's society, money is power.

Still, it is not the idea of profit that is being questioned here; rather, it is the idea of "profit no matter what." A strong case can be made for the introduction of new social values to contain the compulsive pursuit of profit, which in today's society is usually a manifestation of the unrestrained pursuit of power. Such social values would make obsolete incidents like those discussed by Leah Margulies* and by Eddy, Potter, and Page.*

Margulies argues that the Nestlé Company was aggressively marketing its powdered milk in Third World countries even as evidence of resulting malnutrition, illnesses, and death was mounting. As a solution to health problems created by doubtful business practices, she suggests continuous monitoring of corporate activity, cooperation among concerned health professionals, international agencies, and advocacy groups, and development of popular support for the view that business must be held accountable for unethical practices.

Eddy, Potter, and Page discuss the DC-10 aircraft. They contend that many persons, including officials of the Federal Aviation Administration, were aware of design defects in the aircraft but that these defects were not corrected by the builder, McDonnell Douglas, until a Turkish Airlines DC-10 crashed near Paris in 1974, killing 346 persons. When another DC-10 crash occurred—near Chicago's O'Hare International Airport in June 1979, killing 275 persons—the FAA grounded the aircraft. The grounding order declared, according to *Time*, that the engine-and-pylon assembly "may not be of proper design, material, specification, construction and performance for safe operation."[14] Nevertheless, the grounding order was lifted within days amid reports of severe financial losses—about $5 million a day to airlines using the DC-10. If the loss of revenue indeed played a part in the FAA decision to lift the grounding order, one must question the political-economic organization of our society, which risks lives to save a company from bankruptcy.

An even more stunning incident is that of Three Mile Island. In April 1979, the first widely-known nuclear accident took place—in Pennsylvania, at the Three Mile Island nuclear power plant operated by Metropolitan Edison Company. The accident, which threatened at one point to produce the worst possible situation, a core melt-down, caused the plant to shut down. As antinuclear sentiment spread, all similar plants around the nation were shut down also. The accident was charged to human error, but a plant engineer interviewed by *Time* had more to say on this topic. According to him, Unit 2, the site of the accident, had been plagued with problems during the shakedown phase. These problems, though not serious, did indicate a need for a thorough investigation of the cause of the malfunctions. Yet, instead of conducting such an investigation, Metropolitan Edison soon commenced commercial operation of the unit.[15]

Time points out that the unit was pressed into service on December 30, 1978, and that by meeting the year-end deadline, Metropolitan Edison qualified for $17 million to $28 million in 1978 tax investment credits, plus $20 million in depreciation deductions.[16] Thus as David Barasch, attorney for Pennsylvania's State Consumer Advocate Office, noted, "There was no question that there was strong incentive for the company to get that plant on line fast.''[17]

The methods by which Metropolitan Edison attempted to reduce its losses after the incident caused outraged protests among its consumers. Replacing the electricity from its shutdown plant was costing the company about $1 million a day. To recover part of its losses, it announced that it had no alternative but to raise monthly rates by $7.50. Angry consumers could not see why they should pay for the company's mistakes.

In late August 1979, the President's commission investigating the Three Mile Island accident was preparing its recommendations. It had already uncovered ''evidence of faulty design and lax procedures in the operations of nuclear power plants,'' as well as earlier nuclear-plant mishaps elsewhere.[18] As the commission prepared to question Harold R. Denton, the Nuclear Regulatory Commission's chief of reactor regulation, it discovered that Denton had moved to lift a three-month freeze on the licensing of new reactors. Members of the commission were outraged. Arizona Governor Bruce Babbit protested that the licensing decision implied ''a judgement that the system is basically sound and that we ought to proceed with nuclear development.''[19] Another member of the commission, Anna Trunk, who lives near Three Mile Island, asked Denton, ''Could you give me one good reason why I should trust you or the NRC?''[20]

This question hints at even deeper anxieties relating to technological decisions. Of course, we more commonly ask: ''Is nuclear energy safe?'' But there are people who do not care at all about the answer to this question because they reject the use of nuclear energy, even if it were safe. The reason for their opposition stems from ideological considerations. One may say that it is not nuclear technology *per se* that they are opposed to, but rather nuclear technology in the hands of those whose ideology is seen as founded upon an unrestrained Will-to-Power (for more on the Will-to-Power, see the introduction to Part Two).

The basis for such opposition becomes clear if we examine the views of proponents of nuclear energy. For example, Jean-Claud Leny, executive director of Framatone, a company in charge of operating pressurized water reactors, argues that nuclear plants are not dangerous if they are run by competent, and presumably centralized, staff rather than being entrusted to local groups who may or may not be capable of handling the task. He states that ''in my opinion it is essential that few nuclear plants be constructed . . . and [that they be] controlled in a quasi-military way.''[21]

The possibility of a quasi-military organization of nuclear plants is regarded by opponents of nuclear energy as a very disturbing possibility whose significance extends beyond these plants to society as a whole. André Gorz, editor of *Les Temps Modernes,* warns that ''nuclear society assumes the creation of a caste of militarized technicians, obedient like a medieval knighthood, with its own code and its own internal hierarchy, which would be exempt from the common law and invested with extensive powers of control, surveillance, and regulation.''[22] He further cautions that corporations with such technology at their disposal not only will control our society, but will be capable collectively of extending their hegemony over the whole planet.

Therefore, Gorz's rejection of nuclear energy is founded upon his rejection of hegemony and the centralization of power in the hands of corporations which he regards as power-hungry and thus not trustworthy. For Gorz, nuclear energy questions translate primarily not into issues of safety but rather into political and ideological issues concerning democracy.

Harold Ketterer and John R. Schmidhauser* echo these concerns. They argue that a handful of dominant corporations is on its way to controlling nuclear fuel, a major alternative source of energy in the United States. Ketterer and Schmidhauser regard this situation as grave. To underline its gravity they quote former Senator Aiken, who warns that ''when you control energy . . . then you control the nation . . . a very serious threat to political democracy.''

Thus any attempt to defend nuclear energy as clean, efficient, and safe misses the main objections of the anti–nuclear energy groups that are concerned with democracy. These groups regard the political, social, and human rights of the individual as far more important than his material luxury.

One hotly debated issue in business and government circles today ties in directly with these considerations. Many businesses and government officials have questioned the advisability of the transfer of technology from the United States to other countries. In *Foreign Policy* Jack Baranson complains, in "Technology Exports Can Hurt Us," that U.S. corporations have redefined their "self-interest" in such a way as to permit the sale of their sophisticated technological products to noncontrolled foreign enterprises.[23] The problem with such sales is that they sometimes implant in the foreign countries a competitive productive capability. For example, Amdahl corporation, founded by a former IBM engineer, transferred computer technology to Fujitsu Ltd. in Japan in return for successive rounds of finance capital. Now, eighty percent of Amdahl computer manufacturing requirements are based at Fujitsu.

Another example comes from Algeria. In 1972, General Telephone and Electronics signed a $233 million contract to build an electronics plant. Under the agreement Algerian technicians and managers were being trained in the United States and were expected to be capable of managing their entire Sidi-Bel Abbes facility within a few years of the initiation of their training.

Baranson is displeased with such cases of technology transfer. He is concerned that by exporting technology and increasing the technological knowledge of other countries, "U.S. firms may contribute to both the deterioration of the U.S. trade balance and to the loss of U.S. technical leadership."[24] So he urges that in deciding what sort of technology may be exported by the United States, a distinction be drawn between relatively innocuous transfers of technology easily available from other sources in the world and the transfer of sophisticated technology obtainable only in the United States through specific companies. He also urges that a distinction be drawn between technological exports that implant a certain advanced technology in the importing country, and those that do not.

What emerges from Baranson's article is an overriding concern for the loss of overseas profit and control of international markets. The global political and economic effects of pursuing a policy of denying technological know-how to other countries, as a way of protecting overseas profits and maintaining politico-economic control, is discussed by various authors in this book. Daniel Bell* and Zbiginew Brzezinski* are especially vocal about the dangers of technological monopoly to the political stability of the world as a whole. Daniel Bell observes, in "The Future World Disorder," that "the real time bomb in international economic relations is that of industrialization." He adds that effective international means must be designed to achieve a new, more equitable international division of labor—one that can provide economic and perhaps political stability for the world. International agencies could hasten the birth of such a new order by offering technological aid to developing nations.

In "America in the Technetronic Age," Brzezinski voices concerns similar to Bell's. His assessment of potential global political instability leads him to conclude that "international co-operation will be necessary in almost every facet of life" and to suggest "making the massive diffusion of scientific-technological knowledge a principle focus of American involvement in world affairs." Thus, if Bell and Brzezinski had their way, the ideology of unrestrained control would be tempered by the new realities emerging on the international scene.

• • •

So far, this introduction has placed special emphasis on the economic and political aspects of the impact of technology on the professions. But clearly the farmer's plight, the power of corporations, and the development of nuclear technology all have social and ethical implications as well. So the rest of this discussion will emphasize these other considerations by focusing on the controversy surrounding recombinant DNA research. For this controversy provides an excellent example of the ethical and social dilemmas facing today's scientists and technologists.

Some introductory remarks about recombinant DNA research may be helpful at this point. DNA (deoxyribonucleic acid) is the primary genetic material in cells—whether human, animal, or plant. Each DNA molecule is usually a long two-stranded chain that looks like a "double helix." Its chemical and physical properties make it ideal for replication and transfer of information. Indeed, a DNA molecule usually contains a large number of the hereditary units called genes. Thus one of the chief roles of DNA is the transmission of hereditary characteristics, while a related role is the regulation of enzyme synthesis in the cell.

An important point in the development of genetic research was reached when scientists developed techniques by which they could combine a DNA segment from one organism with a DNA segment from another. The resulting DNA strand contained a genetic code which was a composite of the genetic codes contained in the two parent DNA segments. These recombinant DNA techniques made possible the generation of organisms hitherto unknown on earth. The natural barrier that had thus far restricted the genetic combination of different species was surmounted.

Recombinant DNA techniques, then, are techniques by which genes from one organism are spliced to genes of another. These techniques have various applications and consequences. One of the more valuable applications is the creation of an insulin-producing bacterium, which was developed by splicing genes of a host bacterium to human genes having the code appropriate for the synthesis of insulin. As a result, the modified insulin-producing bacterium is an "insulin factory," and when it replicates itself it produces more insulin factories. This achievement has been highly beneficial in the treatment of diabetics, who have had so far to rely on insulin produced by cows and pigs. The new hybrid bacterium produces human insulin, which is far superior because it has a much lower rejection rate among diabetics.

Many scientists are understandably enthusiastic about recombinant DNA research. Molecular biologist Stanley Cohen,* from Stanford University School of Medicine, provides some reasons for his enthusiasm. He explains that within a short period of time the use of recombinant DNA technology has already substantially improved our fundamental knowledge of living organisms. Among other things, it has provided us with a great deal of information concerning the structure, propagation, and gene regulation of some of these organisms. Such knowledge is useful in understanding genetic defects that are the causes of various serious disorders.

Recombinant DNA research has become part of "the high-technology of medicine," defined by Lewis Thomas* as that technology which is based on a genuine understanding of the mechanisms of disease. As such, this technology is not a stopgap measure in the face of incapacitating illnesses, rather, it attempts to cure, if not eradicate, disease altogether. Recently, it was announced that one type of cancer had been traced to a genetic defect in the kidney. Through recombinant DNA research such a defect may become correctable in the future, thus preempting half-way technologies used for kidney cancer victims, such as surgery. Other diseases resulting from genetic defects or disorders seem to be equally accessible to cure—for example, Tay-Sachs disease, cardiovascular disease, diabetes, and hemophilia.

Furthermore, recombinant DNA research promises to provide us with extensive control over various aspects of our lives, and to make possible new sets of choices. Biologist Robert Sinsheimer, at the California Institute of Technology in Pasadena, discusses these prospects in the following passage:

> Would you like to control the sex of your offspring? It will be as you wish. Would you like your son to be six feet tall—seven feet? Eight feet? What troubles you?—allergy, obesity, arthritic pain? These will be easily handled. For cancer, diabetes, phenylketonuria (a metabolic disease) there will be genetic therapy. The appropriate DNA will be provided in the appropriate dose. Viral and microbial disease will be easily met. Even the timeless patterns of growth and maturity and aging will be subject to our design.[25]

But Amatai Etzioni* points out that the ability to control the sex of one's offspring is not without its social repercussions. For example, girls in many cultures are a liability; thus if parents were given a choice in the matter, the result could be a sexual imbalance in society, which could lead ultimately to social dislocations. Similarly, the ability to extend the human life span is not without its social repercussions. Such an ability could ultimately force us to drastically alter our concepts of education, marriage, and reproduction, among others.

Thus, although recombinant DNA research is an extremely exciting field of research, it has the potential of creating enormous ethical and social problems. But this is hardly the most disconcerting aspect of such research. Recently, some scientists have been experimenting with the recombination of genes carrying antibiotic resistance, and with the recombination of genes with oncogenic (that is, cancer-causing) viruses. If any of the hybrid genes were to escape the confines of the laboratory, the effect on the community, and perhaps the nation and the world, could be disastrous.

These fears have spurred an extensive debate in the scientific community concerning the risks involved in recombinant DNA research. In April 1974, a special committee of scientists, chosen by the National Academy of Sciences to evaluate such risks, asked all scientists to honor a temporary moratorium on certain types of potentially dangerous research until the committee had time to assess the situation. The scientists agreed. After several meetings and a major international conference in Asilomar, California, the moratorium was finally ended and some voluntary guidelines for recombinant DNA research were suggested. In July 1976 the National Institutes of Health provided a mandatory set of guidelines for scientists receiving federal funding for such research.

Still, these measures have not succeeded in ending the heated debate among scientists on the risks of recombinant DNA research. Many continue to doubt the wisdom of tinkering with new forms of life in laboratories. Retired Columbia biochemist Erwin Chargaff* asks, "Have we the right to counteract, irreversibly, the evolutionary wisdom of millions of years in order to satisfy the ambition and curiosity of a few scientists?" But Stanley Cohen retorts by reminding Chargaff that this same evolutionary wisdom provided the gene combinations for bubonic plague, small pox, typhoid, polio, diabetes, cancer, and many other diseases that have caused the suffering and death of millions. Clearly, for Cohen, if recombinant DNA is to be opposed, such opposition must rest on something other than the appeal to evolutionary wisdom.

A closer look at Chargaff's position reveals that Chargaff is not opposed in principle to recombinant DNA research. Rather, he is appalled by the apparent haste exhibited by scientists who experiment with the genetic material before fully understanding various features of DNA. He points out that the significance of spacer regions and repetitive sequences in DNA structures is still unknown, and he expresses concern that many of the experiments are being performed without a full appreciation of what is going on. Chargaff worries that such haste in experimentation may result in the introduction of a form of life which is damaging to humans while at the same time resistant to attempts aimed at controlling or destroying it. He warns, "You can stop splitting the atom; you can stop visiting the moon . . . But you cannot recall a new form of life."

The similarity between this debate and the debate between the ecologists and their adversaries (see the introduction to Part Two, as well as the article by William Tucker*) is striking.

So it is not surprising that Chargaff has received support for his position from the ecologists. Francine Robinson Simbring, of the Committee for Genetics, organized by Friends of the Earth (a group of ecologists), argues that there are clear parallels between the recombinant DNA controversy and the nuclear energy controversy.[26] In both cases, Simbring claims, the proponents of the research have defined the problems associated with the research narrowly, and then addressed themselves to these problems, ignoring the other, unquantifiable, ones. In the case of nuclear energy, attention has centered on design criteria, reactor safety, and regulation when it should also have centered on the genetic risk to future generations, human fallibility, and the vulnerability of centralized electric generation, to mention only a few other major problems. Similarly, in the case of recombinant DNA research, attention is focusing now on laboratory containment as the pivotal problem. However, there are other pressing problems to be addressed, and such a narrow definition of the problems at hand must be rejected. Simbring concludes:

> It is therefore essential that open discussion include the entire range of problems in the field of genetic engineering and take into account the biohazards of accidental release of uncontrollable new organisms, the implications of interference with evolution, reduction of diversity in the gene pool, the imposition of complex medical decisions on individuals and society, and the inherent fallibility (not to mention corruptibility) of inspection, enforcement, and regulatory bodies.[27]

What complicates matters is that a new DNA industry has mushroomed. Various corporations, including Standard Oil of Indiana and National Distillers Corporation, have invested about $150 million in recombinant DNA research. Cetus Corporation is engineering a bug to make alcohol from manioc, a starchy vegetable. Such research is expected to yield astronomical profits. Ronald Cape, chairman of Cetus, says "we are talking about billion-dollar possibilities."[28] Given the high stakes, corporations are reportedly exercising an increasing amount of control over the research they fund at various academic institutions.

This new situation has disturbed many scientists and fueled the controversy further. Jonathan King, a biologist at the Massachusetts Institute of Technology, warns that corporations are using their research scientists as front men who claim interest in the expansion of knowledge. The true motives of the corporation, King argues, are pecuniary: "A fortune is going to be made from the cloning of insulin in bacteria . . . They are not going to sell that insulin cheap . . . because it's human insulin. [But] they are going to *produce* it cheap. That is a very, very powerful force behind the scenes . . ."[29]

• • •

Clearly, technology has deeply permeated the professions, giving rise both to outstanding achievements and to complicated problems. In the process, it has left its mark on every aspect of our individual and social existence.

NOTES

1. James Henry Breasted, *A History of Egypt*, 2nd ed. rev. (New York: Charles Scribner's Sons, 1950), p. 118.
2. Lewis Mumford, "Technics and the Nature of Man," in *Philosophy and Technology*, ed. Carl Mitcham and Robert Mackey (New York: The Free Press, 1972), p. 82.
3. John Kenneth Galbraith, *The New Industrial State*, 2nd ed. rev. (New York: The New American Library, Inc., 1972), pp. 74-75.
4. Galbraith, p. 76.
5. Douglas Guthrie, *A History of Medicine* (Philadelphia: J. B. Lippincott Co., 1946), p. 18.
6. "The New American Farmer," *Time*, 6 Nov. 1978, p. 96.
7. "Voices From and About the Farm Protest Movement," *Progressive Farmer*, April 1978, p. 30.
8. Frances Moore Lappé and Joseph Collins with Cary Fowler, *Food First: Beyond the Myth of Scarcity* (Boston: Houghton Mifflin Co., 1977). For example, see pp. 233-241.
9. Susan George, *How the Other Half Dies*. (Montclair, N.J.: Allanheld, Osmun & Co., 1977), pp. 13-14.
10. George, p. 14.

11. George, pp. 14-15.
12. Lappé and Collins, pp. 247, 273.
13. General Foods' 10-K report for the fiscal year ending 4-3-1979, pp. 23, 30.
14. "Debacle of the DC-10," *Time,* 18 June 1979, p. 15.
15. "Now Comes the Fallout," *Time,* 16 April 1979, p. 25.
16. *Time,* p. 25.
17. *Time,* p. 25.
18. "A Scrap Over Nuclear Safety," *Newsweek,* 3 September 1979, p. 32.
19. *Newsweek,* p. 32.
20. *Newsweek,* p. 32.
21. Quoted in André Gorz, "Nuclear Energy and the Logic of Tools," *Radical America,* 13, No. 3 (1979), 15.
22. Gorz, p. 15.
23. Jack Baranson, "Technological Exports Can Hurt Us," *Foreign Policy,* No. 25 (1976-77), p. 184.
24. Baranson, p. 180.
25. Joel Kurtzman and Phillip Gordon, *No More Dying: The Conquest of Aging and the Extension of Human Life* (Los Angeles: J. P. Tarcher, Inc., 1976), p. 153.
26. Francine Robinson Simbring, letter in *Science,* 192 (1976), 940.
27. Simbring, p. 940.
28. Quoted by Sharon Begley with Pamela Abramson, "The DNA Industry," *Newsweek,* 20 August 1979, p. 53.
29. Jonathan King, "New Disease in New Niches," *Nature,* 276 (1978), pp. 6-7.

15

Technology and agriculture

The energy we eat

CAROL and JOHN STEINHART

■ Carol Steinhart *is a biologist, as well as a scientific writer and editor. She has done research on ecology and environmental problems.* John Steinhart *is Professor of Geophysics and Environmental Studies at the University of Wisconsin at Madison. He is also Associate Director of the Marine Studies Center, and has done research in the area of environmental policy.*

Food remains the most basic human need. In this selection from the Steinharts' book Energy: Sources, Use and Role in Human Affairs, *a thorough, energy-oriented analysis of the food system in the United States is developed. Some startling facts—unnoticeable in a conventional economic analysis—are revealed. "In primitive cultures, 5 to 50 calories of food were obtained for each calorie invested. Some highly civilized cultures have done as well and occasionally better. In sharp contrast, industrialized food production requires an input of 5 to 10 calories of fuel to obtain 1 calorie of food." This situation presents a problem immediately. Given the imminent energy shortages, how can the United States reduce its energy expenditures in agriculture and the food system as a whole? The Steinharts suggest several ways of achieving that end.*

The kind of energy that has always been of first importance is food. Throughout most of human history, man has relied on his own labor to provide food. If the energy value of the food obtained had not substantially exceeded the energy expended in obtaining it, our species would not have survived.

Through the development of agriculture, man was able to manipulate the flow of energy in various ecosystems in order to divert an increasingly large fraction of the earth's productivity to his own use. He did this by simplifying the complex natural ecosystem—by decreasing the number of species in the system and by controlling the species that competed with him for the yield. Maintenance of the simplified system required an endless input of energy. At first this was restricted to manpower, as techniques for preparing the soil, planting, weeding, driving off pests, and harvesting were developed. Later, animal power was exploited. Still later, inanimate energy from wind and water was put to work, and finally, energy from the fossil fuels was utilized in food production.

In many parts of the world, agriculture still depends on energy from people, animals, and the sun. If conditions are favorable, such solar agriculture can give very high yields—for example, in wet rice culture, up to 50 calories are returned in harvest for every calorie of human energy investment. Modern agriculture, however, depends on converting fossil fuels into meat and potatoes, and there is an increasingly unfavorable ratio of energy input to food output. It is due to large-scale energy subsidy that agricultural yields have increased manyfold in the United States during this century. In 1900, a single farmer could feed about five people. By 1940 he could feed 10 and in 1960 he could feed 25. Today, the labor of one farmer feeds nearly 50 people, owing to the development of new fungicides, herbicides, rodenticides, insecticides, miticides, nematocides, antibiotics, vaccines, fertilizers, equipment, and specialized varieties of plants and animals. But it is not just one farmer who feeds 50 people. It is one farmer, many tons of coal, many barrels of oil, and an immense food processing and distribution system. In this chapter we will see how the present situation has come about.

ENERGY IN ECOSYSTEMS

Life can be viewed as a ceaseless web of energy conversions which follow well-known principles of energy conservation and transformation. Many ecologists describe an ecosystem in terms of the energy that flows into and out of it and from one trophic level to another (that is, from green plants to herbivores to carnivores or from any of these levels to decomposers). The tropical rain forest is an example of a complex ecosystem characterized by a large mass of living organisms and efficient use of available energy. In the rain forest there are many hundreds of species of plants, each specialized in its structure and function to fill a unique niche, and each accompanied by its cadre of insect predators specialized to deal with it. Armies of microorganisms live in association with roots, soil, and dead matter. Fungus-eating flies, beetles, termites, and other small animals feed on the microorganisms. They in turn are eaten by a variety of amphibians and reptiles. Birds and mammals feed on everything, including each other. Man, the most relentless and least specialized predator of all, is the one creature likely to upset the balance.

A great deal of energy is expended in maintaining the integrity of such a system. In the rain forest, energy flows through an intricate network of feedback loops that regulate all the interrelationships among species. Epidemics are virtually impossible because of the low density of any particular population. If one species increases, others usually increase to counteract it until order is restored. Similarly, should a species decline, its predators also decline or shift their attention to an alternative food source until decreased predation permits recovery of the original population. Even if one species should disappear, the system survives because there are many alternative pathways along which energy and materials can flow. Diversity of species insures that nutrients will be cycled effectively and soil structure preserved.

This complex, stable natural system supports the highest productivity and the greatest mass of living organisms possible under prevailing climatic conditions. Man can glean very little from this system, however, for little remains after respiration, predators, parasites, and decomposers claim their share. The situation in a rain forest is in sharp contrast to that in a field of grass. Energy is stored in a field of grass: it is used primarily for growth, rather than for maintaining the structure of the system. But a field of grass generally lacks stability because it contains relatively few species and few feedback loops which regulate interactions among species. Many ecological niches are empty, inviting other species to invade. Under many conditions, the short-lived, rapidly growing, prolific species in the open field tend eventually to be displaced by a succession of slower growing, longer lived, less prolific species until a system develops which is relatively stable under the prevailing conditions of soil and climate. Left undisturbed, a meadow usually becomes a forest.

The achievement of agriculture is to simplify the system so that, as in a field of grass, the maximum amount of solar energy is channeled into growth. Man supplies the energy for maintenance. He plows and plants. He fights off competitors, predators, and disease. He supplies fertilizer to replace losses from harvesting. He shelters his animals, feeds and vaccinates them, and helps them to breed and to bear and raise their young. Thus, only through a steady input of energy is man able to divert most of the productivity of his fields to his own ends.

It is a difficult struggle to maintain the simplified system however much energy is put into it. Where fields lie barren for part of the year, erosion and deterioration of soil structure claim their toll. Bigger and better pests continually appear. Where irrigation is practiced, water shortages and increasing salinity haunt the farmer. And if a wheat field is abandoned, does a crop of wheat come up the next year? Of course not. This illustrates a major difference between wild plants and the plants that are adapted to modern agricultural methods. Wild plants are hardy, physiologically adaptable, and able to withstand adversity. Man's crops are highly specialized for growth, and consequently they are dependent on man for protection and even their own propagation.

THE PENALTY FOR IMPROVEMENT

The success of modern agriculture depends in part on the development of fast growing, high yielding strains of plants and animals, whose productivity in turn depends on energy-intensive agriculture. For more than ten thousand years man has engaged in selective breeding of domestic plants and animals. One of the most remarkable stories is that of maize, or corn. The first evidence of wild maize is from Mexico, dating from more than 7000 years ago. The plant probably looked similar to any other grass that grows in meadows, and the ear was no larger than your thumbnail. Lurking in the small seed, however, was the potential to become the corn of today, with huge ears of closely packed grain. The yield of modern hybrid corn is more than 90 bushels per acre in the midwestern United States. But corn can no longer survive without man. Even if it weathered a dry season on poor soil, escaping disease and predation, it would perish within a generation or two, for it can no longer shed its seed to propagate itself.

Other domestic plants and animals have also become increasingly dependent on man for their survival. It seems that an organism can be either a generalist or a specialist, but not both. An organism can process just so much energy in its lifetime. If this energy is channeled largely into growth, man must tend to the other needs of the organism. Our improved plants and animals are not truly improved, but are merely specialized. They are specialized in growing wool, fat roots, or giant fruits, in laying eggs or making milk. In diverting their energy into rapid growth, they sacrifice many qualities that allowed their wild forebears to survive. Selective breeding may solve specific problems, but we are fooling ourselves when we expect a plant to resist disease, discourage predators and competitors, withstand unfavorable climatic and nutritional conditions, grow rapidly, reproduce abundantly, and still yield a bountiful harvest that is tasty, nutritious, and beautiful. One manifestation of the penalty for improvement is that, in the plant world, favorite varieties of grapes, roses, and citrus, to name a few, are routinely grafted onto wild rootstocks to increase their vigor.

Another feature of modern crop plants and domestic animals is genetic uniformity. Just as species diversity enhances a natural system's chances for survival, genetic diversity enhances a species' chances. If misfortune befalls a genetic subset of a group, the remaining members can avert disaster for the population as a whole. In addition to being specialized for growth of one sort or another, domestic plants and animals also tend toward genetic uniformity, which implies uniform resistance or susceptibility to adversity. Through modern practices in animal breeding, a dairy bull can spread his genes around the world before some camouflaged weakness makes itself known in his offspring. Vegetative propagation of plants and other common practices insure dependably uniform crops down to the end. For a popular variety of potato known as the lumper, the end was the Irish potato famine of 1846.

It is contradictory to desire uniformity and diversity at the same time. Modern agriculture values uniformity above all else. The risks are well known, but the benefits seem worth the gamble. As a result, there have been some disastrous epidemics in our country, including southern leaf blight of corn, fungus disease of a popular strain of oats called Victory, and the wildfire spread of a virulent new race of wheat stem rust which attacked a variety of wheat bred for resistance to the old race of wheat stem rust. The problem of plant breeders is that while they make hundreds of experimental crosses, nature tries millions. One mutant fungus spore or one fecund insect can undo years of the work of man.

The high-yielding varieties of grain that are being spread throughout the world as the green revolution represent the most extensive experiment in uniformity of crops yet tried. When things go well with the green revolution they go very well indeed; but the miracle grains are susceptible to widespread disaster and they produce no miracles unless grown under the most favorable conditions. But complex as are the problems of the green revolution, ther alternative seems to be starvation for about a billion people.

LAND RESOURCES AND AGRICULTURE

There are three ways in which we can try to feed the growing number of human beings on earth: open up new agricultural and grazing lands, increase the productivity of the lands already in use, and develop new sources of food. The long-term success of any of these depends on the size of the population we are trying to support and on our ability to design self-sustaining systems for which there is a non-depletable source of energy and in which materials are recycled.

We are familiar enough with the surface of our planet to know that no hidden paradise remains to be discovered. The area of the earth's ice-free land surface is about 32 billion acres. Roughly one quarter of this is potentially arable, another quarter is potentially grazable, and half is useless for agriculture, although some of it is forested. Much of the unused potentially arable land is in the tropics, where its potential may be more theoretical than practical. It certainly is not farmable with today's knowledge and techniques and economic restrictions. Practically all the land that can be cultivated under existing social and economic conditions is already under cultivation.

In North America, large amounts of arable land are not being farmed. However, we should not be optimistic about possibilities for expansion, for the best land is already cultivated and what remains has serious deficiencies. Furthermore, buildings and highways are spreading so malignantly over so much of our prime farmland that projections for California, for example, indicate that in less than 50 years half of the state's agricultural land will have been converted into nonagricultural uses. It is doubtful that, in opening new lands, we can run fast enough to stay where we are.

There are sound reasons for not rushing to open up vast new areas for crops or grazing, even if technical and economic obstacles could be overcome. One reason is that agriculture, especially as practiced in industrialized nations, depends on the activities of unmanaged ecosystems for the cycling of wastes and other materials. The capacities of some of these systems are already overtaxed, and further reduction in their size would intensify the problems. Another is that biologists emphasize the importance of maintaining reservoirs of wild plants and animals from which new and valuable domesticates may be developed. In general, the best plan seems to be to increase the productivity of lands already under cultivation, which inevitably requires an energy subsidy. The rest of this chapter analyzes the energy intensive food system in the United States and questions the extent to which it is desirable or even possible to transfer this technology to the developing nations.

ENERGY USE IN AN INDUSTRIAL FOOD SYSTEM

In a modern industrial society, only a tiny fraction of the population is in frequent contact with the soil, and an even smaller fraction of the population raises food on the soil. The proportion of the population engaged in farming halved between 1920 and 1950, and then halved again by 1962. Now it has almost halved again and yet a majority of the remaining farmers hold part-time jobs off the farm.[1] Simultaneously, work animals declined from a peak of more than 22 million in 1920 to very small numbers at present.[2]

In economic terms, the value of food as a portion of the total goods and services of society now amounts to a smaller fraction of the gross national product than it once did. Energy inputs to farming have increased enormously since 1920,[3] and the apparent decrease in farm labor is offset in part by the growth of support industries for the farmer. But with these changes on the farm have come a variety of other changes in the U.S. food system, many of which are now deeply embedded in the fabric of daily life. In the past fifty years, canned, frozen and other processed foods have become principal items in our diet. At present the food processing industry is the fourth largest energy consumer in the Standard Industrial Classification groupings. The use of transportation in the food system has grown apace, and the proliferation of appliances in both numbers and complexity still continues in homes, institutions, and stores. Hardly anyone eats much food as it comes from the fields. Even farmers purchase most of their food from markets in town.

Present energy supply problems make this growth of energy use in the food system worth investigating. It is the purpose of this chapter to do so. But there are larger matters at stake. Georgescu-Roegen notes that "the evidence now before us—of a world which can produce automobiles, television sets, etc., at a greater speed than the increase in population, but is simultaneously menaced by mass starvation—is disturbing."[4] In the search for a solution to the world's food problems, the common attempt to transplant a small piece of a highly industrialized food system to the hungry nations of the world is plausible enough, but so far the outcome is unclear. Perhaps an examination of the energy flow in the U.S. food system as it has been developed can provide some insights that are not available from the usual economic measures.

Measures of food systems

Descriptions of agricultural systems are given most often in economic terms. A wealth of statistics is collected in the United States and in most other countries indicating production amounts, shipments, income, labor, expenses and dollar flow in the agricultural sector of the economy. In what follows, we will make use of these statistics, for these values are of considerable use in determining the economic position of farmers in our society. But agricultural statistics are only a tiny fraction of the story.

Energy flow is another measure available to gauge societies and nations. Only after some nations shifted large portions of the population to manufacturing, specialized tasks, and mechanized food production, and shifted the prime sources of energy to fuels that were transportable and usable for a wide variety of activities could energy flow be a measure of a society's activities. Today it is only in one-fifth of the world that these conditions are sufficiently advanced.

What we would like to know is: how does our food supply system compare in terms of energy use to other societies, and to our own past. Perhaps, knowing this, we can estimate the value of energy flow measures as an adjunct to, but different from, economic measures.

Energy in the United States food system

In the morning, breakfast offers orange juice from Florida by way of the Minute Maid factory, bacon from a midwestern meat packer, cereal from Nebraska and General Mills, eggs from California, milk from not too far away, and coffee from Colombia. All these things are available at the local supermarket (4.7 miles each way in a 300 h.p. automobile), stored in a refrigerator-freezer and cooked on an instant-on gas stove (see appendix D).

The present food system in the United States is complex and the attempt to analyze it in terms of energy use will introduce questions far more perplexing than would the same analysis performed on simpler societies. Such an analysis is worthwhile, however, if only to find out where we stand. We have a food system and most of us get enough to eat from it. If, in addition, one considers the food supply problems, present and future, of societies where a smaller fraction of the people get enough to eat, then our experience with an industrialized food system is even more important. There is simply no gainsaying the fact that most nations are trying to acquire industrialized food systems of their own, whether in whole or in part.

What economics tells us is that food in the United States is expensive by world standards. In 1970 the average annual per capita expenditure for food in the United States was about $600.[3] This is larger than the per capita gross domestic product in more than 30 nations of the world. These 30 nations contain most of the world's people and a vast majority of those who are underfed. It would be convenient to know whether we can put our hands into the workings of our own industrialized food system to extract a piece of it that might mitigate their plight, or whether they must become equally industrialized in order to operate such a food supply system. Even if we consider the diet of a poor resident of India, the annual cost of his food at U.S. prices would be about $200—more than twice his annual income.

The analysis of energy use in the food system begins with an omission. We will neglect that crucial input of energy provided by the sun to the plants upon which the entire food supply depends. Photosynthesis is about 1 percent efficient; thus the maximum solar radiation captured by plants is about 5×10^3 kcal/m^2 per year. Ultimately we can compare the solar input with the energy subsidy supplied by modern technology.

Seven categories of energy use on the farm were considered. The amounts of energy used are shown in [Figure 1]. The values for farm machinery and tractors are for the manufacture of new units only and do not include parts and maintenance for existing units. The amounts shown for direct fuel use and electricity consumption are a bit too high because they include some residential uses of the farmer and his family. On the other hand, some uses in these categories are not reported in the summaries employed to obtain the values for direct fuel and electricity usage. These and similar problems are discussed in the appendix. Note the relatively high energy cost associated with irrigation. In the United States, less than 5 percent of the cropland is irrigated. In some countries where the green revolution is being attempted, the new high yield varieties require irrigation while native crops did not. If that were the case in the United States, irrigation would be the largest single use of energy on the farm.

Little food makes its way from field and farm directly to the table. The vast complex of processing, packaging, and transport has been grouped together in a second major subdivision of the food system. [Figure 2] displays the energy use in food processing and packaging. Energy use for transport of food should be charged to the farm in part, but we have not done so because the calculation of the energy

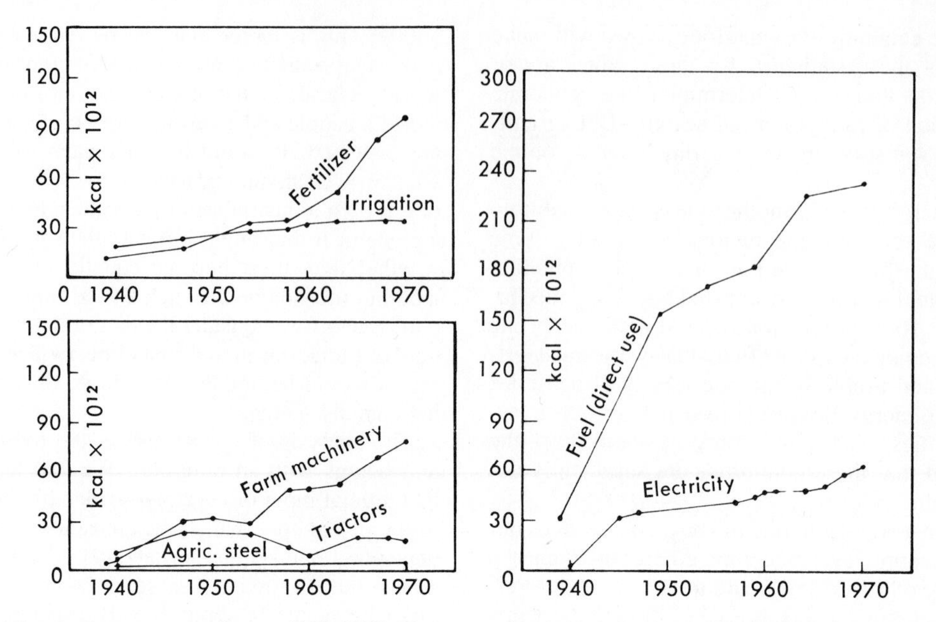

Fig. 1. Energy use on farms, 1940-1970. Transportation is included in the food processing sector.

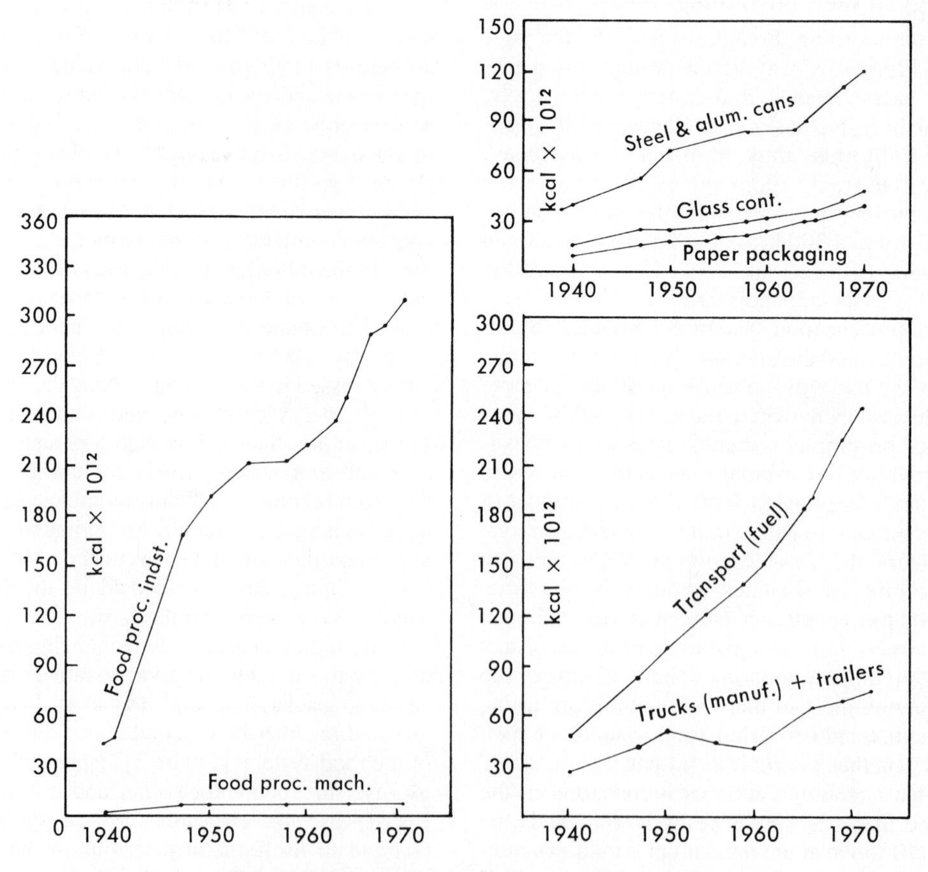

Fig. 2. Energy use in the food processing sector.

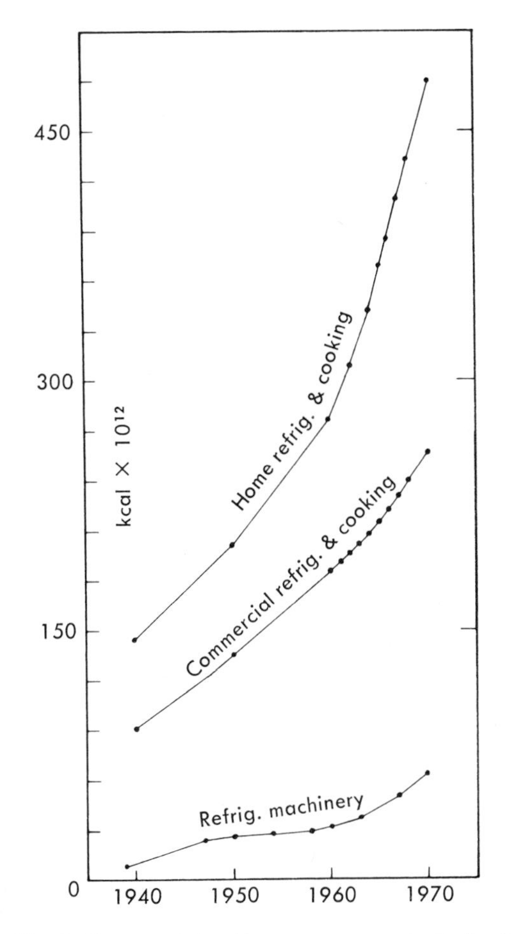

Fig. 3. Commercial and home energy use in the food system. These are selected uses only.

values is easiest (and we believe most accurate) if they are taken for the whole system.

After food is processed there is further energy expenditure. Transportation enters again, and some fraction of the energy used for transportation should be assigned here. But there are also the distributors, wholesalers and retailers, whose freezers, refrigerators and very establishments are an integral part of the food system. There are also the restaurants, schools, universities, prisons, and a host of other institutions engaged in the procurement, preparation, storage, and supply of food. We have chosen to examine only 3 categories: the energy used for home refrigeration and cooking, for commercial refrigeration and cooking, and that used for the manufacture of the refrigeration equipment. [Figure 3] shows energy consumption for these categories. There is no attempt to include the energy used in trips to the store or restaurant.

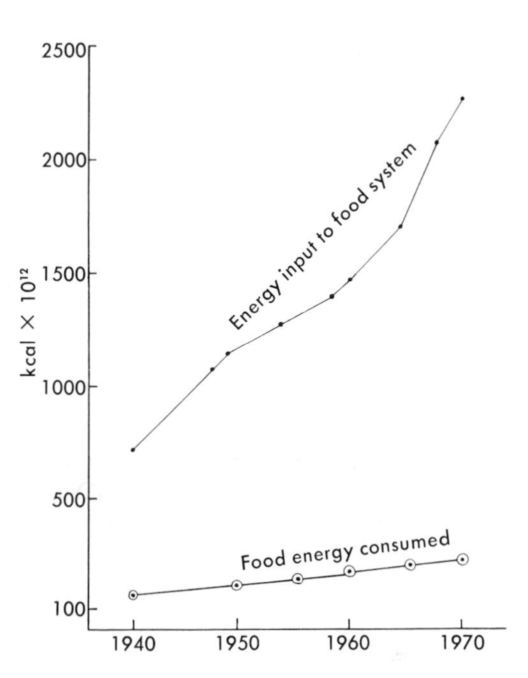

Fig. 4. Energy use in the food system, 1940-1970, compared to caloric energy content of food consumed.

Garbage disposal has also been omitted, although it is a persistent and growing feature of our food system. Twelve percent of the nation's trucks are engaged in waste disposal which is largely, though not entirely, related to food. If there is any lingering doubt that these activities—both the ones included and the ones left out—are an essential feature of our present food system, one need only ask what would happen if everyone should attempt to get on without a refrigerator or freezer or stove? Certainly the food system would change.

. . . As for many activities in the past few decades, the story [of primary energy use by the U.S. food system] is one of continuing increase. The totals [for 1940 to 1970] are displayed in [Figure 4] along with the energy value of the food consumed by the public. The food values were obtained by multiplying the annual caloric intake with the population. The difference in caloric intake over this 30-year period is not significant and the curve mostly indicates the population increase in this period.

PERFORMANCE OF AN INDUSTRIALIZED FOOD SYSTEM

The difficulty with history as a guide for the future or even the present lies not so much in the fact that conditions change—we are at least continually reminded of that fact—but that history is only one experience of the many that might have been. The U.S. food system developed as it did for a variety of reasons, many of them probably not understood. It would do well to examine some of the dimensions of this development before attempting to theorize about how it might have been different, or how parts of this food system can be transplanted elsewhere.

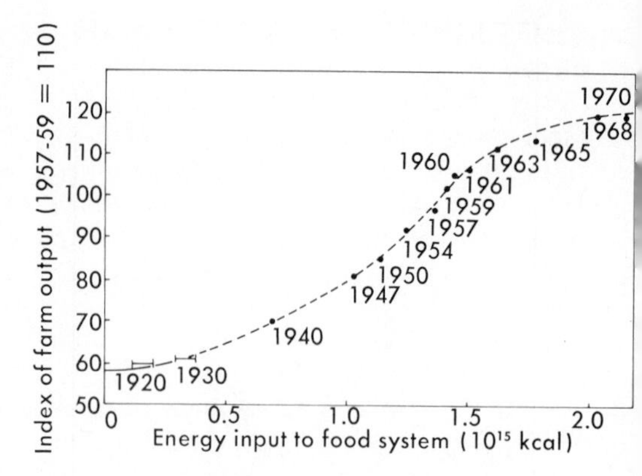

Fig. 5. Farm output as a function of energy input to the U.S. food system, 1920-1970.

Energy and food production

[Figure 5] displays features of our food system not easily seen from economic data. The curve shown has no theoretical basis, but is suggested by the data as a smoothed recounting of the history of increasing food production. It is, however, similar to growth curves of the most general kind, and it suggests that, to the extent that the increasing energy subsidies to farm production have increased that production, we are near the end of an era. . . . there is an exponential phase which began in 1920 or earlier and lasted until 1950 or 1955. Since then the increments in production obtained by the growth in energy use have become smaller. It is likely that further increases in food production from increasing energy inputs will be harder and harder to come by. Of course, a major modification in the food system could change things. However, the argument advanced by the technological optimists—that we can always produce more if we have enough energy and that no other major changes are needed—is not supported by our own history.

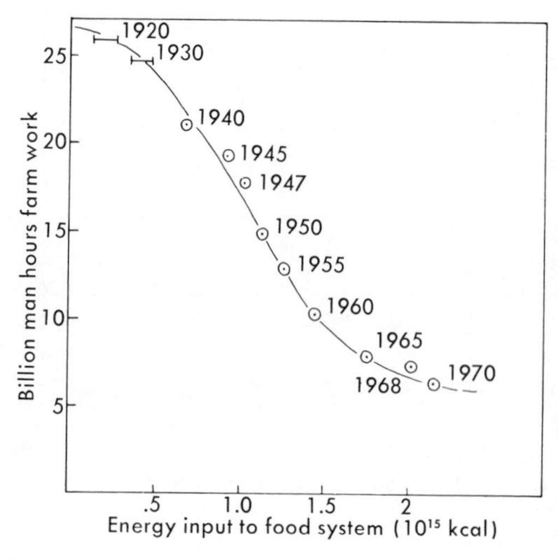

Fig. 6. Labor use on farms as a function of energy use in the food system.

Energy and labor in the food system

One farmer now feeds 50 people, and the common expectation is that labor inputs to farming will continue to decrease in the future. Behind this expectation is the assumption that continued application of technology—and energy—to farming will substitute for labor. [Figure 6] is the substitution curve of energy for labor on the farm. It shows the historic decline in farm labor as a function of the energy subsidy to the food system. Again the familiar "S" shaped curve may be seen. Reduction of farm labor by increasing energy inputs cannot go much further.

The food system that has grown during this period has provided a great deal of employment that did not exist 20, 30 or 40 years ago. Perhaps even the idea of a reduction of labor input is a myth when the food system is viewed as a whole, instead of examining the farm worker only. Pimentel and associates cite an estimate of two farm support workers for each person employed on the farm.[3] To this must be added employment in food processing industries, in food wholesale and retail establishments and in the manufacturing enterprises that support the food system. Yesterday's farmer is today's canner, tractor mechanic and fast food carhop. The process of change has been painful to many ordinary people. The rural poor, who could not quite compete in the industrialization of farming, migrated to the cities. Eventually they found other employment, but one must ask if the change was worthwhile. The answer to that question cannot be provided by energy analysis any more than by economic data, because it raises fundamental questions about how individuals would prefer to spend their lives. But if there is a stark choice of long hours as a farmer, or shorter hours on the assembly line of a meat packing plant, it seems clear that the choice would not be universally in favor of the meat packing plant. Thomas Jefferson dreamed of a nation of independent small farmers. It was a good dream, but society did not develop in that way. Nor can we turn back the clock to recover his dream. But in planning our future, we had better look honestly at our collective history, and then each of us look closely at his own dreams.

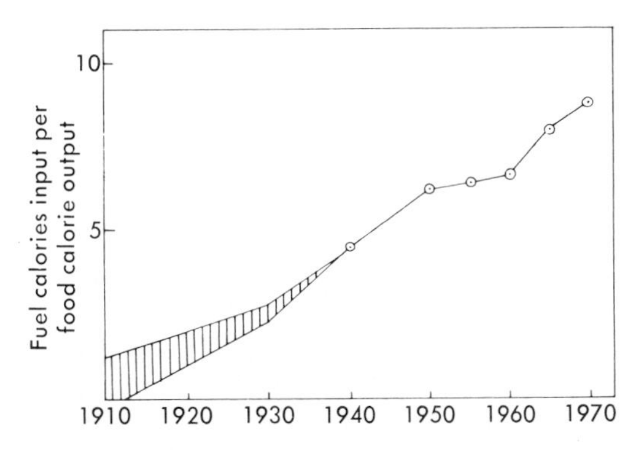

Fig. 7. Energy subsidy to the food system to obtain one food calorie. The values for period 1910-1937 cannot be fully documented, and thus we present a range of values for that period.

The energy subsidy to the food system

The data on [Figure 5] can be combined to show the energy subsidy provided to the food system for the recent past. We take as a measure of the food supplied, the caloric content of the food actually consumed. This is not the only measure of the food supplied, as many protein-poor peoples of the world clearly show. Nevertheless the ratio of caloric input to output is a convenient way to compare our present situation with the past. [Figure 7] shows the history of the U.S. food system in terms of the number of calories of energy supplied to produce one calorie of food for actual consumption. It is interesting and possibly frightening to note that there is no indication that this curve is leveling off. Fragmentary data for 1972 suggest that the increase continued unabated. We appear to be increasing the energy input even more. Note that a graph like [Figure 7] could go to zero. A natural ecosystem has no fuel input at all, and those primitive people who live by hunting and gathering have only the energy of their own work to count as input.

Some economic features of the U.S. food system

The markets for farm commodities in the United States come closer than most to the economist's ideal of a free market. In a free market there are many small sellers, many buyers, and thus no individual is able to affect the price by his own actions in the marketplace. But a market would satisfy these conditions only in the absence of intervention in its function. Government intervention in the prices of agricultural products (and hence of food) has been a prominent feature of the U.S. food system for at least thirty years. Between 1940 and 1970 total farm income has ranged from $4.5 to $16.5 billion, and that part of the National Income having its origin in agriculture (which includes indirect income from agriculture) has ranged from $14.5 to $22.5 billion. Meanwhile government subsidy programs, primarily farm price supports and soil bank payments, have grown from $1.5 billion in 1940 to $6.2 billion in 1970. In 1972 these subsidy programs had grown to $7.3 billion despite foreign demand for U.S. agricultural products. Viewed in a slightly different way, direct government subsidies have accounted for 30 to 40 percent of farm income and they have accounted for 15 to 30 percent of the National Income attributable to agriculture for the years since 1955. The point is important because it emphasizes once again the striking gap between the economic description of society and the economic models used to account for that society's behavior.

The issue of farm price supports is related to energy in this way: first government intervention in the food system is a feature of almost all highly industrialized countries (and, despite the intervention, farm incomes still tend to lag behind national averages); and secondly, because reduction of the energy subsidy to agriculture (even if we could manage it) might reduce farmer's incomes. One reason for this state of affairs is that quantitative demand for food has definite limits and, without farm price supports, the only way to increase farm income is to increase the unit cost of agricultural products. Consumer boycotts and protests in the early 1970s suggest that there is considerable resistance to this course of action.

Government intervention in the functioning of the market for agricultural products has increased with the use of energy in agriculture and the food supply system and we have nothing but theoretical suppositions to suggest that either event could happen alone.

SOME ENERGY IMPLICATIONS FOR THE WORLD FOOD SUPPLY

The food supply system of the United States is complex and interwoven into a highly industrialized economy. We have tried to analyze this system owing to its implications for future energy use. But the world is also short of food. A few years ago it was widely predicted that the world would experience widespread famine in the 1970s. The adoption of new high-yield varieties of rice, wheat, and other grains has caused some experts to predict that the threat of these expected famines can now be averted—perhaps indefinitely. Yet, despite increases in grain production in some areas, the world still seems to be headed towards famine. The adoption of these new varieties of grain—dubbed hopefully the green revolution—is an attempt to export a part of the energy-intensive food system of the highly industrialized countries to non-industrialized countries. It is an experiment, because the whole food system is not being transplanted to new areas, but only a small part of it. The green revolution requires a great deal of energy. Many of the new grain varieties require irrigation in places where traditional crops did not, and almost all the new crops require extensive fertilization. Both irrigation and fertilization require high inputs of energy.

The agricultural surpluses of the 1950s have largely disappeared. Grain shortages in China and the U.S.S.R. have attracted attention because they have brought foreign trade across ideological barriers. There are other countries that would probably import considerable grain if they could afford it. But only four countries may be expected to have any substantial excess agricultural production: Canada, New Zealand, Australia, and the United States. None of these is in a position to give grain away, because they need the foreign trade to avert ruinous balance of payment deficits. Can we then export energy-intensive agricultural methods instead?

Energy-intensive agriculture abroad

It is quite clear that the United States food system cannot be exported intact at present. For example, India has a population of 550×10^6 persons. To feed the people of India at the United States level of about 3,000 kilocalories per day (instead of their present 2,000) would require more energy than India now uses for all purposes. If we wished to feed the entire world with a food system of the U.S. type almost 80 percent of the world's annual energy expenditure would be required.

The recourse most often suggested is to export only methods of increasing crop yield, and to hope for the best. We must repeat that this is an experiment. We know that our food system works (albeit with some difficulties and warnings for the future) but we do not know what will happen if we take a piece of that system and transplant it to a poor country that is lacking the industrial base of supply, transport system, processing industry, appliances for home storage and preparation, and most of all, a level of industrialization permitting higher food costs.

The energy requirements of green revolution agriculture have some important political and social implications. To the extent that the Western, highly industrialized countries must continue research and development for the new strains continually required to respond to new plant diseases and pests that can and do sweep through areas planted with a single variety (consider the recent problem with corn blight in the midwest), the Western countries will possess a hold over the developing countries. Political radicals sometimes dub this state of affairs "technological imperialism," but, whatever the name, the developing countries resent their dependence upon the vagaries of another nation's priorities. In order to avoid this source of friction the improved agriculture must be managed within the developing countries. In many of the developing countries such internal programs have begun. But establishment of anything like the agricultural extension network of the United States will require a significant expenditure of energy. Failure to establish networks of this type has, in some green revolution areas, favored the better-educated farmer against the peasants, who have little access to or knowledge of the new grain varieties. The necessity to fertilize and irrigate also favors the larger, more affluent farms—often with the result of driving more peasants off the land and into the cities, where developing nations face a difficult problem already.

Fertilizers, herbicides, pesticides and in many cases, machinery and irrigation are needed to have any hope of success with the green revolution. Where is the energy for this to come from? Many of the nations with the most serious food problem are also those with scant supplies of fossil fuels. In the industrialized nations, solutions to the energy supply problems are being sought in nuclear energy. This technology-intensive solution, even if successful in advanced countries, poses additional problems for underdeveloped nations. To create the base of industry and technologically sophisticated people within their own country will be beyond many of them. Once again they face the prospect of depending upon the good will and policies of industrialized nations. Since the alternative could be famine, their choices are not pleasant, and their irritation at their benefactors—ourselves among them—could grow to threatening proportions. It would be comfortable to rely on our own good intentions, but our good intentions have often been unresponsive to the needs of others. The matter cannot be glossed over lightly. World peace may depend upon the outcome.

Choices for the future

Application of energy on our farms is now near 10^3 kcal/m^2 per year for corn,[3] and this is more or less typical of intensive agriculture in the United States. With this application of energy we have achieved yields of 2×10^3 kcal/m^2 per year of usable grain—bringing us to almost half of the photosynthetic limit of production. Further applications of energy are likely to yield little or no increase in the level of productivity. In any case research is not likely to improve the efficiency of the photosynthetic process itself. There is a further limitation of improvement of yield. Faith in technology and research has at times blinded us to the basic limitations of the plant and animal material with which we work. We have been able to emphasize desirable features already present in the gene pool, and to suppress others that we find undesirable. At times the cost of increased yield is the loss of desirable characteristics—hardiness, resistance to disease and adverse weather and the like. The further we get from characteristics of the original plant and animal strains, the more care—and energy—is required. Choices must be made in the directions of plant breeding. And the limitations of the plants and animals we use must be kept in mind. We have not been able to alter the photosynthetic process, or to change the gestation period of animals. In order to amplify or change an existing characteristic we will probably have to sacrifice something in the overall performance of the plant or animal. If the change requires more energy, we could end with a solution that is too expensive for people who need it most. These concerns are intensified by the degree to which energy becomes more expensive in the world market.

WHERE NEXT FOR FOOD?

[Figure 8] shows the energy subsidy ratio to energy output for a number of widely used foods in a variety of times and cultures. For comparison the overall behavior of the United States food system is shown, but the comparison is only approximate because, for most of the specific crops, the energy input ends at the farm. As has been pointed out, it is a long way from the farm to the table in industrialized societies. Several things are immediately apparent, and coincide with expectations. High protein foods, such as milk, eggs, and meat, have a far poorer energy return than do plant foods. Because protein is essential for human diets, and because the amino acid balance necessary for good nutrition is not found in most cereal grains, we cannot abandon meat sources altogether. [Figure 8] indicates how unlikely it is that increased fishing or production of fish protein concentrate will solve the world's food problems. Even if we leave aside the question of whether the fish are available—a point on which expert opinions differ—it would be hard to imagine, with rising energy prices, that fish protein concentrate will be anything more than a by-product of the fishing industry, for it requires more than twice the energy of production of grass-fed beef or eggs. Distant fishing is still less likely to solve food problems. On the other hand, coastal fishing is relatively low in energy cost. Unfortunately, however, coastal fisheries are threatened with overfishing as well as pollution.

The position of soybeans may be crucial in [Figure 8]. Soybeans possess the best amino acid balance and protein content of any widely grown crop. This has long been known to the Japanese, who have made soybeans a staple of

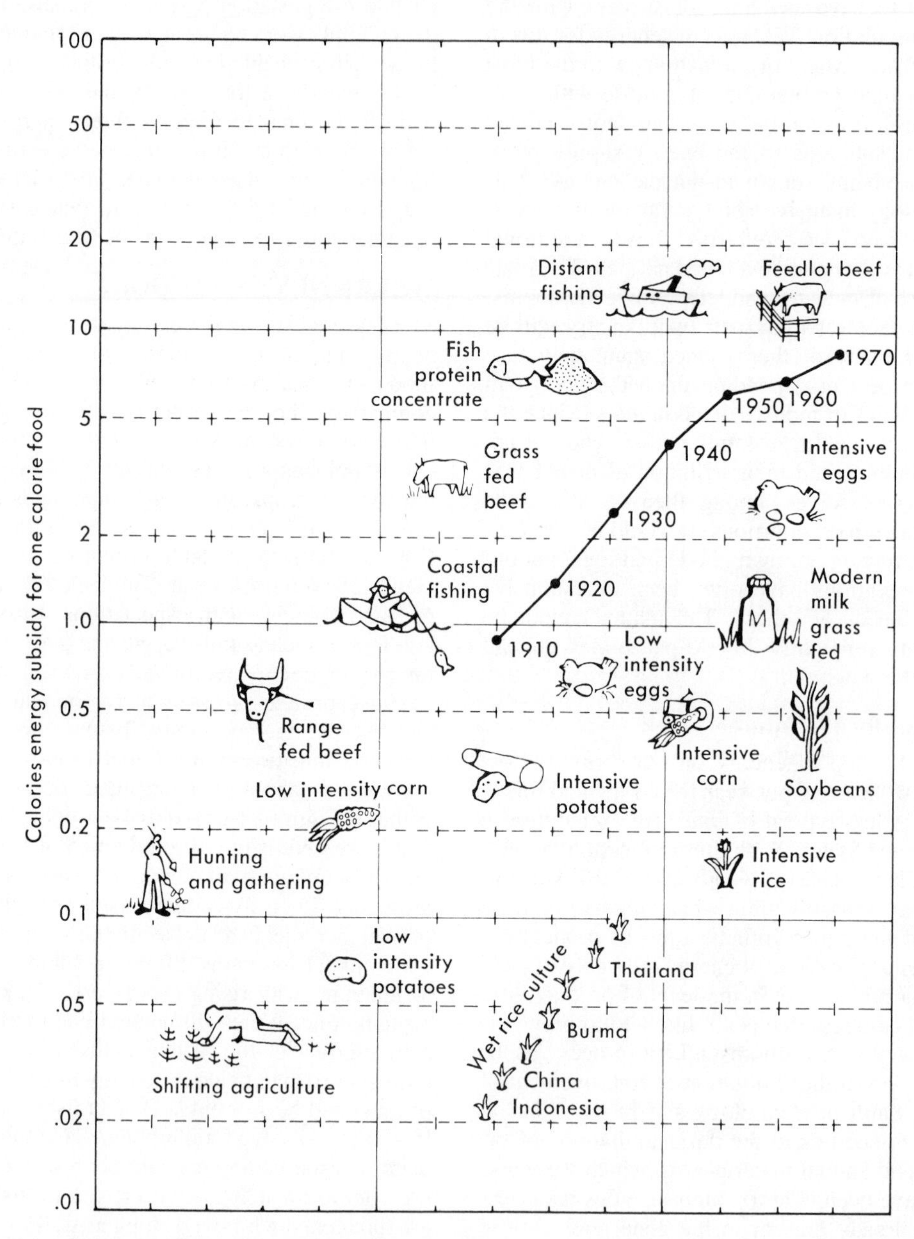

Fig. 8. Energy subsidies for various food crops. The energy history of the U.S. food system is shown for comparison.

their diet, and to beef feedlot operators. Are there other plants, possibly better suited for local climates, which have adequate proportions of amino acids in their proteins? There are about 80,000 edible species of plants, of which only about 50 are actively cultivated on a large scale (and 90 percent of the world's crops come from only 12 species). We may yet be able to find species that can help the world's food supply.

The message of [Figure 8] is simple. In primitive cultures, 5 to 50 calories of food were obtained for each calorie invested. Some highly civilized cultures have done as well and occasionally better. In sharp contrast, industrialized food production requires an input of 5 to 10 calories of fuel to obtain 1 calorie of food. We must pay attention to this difference—especially if energy costs increase. If some of the energy subsidy for food production could be supplied on-site, with renewable sources—primarily sun and wind—we might be able to provide an energy subsidy. Otherwise the choices appear to be less energy-intensive food production or famine for many areas of the world.

Energy reduction in agriculture

It is possible to reduce the amount of energy required for agriculture and the food system. A series of thoughtful proposals by Pimentel and associates deserve wide attention.[3] Many of these proposals mitigate environmental problems, and any reductions in energy use provide direct reduction of the pollutants due to fuel consumption as well as more time to solve our energy supply problems. Among the suggestions made by Pimentel and associates are the following.

First, we should use natural manures. The United States has a pollution problem from runoff from animal feedlots, and yet we apply large amounts of manufactured fertilizer to fields. More than one million kcal per acre could be saved by substituting manure for manufactured fertilizer (and as a side benefit, the condition of the soil would be improved). Widespread use of natural manure will require decentralization of feedlot operations so that manure is generated closer to the point of application. Decentralization would increase feedlot costs, but if energy prices rise, feedlot operations will rapidly become more expensive in any case. Crop rotation is less widely practiced than it was even twenty years ago. Increased use of crop rotation or interplanting winter cover crops of legumes (which fix nitrogen as a green manure) saves 1.5 million kcal per acre compared to commercial fertilizer.

Second, weed and pest control could be accomplished at a much smaller cost in energy. A 10 percent saving of energy in weed control could be obtained by using the rotary hoe twice in cultivation instead of using herbicides (again with pollution abatement as a side benefit). Biologic pest control—that is, the use of sterile males, introduced predators and the like—requires only a tiny fraction of the energy needed for pesticide manufacture and application. A change to a policy of "treat when and where necessary" in pesticide application would bring a 35 to 50 percent reduction in pesticide use. Hand application of pesticides requires more labor than machine or aircraft application, but the reduction of energy is from 18,000 kcal per acre to 300 kcal per acre. Changed cosmetic standards, which in no way affect the taste or edibility of food stuffs, could also bring about a substantial reduction in pesticide use.

Third, the directions in plant breeding might emphasize hardiness, disease and pest resistance, reduced moisture content (to end the wasteful use of natural gas in drying crops), reduced water requirements, and increased protein content—even if it means some reduction in overall yield. In the longer run, plants not now widely cultivated might receive some serious attention and breeding efforts.

The direct use of solar energy on farms, a return to wind power (using the modern windmills now in use in Australia), and the production of methane from manure are all possibilities. These methods require some engineering to be economically attractive, but it should be emphasized that these technologies are now better understood than is the technology of breeder reactors. If energy prices rise, these methods of energy generation would be attractive alternatives even at present costs of implementation.

Energy reduction in the U.S. food system

Beyond the farm, but still far short of the table, many more energy savings could be introduced. The most effective way to reduce the large energy requirements of food processing would be a change in eating habits towards less highly processed foods. The current dissatisfaction with many processed foods from "marshmallow" bread to hydrogenated peanut butter could presage such a change, if it is more than just a fad. Technological changes could reduce energy consumption on an industry by industry basis but the most effective way to encourage the adoption of methods using less energy would be to increase the cost of energy. Such price increases almost certainly await us.

Packaging has long since passed the stage of simply holding a convenient amount of food together and providing it with some minimal protection. Legislative controls may be needed in packaging to end the spiralling competition of manufacturers in amount and expense of packaging. In any case, recycling of metal containers and wider use of returnable bottles could reduce this large energy use.

The trend toward the use of trucks in food transport to the virtual exclusion of trains should be reversed. By simply reducing the direct and indirect subsidies of trucks, we might go a long way toward enabling trains to compete.

Finally, in the home we may have to ask whether the ever larger frostless refrigerators are needed, and whether the host of kitchen appliances really mean less work.

Store delivery routes, even by truck, would require only a fraction of the energy used by private autos for food shopping. Rapid transit, giving some attention to the problems of shoppers with parcels would be even more energy-efficient.

If we insist on a high energy food system, we should consider starting with coal, oil, garbage—or any other source of hydrocarbons—and producing food in factories from bacteria, fungi, and yeasts. These products could be flavored and colored appropriately for cultural tastes. Such a system would be more efficient in use of energy, solve waste problems, and permit much or all of the agricultural land to be returned to its natural state.

Energy, prices, and hunger

If energy prices rise—as they have already begun to do—the rise in the price of food in societies with industrialized agriculture can be expected to be even larger than the energy price increases. Slesser, in examining the case for England, suggests that a quadrupling of energy prices in the next forty years would bring about a sixfold increase in food prices.[5] Even small increases in energy costs may make it profitable to increase labor input to food production. Such a reversal of the fifty year trend toward energy-intensive agriculture would present environmental benefits as a bonus.

We have tried to show how analysis of the energy flow in the food system illustrates features of the food system that are not easily deduced from the usual economic analysis. Despite some suggestions for lower intensity food supply and some frankly speculative suggestions, it would be hard to end this chapter on a note of optimism. The world drawdown in grain stocks which began in the mid 1960s continues, and some food shortages are likely throughout the 1970s and early 1980s. Even if population control measures begin to limit world population, the rising tide of hungry people will be with us for some time.

Food is basically a net product of an ecosystem, however simplified. Food starts with a natural material, however modified later. Injections of energy (and even brains) will carry us only so far. If humankind cannot adjust its wants to the world in which it lives, there is little hope of solving the food problem for mankind. In that case the problem will solve mankind.

NOTES

1. *Statistical Abstract of the United States,* 1973.
2. *Historical Statistics of the United States* (Washington, D.C., 1960).
3. D. Pimentel et al., "Food Production and the Energy Crisis." *Science* 182 (1973): 443-449.
4. N. Georgescu-Roegen, *The Entropy Law and the Economic Process* (Cambridge, Mass.: Harvard University Press, 1971), 301.
5. M. Slesser, "How Many Can We Feed?," *The Ecologist* 3 (1973): 216-220.

Of mites and men

In which scientists and environmentalists argue about the right way to kill insects

WILLIAM TUCKER

■ William Tucker *is a contributing editor of* Harper's *magazine. His article "Environmentalism and the Leisure Class"* (Harper's, *December 1972) earned him an Honorable Mention in the Annual Gerald Loeb Awards. He also was recently chosen a winner of the Annual John Hancock Awards. Both awards are for excellence in business and financial journalism.*

In "Of Mites and Men" Tucker addresses the issue of biological insect control. A case history of a company engaged in the development and sale of biological insecticides is cited. As a result of obstacles placed by the Environmental Protection Agency in the path of biological-control research and the sale of products resulting from such research, the company experienced serious financial difficulties and was subsequently sold. Tucker blames the policies of the EPA on environmentalists, and launches a severe attack against them. In particular he singles out Rachel Carson's book Silent Spring, *in which he sees the "embryonic form" of the later "excesses of environmentalists." Three claims made by environmentalists and defended in Carson's book are assessed and disposed of. They are: (1) that in the good old days, the "pre-pesticide past," the crops were good and the insects were few; (2) that there are "natural" and "unnatural" chemicals, pesticides being of the latter sort; and (3) that "unnatural" chemicals, like pesticides, cause cancer.*

Nothing is easier than to admit in words the truth of the universal struggle for life, or more difficult—at least I found it so—than constantly to bear this conclusion in mind.

Charles Darwin, *The Origin of Species*

In the early 1950s, a small company named Nutrilite Products, Inc., in Buena Park, California, was making a modest income selling a vitamin supplement guaranteed to be made completely from "natural" products. The vitamins were extracted from alfalfa grown on Nutrilite's 1,000-acre farm in the San Jacinto Valley, 100 miles northeast of Los Angeles. The company was using only "organic" humus fertilizers and no chemical pesticides when, in 1953, it discovered an infestation of small, green aphids eating their way through the crop.

Nutrilite felt morally obliged to avoid treating its fields with chemical pesticides, so the late Carl Rehnborg, founder of Nutrilite, consulted agricultural scientists at the University of California at Riverside, who suggested he try spreading an insect-attacking fungal disease among the aphids. "We did it and it worked," Rehnborg wrote later. "It was a great moment in the history of this company.

Two years later, in 1955, when Nutrilite's alfalfa fields were again attacked by the voracious catepillar larvae of a small, mothlike lepidopterous insect, Nutrilite again turned to the universities. This time Rehnborg was directed to Berkeley, where Dr. Edward A. Steinhaus, often called the "father of insect pathology in the United States," introduced him to an insect-attacking bacteria called *Bacillus thuringiensis,* which had been first isolated in Germany in 1911. "BT," as it came to be called, was known to infest a wide variety of lepidopterous (moth and butterfly) larvae, while being completely harmless to humans, animals, and other insect families. Once again, the product worked.

Rehnborg was impressed and began considering marketing BT for use against the dozens of lepidopterous insects that variously attack cotton, vegetables, fruit orchards, forest trees—almost every form of vegetation. He hired an entomologist named Dr. Abdul Chauthani, who went to work in Nutrilite's small laboratory, developing various strains of BT and trying to isolate other insect-attacking bacteria and viruses. In 1960, at a cost of $300,000, Nutrilite was able to register BT with the U.S. Department of Agriculture for use against the cabbage looper. Since the USDA would require similar "efficacy" testing for use of BT against each separate pest on each separate crop, Nutrilite limited its work to cabbages and tried to save enough money to extend registration to other crops. Several larger companies in the pesticide field noted Nutrilite's success and registered BT for other uses. By 1962, when naturalist-author Rachel Carson first publicized the "biological control" of insects as an alternative to toxic chemicals like DDT, she was able to describe *Bacillus thuringiensis* as one of the most promising "alternate" methods, already used successfully against alfalfa pests in California, gypsy moths in Canada and Vermont, and banana-eating insects in Panama.

For the next ten years, the pace of research accelerated at Nutrilite, and by 1970 Dr. Chauthani and other researchers had isolated a wide range of bacteria and viruses that could selectively attack a variety of insects. The company had obtained two "experimental-use" permits from the USDA and had about ten other promising products waiting to go into registration procedures, when it ran into an unusual and unexpected opponent—the environmental movement and the newly formed Environmental Protection Agency (EPA).

In 1972, the EPA, formed after nearly a decade of public agitation about environmental problems, began enforcing the brand-new Federal Environmental Pesticide Control Act, passed that year. The bill had been adopted in response to widespread public fears about DDT and other chemical pesticides, first raised by Carson's *Silent Spring,* published in 1962. Responding to the Congressionally mandated task of reviewing registration of all 30,000 existing pesticides, in addition to enforcing tighter registration requirements against new pesticides, the EPA revoked Nutrilite's two-year-old experimental permits and asked for two more years of data proving that insect bacteria and viruses could be used safely in the environment. Nutrilite would be forced to spend about $200,000 on testing before it could begin *experimenting* with the bacteria again. In addition, there would eventually have to be extensive toxicity testing to prove that the bacteria would not have unintended effects on small mammals, fish, birds, marine life, or farm animals, nor would it leave residues that might produce cancer, mutations, or birth defects in humans. What was worse, the EPA itself seemed unsure about how the strict environmental standards should be applied to such "biological" controls. "The EPA changed its mind so many times, we gave up trying to figure out what they wanted," said Dr. Chauthani when I interviewed him by telephone this spring.

After several years of frustration Nutrilite retrenched its efforts to register new products, and tried to continue making money with BT. By 1971, however, Abbott Laboratories had developed another strain of BT that worked more effectively. Nutrilite would have to switch to the new strain to remain competitive, but company officials soon realized that the EPA was going to require complete re-registration of the new strain even though it was genetically only slightly different from the old one. In desperation, Nutrilite proposed combining its old BT strain with the newly developed pyrethroids, a synthetic version of the pyrethrin chemicals derived from the chrysanthemum flower and used against insects for centuries. The EPA informed Nutrilite that it would still have to go through the $500,000 registration procedure for each separate insect on each separate crop because the new synthetic pyrethrins had not been proved to be safe, even though they are almost the same chemical compounds as the natural pyrethrins that are known to be safe.

With nowhere to go, Nutrilite withdrew its own BT strain from the market in 1975, and has since abandoned all further research on insect bacteria and viruses. The company has decided to continue some research in breeding parasitic insects *simply because this form of biological control has not yet been required to go through registration procedures by the EPA*. The company was financially weakened by its unsuccessful venture into biological controls, and in 1975 most of its stock was bought by the Amway Corporation, a Michigan firm that sells shoe polish and cleaning products door-to-door through a franchising system. Amway officials say they intend to continue spending some money for insect-control research, but are mainly interested in marketing Nutrilite's vitamin supplement.

"We're very bitter," said Dr. James Cupello, manager of insect-control research at Nutrilite, when I talked to him on the phone in March. "But this company is not going to spend another penny trying to develop biological controls as long as we have to go through the EPA. The risks are too great that we'd spend a million dollars on research and four years later we'd find out that the EPA wouldn't let us register the product. We've had the reputation of being the leading marketing company for biological controls in this country, but nobody is going to be able to do anything in this field as long as they have to contend with the EPA. We're going back to making vitamin supplements and trying to stay as far away as possible from the Environmental Protection Agency.''

I first became interested in finding out what happened to Rachel Carson's ''other road'' of biological insect-controls after reading several newspaper stories on the subject in the past two years. Each of these accounts told of the wonders that had been coming out of the laboratories over the past decade—insect chemical mating signals, or ''sex pheromones,'' had been molecularly decoded and synthesized so they could be broadcast on infested fields where they would turn the insects' mating attempts into a three-ring circus; ''juvenile'' and ''anti-juvenile'' hormones had been discovered that could either keep insects forever young and sexually immature or make them try to metamorphose prematurely into adults before they had even had time to grow their larval whiskers; strange bacterial and viral diseases had been isolated that attacked only certain insects and left other species unharmed. Checking back into *Silent Spring,* I found that the early research on all these methods of biological control had been the main substance of Rachel Carson's ''other road'' of biological controls that would lead us away from toxic chemicals like DDT.

But there was a curious footnote in all these stories that usually didn't occur until about the last three paragraphs. For some incomprehensible reason, the Environmental Protection Agency was not allowing any of these new ''third generation'' pesticides to be registered without demanding the enormously expensive testing procedures originally designed to keep chemicals like DDT and the other ''bad'' pesticides off the market. As a result, most of these new methods were still languishing in the laboratories. The situation was always treated as some odd mistake, some bureaucratic foul-up that would be straightened out as soon as the EPA could settle down, stop ''reorganizing,'' and understand the facts clearly. No one seemed willing to consider that the generals at the EPA might still be fighting the last war, and that the broad snare of regulation designed to capture DDT and other ''old-fashioned'' pesticides had now entangled the new generation of pesticides as well. It appears, however, that that is what has happened.

The biological controls that Rachel Carson offered as the other road to pest control have indeed come of age after a decade of brilliant research by American chemists and entomologists. Scientists have discovered all anybody would ever want in an insecticide—carefully isolated chemicals that attack only the ''target'' pests, leave beneficial insects unharmed, and seem to leave no long-term residues that could harm other organisms in the environment. But while these serious research specialists were seeking the answers to environmental problems in the laboratories, another army of enthusiasts was traveling its own other road, which led straight to Washington. This was the environmental movement, a concatenation of glorious amateurs, ''aroused'' citizens with a knack for talking about what they really didn't understand, vocationless aristocrats defending the imagined glories of the past, housewives with a flare for writing publicity releases, lawyers with a talent for histrionics, and ''militant'' scientists and academics with a willingness to shade the truth just a bit in pursuit of a ''good cause.'' This army arrived in the Capital in the early 1970s, quickly routed DDT and its allied devils, occupied offices close to Capitol Hill, and have roamed the halls of Congress ever since. Its major accomplishment has been to build a wall of regulation so solid and insurmountable that almost *no* pesticides should ever be able to scale it again. When the serious scientists, who had attempted a positive approach to the problem, arrived in Washington with the results of their research, they ran up against the brick wall of the Environmental Protection Agency. They have been fruitlessly beating their heads against it ever since. . . .

MYTHS OF ENVIRONMENTALISM

There are three fundamental problems that have caused the current dilemma of environmentalism. First, there is the myth, which environmentalists have fashioned, of an ideal, preindustrial, prepesticide past, when crops were good, living was easy, and insects were few. This is a complete fantasy. Second, there is the false distinction between ''natural'' and ''unnatural'' chemicals, and the implicit assumption that chemicals like pesticides never occur in nature. Third, there is the myth that these "unnatural" chemicals are causing an equally mythical "epidemic" increase in cancer. Unfortunately, the genesis of all three of these ideas can be traced directly to *Silent Spring*.

Silent Spring is a great book, and for the most part has stood the test of time. No one would argue that it was not enormously successful in alerting the public to the dangers of pesticide use and to some of the worst abuses that were then prevalent. It is hard to believe, for example, that whole towns were once sprayed with highly toxic chemicals in an effort to wipe out a single pest species lurking somewhere among the leaves. It also brought to public attention the persistence of some pesticides, and their magnification through the food chain. For this we owe Rachel Carson an enormous debt.

But *Silent Spring* is also a terrible book, and the future excesses of environmentalism appear in embryonic form on every page. In discussing what she calls the system of "deliberately poisoning our food" with pesticides, Carson says:

> But if, as is now the presumable goal, it is possible to use chemicals in such a way that they leave a residue of only 7 parts per million (the tolerance of DDT), or 1 part per million (the tolerance for parathion), or even of only 0.1 part per million as is required for dieldrin on a great variety of fruits and vegetables, then why is it not possible, with only a little more care, to prevent the occurrences of any residues at all?

The constant insistence on "zero pollution levels" has proved to be the most costly and unenforceable aspect in much environmental legislation. More important, however, is the argument that DDT and other pesticides were causing what Carson called "an alarming increase in malignant disease" (cancer), the proof of which is entirely contained in the following sentence: "The monthly report of the Office of Vital Statistics for July, 1959, states that malignant growths, including those of the lymphatic and blood-forming tissues, accounted for 15 percent of the deaths in 1958 compared with only 4 percent in 1900." A high school student would probably blush at the distortion. In 1900, the average American lived to be forty-five and had a good chance of dying of influenza. In 1962, the same citizen could expect to live to seventy and was therefore six times more likely to contract cancer, which is predominantly a disease of old age. The only reason the *percentage* of cancer deaths has increased is because industrial civilization has allowed people to live longer, and bacterial diseases have essentially been eliminated.

The myth of the pest-free past was not explicit in Rachel Carson's book, but was implied by her failure even to *mention* the problems of controlling insects in agriculture. This omission caused one writer, environmentalist LaMont Cole, in reviewing the book for *Scientific American*, to remark: "She does not convey an appreciation of the really great difficulties of the problem [of insect control]. . . . But what I interpret as bias and oversimplification may be just what it takes to write a best-seller." Rather than heeding such warnings, however, the environmental movement has woven an elaborate vision of a mythical, pest-free past against which the problems of current pest-control methods can be contrasted. This fallacy was recently reiterated in the *Washington Post's* front-page Sunday editorial section, in an article entitled "The Pesticide Plague" (March 5, 1978):

> Before synthetic pesticides hit the market in 1946, corn belt farmers didn't have many insect problems. They grew a rich diversity of crops, rotating them from one field to the next. That way the pests attracted to any single crop could not sweep the farm like a plague. But with the birth of the Green Revolution, small, diverse farms were wiped out and massive monocultures, vast tracts of a single crop planted year after year, spread across the corn belt. . . . What have [the farmers] got to show for it? Since pesticides came to the farms, pest damage to corn has not decreased. The latest USDA estimates indicate corn losses from pests have in fact more than tripled. . . ; [meanwhile] the major pesticide producers—petrochemical giants such as Dow, du Pont, Monsanto, American Cyanamid, Standard Oil of California (Chevron), Shell—just celebrated a record year, with $3 billion in sales.

In nineteenth-century America, insect problems were so much a part of life that whole towns were sometimes asked to pray for deliverance. Even the pests themselves have not changed to any great degree. Despite the "rich diversity of crops," the Colorado potato beetle easily spread across the Midwest in the 1860s and eventually made it to Europe, where it became a major pest. After the first gypsy moths escaped from a silkworm experiment in Boston in 1869, the streets of New England were so infested that caterpillars were crawling up the sides of houses and into people's beds. The standard method of protecting crops was to spray them with lead arsenate, a practice that produced its own *Silent Spring,* a book called *100,000,000 Guinea Pigs,* which caused a sensation in the 1930s. The introduction of less toxic DDT in 1946 was regarded as a major advance at the time.

Of course, there is some truth to the statement that "monocultures" of corn have replaced the old diversity, although growing a rich variety of crops in the old days often meant simply having a rich variety of pests. But what farmers in the corn belt also have to show for their efforts, despite the misleading "increase" in pest damage, is contained in the following graph:

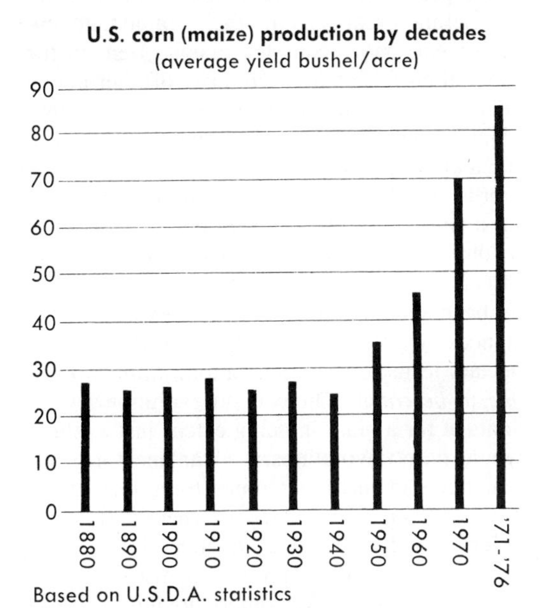

According to the USDA figures, there was never *any* increase in corn productivity in the United States until synthetic pesticides, chemical fertilizers, and new hybrid varieties were introduced after 1945 (pesticides probably account for only 20 percent of the increase, but have ensured the success of other improvements). To produce the same amount of corn under the old methods would mean that an additional area equivalent to Colorado and Wyoming would have to be planted. Nor have farmers and chemical manufacturers been the only beneficiaries. As John Stuart Mill said: "When commerce is spoken of as a source of national wealth, the imagination fixes itself upon the large fortunes acquired by merchants, rather than upon the savings of price to consumers." Most of the corn is used to raise beef cattle, and as a result Americans now consume twice as much beef as they did in 1940, even though they spend a one-third smaller portion of their income on food.

> That which the palmerworm hath left hath the locust eaten; and that which the locust hath left hath the cankerworm eatern; and that which the cankerworm hath left hath the caterpillar eaten.
>
> **The Book of Joel**

The second and more difficult problem of environmentalism is the widely held belief it has fostered that there is an important distinction between "natural" and "man-made" chemicals, and that it is the "synthetic" chemicals manufactured by industrial society that are the cause of all our problems. Rachel Carson played heavily on this distinction in *Silent Spring*. An infinite number of potential chemicals can be made through nature's system of stringing long carbon chains together in various forms to form the "organic hydrocarbons." Only a fraction of the potential number are actually synthesized in nature, but then objects such as shovels, axes, plows, and most of the other implements of our daily lives do not occur in nature either. There is nothing inherently "evil" (Rachel Carson's word) about changing nature by synthesizing new chemicals, and the distinction that "natural" chemicals are "good" and synthetics are "dangerous" is completely meaningless. There are hundreds of highly dangerous "natural" chemicals, just as there are thousands of perfectly harmless "synthetics." Yet environmentalism has managed to establish the doctrine that everything in nature is "good," while things that are made in the laboratory hold the potential for destruction.

The key sentence that expresses this in *Silent Spring* reads as follows:

> The chemicals to which life is asked to make its adjustment are no longer merely the calcium and silica and copper and all the rest of the minerals washed out of the rocks and carried in rivers to the sea; they are the synthetic creations of man's inventive mind, brewed in his laboratories, and having no counterparts in nature.

This statement is so filled with absurdities and errors that it is hard to know where to begin. In the first place, calcium, copper, and other minerals form only the tiniest fraction of the diet of living organisms. Except for certain one-celled creatures, all living things derive all their energy and most of their substance by taking apart large organic molecules (plants make their own carbohydrates using the sun's energy, and then break them down themselves, process called "autotrophism," or self-nourishment). The point that Carson was probably trying to make was that plants and animals never had to deal with special kinds of organic molecules like the "chlorinated hydrocarbons," but if so, she was *completely* wrong.

Practically every schoolchild knows the story of Dr. Alexander Fleming, the British scientist who in 1928 accidentally dropped some cheese in a bacterial culture and later noticed that a few small, sterilized zones had been created. A variety of Penicillium mold was growing on the cheese, and Fleming discovered that the mold excreted small amounts of a substance that killed bacteria. It was soon realized that a wide variety of soil fungi and other organisms produce antibacterial molds that they use in competing for space with other organisms. Penicillin was the result, and since that time our major effort against bacterial diseases has been a process of imitating these soil organisms. What is not generally known, however, is that many of the chemicals in these antibacterial molds are *chlorinated hydrocarbons*. One of the biggest producers of chlorinated hydrocarbons are the long "ray fungi" (actually bacteria) that illustrate one of the opening chapters of *Silent Spring,* and that Carson describes as "growing in long threadlike filaments" at a rate of more than 1,000 pounds per acre! None of these organisms actually make DDT or other common pesticides, but they do use chemicals that are remarkably similar. One Penicillium fungus excretes a chemical that is only one molecule different from a commonly used fungicide "Dowcide 2S," manufactured by Dow Chemical. In fact, it seems quite possible that the presence of these large amounts of chlorinated hydrocarbons in nature may offer an explanation for the enormously large quantities of "pesticide residues" that environmentalists have always been able to find. (In 1970, Frank Graham, Jr., reported without irony in *Silent Spring* that "the amount of DDT in Swedish soils exceeds the total quantities ever used in that country.") To be sure, scientists who developed the pesticides and herbicides from chlorinated hydrocarbons may not have been aware that they were copying nature so closely, but there was a brilliant kind of inductiveness in that we arrived at the same kinds of chemicals that are used for almost the same purposes in nature.

In one of the most beautiful passages in *Silent Spring,* Rachel Carson writes: "Most of us walk unseeing through the world, unaware alike of its beauties, its wonders, and the strange and sometimes terrible intensity of the lives that are being lived about us." She was talking of the insects and their ever-present predators. But what Carson was only peripherally aware of, and what has emerged clearly only in the past decade of research, is that *plants themselves* are also intensely involved in this struggle for existence, and that their form of "warfare" is largely *chemical* warfare. Simple organisms like fungi and bacteria excrete substances that kill competing organisms in their immediate environment. More complex plants often do the same thing. Certain cacti give off herbicidal chemicals that make it impossible for other plants to germinate in their immediate vicinity. In addition, plants are constantly growing thorns, needles, and tough coatings, and synthesizing chemicals to make themselves bitter, inedible, and even poisonous to animals and insects. In a way it seems foolish for us never to have realized it before, but except in instances where the consumption of fruits and nectars leads to seed generation, *plants do not like to be eaten*. There is no evolutionary advantage for a plant in being eaten, just as there would be no evolutionary advantage for a human being in becoming dinner for a lion. Plants have evolved a vast array of chemicals, from chlorinated hydrocarbons to juvenile-hormone mimics, in trying to protect themselves from becoming dinner for other organisms. The simple proof of the matter is that almost everything we eat—wheat, barley, oats, potatoes, corn, carrots, peas, beans, bananas, oranges, lettuce, tomatoes, the list is endless—is a *human invention* that does not exist in nature. They are completely "unnatural" organisms that we have invented for our own purposes through a process of chemical and genetic manipulation that is in no way different from synthesizing a new organic compound in the laboratory. There is no fundamental difference between changing a few atoms in an organic compound and calling it a "pesticide," and manipulating a few genes on a couple of wild plants and calling the result a "carrot." The brilliant realization of the past decade of research in insect control has been that plants, too, are involved in the process of synthesizing chemicals to protect themselves from insect attack, and that the most fruitful path of research may lie in following the trail they have blazed over the last few hundred million years.

> Of all the chemicals in the whole history of the world that have done the most good for humanity, in terms of limiting disease, in terms of providing food, in terms of relieving suffering, the one that has done the most good would have to be DDT.

Dr. William Bowers

The long fight for a complete ban on DDT, and the excesses that were practiced in its pursuit, are what is now haunting the environmental movement in its attempt to replace the chlorinated hydrocarbons and other toxic chemicals with "biological" chemicals. The problem is that there is no basic distinction between the two.

There is no question that there were enormous abuses of DDT and other pesticide chemicals when Rachel Carson wrote *Silent Spring* in 1962. Pesticides were being used with a "shotgun" approach that was having a tremendous impact on wildlife. Carson was on firm ground in voicing these concerns, in part because the same worries had been expressed by scientists for more than fifteen years. Writing a prophetic essay entitled "DDT and the Balance of Nature," published in *Atlantic* magazine in 1945, the same Dr. Wigglesworth who had already identified the juvenile-hormone gland in 1934 wrote:

> DDT is like a blunderbuss, discharging shot in a manner so haphazard that friend and foe alike are killed. . . . Without careful study it is impossible to guess what the ultimate results of this process may be. . . . Some fish . . . are reported to have been killed when they fed on poisoned insects. . . . DDT sprayed on peach trees with the object of killing the caterpillars of the Oriental fruit moth is even more effective in killing the parasite that is controlling this pest. . . . It is obvious enough that DDT is a two-edged sword. . . . Chemicals which upset the balance of nature have been known before. DDT is merely the latest and one of the most violent. . . . We need to know far more about [the insects'] ecology—that is, about their natural history studied scientifically. When the ecology of an insect pest is fully known, it is often possible to modify the conditions in such a way that its world no longer suits it. . . . But when all these so-called cultural or naturalistic methods of control have been developed, there remains a large residue of pests for which insecticides must be used.

Although she essentially ignored the warning in the last sentence, Rachel Carson added two more concerns to this list—the unforeseen development that long-lasting pesticide residues would be "magnified" through the food chain, building up in predators and higher organisms, and the concern that insects eventually would develop resistance to pesticides and that ever-increasing doses would have to be used.

This was all well and good, but neither Carson nor the environmentalists were ever willing to admit that it was precisely DDT's long persistence that had in many ways made it a superior pesticide, and that *any* pesticide would eventually face the same problem of growing insect resistance. The pyrethrins are a classic example. Derived from the chrysanthemum flower, the pyrethrins are a group of natural chemicals whose origins were once held a secret by the Persians until they were ferreted out by the English, who started growing large quantities of chrysanthemums in Kenya in the 1880s. Natural pyrethrins presented two problems, however—the laborious method of production could not supply the world market, and the pyrethrins themselves broke down quickly in sunlight. In the 1950s the problems were finally transferred to the laboratories, where scientists soon synthesized the molecule. In 1962, Rachel Carson could write:

> The ultimate answer [to highly toxic pesticides] is to use less toxic chemicals so that the public hazard from their misuse is greatly reduced. Such chemicals already exist: the pyrethrins, rotenone, ryania, and others derived from plant substances. Synthetic substitutes for the pyrethrins have recently been developed so that an otherwise critical shortage can be averted.

But the problem was that, although they were not very toxic to mammals, the pyrethrins were still fairly dangerous to fish. In addition, there was still no adequate solution to the chemicals' short life. A variety of carriers were tried, but finally it became simpler to change the molecule to create a chemical that would last long enough to be effective. Now the pyrethrins were a useful insecticide—but suddenly they were an environmental problem as well. Because of their new persistence, they posed a danger to fish. In addition, now that they were being used more widely, insects were beginning to develop more resistance. Thus, when the USDA began introducing the synthetic pyrethrins into cotton farms in recent years, the Environmental Defense Fund told the EPA it was opposed to their use, *even though these synthetic chemicals had been specifically approved by Rachel Carson in "Silent Spring."* What Carson failed to realize,

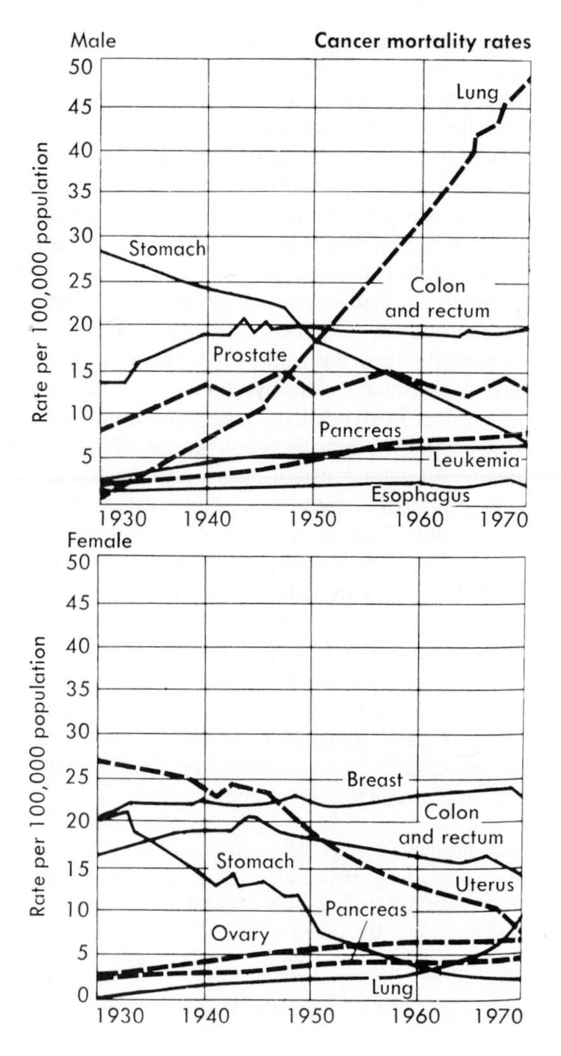

and what the environmental movement has since ignored, is that insects are eventually going to build up resistance to *any* chemical, natural or unnatural, just as bacterial diseases have eventually evolved strains that are resistant to antibiotics. There is already evidence that insects are going to be able to develop resistance to juvenile growth hormones, pheromones, and other "bio-rational" controls as well. In short, the battle with the insects is never going to be over, just as the battle against bacterial infection will never really be over.

Rachel Carson's speculation that residues of DDT in human and animal tissues were causing cancer has mushroomed into a widespread public certainty that it is the products of industrial society that are causing an "epidemic" increase in cancer. . . . [the accompanying] graphs of cancer mortality for an age-adjusted population in the United States indicate that there is no "epidemic" increase in cancer in this country. The only instance of a clear increase in cancer rates is lung cancer among men. Ironically, this is the only instance where people are known to have a personal choice in avoiding the carcinogenic material (the National Cancer Institute estimates that 80 percent of lung cancer incidents are the result of smoking). Among the twenty-four leading industrial nations, the United States is sixteenth in cancer mortalities for an age-adjusted population.

What is perhaps most notable in the graph is the steady *decrease* in the rate of stomach cancer over the past forty years. Stomach cancer is rife in underdeveloped countries in Asia and Africa, and the suspected carcinogen is a completely *natural* substance called "aflatoxin," the excretion of a mold that grows in stored peanuts and grains. Cancers of the digestive system occur in underdeveloped countries at rates up to 200 times their incidence in the United States because of aflatoxins, which are among the most potent carcinogens known. The rate of liver cancer from simply *eating* in East Africa is double the rate of liver cancer found among 25,000 industrial workers exposed to one of the most famous industrial carcinogens, vinyl chloride. Moreover, the aflatoxin mold is known to establish itself best in peanuts and grains that have been damaged by insects! The highest quantity of aflatoxin ever found in the U.S. by the Food and Drug Administration was in a jar of "natural" peanut butter. It would be entirely possible to argue that, rather than causing an increase in cancer, pesticides and fungicides have been partly responsible for the notable *decrease* in cancers of the digestive system in industrialized countries.

Most of the notions on which Rachel Carson based her claim that DDT might be causing cancer were highly speculative at the time, and are now a part of medical history. She suggested that DDT acted on all cells by affecting their ability to use oxygen, causing them to mutate back to a more primitive process of "fermentation" in order to break down carbohydrates. The assumption was that this process would affect the nerve cells of insects, causing nerve dysfunctioning, but would produce cancer in the other human and animal cells as well. This was based on another speculation of the time, that cancer cells were also formed by mutations back to this same primitive fermentation process. All these theories have since been abandoned.

It was generally accepted at the time, and has since been proved, that DDT acts as a "nerve poison" by fitting into certain highly specialized receptacles at the end of all nerve cells. Insects are extremely vulnerable since they have no fat tissues in which to store DDT. Humans and other vertebrate animals avoid the nerve poisoning by storing DDT in fat cells, *but they do not go on building up stored quantities indefinitely*. Numerous tests have shown that a peak level is reached, and all new material is immediately excreted, so there is no danger of "slow poisoning" from DDT. In a single dose, DDT has about the same toxicity as aspirin. On the other hand, parathion, which replaced DDT in many uses, is so highly toxic that a single drop in the eye can kill a person. It is interesting to note that although Rachel Carson said as many bad things about parathion as DDT in *Silent Spring,* the environmental movement chose to concentrate its efforts against DDT because of the "slow poisoning" concerns. The result has been that, while the hysteria has been relieved in suburban living rooms, hundreds and hundreds of farm workers and farm children have been poisoned because of the increased use of parathion, and about twenty-five people die each year.

A PHILOSOPHICAL GAME

While *Silent Spring's* theory for the action of DDT has not held up, neither has its model for the development of cancer. The assumption widely held in 1962, and since increased in stature, is that both a disruption of the genetic material and the intrusion of a cancer virus are involved in the beginning of a cancer "incident." The genetic material temporarily can be disrupted by a "carcinogenic" substance (which is probably the same thing as a "mutagenic" substance), and before the genes can be repaired, a virus (which is really nothing more than a set of "naked genes") become permanently linked into the long genetic molecule. One current theory, widely accepted, is that such cancerous "incidents" occur in the body every day, but most are destroyed by the body's immune system. Once in a great while, however, an invaded cell escapes detection and is able to survive, eventually multiplying into a cancerous growth. The participation of the immune system suggests that the body's general health can play a large part in preventing cancer, and there are many studies linking general malnutrition with the very high rates of cancer among some South African Bushmen and other Third World peoples. This suggests that one way to reduce cancer might be to feed people better, but this is an avenue environmentalism has chosen not to take.

Because of the mutation/virus-intrusion assumption, the hunt for industrial carcinogens has settled upon substances that cause mutations among laboratory organisms. The most recently developed method is the "Ames test," invented by Berkeley biologist Bruce Ames, which uses a highly specialized strain of bacteria that is very susceptible to mutations to measure mutagenic effect. DDT and other chlorinated hydrocarbons have been subjected to the Ames test, and the results show that they do *not* cause mutations. The only exception is toxaphene, which—ironically—is the only chlorinated hydrocarbon pesticide still in use. (Interestingly enough, Ames, who is a staunch environmentalist, disputes the results of his own test and says DDT is "one of the 10 percent of all carcinogens that our test doesn't catch," although it is hard to know how he has decided this. There is one other study, performed at the Epley Institute in Omaha, that showed that DDT causes a slight increase in liver tumors in mice, although not in rats or hamsters. The results of that test are still disputed, because the tumors disappear when DDT dosages are stopped, but Dr. David Clayton, who performed the test, says he is satisfied DDT is a "mild carcinogen.")

In addition, there is one more indication, known for many years, that DDT was not causing any noticeable increase in cancer. This is simply that, among the thousands and thousands of factory hands, pesticide sprayers, farm workers, and people in malaria-prone underdeveloped countries who have been heavily exposed over the course of thirty-five years, there has *never* been any indication of an increase in cancer, even among workers who suffered accidental exposure great enough to put them in the hospital. In the 1950s, volunteers ate large quantities of DDT in a series of tests and never suffered any adverse effects.

Most of these facts were known during the late 1960s when environmentalists were determined to show that DDT was a public health menace. To solve their problem, environmentalists invented a kind of philosophical game which stated that, although there was no evidence to show that DDT did cause cancer, it was *philosophically impossible* for anyone to show that it *couldn't* cause cancer. In part, this argument relied on the fact that many cancers take from thirty to forty years to show up after exposure to carcinogenic substances. But even where the evidence was weakest, the environmentalists maintained that their position was unassailable. William Butler, chief counsel for the Environmental Defense Fund, which led the attack on DDT between 1966 and 1972, repeats the argument today: "You can't prove a negative," he said when I called him in April. "You can't say something doesn't exist because there's always a chance that it does exist but nobody has seen it. Therefore you can't say something doesn't cause cancer because there's always the chance that it does cause cancer but it hasn't showed up yet. You can't prove a negative statement." Does that mean you can't prove that dragons don't exist? I asked him. "That's right, you can't say dragons don't exist."

Butler is absolutely right, of course, in strict logical terms. The problem is that the same argument applies to any other synthetic chemical that is introduced into the environment, including the "biological controls." Like DDT, the blunderbuss of environmental regulation has turned out to be a killer of freind and foe alike. But environmentalism has by no means learned its lesson from the experience. In fact, it is already looking around for new worlds to conquer. Armed with the assurance that only industrial chemicals are causing cancer, the environmental movement and the federal government are now preparing to do for the rest of American industry what they have already done for pesticides by trying to remove all carcinogens from the environment. Speaking like a Puritan schoolmaster calling the class to order, Gus Speth, member of the President's Council on Environmental Quality, recently announced on the *New York Times* op-ed page: "The recent controversy on the proposed Food and Drug Administration ban on saccharin treated us to a dangerous amount of hilarity about the high dosage levels used in animal tests, and demonstrated the prevalence of misunderstanding in this area." The truth is, he announced, that 1) there is "no safe level" of a carcinogen, and 2) laboratory-animals tests are a sure indication of whether a substance causes cancer in humans. "With one or possibly two exceptions, every chemical known to cause cancer in humans also causes it in animals," he concluded. The question, of course, is whether it works the other way around.

The National Academy of Sciences has made an effort to bring some rationality to the notion that we will be able to purge our world of every last trace of carcinogenic material. In 1973, it published a book entitled *Toxicants Occurring Naturally in Foods,* which noted that trace amounts of cancer-inducing chemicals occur naturally in many foods. Another survey of the literature by Dr. Russell S. Adams, Jr., of Penn State University, found that such common foods as rutabagas, tea, cabbage, turnips, peas, strawberries, and milk all contain traces of chemicals that either cause cancer or are closely related to chemicals known to cause cancer. Dr. Julius M. Coon, one of the authors of *Toxicants Occurring Naturally in Foods,* and retired chairman of the Department of Pharmacology at Thomas Jefferson University, Philadelphia, had this to say about the "no safe dose of carcinogens" doctrine when I called him in April:

"When people say there is no safe dose of a carcinogen, what they are saying is that we can't find a safe level so we have to assume that there is no safe level. But the statement that there is no safe dose of a carcinogen is not a valid scientific statement in any sense of the word. We are constantly surrounded by chemicals that may cause cancer, most of them perfectly natural. I think it's reaching for the stars to say we're going to eliminate all carcinogens from our environment. I think the inflation we've seen so far will be a drop in the bucket compared with what we'd see if they try to enforce this new law [the 1976 Toxic Substances Control Act]. We may not have enough well-trained toxicologists to perform the tests."

Yet there seems to be no limit to what the federal government is willing to do to indulge the fanatical concerns about what we eat, drink, and breathe. Not to be outdone by the EPA, the Food and Drug Administration has started enforcing new regulations that apply the same elaborate toxicological standards to *all new hybrid* varieties of crops that are developed in the genetic laboratories. The FDA is no less aware that these human inventions are "synthetics" and that they offer the same dire possibilities that we may at last be poisoning ourselves. This means that the entire centuries-old effort of improving breeds for greater yields and better disease resistance could easily drown in the same sea of red tape that has already suffocated the pesticide industry. The National Academy of Sciences' 1975 report on pest controls voiced considerable alarm about the FDA effort. Yet, fueled by the fanatical concerns about pesticide residues and other toxicant traces, the FDA is moving ahead, and even now the results seem predictable. The hubris of the people who tell us we can wipe the last traces of toxic and carcinogenic materials from our environment is the same hubris of the people who once told us we were going to be able to rid the world of insects by spraying ever-increasing amounts of DDT.

Nature herself has met many of the problems that now beset us, and she has usually solved them in her own successful way. Where man has been intelligent enough to observe and to emulate Nature he, too, is often rewarded with success.

Rachel Carson, *Silent Spring*

The more I examine the environmental movement, the more it seems like a kind of secular religion, with a decidedly Puritan strain. Like all religious movements, it draws its strength from what we *don't* know. It tries to hide in the cracks of our understanding, instilling us with the fear of what we haven't yet been able to learn from nature. Public anxiety about scientific experimentation is nothing new. Louis Pasteur's neighbors in Paris besieged the authorities to put an end to his work. Practically every major medical advance, from autopsies and dissections to vaccination and surgery, has met with suspicion—and sometimes violent opposition—from a large portion of the population. Such misgivings have always existed, and are not always ill-founded. But it is only when some deeply conservative organization such as the Church or environmentalism has orchestrated such fears that these anxieties become institutionalized and *all* scientific advance comes under suspicion. Only then do ordinary human fears about newness and invention start to play a decisive role in history.

I am not foolish enough to think that there will not be a solution to the problem of biological insect controls. The newspapers will discover the situation and soon a new "crisis" will be upon us. But what keeps nagging in the back of my mind in this great Age of Environmentalism is what we are going to look like a few years from now. Somehow it seems we are going to appear as a generation that was so obsessed with misgivings, so afraid of what we didn't—and couldn't—know, so anxious to point hysterical accusing fingers at one another, that we neglected to pick up and use the simple tools we had at hand. I have no doubt that someone will eventually use these tools. I only wonder if we will ever calm down enough to do it ourselves.

Ergot: the taming of a medieval pestilence

LEO VINING

■ Leo Vining, *a native of New Zealand, is a professor of biology at Dalhousie University in Nova Scotia. He has spent most of his professional career researching microbial biosynthesis and chemical microbiology under the auspices of the National Research Council of Canada. This research has earned him several awards.*

In this article, Vining tells us a success story in the history of technology. Ergot is a fungus that grows in the seed-head of grasses, including rye and other plants cultivated for food grains. During the Middle Ages it was the cause of many illnesses and death. In come cases of ergot poisoning "the victims burned with 'holy fire' and the extremities atrophied." In other cases the victims "became demented and violent." Since the true reason for such illness was unknown, it was considered a punishment from heaven.

Slowly, over the years, the benefits of ergot administered in small doses were discovered. That ergot is a fungus was established in 1853. Nevertheless, understanding of the medicinal powers of ergot remained highly fragmentary.

Today ergot is used to treat a wide range of conditions, including migraine, and the "holy fires" of the Middle Ages have been put out once and for all.

In the year 857 A. D., according to German chroniclers of that period, the population around Duisberg was ravaged by "a great plague of swollen blisters that consumed the people by a loathsome rot so that their limbs were loosened and fell off before death." The circumstances and description of this tragedy indicate that it was not another of the bacterial epidemics that we associate with the great plagues of the Dark and Middle Ages, but was, in fact, due to mass food poisoning. The cul-

□ From *Technology Review,* 81, No. 3(1979), 65-74. *Technology Review* is edited at the Massachusetts Institute of Technology.

prit was ergot, a common but unrecognized contaminant in the food grains of those times.

A DESCRIPTION OF ERGOT

Ergot is a fungus. While we have no difficulty in recognizing food spoilage by the ubiquitous molds, which are also fungi, the spoilage caused by ergot is more subtly introduced and, through a combination of circumstances, remained undetected for centuries.

The ergot fungus in its natural habitat is a plant parasite. It grows on and is nourished by a living host, in contrast to the usual refrigerator molds which grow mostly as saprophytes, on dead tissue. Unlike the molds, ergot undergoes several changes in form and appearance during its annual life cycle. It grows in the seed-heads of grasses, including those that man has developed and cultivated for food grains, and replaces a normal seed. At the growth stage coinciding with harvest time for cereal crops, it masquerades as the grain it has replaced. This is the most commonly recognized form of ergot; it is also the form which contains a group of very potent chemical substances, called ergot alkaloids, which cause profound physiological effects if eaten. Without realizing it the farmers of earlier times were harvesting a mixture of grain and ergot with lethal properties.

Those who ate the bread baked from ergot-contaminated flour fell victim to any one of ergot poisoning's many forms. The most widespread symptom was a feeling of cold fire in the limbs—rather like the pins and needles sensation one experiences when circulation has been restricted. We know now that the ergot toxins constrict the blood vessels in the body to cause this effect. The sensation is agonizing and reduced circulation over prolonged periods also has the more severe effects of causing the affected limbs and tissue to atrophy. The contractive action of the toxins on smooth muscle can cause epileptic convulsions and, in addition, ergot has an hallucinogenic effect, inducing mental distortions and dementia. All of these symptoms develop when small quantities of bread containing ergot are eaten over an extended period. Eating large amounts is rapidly fatal, with any one of the symptoms dramatically prominent.

ERGOT: THE WRATH OF GOD

Ergot poisoning was rampant in Europe throughout the Middle Ages and did not disappear until comparatively recent times. Like the great bacterial plagues its real cause was not recognized, but the remarkable symptoms of ergot intoxication soon acquired for it a plausible explanation.

The French archivist Frodoard noted that in an outbreak around Paris in 945 A.D. some of the afflicted, who were cared for in Saint Mary's Church and fed wholesome rations provided by a benevolent Count Hugo, recovered. Those who returned home often found the fire in their bodies rekindled, and were taken again to the chapel to be healed by penitence and prayer. In the religious climate of the times the "fire" was attributed to divine punishment for sinful living, and the recovery to divine forgiveness. If anyone suspected a connection between recovery and the quality of Count Hugo's flour, Frodoard made no note of it.

In the period between 800 and 1500 A.D., historical records describe hundreds of "plagues" in which the symptoms match those of ergot poisoning—usually either the gangrenous form where the victims burned with "holy fire" and the extremities atrophied, or the convulsive kind where the victims became demented and violent. There are frequent references to single epidemics in which 40,000 to 50,000 people died during a year of suffering. The cumulative death toll is staggering: millions of people perished or survived as cripples, and entire regions were decimated. Coincidental events took on significance and the holy fire disease came to be known by many names.

Around Aquitaine in France an epidemic raged for months until the abbé had the bones of Saint Martial dug up and displayed to the sufferers. The pestilence ceased soon afterwards, probably because it was harvest time and the new grain was free of ergot, but in that region "ignis plaga" became "ignis Sancti Martialis." In another area where "holy fire" symptoms in the nose, mouth, and hands were especially prominent it was called "ignis judicialis"—a judgment on the debauchery of the inhabitants. The most common name, however, was "ignis Sancti Antonius"—Saint Anthony's fire. In 1090 the son of a rich nobleman in the Dauphinée region of France was afflicted. Kneeling before the bones of Saint Anthony, the father pledged his riches to help victims of the holy fire if his son were spared. When the son recovered the nobleman kept his promise by founding a series of hospitals throughout the country and Saint Anthony became the patron saint of all "fires," epilepsies and eczemas.

Since food selection has been an important element of survival—only those of us whose ancestors were good at recognizing the poisonous from the non-poisonous are here today—ergot was clearly an insidious and deceptive toxin. In a large measure the deception was grounded in the religious beliefs of the early Middle Ages which provided such satisfying explanations for the symptoms of ergot poisoning. Ergot was widely known to midwives and physicians for its potent and toxic drug action. Historians frequently recounted the sufferings of afflicted people in areas where the crops had been bad and the bread of poor quality. And men of science had been interested in ergot as a disease of grain crops for many years. Yet none of those most familiar with ergot saw the connection between severe ergot infestations and subsequent outbreaks of the fire plague. It was left to an observant Paris lawyer, M. Dodart, whose work took him for several years on visits to the low-lying Sologne district of France where rye was the main cereal crop, to link the prevalence of Saint Anthony's fire with the high ergot content in grain ground by the millers.

THE CATCH IN THE RYE

Rye is a hardy plant. It was not used as a grain by the ancients but was introduced after Roman times into Europe where it thrived on the poorer and damp land unsuitable for wheat. Because of its affinity for infertile land, rye tended to be the grain of the poorer peasants. Reluctant to discard any part of their harvest, even the dark and distorted "seeds" of ergot which flourished in wet seasons, they would often harvest a large amount of ergot with their crop. Although they might set aside the most obviously contaminated grain, often by late winter even this would be used. Thus the severity of the fire plagues often increased thoughout the spring and summer. Interestingly, too, the monks who nursed the plague victims seeking atonement at the shrines of benevolent saints usually farmed the better lands on which wheat could be grown. Wheat is less prone to parasitism by the ergot fungus and the monks could better afford to clean their grain. The bread they provided for the pilgrims was of good quality and the plague victims found physiological as well as psychological relief from their suffering. Meanwhile, their confidence in divine intercession prospered.

Ergot poisoning may also have been mistaken for supernatural intervention on this continent in the early 17th century and triggered the Salem witch hunt. There is some evidence that the young girls who were believed to have been possessed by the devil may have been eating ergot-contaminated rye bread. Their initial behavior coincided in some respects with that of fire plague victims.

When Dodart reported his deductions about the cause of Saint Anthony's fire to the French Academy in 1676, he proposed that the government introduce laws requiring that all rye grains be sieved to remove ergot before being milled into flour. Strict controls were soon adopted in France, but outbreaks of ergot poisoning did not cease immediately. Peasants refused to believe that God would poison their crops and distrusted laws enacted by the authorities. In Russia there were severe epidemics even to the 20th century. During the winter of 1926-1927 11,000 people were stricken and 93 died. In France, bureaucratic control of the milling and distribution of flour eventually became too unwieldy and failed to prevent an outbreak of ergotism as late as 1951.

THE BIRTH OF SCIENTIFIC EVIDENCE

With the increase in scientific observation during the Renaissance, naturalists began to study and speculate about the nature of ergot without being in any way aware of its relationship to the "holy fire" disease. Thalius, a respected authority of the 16th century, considered it to be a malformed plant seed. The suggestion that ergot might be fungus did not surface until 1711, and was hotly disputed. In 1815 the French Academy commissioned one of their members to settle the issue. On the basis of a chemical analysis, Vauquelin declared it to be a seed! It was not until 1853 that the question was finally resolved by Tulasne, a biologist. He showed that the peculiar structure was only one stage in the lifecycle of a fungus—the stage that enables it to survive the winter in a dormant state.

At some point not recorded in medical history it was discovered that eating small quantities of ergot would hasten childbirth. Knowledge of this effect seems to have spread through the folklore of midwives, and the first written prescription for its use appeared in Adam Lonicer's *Krauterbuch* or herbal of 1582. Over the next 300 years ergot gradually found its way into most unofficial medical texts, usually with instructions that one to three "spurs" could be eaten if labor was slow or prolonged. It was introduced into the official pharmacopoeia in the 19th century largely because of the work of John Stearns, a New York state physician who experimented with ergot after hearing of its use from a local midwife. Stearns collected ergot from granaries in the neighborhood and compared the effects of extracts as well as powdered ergots. He found that the substance extracted with water was as active as the ergots themselves—a valuable drug, when properly used. The potency of the extract varied with the source of the ergot and the care with which it was prepared. Moreover, there was no convenient way of testing the activity before use. Overdoses produced severe toxic reactions and, as Stearns had warned, the action of the drug was so powerful and immediate that the uterus would rupture if the child had not been correctly positioned for delivery. By the end of the 19th century the use of ergot extracts to hasten childbirth was considered too dangerous and was discouraged in medical teaching. Instead, it was recommended for postpartum administration to contract the uterus and prevent hemorrhaging. Ergot drugs are routinely used for this purpose today.

THE BIOACTIVE INGREDIENTS IN ERGOT

For 100 years after John Stearns described the effects of an ergot extract chemists tried without success to purify the active principle. Their main difficulty was the lack of a simple, meaningful assay to test the different fractions obtained through purification, and their search tended to concentrate on discovering substances that had chemical resemblance to the higher plant alkaloids already known to be potent natural drugs. In 1918 this effort paid off with the isolation by Alfred Stoll at Sandoz laboratories in Switzerland of ergotamine, a pure alkaloid with many of the properties of the parent ergot. With this discovery a reliable ergot drug preparation became possible, but it took another 33 years before chemists could describe in precise terms the molecular structure of the compound.

One of the difficulties in isolating ergotamine was that it was hidden among several other very similar alkaloids. It soon became apparent that the activity of ergot is due not to a single substance but to a group of alkaloids. Additionally, these are rather unstable compounds which rearrange to inactive forms under some of the treatments routinely used in isolating natural products.

Examining the pharmacological activity of each pure alkaloid of the ergotamine type, chemists failed to match the rapid and powerful uterocontractive activity shown by crude ergot extracts. An English pharmacologist, L. Chassar Moir, devised a very direct way of measuring this activity by placing in the uterus a small water-filled balloon connected by tubing to a barometric recorder. With this as a bioassay and the help of a chemist he then isolated from ergot a new alkaloid. . . . Moir was not alone in his search for this compound: it was described by four different laboratories at about the same time. Thus the compound is variously known as ergometrine, ergobasine, ergonovine and ergostetrine. It is this compound and a semi-synthetic methylergobasine that are now used routinely in childbirth to minimize the chances of postpartum hemorrhage.

The semi-synthetic methylergobasine is prepared by treating the natural alkaloids with alkali to cleave the molecules into their components. The lysergic acid fragment is recovered and attached chemically to a new base (in this case butanolamine) to form methylergobasine. The semi-synthetic drug has similar but not identical uterocontractant activity to its natural homologue, ergobasine.

In the search for other semi-synthetic variants, Albert Hofmann in the Sandoz laboratories prepared many new alkaloids like methylergobasine. One of these produced startling effects, which he discovered in 1943 by accidental self-administration. The compound was a diethylamide of lysergic acid, LSD-25. Hofmann, abandoning laboratory work for the day, set off for home on his bicycle and later recalled his experience: "My field of vision swayed and objects appeared distorted, like images in curved mirrors. I felt fixed to the spot, although my assistant told me afterwards we were cycling at a good speed. I recall the most outstanding symptoms as vertigo and visual disturbance; the faces of those around me appeared as grotesque colored masks. I recognized my condition clearly and sometimes, as if I were an independent neutral observer, saw that I babbled half insanely and incoherently. Occasionally I felt out of my body. When I closed my eyes endless colorful, realistic and fantastic images surged in on me. Acoustic perceptions, such as the noise of a passing car, were transformed into optical effects, every sound evoking a corresponding colored hallucination constantly changing in shape and color."

LSD is an extremely potent substance that produces its hallucinogenic effects by acting directly on the central nervous system. A dose of 1 to 2 μg per kg. body weight will usually elicit symptoms within a few minutes and in the range from 1 to 16 μg per kg. the intensity of the effect is proportional to dose. An immediate dizziness, weakness and nausea is usually followed by inner tension and, in the second or third hour, by visual illusions and sensory distortions. Loss of sensory boundaries creates a need for a supporting environment. Recollections from the past may overlap the present, while moods shift suddenly from elation to fear. If a major panic episode has not occurred after four to five hours, a sense of detachment and supreme control may arise. Although the half-life of the drug is about 3 hours, the entire syndrome usually lasts for about 12 hours. Because of the unpredictable incidence of "bad trips" which can neither be prevented nor treated, other than by reassurance in a supportive atmosphere, the use of LSD is considered by psychologists to be a hazardous undertaking, even though the extreme potency of the drug means that there is little risk of a toxic overdose causing the more clinically dangerous symptoms of ergot poisoning. The drug was used experimentally for a time to treat alcoholism and opiate addiction but caused no permanent change in psychological state. This use has been abandoned and LSD appears now to have no clinical value.

LSD is not a natural constituent of ergot, but related ergot alkaloids do possess similar, if much less intense, abilities. Ergine, the parent lysergic acid amide, is one such compound. Well known to chemists for many years as a degradation product of the more complex alkaloids, its psychotomimetic action was not discovered until 1961 when Sandoz became interested in the drugs used by Central American Indians in their religious ceremonies. *Ololinqui,* the seed of a species of morning glory native to Mexico, contained ergine as the active principle.

THE MODERN ERGOT PHARMACOPOEIA

Ergot alkaloids exhibit a remarkably wide range of physiological effects, all of which stem from actions on the nervous system. To various degrees, they affect the central nervous system directly and cause responses such as a drop in pulse rate, respiration and heart beat by depressing the vasomotor control center of the medulla. LSD, by far the most powerful central nervous system activator, appears to affect the transmission of signals mediated by 5-hydroxytryptamine at much lower concentration than other neurohumoral responses and thus produces a specific psychedelic action at very low doses.

Nerve impulses generated in the brain control the activities of the body through a series of parallel transmission systems. The ergot alkaloids act in the hypothalamic region of the brain which is the principal locus of integration for autonomic functions—those activities that are under involuntary control such as blood pressure, fluid balance, and the less mechanical responses of sleep, emotions and sexual reflexes. The hypothalamus also links the brain with its neural mode of transmission to the endocrine system which regulates body function through the use of chemical messengers (hormones) that travel in the blood stream to act on distant glands and tissues.

A study of pig mortality in Africa pinpointed the capacity of ergot alkaloids to interfere with this process. Death of newborn piglets caused by drying up of the sow's milk supply was traced to a high content of ergot in the millet supplied to the piggery. This effect was in turn traced to the action of agroclavine, a member of the ergot alkaloid group, which interferes with hypothalamic-mediated release of the hormone prolactin from the pituitary. Prolactin is required to induce milk secretion in the mammary glands. The action of agroclavine is mimicked by several semi-synthetic lysergic acid derivatives; such compounds, more active and less toxic than the natural alkaloid, are now in clinical trial for treatment of conditions such as galactorrhea (excessive milk production) and mammary carcinomas that depend on prolactin for continued growth.

During the studies on agroclavine that established its effect on prolactin release, it was observed that mice treated with the drug immediately following copulation failed to conceive. It is now known that the alkaloid and several semi-synthetic members of the group prevent pregnancy by interfering with implantation of the fertilized egg in the uterus. The activity is attributed to inhibition at the hypothalamus of necessary hormone factors essential for the proper development of pregnancy, and offers one potential route to the development of "morning-after" birth control.

The classic uterocontractive activity of ergot alkaloids is due to their effect in the peripheral system where the nerves stimulate or relax smooth muscle. They interfere with the action of noradrenalin, one of the chemical substances that connects nerve activity at synaptic junctions, and the blockade is manifest to varying degrees in alkaloids of different structural types. Most active are the alkaloids, such as ergotamine, which possess a peptide component. Their powerful constrictive action on blood vessels is the basis for their most widespread modern use, the treatment of migraine headaches. Ergotamine is the most effective compound and, so far, the only useful remedy. Migraine is believed to be due to increased amplitude of pulsations in the cranial arteries. The alkaloid, by constricting the arterioles, suppresses these pulsations. Relief is often instantaneous if the drug is injected into the bloodstream, but it is usually given by a less direct route. The dose regimen must be monitored carefully to avoid ergot poisoning.

MANUFACTURING ERGOT ALKALOIDS

Searching out and collecting ergot from naturally infected grasses and cereal crops has long ceased to provide the quantity needed to prepare pure drugs. The immediate solution to the supply problem was to develop ways of artificially infecting crops with the ergot fungus to obtain higher yields. Since ergot is worth 20 times as much as the grain it replaces this is a worthwhile proposition for a farmer, provided the fungus infestation can be contained. Rye is the crop of choice because of the relatively long period when the glumes are open and fungus spores can infect the rye ovary. Ergot cultivation is normally confined to isolated valleys, in fields surrounded by tall trees to prevent spore dispersal by wind and insects. At first, crops were sprayed during flowering with a water suspension of spores, but repeated applications were needed since the glumes do not all open at the same time. The technique was eventually superseded by directly injecting spores by a needle-puncture method. Tractors were fitted at the front with a device that pressed the seed heads between two moving belts, one of which carried a set of needles that had passed through a suspension of spores. A very heavy primary infection can be obtained in this way so that good yields do not depend on secondary dispersal by insects, wind and rain at the honeydew stage.[1]

Along with this kind of agrotechnical research went an empirical selection of genetic strains that gave superior yields of the desired alkaloids and less complicated mixtures to make isolation and purification easier. The overall effort was so successful that, until very recently, all ergot used by the pharmaceutical industry was produced in this way. However concurrent with research into field production methods an alternative approach aimed at achieving production of the alkaloids by cultivating the fungus in tanks of artificial media was developed. This process is now displacing field cultivation as the more economical and reliable method.

The ergot fungus can be grown easily in the laboratory, but most early attempts to persuade it to produce ergot alkaloids in artificial culture were unsuccessful. Where production was achieved the yields were invariably low. To overcome this problem, Matazo Abe in 1948 screened many hundreds of fungus isolates from ergot growing on wild Japanese grasses until he found strains that yielded well when cultivated in a simple nutrient solution. Surprisingly, however, the alkaloids formed by these strains were different from ergotamine and the other alkaloids previously isolated from natural ergot. The main product in his first culture was aproclavine. Over the next ten years

[1]The life-cycle of ergot provides for reproduction and dispersal in appropriate seasons. The ergot body falls to the ground in late fall and germinates in late spring; this sequence of winter cold followed by spring moisture and warmth are necessary to break dormancy. The germinating ergot produces numerous finger-like outgrowths that develop heads packed with threadlike spores. These are dispersed by wind or carried by bees and other insects to the flowering heads of grasses. A fungus spore landing in the open floret grows on the ovary, eventually consuming it. Filamentous cells proliferate, followed by a "honeydew" stage when filaments exude a sweet, sticky liquid containing masses of small round asexual spores. These are carried by the many kind of insects that feed on honeydew to other plants, spreading the infection widely. After the honeydew stage, the fungus produces a core of closely packed cells which expands to form the ergot body. This core preserves the fungus in a dormant state during the winter and it alone contains the poisonous alkaloids.

dozens of similar alkaloids were found, but none of the high-yielding cultures produced clinically useful compounds.

The first break came in 1960 when an Italian group working with Ernest Chain found a strain of *Clavicaps paspali,* a species of the ergot fungus growing on paspalum grass, that produced large amounts of the hitherto unknown alkaloid, lysergic acid α-hydroxyethyl amide, in culture. This can be easily hydrolyzed chemically to lysergic acid and converted to the clinically useful ergometrine or methylergobasine. A few years later workers at the Sandoz laboratories in Basel discovered a subspecies of *Claviceps paspali* that produced high yields of paspalic acid, an alkaloid easily rearranged to lysergic acid. This process, too, became a starting point for the semi-synthetic production of uterocontractant alkaloids. Persistent research has gradually solved the riddles of ergot fungus physiology and even the recalcitrant *Claviceps purpurea* has now been persuaded to make large quantities of the valuable ergotamine in culture. Once reasonable yields are obtained the inherent advantages of fermentations over field cultivation make the choice between these processes a simple one.

THE RAISON D'ETRE FOR ERGOT ALKALOIDS

With Saint Anthony's fire eradicated and the scourge of medieval peasants and villagers now producing useful drugs for mankind, one intriguing question remains: what possible use can the ergot alkaloids have in the fungus that makes them? We cannot arrive at a decisive answer since proof is hard to come by, but consider these facts:

The genetic and biochemical machinery needed to make these compounds is complex, and therefore expensive for the fungus to maintain.

Unnecessary characters are normally lost during the evolutionary struggle for survival.

There seems to be a link between the ability of the fungus to parasitize plants and to produce alkaloids.

The ergot bodies are dark-colored and therefore visible as well as toxic, so that predators, especially birds, would learn to avoid them.

It seems reasonable to believe that formation of alkaloids is a protection for the unusual lifestyle of the fungus. Unfortunately man, through reasons of his own making, took an exceptionally long time to discover that ergot was not edible and should be strictly avoided.

Ergot is a word of French origin meaning "cockspur." The German name is *Mutterkorn,* a folklore term that was adopted when the cornfields were believed to be visited by a demon spirit, the corn mother. The corn could be seen to sway and part as she passed through; where she touched the seed heads *Wolfzähne,* sprang up, intended for her children, the rye-wolves. It seems we have always tried as best we can to explain away the things we don't quite understand. We may no longer believe that ergot is made by a demon spirit to provide teeth for her children but perhaps our present attempt to explain its existence may prove to be no less fanciful.

16

Technology and the biomedical sciences

The future impact of science and technology on medicine

LEWIS THOMAS

▪ Lewis Thomas, *a physician and an educator, is president of Memorial Sloan-Kettering Cancer Center. He is also Professor of Medicine and Pathology at the Medical School, Cornell University, and co-director of the Graduate School of Medical Science. He is the author of* Lives of a Cell.

In the following article Thomas discusses the three levels of technology involved in modern medicine. He calls the first and highest level "high technology." "This is the genuinely decisive technology of modern medicine exemplified best by methods of immunization against . . . various virus diseases, and the contemporary use of antibiotics and chemotherapy for bacterial infections." The important feature of high technology is that it is the result of a genuine understanding of disease mechanisms. The second level of technology he calls "halfway technology." Exemplified by transplants and the invention of artificial organs, this kind of technology is an attempt to compensate for the effect of disease which we do not yet know how to control. The third and lowest level of technology is best termed "non-technology." It is basically "caring for" and "standing by" the patient when no known technology can be used to help him.

Lewis points out that the last two levels are the most expensive. In times when money is becoming scarce, he suggests that it would be wise to put most of the money where it can produce the best results, namely in basic science research. If the mechanism of a disease becomes understood, the expensive units designed to compensate for the damage resulting from the disease become unnecessary. The need for non-technology would also be substantially reduced.

It is said that we are spending this year something like $85 billion on health in this country. Last year the figure was $70 billion; the year before around 60. Nobody can vouch with certainty for the accuracy of these figures, nor even count up all the things the dollars are presumably buying. But no matter; they are socking great sums, enough to warrant the term Health Industry for the whole enterprise.

With an investment of this size, much of it representing public funds, it is surprising that there is so little analytical information concerning the enterprise; there is really no such thing as a Health Policy for the country in the sense that the term Policy is used for other major public ventures and certainly nothing like real Policy Planning. There is only an intense public anxiety that it is costing much too much money and we cannot afford to put in more; also, there is a spreading doubt that we are getting anything like our money's worth.

In this climate, it is no wonder that the general support for scientific research and training in the biomedical sciences has come upon such hard times in the past several years. This, by the way, is not a special bias of the present administration. The cutting back on funds to support medical research began in 1967 and would probably have continued regardless of what administration took power. There is an unmistakable loss of confidence in the value and effectiveness of science. It applies to science in general, not just medicine. It derives in part from the anxiety about our mixed technological blessings and some general apprehension about the future. At the same time, doubts have arisen about the capacity of science to solve our health

☐ Reprinted, with permission, from *BioScience,* 24, No. 2 (1974), 99-105, published by the American Institute of Biological Sciences.

problems and there are new fears concerning the harmful effects of science in medicine. We are suspected of busying our laboratories devising hideous new technologies to engineer ordinary, friendly, everyday man out of existence. We have special basement laboratories where we invent ways to control human behavior to our liking, transplant heads, raise identical parthenogenetic babies in plastic test tubes, clone prominent political figures, and teach computers to think rings around us. Scientists, in short, are suspected of having gone crazy—partly, it is also suspected, under the influence of money.

These are bad times for reason, all around. Suddenly, all of the major ills are being treated by acupuncture. If not acupuncture, it is apricot pits, astrology, or transcendental meditation.

We are in danger of losing sight of our genuine assets and in even more danger of failing to take advantage of the opportunities in science which, in the long history of science, have never been so real. There is, not too far ahead, the solid possibility that man may be rid of the worst of his diseases once and for all. This is a time when we ought to be making very careful plans for the long-term future. It is the unlikeliest of all times for us to be getting ourselves into a depression about medical science.

What we need today in medicine is a better technology assessment—the same kind of periodic hard-headed assessment of technology that has become a routine and profitable exercise for most of the large industrial enterprises of the nation. We need an inventory of our information, our methods, and our real prospects for the future. Up to now medicine has received very little of this kind of treatment. The technology of medicine has simply been accepted as given—as something there, for better or worse, to be taken for granted. Even the term "technology" means something different to most people from what it means to the rest of science. We are not accustomed to thinking of antibiotics, insulin, or coronary bypass surgery as the items of technology they obviously are, and we tend to speak of technology only in the sense—and a very limited sense it is—of the methods involved in the distribution and delivery of health care.

It would be useful, I think, to take a careful look at the actual measures employed in the management of disease and the preservation of health, and to make, periodically, the same objective appraisals of their effectiveness as are made by other industries. It would be a lot easier to work out long-range policies, set future goals, and assign budgets if we had some system in operation to tell us, at any given moment, where we stand with our technology, what these measures are costing us, and whether there are options in sight, especially whether there is new information that might lead to basic changes in technology just ahead.

We have no such system for appraisal. There are good reasons for this although I believe the time has come for us to find our way around the difficulties. The technology of medicine has certain features that distinguish this field from the rest. One difference has to do with the economics which seem to govern all other technological advances but have no discernible influence on the kinds of things we do, or *think* we do, in medicine. For example, we do not build new bridges at great cost without knowing in advance, quite precisely, what the transport requirements will be in the future, and having some kind of assurance that the bridges will bear the traffic, meet all forseeable demands and stand up to all forseeable stresses. But we will undertake the development of an artificial heart at the cost of many bridges without going through any sort of cost-benefit, logistic, or even moral analysis of what it is that we are making. Indeed, in medicine, it is characteristic of our technology that we do not count the cost, ever, even when the bills begin coming.

This is plainly a defect in our system—if we can be said to have a system. It is, in part, explainable by our history, by the brand newness of any kind of technology at all in this field, and our consequent unfamiliarity with any methods, or indeed, any incentive in the first place, for technology assessment in medicine. We have had almost no genuine science to tap into for our technology until just the past three decades. As a profession, we go back a very much longer time, probably thousands of years. During most of our history, therefore, we have become accustomed to no technology, or to pseudotechnologies without science. We have long since acquired the habit of improvising, of trying whatever came to hand; and in this way, have gone through our cyclical fads and fashions, generation after generation, ranging from bleeding, cupping, and purging, through incantations and the reading of omens, to prefrontal lobotomy and metrazol convulsions; we have all gotten quite used to this kind of thing, whether we will admit it or not. Early on we became accustomed to the demand that a doctor must *do* something; doctors who didn't *do* something, no matter what, were not real doctors. During the long period when we knew of nothing to do about typhoid fever except to stand by and wait for the patient to struggle through, keeping an eye out for the hemorrhages and perforations that might kill him at any time, the highest level of technology was the turpentine stupe. This was an elaborate kind of fomentation applied to the belly, very difficult to make without ending up with a messy shambles, and capable, I believe, of doing absolutely no good whatever beyond making everyone feel that the doctor was *doing* something. This, by the way, is not a baroque item of our distant history. I learned to make a turpentine stupe at the Peter Bent Brigham Hospital in 1937; it is, therefore, in my view, a relatively recent, almost modern example of the way we develop technology, and it is not yet all behind us as we shall see. We still have our equivalents of bleeding, cupping, and turpentine stupes, and they are all around us.

The trouble with this kind of pseudo-technology or magical technology is that it has become unbelievably expensive in its more modern forms, and, at times, is dangerous. It is particularly dangerous and expensive when it takes the form of strong drugs, bizarre diets, or surgery.

But now that science has entered medicine in full force, we must begin to sort out our affairs. From now on we will need, as never before, to keep the three central enterprises of medicine—"to cure, to relieve, to comfort"—clearly separated from each other in our minds. They do not really overlap, but we tend to view them, and the public takes the same view, as though they were all of a piece, all the same body of technology, all derived from science, all modern. I think perhaps one reason we do this is because of an unconscious conviction that dollar values must be placed on all human enterprises, and we do not like to confess to ourselves that so many of the things that we do are simply provided for comfort and reassurance. Somehow this has come to seem a less significant product than a cure; so we try, consciously or unconsciously, to pretend that there is more continuity than is really there, that everything we do is directed toward the same end.

In fact we are engaged in three entirely different kinds of technology in medicine. I have an idea that if we could conduct a sort of technology assessment on ourselves, and come to some sort of general agreement about which technology belongs in which category, we might be in a better position to make intelligent plans and forecasts for the future. We would almost surely be clearer in our minds about how to set priorities for the investment of scientific resources for the future.

Before beginning on my own version of a classification, I would like to make a general declaration of faith, and a general confession of optimism. My dogmas are as follows: I do not believe in the inevitability of disease. I concede the inevitability of the *risk* of disease, but I cannot imagine any category of human disease that we are precluded, by nature, from thinking our way around. Moreover, I do not believe that when we succeed in controlling or curing one kind of disease we will necessarily, automatically, find that it has been replaced by another.

Even if I am wrong, and it should turn out that there is some law of nature that mandates the doling out of new diseases up to some optimal number whenever old ones disappear (which strikes me as a piece of illogic as well as high improbability), I still cannot imagine remaining helpless before all the new ones. Nature is inventive but not so inventive as to continue elaborating endless successions of new, impenetrable disease mechanisms. After we have learned enough to penetrate and control the mechanisms of today's disease, I believe we will be automatically well-equipped to deal with whatever new ones turn up. I do not say this in any arrogance; it just seems reasonable.

I have no more difficulty in imagining a disease-free human society, or at least a society in which major diseases are held under control, than I do with the idea that valuable stocks of animals, or varieties of plants, can be maintained *relatively* free of disease.

I believe that disease is fundamentally unnatural. It is not a normal or natural part of the human condition for aging human beings to become paralyzed and idiotic for long years before they finally die any more than it is for young people to develop acute leukemia. I believe that disease comes generally as the result of biological mistakes: misinterpretations of signals on the part of cells and tissues; misuse of information. I believe that the mechanisms of disease are quite open to intelligent intervention and reversal whenever we learn more about how they operate.

To say it another way, I do not consider that

the ambition to control or eliminate disease, which is an ambition shared by everyone in biomedical science, is either unthinkable or any distance beyond imagining. What makes it seem to many people like an outlandish, even outrageous way to be talking, is that it becomes assumed that we are talking about human happiness, which is really quite another matter, or about human mortality, which is also quite another matter. As to the first, it is of course true that disease has long been a major cause of human despair and wretchedness, but this is no reason to believe that we will all become happy, well-adjusted people by being rid of it. We will still be left with our share of worrisome problems, and we will still have more than our share of ample reasons for despair, and no medical science—not even psychiatry—has any forseeable contribution to make to these matters. War and bombs, failure and anomie, clouding of the sun by particles of our own waste, the shutting off of oxygen, the loss of room to move around—these are problems still to be with us for some time to come, healthy or ailing, and I hope that no one will suggest that these are in any sense problems for medicine—or we will never get any of them solved. But perhaps human society will be better equipped to think its way through these imponderables if, at least, we no longer have today's roster of diseases to worry us at the same time.

As to mortality, I have a hunch that we will discover, someday, that disease and death are not as inextricably interrelated as we tend to view them today. All the rest of nature undergoes, in its variable cycles, the physiological process of death by the clock; all creatures, all plants, age finally, and, at the end, they all die. Diploid cells in tissue culture have finite life spans which are different for different lines of cells, and characteristic of particular cell stocks. Some live for 40 generations and then die; others for 70. They do not develop fatal diseases; it is not a catastrophe—they simply reach the end of a life-span programmed for them in their own genomes, and at the end of that span, they die.

I believe that we are also like this. If we are not struck down prematurely by one or another of today's diseases, we live a certain length of time and then we die, and I doubt that medicine will ever gain a capacity to do anything much to modify this. I can see no reason for trying, and no hope of success anyway. At a certain age, it is in our nature to wear out, to come unhinged and to die, and that is that. My point here is that I very much doubt that the age at which this happens will be drastically changed for most of us when we have learned more about how to control disease. The main difference will be that many will die in relatively good health, in a manner of speaking. Rather after the fashion of Bertrand Russell, we may simply dry up and blow away.

Even if our technology were to become so dazzlingly effective as to rid us of all the major diseases that now kill many of us before our time of wearing out, I doubt that the resulting population increase would make more than a marginal difference to the general problem of world overpopulation. Indeed, it might help some since there would be smaller numbers of us in hospitals, or living out their lives in various degrees of incapacitation and suffering. Being overpopulated is bad enough as social problems go, but to be overpopulated with so many disabled by disease, especially by the chronic diseases of the elderly let alone schizophrenia, presents an unthinkable prospect for the approaching century.

In any case, we do not really owe much of today's population problem to the technology of medicine. Overpopulation has been coming for several centuries and the alarming upward slope began long before we had developed a genuine capacity to change the outcome of disease. Modern medical science is a recent arrival and the world population had already been set on what seems to be its irreversible course by the civilizing technologies of agriculture, engineering, and sanitation—most especially the latter. From here on the potential benefits of medicine greatly outweigh any conceivable hazard. We will perhaps change slightly the numbers living at any moment in time, but it lies within our capacity to change very greatly the quality of life.

Well, where do we stand today as a science? This is not the same question, of course, as the one concerning the state of our technology. Our science is the science of the biological revolution, and we have scarcely begun to apply any of it. We do not yet, in fact, know where to begin. In contrast with today's genuinely high technologies of molecular biology, neurobiology, or cellular biology, with the immense power of their instruments for exploring the most fundamental questions about the processes of life, the condition of our knowledge of disease mechanisms has a primitive, 19th century look, and our capacity to intervene in disease is not much better. This is the general shape of things today, but tomorrow will be very different indeed. I simply cannot imagine any long persistence of our ignorance about disease mechanisms in the face of all that is being learned about normal cells and tissues. Our time for the application of science on a major scale is approaching rapidly, and medicine will be totally transformed when it happens. The hard problem just ahead will be setting priorities and making choices between options. We will be obliged, as never before in our history, to select alternative possibilities in technology; we will be compelled to make long range predictions as to the outcome of this course or that. In short, we will be thrust into the business of technology assessment just like all the other great national enterprises.

It is a curious position that we are in today, poised as we are between the old world of trial-and-error empiricism, superstition, hunch, and resignation to defeat, and the new world, just ahead, of hard information and applied science. We seem to work, as of now, with three different levels of technology.

1. First, and necessarily foremost, is what might be termed the *high technology* of medicine, equivalent in its sophistication and effectiveness to the high technologies of the physical sciences. It is a curious fact that although the accomplishments here represent the major triumphs of medicine to date, most of us tend to take them for granted. We often forget what they mean for the quality of life in modern society. This is the genuinely decisive technology of modern medicine exemplified best by methods for immunization against diphtheria, pertussis, and various virus diseases, and the contemporary use of antibiotics and chemotherapy for bacterial infections. The capacity to deal effectively with syphilis and tuberculosis represents a milestone in human endeavor, even though full use of this potential has not been made. And there are, of course, other examples: the treatment of endocrinologic disorders with appropriate hormones, the prevention of hemolytic disease of the newborn, the treatment and prevention of various nutritional disorders, and perhaps just around the corner the management of Parkinsonism and sickle-cell anemia. There are other examples and everyone will have his favorite candidates for the list; but the truth is that there are not nearly as many as the public has been led to believe.

The point to be made about this kind of technology—the real high technology of medicine—is that it comes as the result of a genuine understanding of disease mechanisms and, when it becomes available, it is relatively inexpensive, relatively simple, and relatively easy to deliver.

Offhand, I cannot think of any important human disease for which medicine possesses the capacity to prevent or cure where the cost of the technology is itself a major problem. The price is never as high as the cost of managing the same disease during the earlier stages of ineffective technology. If a case of typhoid fever had to be managed today by the best methods of 1935, it would run to a staggering expense. At say around 50 days of hospitalization requiring the most demanding nursing care, with the obsessive concern for details of diet that characterized the therapy of the time, with daily laboratory monitoring and, on occasion, surgical intervention for abdominal catastrophe, I should think $10,000 would be a conservative estimate for the illness as contrasted with today's cost of a bottle of Chloramphenicol and a day or two of fever. The technology that was evolving for poliomyelitis in the early 1950's, just before the emergence of the basic research that made the vaccine possible, provides another illustration. It is the cost of those kinds of technology and their relative effectiveness that must be compared with the cost and effectiveness of the vaccine.

Pulmonary tuberculosis had similar episodes in its history. There was a sudden enthusiasm for the surgical removal of the infected lung tissue in the early 1950's, and elaborate plans were being made for new and expensive installations for major pulmonary surgery in tuberculosis hospitals. Then, the drug isoniazid and the antibiotic streptomycin were discovered and the hospitals were closed.

It is when physicians are bogged down by their incomplete technologies and by the innumerable tasks they are obliged to do in medicine when they lack a clear understanding of disease mechanisms, that the deficiencies of the health care system are most conspicuous.

2. This brings me to the second level of technology in this classification, which I have termed the *halfway technology* of medicine. This represents what must be done after the fact in efforts to compensate for the incapacitating effects of certain diseases whose courses about which we are unable to do very much. It is a technology designed to make up for disease or to postpone death.

The outstanding examples in recent years are the transplantations of hearts, kidneys, livers, and other organs, and the equally spectacular inventions of artificial organs. In the public mind, this kind of technology now seems like the equivalent of the high technologies in the physical sciences. The media tend to present each new procedure as though it represented a breakthrough and therapeutic triumph instead of the makeshift that it really is.

In fact, this level of technology is, by its nature, at the same time highly sophisticated and profoundly primitive. It is the kind of thing that we must continue to do until there is a genuine understanding of the mechanisms involved in disease. In chronic glomerulonephritis, for example, a much clearer insight will be needed into the events leading to the destruction of capillaries in the kidneys. There is solid evidence that abnormal immunologic reactions are the basis for this destruction. If more information can be obtained, it should become possible to intervene intelligently to prevent the process, or turn it around. When this level of understanding has been reached, the technology of kidney replacement will not be much needed and should no longer pose the huge problems of logistics, cost, and ethics that it poses today.

An extremely complex and costly technology for the management of coronary heart disease has evolved involving specialized ambulances and hospital units, all kinds of electronic gadgetry, and whole platoons of new professional personnel to deal with the end results of coronary thrombosis. Almost everything offered today for the treatment of heart disease is at this level of technology, with the transplanted and artificial hearts as ultimate examples. When enough has been learned to know what really goes wrong in heart disease, we ought to be in a position to figure out ways to prevent or reverse the process, and when this happens the current elaborate technology will be set to one side.

The impending development of an artificial heart illustrates the kind of dilemma we are placed in by today's emphasis on halfway technology. Let us assume that heart disease, for all its manifold origins and its complexity, does represent an approachable scientific problem—that if we study the matter with sufficient imagination and energy, making use of all the new information about muscle structure and function and blood coagulation and lipid metabolism, and making capital use of new information along other lines as yet unguessed by any of us, we will eventually solve this problem and we will then learn how to intervene before the onset of irreversible muscle or valve disease, to prevent the process, or to turn it around. As a non-cardiologist, an outsider, I have total confidence that this can be done, that sooner or later it will be done, and my colleagues who know a lot about heart disease have, I sense, this same kind of confidence for the long term. This, then, is one option, and an altogether wise one to adopt. But the artificial heart represents a completely different attitude, basically opposed. To be willing to invest the many millions of dollars that will probably be necessary for this one piece of new technology almost demands of its proponents the conviction that heart disease represents an unapproachable, insoluble biological problem. It assumes that the best we will be able to do, within the next few decades, is to wait until the underlying mechanisms of heart disease have had their free run, until the organ has been demolished, and then to put into the chest this nuclear-powered, plastic-and-metal, essentially hideous engine. Even if it works, which I am afraid is not at all unlikely, I cannot imagine how society will solve the problems of cost, distribution, and priority. Who will be entitled to buy and have installed these engines—those with enough wealth to pay for them?—those who strike the rest of us, or our committees, as potentially useful citizens? Once we have started on this endless line of unsoluble problems, there may be no turning back. If ever there were an urgent, overwhelmingly important problem in biomedical science, it is with us now: someone simply *must* provide us with a quick solution to the problem of coronary atherosclerosis. If this can be done, the artificial heart can become, overnight, an interesting and ingenious contraption, something clever and decorative but no longer a practical thing, with some of the charm of a Tiffany lamp—a sort of instant antique, and we will all be the better off for this transformation. Otherwise, we are in for real trouble, just ahead, and I'm not sure we have the collective intelligence in medicine to deal with it.

Much of what is done in the treatment of cancer, by surgery, irradiation and chemotherapy, represents halfway technology, in the sense that these measures are directed at the existence of already established cancer cells, but not at the mechanisms by which cells become neoplastic. The policy problems that confront us now, with the nation's declared commitment to conquer cancer, are somewhat like those involved in the artificial heart question. There will be, for a while, anyway, a running argument between two opposing forces. There will be, on one side, those who believe that cancer is a still unsolved but eminently approachable scientific puzzle, requiring only enough good research by imaginative investigators, on a broad enough biological base. Provided with enough financial support and enough time, we will find ourselves home and dry. On the other side, there will be those who believe themselves to be more practical men of the real world, who feel that we have already come as great a distance toward understanding cancer as we are likely to move for some time. These men think we should give the highest priority to applying, on a much larger scale, what we know today about this disease—that with surgery, chemotherapy, and radiation we can now cure or palliate a considerable number of patients, and what we need at this time is more and better technology of essentially today's model. I do not know how this argument will come out, but I believe it is to be an issue of crucial, symbolic significance; whichever way it goes, it is possible that this will be the drift of biomedical science for the next decade. Personally, I would prefer the middle ground, for I like a comfortable position, but I am afraid that I belong with the first group of extremists in this one—for I regard cancer as an entirely open, entirely unsolved problem, wide open to research, and soluble, and I regard the technology of today's forms of therapy as paradigms of halfway technology, directed at the end-results of the disease rather than at underlying mechanisms.

It is characteristic of this kind of technology that it costs an enormous amount of money and requires a continuing expansion of hospital facilities. There is no end to the need for new, highly trained people to run the enterprise. And there is really no way out of this, at the present state of knowledge. If the installation of specialized coronary care units results in the extension of life for only a few patients with coronary disease (and there is no question that this technology is effective in a few cases), it seems to me an inevitable fact that as many of these as can be built will be put together, and as much money as can be found will be spent. I do not see that anyone has much choice in this. We are obliged, by the very nature of our professional responsibility, to adopt a new technology that will benefit patients with otherwise untreatable diseases, even when only a very small percentage will be benefitted, even when the costs are very high. Neither we, nor any other sector of our society, control this aspect of our economy. We cannot, like other industries, withhold a technology from the marketplace because it costs too much money or benefits too small a percentage of patients. The only thing that can move medicine away from this level of technology is new information, and the only imaginable source of this information is research.

The best we can do when the economic or logistic problems associated with our technology verge on the unsupportable, or when the odds are too high against the success of our procedures, is to try to improve the technology or to discover an altogether new technology as quickly as possible. Meanwhile, however, we must continue to employ the less than satisfactory ones.

3. This brings me to the third level of technology—the large body best termed *non-technology*. It is, in effect, the substitute for technology which medicine has always been compelled to use when we are unable to alter either the natural course of disease or its eventual outcome. A great deal of money is spent on this. It is valued highly by the professionals as well as the patients and consists of what is sometimes called "supportive therapy." It tides patients over through diseases that are not understood by and large. It is what is meant by "caring for" and "standing by," and is absolutely indispensable. It is not, however, a technology in any real sense.

It includes the large part of any good doctor's time that is taken up with simply providing reassurance, explaining to patients who fear that they have contracted one or another lethal disease that they are, in fact, quite healthy.

It is what physicians used to be engaged in at the bedside of patients with diphtheria, meningitis, poliomyelitis, lobar pneumonia, and all the rest of the infectious diseases that have since come under control.

It is what physicians must now do for patients with intractable cancer, severe rheumatoid arthritis, multiple sclerosis, stroke, and advanced cirrhosis. One can think of at least 20 major diseases that require this kind of supportive medical care because of the absence of an effective technology. In this category I would include a large amount of what is called mental disease and most varieties of cancer.

The cost of this nontechnology is very high and getting higher all the time requiring not only a great deal of time but also very hard effort and skill on the part of physicians. Only the very best of doctors are good at coping with this kind of defeat. It also involves long periods of hospitalization, a great deal of nursing, and involvement of non-medical professionals in and out of the hospital. It represents, in short, a substantial segment of today's expenditures for health. It is not as great a financial problem for the future as halfway technology, but between them, non-technology and halfway technology will sooner or later drive any system of health care that we may devise into bankruptcy.

If I were a policy-maker interested in saving money for health care over the long haul, I would regard it as an act of prudence to give high priority to a lot more basic research in biological science. This is the only way to get the full mileage that biology owes to the science of medicine even though it seems, as it used to be said in the days when the phrase still had some meaning, like asking for the moon.

Finally, I'd like to make a brief comment on biomedical science planning. This is an especially lively topic at the moment because of the immediate implications for national science policy. It is administratively fashionable in Washington to attribute the delay of applied science in medicine to a lack of systematic planning. Under a new kind of management, it is said that with more businesslike attention to the invention of practical applications we should arrive at our targets more quickly and, it is claimed as a bonus, more economically. Targeting is the new word. We need more targeted research, more mission-oriented science. And maybe less basic research—maybe considerably less. This is said to be the new drift.

One trouble with this view is that it attributes to biology and medicine a much greater store of usable information with coherence and connectedness than actually exists. In real life, the biomedical sciences have not yet reached the stage of any kind of general applicability to disease mechanisms. In some respects we are like the physical science of the early 20th century, booming along into new territory but without an equivalent for the engineering of that time. It is possible that we may be on the verge of developing a proper applied science, but it has to be said that we don't have one yet. The important question before the policy-makers is whether this should be allowed to occur naturally as a matter of course, or whether it can be ordered up more quickly under the influence of management and money.

There are risks. We may be asking for more of the kind of trouble with which we are already too familiar. There is a trap here that has enmeshed medicine for all the millenia of its professional existence. It has been our perpetual habit to try anything on the slimmest of chances, the thinnest hopes, empirically and wishfully. We have proven to ourselves over and over again that the approach doesn't work well. There is no question about our good intentions in this matter; we all hanker, collectively, to become applied scientists as soon as we can—overnight, if possible.

It takes some doing, however. Everyone forgets how long and hard the work must be before the really important applications become applicable. The great contemporary achievement of modern medicine is the technology of controlling and preventing bacterial infection, but this did not fall into our laps with the appearance of penicillin and the sulfonamides. It had the beginnings in the final quarter of last century, and decades of the most painstaking and demanding research were required before the etiology of pneumonia, scarlet fever, meningitis, and the rest could be worked out. Generations of energetic and imaginative investigators exhausted their entire lives on the problems. It overlooks a staggering amount of basic research to say that modern medicine began with the era of antibiotics.

We have to face, in whatever discomfort, the real possibility that the level of insight into the mechanisms of today's unsolved diseases—schizophrenia, for instance, cancer, or stroke—may be comparable to the situation for infectious disease in 1875, with similarly crucial bits of information still unencountered. We could be that far away in the work to be done, if not in the years to be lived through. If this is the prospect, or anything like this, all ideas about better ways to speed things up should be given open-minded, close scrutiny.

Long-range planning and organization on a national scale are obviously essential. There is nothing unfamiliar about this; indeed, we've been engaged in a coordinated national effort for over two decades through the established processes of the National Institutes of Health. Today's question is whether the plans are sharply focussed enough, the organization sufficiently tight. Do we need a new system of research management, with all the targets in clear display, at which we should aim?

This would seem reassuring and tidy. There are some important disease problems where it has already been done effectively demonstrating that the direct, frontal approach does work. Polio is the most spectacular example. Once it had been learned (from basic research) that there were three antigenic types of virus and that they could be abundantly grown in tissue culture, it became a certainty that a vaccine could be made. Not to say that the job would be easy, or in need of any less rigor and sophistication than the previous research; simply that it could be done. Given the assumption that experiments would be carried out with technical perfection, the vaccine was a sure thing. It was an elegant demonstration of how to organize applied science and for this reason it would have been a surprise if it had not succeeded.

This is the element that distinguishes applied science from basic. Surprise is what makes the difference. When you are organized to apply knowledge, set up targets, and produce a usable product, you require a high degree of *certainty* from the outset. All the facts on which you base protocol must be recognisably hard facts with unambiguous meaning. The challenge is to plan the work and organize the workers so that it will come out precisely as predicted. For this, you need centralized authority, elaborately detailed time schedules, and some sort of reward system based on speed and perfection. But most of all you need the intelligible basic facts to begin with, and these must come from basic research. There is no other source.

In basic research, everything is just the opposite. What you need at the outset is a high degree of *uncertainty;* otherwise it isn't likely to be an important problem. You start with an incomplete roster of facts, characterized by their ambiguity; often the problem consists of discovering the connections between unrelated pieces of information. You must plan experiments on the basis of probability, even bare possibility, rather than certainty. If an experiment turns out precisely as predicted, this can be very nice, but it is only a great event if at the same time it is a surprise. You can measure the quality of the work by the intensity of astonishment. The surprise can be because it *did* turn out as predicted (in some lines of research, one percent is accepted as a high yield), or it can be confoundment because the prediction was wrong and something totally unexpected turned up, changing the look of the problem and requiring a new kind of protocol. Either way you win.

I believe, on hunch, that an inventory of our major disease problems based on this sort of classification would show a limited number of significant questions for which the predictable answers carry certainty. It might be a good idea, when Commissions go to work laying out long-range plans for disease-oriented research, for these questions to be identified and segregated from all the rest, and the logic of operations research should be invaluable for this. There will be disputing among the experts as to what is certain and what is not; perhaps the heat and duration of dispute could be adapted for the measurement of uncertainty. In any case, once a set of suitable questions becomes agreed upon, these can be approached by the most systematic methods of applied science.

However, I have a stronger hunch that the greatest part of the important biomedical research waiting to be done is in the class of basic science. There is an abundance of interesting facts relating to all of our major diseases, and more items of information are coming in steadily from all quarters of biology. The new mass of knowledge is still formless, incomplete, lacking the essential threads of connection, displaying misleading signals at every turn, riddled with blind alleys. There are fascinating ideas all over the place, irresistible experiments beyond numbering, all sorts of new ways into the maze of problems. But every next move is unpredictable, every outcome uncertain. It is a puzzling time, but a very good time.

I am, as I've indicated, an unqualified optimist about the future of medicine provided that we can keep the science going, and going in the right directions. One mistake which we could make, if we are unlucky, is to cut back the financial support of research to such an extent that we begin to lose the critical mass of good minds required for the job. Somehow, I doubt that this is going to happen. I believe that the biological revolution of the past 20 years has launched us on one of the really great events in human history, and I do not see how this can be turned off or turned back even though the pace can be slowed by lack of adequate support. The events that lie ahead are, it seems to me, absolutely inevitable. You cannot accumulate information of such power and profundity about the life of cells and tissues without uncovering the mechanisms of disease at the same time, and this is what I believe is beginning to happen. We are going to learn our way around disease, sooner or later, and this is a new fact of life.

We can make a worse mistake, and delay things for a longer time, by planning the science in the wrong way. If you begin by making the assumption that we know more than we really do, and are ready for full-scale applied science across the board, you can turn off all progress. I hope this will not happen in cancer research where almost all of the really important and interesting problems are matters of high uncertainty awaiting surprise answers. To be sure, there are a few areas of cancer ready for centrally planned applied science—the chemotherapy and radiation treatment of Hodgkin's disease and lymphomas provide good examples of this—but the major part of cancer research is still at the frontier of the unknown and has to be regarded as basic science. The same thing can be said for the problem of chronic nephritis and renal failure, for heart disease, stroke, multiple sclerosis, rheumatoid arthritis, schizophrenia, and all the rest.

The huge difference between our situation today and that of 10 years ago, or even 5 years ago, is that these all seemed then to be impenetrable mysteries and today we can see paths leading to what we think might turn out to be the center in each of them. They are now approachable and are soluble. That is the great challenge that has occurred in these last few years. The mechanisms of disease are soluble problems and now it is a matter of time.

Erwin Chargaff *was born in Austria in 1905 and moved to the United States in 1928. He has made significant contributions to the field of biochemistry. Among them are the discovery of the base pairing regularities in DNA and the demonstration of the existence of different deoxyribonucleic acids in different biological species. Chargaff, who has received many honors, is currently a professor emeritus at Columbia University. He co-edited the three-volume work* The Nucleic Acids.

Stanley Cohen, *a molecular geneticist, is professor and chairman of the Department of Genetics at the Stanford University School of Medicine. In 1973, he and his colleagues reported the first successful gene transplantation experiments. Cohen was also a member of the National Academy of Sciences' committee that first called for a pause in certain types of recombinant DNA studies.*

The following two short articles reflect the depth of the disagreement in the scientific community surrounding recombinant DNA research. Chargaff finds such research dangerous, and laments that "our time is cursed with the necessity for feeble men, masquerading as experts, to make enormously far-reaching decisions." He asks, "Is there anything more far-reaching than the creation of new forms of life?" He questions the wisdom of counteracting irreversibly through certain types of recombinant DNA research "the evolutionary wisdom of millions of years, in order to satisfy the ambition and the curiosity of a few scientists." "My generation," he concludes, "or perhaps the one preceding mine, has been the first to engage, under the leadership of the exact sciences, in a destructive colonial warfare against nature. The future will curse us for it."

Cohen, on the other hand, is quite optimistic about the benefits to be reaped from recombinant DNA research. "Use of recombinant DNA techniques has provided knowledge about how genes are organized into chromosomes and how expression is controlled. With such knowledge we can begin to learn how defects in the structure of such genes alter their function." To Chargaff's warning about counteracting the evolutionary wisdom of nature, Cohen retorts that "it is this so-called evolutionary wisdom that gave us the gene combinations for bubonic plague, smallpox, yellow fever, typhoid, polio, diabetes, and cancer." He concludes that "we must then examine the 'benefit' side of the picture—against the vague fear of the unknown that has in my opinion been the focal point of this controversy."

On the dangers of genetic meddling

ERWIN CHARGAFF

A bizarre problem is posed by recent attempts to make so-called genetic engineering palatable to the public. Presumably because they were asked to establish "guidelines," the National Institutes of Health have permitted themselves to be dragged into a controversy with which they should not have had anything to do. Perhaps such a request should have been addressed to the Department of Justice. But I doubt that they would have wanted to become involved with second-degree molecular biology.

Although I do not think that a terrorist organization ever asked the Federal Bureau of Investigation to establish guidelines on the proper conduct of bombing experiments. I do not doubt what the answer would have been; namely, that they ought to refrain from doing anything unlawful. This also applies to the case under discussion: no smokescreen, neither P3 nor P4 containment facilities, can absolve an experimenter from having injured a fellow being. I set my hope in the cleaning women and the animal attendants employed in laboratories playing games with "recombinant DNA"; in the law profession, which ought to recognize a golden opportunity for biological malpractice suits; and in the juries that dislike all forms of doctors.

In pursuing my quixotic undertaking—fighting windmills with an M.D. degree—I shall start with the cardinal folly, namely, the choice of *Escherichia coli* as the host. Permit me to quote from a respected textbook of microbiology: "*E. coli* is referred to as the 'colon bacillus' because it is the predominant facultative species in the large bowel."[1] In fact, we harbor several hundred different varieties of this useful microorganism. It is responsible for few infections but probably for more scientific papers than any other living organism. If our time feels called upon to create new forms of living cells—forms that the world has presumably not seen since its onset—why choose a microbe

[1] B. D. Davis, R. Dulbecco, H. N. Eisen, et al., *Microbiology* (New York: Harper & Row, Publishers, Inc., 1967), p. 769.

that has cohabited, more or less happily, with us for a very long time indeed? The answer is that we know so much more about *E. coli* than about anything else, including ourselves. But is this a valid answer? Take your time, study diligently, and you will eventually learn a great deal about organisms that cannot live in men or animals. There is no hurry, there is no hurry whatever.

Here I shall be interrupted by many colleagues who assure me that they cannot wait any longer, that they are in a tremendous hurry to help suffering humanity. Without doubting the purity of their motives, I must say that nobody has, to my knowledge, set out clearly how he plans to go about curing everything from alkaptonuria to Zenker's degeneration, let alone replacing or repairing our genes. But screams and empty promises fill the air. "Don't you want cheap insulin? Would you not like to have cereals get their nitrogen from the air? And how about green man photosynthesizing his nourishment: 10 minutes in the sun for breakfast, 30 minutes for lunch, and 1 hour for dinner?" Well, maybe Yes, maybe No.

If Dr. Frankenstein must go on producing his little biological monsters—and I deny the urgency and even the compulsion—why pick *E. coli* as the womb? This is a field where every experiment is a "shotgun experiment," not only those so designated; and who knows what is really being implanted into the DNA of the plasmids which the bacillus will continue multiplying to the end of time? And it will eventually get into human beings and animals despite all the precautions of containment. What is inside will be outside. Here I am given the assurance that the work will be done with enfeebled lambda and with modified, defective *E. coli* strains that cannot live in the intestine. But how about the exchange of genetic material in the gut? How can we be sure what would happen once the little beasts escaped from the laboratory? Let me quote once more from the respected textbook: "Indeed, the possibility cannot be dismissed that genetic recombination in the intestinal tract may even cause harmless enteric bacilli occasionally to become virulent."[2] I am thinking, however, of something much worse than virulence. We are playing with hotter fires.

It is not surprising, but it is regrettable that the groups that entrusted themselves with the formulation of "guidelines," as well as the several advisory committees, consisted exclusively, or almost exclusively, of advocates of this form of genetic experimentation. What seems to have been disregarded completely is that we are dealing here much more with an ethical problem than with one in public health, and that the principal question to be answered is whether we have the right to put an additional fearful load on generations that are not yet born. I use the adjective "additional" in view of the unresolved and equally fearful problem of the disposal of nuclear waste. Our time is cursed with the necessity for feeble men, masquerading as experts, to make enormously far-reaching decisions. Is there anything more far-reaching than the creation of new forms of life?

Recognizing that the National Institutes of Health are not equipped to deal with a dilemma of such import, I can only hope against hope for congressional action. One could, for instance, envision the following steps: (i) a complete prohibition of the use of bacterial hosts that are indigenous to man; (ii) the creation of an authority, truly representative of the population of this country, that would support and license research on less objectionable hosts and procedures; (iii) all forms of "genetic engineering" remaining a federal monopoly; (iv) all research eventually being carried out in one place, such as Fort Detrick. It is clear that a moratorium of some sort will have to precede the erection of legal safeguards.

But beyond all this, there arises a general problem of the greatest significance, namely, the awesome irreversibility of what is being contemplated. You can stop splitting the atom; you can stop visiting the moon; you can stop using aerosols; you may even decide not to kill entire populations by the use of a few bombs. But you cannot recall a new form of life. Once you have constructed a viable *E. coli* cell carrying a plasmid DNA into which a piece of eukaryotic DNA has been spliced, it will survive you and your children and your children's children. An irreversible attack on the biosphere is something so unheard-of, so unthinkable to previous generations, that I could only wish that mine had not been guilty of it. The hybridization of Prometheus with Herostratus is bound to give evil results.

Most of the experimental results published so far in this field are actually quite unconvincing. We understand very little about eukaryotic

2. Davis, Dulbecco, Eisen, et al., p. 769.

DNA.[3] The significance of spacer regions, repetitive sequences, and, for that matter, of heterochromatin[4] is not yet fully understood. It appears that the recombination experiments in which a piece of animal DNA is incorporated into the DNA of a microbial plasmid are being performed without a full appreciation of what is going on. Is the position of one gene with respect to its neighbors on the DNA chain accidental or do they control and regulate each other? Can we be sure—to mention one fantastic improbability—that the gene for a given protein hormone, operative only in certain specialized cells, does not become carcinogenic when introduced naked into the intestine? Are we wise in getting ready to mix up what nature has kept apart, namely the genomes of eukaryotic and prokaryotic cells?[5]

3. Eukaryotic DNA is the DNA of higher organisms which are composed of cells with nuclei—for example, animal DNA. [ed. note]

4. Heterochromatin is a kind of chromatin. Chromatin is the material of which chromosomes are made. [ed. note]

5. Prokaryotic cells are cells with no nuclei. All bacteria are prokaryotes. [ed. note]

The worst is that we shall never know. Bacteria and viruses have always formed a most effective biological underground. The guerilla warfare through which they act on higher forms of life is only imperfectly understood. By adding to this arsenal freakish forms of life—prokaryotes propagating eukaryotic genes—we shall be throwing a veil of uncertainties over the life of coming generations. Have we the right to counteract, irreversibly, the evolutionary wisdom of millions of years, in order to satisfy the ambition and the curiosity of a few scientists?

This world is given to us on loan. We come and we go; and after a time we leave earth and air and water to others who come after us. My generation, or perhaps the one preceding mine, has been the first to engage, under the leadership of the exact sciences, in a destructive colonial warfare against nature. The future will curse us for it.

Recombinant DNA: fact and fiction

STANLEY N. COHEN

Almost 3 years ago, I joined with a group of scientific colleagues in publicly calling attention to possible biohazards of certain kinds of experiments that could be carried out with newly developed techniques for the propagation of genes from diverse sources in bacteria *(1)*. Because of the newness and relative simplicity of these techniques *(2)*, we were concerned that experiments involving certain genetic combinations that seemed to us to be hazardous might be performed before adequate consideration had been given to the potential dangers. Contrary to what was believed by many observers, our concerns pertained to a few very specific types of experiments that could be carried out with the new techniques, not to the techniques themselves.

□ From *Science,* 195, 18 February 1977, pp. 654-657.

Guidelines have long been available to protect laboratory workers and the general public against known hazards associated with the handling of certain chemicals, radioisotopes, and pathogenic microorganisms; but because of the newness of recombinant DNA techniques, no guidelines were yet available for this research. My colleagues and I wanted to be sure that these new techniques would not be used, for example, for the construction of streptococci or pneumococci resistant to penicillin, or for the creation of *Escherichia coli* capable of synthesizing botulinum toxin or diphtheria toxin. We asked that these experiments not be done, and also called for deferral of construction of bacterial recombinants containing tumor virus genes until the implications of such experiments could be given further consideration.

During the past 2 years, much fiction has been written about "recombinant DNA research." What began as an act of responsibility by scientists, including a number of those involved in the development of the new techniques, has become the breeding ground for a horde of publicists—most poorly informed, some well-meaning, some self-serving. In this article I attempt to inject some relevant facts into the extensive public discussion of recombinant DNA research.

SOME BASIC INFORMATION

Recombinant DNA research is not a single entity, but rather it is a group of techniques that can be used for a wide variety of experiments. Much confusion has resulted from a lack of understanding of this point by many who have written about the subject. Recombinant DNA techniques, like chemicals on a shelf, are neither good nor bad per se. Certain experiments that can be done with these techniques are likely to be hazardous (just as certain experiments done with combinations of chemicals taken from the shelf will be hazardous), and there is universal agreement that such recombinant DNA experiments should not be done. Other experiments in which the very same techniques are used—such as taking apart a DNA molecule and putting segments of it back together again—are without conceivable hazard, and anyone who has looked into the matter has concluded that these experiments can be done without concern.

Then, there is the area "in between." For many experiments, there is no evidence of biohazard, but there is also no certainty that there is not a hazard. For these experiments, guidelines have been developed in an attempt to match a level of containment with a degree of hypothetical risk. Perhaps the single point that has been most misunderstood in the controversy about recombinant DNA research, is that discussion of "risk" in the middle category of experiments relates entirely to hypothetical and speculative possibilities, not expected consequences or even phenomena that seem likely to occur on the basis of what is known. Unfortunately, much of the speculation has been interpreted as fact.

There is nothing novel about the principle of matching a level of containment with the level of anticipated hazard; the containment procedures used for pathogenic bacteria, toxic substances, and radioisotopes attempt to do this. However, the containment measures used in these areas address themselves only to known hazards and do not attempt to protect against the unknown. If the same principle of protecting only against known or expected hazards were followed in recombinant DNA research, there would be no containment whatsoever except for a very few experiments. In this instance, we are asking not only that there be no evidence of hazard, but that there be positive evidence that there is no hazard. In developing guidelines for recombinant DNA research, we have attempted to take precautionary steps to protect ourselves against hazards that are not known to exist—and this unprecedented act of caution is so novel that it has been widely misinterpreted as implying the imminence or at least the likelihood of danger.

Much has been made of the fact that even if a particular recombinant DNA molecule shows no evidence of being hazardous at the present time, we are unable to say for certain that it will not devastate our planet some years hence. Of course this view is correct; similarly, we are unable to say for certain that the vaccines we are administering to millions of children do not contain agents that will produce contagious cancer some years hence, we are unable to say for certain that a virulent virus will not be brought to the United States next winter by a traveler from abroad, causing a nationwide fatal epidemic of a hitherto unknown disease—and we are unable to say for certain that novel hybrid plants being bred around the world will not suddenly become weeds that will overcome our major food crops and cause worldwide famine.

The statement that potential hazards could result from certain experiments involving recombinant DNA techniques is akin to the statement that a vaccine injected today into millions of people *could* lead to infectious cancer in 30 years, a pandemic caused by a traveler-borne virus *could* devastate the United States, or a new plant species *could* uncontrollably destroy the world's food supply. We have no reason to expect that any of these things will happen, but we are unable to say for certain that they will not happen. Similarly, we are unable to guarantee that any of man's efforts to influence the earth's weather, explore space, modify crops, or cure disease will not carry with them the seeds for the ultimate destruction of civilization. Can we in fact point to one major area of human activity where one can say *for certain* that there is zero risk? Potentially, we could respond to such risks by taking measures such as prohibiting foreign travel to reduce the hazard of deadly virus importation and stopping experimentation with hybrid plants. It is possible to develop plausible "scare scenarios" involving virtually any activity or process, and these would have as much (or as little) basis in fact as most of the scenarios involving recombinant DNA. But we must distinguish fear of the unknown from fear that has some basis in fact; this appears to be the crux of the controversy surrounding recombinant DNA.

Unfortunately, the public has been led to believe that the biohazards described in various scenarios are likely or probably outcomes of recombinant DNA research. "If the scientists themselves are concerned enough to raise the issue," goes the fiction, "the problem is probably much worse than anyone will admit." However, the simple fact is that there is no evidence that a bacterium carrying any recombinant DNA molecule poses a hazard beyond the hazard that can be anticipated from the known properties of the components of the recombinant. And experiments involving genes that produce toxic substances or pose other known hazards are prohibited.

FREEDOM OF SCIENTIFIC INQUIRY

This issue has been raised repeatedly during discussions of recombinant DNA research. "The time has come," the critics charge, "for scientists to abandon their long-held belief that they should be free to pursue the acquisition of new knowledge regardless of the consequences." The fact is that no one has proposed that freedom of inquiry should extend to scientific experiments that endanger public safety. Yet, "freedom of scientific inquiry" is repeatedly raised as a straw-man issue by critics who imply that somewhere there are those who argue that there should be no restraint whatsoever on research.

Instead, the history of this issue is one of self-imposed restraint by scientists from the very start. The scientific group that first raised the question of possible hazard in some kinds of recombinant DNA experiments included most of the scientists involved in the development of the techniques—and their concern was made public so that other investigators who might not have adequately considered the possibility of hazard could exercise appropriate restraint. While most scientists would defend their right to freedom of scientific thought and discourse, I do not know of anyone who has proposed that scientists should be free to do whatever experiments they choose regardless of the consequences.

INTERFERENCE WITH "EVOLUTIONARY WISDOM"

Some critics of recombinant DNA research ask us to believe that the process of evolution of plants, animals, and microbes has remained delicately controlled for millions of years, and that the construction of recombinant DNA molecules now threatens the master plan of evolution. Such thinking, which requires a belief that nature is endowed with wisdom, intent, and foresight, is alien to most post-Darwinian biologists *(3)*. Moreover, there is no evidence that the evolutionary process is delicately controlled by nature. To the contrary, man has long ago modified the process of evolution, and biological evolution continues to be influenced by man. Primitive man's domestication of animals and cultivation of crops provided an "unnatural" advantage to certain biological species and a consequent perturbation of evolution. The later creation by man of hybrid plants and animals has resulted in the propagation of new genetic combinations that are not the products of natural evolution. In the microbiological world, the use of antimicrobial agents to treat bacterial infections and the advent of mass immunization programs against viral disease has made untenable the thesis of delicate evolutionary control.

A recent letter *(4)* that has been widely quoted by critics of recombinant DNA research asks, "Have we the right to counteract irreversibly the evolutionary wisdom of millions of years. . . ?" It is this so-called evolutionary wisdom that gave us the gene combinations for bubonic plague, smallpox, yellow fever, typhoid, polio, diabetes, and cancer. It is this wisdom that continues to give us uncontrollable diseases such as Lassa fever, Marburg virus, and very recently the Marburg-related hemorrhagic fever virus, which has resulted in nearly 100 percent mortality in infected individuals in Zaire and the Sudan. The acquisition and use of all biological and medical knowledge constitutes an intentional and continuing assault on evolutionary wisdom. Is this the "warfare against nature" that some critics fear from recombinant DNA?

HOW ABOUT THE BENEFITS?

For all but a very few experiments, the risks of recombinant DNA research are speculative. Are the benefits equally speculative or is there some factual basis for expecting that benefits will occur from this technique? I believe that the anticipation of benefits has a substantial basis in fact, and that the benefits fall into two principal categories: (i) advancement of fundamental scientific and medical knowledge, and (ii) possible practical applications.

In the short space of 3½ years, the use of the recombinant DNA technology has already been of major importance in the advancement of fundamental knowledge. We need to understand the structure and function of genes, and this methodology provides a way to isolate large quantities of specific segments of DNA in pure form. For example, recombinant DNA methodology has provided us with much information about the structure of plasmids that cause antibiotic resistance in bacteria, and has given us insights into how these elements propagate themselves, how they evolve, and how their genes are regulated. In the past, our inability to isolate specific genetic regions of the chromosomes of higher organisms has limited our understanding of the genes of complex cells. Now use of recombinant DNA techniques has provided knowledge about how genes are organized into chromosomes and how gene expression is controlled. With such knowledge we can begin to learn how defects in the structure of such genes alter their function.

On a more practical level, recombinant DNA techniques potentially permit the construction of bacterial strains that can produce biologically important substances such as antibodies and hormones. Although the full expression of higher organism DNA that is necessary to accomplish such production has not yet been achieved in bacteria, the steps that need to be taken to reach this goal are defined, and we can reasonably expect that the introduction of appropriate "start" and "stop" control signals into recombinant DNA molecules will enable the expression of animal cell genes. On an even shorter time scale, we can expect recombinant DNA techniques to revolutionize the production of antibiotics, vitamins, and medically and industrially useful chemicals by eliminating the need to grow and process the often exotic bacterial and fungal strains currently used as sources for such agents. We can anticipate the construction of modified antimicrobial agents that are not destroyed by the antibiotic inactivating enzymes responsible for drug resistance in bacteria.

In the area of vaccine production, we can anticipate the construction of specific bacterial strains able to produce desired antigenic products, eliminating the present need for immunization with killed or attenuated specimens of disease-causing viruses.

One practical application of recombinant DNA technology in the area of vaccine production is already close to being realized. An *E. coli* plasmid coding for an enteric toxin fatal to livestock has been taken apart, and the toxin gene has been separated from the remainder of the plasmid. The next step is to cut away a small segment of the toxin-producing gene so that the substance produced by the resulting gene in *E. coli* will not have toxic properties but will be immunologically active in stimulating antibody production.

Other benefits from recombinant DNA research in the areas of food and energy production are more speculative. However, even in these areas there is a scientific basis for expecting that the benefits will someday be realized. The limited availability of fertilizers and the potential hazards associated with excessive use of nitrogen fertilizers now limits the yields of grain and other crops, but agricultural experts suggest that transplantation of the nitrogenase system from the chromosomes of certain bacteria into plants or into other bacteria that live symbiotically with food crop plants may eliminate the need for fertilizers. For many years, scientists have modified the heredity of plants by comparatively primitive techniques. Now there is a means of doing this with greater precision than has been possible previously.

Certain algae are known to produce hydrogen from water, using sunlight as energy. This process potentially can yield a virtually limitless source of pollution-free energy if technical and biochemical problems indigenous to the known hydrogen-producing organisms can be solved. Recombinant DNA techniques offer a possible means of solution to these problems.

It is ironic that some of the most vocal opposition to recombinant DNA research has come from those most concerned about the environment. The ability to manipulate microbial genes offers the promise of more effective utilization of renewable resources for mankind's food and energy needs; the status quo offers the prospect of progressive and continuing devastation of the environment. Yet, some environmentalists have been misled into taking what I believe to be an antienvironmental position on the issue of recombinant DNA.

THE NIH GUIDELINES

Even if hazards are speculative and the potential benefits are significant and convincing, wouldn't it still be better to carry out recombinant DNA experiments under conditions that provide an added measure of safety—just in case some of the conjectural hazards prove to be real?

This is exactly what is required under the NIH (National Institutes of Health) guidelines *(5)* for recombinant DNA research:

1) These guidelines prohibit experiments in which there is some scientific basis for anticipating that a hazard will occur. In addition, they prohibit experiments in which a hazard, although it might be entirely speculative, was judged by NIH to be potentially serious enough to warrant prohibition of the experiment. The types of experiment that were the basis of the initial "moratorium" are included in this category; contrary to the statements of some who have written about recombinant DNA research, there has in fact been no lifting of the original restrictions on such experiments.

2) The NIH guidelines require that a large class of other experiments be carried out in P4 (high level) containment facilities of the type designed for work with the most hazardous naturally occurring microorganisms known to man (such as Lassa fever virus, Marburg virus, and Zaire hemorrhagic fever virus). It is difficult to imagine more hazardous self-propagating biological agents than such viruses, some of which lead to nearly 100 percent mortality in infected individuals. The P4 containment requires a specially built laboratory with airlocks and filters, biological safety cabinets, clothing changes for personnel, autoclaves within the facility, and the like. This level of containment is required for recombinant DNA experiments for which there is at present no evidence of hazard, but for which it is perceived that the hazard might be potentially serious if conjectural fears prove to be real. There are at present only four or five installations in the United States where P4 experiments could be carried out.

3) Experiments associated with a still lesser degree of hypothetical risk can be conducted in P3 containment facilities. These are also specially constructed laboratories requiring double door entrances, negative air pressure, and special air filtration devices. Facilities where P3 experiments can be performed are limited in number, but they exist at some universities.

4) Experiments in which the hazard is considered unlikely to be serious even if it occurs still require laboratory procedures (P2 containment) that have for years been considered sufficient for research with such pathogenic bacteria as *Salmonella typhosa, Clostridium botulinum,* and *Cholera vibrio*. The NIH guidelines require that P2 facilities be used for work with bacteria carrying interspecies recombinant DNA molecules that have shown no evidence of being hazardous—and even for some recombinant DNA experiments in which there is substantial evidence of lack of hazard.

5) The P1 (lowest) level of containment can be used only for recombinant DNA molecules that potentially can be made by ordinary biological gene exchange in bacteria. Conformity to even this lowest level of containment in the laboratory requires decontamination of work surfaces daily and after spills of biological materials, the use of mechanical pipetting devices or cotton plugged pipettes by workers, a pest control program, and decontamination of liquid and solid waste leaving the laboratory.

In other areas of actual or potential biological hazard, physical containment is all that microbiologists have had to rely upon; if the Lassa fever virus were to be released inadvertently from a P4 facility, there would be no further barrier to prevent the propagation of this virus which is known to be deadly and for which no specific therapy exists. However, the NIH guidelines for recombinant DNA research have provided for an additional level of safety for workers and the public: This is a system of biological containment that is designed to reduce by many orders of magnitude the chance of propagation outside the laboratory of microorganisms used as hosts for recombinant DNA molecules.

An inevitable consequence of these containment procedures is that they have made it difficult for the public to appreciate that most of the hazards under discussion are conjectural. Because in the past, governmental agencies have often been slow to respond to clear and definite dangers in other areas of technology, it has been inconceivable to scientists working in other fields and to the public at large that an extensive and costly federal machinery would have been established to provide protection in this area of research unless severe hazards were known to exist. The fact that recombinant DNA research has prompted international meetings, extensive coverage in the news media, and governmental intervention at the federal level has been perceived by the public as prima facie evidence that this research must be more dangerous than all the rest. The scientific community's response has been to establish increasingly elaborate procedures to police itself—but these very acts of scientific caution and responsibility have only served to perpetuate and strengthen the general belief that the hazards under discussion must be clear-cut and imminent in order for such steps to be necessary.

It is worth pointing out that despite predictions of imminent disaster from recombinant DNA experiments, the fact remains that during the past 3½ years, many billions of bacteria containing a wide variety of recombinant DNA molecules have been grown and propagated in the United States and abroad, incorporating DNA from viruses, protozoa, insects, sea urchins, frogs, yeast, mammals, and unrelated bacterial species into *E. coli,* without hazardous consequences so far as I am aware. And the majority of these experiments were carried out prior to the strict containment procedures specified in the current federal guidelines.

Despite the experience thus far, it will always be valid to argue that recombinant DNA molecules that seem safe today may prove hazardous tomorrow. One can no more prove the safety of a particular genetic combination under all imaginable circumstances than one can prove that currently administered vaccines do not contain an undetected self-propagating agent capable of producing cancer in the future, or that a hybrid plant created today will not lead to disastrous consequences some years hence. No matter what evidence is collected to document the safety of a new therapeutic agent, a vaccine, a process, or a particular kind of recombinant DNA molecule, one can always conjure up the possibility of future hazards that cannot be disproved. When one deals with conjecture, the number of possible hazards is unlimited; the experiments that can be done to establish the absence of hazard are finite in number.

Those who argue that we should not use recombinant DNA techniques until or unless we are absolutely certain that there is zero risk fail to recognize that no one will ever be able to guarantee total freedom from risk in any significant human activity. All that we can reasonably expect is a mechanism for dealing responsibly with hazards that are known to exist or which appear likely on the basis of information that is known. Beyond this, we can and should exercise caution in any activity that carries us into previously uncharted territory, whether it is recombinant DNA research, creation of a new drug or vaccine, or bringing a spaceship back to Earth from the moon.

Today, as in the past, there are those who would like to think that there is freedom from risk in the status quo. However, humanity continues to be buffeted by ancient and new diseases, and by malnutrition and pollution; recombinant DNA techniques offer a reasonable expectation for a partial solution to some of these problems. Thus, we must ask whether we can afford to allow preoccupation with and conjecture about hazards that are not known to exist, to limit our ability to deal with hazards that do exist. Is there in fact greater risk in proceeding judiciously, or in not proceeding at all? We must ask whether there is any rational basis for predicting the dire consequences of recombinant DNA research portrayed in the scenarios proposed by some. We must then examine the "benefit" side of the picture and weigh the already realized benefits and the reasonable expectation of additional benefits, against the vague fear of the unknown that has in my opinion been the focal point of this controversy.

REFERENCES AND NOTES

1. P. Berg, D. Baltimore, H. W. Boyer, S. N. Cohen, R. W. Davis, D. S. Hogness, D. Nathans, R. Roblin, J. D. Watson, S. Weissman, N. D. Zinder, *Proc. Natl. Acad. Sci. U.S.A.* **71,** 2593 (1974).
2. S. N. Cohen, A. C. Y. Chang, H. W. Boyer, R. B. Helling, *ibid.* **70,** 3240 (1973); S. N. Cohen, *Sci. Am.* **233** (No. 7), 24 (1975).
3. If we accept the view that any natural barriers to the propagation of genetic material derived from unrelated species do not owe their existence to the intent of nature, we can reason that evolution has created and maintained such barriers because opportunities for genetic mixing occur in nature. Furthermore, we must conclude that limitations to gene exchange have evolved because the mixing of genes from diverse organisms is biologically undesirable—not in a moral or theological sense as some observers would have us believe—but to those organisms involved.
4. E. Chargaff, *Science* **192,** 938 (1976).
5. *Fed. Reg.* **41**(176) (9 September 1976), pp. 38426-38483.

Sex control, science and society

AMITAI ETZIONI

■ Amitai Etzioni *is an educator and a sociologist. He developed organizational analysis, a typology based on means used to control participants in organizations and how organizations change, survive, and are integrated into larger social units. He is Professor of Sociology, and chairman of the sociology department, at Columbia University. Among his books are* The Active Society *and* Social Problems.

In the following selection from Genetic Fix, *another of his books, Etzioni is concerned about the freedom of scientific research in this age of technological innovations. He asks: "Are there any circumstances under which the societal well-being justifies some limitation on the freedom of research?" To make this issue more concrete, he focuses on the problem of sex control. Technology may in the future afford parents the choice of the sex of their offspring. The benefits of such an opportunity are obvious. The harmful effects are numerous. Given our societal values, parents would tend to favor having boys. The accumulative effect of the resulting sex imbalance in society would have social, ethical, and political ramifications of great significance. Society would not collapse, but, asks Etzioni, "Are the costs justified?"*

He does not think so. He concludes that "what may have to be considered now is a more preventive and more national effective guidance, one that would discourage the development of those technologies which, studies would suggest, are likely to cause significantly more damage than payoffs."

Using various techniques developed as a result of fertility research, scientists are experimenting with the possibility of sex control, the ability to determine whether a newborn infant will be a male or a female. So far, they have reported considerable success in their experiments with frogs and rabbits, whereas the success of experiments with human sperm appears to be quite limited, and the few optimistic reports seem to be unconfirmed. Before this new scientific potentiality becomes a reality, several important questions must be considered. What would be the societal consequences of sex control? If they are, on balance, undesirable, can sex control be prevented without curbing the freedoms essential for scientific work? The scientific ethics already impose some restraints on research to safeguard the welfare and privacy of the researched population. Sex control, however, might affect the whole society. Are there any circumstances under which the societal well-being justifies some limitation on the freedom of research? These questions apply, of course, to many other areas of scientific inquiry, such as work on the biological code and the experimental use of behavior and thought-modifying drugs. Sex control provides a useful opportunity for discussion of these issues because it presents a relatively "low-key" problem. Success seems fairly remote, and, as we shall see, the deleterious effects of widespread sex control would probably not be very great. Before dealing with the possible societal effects of sex control, and the ways they may be curbed, I describe the work that has already been done in this area.

☐ From *Science*, 161, 13 September 1968, pp. 1107-1112.

THE STATE OF THE ART

Differential centrifugation provided one major approach to sex control. It was supposed that since X and Y chromosomes differ in size (Y is considerably smaller), the sperm carrying the two different types would also be of two different weights; the Y-carrying sperm would be smaller and lighter, and the X-carrying sperm would be larger and heavier. Thus, the two kinds could be separated by centrifugation and then be used in artificial insemination. Early experiments, however, did not bear out this theory. And, Witschi pointed out that, in all likelihood, the force to be used in centrifugation would have to be of such magnitude that the sperm may well be damaged *(1)*.

In the 1950s a Swedish investigator, Lindahl *(2)*, published accounts of his results with the use of counterstreaming techniques of centrifugation. He found that by using the more readily sedimenting portion of bull spermatozoa that had undergone centrifugation, fertility was decreased but the number of male calves among the offspring was relatively high. His conclusion was that the female-determining spermatozoa are more sensitive than the male and are damaged due to mechanical stress in the centrifuging process.

Electrophoresis of spermatozoa is reported to have been successfully carried out by a Soviet biochemist, V. N. Schröder, in 1932 *(3)*. She placed the cells in a solution in which the *p*H could be controlled. As the *p*H of the solution changed, the sperm moved with different speeds and separated into three groups: some concentrated next to the anode, some next to the cathode, and some were bunched in the middle. In tests conducted by Schröder and N. K. Kolstov *(3)*, sperm which collected next to the anode produced six offspring, all females; those next to the cathode—four males and one female; and those which bunched in the center—two males and two females. Experiments with rabbits over the subsequent 10 years were reported as successful in controlling the sex of the offspring in 80 percent of the cases. Similar success with other mammals is reported.

At the Animal Reproduction Laboratory of Michigan State University, Gordon replicated these findings, although with a lower rate of success *(4)*. Of 167 births studied, in 31 litters, he predicted correctly the sex of 113 offspring, for an average of 67.7 percent. Success was higher for females (62 out of 87, or 71.3 percent) than for males (51 out of 80, or 63.7 percent).

From 1932 to 1942, emphasis in sex control was on the acid-alkali method. In Germany, Unterberger reported in 1932 that in treating women with highly acidic vaginal secretions for sterility by use of alkaline douches, he had observed a high correlation between alkalinity and male offspring. Specifically, over a 10-year period, 53 out of 54 treated females are reported to have had babies, and all of the babies were male. In the one exception, the woman did not follow the doctor's prescription, Unterberger reported *(5)*. In 1942, after repeated tests and experiments had not borne out the earlier results, interest in the acid-alkali method faded *(6)*.

It is difficult to determine the length of time it will take to establish routine control of the sex of animals (of great interest, for instance, to cattle breeders); it is even more difficult to make such an estimate with regard to the sex control of human beings. In interviewing scientists who work on this matter, we heard conflicting reports about how close such a breakthrough was. It appeared that both optimistic and pessimistic estimates were vague—"between 7 to 15 years"—and were not based on any hard evidence but were the researchers' way of saying, "don't know" and "probably not very soon." No specific roadblocks which seemed unusually difficult were cited, nor did they indicate that we have to await other developments before current obstacles can be removed. Fertility is a study area in which large funds are invested these days, and we know there is a correlation between increased investment and findings *(7)*. Although most of the money is allocated to birth control rather than sex-control studies, information needed for sex-control research has been in the past a by-product of the originally sponsored work. Schröder's findings, for example, were an accidental result of a fertility study she was conducting (*4*, p. 90). Nothing we heard from scientists working in this area would lead one to conclude that there is any specific reason we could not have sex control 5 years from now or sooner.

In addition to our uncertainty about when sex control might be possible, the question of how it would be effected is significant and also one on which there are differences of opinion. The mechanism for practicing sex control is important because certain techniques have greater psychic costs than others. We can see today, for example, that some methods of contraception are preferred by some classes of people because they involve less psychic "discomfort" for them; for example, the intrauterine device is preferred over sterilization by most women. In the same way, although electrophoresis now seems to offer a promising approach to sex control, its use would entail artificial insemination. And whereas the objections to artificial insemination are probably decreasing, the resistance to it is still considerable *(8)*. (Possibly, the opposition to artificial insemination would not be as great in a sex-control situation because the husband's own sperm could be used.) If drugs taken orally or douches could be relied upon, sex control would probably be much less expensive (artificial insemination requires a doctor's help), much less objectionable emotionally, and significantly more widely used.

In any event both professional forecasters of the future and leading scientists see sex control as a mass practice in the foreseeable future. Kahn and Wiener, in their discussion of the year 2000, suggest that one of the "one hundred technical innovations likely in the next thirty-three years" is the "capability to choose the sex of unborn children'' *(9)*. Muller takes a similar position about gene control in general *(10)*.

SOCIETAL USE OF SEX CONTROL

If a simple and safe method of sex control were available, there would probably be no difficulty in finding the investors to promote it because there is a mass-market potential. The demand for the new freedom to choose seems well established. Couples have preferences on whether they want boys or girls. In many cultures boys provide an economic advantage (as workhorses) or as a form of old-age insurance (where the state has not established it). Girls in many cultures are a liability; a dowry which may be a sizeable economic burden must be provided to marry them off. (A working-class American who has to provide for the weddings of three or four daughters may appreciate the problem.) In other cultures, girls are profitably sold. In our own culture, prestige differences are attached to the sex of one's children, which seem to vary among ethnic groups and classes *(11*, pp. 6-7).

Our expectations as to what use sex control might be put in our society are not a matter of idle speculation. Findings on sex preferences are based on both direct ''soft'' and indirect ''hard'' evidence. For soft evidence, we have data on preferences parents expressed in terms of the number of boys and girls to be conceived in a hypothetical situation in which parents would have a choice in the matter. Winston studied 55 upperclassmen, recording anonymously their desire for marriage and children. Fifty-two expected to be married some day; all but one of these desired children; expectations of two or three children were common. In total, 86 boys were desired as compared to 52 girls, which amounts to a 65 percent greater demand for males than for females *(12)*.

A second study of attitudes, this one conducted on an Indianapolis sample in 1941, found similar preferences for boys. Here, while about half of the parents had no preferences (52.8 percent of the wives and 42.3 percent of the husbands), and whereas the wives with a preference tended to favor having about as many boys as girls (21.8 percent to 25.4 percent), many more husbands wished for boys (47.7 percent as compared to 9.9 percent) *(13)*.

Such expressions of preference are not necessarily good indicators of actual behavior. Hence of particular interest is ''hard'' evidence of what parents actually did—in the limited area of choice they already have: the sex composition of the family at the point they decided to stop having children. Many other and more powerful factors affect a couple's decision to curb further births, and the sex composition of their children is one of them. That is, if a couple has three girls and it strongly desires a boy, this is one reason it will try ''once more.'' By comparing the number of families which had only or mainly girls and ''tried once more'' to those which had only or mainly boys, we gain some data as to which is considered a less desirable condition. A somewhat different line was followed in an early study. Winston studied 5466 completed families and found that there were 8329 males born alive as compared to 7434 females, which gives a sex ratio at birth of 112.0. The sex ratio of the last child, which is of course much more indicative, was 117.4 (2952 males to 2514 females). That is, significantly more families stopped having children after they had a boy than after they had a girl.

The actual preference for boys, once sex control is available, is likely to be larger than these studies suggest for the following reasons. Attitudes, especially where there is no actual choice, reflect what people believe they ought to believe in, which, in our culture, is equality of the sexes. To prefer to produce boys is lower class and discriminatory. Many middle-class parents might entertain such preferences but be either unaware of them or unwilling to express them to an interviewer, especially since at present there is no possibility of determining whether a child will be a boy or a girl.

Also, in the situations studied so far, attempts to change the sex composition of a family involved having more children than the couple wanted, and the chances of achieving the desired composition were 50 percent or lower. Thus, for instance, if parents wanted, let us say, three children including at least one boy, and they had tried three times and were blessed with girls, they would now desire a boy strongly enough to overcome whatever resistance they had to have additional children before they would try again. This is much less practical than taking a medication which is, let us say, 99.8 percent effective and having the number of children you actually want and are able to support. That is, sex control by a medication is to be expected to be significantly more widely practiced than conceiving more children and gambling on what their sex will be.

Finally, and most importantly, such decisions are not made in the abstract, but affected by the social milieu. For instance, in small *kibbutzim* many more children used to be born in October and November each year than any other months because the community used to consider it undesirable for the children to enter classes in the middle of the school year, which in Israel begins after the high holidays, in October. Similarly, sex control—even if it were taboo or unpopular at first—could become quite widely practiced once it became fashionable.

In the following discussion we bend over backward by assuming that actual behavior would reveal a smaller preference than the existing data and preceding analysis would lead one to expect. We shall assume only a 7 percent difference between the number of boys and girls to be born alive due to sex control, coming on top of the 51.25 to 48.75 existing biological pattern, thus making for 54.75 boys to 45.25 girls, or a surplus of 9.5 boys out of every hundred. This would amount to a surplus of 357,234 in the United States, if sex control were practiced in a 1965-like population *(14)*.

The extent to which such a sex imbalance will cause social dislocations is in part a matter of the degree to which the effect will be cumulative. It is one thing to have an unbalanced baby crop one year, and quite another to produce such a crop several years in a row. Accumulation would reduce the extent to which girl shortages can be overcome by one age group raiding older and younger ones.

Some demographers seem to believe in an invisible hand (as it once was popular to expect in economics), and suggest that overproduction of boys will increase the value of girls and hence increase their production, until a balance is attained under controlled conditions which will be similar to the natural one. We need not repeat here the reasons such invisible arrangements frequently do not work; the fact is they simply cannot be relied upon, as recurrent economics crisis in pre-Keynesian days or overpopulation show.

Second, one ought to note the deep-seated roots of the boy-favoring factors. Although there is no complete agreement on what these factors are, and there is little research, we do know that they are difficult and slow to change. For instance, Winston argued that mothers prefer boys as a substitute for their own fathers, out of search for security or Freudian considerations. Fathers prefer boys because boys can more readily achieve success in our society (and in most others). Neither of these factors is likely to change rapidly if the percentage of boys born increases a few percentage points. We do not need to turn to alarmist conclusions, but we ought to consider what the societal effects of sex control might be under conditions of relatively small imbalance which, as we see it, will cause a significant (although not necessarily very high) male surplus, and a surplus which will be cumulative.

SOCIETAL CONSEQUENCES

In exploring what the societal consequences may be, we again need not rely on the speculation of what such a society would be like; we have much experience and some data on societies whose sex ratio was thrown off balance by war or immigration. For example, in 1960 New York City had 343,470 more females than males, a surplus of 68,366 in the 20- to 34-age category alone *(15)*.

We note, first, that most forms of social behavior are sex correlated, and hence that changes in sex composition are very likely to affect most aspects of social life. For instance, women read more books, see more plays, and in general consume more culture than men in the contemporary United States. Also, women attend church more often and are typically charged with the moral education of children. Males, by contrast, account for a much higher proportion of crime than females. A significant and cumulative male surplus will thus produce a society with some of the rougher features of a frontier town. And, it should be noted, the diminution of the number of agents of moral education and the increase in the number of criminals would accentuate already existing tendencies which point in these directions, thus magnifying social problems which are already overburdening our society.

Interracial and interclass tensions are likely to be intensified because some groups, lower classes and minorities specifically *(16)*, seem to be more male-oriented than the rest of the society. Hence while the sex imbalance in a society-wide average may be only a few percentage points, that of some groups is likely to be much higher. This may produce an especially high boy surplus in lower status groups. These extra boys would seek girls in higher status groups (or in some other religious group than their own) *(11)*—in which they also will be scarce.

On the lighter side, men vote systematically

and significantly more Democratic than women; as the Republican party has been losing consistently in the number of supporters over the last generation anyhow, another 5-point loss could undermine the two-party system to a point where Democratic control would be uninterrupted. (It is already the norm, with Republicans having occupied the White House for 8 years over the last 36.) Other forms of imbalance which cannot be predicted are to be expected. "All social life is affected by the proportions of the sexes. Wherever there exists a considerable predominance of one sex over the other, in point of numbers, there is less prospect of a well-ordered social life. . . . Unbalanced numbers inexorably produce unbalanced behavior *(17)*."

Society would be very unlikely to collapse even if the sex ratio were to be much more seriously imbalanced than we expect. Societies are surprisingly flexible and adaptive entities. When asked what would be expected to happen if sex control were available on a mass basis, Davis, the well-known demographer, stated that some delay in the age of marriage of the male, some rise in prostitution and in homosexuality, and some increase in the number of males who will never marry are likely to result. Thus, all of the "costs" that would be generated by sex control will probably not be charged against one societal sector, that is, would not entail only, let us say, a sharp rise in prostitution, but would be distributed among several sectors and would therefore be more readily absorbed. An informal examination of the situation in the USSR and Germany after World War II (sex ratio was 77.7 in the latter) as well as Israel in early immigration periods, support Davis's nonalarmist position. We must ask, though, are the costs justified? The dangers are not apocalyptical; but are they worth the gains to be made?

A BALANCE OF VALUES

We deliberately chose a low-key example of the effects of science on society. One can provide much more dramatic ones; for example, the invention of new "psychedelic" drugs whose damage to genes will become known only much later (LSD was reported to have such effects), drugs which cripple the fetus (which has already occurred with the marketing of thalidomide), and the attempts to control birth with devices which may produce cancer (early versions of the intrauterine device were held to have such an effect). But let us stay with a finding which generates only relatively small amounts of human misery, relatively well distributed among various sectors, so as not to severely undermine society but only add, maybe only marginally, to the considerable social problems we already face. Let us assume that we only add to the unhappiness of seven out of every 100 born (what we consider minimum imbalance to be generated), who will not find mates and will have to avail themselves of prostitution, homosexuality, or be condemned to enforced bachelorhood. (If you know someone who is desperate to be married but cannot find a mate, this discussion will be less abstract for you; now multiply this by 357,234 per annum.) Actually, to be fair, one must subtract from the unhappiness that sex control almost surely will produce, the joy it will bring to parents who will be able to order the sex of their children; but as of now, this is for most, not an intensely felt need, and it seems a much smaller joy compared to the sorrows of the unmatable mates.

We already recognize some rights of human guinea pigs. Their safety and privacy are not to be violated even if this means delaying the progress of science. The "rest" of the society, those who are not the subjects of research, and who are nowadays as much affected as those in the laboratory, have been accorded fewer rights. Theoretically, new knowledge, the basis of new devices and drugs, is not supposed to leave the inner circles of science before its safety has been tested on animals or volunteers, and in some instances approved by a government agency, mainly the Federal Drug Administration. But as the case of lysergic acid diethylamide (LSD) shows, the trip from the reporting of a finding in a scientific journal to the bloodstream of thousands of citizens may be an extremely short one. The transition did take quite a number of years, from the days in 1943 when Hoffman, one of the two men who synthesized LSD-25 at Sandoz Research Laboratories, first felt its hallucinogenic effect, until the early 1960s, when it "spilled" into illicit campus use. (The trip from legitimate research, its use at Harvard, to illicit unsupervised use was much shorter.) The point is that no additional technologies had to be developed; the distance from the chemical formula to illicit composition required in effect no additional steps.

More generally, Western civilization, ever since the invention of the steam engine, has proceeded on the assumption that society must adjust to new technologies. This is a central meaning of what we refer to when we speak about an industrial revolution; we think about a society being transformed and not just a new technology being introduced into a society which continues to sustain its prior values and institutions. Although the results are not an unmixed blessing (for instance, pollution and traffic casualties), on balance the benefits in terms of gains in standards of living and life expectancy much outweigh the costs. (Whether the same gains could be made with fewer costs if society would more effectively guide its transformation and technology inputs, is a question less often discussed [*18*].) Nevertheless we must ask, especially with the advent of nuclear arms, if we can expect such a favorable balance in the future. We are aware that single innovations may literally blow up societies or civilization; we must also realize that the rate of social changes required by the accelerating stream of technological innovations, each less dramatic by itself, may supersede the rate at which society can absorb. Could we not regulate to some extent the pace and impact of the technological inputs and select among them without, by every such act, killing the goose that lays the golden eggs?

Scientists often retort with two arguments. Science is in the business of searching for truths, not that of manufacturing technologies. The applications of scientific findings are not determined by the scientists, but by society, politicians, corporations, and the citizens. Two scientists discovered the formula which led to the composition of LSD, but chemists do not determine whether it is used to accelerate psychotherapy or to create psychoses, or, indeed, whether it is used at all, or whether, like thousands of other studies and formulas, it is ignored. Scientists split the atom, but they did not decide whether particles would be used to produce energy to water deserts or superbombs.

Second, the course of science is unpredictable, and any new lead, if followed, may produce unexpected bounties: to curb some lines of inquiry—because they may have dangerous outcomes—may well force us to forego some major payoffs; for example, if one were to forbid the study of sex control one might retard the study of birth control. Moreover, leads which seem "safe" may have dangerous outcomes. Hence, ultimately, only if science were stopped altogether, might findings which are potentially dangerous be avoided.

These arguments are often presented as if they themselves were empirically verified or logically true statements. Actually they are a formula which enables the scientific community to protect itself from external intervention and control. An empirical study of the matter may well show that science does thrive in societies where scientists are given less freedom than the preceding model implies science must have—for example, in the Soviet Union. Even in the West in science some limitations on work are recognized and the freedom to study is not always seen as the ultimate value. Whereas some scientists are irritated when the health or privacy of their subject curbs the progress of their work, most scientists seem to recognize the priority of these other considerations. (Normative considerations also much affect the areas studied; compare, for instance, the high concern with a cancer cure to the almost complete unwillingness of sociologists, since 1954, to retest the finding that separate but equal education is not feasible.)

One may suggest that the society at large deserves the same protection as human subjects do from research. That is, the scientific community cannot be excused from the responsibility of asking what effects its endeavors have on the community. On the contrary, only an extension of the existing codes and mechanisms of self-control will ultimately protect science from a societal backlash and the heavy hands of external regulation. The intensification of the debate over the scientists' responsibilities with regard to the impacts of their findings is by itself one way of exercising it, because it alerts more scientists to the fact that the areas they choose to study, the ways they communicate their findings (to each other and to the community), the alliances they form or avoid with corporate and governmental interests—all these affect the use to which their work is put. It is simply not true that a scientist working on cancer research and one working on biological warfare are equally likely to come up with a new weapon and a new vaccine. Leads are not that random, and applications are not that readily transferable from one area of application to another.

Additional research on the societal impact of various kinds of research may help to clarify the issues. Such research even has some regulatory impact. For instance, frequently when a drug is shown to have been released prematurely, standards governing release of experimental drugs to mass production are tightened *(19),* which in effect means fewer, more carefully supervised technological inputs into society; at least society does not have to cope with dubious findings. Additional progress may be achieved by studying empirically the effects that various mechanisms of self-regulation actually have on the work of scientists. For example, urging the scientific community to limit its study of some topics and focus on others may not retard science; for instance, sociology is unlikely to suffer from being now much more reluctant to concern itself with how the U.S. Army may stabilize or undermine foreign governments than it was before the blowup of Project Camelot *(20)*.

In this context, it may be noted that the systematic attempt to bridge the "two cultures" and to popularize science has undesirable side effects which aggravate the problem at hand. Mathematical formulas, Greek or Latin terminology, and jargon were major filters which allowed scientists in the past to discuss findings with each other without the nonprofessionals listening in. Now, often even preliminary findings are reported in the mass media and lead to policy adaptations, mass use, even legislation *(21)*, long before scientists have had a chance to double-check the findings themselves and their implications. True, even in the days when science was much more esoteric, one could find someone who could translate its findings into lay language and abuse it; but the process is much accelerated by well-meaning men (and foundations) who feel that although science ought to be isolated from society, society should keep up with science as much as possible. Perhaps the public relations efforts on behalf of science ought to be reviewed and regulated so that science may remain free.

A system of regulation which builds on the difference between science and technology, with some kind of limitations on the technocrats serving to protect societies, coupled with little curbing of scientists themselves, may turn out to be much more crucial. The societal application of most new scientific findings and principles advances through a sequence of steps, sometimes referred to as the R & D process. An abstract finding or insight frequently must be translated into a technique, procedure, or hardware, which in turn must be developed, tested, and mass-produced, before it affects society. While in some instances, like that of LSD, the process is extremely short in that it requires few if any steps in terms of further development of the idea, tools, and procedures, in most instances the process is long and expensive. It took, for instance, about $2 billion and several thousand applied scientists and technicians to make the first atomic weapons after the basic principles of atomic fission were discovered. Moreover, technologies often have a life of their own; for example, the intrauterine device did not spring out of any application of a new finding in fertility research but grew out of the evolution of earlier technologies.

The significance of the distinction between the basic research ("real" science) and later stages of research is that, first, the damage caused (if any) seems usually to be caused by the technologies and not by the science applied in their development. Hence if there were ways to curb damaging technologies, scientific research could maintain its almost absolute, follow-any-lead autonomy and society would be protected.

Second, and most important, the norms to which applied researchers and technicians subscribe and the supervisory practices, which already prevail, are very different than those which guide basic research. Applied research and technological work are already intensively guided by societal, even political, preferences. Thus, while about $2 billion a year of R & D money are spent on basic research more or less in ways the scientists see fit, the other $13 billion or so are spent on projects specifically ordered, often in great detail, by government authorities, for example, the development of a later version of a missile or a "spiced-up" tear gas. Studies of R & D corporations—in which much of this work is carried out, using thousands of professionals organized in supervised teams which are given specific assignments—pointed out that wide freedom of research simply does not exist here. A team assigned to cover a nose cone with many different alloys and to test which is the most heat-resistant is currently unlikely to stumble upon, let us say, a new heart pump, and if it were to come upon almost any other lead, the boss would refuse to allow the team to pursue the lead, using the corporation's time and funds specifically contracted for other purposes.

Not only are applied research and technological developments guided by economic and political considerations but also there is no evidence that they suffer from such guidance. Of course, one can overdirect any human activity, even the carrying of logs, and thus undermine morale, satisfaction of the workers, and their productivity; but such tight direction is usually not exercised in R & D work nor is it required for our purposes. So far guidance has been largely to direct efforts toward specific goals, and it has been largely corporate, in the sense that the goals have been chiefly set by the industry (for example, building flatter TV sets) or mission-oriented government agencies (for instance, hit the moon before the Russians). Some "preventive" control, like the suppression of run-proof nylon stockings, is believed to have taken place and to have been quite effective.

I am not suggesting that the direction given to technology by society has been a wise one. Frankly, I would like to see much less concern with military hardware and outer space and much more investment in domestic matters; less in developing new consumer gadgets and more in advancing the technologies of the public sector (education, welfare, and health); less concern with nature and more with society. The point though is that, for good or bad, technology is largely already socially guided, and hence the argument that its undesirable effects cannot be curbed because it cannot take guidance and survive is a false one.

What may have to be considered now is a more preventive and more national effective guidance, one that would discourage the development of those technologies which, studies would suggest, are likely to cause significantly more damage than payoffs. Special bodies, preferably to be set up and controlled by the scientific community itself, could be charged with such regulation, although their decrees might have to be as enforceable as those of the Federal Drug Administration. (The Federal Drug Administration, which itself is overworked and understaffed, deals mainly with medical and not societal effects of new technologies.) Such bodies could rule, for instance, that whereas fertility research ought to go on uncurbed, sex-control procedures for human beings are not to be developed.

One cannot be sure that such bodies would come up with the right decisions. But they would have several features which make it likely that they would come up with better decisions than the present system for the following reasons: (i) they would be responsible for protecting society, a responsibility which so far is not institutionalized; (ii) if they act irresponsibly, the staff might be replaced, let us say by a vote of the appropriate scientific associations; and (iii) they would draw on data as to the societal effects of new (or anticipated) technologies, in part to be generated at their initiative, while at present—to the extent such supervisory decisions are made at all—they are frequently based on folk knowledge.

Most of us recoil at any such notion of regulating science, if only at the implementation (or technological) end of it, which actually is not science at all. We are inclined to see in such control an opening wedge which may lead to deeper and deeper penetration of society into the scientific activity. Actually, one may hold the opposite view—that unless societal costs are diminished by some acts of self-regulation at the stage in the R & D process where it hurts least, the society may "backlash" and with a much heavier hand slap on much more encompassing and throttling controls.

The efficacy of increased education of scientists to their responsibilities, of strengthening the barriers between intrascientific communications and the community at large, and of self-imposed, late-phase controls may not suffice. Full solution requires considerable international cooperation, at least among the top technology-producing countries. The various lines of approach to protecting society discussed here may be unacceptable to the reader. The problem though must be faced, and it requires greater attention as we are affected by an accelerating technological output with ever-increasing societal ramifications, which jointly may overload society's capacity to adapt and individually cause more unhappiness than any group of men has a right to inflict on others, however noble their intentions.

REFERENCES AND NOTES

1. E. Witschi, personal communication.
2. P. E. Lindahl, *Nature* 181, 784 (1958).
3. V. N. Schröder and N. K. Koltsov, *ibid.* 131, 329 (1933).
4. M. J. Gordon, *Sci. Amer.* 199, 87-94 (1958).
5. F. Unterberger, *Deutsche Med. Wochenschr.* 56, 304 (1931).
6. R. C. Cook, *J. Hered.* 31, 270 (1940).
7. J. Schmookler, *Invention and Economic Growth* (Harvard Univ. Press, Cambridge, Mass., 1966).
8. Many people prefer adoption to artificial insemination. See G. M. Vernon and J. A. Boadway, *Marriage Family Liv.* 21, 43 (1959).
9. H. Kahn and A. J. Wiener, *The Year 2000: A Frame-*

work for Speculation on the Next Thirty-Three Years (Macmillan, New York, 1967), p. 53.
10. H. J. Muller, *Science* 134, 643 (1961).
11. C. F. Westoff, "The social-psychological structure of fertility," in *International Population Conference* (International Union for Scientific Study of Population, Vienna, 1959).
12. S. Winston, *Amer. J. Sociol.* 38, 226 (1932). For a critical comment which does not affect the point made above, see H. Weiler, *ibid.* 65, 298 (1959).
13. J. E. Clare and C. V. Kiser, *Millbank Mem. Fund Quart.* 29, 441 (1951). See also D. S. Freedman, R. Freedman, P. K. Whelpton, *Amer. J. Sociol.* 66, 141 (1960).
14. Based on the figure for 1965 registered births (adjusted for those unreported) of 3,760,358 from *Vital Statistics of the United States 1965* (U.S. Government Printing Office, Washington, D.C., 1965), vol. 1, pp. 1-4, section 1, table 1-2. If there is a "surplus" of 9.5 boys out of every hundred, there would have been 3,760, 358/100 × 9.5 = 357,234 surplus in 1965.
15. Calculated from C. Winkler, Ed., *Statistical Guide 1965 for New York City* (Department of Commerce and Industrial Development, New York, 1965), p. 17.
16. Winston suggests the opposite but he refers to sex control produced through birth control which is more widely practiced in higher classes, especially in the period in which his study was conducted, more than a generation ago.
17. Quoted in J. H. Greenberg, *Numerical Sex Disproportion: A Study in Demographic Determinism* (Univ. of Colorado Press, Boulder, 1950), p. 1. The sources indicated are A. F. Weber. *The Growth of Cities in the Nineteenth Century,* Studies in History, Economics, and Public Law, vol. 11, p. 85, and H. von Hentig, *Crime: Causes and Conditions* (McGraw-Hill, New York, 1947), p. 121.
18. For one of the best discussions, see E. E. Morison, *Men, Machines and Modern Times* (M.I.T. Press, Cambridge, Mass., 1966). See also A. Etzioni, *The Active Society: A Theory of Societal and Political Processes* (Free Press, New York, 1968), chaps. 1 and 21.
19. See reports in The *New York Times:* "Tranquilizer is put under U.S. curbs; side effects noted," 6 December 1967; "F.D.A. is studying reported reactions to arthritis drug," 19 March 1967; "F.D.A. adds 2 drugs to birth defect list," 3 January 1967. On 24 May 1966, Dr. S. F. Yolles, director of the National Institute of Mental Health, predicted in testimony before a Senate subcommittee: "The next 5 to 10 years . . . will see a hundredfold increase in the number and types of drugs capable of affecting the mind."
20. I. L. Horowitz, *The Rise and Fall of Project Camelot* (M.I.T. Press, Cambridge, Mass., 1967).
21. For a detailed report, see testimony by J. D. Cooper, on 28 February 1967, before the subcommittee on government research of the committee on government operations, United States Senate, 90th Congress (First session on Biomedical Development, Evaluation of Existing Federal Institutions), pp. 46-61.

17

Technology, business, and political economy

The future world disorder: the structural context of crises

DANIEL BELL

In the following article Daniel Bell (who was introduced in Part Two) focuses on two "extraordinary sociological and geopolitical transformations in the social structures of the world" which occurred between 1948 and 1973. The first transformation took place, according to Bell, in Western advanced industrial nations, where there was a transition to a more open and egalitarian society. Two aspects of this transformation are the growth of union power and the spread of the Women's Liberation Movement. The second transformation occurred when the old international order collapsed and a large number of new states emerged. With their emergence new dichotomies in world politics superseded the old East-West dichotomy, and created new challenges for international stability. Underlying both these transformations were two "extraordinary technological revolutions: the revolution in transportation and communication . . . and the rise of the new science-based industries." The first tied the world closer together, while the second brought about "the postindustrial society."

Bell discusses four structural problems that the advanced industrial societies will face in the next decade as a result of the two transformations mentioned earlier. He draws a bleak picture of the situation, concluding that "the existing political structures no longer match the underlying economic and social realities," and that this mismatch may be the source of disintegration.

I

Historians now understand that Metternich—the other one, that is—made a strategic mistake at the Congress of Vienna. His policy was based on the premise that France, which had overrun almost all of Europe with Napoleon's armies, should not have the power to do so again. What he did not see was that in his backyard there would be looming a new and more powerful threat—that of an industrializing Germany.

The lack of foresight was understandable. Germany—to the extent that there was such an entity—had been disunited for almost a thousand years and the existing loose federation showed little promise of uniting. Indeed, if only for reasons of river-valley geography, the centralization of Germany, in any effective form, was not possible before the invention of the railroad.[1]

The cautionary moral of this tale is that today's policy-makers, in their understandable preoccupation with Great Power strategies, the rivalries of ideologies and national passions, the problems of nuclear proliferation and the like—all of which are their more immediate concerns—risk losing sight of changes in underlying contexts. These contexts are today necessarily more sociological than technological, more diffuse and difficult to define. And the issues to which they give rise are on a very different time scale from the crisis situations which flare up in the Middle East or in southern Africa, for example. But they nonetheless

☐ Reprinted with permission from *Foreign Policy*, No. 27(Summer 1977).

shape the problems that decision-makers will have to deal with in the next decade. Any attempts to deal with these issues require the redesign of political and social institutions and so confront both the inadequacies of economic and social knowledge and the resistance of traditions (which have their own justifications) and vested interests and privileged groups (which have great power).

What follows is thus not a forecast of the next decade—it could not be, for it eschews the overt political rivalries of the different powers, as well as such explosive questions as nuclear proliferation—but an effort to sketch the broad socio-economic context which, at its loosest, will constrain policy-makers and pose, in direct form, as yet unresolved dilemmas.

II

From 1948 to 1973, there was a 25-year boom in the world economy which was greater than that of any previous period in economic history. Gross Domestic Product, in real terms, increased by more than three and a half times, a world rate of over 5 per cent a year. Japan's growth was almost double that rate; Britain's was half.[2] This real per capita growth was shared almost equally by about half the world (the middle-income countries—e.g., Brazil and Mexico—being slightly the largest gainers). The very poor countries grew at an annual rate of 1.8 per cent, small in comparison with the others, respectable on the basis of their own past.[3]

The same period saw two extraordinary sociological and geopolitical transformations in the social structures of the world. Within the Western advanced industrial societies, there was the transition to a more open and egalitarian society: the inclusion of disadvantaged groups into the society, the expansion of educational opportunities, the growth of union power, the spread of Social Democratic governments,[4] the enlargment of personal liberties and the tolerance of diverse lifestyles, the spread of Women's Liberation, the increase in public spending on social services—in short, that complex of new social rights which is summed up in the ideas of the Welfare State, what I have called the "revolution of rising entitlements," and the greater freedom in culture and morals. With it came the cultural shocks to the older middle classes and the challenges to authority that arose first in the universities, with the student uprisings, and have spread to many other institutions in the society.

The second transformation, which in historical perspective is of greater import for the future, was the end of the old international order with a rapidity that had been almost entirely unforeseen,[5] and the emergence of a bewilderingly large number of new states of vastly diverse size, heterogeneity, and unevenly distributed resources. As a result of this development, the problem of international stability in the next 20 years will be the most difficult challenge for those responsible for the world polity. Some of the consequences of this transformation have been conceptualized as new North-South divisions, cutting across the East-West divisions which have been the axis of Great Power conflicts for almost all of modern times. Whether this is a useful conceptualization, or as vague and tendentious as the phrase "the Third World," is moot. (As Jean-Francois Revel has wryly observed: most of the South is East, but not all the North is West.) The fact remains that, just as within the advanced industrial societies of the West, so in the world at large, there has been a vast multiplication of new actors, new constituencies, new claimants in the political arenas of the world.

Underlying both these changes (though not determining them) have been two extraordinary technological revolutions: the revolution in transportation and communication which has tied the world together in almost real time[6] and the rise of the new science-based industries of what I have called the postindustrial society. The revolutions have given the Western countries an extraordinary advantage in high technology, and paved the way (if one can handle the huge problems of economic dislocation and displacement) for the transfer of a large part of the routinized manufacturing activities of the world to the less-developed countries.[7]

III

These structural changes, which have been taking place within each advanced industrial society and in the world economy, have created a new kind of "class struggle," with a greater potential for social instability and difficulties of governance than those characteristic of the old industrial order. The expansion everywhere of state-managed or state-directed societies—the most crucial political fact about the third quarter of the twentieth century—has meant the emergence of what Schumpeter years ago ironically called ''fiscal sociology.''

In this situation, the salient social struggles in the advanced—and, one must also say, open and democratic—industrial societies are less between employer and worker, as in the nineteenth and early twentieth centuries, than between organized social groups—syndicalist (such as trade unions), professional (such as academic, medical, scientific research complexes), corporate (business and even nonprofit economic enterprises), and intergovernmental units (states, cities, and counties)—for the allocation of the state budget.[8] And as state tax policy and direct state disbursements become central to the economic well-being of these groups, and as political decision-making rather than the market becomes decisive for a whole slew of economic questions (energy policy, land use, communications policy, product regulation and the like), the control and direction of the political system, not market power, becomes the central question for the society.

The corollary fact, that economic dealings between nations become more subject to national political controls, means that the international political arena becomes the cockpit for overt economic demands by the ''external proletariat'' (to use Toynbee's phrase) of the world against the richer industrial nations. Lin Piao may have perished in the plane crash in outer Mongolia, and China may, in the coming decades, be preoccupied with the building of ''socialism in one country,'' but the call that Lin uttered a decade ago for the periphery of the world system to crush the core is a seismic force that could yet be released.

It is in this context that the worldwide recession which began in 1973 acquires such brutal significance. If the economic growth which has been the means of raising a large portion of the world into the middle class—and also a political solvent to meet the rising expectations of people and finance social welfare expenditures—cannot continue, then the tensions which are being generated will wrack every advanced industrial society and polarize the confrontation between the ''south'' (in all probability tied more and more to the ''east'') and the advanced industrialized, capitalist societies of the West.

The current recession can be interpreted in many ways. From a Marxist point of view, it is one more long swing in the inevitable fluctuations of the business cycle. For an economic historian like W.W. Rostow, we may be entering a new downward turn in the Kondratieff cycle, indicating an exhaustion of technological and investment possibilities. The difficulty with these statements is that they are so general and even contradictory. They do not take into account the structural changes in the character of contemporary capitalism, in particular the key role of the state. They are not responsive to what is the unique and different fact about the 1970s' recession, namely that it arose out of a worldwide inflation and that, as much as anything, it has been the deflationary actions of governments that have been responsible for the drop in industrial production and the rise of unemployment.

If one assembles the evidence about the 1970s' recession, one can see the conjoining of a number of short-term cyclical and longterm factors, with two wholly new elements—the surprise ability of the Organization of Petroleum Exporting Countries (OPEC) to create an effective cartel and to quadruple oil prices, and, ultimately more important, the worldwide synchronization of demand, indicating the emergence of a genuine world economy, which led to the inflationary pressures that brought about the end of the boom.

In one sense, the OPEC oil price rises imply a large-scale international redistribution of income which may continue for many years. This is a factor which every dependent economy has to take into account in estimating its costs and rate of possible growth. It is structural in a narrow political sense. But the synchronization of world-wide demand is a new structural feature of the world economy.

In a crucial sense, the modern era is defined as the shift in the character of economies—and in the nature of modern economic thinking—from supply to demand. For thousands of years, the level of supply (and its low technological foundation) dictated the standard of living. What has been singular about modern life is the emphasis on demand, and the fact that demand has become the engine of economic advance, moving entrepreneurs and inventors into the search for new modes of productivity, new combinations of materials and markets, new sources of supply, and new modes of innovation. The re-entry of a destroyed Germany and Japan into the world economy; the rapid industrialization of Brazil, Mexico, Taiwan,

Korea, Algeria, South Africa, and similar countries; the expanding world trade of the Soviet-bloc countries—the revolution of rising expectations and the urge to get into the middle class—have all produced this extraordinary synchronization. Yet, while we have the genuine foundations of a world economy, we evidentally lack those cooperative mechanisms which can adjust these different pressures, create a necessary degree of stabilization in commodity prices, and smooth the transition to a new international division of labor that would benefit the world economy as a whole. We shall return to this below.

IV

If one looks ahead to the next decade, there are four structural problems that will confront the advanced industrial societies in the effort to maintain political stability and economic advance.

1. *The double bind of advanced economies.* The facts that every society has become so interconnected and interdependent and that the political system has taken on the task of managing, if not directing, the economy mean that, increasingly, "someone" has to undertake the obligation of thinking about the system "as a whole." When the economic realm had greater autonomy, the shocks and dislocations generated through the market could be walled off, or even ignored—though the social consequences were often enormous. But now all major shocks are increasingly *systemic,* and the political controllers must make decisions not for or against particular interests, powerful as these may be, but for the consequences to the system itself.

Yet that very fact increases the inherent double bind in the nature of a democratic or responsive polity. For the state increasingly has the double problem of aiding capital formation and growth (*accumulation,* in the Marxist jargon) and meeting the rising claims of citizens for income security, social services, social amenities, and the like (the problem of *legitimation,* in Max Weber's terminology).

In one sense, this is the fulfillment of a different kind of prediction made by Marx. Already in 1848 to 1849, when he was engaged in political activities in Cologne, he said that once the "democratic revolution" (i.e., the achievement of the franchise and other civil rights) was achieved, the "social revolution" (the transformation of society) would follow. This was the basis for his "right-wing" and "coalitionist" tactics toward the democratic (i.e., bourgeois) groups at the time. What is striking is how long it took for Marx's prediction to come true. The electoral franchises were secured, in most Western European countries, only by the end of the century, and it took 50 years beyond that (facilitated by the structural changes in the economy) for democratic pressure to be turned into social leverage.

In practical fact, this major change has resulted in the sharp rise in governmental expenditures over the last 40 years and in social expenditures in the last decade and a half. Since 1950, the growth in public expenditure, per year, has been between 4.3 per cent in Great Britain, at the low end of the scale, to 11.6 per cent for Italy, at the high end. In these years, the growth in GNP has been from 2.8 per cent a year in Britain to 5.7 per cent in Germany. (Italy was growing at 5.3 per cent a year.) As a share of GNP, public expenditure varies from 30 per cent of GNP in France to 64 per cent in Sweden, which has experienced the highest growth in the 25-year period. (Italy's public expenditure is 58 per cent of GNP, Britain's is 53 per cent, and the United States' is 38 per cent.)

These rates of growth of public expenditures over a quarter of a century, in countries such as Great Britain and Sweden—almost 50 to 75 per cent greater than the growth of GNP—raise some complex economic and social questions. Direct comparisons on the basis of *growth rates* are difficult, since some nations started from a low absolute base of public expenditure. It is too easy to say, as some conservative economists do, that public expenditure is eating up the national patrimony. And it is hard to calculate how much the expenditures on education and health increase the skills and capabilities of individuals in the society. With all that, some questions remain. The Oxford economists Bacon and Eltis have argued that expenditures in the public sector are, inevitably, of lower productivity than a comparable amount spent in the private sector, and that these differential rates account for the slowdown of the British economy. And if these rates of growth of public expenditure continue, who will pay for them? If they cannot be financed from economic growth, then they have to be financed by higher taxes, by inflation (a disguised form of taxation), or by external borrowing.

A recent group of theorists has sought to draw some larger consequences from this state of affairs. Richard Rose calls it "over-load"—the condition in which expectations are greater than the system can produce—and speculates whether nations can go bankrupt. Jurgen Habermas calls it a "legitimation crisis," putting it into the larger philosophical context of political justifications. Under the prevailing tenets of the liberal theory of society, each individual is free to pursue his own interests and the rule of law is only formal and procedural, establishing the rules of the game without being interventionary. But the emerging system of state capitalism lacks the kind of philosophical legitimation that liberalism has provided. Samuel P. Huntington and Samuel Brittan have argued that democracies are becoming increasingly ungovernable, because the "democratization of political demands," in the Schumpeterian sense of the term, is subject to few constraints, or fewer than those represented by the limited credit available to individuals or firms that at some point would have to pay their debts, rather than "postpone" them by increasing the public debt.[9]

I think these diagnoses are all accurate but partial. For the issue concerns not only the democracies but *any* society which seeks economic growth, yet has to balance the needs (if not the public demands) of its citizens for satisfactions and security. The Soviet Union could emphasize growth (a naked "primitive accumulation," in Marx's very sense of the term) by promises of a utopian tomorrow, the brutal repression of its peasants, and the direct and indirect coercion of its workers. But how long could this go on? It is evident that the next generation of Soviet rulers will face more and more demands, open or disguised, for the expansion of social claims, as well as for some influence (particularly among the managerial elites) over the allocation of state budgets.

The problem already exists in Poland, where Gierek—who in that sense faces the same problems as Denis Healey—has to worry about capital formation for the renovation of Polish industry, yet maintains high prices for peasants as inducements to produce, and food subsidies for workers to keep *their* prices down. When he sought to realign the system by raising food prices, as economic logic compelled him to, he had almost a full-scale revolt by the workers on his hands. In fact, one can say that Poland is probably the only real Socialist government in Europe since it is the government most afraid of its working class.

If one searches for a solution, the double bind manifests itself in the fact that inflation or unemployment have become the virtual trade-offs of government policy, and governments are in the difficult position of constantly redefining what is an "acceptable" level of unemployment and an "acceptable" level of inflation. It is compounded by the fact that where there are deflationary pressures, particularly within declining economies, every group seeks to escape the necessary cut in its standard of living or its wealth, so that the pressures toward a greater corporate organization of society (and the ability to use that corporate power for wage indexing or tax advantages) increase, and the heaviest burdens fall on the unorganized sections of the society, largely sections of the poor and the middle classes. The final irony is that with all the money being spent on social expenditures there is an evident sense that the quality of the services is poor, that the social-science knowledge to design a proper health system, or a housing environment, or a good educational curriculum, is inadequate, and that large portions of these moneys are increasingly spent on administrative and bureaucratic costs.

2. *Debt and protectionism.* Almost every Western society, as a result of Keynesian thinking, has stimulated its economy in the last 40 years by means of deficit financing and pump priming (or in the newer, fashionable phrase, "demand management"), with the result that it has incurred ever deeper debts.

According to the earlier theorists of "functional finance," such as A.P. Lerner, debt meant very little in economic terms so long as (a) the amount of debt service was manageable and did not become too large a lien on the society, and (b) a nation could not go bankrupt since it owed the money, really, to "itself" and could always reduce the debt if necessary, so long as it had effective taxing power. In fact, the theory went, a nation, like a giant utility company, would never even "redeem" its debt but continue to roll it over in new borrowings, so long as the debt management level was within "reasonable" limits—an "acceptable" level which, like that of inflation and unemployment, was constantly being redefined.

The difficulty in most countries today is that not only has the "internal" debt level been mounting steadily, but there is also a rising "external" debt which presumably has to be repaid at some point. And it is the combination of the two which seems so threatening to the stability of the international monetary system.

The major problem is the growth of external debt. To meet its obligations, Great Britain has now borrowed about $20 billion dollars, quite a low figure compared to its internal debt. Yet that money has to be repaid. To obtain money from the International Monetary Fund (IMF), Britain (like Italy, which is in a similar situation) has had to comply with various stringencies imposed by the IMF as its "price" for the loan, one of these being even larger cuts in public expenditures than the Labour party had planned.

But the question of external debt is a minor one, as yet, for the advanced industrial societies. The heaviest burdens fall on the non-oil-producing less-developed countries, about a hundred in number. A conservative estimate by the Organization of Economic Cooperation and Development (OECD) in its Economic Outlook of December 1976 puts the figure at roughly $186 billion (some estimates go as high as $220 billion), most of this incurred in recent years as a result of the rise in oil prices. Projections of that debt in 1985 go from nearly $350 billion to $500 billion. For these countries, the ratio of *external* debt to GNP is about 25 per cent; by 1985 it would rise to 45 per cent.

If one takes the conservative figure of the aggregate external debt in 1976 as $190 billion, the deficit trade balance (imports over exports) is about $34 billion, and the debt service about $13 billion. This makes, for 1976, a total of $47 billion as the amount of *additional* external borrowing required. If one takes the scenario to 1985, and an external debt of $500 billion, the projected trade deficit would be $52 billion and the debt service $34 billion, or a requirement of $86 billion in that year from the "richer" countries.

How can this be done? In 1974 to 1976, two-thirds of the Third World's borrowing (of $78 billion) was financed by the recycling of petrodollars through the Western banks. But how long can this continue? Any new loans would have to come from international agencies such as the IMF. But one of the conditions that the IMF usually imposes is that debtor countries reduce or eliminate their payments deficits—and this can be done only by the sizable reduction of imports.

In effect, the very discipline that an IMF would impose could only lead to a heightened economic nationalism and protectionism. This is the very prescription that the British Labour Left (aided by the thinking of the "new" Cambridge school of economists, Wynne Godley, Michale Posner, and Robert Neild) has put forward. Import restrictions, they argue, are preferable to cuts in public expenditure. Too many of the "wrong" things are being imported and, besides, if import controls were being established, domestic industry would take up the slack and produce the necessary items that are now being imported (such as more British cars).

The British Left is advocating a "siege economy." But the pressures for protectionism are evident in almost every country that is feeling the shock of dislocations under competitive pressures. Japan, as every country knows, has subtly kept many foreign products outside its home market, while allegedly "dumping" various products onto other markets. The United States has begun retaliating by raising the tariff on Japanese television sets. American trade unions, once largely for free trade, are now completely protectionist, and the maritime unions have often been successful in their demands that various subsidized exports be carried in American bottoms.

The 1929 world depression came when Britain decamped from international free trade and instituted "imperial preference"; actions soon followed by other countries, such as the United States, going off the gold standard and imposing export controls on capital. None of the present-day pressures exist on the same scale. But there is a great temptation for many countries, Britain included, to have a go at the game of protectionism. As *The Economist* (February 26, 1977) recently commented:

> Economic nationalism will develop first among the poor and the weak, the countries with the largest trade deficits which have least to fear from retaliation. Their governments will put on import restrictions, because they fear to impose socially disastrous and politically dangerous austerity measures at home. The first in this field will gain. But for the world as a whole this will be a negative-sum game. The result will be a further period of serious international recession until inflationary pressures have been purged from the system. When, where and how quickly will this happen?

3. *The demographic tidal wave*. The third structural problem derives from demographic change, particularly in Latin America and Asia. Most demographic discussions have focused on the problem of the size of the world's population by the year 2000—whether it would be six

Area or Country	Population 1975	Population growth rate	% Urban	Inflation rate	Population under 15 (%)
Latin America	327.6	2.9	60.4		43
Mexico	59.3	2.4	63.2	22.5	48
Brazil	113.8	2.9	59.5	32.7	42
Columbia	24.7	3.2	61.8	31	47
Venezuela	12.0	2.9	82.4	11.9	45
Chile	10.7	1.9	83.0	365	40
Argentina	25	1.5	80		29
Asia	2,407.4	2.5			
India	636.2	2.6	21.5	31	42
Bangladesh	79.6	3.0	6.8	100	45
Pakistan	71.6	3.6	26.9	*	44
Indonesia	137.9	2.7	19.3	34.4	45
Philippines	44.7	3.2	36	30	43
Thailand	42.3	3.1	16.5	21.3	46
China	942	2.4	23.5	*	36
Japan	111.9	1.3	75.2		24
Africa	420.1	2.8	24.5		44
Nigeria	81.8	2.5	23.1	12	45
Ethiopia	28.8	2.6	11.2		45
Zaire	24.9	2.8	26.2	29.3	44
Egypt	37.2	2.2	47.7		42
Algeria	16.8	3.3	49.9		47
Europe (excluding USSR)	474.2	0.8	67.2		26
United Kingdom	56.2	0.2	78.2		24
France	53	0.8	76.1		25
W. Germany	62.6	0.5	83.4		23
E. Germany	16.8	−0.4	74.9		21
Poland	34.0	0.9	56.5		25
USSR	254.3	0.9	60.5		29
USA	219.7	1.0	76.3		27

*Not available.

billion or seven billion, and whether the world could sustain those numbers. But in any immediate sense, the year 2000 is not the issue. A scrutiny of the accompanying table shows what *is* urgent: *the percentage of the age cohort now under 15 years of age*. This is a group already alive, which within the next decade will flood the schools and labor markets of the less-developed countries.

If one recalls the events of the 1960s in the West, much of the student unrest was due (not as a cause, but as a condition) to the tidal wave of young people that rolled through the universities in the middle and late 1960s. In the United States, for example, there was no increase at all in the proportion of young people between 17 and 22 in the 1940 to 1950 decade, and no increase at all in the proportion of young people in the following decade. Yet from 1960 to 1968, reflecting the "baby boom" of the early postwar years, the proportion of young people jumped more than 50 per cent. What one found was an increasing self-awareness of the group as a separate "youth culture" (and youth market), an increasing competitiveness to get into the good schools, and, owing to the draft, into graduate and professional schools. This large expansion of an age cohort, combined with the moral ambiguity of the Vietnam war, turned a large part of this generation, particularly its elites, against the society. And a similar process occurred in Western Europe.

If one looks ahead to the next decade, what is striking is the extraordinarily high proportion of young people in Latin America (with the exception of Argentina), Asia (except Japan), and Africa. In Europe, during the 1960s, the large number of "surplus" workers in Turkey, Yugoslavia, Greece, and southern Italy could be drawn "north" by the expanding economies of the Western European tier. (Now large pockets of such workers remain, creating a growing problem for these countries, such as the Turkish knots in West Berlin.) But where will the "surplus" populations of the developing world go in the coming years? The problem is compounded by the fact that there already exists in Latin America a high degree of urbanization, high inflation rates, and high unemployment or underemployment rates. Both Mexico and Brazil, whose industrial production have been growing at the astounding rates of between 12 and 15 per cent a year, are by now almost at the peak of their potential. Yet both face a doubling of the entry rate into schools and the labor force in the next decade.

Mexico, with its highly concentrated population in the Federal District of Mexico City—which contains about a fourth of the entire population of Mexico—is an especially sensitive case. In 1920, Mexico had a population of little more than 14 million persons. Fifty years later, it was more than 60 million (or more than almost every country in Western Eurpoe), and by the end of the century it will probably have at least 100 million persons. The United States is belatedly waking up to the problem of millions of illegal aliens flowing across the border and finding sleazy jobs in small service and manufacturing establishments whose owners welcome the cheap, exploitable labor, since they need not pay large social fringe benefits, and the workers have to be docile lest they be deported. But what is the solution? Is one to string barbed wire across two thousand miles of border? Or engage in periodic dragnets in the major cities of the country? And can Mexico itself, facing these explosive problems of population, escape the risks of military dictatorship when its problems become "unmanageable"? What will foreign capital do under those circumstances? Can any of these questions be met without some form of international migration policy?

4. *Rich and Poor Nations*. The rich and the poor may always be with us, but in what proportions? One of the most striking facts about the period since World War II, in terms of its psychological impact, has been the growth in the world's middle class—using the term, crudely, to mean those who could purchase domestic electrical appliances, have a telephone, buy a car, use a stated amount of energy per capita, etc. According to the calculations of Nathan Keyfitz, between 1950 and 1970, the middle class grew from 200 million to 500 million persons—to about 12.5 per cent of the world's population, or more than 40 per cent if we assume that this growth was largely within the rich and middle income countries.[10] If we were able, in the next 20 years, to maintain that rate—4.7 per cent a year achieved in the best period we have seen in world economic history—about 15 million of the 75 million persons who are being added to the world's population each year would be added to the middle class. But the remaining 60 million would be poor.

Of the many important issues between the rich and the poor nations, perhaps the most sticky, and the real time bomb in international economic relations, is that of industrialization. The goal of the developing countries, stated in the UNIDO Declaration and Plan of Action on Industrial Development and Co-operation issues agreed in Lima in 1975, is that *by the year 2000, the developing countries should account for at least 25 per cent of the world's industrial production*. It is typical of the rhetoric that in the Lima Declaration the term "industry" was not defined, nor was it specified whether "industrial production" means *gross* or *net* industrial output, nor was there even an unambiguous definition of what constituted the group of developing countries!

However, at the United Nations Conference on Trade and Development (UNCTAD) meeting in Nairobi in May 1976, a more serious and specific effort was made to spell out the implications of that target. The paper presented to UNCTAD considers manufacturing only (excluding mining, electricity, gas, and water), defines production as *net* output (value added, or the sector's contribution to Gross Domestic Product), and includes Yugoslavia and Israel within the definition of developing countries.[11]

Taking the growth rates of manufacturing output for the developed-market economies and for the countries of Eastern Europe for 1960 to 1972, the UNCTAD document projects the estimated production values from 1972 to the year 2000 *at those growth rates,* and reaches a figure of $6,500 billion in 1972 dollars. "The Lima Target," declared the document, "postulates that the share of the developing countries in world manufacturing output will increase from a share of 9.3 per cent in 1972 to 25 per cent by the year 2000 which, when applied to the figure given above, yields a value of $2,165 billion. To reach this output volume, *manufacturing output in the developing countries would have to maintain an annual growth rate of over 11 per cent per year—compared with the growth of 6.6 per cent attained during the period 1960-1972—or in other words their manufacturing output would have to be 20 times the output achieved in 1972.*" (emphasis added)

To put that figure in meaningful perspective, the growth rates of manufacturing output in the developed "market-economy countries" from 1960 to 1972 was 5.6 percent and for the "Socialist countries" 9.0 per cent a year. The prospect of reaching the UNCTAD target, even by radical restructuring of the composition of the manufacturing output (i.e., a shift from light to heavy industry), is clearly improbable. The UNCTAD document then draws upon another report, prepared for the International Labor Organization (ILO) conference in June 1976 on Income Distribution and Social Progress and the International Division of Labor. This document deals with the "eradication of absolute poverty" among the hard core of the poor, defined as the poorest 20 per cent of the world's population, and points out that to achieve this target by the year 2000 by economic growth alone would require a "doubling of the already rapid rates of GNP growth in developing countries, a contingency that is considered unlikely."

What, then, is the answer? The ILO report, echoed by the UNCTAD document, states that "*if substantial income redistribution policies were introduced,* most developing countries would appear to achieve the basic needs objective by growth at an annual rate of approximately 7 to 8 per cent," and that "the proposed strategy implies quite high levels of investment, without which there would be neither growth nor meaningful redistribution." The rhetoric is not that of the *Communist Manifesto*. Given the platforms, those of United Nations' agencies, the language is stiff and bureaucratic. Given the proponents, however, the key terms "substantial income redistribution" and "high levels of investment" have a menacing ambiguity. The point, however, is clear. Here is the agenda of international politics for the rest of the century. Whether the proponents of the "new international economic order" have the political or economic strength to enforce these demands, is another question.

V

If one reviews the nature of the structural situations facing the advanced industrial societies in the 1970s, what is striking are the parallels to the 1920s and 1930s. If one looks at the period not in terms of the character of the extremist movements of the time, but to understand why the Center could not hold—from the vantage point of the governments, so to speak—there were four factors that, conjoined, served to reduce the authority of the governments, imperil their legitimacy, and facilitate the destruction of these regimes. These were:

The existence of an "insoluble" problem.

The presence of a parliamentary impasse with no group being able to command a majority.

The growth of an unemployed educated intelligentsia.

The spread of private violence which the ruling regimes were unable to check.

In that period, the "insoluble" problem was unemployment. No government had an answer. The Socialists, when in office, as in Germany in 1930 or England in 1931 could only say (as did Rudolf Hilferding, the most eminent Marxist economist of the time, who served as a minister in the Müller cabinet in 1930) that under capitalism the state could not intervene and one had to let the depression run its course. In England, as Tom Jones, the friend of Ramsay Macdonald, confidant of Stanley Baldwin, and a member of the key Unemployment Board with Sir William Beveridge, noted in his *A Diary with Letters,* no one at the time knew what to do.

The parliamentary impasse arose out of the polarization of parties and, in the Latin countries, the unwillingness of the Socialist parties to enter "bourgeois governments" lest they be co-opted (as a large number of French Socialists from Briand to Mitterand have been) and leave the Socialist movement. Thus in Spain, in Italy, in France, the parliaments were in shambles.

The unemployed intelligentsia consisted of lawyers without clients, doctors without patients, teachers without jobs, the group that Konrad Heiden, the first historian of National Socialism, was to call "the armed Bohemians." The entire first layer of the Nazi party leadership, Goebbels, Rosenberg, Strasser, were of this stripe.

The spread of private violence arose out of the private armies of the extremist groups—the Black Shirts, the Brown Shirts, the Communists, with their own grey and red uniformed detachments, and even the Socialists with their Schutzbund in Austria—and the efforts of these groups to control the "streets" and carry out their demonstrations.

The result, of course, was the rise of authoritarian and Fascist regimes in Portugal, Italy, Germany, Austria, and Spain, and the menacing threat of Fascist movements in France (de la Roque and the Cagoulards), in Belgium (Degrelle), and the Great Britain (Mosley). In these instances, the decisive support came from the middle class, which feared being declassed, and the traditionalist elements, which feared the rising disorder. When Hans Fallada asked, in the famous title of his novel, "Little Man, What Now?" the answer was a right-wing reaction as preferable to left-wing Bolshevism. The Center no longer had a chance in most of these countries.

If one looks at the situation in the 1970s, there are some sinister parallels. The insoluble problem is inflation. Few of the economists, once so sure of their mastery of policy, now can agree upon an answer; and to the extent that there is one, it is reminiscent of the old answer of Hilferding: a deflationary policy that takes its toll by unemployment. To reduce the fever, one resorts to amputation. With continuing or a yo-yo inflation, there is rising anxiety, especially in the middle classes. With high levels of unemployment, the young, the blacks, and the poor suffer most.

The parliamentary impasse is reflected in the fact that there is not a single majority government in Western Europe. Every country is ruled by a coalition of parties, no single one of which commands a majority on its own. In England, France, and Italy the ruling governments are led by minority parties that often dare not act, or cannot govern effectively.

The increase in the educated intelligentsia is an obvious fact in every Western country, a product of demographic idiosyncracy and deflationary cuts in public expenditures, but an explosive force no less, as is being shown in Italy today.

The private violence of the 1920s and 1930s is replaced by urban terrorism, fitful and sporadic in most cases, yet sufficiently menacing in Northern Ireland to turn that country into a garrison state.

No parallels are ever historically exact, and they can mislead as often as help, as we have seen by the occasions when words like "Munich" or the "betrayal of Ramsay Macdonald" are invoked. Yet, distorting mirrors though they may be, they allow us to see what may be similar and what may be different.

Even with the growing anxieties of the middle class, as in Denmark and Sweden and England, and, less obviously, in France and Italy, it is highly unlikely that any of the European countries will go Fascist, or see a strong right-wing reaction. These movements are too discredited politically and would lack any historical legitimacy. What is more likely to happen in Europe, as well as in many other countries, is *fragmentation*—both in geographical terms and as a result of the unraveling of the society in functional terms.

There are two reasons for the greater possibility of fragmentation as the likely response in the coming decade, and they are clearly visible. One is that most societies have become more self-consciously *plural* societies (defined in ethnic terms) as well as *class* societies. The resurgence of minority-group consciousness in almost every section of the world—in national, linguistic, religious, and communal terms—shows that ethnicity has become a salient political mechanism for hitherto disadvantaged groups to assert themselves. The second reason is that in a world marked by greater economic interdependence, yet also by a growing desire of people to participate at a local level in the decisions that affect their lives, *the national state has become too small for the big problems in life, and too big for the small problems.* In economic terms, enterprises seek regional or transnational locations, moving their capital and often their plants where there is the greatest comparative advantage. In sociological terms, ethnic and other groups want more direct control over decisions and seek to reduce government to a size that is more manageable for them.

The threat of *geographical* fragmentation can be seen in the United Kingdom, with possible devolution for Scotland and Wales; in Northern Ireland, with the bitter religious fratricide; in Belgium, with the traditional enmity of the Flemish and the Walloons; in Canada, on the linguistic issue between the French in Quebec and the English-speaking groups in the other provinces; in France, where there are small separatist movements in Corsica and Brittany; in Spain, with the traditional claims for Catalonian and Basque autonomy; in Yugoslavia, where there are the smoldering rivalries of the Serbs, Croats, Slovenes, and Montenegrins; in

Lebanon, where the binational state has fallen apart and become a client of Syria. Pakistan split apart into West Pakistan and Bangladesh. Nigeria has just survived a civil war, overcoming the threat of Biafran succession. In various African countries, in the landlocked areas of the Sudan, and Rwanda-Burundi, whose tribes and peoples are being quitely slaughtered, almost unnoticed.

Nor is the Soviet bloc immune. Politically, there has been a very real fragmentation in the loss of the earlier Stalinist hegemony over the countries of Eastern Europe and the European Communist parties. The unrest is ever latent in Poland and in Czechoslovakia. Within the Soviet Union, there is the evident unease at the shifting demographic balances that, by the year 2000, will make the Great Russians a minority in the Soviet world, and will produce a piquant situation where three of every ten recruits for the Soviety army will be Muslim.

Functionally, fragmentation consists of the effort of organized corporate groups to exempt themselves from the incomes policies that regimes inevitably have to resort to, in one way or another—through an overt social contract or through the tax mechanism—in order to reduce inflation. There is the likelihood in many countries of the breakup of the party systems. Though such structures have a powerful life of their own, in many countries they evidentiy do not reflect underlying voter sentiment. In Britain, the majority of people are for the "center," yet the party machines fall into the hands of the more extreme right-wing, as in the Conservative party, or in the hands of the left-wing, as is almost the case in the Labour party. Where the party system does not break up, there is a greater likelihood of volatility, with individuals arising—as did Jimmy Carter—to present themselves as "protest" candidates, and, using the mechanisms of primaries, direct elections, and the visibility generated by the media, catapult themselves into office.

VI

Is there a way out? In principle, there is an answer. It is the principle of "appropriate scale." What is quite clear is that the existing political structures no longer match the underlying economic and social realities, and just as disparities of status and power may be a cause for revolution, so the mismatch of scales may be the source of disintegration.

What was evident in the 1930s, in a wide variety of political circumstances, was that the national state became the means to pull the economy and society together. If one looks back at the New Deal of Franklin D. Roosevelt, it was not "creeping socialism" or "shoring up capitalism" that characterized his reforms (though there were elements of both in his measures), but the effort to create national political institutions to manage the national economy that had arisen between 1910 and 1930. By shifting the locus of policy from the states to the federal government, Roosevelt was able to carry out macroeconomic measures which later became more self-conscious, particularly as the tools of macroeconomic analysis (the ideas of national income accounts and GNP, both of which were only invented in the 1940s and were introduced in the Roosevelt budget message of 1945) came to hand.

But the national state is an ineffective instrument for dealing with the scale of major economic problems and decisions which will be necessary in the new world economy that has grown up, though national interests will always remain. The problem, then, is to design effective international instruments—in the monetary, commodity, trade, and technological areas—to effect the necessary transitions to a new international division of labor that can provide for economic and, perhaps, political stability. (It would be foolish, these days, to assert that economics determines politics; but the economic context is the necessary arena for political decisions to be effective.) Such international agencies, whether they deal with commodity buffer stocks or technological aid, are necessarily "technical," though political considerations will always intrude. Yet the creation of such mechanisms is necessary for the play of politics to proceed more smoothly, so that when some coordinated decisions are taken for political reasons, there is an effective agency to carry them out.

At the other end of the scale, the problem of decentralization becomes ever more urgent. The multiplication of political decisions and their centralization at the national level only highlight more nakedly the inadequacies of the administrative structures of the society. The United States, as Samuel P. Huntington once remarked, still resembles a Tudor polity in its multiplication of townships, counties, incorporated or unincorporated villages. With such overlapping jurisdictions and inefficiencies not only are costs—and taxes—multiplied, but services continue to decline. We have little sense of what is the appropriate size and scope of what unit of government to handle what level of problem. What is evident is that the overwhelming majority of people are increasingly weary of the large bureaucracies that now expand into all areas of social life—an expansion created, not so paradoxically, by the increased demand for social benefits. The double bind of democracy wreaks its contradictory havoc in the simultaneous desire for more spending (for one's own projects) and lower taxes and less interference in one's life.

Yet here, too, there is the possibility of a way out: the use of the market principle—the price mechanism—for social purposes. As against the ritualistic liberal, whose first reaction regarding any problem is to call for a new government agency or regulation, or the hoary conservative who argues that the private enterprise system can take care of the problems (it often cannot, for some coordinated action by a communal agency is necessary), one can use the market for social purposes—by giving people money and letting them buy the services they need in accordance with their diverse needs, rather than through some categorical program.

. . . In a world where, at the large and small ends of the scale, social stability is threatened and governance becomes difficult, questions of domestic and foreign policy quickly intertwine. For if the national state is too small for the big problems of life and too big for the small problems, we have to begin to think—and, given the shortness of time and the specter in the streets, to concentrate the mind, as Dr. Johnson would have said—about what other political arrangements may be necessary to give us stability and freedom in this shrinking world.

NOTES

[1]This, as well as many other striking insights, is to be found in the neglected book of Brooks Adams, *The New Empire* (New York: MacMillan, 1902), a powerful history of the rise and fall of empires in response to the changing trade routes, the exhaustion of metals and resources, and the intersecting influences of geography and technology.

[2]UNCTAD *Statistical Handbook,* 1973.

[3]See Richard Jolly, "International Dimension," in Hollis Chenery, et. al., *Redistribution with Growth* (New York: Oxford University Press, 1974).

[4]So rapid was this political change that most persons do not know that the issue which threatened to split the Socialist International in the 1930s was the question (summed up in the so-called Bauer-Dan Zyromski theses) of entering "bourgeois coalitions."

[5]If one reviews the sociological and political literature of the 1930s, it is striking that almost none of the major works dealing with contemporary crises foresaw the change in the international system. The only country that had a "visible" independence movement was India, and it was assumed that, someday, it would achieve a greater degree of self-government within the Commonwealth framework. Almost all the preoccupations were with the threat of fascism and the breakup of the liberal bourgeois states. For a representative book of those times, see Karl Mannheim, *Man and Society in an Age of Reconstruction* (New York: Harcourt Brace Jovanovich, 1967).

[6]International money markets are now so sensitive that—as the *London Times* of November 1, 1976 reported—some 800 banks and 250 corporations, from Hong Kong to Europe and across the United States, pay £7,000 a year to be plugged into the Reuters Money Market service, a computerized electonric monitoring service on exchange rates in different world centers.

[7]I have tried to deal with the social consequences of the impact of each on the other in a monograph, "The Social Framework of the Information Society," for the Laboratory for Computer Science at M.I.T. Part of that study will be included in a volume on the future of computer technology, edited by Michael Dertouzos and Joel Moses, to be published by the M.I.T. Press in spring of 1978. A different section appears in *Encounter,* June 1977. See also Daniel Bell, *The Coming of Post-Industrial Society* (New York: Basic Books, 1973); also a paperback version (New York: Basic Books, 1976) with a new introduction.

[8]For the origin and development of Schumpeter's idea, see Daniel Bell, "The Public Household—on Fiscal Sociology and the Liberal Society," in *The Public Interest,* Fall 1974. A variant version of that essay is included in Daniel Bell, *The Cultural Contradictions of Capitalism* (New York: Basic Books, 1976).

[9]See Richard Rose and Guy Peter, "Can Government Go Bankrupt," unpublished paper, December 1976, to appear in a volume of the same title by Basic Books in spring 1978; Jurgen Habermas, *Legitimationsprobleme in Spätkapitalismus* (Frankfurt, 1976. English translation, *Legitimation Crisis,* by Beacon Press, 1976); Samuel P. Huntington, "The Democratic Distemper," in *The Public Interest,* Fall 1975; and Samuel Brittan, "The Economic Contradictions on Democracy," in the *British Journal of Political Science,* 1975, no. 2. For a neo-Marxist view, see James O'Connor, *The Fiscal Crisis of the State* (New York: St. Martin's Press, 1973), and, for an effort to put the economic issues in a cultural as well as political context, Daniel Bell, *Cultural Contradictions of Capitalism* (New York: Basic Books, 1976).

[10]See Nathan Keyfitz, "World Resources and the World Middle Class," *Scientific American,* July 1976.

[11]United Nations, Secretariat, Conference on Trade and Development, *The Dimensions of the Required Restructuring of World Manufacturing Output and Trade in order to Reach the Lima Target,* Supp. 1 (TD/185), April 12, 1976.

Bottle babies: death and business get their market

LEAH MARGULIES

■ Leah Margulies *has been a coordinator of the Infant Formula Campaign since 1975. She has long been interested in the dynamics of market expansion by Western corporations into Third World countries, and especially in the marketing of Western middle-class products to poor people in the Third World. She is now co-authoring a book about the baby-formula controversy.*

In "Bottle Babies: Death and Business Get Their Market," *Margulies provides a bleak account of the attempts made by Nestlé and other corporations to replace breast feeding with bottle feeding among the poor of the Third World. She describes the suffering, malnutrition, and death caused by the Nestlé infant formula campaign, as well as the dilemma of the hospitals and physicians who co-operate with Nestlé in exchange for sorely needed medical equipment. "Because of . . . growing condemnation of industry practices, the companies have made some attempts to deal with their critics. In most cases however, the concessions do not significantly alter the outcome of formula promotion." She points out that the profits from infant formula sales in the Third World run quite high.*

Caracas, Venezuela, July 1977: In the emergency room of the Hospital de Niños, a large facility in the center of the city, lie 52 infants. All are suffering from gastroenteritis, a serious inflammation of the stomach and intestines. Many also suffer from pneumonia. According to the doctor in charge, 5,000 Venezuelan babies die each year from gastroenteritis, and an equal number die from pneumonia. The doctor further explains that these babies, like many who preceded them and those who would follow, have all been bottle-fed. He remarks, "A totally breast-fed baby just does not get sick like this."

Poverty, inadequate medical care, and unsanitary conditions make bottle feeding, to quote a government nurse in Peru, "poison" for babies in the developing countries. Yet bottle feeding is rapidly becoming the norm in Third World countries. In 1951, almost 80 percent of all three-month-old babies in Singapore were being breast-fed at the age of three months; twenty years later, only 5 percent of them were at the breast. In 1966, 40 percent fewer mothers in Mexico nursed their six-month-old babies than had done so six years earlier.

The end result of this significant change in human behavior is higher morbidity and mortality rates among bottle-fed babies. Many well-known studies provide evidence of the relation between bottle feeding and infant malnutrition, disease, and death. Of course, it is impossible to know how many babies are getting sick or dying because of bottle feeding, but the number is large and growing throughout the developing world. Dr. Derrick Jelliffe, head of the Department of Population, Health, and Family Planning at the UCLA School of Public Health, conservatively estimates that about 10 million babies a year suffer from malnutrition related to bottle feeding. The phenomenon is literally worldwide. According to medical reports of malnutrition among Eskimo children in the Baffin Zone of Canada, almost 5 percent of the infants born there in 1973-74 had to be flown to Montreal for emergency treatment, and doctors believe that one of the major causes of this tragic development was bottle feeding.

At the center of the bottle-feeding controversy are the promotional practices of the corporations who sell bottles and powdered baby milks in the Third World. Critics believe that promotion of these powders to mothers who do not have the facilities to properly prepare the feeds is a deadly way to make a profit. However, despite the increased activity of critics and acknowledgments by industry that improper bottle feeding can be dangerous, sales of infant formulas in poor countries are still escalating.

The corporations that sell infant formula in the Third World run the gamut from prestigious American, Swiss, British, and Japanese multinational corporations—like Abbott, American Home products, Bristol-Myers, Nestlé's, and Cow and Gate—to local fly-by-night manufacturers trying to cash in. The concentrated campaign to attract Third World consumers began in the late 1950s. Soon a body of literature arose to help business conquer this almost virgin territory. For example, various articles advised foreign marketers that, in the absence of a middle class, they should consider the urban poor as an important potential market.

Business began to understand the market potential of a poor population with many unfulfilled needs. Often the real needs of the poor could be obscured by a corporate sales strategy which promised the satisfaction of newly created needs. Mass media—TV, radio, and newspapers—could convey the promise that new products would meet these new needs. *Fortune* magazine heralded this new age with an article entitled, "Welcome to the Consumption Community." It was therefore not surprising that when the "Community" of infant formula consumers in the United States began to shrink as postwar birth rates declined and middle-class women in the developed countries decided they had been deprived of the experience of breast-feeding and began turning to the more natural way, the corporations turned to the ripe Third World market.

For the companies, baby formula sales strategies have paid off. Unfortunately no reliable statistics on infant formula sales are publicly available, although sometimes companies have inadvertently revealed the extent of their commitment to the product. World-wide sales of formula are estimated to total around $1 billion, with Nestlé's figure at roughly $300-400 million. Nestlé reportedly controls approximately half of the formula market in developing countries.

Whatever the sales figures at present, they will undoubtedly increase in the future. Bristol-Myers, for instance, has consistently reported sales gains for its Enfamil infant formula. Moreover, the upward trend, for the other companies as well as for Bristol-Myers, shows few signs of abating. Of course, sales figures do not tell the full story. Profit rates for infant formulas are also thought to be quite high. According to a 1977 supermarket sales printout from Brazil, commercial formula enjoyed a 72 percent profit margin, while all other supermarket products ranged between 15 percent and 25 percent.

WHAT IS IT?

What kind of product is infant formula? It is a highly processed food, based primarily on cow's milk. While the fat content and sugar source are patterned after mothers' milk, the company's claim that it is "nearly identical to mother's milk" is ridiculous. Maternal milk is a living substance, unique in many ways. Besides supplying the proper quantities of protein, fats, and other nutrients, it protects the infant from disease by providing antibodies important to the development of the immunization system. Formula does not have the digestibility of mothers' milk. Sometimes the product is sold premixed, but in the Third World it is more often sold as a powder that requires measured amounts of *pure* water for the proper reconstitutions. Sterilized bottles and nipples are also necessary.

There are a number of reasons why infant formula sells so well in the Third World. A mother in a developing country often finds herself in situations totally unlike those her mother ever experienced. She may, for instance, work outside the home, listen to the radio, or watch TV. These situations can be disorienting, and new values and attitudes must be formed in order to deal with them. Newly acquired values such as social mobility, as well as a high regard for modern products and medical expertise, make her a particularly vulnerable target for sophisticated formula marketing campaigns. The smiling white babies pictured on the front of formula tins can lead her to think that rich, white mothers feed their baby this product and that therefore it must be better.

Going into a hospital to give birth can be an especially frightening situation for a young Third World woman. Since in many countries only a small proportion of women attend the prenatal clinic (if there is one), a mother's maternity stay may be one of the few times in her life that she will go into a hospital. Any products given to her in this environment will seem to carry medical endorsement.

Imagine the reaction of a Third World mother in her home, or a group of mothers in a clinic or hospital attending a class, to a woman in a crisp nurse's uniform. The woman may or may not be a nurse. She begins her speech, tactfully enough, by reassuring them that "breast is best," but she ends by extolling the virtues of her company's product over the natural method. Capitalizing on the respect given a nurse, the use of a "milk nurse" implies a connection between the health care profession and the commercial product.

In developed and developing countries alike, one of the hospital practices most damaging to breast-feeding efforts—and one implicitly supported by company promotional practices—is the separation of mothers and infants shortly after birth. During the twelve to forty-eight hours of separation, the infants are bottle fed in the nursery. Mothers are sometimes given antilactation shots during this period. Thus when a mother is finally reunited with her baby, switching from bottle to breast is made more difficult.

Furthermore, if the hospital has no incentive to teach her, the woman is even less likely to breast-feed. Formula companies create a strong climate for their products with their constant offers to set up bottle sterilization and preparation facilities, to equip nurseries, and to provide free supplies of formula. Busy doctors and nurses are led to adopt the postnatal separation strategy by the willingness of formula companies to make this approach easier than breast-feeding.

Medical personnel are a prime target for promotion because they are the direct link to mothers. Although it is the patient who ultimately pays for the product, doctors tell her what to buy, and the difference in backgrounds of doctor and patient may well lead to an inappropriate choice. As Dr. John Knowles, president of the Rockefeller Foundation, stated in a letter to the chairman of Bristol-Myers:

> The problem is not a 'scientific' one. The problem is poverty and the inadequate home environment which makes the use of prepared formulae so lethal. This the physician is *not* uniquely qualified to understand. In fact, he may be precisely the most unqualified to understand, since he undoubtedly comes from a different socio-economic background and may have no idea of the home conditions of the poorest mothers of his own society.

Many dedicated physicians face a real dilemma when dealing with the promotion efforts of formula companies. Their hospitals and clinics are often woefully short of medical equipment and supplies. Under such circumstances, it may seem harmless, indeed charitable to agree to give away free samples of infant formula to mothers in exchange for the company's gift of medical stocks or a new nursery. One hospital administrator in Malaysia has explained. "It is a very corrupting influence. You are always aware that you could have virtually anything you ask for."

MARKETING FOR BABIES

These marketing strategies are consciously decided upon and implemented through instructions to sales personnel, milk nurses, and distributors. Note the following extract from American Home Products selling instructions for 1975:

> **Selected doctors:** 40-50 doctors per territory including 5 or 6 VIP's. These doctors should all be selected on the basis of their known influence on the selection of formula by mothers and by hospital or clinic maternity services.
>
> **Sampling:** . . . Maternity services should be given primary allocation of free samples, geared to producing potential sales.

Companies believe, and with good reason, that the product a mother goes home with is the product she will be loyal to. A 1969 study of 120 mothers in Barbados showed that 82 percent of the mothers given free samples, whether in a hospital or at home, later purchased the same brand. Thirty-two percent of them admitted that they were influenced by the free sample.

This aggressive market penetration and consumer creation are particularly destructive because they affect the most important resource developing countries have—people. In Chile in 1973 three times as many deaths occurred among infants who were bottle fed before three months old than among wholly breast-fed infants. A research team inspecting feeding bottles there discovered a bacterial contamination rate of 80 percent. Poverty and underdevelopment lead to abuse of even legitimate baby milk substitutes. Poor mothers cannot afford them in the quantities needed. Water is often contaminated, and the necessary boiling is rarely possible. Illiteracy makes it difficult to follow proper directions. Early weaning of infants from the breast to bottled infant formula is accompanied by increasing cases of diarrhea and gastroenteritis. Improperly attended—as they are likely to be due to inadequate medical care—these disorders result in many deaths.

Malnutrition is another common result and has been described as "commerciogenic malnutrition." This is not meant to imply that the manufacturers are solely responsible but simply that this type of malnutrition has nothing directly to do with underdevelopment and lack of food resources. As Dr. Michael Latham, a pediatrician and Cornell University professor of nutrition stated, "Placing a baby on the bottle in the Third World might be tantamount to signing that baby's death certificate."

A 1975 Pan American Health Organization study found that childhood deaths from malnutrition peaked in the third and fourth months of life, because of the early abandonment of breast-feeding. The study covered some 35,000 deaths in fifteen countries. Medical studies linking bottle feeding with infant mortality and morbidity cover practically all areas of the Third World and some developed countries as well. A 1977 study in Cooperstown, New York, compared 164 breast-fed infants with 162 formula-fed infants; significant illnesses increased as breast-feeding declined. In 1970 a study in Jamaica, West Indies, revealed a higher incidence of gastroenteritis in the first four months of life among partly or wholly bottle-fed babies than among breast-fed babies. Other studies have reported similar results from Chile, Lebanon, Israel, Lagos, and others.

Hospital reports and personal testimony from doctors and nurses confirm these findings. Doctors in Jamaica have reviewed the records of thirty-seven seriously ill infants admitted in 1975 into their hospital, the Tropical Metabolism Research Unit in 1975. Twenty-five of the thirty-seven patients had been fed a brand-name infant formula. The average body weight of the babies was only 58 percent of the normal value. Their families were simply not equipped to safely bottle feed. About one fifth of the mothers were illiterate. The remainder were able to sign their names but were functionally illiterate. It was highly unlikely that they would be able to read, no less understand written directions.

Nearly all the families lived in cramped, overcrowded, and unsanitary conditions, with an average weekly income of sixteen dollars. A tin of baby formula costs approximately two dollars and a baby needs two cans a week if exclusively formula-fed. Despite optimal medical care, five of these babies died. The case studies graphically show the inevitability of bottle contamination and dilution—the key culprits leading to illness and malnutrition.

Since 1970 when the Protein Advisory Group (recently dismantled) of the United Nations first met with the baby formula industry, there has been a growing international campaign aimed at stopping unethical promotional practices. In 1973, the Protein Advisory Group published guidelines for promoting infant nutrition and included the need for restrictions in advertising. In 1974, the World Health Assembly called for a critical review of company promotion, and the issue has been discussed extensively at medical conferences, international seminars, in U.N. papers, etc. Most recently, on January 31, 1978, the World Health Organization, announced, "The advertising of food for nursing infants or older babies and young children is of particular importance and should be prohibited on radio and television . . . finally, the distribution of free samples and other sales promotion practices for baby foods should be generally prohibited."

In 1975 the International Pediatrics Association issued a series of recommendations to encourage breast-feeding. The section entitled "Curtailing Promotion of Artificial Feeding" reads:

1. Sales promotion activities of organizations marketing baby milks and feeding bottles, that run counter to the general intent expressed in this document, must be curtailed by every means available to the profession, including, where necessary and feasible, legislation to control unethical practice.
2. Dissemination of propaganda about artificial feeding and distribution of samples of artificial baby foods in maternity units should be banned immediately.

In the U.S. recently, considerable interest has centered around the stockholder lawsuit against Bristol-Myers (Mead Johnson Division). The Sisters of the Precious Blood have charged the company with making "false and misleading statements" about their overseas promotion and sales of infant formula. The statements appeared in a proxy report to stockholders, which is required by law to be accurate. In May 1977, a U.S. district court judge dismissed the case, stating that the Sisters had not shown that they, as shareholders, had been caused "irreparable harm" by the alleged misstatements. The judge declined to comment on the accuracy of the company's proxy report. The nuns appealed this decision.

Then in the first weeks of 1978, the Sisters signed an out-of-court settlement with Bristol-Myers. The settlement stipulates that a report be sent to all shareholders of the company, outlining the legal action and the positions of both parties. The Sisters' statement in the report contains affidavits from five countries and an analysis of their current criticisms of company practices. The company's statement announces a more stringent interpretation of its Code of Policies and Practices and the fact that it has discontinued the use of milk nurses in Jamaica. Industry critics view the settlement as an important step toward convincing the companies that public opinion has changed the social climate in which marketing takes place: What was at one time an "acceptable" social cost no longer is the case, primarily because of increased public knowledge and protest.

The findings of the lawsuit have prompted local consumer advocacy groups in the United States to join forces in a coalition called INFACT (Infant Formula Action). These groups believe it vital to keep pressuring Nestlé—the largest manufacturer of baby formula in the Third World—to desist from its promotion tactics. The Minnesota-based Third World Institute has initiated a consumer boycott which is quietly spreading throughout the U.S. In addition, church groups, acting in their capacity as stockholders in the American companies, are continuing their efforts to further restrict the promotion these companies engage in. This year two new shareholder resolutions were filed with American Home Products and Carnation, both of whom widely advertise their condensed milk in the Third World.

Because of this growing condemnation of industry practices, the companies have made some attempts to deal with their critics. In most cases however, the concessions do not significantly alter the outcome of formula promotion. There have been a number of changes:

• After blatant advertising, especially mass-media promotion had made some of the companies highly vulnerable to criticism, these companies switched the focus of their promotion efforts to the medical profession. This new marketing approach is more sophisticated, less risky, and far more effective. Via mass media, everyone heard the message, whether they were potential customers or not. Now marketing focuses more directly on the consumer through the use of health workers. For example, in a poverty hospital in the Philippines, name tags with a prominent brand-name logo are found on each crib in the nursery. Nestlé wrist labels have also been provided. There and elsewhere, while the most blatant ads have been curtailed, direct consumer promotion continues in the hospitals themselves and appears to be sanctioned by the medical authorities.

• In the past, critics charged that companies encouraged the abandonment of breast-feeding. Now the companies agree that bottle feeding to the exclusion of breast-feeding is not desirable. They talk about "supplementation." However, mixed feeding has also been shown to be quite dangerous. Consuming smaller amounts of contaminated and diluted formula is preferable, one assumes, but it is not the answer. Furthermore, the encouragement of supplementation in fact undermines breast-feeding. According to most medical experts, supplementation negatively affects the production of human milk.

• Critics have also complained about milk nurses and the ethics involved in employing nurses as a company sales force. Again, the companies have adapted. They often change the colors of the uniforms, add belts, call them "company representatives," and may even agree to alter somewhat milk-nurse sales techniques. But visits to hospitals and homes continue, and the nurses are still being lured away from government health services.

A more significant adaptive technique is that of employing nutritionists and other highly trained professionals. In Venezuela, for example, Nestlé employs no milk nurses but several nutritionists. These nutritionists interact on a regular basis with Ministry of Health, nutrition, and hospital personnel. One Nestlé nutritionist in Caracas appears to have been totally integrated into the health care team at Maternidad Hospital as she made her rounds with the paid hospital staff. This type of interaction between government and business personnel raises serious ethical questions about the extent to which industry's point of view should be institutionalized within government health services.

• When critics argued that formula was being promoted to the poor, the companies responded that formula is priced above the income of poor people and is purchased almost exclusively by upper-income groups. But the companies have provided no evidence to confirm this argument. Indeed, there is more than adequate proof that the products are being promoted and sold indiscriminately to mothers who have neither the financial nor the sanitary facilities to use the products safely. Since July 1977 alone, documentation confirms the presence of promotional displays in markets, pharmacies, and grocery stores in the mountain villages of India, the barrios of Caracas, and the slums of Manila.

• In response to these kinds of intense promotion efforts, the critics finally called for regulation of the formula industry. The industry, in turn, has responded with "self-regulation," which mainly consists of business codes. There are now several codes of ethics, some more stringent than others. All, however, share two inherent weaknesses.

First, the codes legitimize promotion to the medical profession and characterize the latter as "intermediaries" between the baby food industry and the mother. However, given the desperate shortages of medical personnel in developing countries and the constant pressure exerted on existing workers by the companies, it is very difficult for these intermediaries to be impartial.

Second, insuring that the companies will adhere to their self-imposed restrictions is virtually impossible in the absence of regular scrutiny by an independent body. In August 1977, a Bristol-Myers milk nurse was interviewed by this author on the ward of the largest public hospital in Jamaica. The milk nurse had in her hand a list of mothers she intended to visit in their homes. She had copied the names from ward lists. In an interview just two days before, the chief medical officer of Jamaica had explained that government policy prohibited milk nurses from entering public hospitals. The milk nurse's actions were therefore doubly in violation of Bristol-Myers's code of ethics which specifically requires cooperation with government health policies as well as the solicitation of references from medical professionals for all home visits. The publicity surrounding this incident most likely influenced Bristol-Myers's decision to discontinue milk nurses in Jamaica.

Stopping the promotion of infant formula products will not, in and of itself, eliminate malnutrition. Infant formula products could still be sold under carefully controlled and supervised conditions and still be misused because the existing social and economic conditions make proper usage virtually impossible. An end to malnutrition will ultimately require massive changes in the distribution of wealth, land, and power. But that is no reason not to take intermediate steps. The shifts in promotion thus far are adaptations to a new business climate and clearly prove that the formula industry is vulnerable to pressure.

If promotion could be eliminated entirely, health care institutions and governments would be freer to develop their own capacity to handle the monumental health problems that face Third World countries. To accomplish this, the public needs a strategy. It must include the continuous monitoring and disclosure of corporate activity; cooperation between concerned health professionals, international agencies, and advocacy groups; and the development of an increasingly larger audience of people who share the belief that business must be held accountable for unethical practices, however costly and inconvenient. As Dr. Alan Jackson of the Tropical Metabolism Research Unit in Jamaica stated in a recent interview:

> When you spend your time working with children who are malnourished and you see children dying because they are either getting wrong food or food prepared improperly, it has a devastating effect on you. It's very hard to think that people who are involved in selling, encouraging people to buy infant preparation, can carry on in this kind of a way, and at the same time pretend that they are not involved in the end results, which is malnutrition, malnourished children.

Industry's new frontier in space

GENE BYLINSKY

▪ Gene Bylinsky *is associate editor of* Fortune *magazine. He is the recipient of various journalism awards, among them the twenty-first Albert Lasker Medical Journalism Award. He wrote* The Innovation Millionaires.

In the following article Bylinsky discusses the benefits that can be reaped by industry and society from locating some industries in outer space. He points out that the virtually gravity-free environment of outer space can be used to produce "no fewer than 400 alloys that cannot be made on earth because of the gravitational pull." These alloys can be used to produce, among other things, lighter cars and "featherweight metal furniture." He also notes that the gravity-free environment is more suitable for growing crystals, the material from which chips are made in electronic industries. A crystal grown in outer space is larger and more uniform. The benefits of outer space also extend to biology and medicine. "Vaccines may attain a purity not possible on earth," and hundreds of biological products that "simply cannot be synthesized or separated on earth" can be so treated in outer space.

All this, Bylinsky says, has increased the interest of industry in outer space research and has led to the allocation of huge sums of money in that area by corporations, which see an opportunity for substantial profits.

The new battle cry at NASA, only slightly amplified, is: Thar's gold in them thar stars. In the wide and starry band of near-earth space, beginning about 200 miles up and extending to 22,500 miles, where a satellite can be placed in stationary orbit rotating in unison with the earth, the National Aeronautics and Space Administration sees the possibility of an industrial bonanza. Operating in this pure and virtually gravity-free environment, factories could produce novel materials worth as much as $30,000 a pound back here on earth. Spidery, dreamlike

☐ From *Fortune*, 29 Jan. 1979, pp. 77-83.

power stations could collect energy from sunlight and beam it to the planet below. The freighter servicing these new industries will be the Space Shuttle, which is scheduled to make its first orbital flight this September.

So far, the reaction of earthling industrialists to all these glittering promises has been mixed—and rather muted on the whole. High-technology companies that have contracts to develop equipment and experimental manufacturing processes for NASA are naturally enthusiastic. Other corporations are apt to be poorly informed about the possibilities, and skeptical to boot. They are well aware that the space agency itself, which desperately needs a post-Apollo mission that can command broad public support, has a great deal riding on its fledgling industries in space and has been promoting them with gusto.

There is also the rather basic matter of costs. Even a simple experiment aboard the Shuttle can cost several hundred thousand dollars, while a small, automated production plant, designed to be left in orbit and serviced periodically by the Shuttle crew, would run into tens of millions. The military, or scientific researchers backed by the government, might be willing to pay fabulous sums for materials that can't be made on earth—a perfect lens for a spy satellite, say, or perfect spheres of hydrogen isotopes for use in laser-fusion research—but NASA acknowledges that the costs and risks of space manufacture are too high and too ill defined at present to interest most corporations. Robert A. Frosch, NASA's administrator, says that his job right now is "to provide access to space and to develop basic technologies, which eventual users will need to evaluate before making investment decisions."

But skepticism on the part of profit-seeking corporations can be overdone. NASA is not exactly starting from scratch out there in space; it is building on promising experiments done on prior space flights. Those tests, mainly on the Skylab and Apollo-Soyuz flights, showed that beyond the pull of the earth's gravity remarkable things happen to materials. Crystals grow more uniformly—and in some cases ten times bigger than on earth. Biological substances can be separated and sorted out much more easily, suggesting the possibility of purer vaccines and brand-new drugs. Furthermore, those earlier flights established that it is at least technically possible to create new types of glass, "super" alloys of various sorts, and materials of variable density, with properties never seen on earth. In fact, some scientists believe that the Shuttle flights will mark a milestone for human invention, comparable to the development of the vacuum pump back in the seventeenth century.

NASA considers it significant that West German and Japanese companies are more excited about the Shuttle than their American counterparts, and that the European Space Agency, an active and enthusiastic partner, has budgeted $600 million for the design and construction of the Shuttle's spacelab—more than twice NASA's own current budget for this work. At this point, in a field involving so many unknowns, perhaps the best judgment that can be made is that while few corporations will care to take a plunge into space manufacturing, no corporation affected by changes in technology can afford to ignore the new era of innovation that is about to begin.

The advantages of manufacturing in space can best be understood as the flip side of various *disadvantages* here on earth, the most important of which is gravity. Most solid materials go through a liquid, or molten, stage at some point during their creation or processing, and, where gravity exists, they must be supported by a container—a source of contaminants.

More important, gravity induces convection currents, which flow along the thermal gradients, or temperature differences, in layers of the liquid. Convection currents, being chaotic and unpredictable, often lead to unpredictable and undesirable structural and compositional differences in the solid material. Convection can create soft and mushy zones. Gravity also pulls molecules apart, leaving holes where impurities collect. If the liquid contains more than one type of material, gravity tends to separate these different materials, and the resulting solid lacks uniformity.

These adverse effects of gravity have bedeviled materials manufacturers ever since man cast the first bronze figurine, and because of them metals have never achieved the strength and other mechanical properties that theory predicts. Steels, for instance, could be anywhere from 100 to 1,000 times stronger than they are today. Blades of jet engines now fall apart at temperatures where the efficiency of engine operations would be appreciably higher. The wires in heart pacemakers and the pins in bone prosthetics—extremely expensive, not to mention traumatic to replace—fail much sooner than they should.

In the weightlessness of outer space, most of these problems in the processing of materials disappear. Strictly speaking, of course, there is no such thing as "zero gravity"—every particle and atom has an attraction for every other one. But weightlessness aboard the Shuttle will come close to that unattainable standard. When things are quiet on the Shuttle, the pull of gravity will be only a millionth as great as that on earth. When the astronauts fire small rockets to correct their course—or merely clump around in their suction-cup boots—the pull could shoot up to a thousandth of the usual earthly value. Some scientists call these fractional conditions "microgravity."

TRW, in a major study for NASA, has identified no fewer than 400 alloys that cannot be made on earth because of the gravitational pull. Many of them are metallic combinations that, like oil and water, will not mix on earth. When allowed to solidify in the weightlessness of space, they would mix down to microscale, yielding unusual strength or hitherto unrealized mechanical, electrical, and magnetic properties. Light but sturdy vehicles could be built out of such metals—tanks that would weigh no more than a car, for example, and featherweight metal furniture. A lively topic of interest to the utilities are superconducting metals that could transmit electrical energy at low temperatures with virtually no loss of power.

In certain compositional ranges, metals such as copper and lead, or aluminum and lead, would display self-lubricating properties, possibly leading to automobile engines that could last 500,000 miles and more. BMW, the West German automaker, has shown an interest in financing some experiments with aluminum-lead combinations.

Many of these materials could be produced in a mode unique to the space environment—containerless processing. This method is possible because levitation, which may have magical overtones on earth, is the natural mode of behavior for objects in space. A blob of liquid or a solid can be positioned easily with a minimum application of force in an acoustic, electromagnetic, or electrostatic field. Since second-order forces, such as surface tension, take over in space, a blob of molten material will automatically assume a spherical shape. It can then be changed into the desired shape by applying slight outside forces. Containerless processing hasn't got very far on earth because of the great forces needed. In space, even the sound from a good commercial hi-fi set would levitate a blob of steel. This conjures up visions of a with-it generation of astronauts laying back and letting the Rolling Stones do the work.

Dispensing with containers could lead to important improvements in the microstructure of tungsten, which has a melting point so high (6,170° F.) that it is particularly prone to contamination when a melt is achieved. Impurities from the crucible also have prevented the manufacture of truly pure optical glass and have greatly increased the cost of producing the high-quality glass fibers needed for the novel transmission lines being developed by A.T.&T. and others. Glasses from space, with unique refraction and dispersion qualities, have endless possibilities in lasers and other high-technology optical systems. Ralph A. Happe, a glass specialist at Rockwell International, predicts: "We'll be doubling the catalogue of the optical designer."

But the most immediately promising field for materials processing in space is the culture of crystals, which have become the sum and substance of modern electronics and electro-optics. In electronics, the principal virtue of a crystal is its ability to transmit electrons under precisely defined and controllable conditions. In optics, crystals offer better transparency than even the best glasses because variations in the amorphous structure of glass will scatter some of the light.

Crystal culture here on earth is generally not a science but an art. The specialists who grow the large, carrot-shaped crystals used to make semiconductor chips are given to bragging about their "green thumbs"—and the metaphor is not farfetched. Although crystals are not living things, they grow in a manner roughly similar to plants. They demand nourishment for instance, and will reach toward the source of food. A crystal grower, says one specialist, "adds a little bit of this, a little bit of that—it's a recipe-type operation." Those all-important impurities, called dopants, which impart the desired electronic properties to a semiconductor crystal, are difficult to distribute evenly on earth because of convection currents induced by gravity. Consequently, the yield of usable chips from a crystal is low.

What can be accomplished in space was shown dramatically aboard Skylab, where an experiment designed by Dr. Harry C. Gatos, professor of materials science and engineering at M.I.T., produced a remarkably smooth sample of indium-antimonide crystal. Measuring the conductivity of the crystal along its length, Gatos found that the electrical properties were constant. In a similar crystal grown on earth, these properties vary continuously from one end to the other. On the Apollo-Soyuz flight, he grew an equally perfect germanium crystal. Although the experiments were of necessity somewhat primitive, the results, in Gatos's words, were "way beyond expectations."

Crystal culture in space will resume with the first materials-processing flights, which are scheduled to start in 1981, and Gatos declares that he can already see profits ahead. He cites the case of gallium arsenide, which is widely used in light-emitting diodes, lasers, microwave devices, and in other high-technology products. Gallium arsenide of not very high quality sells today for $15,000 a pound. "The cost of processing it in space," says Gatos, "will eventually be a small fraction of its selling price." The space-made crystals will have a much higher yield of usable chips, he explains, justifying a much higher price for the crystal. If, as he expects, the higher quality gives birth to new applications, then the value becomes incalculable.

Another space product that may turn a profit is a tiny sphere of a rather ordinary plastic, polystyrene latex. Spheres smaller than two microns in diameter (that is, two-millionths of a meter) or larger than forty microns can be made on earth. But for complicated technical reasons, spheres in the intermediate ranges are unstable and can't be mass produced. And it so happens that scientists crave those particular sizes. If the spheres could be sprinkled into a culture before it is exposed to electron microscopy, for example, the known sizes of the spheres would permit researchers to take the exact measure of many things, from viruses to the apertures in membranes. The tiny spheres would also be useful for calibrating the electron microscope itself, as well as medical filters and other devices.

Eager to demonstrate the advantages of materials processing in space, NASA will try to produce these latex spheres right away, possibly on the first orbital test flight this September. By the third flight, the experimenters, led by John W. Vanderhoff of Lehigh University, hope to be making the spheres in batches of up to four ounces. Waiting in the wings is Accupart Laboratories, a small company in Huntsville, Alabama, founded by a retired NASA executive named Brian Montgomery, who says that he has venture capitalists lined up willing to invest $5.6 million in the making of the microspheres, which presumably will fetch a substantial premium over the $30,000 a pound that the smaller sizes sell for today.

Space holds vast possibilities for biology and medicine. Microgravity should greatly improve man's ability to separate specific cell types, cell components, cell products, and proteins. Vaccines may attain a purity not possible on earth. Earlier space flights have already yielded some clues—and some cautionary lessons, as when bacteria got into a test on salmon-sperm DNA and ate it all up.

The basic opportunity lies in the fact that hundreds of biological products simply cannot be synthesized or separated on earth—once again because convective flows produce irregular and unpredictable mixtures. Many of these desirable products are substances that the body itself makes and packages in a very complex soup of other ingredients. Urokinase, for example, is an immensely useful chemical that activates an enzyme which dissolves blood clots. Yet urokinase is manufactured by only 5 percent of all kidney cells. The Shuttle's goal is to separate these specialized cells and then establish a culture on earth, thus increasing the yield. In fact, kidney cells separated on the Apollo-Soyuz flight did produce about seven times more urokinase than usual, but for some reason, which the researchers are naturally curious to pin down, the cells stopped making urokinase when they were cultured back here on earth.

Similarly, hormones and other substances made by the body in minute amounts—the antiviral agent, interferon, for example, and the brain's own painkillers, the endorphins—could be purified in orbit. Still another key candidate for space processing is erythropoietin, a kidney hormone that stimulates bone-marrow cells into producing red blood corpuscles. So far, nobody has succeeded in extracting pure erythropoietin on earth.

Earthbound researchers have already made great progress with white blood cells, which have been found to contain whole subpopulations of substances that act as the body's immunological defenders. In the felicitous absence of gravity, scientists think they might be able to isolate new drugs that could combat such imbalances of the immune system as rheumatoid arthritis. John C. Carruthers, the director of materials processing, who spent fifteen years at Bell Labs before joining NASA last year, predicts that "one day we'll be making pharmaceuticals in space."

If the greatest advantage of space is its lack of gravity, the second most important is the purity and thinness of the atmosphere 200 miles up. Robert T. Frost of G.E.'s space division refers to these upper regions as "the world's greatest vacuum chamber." Once again, qualifications are necessary. The space around the Shuttle won't be as clean as researchers would like to have it, because trace amounts of gases from the rocket engines and debris dumped from the cargo bay will trail the spacecraft in its orbit. And even that high up there is still an atmosphere, composed of widely dispersed atoms of oxygen, which create a pressure amounting to only ten-billionths of that at sea level on earth. So NASA is thinking of building a flying shield that could be deployed at the end of a boom. As the "air" rushed past the outer edges of the shield at the tremendous speeds of space, it would create a nearly perfect vacuum behind it. Frost has suggested that in this ultra-clean environment the thin film used in solar cells might be manufactured for 1 percent of the cost on earth.

All these marvels won't be accomplished overnight. The first flight in September will be devoted mainly to checking out the systems. Next year the Shuttle will be used to launch communications and other satellites. The first spacelab, equipped for experiments in materials processing, won't go up until the twelfth flight, in mid-1981. (The public will have to wait until about that time to see the first American woman in space.) For most people, the main excitement this fall will come when the astronauts try to jockey their stubby, ungraceful craft back to a safe landing on the salt flats around Edwards Air Force Base in California.

Around 1984, the Shuttle will take up packages containing automated experiments in biology and materials processing. The astronauts will stay in the cockpit, while on-board computers, monitored by scientists in Houston, run the show. A bit later, the Shuttle is expected to take up and release a free-flying automated laboratory powered by its own solar module. Some of the scientists seem to think that the best work will be done when the machines are left alone to do their thing.

"I doubt very much that it will be optimal to have people operating any of our experiments in space," says James H. Bredt, manager of space processing applications at NASA. "The function of man in space is not routine operations but troubleshooting. When you don't have astronauts walking around in their suction-cup boots, more perfect microgravity can be maintained."

In the fourth phase of this great adventure, however, the astronauts will come into their own again. They will be needed to construct the huge stations that will beam solar power back to earth. . . . By then, NASA fervently hopes it will be less of a drain on the taxpayers. It is spending $4.3 billion this year, of which $1.4 billion can be attributed to the Shuttle. By mid-1980's, it may have become a new sort of public utility—exercising a near-monopoly on space and all the wonders that it holds, selling its services to corporations around the world. Or it might even have turned this growing business over to a private corporation. Boeing, for one, thinks that it could run the Shuttle profitably as a commercial enterprise.

18

Technology and engineering

Are engineers responsible for the uses and effects of technology?

AARON ASHKINAZY

Aaron Ashkinazy *is a member of the technical staff in the Solid State Division of RCA in Somerville, New Jersey. In the early 1970's he was a member of the now defunct Committee for Social Responsibility in Engineering. The aim of this committee was to enlighten engineers as to the consequences of the misuse of technology, especially in the area of weaponry.*

In this article, Ashkinazy rejects the view that engineers are technologists "whose role does not include the evaluation of the product's social or environmental desirability." He argues that the concept of professionalism as applied to engineers "implies a code of ethics, to be promoted and enforced by the mutual consent of all members of that profession." Ashkinazy admits that if responsible engineers were to live by such a code and refuse to work on certain projects, they could run the risk of being fired. "Until the numbers change, every concerned engineer must individually resolve this conflict between principle and personal security." Ashkinazy provides several suggestions for effecting "a positive change in our technological priorities" so that they better serve society and the rights of future generations.

How can the individual engineer help to redirect the utilization of technology toward solving society's problems rather than creating more problems? In this article, we consider the efficacy of two fronts of action—individual action at the place of one's employment and public action (i.e., dialogue with and information dissemination to government officials, the public, students), especially by the academic community and the professional societies, with the full support of individual engineers in industry.

Most corporations exist primarily to show a profit. Nonprofit foundations are notable exceptions. As gross as this assertion seems, given the interest that many corporations have been showing in community affairs and in the welfare of their employees, the fact remains that economic gain is the crucial contribution to the future fate of private corporations. The primary goal of an employee is to earn a salary that will enable that employee to achieve an "acceptable" standard of living within society. Unless a crisis situation occurs requiring emergency action, any request to the corporation or to the employee to jeopardize their respective primary goals would not only be considered unreasonable but would also be ignored.

Over 90 percent of all engineers are employees of industrial corporations. The decision-making responsibility of the engineer lies primarily in the realm of how to design a particular product or perform a particular service. If the engineer must also decide whether the product should be developed, the relevant factors considered are almost always technical or economic feasibility; if the product will not ultimately return a profit to the employer, it should not be developed.

This view of the engineer—as a technologist whose role does not include the evaluation of the product's social or environmental desirability—is held not only by the management but by most engineers as well. It should be men-

☐ From *Professional Engineer,* 42, No. 8(1972), 46-47.

tioned that some changes are occurring, primarily due to public and governmental pressure (e.g., compliance with federal auto-emission standards, attention to pollution caused by dumping industrial wastes, and the like). However, if there are no overriding external pressures, the decision-making function relegated to the engineer often excludes social consequences as a major consideration.

As a group, engineers might claim a passive role in contributing to societal problems; although it was the engineer who developed the technology, the priorities and applications of this technology were not under the engineer's control. As any employee must do in order to remain among the ranks of the employed, one does what one's employer wishes. However, it is doubtful that many engineers grudgingly perform their jobs, feeling economically pressured to do the bidding of an immoral and unethical employer. Rather, engineers have been the willing associates (if not accomplices) of industrial and governmental leaders in constructing our technological society, and therefore are equally accountable for the problems that technology has wrought.

The assertion that engineers as a group are responsible for the effects of technology might appear as an unwarranted generalization. What about all the engineers who have never developed anything of detrimental socioecological consequence? Why should they be held accountable for their colleagues' actions? Because the concept of professionalism, if it applies to engineers, implies a code of ethics, to be promoted and enforced by the mutual consent of all members of that profession. Because, as rational, compassionate human beings, everybody (including engineers) has the obligation of ensuring that future generations not only have the right to exist but also have the best possible opportunity of achieving a healthy and untroubled existence.

If the current situation were to be perpetuated, nobody may be around to receive anything from us, and the legacy we leave could be annihilation. Engineers, through personal contacts and through professional societies, are in an excellent position to influence the attitudes of their colleagues. To do nothing, to not communicate with engineers who are working on a socially detrimental project, constitutes implicit support for their actions.

A growing number of engineers are beginning to evaluate their jobs on the basis of the consequent social implications. If an engineer refuses to work on certain projects, this action could serve to inspire others to do the same thing. If enough engineers were to be involved, the risk to each individual would be reduced. Since, however, relatively few engineers would currently adopt such a stand, any individual who attempts this tactic should anticipate the possibility of being fired. Until the numbers change, every concerned engineer must individually resolve this conflict between principle and personal security.

Even if the engineer working in industry were free to follow the dictates of conscience without penalty, it is doubtful that his action, by itself, would cause the future effects of technology to be any less devastating than the technologically induced social maladies we must currently contend with. The key phrase in this assertion (or denial) is "by itself;" that is, although it is absolutely necessary for the engineer to exercise, and be allowed to exercise, judgment about the projects undertaken on the job, significant changes in our technological priorities must ultimately be achieved by engineers working on other fronts. This is primarily due to three factors:

1) The engineer's position in industry is such that it is very far from the corporate executive's ear. That is, the engineer is only expected to decide on how to produce, not what to produce. The only exception is the rarity in which the engineer proposes a product that is obviously a big money-maker.
2) Value judgments are subjective and vary from individual to individual. Thus, individuals will pull in random directions, and no net motion can take place.
3) Technological knowledge is necessarily fragmented and specialized. Hence no individual engineer can be expected to foresee all the undesirable end effects of a particular project.

The first factor is an obvious statement of the engineer's impotent persuasive position within the corporation. In effect, it asserts that if one really wants to see the business community change its directions, he must try to convince the person in marketing that he is a professional.

Most people realize that society faces serious problems, but still, for most people, life goes on as usual. The reason is probably due to the fact that very few people believe that this time the problems may actually lead to our destruction. As long as individuals in both labor and management believe that we will somehow survive, that a crisis situation does not exist, why should they abandon their primary goals and adopt a new priority of saving society? In short, very few individuals will abandon the conditioned drives acquired over a lifetime unless they feel that their own personal survival is being threatened.

Well, is our survival threatened? Will we manage to limp along in the future even if we don't take emergency action now? The academic community is the only group possessing the capability of researching the multitude of problems facing society. Conceivably, academia could not only answer these questions, but it could also recommend actions in the event that we indeed face a serious survival crisis. This is no small task.

If it is possible at all, it will require a coordinated interdisciplinary effort of sociologists, political scientists, physicists, chemists, engineers, and others. All aspects of our socio-ecological problems must be analyzed. Clearly, no other institution has such talents so readily available as do the universities. The professional societies could be of great assistance by providing a medium of communication. Engineers in industry could also provide inputs to this study.

If it is found that we do not face imminent disaster, we can all breathe a sigh of relief. If, however, it is found that we are at the point where irreversible damage has occurred or will shortly occur, then this knowledge must be made public. Every avenue should be pursued to achieve this end, including lobbying, expert testimony before Congressional committees, mass-media appearances, and magazine articles.

Hopefully, this will cause individuals, including many engineers, to pressure their congressmen and senators. Government officials could no longer overlook the political fact that the solution of socio-ecological problems is not only necessary but should be made a national priority. Many corporate executives will be influenced by the media, but with the encouragement of Government, industry will quickly respond. A positive change in our technological priorities would then have been effected.

From *Destination Disaster: From the Tri-Motor to the DC-10, The Risk of Flying*

PAUL EDDY, ELAINE POTTER, and BRUCE PAGE

■ Paul Eddy, *an English journalist, has worked for* The Sunday Times *since 1972. He co-authored* The Hughes Papers, *an investigation of the Howard Hughes empire in Las Vegas, and* The Plumbat Affair, *an investigation into how Israel acquired the means to manufacture nuclear weapons. He has just completed a book on the French underworld, and is co-leader of the* Sunday Times Insight Team.

Elaine Potter, *who was born in South Africa, received a doctorate in politics from Oxford University. She has co-authored* The Press as Opposition: The Political Role of South African Newspapers *and* Suffer the Children, *an investigation into the thalidomide tragedy. She is a specialist writer for* The Sunday Times.

Bruce Page *was educated in Australia and worked on the* Melbourne Herald *and other newspapers before joining* The Sunday Times, *where he became Managing Editor in Charge of Special Projects. He has co-authored several bestsellers, among which are* Philby: The Spy Who Betrayed a Generation *and* American Melodrama *(about the American presidential election of 1968). He left the* Times *in 1976 to become editor of the* London New Statesman.

The following is a passage from Destination Disaster: From the Tri-Motor to the DC-10, The Risk of Flying, *which was a cooperative effort of Eddy, Potter, and Page. The book provides a thorough report on current commercial aviation, with a focus on the crash of a Turkish Airlines DC-10 near Paris in March 1974 which killed 346 persons. The passage informs us of serious design defects in the DC-10, which were known long before the crash but which were ignored. The passage also points out serious defects in the Federal Aviation Administration's certification process. The FAA is responsible for certifying every new airplane. Since it lacks the necessary manpower, the FAA appoints at every manufacturing plant "company men, paid by the manufacturer," to carry out the certification process. This clearly creates a conflict of interest.*

We have already shown that there were inherent weaknesses in the design of the DC-10 and that given the vulnerability of the cabin floor, and its crucial role in carrying the flight controls, the cargo-door locking system was quite inadequate. Often, such perceptions are available only in hindsight. But what the FAA seemingly did not know was that some well-qualified members of the DC-10 design teams were aware almost from the first that the cargo-door system was suspect.

The detail design of the DC-10 fuselage and its doors was largely done by engineers from the Convair division of General Dynamics in San Diego, California. This was for financial as well as industrial reasons: McDonnell Douglas was naturally looking for subcontractors capable of sharing the enormous load of financing a program which could not run into profit for many years. (This is standard procedure in the U.S. aerospace industry. Subcontractors pay their own start-up and engineering costs, getting the money back bit by bit over a predetermined period as they deliver units to the main contractor.)

Convair and its unsuccessful rivals for the contract (North American Rockwell, Rohr Aircraft, and Aerfer) were given, early in 1968, a weighty Subcontractor Bid Document which set out Douglas's requirements for the DC-10. The passenger doors were to be plugs, but the lower cargo doors were to be outward-hinging tension-latch doors. They were to have over-center latches driven by *hydraulic cylinders,* a system already used on some DC-8 and DC-9 doors. In addition to hydraulic latches, each cargo door was to have a manual locking system "designed so that the handle or latch lever cannot be stowed unless the door is properly closed and latched." The bid document was insistent that weight should be saved wherever possible (paint was to be used "to a minimum"), and the subcontractors were told that one pound of weight should be thought of as costing $100.

On August 7, 1968, McDonnell Douglas signed a subcontract with Convair, William Gross of Douglas later said that the reason for choosing Convair, in addition to the financial strength of the parent company, General Dynamics, was the excellent reputation Convair had for structural design. If that was so, it is surprising that Douglas did not take more notice of Convair's reservations about the DC-10 cargo-door design. These began to emerge in November 1968, when Douglas told the San Diego engineers that instead of hydraulic cylinders

□ New York: Quadrangle, 1976, pp. 175-188.

they must use *electric* actuators to drive the cargo-door latches.

The principles are very different. . . . A hydraulic actuator uses high-pressure fluid to drive a piston along a cylinder. An electric actuator, on the other hand, works through gears, and is driven by an electric motor.

Douglas gave two reasons for changing from hydraulic actuators: to do so would save twenty-eight pounds of weight per door and also would "conform to airline practice." The fact was that American Airlines had asked for electric actuators, saying that as they would have fewer moving parts they would be easier to maintain. And in the competitive conditions of the airbus market, requests of this kind from airlines were not likely to be rejected.

Some Convair engineers, and in particular their Director of Product Engineering, F. D. "Dan" Applegate, were never fully reconciled to the change. The hydraulic system was, in Applegate's judgment at least, better because it was more "positive." When engineers say that one system is more positive than another, they are making a value judgment rather than a precise, mathematical statement just as they do when they say that one system is "simpler" than another. Nonetheless, the Convair argument has much substance to it.

Airplane hydraulic systems make power—pressure—available *continuously* to the mechanisms they serve. Fluid pressure from a hydraulic system is pumped through valves into an actuator cylinder. It then pushes the piston through its travel and continues to exert pressure against it. Piston travel is reversed by adjusting the valves so as to allow fluid to flow in from the other end of the cylinder, and thus press upon the other surface of the piston. The point is that whatever the position of the piston, hydraulic fluid is always working on it.

Electric power, on the other hand, is not used continuously. It is switched on to produce movement, and then switched off again. As there is no continuous pressure to maintain the position that an actuator has reached, some kind of mechanical device must be used to prevent the electrically driven piston from slipping back. Most electric actuators, and certainly the type fitted to the DC-10 doors, are therefore *irreversible* in action. Once they have achieved the maximum travel that they can achieve in any particular direction, they remain fixed until the travel is specifically reversed by the application of an opposite electric impulse.

Therefore, an electric latch will behave very differently from an hydraulic one. An hydraulic latch, though positive, is *not* irreversible. If it fails to go over-center, it will in the nature of things "stall" at a point where the pressure inside the cylinder has reached equilibrium with the friction which is obstructing the travel of the latch. Thereafter, quite a small opposite pressure will move it in the reverse direction—and what this means in a pressure-hull door is that if the latches have not gone quite "over," they will slide open quite smoothly as soon as a little pressure develops inside the hull and starts pushing at the inside of the door. Thus, they will slide back and the door will open, well before the pressure inside the aircraft hull is high enough to cause a dangerous decompression. The door will undoubtedly be ripped from its hinges by the force of the slip stream but, at low altitude, that poses no threat of structural damage to the plane and no danger to its passengers. The crew will immediately become aware of the problem, because the aircraft cannot be pressurized and can simply return to the airport.

However, if an irreversible electric latch fails to go over-center, the result will usually be quite different. Once current is switched off, the attitude of the latch is fixed, and if it has gone quite a long way over the spool, there will be considerable frictional forces between the two metal surfaces, holding the latch in place. Pressure building up inside the door cannot *slide* the latches open. It can only force the fixed, part-closed latches off their spools. This, typically, will happen in a swift and violent movement, occurring only when pressure inside the airplane has built up to a level when sudden depressurization will be structurally dangerous.

Hydraulic latches would have been intrinsically safer with even a mediocre manual-lock system to back them up. But once the changeover was made to electric power, it became essential to provide a totally foolproof checking-and-locking backup.

In the summer of 1969 Douglas asked Convair to start drafting a Failure Mode and Effects Analysis (FMEA) for the lower cargo-door system of the DC-10. The purpose of the FMEA is, as the name suggests, to assess the likelihood of failure in a particular system, and the consequences of failure should it occur. Before an airplane can be certificated, the FAA must be given an FMEA for those major systems which are critical to safety.

Convair submitted a draft FMEA for the door

system in August 1969. The design examined was an early one, in which the back-up locking system consisted simply of spring-loaded locking pins (later, of course, the spring-loaded pins were replaced by the manual locking handle and its complex linkages). Convair apologized for having taken two months over the work, due to the fact that their engineers had been working lately on military programs and were not familiar with current FMEA procedures for civil aircraft. In spite of this, and the fact that the door design analyzed was a relatively early one, Convair produced a document which accurately foresaw the deadly consequences of cargo-door failure.

Among the "ground rules and assumptions" of the FMEA, those dealing with failure-warning systems stand up especially well to hindsight. First, said Convair, no great reliance was to be given to warning lights on the flight deck because "failures in the indicator circuit, which result in incorrect indication (i.e., 'lights out') of door locked and/or closed, may not be discovered during the checkout prior to take-off."

Convair claimed that even less reliance should be placed on warning systems which relied on the alertness of ground crews. In this early design, the only way of telling from the outside whether the latches had gone home was to look at the "manual override" handle provided to wind them shut by hand in case of electrical failure. It if had moved through its full travel, the latches must be safe. The Convair FMEA found:

> . . . That the ground crew requirement to visually check the angular position of the manual override handle, to detect an "unlocked" condition, to be subject to human error. It is assured that routine handling of repetitive aircraft could result in the omission of this check or visual error due to the location of the handle on a curved surface under the lower fuselage.

The substance of the FMEA then went on to show that there were nine possible failure-sequences which could lead to a "Class IV hazard"—that is, a hazard involving danger to life. Five of these involved danger to ground crew by doors falling suddenly shut or coming open with undue violence, and these do not concern our narrative. But there were four sequences shown as capable of producing sudden depressurization in flight: also a "Class IV hazard," and meaning in this context the likely loss of the airplane. One of these sequences was, in principle, remarkably similar to what actually occurred over Windsor and later outside Paris.

The starting point was seen as a failure of the locking-pin system, due to the jamming of the locking tube or of one or more of the locking pins. In that case, said the FMEA: "Door will close and latch, but will not safety lock." There should of course be a warning against this, and if it works properly: "Indicator light [in the cockpit] will indicate door is unlocked and/or open." But one of the ground rules of the FMEA was that circuit failures in the indicator system might well go undetected, in which case: "Indicator light will indicate normal position." If that happened, malfunction of the electric latch actuators could produce a situation in which the "door will open in flight—resulting in sudden depressurization and possibly: structural failure of floor; also damage to empennage* by expelled cargo and/or detached door. *Class IV hazard* in flight."

The difference between this and what actually happened over Windsor is: first, the FMEA envisaged failure in a spring-loaded, rather than a hand-driven, locking-pins system; second, the failure mentioned was one of inadvertent electrical reversal of the latches rather than a failure of the latches to go "over-center" in the first place. But it was a powerful demonstration that the door design was potentially dangerous without a totally reliable fail-safe locking system.

The other three depressurization sequences were rather different in that they envisaged total latch failure due to electrical faults, with the door being held shut until danger point by just the electric system which closes the door. These were not relevant to the accidents which ultimately occurred; nonetheless, together with the FMEA's general skepticism about warning lights and ground-crew assessments, they helped to produce a document that spelled out very clearly the terrible consequences that could follow from ill-thought-out door design.

But neither this FMEA draft nor anything seriously resembling it was shown to the FAA by Douglas, who, as lead manufacturers, made themselves entirely responsible for certification of the airplane. (Indeed, under the terms of the subcontract, General Dynamics was forbidden from contacting the FAA about the DC-10.) Our evidence is drawn from documents pro-

*Tail flying surfaces.

duced by Douglas and testimony given in the complex of compensation lawsuits which resulted from the Paris crash.[1] Evidence given by J. B. Hurt, Convair's DC-10 support program manager during the litigation, was that Douglas never replied to the Convair FMEA.

FMEAs submitted by Douglas to the FAA, leading up to certification of the DC-10, do not mention the possibility of Class IV hazards arising from malfunction of lower cargo doors.

The documentary warning of the dangers of depressurization were followed in 1970 by a physical manifestation at Long Beach. But even this it seems could not dent the self-assurance of the Douglas design team.

By May 1970, the first DC-10 (Ship1) had been assembled at Long Beach and was going through ground tests to prepare for the maiden flight scheduled for August. On May 29, outside Building 54 the air-conditioning system was being tested, which involved building up a pressure-differential inside the hull of four to five pounds per square inch. Suddenly the forward lower cargo door blew open. Inside, a large section of the cabin floor collapsed into the hold.

The Douglas response to this foreshadowed the company's response to later and more serious accidents. It was simply blamed on the "human failure" of the mechanic who had closed the door. This explanation was still adhered to by William Gross when he gave evidence during the Paris crash litigation in 1974-75. He gave no sign of thinking that there might be something basically dubious about a system that could become dangerous simply because one man, fairly low down in the engineering hierarchy, failed to perform exactly to plan.

However, Douglas did at the time acknowledge that some modification of the door was required before presenting the whole system for FAA certification. It had already been decided before the Ship One accident that the spring-loaded locking-pin system should be replaced by hand-driven linkages, and now it was decided to try and build some extra safeguards into that system. Ship One took off on its maiden flight on schedule with an unmodified door. But by the autumn of 1970 the "vent door" concept had been adumbrated.

The essence of this system has already been described in Chapter 8. The miniature plug door was let into the main door above and to the right of the locking handle. This was supposed to stand conspicuously open until the main door was latched, and the locking handle was pulled down to drive the locking pins home.

In truth, it added little or nothing to the safety of the system. Such a vent door can only provide a check on the position of locking pins if its closure is a *consequence* of the pins having gone home. This is the case with the Boeing 747 tension latch cargo door, which was already flying by 1970. But in the Douglas scheme, the closure of the vent door was merely *coincidental* to the action of the locking pins. If the rod transmitting the locking-handle's movement were to break, or to be absent altogether, then the Douglas vent door would still close.

On the face of it it seems especially remarkable that such an obvious flaw should have been overlooked. The certification process which the FAA imposes on every new airplane is supposed to identify and reject the offspring of dubious design philosophy. But the process, long and exhaustive though it undoubtedly is, suffers from a fundamental weakness: Although it is carried out in the name of the FAA, much of the work involved is actually done by the manufacturers themselves.

The FAA says it has neither the manpower nor, in some instances, the specialized expertise to inspect every one of the thousands of parts and systems that go to make up a modern airliner. It therefore appoints at every plant Designated Engineering Representatives (DERs)—company men, paid by the manufacturer, who spend part of their working lives wearing, as it were, an FAA hat.[2] Their job during the certification process is to carry out "conformity inspections" of the plane's bits and pieces to insure that they comply with the Federal Airworthiness Regulations. In the case of the DC-10, there were 42,950 inspections. Only 11,055 were carried out by FAA personnel. The rest were done by McDonnell Douglas DERs.

Designated representatives are chosed by the FAA with a careful eye to their experience and integrity, but inevitably conflicts of interest can arise when manufacturers are called upon, in effect, to police themselves. The system also reduces the chance of mistakes being spotted, and the DC-10 vent door system stands as a classic illustration of what can happen. Before certification the vent door system was submitted to a series of tests by McDonnell Douglas and the results were approved, on behalf of the company, by a senior engineer. *Later, this time wearing his DER hat, the same engineer ap-*

proved the report of the tests as acceptable documentation for showing that the DC-10 cargo door complied with the airworthiness regulations.

There were other faults in the design. The linkages had not been stressed correctly, so that they later turned out to be capable of flexing out of shape when submitted to pressure. And their various degrees of travel were all adjustable, so that the whole system was equivocal. Its validity as a check on the function of the latches depended upon whether the linkages in any given door happened to be correctly rigged.

Conceivably such design faults resulted from inexperience with doors of this kind. (Incidentally, if Douglas needed proof of the efficacy of the hydraulic system they had abandoned in the DC-10 cargo doors, it was abundantly available during 1970. There were five examples that year of hydraulic tension-latch doors in DC-8s and DC-9s opening in flight *before* pressure had built up to a dangerous level. All the aircraft landed safely.) But inexperience is no excuse, because in safety matters most airplane designers are willing to render assistance across competitive commercial boundaries.

Indeed, during November 1970, Convair was able to obtain via American Airlines considerable detail on Boeing vent systems for tension-latch doors. Not only did they discover that the two vent doors in the 747's cargo door were driven off the locking tube itself: they also learned that the locking-system consisted of nonadjustable, and so unequivocal, linkages.

Not that everyone at Douglas's Long Beach plant thought that things were going the right way. In November a Convair engineer named H. B. ("Spud") Riggs, who was attached to the Douglas design team, wrote an internal memo headed: "Approaches to Eliminate Possibility of Cabin Pressurization with Door Unsafe." Riggs wrote that the design conception of the vent door was so far "less than desirable." He canvassed other ways of dealing with the problem: going back to hydraulic actuation; adding redundant electric circuits as back-up on the electric actuators; interlinking the door-closing system with the pressurization system; increasing the floor strength sufficiently to make the floor resist any possible pressure differential after a door blowout; providing vent space in the floor to enable high-pressure air to escape without doing damage.

None of these possibilities was incorporated in the DC-10—nor was Rigg's memo given to the FAA. By the following month Douglas as a corporation seemed to be more concerned about the financial consequences of the May blowout than the engineering ones. As with the doors, Convair was responsible for the detailed design of the DC-10 floor—albeit to Douglas's specifications. On December 4 an internal Convair memorandum recorded a negotiation with Douglas officials:

> Douglas indicated that there was nothing defective in the door but that the passenger floor, having failed in the pressure test, was defective, and Convair owed Douglas a new floor. However, since it was not practical to change the floor Douglas wanted Convair to install the blow out door [vent door] in the 300 aircraft to satisfy its obligation on the floor.

There had always been a case for making the floor of the DC-10, and all other wide-bodied jets, strong enough to withstand full pressure differential. To do so would have cost some three thousand pounds in the DC-10 and the TriStar: say, one dozen passengers and their luggage, which in terms of present load factors might seem quite tolerable.

But that has been rejected—by Lockheed and Boeing, as well as Douglas—and the floor which Convair had built was by this time exactly as strong as Douglas had specified. Now Douglas was trying to say that the floor was not really strong enough ("defective"), so the door needed to be more reliable; therefore, the insertion of the vent door should be paid for by Convair.

On December 15, 1970, another Convair memo noted that Douglas had decided that in all cargo versions of the DC-10 the *upper* cargo doors would not have electric actuators and vents, but would go back to hydraulic actuation. M. R. Yale, Convair's Manager of DC-10 Engineering, wrote: "[We] asked Douglas why this approach would not be a better solution to lower cargo doors than vent doors. Only answer received was that Douglas had considered these factors and concluded that the vent door was the appropriate solution."

In the New Year of 1971, with seven months to go before the scheduled date for the certification of the DC-10 as a commercial airliner, Douglas formally directed the installation of vent doors on all lower cargo doors, and although they were delayed by tooling problems and shortages of parts, deliveries were getting under way in June.

On July 6, 1971, a Convair memo gave a résumé of the situation:

(1) Design criteria and design features of operating, latching, and locking mechanisms were specified by Douglas for Convair.

(2) Basic design work was done by Convair engineers working at Long Beach under supervision of Douglas Engineering Department. Douglas retained total responsibility for obtaining FAA approval of DC-10 and prohibited Convair from discussing any design feature with FAA.

(3) After the 1970 incident Douglas unilaterally directed incorporation of the vent door, even though there were in Convair's opinion several simpler, less costly alternative methods of making the failure more remote.

On July 29, 1971, the DC-10 was certificated by the FAA. Less than one year later came the blowout over Windsor, Ontario.

Fifteen days after Windsor, after the gentleman's agreement had been struck, and after Jack Shaffer had relaxed in the knowledge that Jack McGowen had fixed "his goddam airplane," Dan Applegate of Convair wrote a remarkable memorandum, which demands to be quoted in full. It expresses, with a vehemence not commonly found in engineering documents, all the doubts and fears that some of the Convair team felt about the airplane they were working on.

27 June 1972
Subject: DC-10 Future Accident Liability.

The potential for long-term Convair liability on the DC-10 has caused me increasing concern for several reasons.

1. The fundamental safety of the cargo door latching system has been progressively degraded since the program began in 1968.
2. The airplane demonstrated an inherent susceptibility to catastrophic failure when exposed to explosive decompression of the cargo compartment in 1970 ground tests.
3. Douglas has taken an increasingly "hard-line" with regards to the relative division of design responsibility between Douglas and Convair during change cost negotiations.
4. The growing "consumerism" environment indicates increasing Convair exposure to accident liability claims in the years ahead.

Let me expand my thoughts in more detail. At the beginning of the DC-10 program it was Douglas' declared intention to design the DC-10 cargo doors and door latch systems much like the DC-8s and-9s. Documentation in April 1968 said that they would be hydraulically operated. In October and November of 1968 they changed to electrical actuation which is fundamentally less positive.

At that time we discussed internally the wisdom of this change and recognized the degradation of safety. However, we also recognized that it was Douglas' prerogative to make such conceptual system design decisions whereas it was our responsibility as a sub-contractor to carry out the detail design within the framework of their decision. It never occurred to us at that point that Douglas would attempt to shift the responsibility for these kinds of conceptual system decisions to Convair as they appear to be now doing in our change negotiations, since we did not then nor at any later date have any voice in such decisions. The lines of authority and responsibility between Douglas and Convair engineering were clearly defined and understood by both of us at that time.

In July 1970 DC-10 Number Two[3] was being pressure-tested in the "hangar" by Douglas, on the second shift, without electrical power in the airplane. This meant that the electrically powered cargo door actuators and latch position warning switches were inoperative. The "green" second shift test crew manually cranked the latching system closed but failed to fully engage the latches on the forward door. They also failed to note that the external latch "lock" position indicator showed that the latches were not fully engaged. Subsequently, when the increasing cabin pressure reached about 3 psi (pounds per square inch) the forward door blew open. The resulting explosive decompression failed the cabin floor downward rendering tail controls, plumbing, wiring, etc. which passed through the floor, inoperative. This inherent failure mode is catastrophic, since it results in the loss of control of the horizontal and vertical tail and the aft center engine. We informally studied and discussed with Douglas alternative corrective actions including blow out panels in the cabin floor which would provide a predictable cabin floor failure mode which would accommodate the "explosive" loss of cargo compartment pressure without loss of tail surface and aft center engine control. It seemed to us then prudent that such a change was indicated since "Murphy's Law"[4] being what it is, cargo doors will come open sometime during the twenty years of use ahead for the DC-10.

Douglas concurrently studied alternative corrective actions, inhouse, and made a unilateral decision to incorporate vent doors in the cargo doors. This "bandaid fix" not only failed to correct the inherent DC-10 catastrophic failure mode of cabin floor collapse, but the detail design of the vent door change further degraded the safety of the original door latch system by replacing the direct, short-coupled and stiff latch "lock" indicator system with a complex and relatively flexible linkage. (This change was accomplished entirely by Douglas with the exception of the assistance of one Convair engineer who was sent to Long Beach at their request to help their vent door system design team.)

This progressive degradation of the fundamental safety of the cargo door latch system since 1968 has exposed us to increasing liability claims. On June 12, 1972 in Detroit, the cargo door latch electrical actuator system in DC-10 number 5 failed to fully engage the latches of the left rear cargo door and the complex and relatively flexible latch "lock" system failed to make it impossible to close the vent door. When the door blew open before the DC-10 reached 12,000 feet altitude the cabin floor collapsed disabling most of the control to the tail surfaces and aft center engine. It is only chance that the airplane was not lost. Douglas has again studied alternative corrective actions and appears to be applying more "band-aids." So far they have directed us to install small one-inch diameter, transparent inspection windows through which you can view latch "lock-pin" position, they are revising the rigging instructions to increase "lock-pin" engagement and they plan to reinforce and stiffen the flexible linkage.

It might well be asked why not make the cargo door latch system really "fool-proof" and leave the cabin floor alone. Assuming it is possible to make the latch "fool-proof" this doesn't solve the fundamental deficiency in the airplane. A cargo compartment can experience explosive decompression from a number of causes such as: sabotage, mid-air collision, explosion of combustibles in the compartment and perhaps others, any one of which may result in damage which would not be fatal to the DC-10 were it not for the tendency of the cabin floor to collapse. The responsibility for primary damage from these kinds of causes would clearly not be our responsibility, however, we might very well be held responsible for the secondary damage, that is the floor collapse which could cause the loss of the aircraft. It might be asked why we did not originally detail design the cabin floor to withstand the loads of cargo compartment explosive decompression or design blow out panels in the cabin floors to fail in a safe and predictable way.

I can only say that our contract with Douglas provided that Douglas would furnish all design criteria and loads (which in fact they did) and that we would design to satisfy these design criteria and loads (which in fact we did).[5] There is nothing in our experience history which would have led us to expect that the DC-10 cabin floor would be inherently susceptible to catastrophic failure when exposed to explosive decompression of the cargo compartment, and I must presume that there is nothing in Douglas's experience history which would have led them to expect that the airplane would have this inherent characteristic or they would have provided for this in their loads and criteria which they furnished to us.

My only criticism of Douglas in this regard is that once this inherent weakness was demonstrated by the July 1970 test failure, they did not take immediate steps to correct it. It seems to me inevitable that, in the twenty years ahead of us, DC-10 cargo doors will come open and I would expect this to usually result in the loss of the airplane. (Emphasis added.) This fundamental failure mode has been discussed in the past and is being discussed again in the bowels of both the Douglas and Convair organizations. It appears however that Douglas is waiting and hoping for government direction or regulations in the hope of passing costs on to us or their customers.

If you can judge from Douglas's position during ongoing contract change negotiations they may feel that any liability incurred in the meantime for loss of life, property and equipment may be legally passed on to us.

It is recommended that overtures be made at the highest management level to persuade Douglas to immediately make a decision to incorporate changes in the DC-10 which will correct the fundamental cabin floor catastrophic failure mode. Correction will take a good bit of time, hopefully there is time before the National Transportation Safety Board (NTSB) or the FAA ground the airplane which would have disastrous effects upon sales and production both near and long term. This corrective action becomes more expensive than the cost of damages resulting from the loss of one plane load of people.

F. D. Applegate
Director of Product Engineering

Although this was not a formally set-out safety analysis, like the Convair-drafted FMEA which Douglas did not give to the FAA, the Applegate memorandum was in some respects an even more disturbing document. Yet not only did its contents not reach the FAA, where surely they would have eroded even John Shaffer's durable complacency, they were not even put by Convair to Douglas.

The immediate fault in this must, of course, lie with Convair. But something must also be said about motivation: Briefly, Convair's experience over the previous three years had led to the belief that to raise such major safety questions with Douglas was chiefly to give away points in an ongoing financial contest. Certainly the record appeared to be one in which Douglas had been unimpressed by the safety propositions argued by Convair in its draft FMEA, and by the middle of 1972, it seems clear the position of the two companies was essentially an adversary one. Five days after Applegate wrote his memorandum his immediate superior, Mr. Hurt, wrote an equally revealing comment upon it:

3 July 1972.

From: J. B. Hurt.
Subject: DC-10 Future Accident Liability.
Reference: F. D. Applegate's Memo, same subject, date 27 June 1972.

I do not take issue with the facts or the concern expressed in the referenced memo. However, we should look at the "other side of the coin" in considering the subject. Other considerations include:

1. We did not take exception to the design philosophy established originally by Douglas and by not taking exception, we, in effect, agreed that a proper and safe philosophy was to incorporate inherent and proper safety and reliability in the cargo doors in lieu of designing the floor structure for decompression or providing pressure relief structure for decompressions or providing pressure relief provisions in the floor. The Reliance clause in our contract obligates us in essence to take exception to design philosophy that we know or feel is incorrect or improper and if we do not express such concern, we have in effect shared with Douglas the responsibility for the design philosophy.
2. In the opinion of our Engineering and FAA experts, this design philosophy and the cargo door structures and it original latch mechanism design satisfied FAA requirements and therefore the airplane was theoretically safe and certifiable.
3. In redesigning the cargo door latch mechanism as a result of the first "blowout" experience, Douglas unilaterally considered and rejected the installation of venting provisions in the floor in favor of a "safer" latch mechanism.[6] Convair engineers did discuss the possibility of floor relief provisions with Douglas shortly after the incident, but were told in effect, "We will decide and tell you what changes we feel are necessary and you are to await our directions on redesign." This same attitude is being applied by Douglas today and they are again making unilateral decisions on required corrections as a result of the AAL Detroit incident.[7]
4. We have been informally advised that while Douglas is making near-term corrections to the door mechanism, they are reconsidering the desirability of following-up with venting provisions in the floor.

I have considered recommending to Douglas Major Subcontracts the serious consideration of floor venting provisions based on the concern aptly described by the reference memo, but have not because:

1. I am sure Douglas would immediately interpret such recommendation as a tacit admission on Convair's part that the original concurrence by Convair of the design philosophy was in error and that therefore Convair was liable for all problems and corrections that have subsequently occurred.
2. Introducing such expression at this time while the negotiations of SECP 297 and discussion on its contractual justification are being conducted would introduce confusion and negate any progress that had been made by Convair in establishing a position on the subject.[8] I am not sure that discussion on this subject at the "highest management level" recommended by the referenced memo would produce a different reaction from the one anticipated above. We have an interesting legal and moral problem, and I feel that any direct conversation on this subject with Douglas should be based on the assumption that as a result Convair may subsequently find itself in a position where it must assume all or a significant portion of the costs that are involved.

J. B. Hurt
Program Manager, DC-10 Support Program.

On July 5, 1972, Mr. M. C. Curtis, the Convair vice-president who was in overall charge of the DC-10 project, called a meeting to decide corporate policy in the light of the Applegate memorandum. (Hurt's memo was chiefly a briefing to Curtis for the meeting.) Convair's chief counsel, director of operations and director of contracts, attended along with Applegate and Hurt. (It seems that one person not consulted was David Lewis, the one-time heir apparent of "Mister Mac" who became president of General Dynamics in 1970. Ironically, it was Lewis as president of the Douglas division who, in 1968, negotiated the financial details of the contract with Convair.)

It was acknowledged that Applegate was closer than Hurt to the engineering of the DC-10 and had a better knowledge of the safety factors involved. But Mr. Curtis and his colleagues preferred the reasoning of Hurt's memo and resolved the "interesting legal and moral problem" by deciding that Convair must not risk an approach to Douglas. According to Hurt's testimony, two-and-a-half years later in Hope versus McDonnell Douglas, the meeting came up with a rationalization which, though touched with cynicism, had justification of a sort. "After all," said Hurt, "most of the statements made by Applegate were considered to be well-known to Douglas and there was nothing new in them that was not known to Douglas." And it is certainly hard to believe that the Douglas design team could have claimed, after three years of close collaboration with Convair, that the arguments of the Applegate memorandum were unknown to them.

Both to Douglas and to Convair the dangers were, or should have been, obvious. The determination of Convair, as subcontractors, was not to take upon itself the duty of pointing them out.

And so, because of the interrelated failures of McDonnell Douglas, Convair, and the Federal Aviation Administration, a fundamentally defective airplane continued on its way through the "stream of commerce" (as plaintiff lawyers like to call it). Unhappily, some examples of the DC-10 were soon to find their way into places where that stream flowed in a somewhat troubled fashion. Just as the aviation industry likes to pretend that all airplanes are created equal, so the same superstition is attached, at least in public, to the airlines that operate them. But, as a little investigation will show, the fact is that some airlines are quite strikingly less equal than others.

NOTES

1. *Hope v. McDonnell Douglas et al.:* Civ. No. 17631, Federal District Court, Los Angeles, California.
2. The FAA also appoints Designated Manufacturing Inspection Representatives (DMIRs) to assist the agency in monitoring production.
3. We have been unable to establish whether Applegate's reference to an accident involving Ship Two, in July 1970, is a mistake on his part or whether there were *two* blowout incidents. Certainly Ship One was damaged on May 29, 1970, in circumstances very similar to those described by Applegate.
4. Murphy's Law: "If it can happen, it will." Also known sometimes as the totalitarian or Hegelian law of physics, from Hegel's view that all that is rational is real: that is, if you can think of it, it must exist. This is a case where implausible philosophy makes for good engineering.
5. Douglas's design criteria called for the floor to withstand a pressure of 3 psi, and it eventually did so—although not until Douglas had challenged Convair's original stress analysis, and a stronger kind of aluminum alloy had been introduced to the floor beams.
6. Vents built into the floor would allow pressurized air to escape into the hold without buckling the floor.
7. American Airlines. Detroit is where the DC-10 landed after the Windsor incident.
8. A reference to negotiations over the cost of fitting three vent doors in each of 300 projected aircraft: a total cost of $3 million.

Solar energy—a practical alternative to fossil and nuclear fuels

HAROLD E. KETTERER and JOHN R. SCHMIDHAUSER

■ Harold Ketterer, *a physicist, is currently a Developmental Associate at the Fusion Energy Division of Oak Ridge National Laboratory. He co-authored the following article while at the University of Iowa.*

John Schmidhauser is a professor of political science and chairman of the political science department at the University of Southern California. He was a contributor to the monograph report on ocean thermal energy that was made to the National Science Foundation in 1977. He has published extensively in the area of judicial behavior, and has represented the First Congressional District of Iowa in the Eighty-ninth Congress (1965 to 1966). While in Congress he introduced a bill that would provide funds for the development of solar energy.

This article was published with the following abstract prepared by the authors.

The issue of whether the scientific and engineering communities of highly industrialized societies adequately evaluate and investigate all alternatives to fossil fuels is exceedingly serious with relation to the future development of energy sources. The economic restraints on the developing of solar and solar related energy sources have inhibited its development. The implications of alternative source development for diminishing environmental pollution are exceedingly great. But significant political and economic consequences are also likely to result from low cost development of solar energy with special implications for the domestic poor of America and the underdeveloped nations.

Many years ago it was predicted that there would be a time when the available resources for power production would dwindle to the point that rationing would occur, and eventually these resources would simply run out. At the time, few took notice of these gloomy predictions because there was always more power available than anyone needed. Suddenly it is upon us, and every individual in the country who drives a car or prefers an air-conditioned home in the heat of summer is made aware of these previously meaningless predictions. It is being referred to as the energy crisis, but the current problem is more aptly termed an energy shortage when compared with what will occur in the future if the existing trends are not drastically changed. The future holds the real crisis, but only if we ignore the warning that is being presented to us currently in the form of gasoline shortages and power brown-outs.

Although very difficult to predict because of the number and uncertainty of the variables involved, estimates of the length of time that fossil fuels (coal and petroleum) will be available to supply required energy demands range from 50 to 100 years for petroleum and from 100 to 300 years for coal.

The issue of whether the scientific and engineering communities of the highly industrialized societies adequately evaluate and investigate *all* of the alternatives to fossil fuels as energy sources was raised cogently by Reynhart[1]. He first defined the energy bases for modern civilization in the context of Western industrialization and the relationship of these bases to prosperity. He summed this up as follows:

> . . . the generally accepted standard states that the level of civilization is defined by the *use of energy* by the people. It is tacitly accepted that civilization goes together with prosperity.

He then posed the question: "How long will atomic energy and fossil-fuel energy last?"

> . . . the reserves of oil and coal will be used with an increase of 1% per year after about 200 years and with an increase of 4% per year after 80 years. Taking the quantities of oil and coal on an equal basis . . . then the oil will be used up after 90 years with 1% increase per year, or after 50 years with 4%, i.e., in about the year 2010. Nuclear energy will last quite a little longer, but after a rather short time this energy, too, will be exhausted. Then the last source will be deuterium, if, in the meantime, we succeed in converting this material into usable energy.

He concluded:

> An increase in the consumption of energy of 4% per year means that within 100 years the resources of oil and coal will be exhausted. Continuing with the exploitation of uranium, even if deuterium proves usable, will produce a gigantic scar in the earth's crust within only a couple of centuries, and posterity will have to face the disastrous consequences. The only form of energy mankind will have at his disposal, then, is the energy of the sun.

☐ From *Energy Sources,* 1, No. 3(1974), 249-269.

However, this does not represent the entire picture. In the past few years everyone has become more concerned with the protection of our environment and sustaining a viable ecology. This problem, too, was predicted far in advance of the time when it finally started to affect enough individuals that something has started to be done about it. Thus it is necessary when considering the energy problem that faces us to include in our thinking any effects that our proposed solutions will have on the environment. This means that we must start now to develop methods of power production which will meet the demands of the future on a long-term basis and which will not be a growing hazard to our ecological system. When considering the environmental impact of any proposed energy source the ultimate objective must be to maintain or improve living conditions for all human beings. Thus, when considering potential alternatives, the problems of those individuals or groups who lack access to decision makers or the economic resources to hire lawyer advocates must be considered as thoroughly as the problems of the influential.

Profound understanding of the fundamentals of the human condition in society has, more often than not, been the hallmark of the artist or writer rather than the corporate-financed public-relations-oriented politician. George Orwell[2], in his generally ignored essay *Marrakech,* once described the essence of human relations in colonial Morocco in a manner never equalled by the strutting generals or nervous premiers of the France of that bygone era. His theme was summed up in a moving description of a group of very old Moroccan women who filed singly past his house every afternoon. The essence of colonialism in French Morocco was underscored in his comment: ". . . what is strange about these people is their invisibility." Orwell's observation applies with telling directness to the condition of America's poverty-stricken people. This is especially true when decisions concerning environmental pollution are being made.

It is frequently argued that the rich as well as the poor suffer from environmental pollution because the essence of an "advanced" industrial society is its heavy dependence upon fossil fuels for the production of energy. The very act of extracting fossil fuels may despoil the environment of rich and poor alike. Oil seepage off Santa Barbara brought even-handed desolation to the yachts of the wealthy and the beaches of the poor, and decimated with awful and awesome completeness the multitudes of shore birds of that region. The Santa Barbara incident undoubtedly contributed a great deal to public awareness of the pollution problems inherent in underwater mineral extraction, and assuredly made an indelible impression upon an upper middle class traditionally immune to the more overt manifestations of environmental pollution. However, it did not change significantly either the demography of intensive exposure to the most serious forms of pollution, that which is concentrated in overpopulated and heavily industrialized metropolitan centers, or the conventional mode of environmental policy making and decision making by those who own, manage, or are influenced by the corporations which extract and consume fossil fuels.

As Harry Caudill[3] so eloquently chronicled, few of the mine owners who despoiled the Appalachian region live in it. Neither do the industrialists who pollute the air and water of congested cities live in them. Instead they reside in pleasant suburban or rural enclaves conveniently shielded from the unpleasantness associated with the environmental wrongs they have committed. Although the universal impact of energy use in urban industrial societies ultimately will affect every segment of their population, the particular vulnerability of the poor must receive special attention.

ECONOMIC AND POLITICAL POWER: WHY SOME ALTERNATIVES RECEIVE GREATER CONSIDERATION THAN OTHERS

It is clear that the greatest single source of air pollution is energy conversion. The problem is generally subdivided into four segments, industry, utility electric power generation, transportation (particularly automotive), and commercial and households, and every portion, except households, is dominated by corporate decision making. The poverty stricken and, indeed, the public at large have little direct influence over corporate decision making. In fact decision making concerning fundamentals is becoming concentrated in fewer hands. The production and sale of electrical energy provides an excellent illustration.

The Senate Antitrust Subcommittee accumulated sobering evidence that a massive and, to date, highly successful program to acquire competing fuel sources for the production of electrical energy had been organized by the corporate giants of the petroleum industry. After gathering data on oil industry penetration of competing fuel sources for the Senate Antitrust Subcommittee, the National Economic Research Association (NERA) warned:

The acquisition of the oil companies across the energy market spectrum . . . may be viewed as classic horizontal integration on a scale comparable to the formation of the trusts in the latter decades of the nineteenth century. In short, the oil companies, themselves portraying their activities as efforts at diversification, are in fact systematically acquiring their competition.[4]

The factual record bears out NERA's charge. Twenty-five of the largest oil corporations have penetrated into other fuel industries. Nine of the oil corporations have moved into all of the competing fuel sources, coal, gas, uranium, and shale. NERA and some Senate Antitrust Committee members are seriously concerned about the implications of this penetration by petroleum corporations for electric utility costs to the consumer. This is a very serious threat in and of itself, but the concentration of oil company control has two broader-ranging consequences. One was highlighted decisively by Senator George Aiken of Vermont when, in August of 1970, he warned: "There is some group determined to get control of electrical energy in this nation"[5]. In addition to the possible domination of energy production by a monopolistic group of petroleum corporations, there is a second, more subtle implication. The immense economic and political influence of domestic and international petroleum corporations (as well as other fossil-fuel-oriented companies) inhibits full-blown efforts at finding and developing practical alternatives to fossil fuels as sources of energy production.

The necessity for a reordering of national priorities is starkly evident in the tragic regional sprawl of environmental pollution. Yet change can only come from communities often conditioned by monopolistic energy and fossil-fuel-oriented corporations to consider what are fundamentally nonalternatives to a fossil-fuel basis for the energy and power that is basic to a modern industrial society. The nature of the energy sources for particular societies is directly related to the quality of life in such a society. Consequently, the issue has deep significance for America's poverty stricken and ultimately for the multitude of poor throughout the world, particularly those in the underdeveloped nations.

Michael Tanzer[6] urged recognition of the indispensible role of energy in modern societies:

. . . First, energy fuels are indispensible for modern industry and agriculture. In fact any economist who like this author has studied the relationship between energy and the total economy in a variety of countries would be tempted to add Energy as the fourth factor of production to the classic ones of Land, Labor, and Capital. For just as capital without labor is useless, so too is sophisticated capital without energy.

A United Nations study[7] of energy development in Latin America underscored Tanzer's argument by concluding:

Although in many cases the cost of energy, especially electricity, represents but a minor percentage of total costs, energy exercises great influence because of its qualitative effects. It is the key element without which the production process cannot operate adequately, and the lack or shortage of energy may cause serious difficulties. It stands in the same position as other tangible or intangible factors of industrial production, the economic effect of which is more important than their net cost.

Electric power consumption rises steadily as the appliances and gadgets of modern life multiply. Cities are ringed by fuel-burning steam-electric generators. Even the most carefully engineered and operated power plants cannot help but add great tonnages of combustion wastes, sulfur dioxide, and ash to the surrounding air. Smoke stacks rise as high as 800 ft to aid in dispersion of the pollutants. Few are really willing to reduce the use of electric power or of the modern conveniences utilizing such power. As air pollution laws are tightened, we are still faced with the irreducible minimum pollution load for every kilowatt generated by burning fossil fuel.

In short, the need for new energy sources to replace and supplement carbonaceous fuels is reinforced by the growing environmental pollution problem.

WHAT ARE THE ALTERNATIVES?

We can broadly classify all the energy resources of the earth into five categories: (1) the tides, (2) the earth's heat, (3) fission fuels, (4) fusion fuels, and (5) solar energy.

The tides are produced by the gravitational forces of the earth-moon-sun system. Power from the tides can be obtained by the filling and emptying of a bay or estuary which can be closed by a dam. The enclosed basin is allowed to fill at high tide, and power is extracted via hydraulic generators (as are currently used with river dams) as the water flows out. One such installation exists of the Rance estuary on the Channel Island coast of France. Its operating capacity is 240 MW (electric). Other such projects have been proposed, but the total available power if all favorable inlets were used will, with reasonable estimates, provide only a small fraction of the projected needs of the future. Environmentally, the tidal-power scheme is totally clean in every way, so that this approach can be considered a good method for producing at least some of the power required in the regions where it is applicable.

The earth's heat is derived primarily from radioactive decay within the earth, and is the source of volcanoes and geysers. This heat can be harnessed to produce power in various ways. Hot steam geysers, where they exist, can run generators essentially identical to those used in fossil-fuel plants. Three such facilities currently exist: in the Lardello area of Italy, capacity 370 MW; in the Geysers area of northern California, current capacity 82 MW, projected capacity 400 MW; in the Wairakei area in New Zealand, capacity 290 MW. Lower-temperature steam is available in other areas, but it is a wetter steam and new methods would have to be devised to apply it to power production. Finally, it may be possible to tap the heat of the earth directly by drilling or causing a fissure such that water can be heated by the hot rocks, and applying this hot water to power generators. In terms of the environment the only possible pollution is thermal, except in those areas where the steam geysers had already existed. Again, though, this method was fairly limited in that only certain areas currently have geysers, and only certain areas which would yield heat by drilling or by fissures can be located.

Fission involves the splitting of nuclei of heavy elements such as uranium, and power is obtained by heating water for steam generators with the heat released in the process. Currently, uranium-235 is used as the basic fuel for all fission reactors because it is the only atomic species capable of fissioning under relatively mild conditions. Uranium-235 is rare (0.7% of naturally occurring uranium), and the projected length of time that power could be supplied by this alone is very short (on the order of 50-100 years). Thus the current trend in nuclear power research is toward the breeder reactor, wherein, by absorbing neutrons in a reactor, uranium-238 (99% of naturally occurring uranium) is transformed into fissionable plutonium-239, or thorium-232 becomes fissionable uranium-233. In this case the entire supply of natural uranium and thorium becomes available as fuel for fission reactors and the projected duration increases considerably to the point of no longer being as great a threat. The problem to consider with fission power sources, however, is that of the environment. As the popular advertisements proclaim, nuclear energy is clean in terms of atmospheric pollutants, that is, there is no smoke since there is no fire in the ordinary sense. What the advertisements do not mention, however, are the thermal and radioactive pollution problems.

Thermal pollution occurs because not all of the heat produced to make steam for the electric generators is used and therefore must be disposed of in some manner. Any process that converts one form of energy to another must conform to the laws of thermodynamics. One of these laws states that no conversion process can be 100% efficient. The average efficiency of a fossil-fuel plant is 40% compared with about 30% for nuclear power plants. This is to say that for the nuclear power plant 70% of the energy contained in the fuel appears as waste heat to be disposed of in the environment. In comparing the fossil-fuel and nuclear plant figures it must also be noted that, to produce the same electric output as a fossil-fuel plant, a nuclear plant requires about a third more fuel. Thus the nuclear plant produces about 50% more waste heat than the fossil-fuel plant of the same watt rating. At one time this waste heat was simply dumped into an adjacent body of water. Thanks to ardent environmentalists this practice is being phased out, and power plants are required to reduce the heat input to rivers or lakes usually by the methods of cooling ponds or cooling towers which give up most of the heat to the surrounding atmosphere. Although at this time these are only local heating effects, the overall picture for the future indicates that such amounts of waste heat occurring at numerous power plants will start to produce large-scale effects.

The effects of radiation upon the environment are difficult to assess. Of course, large doses in short periods of time cause effects that are relatively easy to predict. However, the problem with nuclear reactor power plants (barring large-scale disasters) is that the radioactivity released to the surrounding environment is a small amount. For some time now arguments have been taking place as to exactly what level of radiation can be considered harmless over a period of time. This matter has yet to be settled, and, until it is, should we allow the unknown and possibly injurious effects to continue and grow in magnitude as more nuclear power plants are constructed? It is worth noting that the radiation difficulties are more serious with the breeder reactors which would solve any nuclear fuel shortage problems which might occur. The breeder with the most promise is the liqiud metal fast breeder reactor, and it works on the conversion of uranium-238 to plutonium-239. Plutonium-239 is the fissionable material which goes into making the atom bomb, and when this material is available and used in large quantities the probability for a disaster increases to a great extent. One more item to be considered for both types of reactor is the waste-fuel problem. After a time, the fuel elements used to generate the necessary reaction become fatigued to the point of requiring replacement. These spent elements are highly radioactive, and, after processing to save some of that material which is still useful, there remains a certain amount of waste which must somehow be stored in such a manner that it does not affect the environment. Most of these wastes must be safely contained for a period of hundreds of thousands of years before they become degraded to the point of being harmless. Where can we store this amount of radioactive waste such that it will be safe for the length of time? Finally, we must consider what happens with the reactor itself after its useful life of 30-40 years. At the present time it is considered uneconomical to dismantle the facility because of the intense radioactivity at the core. Thus the current approach to this difficulty is to entomb the reactor in concrete and leave this as another problem for future generations to maintain as safely as possible, if possible!

It is interesting to note here that the fuel which dominates current economic and scientific research is fission fuel. Significantly, the same corporate petroleum interests which have captured major segments of the sources of competing fossil fuels are rapidly gaining control of uranium production. The transition has been exceedingly swift. Before 1966 only two oil companies were heavily involved in uranium development. By 1968, 44% of the drilling and exploration for uranium were conducted by petroleum corporations. According to testimony before the Senate Antitrust Subcommittee given in May, 1970, petroleum corporations controlled 45% of the known uranium reserves (as of January, 1970). The interaction of accelerating fuel needs and growing monopolistic control of resources by the oil industry has created serious problems affecting the pocketbooks of consumers, the growing environmental crisis, and the ability or will of government to protect the public from the consequences of unregulated monopolistic control by one sector of corporate enterprise. The Federal Power Commission (FPC) estimated in 1970 that electrical energy needs were being met as follows (percentages rounded): coal, 49%; natural gas, 23%; hydroelectric power, 17%; oil, 9%; nuclear power, 4%. The FPC estimated that by 1980 nuclear energy would total 29%, while by 1990 it would reach 52%. Clearly, the United States has made its commitment to nuclear energy as the auxiliary and/or alternative to fossil fuels. It seems equally clear that in its economic and political aspects, the choice of nuclear energy as the alternative perpetuates and solidifies the control of energy in the hands of a handful of dominant corporations. Indeed, the one major element still outside the control of these corporations, the nuclear enrichment plants under the authority of the Atomic Energy Commission, is scheduled for sale to private enterprise, in President Nixon's announcement of intent, "at such time as various national interests will be best served". Senator George Aiken terms the proposed sale:

> . . . the most advanced step toward private control of the nation that has ever occurred. When you control energy—and the oil interests control coal and are on their way to controlling nuclear fuel—then you control the nation. I see this as a very serious threat to political democracy.

The political implications are compounded by the economic impact of potential control by private corporations. Philip Sporn, retired president of the largest American utility holding company, American Electric Power, Inc., estimated that private ownership of the crucial enrichment plants would generate a 50% increase in the cost of nuclear fuel.[8]

Despite the serious economic and political problems which confront the nation because of the commitment to nuclear fuel under private control, the commitment to nuclear energy was summed up accurately in the *Washington Post;* "When it comes to generating electricity King Coal is losing power: the heir apparent is uranium."[9]

In contrast to an accelerating national commitment to nuclear energy, scientist Peter E. Glaser of Arthur D. Little, Inc., argued in 1967 that "we should examine alternate means to generate power *before* the nations of the world commit themselves irrevocably to the large-scale production of electrical power with nuclear energy"[10] (our italics). He specifically recommended utilization of the technology developed for the space program in projects for the development of satellite solar power stations.[10] Citing the 2nd Progress Report of the House Subcommittee on Science, Research, and Development, Glaser argued that it is not enough to plan to close the widening gap between known energy sources and growing anticipated needs. It is necessary "to learn how to control the deteriorating environment which we create by using presently known energy sources." He quoted the committee report:

> We have grown accustomed to the thought that, at some point in the future when our fossil fuels have been used up (or perhaps long before), atomic energy will be available as an endless source of efficient power. But will it? Since the fission of the first atom in 1938 there has been little progress in devising a way to rid ourselves of the toxic byproducts. The best we can do with radioactive waste is what we first thought of—bury it. So we store it in containers underground. But some day that system will no longer be feasible. Then what? Will we have wasted precious time, money, and effort on an energy source that cannot be made practical for widespread use unless we can break down the byproducts for quick and safe disposal or reuse. Should we have been working much harder on this problem as well as on alternatives such as the fuel cell or magneto hydrodynamics? At this point there is no convincing evidence that anyone really knows—but we should be finding out.[11]

Glaser's challenge and the basic questions raised by the House subcommittee deserve examination.

In summarizing the potential for fission fuels we must try to balance the benefits with all the costs. The only benefits are that it works at present and the availability of recources with the breeder reactor are relatively long term. However, if the nation were to depend in a large part on fission power the following problems will have to be dealt with.

(1) Waste heat, approximately 50% more per watt rating than fossil-fuel plants. If large numbers of nuclear power plants are built this problem will become widespread.

(2) Local effects due to small doses of radiation over long periods of time. This is one more case where controlling corporate interests have been able to dupe the public into accepting terms which could easily be deleterious to present and future generations.[12] This item deserves immediate and extensive attention.

(3) Transportation and security for large quantities of radioactive wastes and fuels. If the plutonium-239, which will become more readily available (when breeder reactors increase in number), were to fall into the wrong hands, any number of disasters might occur. As more radioactive materials are moved from place to place the probability for accidents in this area increases.

(4) Storage of wastes for hundreds of thousands of years. As the use of fission power increases the waste storage problems increases. Imagine trying to find some place which will be dry and free from any natural or man-made disaster for hundreds of thousands of years! This problem deserves a great deal of consideration.

(5) The possible political and economic impacts of permitting an increasing proportion of the energy industry to be controlled by a few corporations.

The fourth category of energy resources to be considered is that of fusion fuels. Fusion involves combining light nuclei; for example, the reactions considered to be most promising are deuterium-tritium and deuterium-deuterium (the hydrogen nucleus contains one proton and no neutrons, the deuterium nucleus contains one proton and one neutron, and the tritium nucleus contains one proton and two neutrons). A well-known fusion reaction is that of the hydrogen bomb. Unfortunately, at the present time, fusion reactions capable of being harnessed for power production are unattainable. Presumably, in view of the research that is taking place at present, this type of power generation will be available some day. If ever it is available, it will be a very useful source of power. Resource supplies for fusion processes are virtually unlimited so this will be no problem. Environmentally, the fusion reactor should be less of a difficulty than current fossil-fuel or nuclear fission reactors if the proposed method or direct power extraction can be attained.

Thus fusion energy does hold great promise for the future, but exactly how long we must wait no one knows because of the huge technological and scientific problems that must be overcome before this process becomes practical. Although no cost estimates are currently available, it would not be too pessimistic to assume that the cost will be a good deal higher, at least initially, for fusion power because of the complex and intricate mechanisms involved.

The last category of energy resources is solar energy. This category is considered in two parts: current and stored solar energy. The term stored solar energy refers to the fossil fuels which at one time chemically stored energy from the sun and are currently being withdrawn from the earth and consumed. As previously stated, the fossil-fuel resources are finite and will eventually be depleted so we need not consider this resource further here. The current solar energy sources include the winds, the hydrologic cycle, and, of course, the direct radiation of the sun.

Various methods have been attempted to harness the energy of the winds and, though successful on a small scale, this source is considered too variable to be applied on any large-scale electric production basis. This source does deserve more research, though, for in some areas it could be used as a supplement to other sources and save on other fuels in these areas. Power production from the wind as currently proposed would be environmentally clean.

The hydrologic cycle encompasses the process by which water is evaporated from the oceans and lakes by the heat of the sun and later deposited as rain, snow, and so on. Power applications from this cycle are the hydrogeneration power plants currently in use, such as those of the Tennessee Valley Authority. This method of power production causes little effect on the environment in that there is no thermal or atmospheric pollution. However, as with tidal and geothermal sources, this method is only applicable at certain locations and is thus limited.

The final source of energy to be considered is the direct radiation from the sun. Solar energy is free and available everywhere. Technical developments have already indicated a variety of possible uses. Operational units are converting this energy directly to electricity, as in the familiar solar cells which power space vehicles and operate telephone line amplifiers. Solar energy converted to heat energy is being tested to warm homes in the winter and cool them in the summer. The energy of the sun is employed in photosynthesis not only in growing crops, but in treating sewage. Sunlight can be harnessed to recover fresh water from the sea by distillation. All these applications, and many more, are instances of substituting solar energy for fuel energy, giving opportunities to cut down pollution.

Putting sunpower to work has been stimulated in fuel-power areas and the developing nations where electric power distribution is nonexistent. Highly industrialized countries can also benefit. For example, Russia is reported to have a larger program than the US for developing solar energy (exclusive of its uses in space technology). Baum[13] provided a highly informative report of the practical progress made in desert areas in the development of solar-energized air-conditioning systems in apartment houses, and in the development of sunpowered thermoelectric water hoists. The report indicates that two major scientific institutions in the Soviet Union, the Turkmenian Academy of Sciences and the Krzhizhanovsky Power Institute, are developing full-scale programs for the utilization of solar energy.

In the United States, one experimental dwelling, developed in 1959 in the Washington, D.C. area, was built at a cost of $15,000. This cost included the solar heating system, plus a conventional oil burner for use in excessively cloudy periods. For three consecutive years this six-room house was heated for a total cost of $18.90, averaging $6.30 a year. The overall size of the collector used was 38 ft by 22 ft to obtain approximately 700 ft^2 of useful collection area. Every attempt was made to keep the cost low, and in 1959 the cost of the collector came to about $1 per ft^2. Rough measurements were made in December 1959 and the efficiency of the collector was found to be approximately 47%. Heat obtained by water flowing through the collector was stored by dissipation into 50 tons of fist-sized stones which surrounded a 1600-gallon drum which held the water long enough for the heat to dissipate before being recycled to the collector. The entire storage bin, located in part of the basement of the house, was 10 ft wide, 25 ft long, and 7 ft deep. The bin was waterproofed and insulated, and ducting for forced air circulation throughout the house was provided to it in much the same manner as with ordinary household furnaces. The heat storage capacity was adequate for about five overcast days. In addition, the solar unit reduced air-conditioning bills by 40%, since it could be modified for air-conditioning purposes during the summer months. The total cost for the solar space heating and air-conditioning system was between $1500 and $2000. With added research and mass production of the materials used, this could possibly be lowered in terms of the present-day dollar.[14-16]

An article by Altman, Telkes, and Wolf[17] indicates that space heating of individual homes could contribute significantly to a reduction in use of other fuels. Taking into consideration all the averaged effects due to poor weather conditions, but applying technology that is currently available, the net result of this study was that if every single family dwelling in the US were to use solar space heating, the reduction in power for this application would be 10%. This seems a small reduction, but should be compared with the fact that in 1967 only 10% of all energy consumed in the US was electrical. Thus a savings in fuel comparable to that used by all the electric utilities could be achieved by the simple method of solar space heating. Another practical use of solar energy applied on an individual basis has been developed experimentally on an Illinois farm. In Hamilton, Illinois, a solar heating unit has been developed for a swine finishing house designed to hold 1200 hogs. Experimental use of solar energy for heating homes and farm animal sheds has indicated that considerable economies are possible.

A method of using solar energy for central power applications was suggested by Meinel and Meinel[18]. Utilizing some of the latest developments in special optical coatings on collectors, the system proposed would generate electricity by solar heating of a brine solution to 540° C. This is a sufficiently high temperature to drive conventional steam turbines currently in existence. Roughly three square miles of land covered by these collectors would be required for a 1000-MW generating station, a size comparable to many nuclear power plants under construction today. The system would attain approximately 30% overall efficiency of conversion of sunlight to electricity. Although Meinel and Meinel estimate an initial cost that could be as high as three times that of an equivalent nuclear power plant, there would be no added fuel costs after construction. Environmentally, the system is attractive as there would be no air pollution and no radioactivity. However, since the steam turbines would require cooling, there would be some possibility of thermal pollution. However, utilization of the waste heat for agricultural or industrial purposes, or possibly for running evaporators to produce freshwater from seawater by desalination, is envisaged.

Despite these breakthroughs, much needs to be learned before widespread employment of solar energy will be economical. Some of the problems are inherent in the nature of sunlight and are facts of life to be faced by the solar scientists and engineers; they include night-to-day ratios, clouds, summer-winter variations, and a low concentration of energy per unit area.

Other problems are more susceptible to research and development. These include reflection and absorption surfaces, energy storage devices, sun tracking and focusing, and more efficient conversion of electric heat or heat energy. This research could benefit directly from the technology resulting from our space effort. The problems are challenging, but not overwhelming. The potential of solar energy is so enormous that development seems mandatory even in the face of admitted difficulties. Each year, the world receives in sunlight 23,000 times more energy than it uses. Any significant conversion of this energy would drastically lower the demand on fossil fuels or nuclear energy.

Conversely, what are some of the present limitations and most serious problems relating to the utilization of solar energy?

> Tapping the vast amount of solar energy that impinges on the earth's surface is made exceedingly difficult because the supply is intermittent, comes in low concentration, has only a limited utility in its directly available form, and its storage is extremely difficult and of brief duration . . .[19]

Yet the scientists who made this comment did so in a report of a successful practical utilization of solar energy through algae-produced methane developed in an urban sewage treatment plant. In their report, Golueke and Oswald detailed these promising findings:

> A promising solution is the duplication and miniaturization of nature's method of utilizing solar energy; for nature solved the problem long ago, as is attested by the vast array of life that has developed whose sole energy source is the sun. This feat was and is accomplished by fixing solar energy into the cellular energy of green plants. In nature's method,

therefore, green plants serve as the primary storers of solar energy . . .

At present, the biological conversion plant (here algae raised in a sewage treatment system) certainly cannot compete economically as a power producer in the United States with plants using fossil fuels . . . On the other hand, the biological conversion plant not only must produce power, it must also produce its own fuel, whereas the plants using fossil fuel need only to produce power. However, the indicated power cost approaches that obtained from nuclear reaction, although the latter has the advantage of being accompanied by technical know-how, while the former has only recently passed from the small pilot plant stage in some of its phases.

In its favor the requirements for a pond-digester plant are much less complete than those for a nuclear reactor. Hence, in certain areas where fossil fuels are in short supply, it may be advantageous to seriously study the system as a possible alternative to presently available sources of power.[19]

Experiments such as those conducted by Golueke and Oswald suggest that the potential of solar energy is not some sort of Buck Rogers fantasy. Indeed American scientific developments in the field engaged the active interest of men such as John Ericcson and Robert H. Goddard. Goddard pioneered several solar energy investigations while pursuing his now widely acclaimed rocket experiment, and he predicted great advances in solar energy and directly related this development to plans for interplanetary travel. He recorded:

In 1919, I suggested the possibility of producing a rocket so powerful that it would leave the earth's surface never to return. Since then, there has been considerable speculation regarding the use of such a rocket in interplanetary space. It is in this connection that I have been interested in developing a very light and efficient solar engine . . .[20]

He insisted, however, that the most obvious first use of such an engine must be "to supply abundant and cheap power for mankind".

France, Japan, and Israel are among the nations currently engaged in projects designed to harness solar energy. Hagemann reported in 1962 that the design of some of their solar devices are "oddly similar to" Goddard's models and plans.

There is a solar furnace on Mont Louis in the French Pyrenees that employs 3500 pieces of glass as a 30 foot parabolic reflector and creates heat of 6000° F. In Israel, scientists and technicians are concentrating on development of electric power. One installation has a parabolic mirror covering more than 3000 square feet, a boiler, heat exchanger, and turbine are employed . . .

Here in the United States, the Ryan Aeronautical Company has recently developed a solar collection model, in size or scale "to evaluate methods for fabricating an operational space system". The total design package is remarkably similar to Goddard's 1929 device.[20]

Goddard's prophetic comments concerning the potentiality of solar energy have gained support not only from analysts such as Glaser, but from experimentalists such as Zahl and Ziegler who, when reporting on potential power sources for satellites and space vehicles, indicated:

Actually, the solar radiation band from the near ultraviolet through infrared appears to be the only worthwhile source of natural energy in outer space which lends itself to profitable utilization.[21]

It is most significant that a long-range appraisal of future energy sources by Texaco Petroleum Corporation reocognized the potentiality of solar energy. L. P. Gaucher, assistant to the manager of Scientific Planning, Research and Technical Department of Texaco, made the candid admission:

If man's ingenuity through the years had been directed to the utilization of solar energy instead of to the development of devices to consume fossil fuels, it is quite conceivable that we might today have a solar economy just as effective as our present fossil-fuel economy.[22]

THE CONTROL OF ENERGY PRODUCTION: THE POLITICAL ECONOMY OF SOLAR ENERGY *versus* FOSSIL AND NUCLEAR FUELS

A crucial question confronting protagonists of the development of solar energy is whether large-scale energy producers and corporate developers have a vested interest in fossil fuels and nuclear energy. Similarly, does this inhibit modest efforts at developing the long-range potential and immediate prospects of solar energy as a partial alternative to fossil fuels and nuclear power? The answer, provided guardedly by solar energy scientists, is in the affirmative. For example, Professor Farrington Daniels of the University of Wisconsin touched upon the lack of financial support for solar energy research in his introductory remarks at the 1959 Conference of the Advisory Council of the Association of Applied Solar Energy as follows:

In the absence of large activities in the universities and in the absence of large financial incentives for industry in the field of solar energy research . . .[23]

One of the most candid appraisals of the situation was made by Löf[24] concerning the governmental subsidy policy, corporate profits, and the allocation of research resource money. He pointed out:

There are no owners of solar energy, so its use cannot be expected to receive the sort of promotion that natural gas does, for example. Thus one incentive for solar energy development is lacking. In the category of converters and sellers of energy, there is a corresponding lack of incentive to develop solar energy because there are ample and, in most cases, cheap supplies of fuel and water power for conversion. An alternate basic supply is not yet needed, and the time when it may be needed is too distant to justify research and development expenditures by these organizations now. To some extent, *the same might be said of atomic energy, but here there is a heavy government subsidy to the developers of this source which is absent in solar energy.* . . . If public utilities will not be using solar energy equipment for many years to come, who are the buyers for solar conversion equipment? The purchase for them appears to be largely built around the individual user—the home owner, the business man, the industrial firm, and the farmer. These individuals and groups have solar energy available to them. They will purchase the equipment to convert this energy to useful forms, and they will utilize the heat, electricity, and other products derived therefrom . . .

One of the most intriguing potentialities of solar energy is in combined energy absorption and storage by means of photosynthetic chemical reactions. For example, in the presence of certain catalysts, water can be decomposed into hydrogen and oxygen by the absorption of energy in the ultraviolet portion of the solar spectrum. These gases can be stored for subsequent combustion and power generation. Certain other aqueous reactions could possibly be utilized whereby absorption of energy would cause a change in one direction which could then be reversed when desired to liberate the absorbed energy as heat or, more ideally, as electrical energy. Limited progress has been made along some of these lines, but, as yet, efficiencies in converting solar to chemical energy of only small fractions of 1% have been achieved. *If a substantial technical breakthrough should occur along some such line, the whole economic picture of large-scale solar energy utilization could be affected.* Mechanisms of suitable reactions, a complete understanding of natural photosynthesis and other basic problems will probably have to be solved before substantial headway is made into practical application of these principles. (Our italics).

One of the most cogent presentations of the rationale for developing solar energy in the context of the generally rural economic societies of the underdeveloped nations was made by J. C. Kapur[25], a scientist employed by private industry in Calcutta, India:

In the march of civilization man has faced no challenge as great in scope or dimensions as confronting him today: the amelioration of the living conditions of more than two billion people. The disparities in the effort are emphasized by the fact that while even the more advanced of the underdeveloped areas are still struggling to comprehend some of the more basic problems of an industrial society, others are making ready to explore space.

In an analysis of the task of raising the living standards of the world, the great differences in energy available to peoples stands as a barrier. Furthermore, the prospect for any major improvement in energy for utilization by the less-developed nations is slight. At present the nations of advanced industrialization and mass consumption, with 31.3% of the world population, consume 84% of the world energy used. It is estimated that this same group of nations will, by the year 2000, comprise 27% of the total population and consume 68.4% of the energy. At the other end of the scale, the least developed nations, comprising now about 19.7% of the people, have at their disposal only 1% of the world energy. At the end of the century it is expected that this group, with 22% of world population, will consume no more than 2.3% of the world total. (Considering projected increases in population, energy use will increase significantly.)

If this projected pattern holds, the known world deposits of fossil fuels (estimated at 30×10^{11} tons of coal) at the anticipated rate of consumption, in the year 2000 would last less than 200 years. With the possibility of a still greater rate of consumption after the year 2000 for the underdeveloped areas, the known deposits may not last 100 years from now. Therefore, by the time the underdeveloped areas succeed in building up an industrial complex around the existing source of energy, the resources might reach a stage of near depletion. This would become particularly alarming if the advances in technology enable us to accelerate the rate of economic growth beyond what appears to be possible under present conditions.

Another factor deserves consideration. This is the nature and distribution of energy demand in underdeveloped areas. (In India, for example) by 1966, only 29,500 of the 557,000 villages and towns with populations of less than 5000 would have electric power.

In the context of most of the underdeveloped countries, solar energy is more a part of the life than any other source of energy. The advanced technology of energy transformations known to us today has no associations with their energy concepts. The utilization of solar energy is most likely to cut across the mental barriers that often resist the introduction of advanced technologies . . .

The availability, intermittence, and storage are a few of the important considerations that have puzzled the scientists and engineers concerning the utilization of the new source of energy. In rural communities in underdeveloped areas, where important sections of the population remain partially or wholly unemployed, the problems of intermittence and storage lose most of their significance. It is really important to have sunshine, or work every day except Sundays and festival days? Why must we apply the concept and economies of the advanced societies in entirely unwarranted situations? . . .
Economics apart, even under the existing conditions it should normally be possible to create employment opportunities by the use of new sources of energy from 200 to 300 days in a year in most underdeveloped areas . . ."

(Noting that both cultural and technological problems remain unsolved, Kapur suggested that)

A practical approach, therefore, would be to organize an experimental station for every 50 million of the population in different regions of the world. This would enable us to resolve some of the outstanding problems of technology under actual conditions of operation, and to study the impact of these changes on the life of these communities.

CONCLUSION

Kapur's analysis sets challenging goals not only for the nonindustrial underdeveloped nations. It provides potentially significant goals which can be adapted to the needs of the urban and rural poor of America in matters as practical and compelling as cheaper heating and cooking costs, low-cost air conditioning, and lower electrical energy costs for both industry and home consumers.

There is an additional consideration of broad intellectual concern. Particularly since World War II, our people have lived through the great transition of American national life which has seen an increasing tension between the world of natural science and other sectors of our society. In the generation of Leonardo da Vinci, or for that matter the generation of Thomas Jefferson, science was viewed as the great benefactor of mankind rather than a source of disquiet and a possible threat to the future of mankind. Yet, in our own generation we have seen not only the development and application of atomic energy, but the development of great tensions within our society concerning the directions in which science may lead mankind. Thoughtful members of the scientific community themselves have been troubled by this, as exemplified in the development of special publications, such as the *Bulletin of the Atomic Scientists*.

In literary fields the same tension and sense of alienation between the community of natural scientists and public policy makers has been dramatically portrayed in the writings of C. P. Snow. Yet, many have felt that there is no necessary conflict between the world of science and the world of public policy making. The very genius of American civilization, our talent for intelligent adaptation to our natural environment, is best exemplified in the bold and inquiring spirit of one of our greatest Presidents, Thomas Jefferson, who combined superb qualities of political leadership with a sensitive and impressive grasp of science, the arts, and the humanities.

The most sensible method of re-establishing the unity between our talented natural scientists and the world of public policy makers is through those measures which illustrate and in practical fashion apply the great contributions science can make to improve the conditions of mankind. Practical and economic use of solar energy represents one of the great frontiers in our scientific endeavors. At the same time, solar energy can provide practical and direct evidence of the compatibility of the world of natural sciences and the compassion so frequently associated with the world of humanities.

REFERENCES

1. A. F. A. Reynhart, "Mankind, civilization, and prosperity", *Solar Energy*, 3:23-25, April 1959.
2. G. Orwell, "Marrakech", *in* "The Collected Essays, Journalism, and Letters of George Orwell, Vol. 1, An Age Like This, 1920-1940", Secker and Warburg, London, 1968, p. 387-393.
3. H. Caudill, "Night Comes to the Cumberland", Little Brown and Co., Boston, 1963.
4. National Economic Research Association, Testimony before Senate Anti Trust Committee, U.S. Senate, May, 1970.
5. M. Mintz and R. K. Warner, "Big oil companies acquire grip on competing fuels", *The Washington Post*, 23 August 1970, pp. A1-A2.
6. M. Tanzer, *The Political Economy by International Oil and the Underdeveloped Countries*, Beacon Press, Boston, 1969, p. 3.
7. United Nations, Economic Commission for Latin America, *Energy Development in Latin America*, UN Publication, Geneva, 1957, p. 4.

8. M. Mintz and R. K. Warner, *op. cit.*, p. A2.
9. T. O'Toole, "Electricity for the Future," *Washington Post*, 24 August 1970, p. A1.
10. P. E. Glaser, "Satellite solar power station," *Solar Energy*, 12:353-360, May 1969.
11. 2nd Progress Report of the Subcommittee on Science Research and Development to the Committee on Science and Astronautics, U.S. House of Representatives, 89th Congress, 2nd Session, Serial R, U.S. Government Printing Office, Washington, D.C., 1966, p. 25.
12. See R. S. Lewis (ed.), *The Environmental Revolution*, Educational Foundation for Nuclear Science, Chicago, 1973, pp. 126-133.
13. V. Baum, *Soviet Weekly*, 26 March 1966.
14. H. E. Thomason, "Solar space heating and air-conditioning in the Thomason home", *Solar Energy*, 4:11, October-December, 1960.
15. H. E. Thomason, "7 Dollars a year heats the house with sunshine in the basement", *Popular Mechanics*, pp. 83-93, 234, February 1965.
16. H. E. Thomason, "Experience with solar houses", *Solar Energy*, 10:17-22, January-March 1966.
17. M. Altman, M. Telkes, and M. Wolf, "The energy resources and electric power situation in the United States", *Energy Conversion*, 12:53-64, 1972.
18. A. B. Meinel and M. P. Meinel, "Physics looks at solar energy", *Phys. Today*, 25(2):44-50, February 1972.
19. C. G. Golueke and W. J. Oswald, "Power from solar energy—via algae-produced methane", *Solar Energy*, 7:86-92, July-September 1963.
20. E. R. Hagemann, "R. H. Goddard and solar power, 1924-1934", *Solar Energy*, 6:47-54, April-June 1962.
21. H. A. Zahl and H. K. Ziegler, "Power sources for satellites and space vehicles", *Solar Energy*, 4:34, January 1960.
22. L. P. Gaucher, "Energy sources of the future for the United States", *Solar Energy*, 9:125, July-September 1965.
23. See *Solar Energy*, 3:7, October 1959.
24. G. O. G. Löf, "Profits in solar energy", *Solar Energy*, 4:9-15, April 1960.
25. J. C. Kapur, "Socio-economic considerations in the utilization of solar energy in underdeveloped areas", *Solar Energy*, 6:99-103, July-September 1962.

BIBLIOGRAPHY

BOOKS

Adler, Alfred. *Superiority and Social Interest: A Collection of Later Writings*. Biographical essay by Carl Furtmuller. Ed. Heinz L. Ansbacher and Rowena R. Ansbacher. Evanston, Ill.: Northwestern University Press, 1964.

Akin, William E. *Technology and The American Dream: The Technocrat Movement, 1900-1941*. Berkeley: University of California Press, 1977.

Allen, Francis R., et al. *Technology and Social Change*. New York: Appleton-Century-Crofts, 1957.

Anderson, Alan Ross, ed. *Minds and Machines*. Englewood Cliffs, N.J.: Prentice-Hall, Inc., 1964.

Ann Arbor Science for the People Editorial Collective. Ann Arbor, Michigan. *Biology as a Social Weapon*. Minneapolis: Burgess Publishing Co., 1977.

Arendt, Hannah. *Between Past and Future: Six Exercises in Political Thought*. New York: The Viking Press, 1961.

Armytage, W. H. G. *The Rise of the Technocrats*. London: Routledge & Kegan Paul Ltd., 1965.

Armytage, W. H. G. *A Social History of Engineering*. 4th ed. Boulder, Colo.: Westview Press, Inc., 1976.

Aron, Raymond. *Progress and Disillusion: The Dialectic of Modern Society*. New York: Praeger Publishers, 1968.

Aron, Raymond, ed. *World Technology and Human Destiny*. Trans. Robert Seaver. Ann Arbor, Mich.: The University of Michigan Press, 1963.

Asimov, Isaac. *The Human Body: Its Structure and Operation*. Boston: Houghton Mifflin Co., 1963.

Augenstein, Leroy G. *Come, Let Us Play God*. New York: Harper & Row, Publishers, Inc., 1969.

Axelos, Kostas. *Alienation, Praxis and the Techne in the Thought of Karl Marx*. Trans. Ronald Bruzina. Austin: University of Texas Press, 1976.

Baker, Elizabeth (Faulkner). *Technology and Woman's Work*. New York: Columbia University Press, 1964.

Baran, Paul A., and Paul M. Sweezy. *Monopoly Capital*. New York: Monthly Review Press, 1966.

Barbour, Ian G. *Science and Secularity: The Ethics of Technology*. New York: Harper & Row, Publishers, Inc., 1970.

Barbour, Ian G., ed. *Western Man and Environmental Ethics: Attitudes Towards Nature and Technology*. Reading, Mass.: Addison-Wesley Publishing Co., Inc., 1973.

Barnet, Richard J., and Ronald E. Muller. *Global Reach: The Power of the Multinational Corporations*. New York: Simon & Schuster, Inc., 1974.

Barrett, William. *The Illusion of Technique: A Search for Meaning in a Technological Civilization*. Garden City, N.Y.: Anchor Press, 1978.

Baum, Robert J., and Albert Flores, eds. *Ethical Problems in Engineering*. Troy, N.Y.: Center for the Study of Human Dimensions of Science and Technology, 1978.

Bell, Daniel. *The Coming of Post-Industrial Society: A Venture in Social Forecasting*. New York: Basic Books, Inc., Publishers, 1973.

Bell, Daniel. *The Cultural Contradictions of Capitalism*. New York: Basic Books, Inc., Publishers, 1976.

Bell, Daniel, ed. *Toward the Year 2000*. Boston: Houghton Mifflin Co., 1968.

Benedict, Murray. *Farm Policies of the United States, 1790-1950: A Study of Their Origins and Development*. New York: Twentieth Century Fund, 1953.

Benthall, Jonathan. *The Body Electric: Patterns of Western Industrial Culture*. London: Thames & Hudson Ltd., 1976.

Berg, Alan S. *The Nutrition Factor: Its Role in National Development*. Washington, D.C.: The Brookings Institution, 1973.

Berger, Peter L. *Pyramids of Sacrifice: Political Ethics and Social Change*. New York: Basic Books, Inc., Publishers, 1974.

Berger, Peter L., et al. *The Homeless Mind: Modernization and Consciousness*. New York: Random House, Inc., 1973.

Berry, Wendell. *The Unsettling of America*. New York: Avon Books, 1978.

Bertalanffy, Ludwig von. *Robots, Men and Minds:*

Psychology in the Modern World. New York: George Braziller, Inc., 1967.

Bettelheim, Bruno. *The Informed Heart: Autonomy in a Mass Age*. New York: The Free Press, 1960.

Boorstin, Daniel J. *The Republic of Technology*. 1st ed. New York: Harper and Row, 1978.

Boulding, Kenneth. *The Meaning of the Twentieth Century*. New York: Harper & Row, Publishers, Inc., 1965.

Bowers, Raymond, Alfred M. Lee, and Carey Hershey, eds. *Communications for a Mobile Society: An Assessment of New Technology*. Beverly Hills, Calif.: Sage Publications, Inc., 1978.

Breasted, James Henry. *The Conquest of Civilization*. Ed. Edith Williams Wane. New York: Harper & Row, Publishers, Inc., 1954.

Breasted, James Henry. *A History of Egypt from the Earliest Times to the Persian Conquest*. 2nd ed. 1909; rpt. New York: Charles Scribner's Sons, 1950.

Brickman, William, and Stanley Lehrer, eds. *Automation, Education and Human Values*. New York: Thomas Y. Crowell Co., 1969.

Brown, Martin, ed. *The Social Responsibility of the Scientist*. New York: The Free Press, 1971.

Brown, Norman O. *Life Against Death: The Psychoanalytical Meaning of History*. Middletown, Conn.: Wesleyan University Press, 1959.

Brown, Norman O. *Love's Body*. New York: Random House, Inc., 1960.

Brunner, John. *Stand on Zanzibar*. New York: Ballantine Books, Inc., 1969.

Brzezinski, Zbigniew. *Between Two Ages: America's Role in the Technetronic Era*. New York: The Viking Press, 1970.

Buchanan, Robert A. *Technology and Social Progress*. Oxford: Pergamon Press Ltd., 1965.

Burke, James. *Connections*. London: Macmillan Publishers Ltd., 1978.

Burke, John G., ed. *The New Technology and Human values*. 2nd ed. Belmont, Calif.: Wadsworth Publishing Co., 1972.

Buttcock, Gregory, ed. *New Artist's Video*. New York: E. P. Dutton, 1978.

Calabresi, Guido, and Philip Bobbitt. *Tragic Choices*. New York: W. W. Norton & Co., Inc., 1978.

Calder, Nigel. *Technopolis: Social Control of the Uses of Science*. New York: Simon & Schuster, Inc., 1970.

Calder, Nigel. *The Violent Universe: An Eye-Witness Account of the New Astronomy*. New York: The Viking Press, 1970.

Callahan, Daniel. *The Tyranny of Survival; And Other Pathologies of Civilized Life*. New York: Macmillan, Inc., 1973.

Cappon, Daniel. *Technology and Perception*. Springfield, Ill.: Charles C Thomas, Publisher, 1971.

Carpenter, Edmund Snow, and H. Marshall McLuhan, eds. *Explorations in Communication: An Anthology*. Boston: Beacon Press, 1960.

Carson, Rachel Louise. *Silent Spring*. Boston: Houghton Mifflin Co., 1962.

Castro, Josué de. *The Geography of Hunger*. Boston: Little, Brown & Co., 1952.

Chargaff, Erwin. *Heraclitean Fire*. New York: Rockefeller University Press, 1978.

Clarke, Arthur C. *Profiles of the Future: An Inquiry into the Limits of the Possible*. New York: Bantam Books, Inc., 1964.

Cobb, John B. *Is It Too Late? A Theology of Ecology*. Beverly Hills, Calif.: Bruce, 1972.

Collins, Joseph, and Frances Moore-Lappé, with Gary Fowler. *Food First: Beyond the Myth of Scarcity*. Boston: Houghton-Mifflin Co., 1977.

Commoner, Barry. *The Closing Circle: Nature, Man and Technology*. New York: Alfred A. Knopf, Inc., 1971.

Commoner, Barry. *The Poverty of Power: Energy and the Economic Crisis*. New York: Alfred A. Knopf, Inc., 1976.

Commoner, Barry. *Science and Survival*. New York: The Viking Press, 1966.

Conference on the Ecological Aspects of International Development. Airlie House, 1968. *The Careless Technology: Ecology and International Development: The Record*. Garden City, N.Y.: The Natural History Press, 1972.

Cooke, E. Mary. *Escherichia Coli and Man*. Edinburgh: Churchill Livingstone, 1974.

Coulson, Charles A. *Science, Technology and the Christian*. New York: Abingdon Press, 1960.

Cox, Harvey. *The Secular City*. New York: Macmillan, Inc., 1965.

Crosson, Frederick J., and Kenneth Sayre, eds. *Philosophy and Cybernetics*. Notre Dame, Ind.: University of Notre Dame Press, 1967.

Curran, Charles E. *Politics, Medicine, and Christian Ethics: A Dialogue with Paul Ramsey*. Philadelphia: Fortress Press, 1973.

Curtis, James M. *Culture as Polyphony: An Essay on the Nature of Paradigms*. Columbia, Mo.: University of Missouri Press, 1978.

Daly, Mary. *Beyond God the Father: Toward a Philosophy of Women's Liberation*. Boston: Beacon Press, 1973.

Davis, Elizabeth Gould. *The First Sex*. New York: G. P. Putnam's Sons, 1971.

Deane, Phyllis. *The First Industrial Revolution*. Cambridge: Cambridge University Press, 1965.

Delgado, José M. R. *Physical Control of the Mind*. New York: Harper Colophon Books, 1971.

Dewey, John. *Freedom and Culture*. New York: G. P. Putnam's Sons, 1939.

Dewey, John. *Philosophy and Civilization*. New York: Minton, Balch & Co., 1931.

Dewey, John. *The Public and Its Problems: An Essay in Political Inquiry*. Chicago: Gateway Books, 1946.

Diebold, John. *Man and the Computer: Technology as an Agent of Social Change*. New York: F. A. Praeger, 1969.

Dobzhansky, Theodosius. *Genetic Diversity and Human Equality*. New York: Basic Books, Inc., Publishers, 1973.

Drucker, Peter F. *The Age of Discontinuity: Guidelines to Our Changing Society*. New York: Harper & Row, Publishers, Inc., 1969.

Drucker, Peter F. *Technology, Management, and Society: Essays*. New York; Harper & Row, 1970.

Dubos, René. *A God Within*. New York: Charles Scribner's Sons, 1972.

Dubos, René. *So Human An Animal*. New York: Charles Scribner's Sons, 1968.

Dumont, René. *Lands Alive*. New York: Monthly Review Press, 1966.

Durbin, Paul T., ed. *Research in Philosophy and Technology*. Vol. 1. Greenwich, Conn.: JAI Press, 1978.

Edge, D. O., and J. N. Wolfe, eds. *Meaning and Control: Essays in the Social Aspects of Science and Technology*. London: Tavistock Publications Ltd., 1973.

Efron, Edith. *The Newstwisters*. New York: Manor Books, Inc., 1972.

Eisenstein, Elizabeth L. *The Printing Press as an Agent of Change: Communications and Cultural Transformations in Early Modern Europe*. Cambridge: Cambridge University Press, 1979.

Ellison, Craig W., ed. *Modifying Man: Implications and Ethics*. Washington, D.C.: University Press of America, 1977.

Ellul, Jacques. *The Political Illusion*. Trans. Konrad Kellen. New York: Alfred A. Knopf, Inc., 1967.

Ellul, Jacques. *Propaganda: The Formation of Man's Attitudes*. Introd. Konrad Kellen. Trans. Konrad Kellen and Jean Lerner. New York: Alfred A. Knopf, Inc., 1965.

Ellul, Jacques. *The Technological Society*. Introd. Robert K. Merton. Trans. J. Wilkinson. New York: Alfred A. Knopf, Inc., 1964.

Ettinger, Robert C. *Man into Superman: The Startling Potential of Human Evolution —and How to be Part of It*. New York: St. Martin's Press, Inc., 1972.

Falk, Richard A. *This Endangered Planet: Prospects and Proposals for Human Survival*. New York: Random House, Inc., 1971.

Fanon, Frantz. *The Wretched of the Earth*. Pref. Jean-Paul Sartre. Trans. Constance Farrington. New York: Grove Press, Inc., 1966.

Feibleman, James K. *The Reach of Politics: A New Look at Government*. New York: Horizon Press, 1969.

Ferkiss, Victor C. *The Future of Technological Civilization*. New York: George Braziller, Inc., 1974.

Ferkiss, Victor C. *Technological Man: The Myth and The Reality*. New York: Mentor Books, 1970.

Ferre, Frederick. *Shaping the Future: Resources for the Post-Modern World*. New York: Harper & Row, Publishers, Inc., 1976.

Fink, Donald G. *Computers and the Human Mind*. New York: Doubleday & Co., Inc., 1966.

Fletcher, Joseph F. *The Ethics of Genetic Control: Ending Reproductive Roulette*. Garden City, N.Y.: Doubleday & Co., Inc., 1974.

Foster, George M. *Traditional Societies and Technological Change*. 2nd ed. New York: Harper & Row, Publishers, Inc., 1973.

Frank, Andre Gunder. *Capitalism and Underdevelopment in Latin America*. New York: Monthly Review Press, 1967.

Friedmann, Georges. *Industrial Society: the Emergence of the Human Problems of Automation*. Trans. and ed. H. L. Sheppard. New York: The Free Press, 1955.

Fromm, Erich, ed. *Socialist Humanism: An International Symposium*. Garden City, N.Y.: Doubleday & Co., Inc., 1965.

Fuller, R. Buckminster. *Earth, inc*. Garden City, N.Y.: Anchor Press, 1973.

Fuller, R. Buckminster. *Ideas and Integrities: A Spontaneous Autobiographical Disclosure*. Ed. Robert W. Marks. Englewood Cliffs, N.J.: Prentice-Hall, Inc., 1963.

Fuller, R. Buckminster. *Operating Manual for Spaceship Earth*. Carbondale, Ill.: Southern Illinois University Press, 1969.

Furtado, Celso. *The Development of Underdevelopment*. Berkeley: the University of California Press, 1964.

Gabor, Dennis. *Inventing the Future*. 1st Amer. ed. New York: Alfred A. Knopf, Inc., 1963.

Gabor, Dennis, *Innovations: Scientific, Technological, and Social*. London: Oxford University Press, 1970.

Galbraith, John Kenneth. *The New Industrial State*. Boston: Houghton Mifflin Co., 1967.

Galbraith, John Kenneth. *The New Industrial State*. 3rd. ed. rev. Boston: Houghton Mifflin Co., 1978.

García, Márquez Gabriel. *One Hundred Years of Solitude*. Trans. Gregory Rabassa. New York: Harper & Row, Publishers, Inc., 1970.

Garrison, Fielding H. *An Introduction to the History of Medicine, with Medical Chronology, Suggestions for Study, and Bibliographic Data*. 4th ed. Philadelphia: W. B. Saunders Co., 1929.

Gendron, Bernard. *Technology and the Human Condition*. New York: St. Martin's Press, Inc., 1977.

George, Frank. *Machine Takeover: The Growing Threat to Human Freedom in a Computer-Controlled Society*. Oxford: Pergamon Press Ltd., 1977.

George, Henry. *Social Problems*. New York: Robert Schalkenbach Foundation, 1953.

George, Susan. *How the Other Half Dies: The Real Reasons for World Hunger*. Harmondsworth, N.Y.: Penguin Books, 1976.

Ghali, Mirrit Boutros. *Tradition for the Future: Human Values and Social Purpose*. Oxford: Alden Press, 1972.

Giedion, Sigfried. *Mechanization Takes Command*. 1948; rpt. New York: W. W. Norton & Co., Inc., 1969.

Gillouin, René. *Man's Hangman Is Man*. Trans. D. D. Lachman. Mundelein, Ill.: Island Press, 1957.

Gimpel, Jean. *The Medieval Machines: The Industrial Revolution of the Middle Ages*. New York: Penguin Books, 1977.

Goodfield, June. *Playing God: Genetic Engineering and the Manipulation of Life*. New York: Random House, Inc., 1977.

Goodman, Paul. *Like a Conquered Province: The Moral Ambiguity of America*. New York: Random House, Inc., 1967.

Gouldner, Alvin W. *The Dialectic of Ideology and Technology: The Origins, Grammar, and Future of Ideology*. New York: The Seabury Press, Inc., 1976.

Goulet, Denis. *The Cruel Choice: A New Concept in the Theory of Development*. New York: Atheneum Publishers, 1971.

Griffin, Keith. *Land Concentration and Rural Poverty*. London: The Macmillan Press Ltd., 1976.

Griffin, Susan. *Woman and Nature*. New York: Harper & Row, Publishers, Inc., 1978.

Griswold, Alfred Whitney. *Farming and Democracy*. New York: Harcourt, Brace, 1948.

Gross, Bertram M., ed. *Action Under Planning*. New York: McGraw-Hill Book Co., 1967.

Gross, Bertram M., ed. *A Great Society?* New York: Basic Books, Inc., Publishers, 1968.

Guardini, Romano. *The End of the Modern World: A Search for Orientation*. Trans. J. Theman and H. Burke. New York: Sheed and Ward, 1956.

Guardini, Romano. *Power and Responsibility: A Course of Action for the New Age*. Trans. E. C. Briefs. Chicago: Regnery, 1961.

Guthrie, Douglas J. *A History of Medicine*. Introd. Samuel C. Harvey. Philadelphia: J. B. Lippincott Co., 1946.

Gutman, Herbert. *Work, Culture and Society in Industrializing America*. New York: Vintage Books, 1977.

Haas, Ernst B., et al. *Scientists and World Order: The Uses of Technical Knowledge in International Organizations*. Berkeley: University of California Press, 1977.

Habermas, Jürgen. *Toward a Rational Society: Student Protest, Science, and Politics*. Trans. Jeremy J. Shapiro. Boston: Beacon Press, 1970.

Halacy, Daniel S. *Genetic Revolution: Shaping Life for Tomorrow*. New York: Harper & Row, Publishers, Inc., 1974.

Hamilton, David. *Technology, Man, and the Environment*. New York: Charles Scribner's Sons, 1973.

Hamilton, Michael R., ed. *The New Genetics and The Future of Man: A Book Produced Through the Cooperation of the National Presbyterian Center* [and others]. Grand Rapids, Mich.: Wm. B. Eerdmans Publishing Co., 1972.

Hardin, Garrett J. *Exploring New Ethics for Survival: The Voyage of the Spaceship Beagle*. New York: The Viking Press, 1972.

Harrington, Alan. *The Immortalist: An Approach to the Engineering of Man's Divinity*. New York: Avon Books, 1970.

Heidegger, Martin. *The Question Concerning Technology, and Other Essays*. Introd. and trans. William Lovitt. New York: Garland Publishing, Inc., 1977.

Heilbroner, Robert L. *An Inquiry into the Human Prospect*. New York: W. W. Norton & Co., Inc., 1974.

Heinemann, F. H. *Existentialism and the Modern Predicament*. New York: Harper & Row, Publishers, Inc., 1953.

Henderson, Hazel. *Creating Alternative Futures: The End of Economics*. Foreword E. F. Schumacher. New York: Berkeley Publishing Corp., 1978.

Henry, Jules. *Culture Against Man*. New York: Random House, Inc., 1965.

Hightower, Jim. *Eat Your Heart Out: Food Profiteering in America*. New York: Crown Publishers, Inc., 1975.

Hill, Johnson D., and Walker E. Stuermann. *Roots in the Soil: An Introduction to Philosophy of Agriculture*. New York: Philosophical Library, Inc., 1964.

Hillegas, Mark R. *The Future as Nightmare: H. G. Wells and the Anti-Utopians*. Carbondale: Southern Illinois University Press, 1967.

Horvat, Branko, et al., eds. *Self-Governing Socialism: A Reader*. 2 vols. White Plains, N.Y.: International Arts and Science Press, 1975.

Houston Conference on Ethics in Medicine and Technology. *Who Shall Live? Medicine, Technology, Ethics*. Ed. Kenneth Vaux. Philadelphia: Fortress Press, 1970.

Hovenkamp, Herbert. *Science and Religion in America, 1800-1860*. Philadelphia: University of Pennsylvania Press, 1978.

Hughes, Thomas Parke, ed. *Changing Attitudes Toward American Technology*. New York: Harper & Row, Publishers, Inc., 1975.

Hughes, Thomas Parke, ed. *The Development of Western Technology Since 1500*. New York: Macmillan, Inc., 1964.

Ihde, Don. *Technics and Praxis*. Boston: D. Reidel Publishing Co., Inc., 1979.

Jalée, Pierre. *The Pillage of the Third World*. Trans. Mary Klopper. New York: Monthly Review Press, 1968.

Jaspers, Karl. *The Future of Mankind*. Trans. E. B. Ashton. Chicago: University of Chicago Press, 1961.

Jaspers, Karl. *Man in the Modern Age*. Trans. Eden

and Cedar Paul. (New ed.) London: Routledge & Kegan Paul Ltd., 1951.

Jaspers, Karl. *The Origin and Goal of History*. Trans. M. Bullock. New Haven: Yale University Press, 1953.

Jequier, Nicholas, ed. *Appropriate Technology: Problems and Promises*. Paris: Development Centre of the Organization for Economic Co-operation and Development, 1976.

Johnson, Michael. *Holistic Technology*. Roslyn Heights, N.Y.: Libra Publishers, 1977.

Jonas, Hans. *Philosophical Essays: From Ancient Creed to Technological Man*. Englewood Cliffs, N.J.: Prentice-Hall, Inc., 1974.

Josephson, Eric and Mary, eds. *Man Alone: Alienation in Modern Society*. Introd. Eric and Mary Josephson. New York: Dell Publishing Co., Inc., 1962.

Julien, Claude. *America's Empire*. 1st. Amer. ed. Trans. Renaud Bruce. New York: Pantheon Books, Inc., 1971.

Kahn, Herman, and Anthony J. Wiener. *The Year 2000*. New York: Macmillan, Inc., 1967.

Kaplan, Max, and Phillip Bosserman, eds. *Technology, Human Values and Leisure*. Nashville: Abingdon Press, 1971.

Kasson, John F. *Civilizing the Machine: Technology and Republican Values in America, 1776-1900*. New York: Penguin Books, 1977.

Klemm, Frederich. *A History of Western Technology*. Trans. Dorothea Waley Singer. Cambridge: The MIT Press, 1964.

Klibansky, Raymond, ed. *Contemporary Philosophy: A Survey*. Florence: La Nuova Italia, Editice, 1968.

Koestler, Arthur. *The Act of Creation*. New York: Dell Publishing Co., Inc., 1966.

Koestler, Arthur. *The Ghost in the Machine*. Chicago: Regnery, 1971.

Kormondy, E. *Concepts of Ecology*. Englewood Cliffs, N.J.: Prentice-Hall, Inc., 1967.

Kranzberg, Melvin, and William H. Davenport, eds. *Technology and Culture: An Anthology*. New York: Schocken Books, Inc., 1972.

Krohn, Wolfgang, et al., eds. *The Dynamics of Science and Technology: Social Values, Technical Norms and Scientific Criteria in the Development of Knowledge*. Boston: D. Reidel Publishing Co., Inc., 1978.

Kugler, Hans J. *Slowing Down the Aging Process*. New York: Pyramid Books, 1973.

Kuhns, William. *Environmental Man*. New York: Harper & Row, Publishers, Inc., 1969.

Kuhns, William. *The Post Industrial Prophets: Interpretations of Technology*. New York: Weybright and Talley, 1971.

Kurtzman, Joel, and Phillip Gordon. *No More Dying: The Conquest of Aging and the Extension of Human Life*. Los Angeles: J. P. Tarcher, Inc.; New York: Hawthorn Books, Inc., 1976.

Landes, David S. *The Unbound Prometheus: Technological Change and Industrial Development in Western Europe from 1750 to the Present*. Cambridge: Cambridge University Press, 1969.

Laslett, Peter. *The World We Have Lost*. 2nd ed. New York: Charles Scribner's Sons, 1973.

Laszlo, Ervin. *A Strategy for the Future: The Systems Approach to World Order*. New York: George Braziller, Inc., 1974.

Laszlo, Ervin. *The Systems View of the World: The Natural Philosophy of the New Developments in the Sciences*. New York: George Braziller, Inc., 1972.

Layton, Edwin T. *The Revolt of the Engineers: Social Responsibility and the American Engineering Profession*. Cleveland: Press of Case Western Reserve University, 1971.

Lehmann, Phyllis. *Cancer and the Worker*. New York: New York Academy of Sciences, 1977.

Leiss, William. *The Domination of Nature*. New York: George Braziller, Inc., 1972.

Lem, Stanislaw. *The Cyberiad*. New York: The Seabury Press, Inc., 1974.

Lenin, V. I. *Imperialism: The Highest Stage of Capitalism; A Popular Outline*. New, rev. trans. New York: International Publishers Co., Inc., 1939.

Lerza, Catherine, and Michael Jacobson, eds. *Food for People Not for Profit: A Source Book on the Food Crisis*. Preface Ralph Nader. New York: Ballantine Books, Inc., 1975.

Lifton, Robert Jay. *Death in Life: Survivors of Hiroshima*. New York: Random House, Inc., 1968.

Lilley, Samuel. *Men, Machines, and History: A Short History of Tools and Machines in Relation to Social Progress*. London: Cobbett Press, 1948.

McHale, John. *The Future of the Future*. New York: George Braziller, Inc., 1969.

McLuhan, Herbert Marshall. *Understanding Media: The Extensions of Man*. New York: McGraw-Hill Book Co., 1964.

McLuhan, Herbert Marshall, and Quentin Fiore. *War and Peace in the Global Village: An Inventory of Some of the Current Spastic Situations That Could Be Eliminated by More Feedforward*. Co-ordinated Jerome Azel. New York: McGraw-Hill Book Co., 1968.

McLuhan, Marshall. *The Gutenberg Galaxy*. New York: The New American Library, Inc., 1969.

Magdoff, Harry. *The Age of Imperialism: The Economics of U.S. Foreign Policy*. New York: Monthly Review Press, 1969.

Mander, Jerry. *Four Arguments for the Elimination of Television*. New York: William Morrow & Co., Inc., 1977.

Marcel, Gabriel. *The Decline of Wisdom*. Trans. M. Hanari. London: Harvill Press Ltd., 1954.

Marcuse, Herbert. *An Essay on Liberation*. Boston: Beacon Press, 1969

Marcuse, Herbert. *One Dimensional Man: Studies of Advanced Industrial Society*. Boston: Beacon Press, 1964.

Marx, Karl. *Capital*. Ed. Frederick Engels. Trans.

from the 3rd German ed. by Samuel Moore and Edward Aveling. London: Lawrence & Wishart Ltd., 1974.

Marx, Leo. *The Machine in the Garden: Technology and the Pastoral Ideal in America*. New York: Oxford University Press, Inc., 1964.

Mead, George Herbert. *Mind, Self, and Society: From the Standpoint of a Social Behaviorist*. Chicago: University of Chicago Press, 1934.

Mead, Margaret, ed. *Cultural Patterns and Technical Change*. (From the Tensions and Technology series, World Federation for Mental Health.) New York: The New American Library, Inc., 1955.

Mead, Margaret. *Culture and Commitment: A Study of the Generation Gap*. London: The Bodley Head, Ltd., 1970.

Mesthene, Emmanuel G., ed. *Technological Change: Its Impact on Man and Society*. Cambridge: Harvard University Press, 1970.

Mesthene, Emmanuel G., ed. *Technology and Social Change*. Indianapolis: The Bobbs-Merrill Co., Inc., 1967.

Michener, James A. *Hawaii*. New York: Random House, Inc., 1959.

Mill, John Stuart. *On Liberty; Annotated Text, Sources and Background, Criticism*. Ed. David Spitz. New York: W. W. Norton & Co., Inc., 1975.

Miller, George W. *Moral and Ethical Implications of Human Organ Transplants*. Foreword Dwight E. Harkin. Springfield, Ill.: Charles C Thomas, Publisher, 1971.

Mitcham, Carl, and Robert Mackey, eds. *Philosophy and Technology*, New York: The Free Press, 1972.

Montagu, Ashley, ed. *Man and Aggression*. New York: Oxford University Press, Inc., 1968.

Montejo, Estaban. *The Autobiography of a Runaway Slave*. Ed. Miguel Barnet. Trans. Jocasta Innes. New York: Pantheon Books, Inc., 1968.

Montgomery, John D. *Technology and Civic Life: Making and Implementing Development Decisions*. Cambridge, Mass.: The MIT Press, 1974.

Morgan, Elaine. *The Descent of Woman*. New York: Stein & Day, Publishers, 1972.

Morison, E. E. *Men, Machines, and Modern Times*. Cambridge, Mass.: The MIT Press, 1966.

Morris, Desmond. *The Naked Ape: A Zoologist's Study of the Human Animal*. New York: Dell Publishing Co., Inc., 1969.

Muller, Herbert J. *The Children of Frankenstein: A Primer on Modern Technology and Human Values*. Bloomington: Indiana University Press, 1970.

Mumford, Lewis. *The City in History: Its Origins, Its Transformations, and Its Prospects*. New York: Harcourt, Brace and World, 1961.

Mumford, Lewis. *The Myth of the Machine: The Pentagon of Power*. New York: Harcourt Brace Jovanovich, Inc., 1970.

Nasr, Seyyed Hossein. *The Encounter of Man and Nature: The Spiritual Crisis of Modern Man*. London: George Allen and Unwin Ltd., 1968.

Nasr, Seyyed Hossein. *An Introduction to Islamic Cosmological Doctrines: Conceptions of Nature and Methods Used for its Study by the Ikhwan al-Safa, al-Birūnī, and Ibn Sīna*. Cambridge, Mass.: Belknap Press, 1964.

Nathan, Otto, and Heinz Norden, eds. *Einstein On Peace*. Pref. Bertrand Russell. New York: Simon & Schuster, Inc., 1960.

Nef, John U. *The Conquest of the Material World*. Chicago: University of Chicago Press, 1964.

Nef, John U. *Cultural Foundations of Industrial Civilization*. Cambridge: Cambridge University Press, 1958.

Nef, John U. *War and Human Progress: An Essay on the Rise of Industrial Civilization*. Cambridge: Harvard University Press, 1950.

Nevins, Allan, and Frank Ernest Hill. *Ford: The Times, the Man, the Company*. New York: Charles Scribner's Sons, 1954.

New York University Institute of Philosophy. *Dimensions of Mind: A Symposium*. Ed. Sidney Hook. New York: New York University Press, 1960.

Niebuhr, Reinhold. *Moral Man and Immoral Society: A Study in Ethics and Politics*. 1932; rpt. New York: Charles Scribner's Sons, 1952.

Noakes, Jeremy, and Geoffrey Pridham, eds. Introd. Jeremy Noakes and Geoffrey Pridham. *Documents on Nazism 1919-1945*. New York: The Viking Press, 1974.

O'Connor, Harvey. *World Crisis in Oil*. New York: Monthly Review Press, 1962.

Odum, Howard T. *Environment, Power, and Society*. New York: John Wiley and Sons, Inc., 1970.

Ogburn, William F. *The Social Effects of Aviation*. Boston: Houghton Mifflin Co., 1946.

Ong, Walter. *Knowledge and the Future*. New York: Simon and Schuster, Inc., 1968.

Ong, Walter J. *Frontiers in American Catholicism: Essays on Ideology and Culture*. New York: The Macmillan Co., 1957.

Papanek, Victor. *Design for the Real World: Human Ecology and Social Change*. New York: Bantam Books, Inc., 1973.

Passmore, John A. *Man's Responsibility for Nature: Ecological Problems and Western Traditions*. New York: Charles Scribner's Sons, 1974.

Piel, Gerard. *The Acceleration of History*. New York: Alfred A. Knopf, Inc., 1972.

Pierce, Harry H. *The Railroads of New York: A Study of Government Aid, 1826-1875*. Cambridge: Harvard University Press, 1953.

Pytlik, Edwards C., et al. *Technology, Change and Society*. Worcester, Mass.: Davis Publications, Inc., 1978.

Ramsey, Paul. *Ethics at the Edges of Life: Medical and Legal Intersections*. New Haven: Yale University Press, 1978.

Ramsey, Paul. *Fabricated Man: The Ethics of Genetic Control*. New Haven: Yale University Press, 1970.

Raskin, Marcus. *Being and Doing*. New York: Random House, Inc., 1971.

Reich, Charles A. *The Greening of America*. New York: Random House, Inc., 1970.

Reich, Wilhelm. *The Mass Psychology of Fascism*. Trans. Vincent R. Carfagno. New York: Farrar, Straus & Giroux, Inc., 1970.

Reiser, Stanley Joel. *Medicine and the Reign of Technology*. Cambridge: Cambridge University Press, 1978.

Ribeiro, Darcy. *The Americas and Civilization*. Trans. Linton Lomas Barrett and Marie McDavid Barrett. New York: E. P. Dutton, 1971.

Rich, Adrienne C. *Of Woman Born: Motherhood as Experience and Institution*. New York: W. W. Norton & Co., Inc., 1976.

Rose, John, ed. *Technological Injury: The Effect of Technological Advances on Environment, Life and Society*. New York: Gordon and Breach, Science Publishers, Inc., 1973.

Rossi, Paolo. *Philosophy, Technology and the Arts in the Early Modern Era*. Trans. Salvator Attanasio. Ed. Benjamin Nelson. New York: Harper and Row, Publishers, Inc., 1970.

Roszak, Theodore. *The Making of a Counter Culture*. New York: Doubleday & Co., Inc., 1969.

Roszak, Theodore. *Where the Wasteland Ends: Politics and Transcendence in Postindustrial Society*. Garden City, N.Y.: Anchor Books, 1973.

Rowbotham, Sheila. *Woman's Consciousness, Man's World*. Harmondsworth, England: Penguin Books Ltd., 1973.

Schiller, Herbert I. *Communication and Cultural Domination*. White Plains, N.Y.: International Arts and Sciences Press, 1976.

Schiller, Herbert I. *Mass Communications and American Empire*. Boston: Beacon Press, 1971.

Schramm, Wilbur L. *Men, Messages and Media: A Look at Human Communication*. New York: Harper & Row, Publishers, Inc., 1973.

Schwartz, Eugene. *Overskill: The Decline of Technology in Modern Civilization*. Chicago: Quadrangle Books, 1971.

Servan-Schreiber, Jean J. *The American Challenge*. Foreword Arthur Schlesinger, Jr. Trans. Ronald Steel. New York: Atheneum Publishers, 1968.

Singer, Charles J. *Technology and History*. London: Oxford University Press, 1952.

Siu, Ralph G. H. *The Tao of Science: An Essay on Western Knowledge and Eastern Wisdom*. Cambridge, Mass.: The MIT Press, 1957.

Slater, Philip E. *Earthwalk*. Garden City, N.Y.: Doubleday & Co., Inc., 1974.

Smith, Adam. *An Inquiry into the Nature and Causes of the Wealth of Nations*. Ed. Edwin Cannan. New York: Modern Library, 1937.

Snow, C. P. *The Two Cultures and the Scientific Revolution*. New York: Cambridge University Press, 1959.

Speer, Albert. *Inside The Third Reich*. Trans. Richard and Clara Winston. New York: Macmillan, Inc., 1970.

Spengler, Oswald. *Man and Technics: A Contribution to a Philosophy of Life*. Trans. Charles Francis Atkinson. Westport, Conn.: Greenwood Press, Inc., 1976.

Stanley, Manfred. *The Technological Conscience: Survival and Dignity in an Age of Expertise*. New York: The Free Press, 1978.

Stobaugh, Robert, and Daniel Yergin, eds. *Energy Future: Report of the Energy Project at the Harvard Business School*. New York: Random House, Inc., 1979.

Susskind, Charles. *Understanding Technology*. Baltimore: The Johns Hopkins University Press, 1973.

Sypher, Wylie. *Literature and Technology: The Alien Vision*. New York: Vintage Books, 1971.

Szasz, Thomas S. *Ideology and Insanity: Essays on the Psychiatric Dehumanization of Man*. New York: Anchor Books, 1970.

Tanger, Michael. *The Political Economy of International Oil and the Underdeveloped Countries*. Boston: Beacon Press, 1969.

Teich, Albert H., ed. *Technology and Man's Future*. 2nd ed. New York: St. Martin's Press, Inc., 1977.

Teilhard de Chardin, Pierre. *The Future of Man*. Trans. Norman Denny. New York: Harper & Row, Publishers, Inc., 1964.

Thrall, Charles A., and Jerold M. Starr. *Technology, Power and Social Change*. Lexington, Mass.: Lexington Books, 1972.

Toffler, Alvin. *Future Shock*. New York: Random House, Inc., 1970.

Toffler, Alvin, ed. *The Futurists*. New York: Random House, Inc., 1972.

Toynbee, Arnold J. *Civilization on Trial and the World and the West*. New York: Meridian Books, 1958.

Toynbee, Arnold J. *A Study of History*. 12 vols. London: Oxford University Press, 1934.

Transitional Institute. *World Hunger, Causes and Remedies: A Transitional Institute Report*. Washington, D.C.: Transitional Institute, 1974.

Truitt, Willis H., et al. *Science, Technology and Freedom*. Boston: Houghton Mifflin Co., 1974.

U.S. Department of Agriculture. *Farmers in a Changing World: The Yearbook of Agriculture: 1940*. Washington, D.C.: U.S. Government Printing Office, 1940.

Vaux, Kenneth. *Subduing the Cosmos: Cybernetics and Man's Future*. Richmond, Va.: John Knox Press, 1970.

Veblen, Thorstein B. *The Theory of the Leisure Class: An Economic Study of Institutions*. New York: Macmillan, Inc., 1972.

Volkov, Genrikh. *Man and the Challenge of Tech-*

nology. Moscow: Novosti Press Agency Publishing House, 1972.

Walker, Charles, and A. G. Walker. *Technology, Industry, and Man: The Age of Acceleration*. New York: McGraw-Hill Book Co., 1968.

Wallia, C. S., ed. *Toward Century 21: Technology, Society, and Human Values*. New York: Basic Books, Inc., Publishers, 1970.

Watts, Alan W. *Nature, Man, and Woman*. New York: Pantheon Books, Inc., 1958.

Watts, Alan W. *The Way of Zen*. New York: Pantheon Books, Inc., 1957.

Weber, Max. *The Protestant Ethic and the Spirit of Capitalism*. Trans. Talcott Parsons. New York: Charles Scribner's Sons, 1958.

Weissaman, Steve, et al. *The Trojan Horse*. San Francisco: Ramparts Press, Inc., 1974.

White, Lynn T. *Machine ex Deo: Essay on the Dynamism of Western Culture*. Cambridge, Mass.: The MIT Press, 1968.

White, Lynn T. *Medieval Technology and Social Change*. New York: Oxford University Press, Inc., 1962.

Whitehead, Alfred North. *Nature and Life*. New York: Greenwood Publishers, 1968.

Whitehead, Alfred North. *Science and the Modern World*. New York: Macmillan, Inc., 1925.

Wiener, Norbert. *God and Golem, Inc.: A Comment on Certain Points Where Cybernetics Impinges on Religion*. Cambridge, Mass.: The MIT Press, 1964.

Wiener, Norbert. *The Human Use of Human Beings: Cybernetics and Society*. Boston: Houghton Mifflin Co., 1950.

Williams, Eric E. *Capitalism and Slavery*. Chapel Hill, N.C.: The University of North Carolina Press, 1944.

Williams, Preston N., ed. *Ethical Issues in Biology and Medicine*. Cambridge, Mass.: Schenkman Publishing Co.; Morristown, N.J.: General Learning Press, 1973.

Williams, Raymond. *Television*. New York: Schocken Books, Inc., 1975.

Williams, Roger. *Politics and Technology*. London: Macmillan, Inc., 1971.

Winner, Langdon. *Autonomous Technology*. Cambridge, Mass.: The MIT Press, 1977.

Winter, Gibson. *Being Free: Reflections on America's Cultural Revolution*. New York: Macmillan, Inc., 1970.

Worster, Donald E. *Nature's Economy: A History of the Subversive Science*. San Francisco: Sierra Club Books, 1977.

Woztinsky, W. S., and E. S. Woytinsky. *World Population and Production: Trends and Outlook*. New York: Twentieth Century Fund, 1953.

Yablonsky, Lewis. *Robopaths*. Indianapolis: The Bobbs-Merrill Co., Inc., 1972.

Young, John F. *Cybernetics*. London: Iliffe, 1969.

Youngblood, Gene. *Expanded Cinema*. New York: E. P. Dutton, 1970.

ARTICLES

Baranson, Jack. "Technology Exports Can Hurt Us." *Foreign Policy*, No. 25 (1976-1977), pp. 180-194.

Barber, Bernard, ed. "Perspectives on Medical-Ethics and Social-Change." *The Annals of the American Academy of Political and Social Science*, Special Issue, No. 437 (May 1978), pp. 1-7.

Beres, Louis René. "The Errors of Cosmopolis: World Order and the Vision of Human Oneness." *Philosophy Today*, 18, No. 3 (1974), 234-247.

Blackstone, William J. "Ethics and Ecology." *Southern Journal of Philosophy*, 11, Nos. 1-2 (1973), 55-71.

Brooks, Harvey. "Technology and Values: New Ethical Issues Raised by Technological Progress." *Zygon*, 8, No. 1 (1973), 17-35.

Callahan, Daniel. "Bioethics as a Discipline." *Hastings Center Studies*, 1, No. 1 (1973), 66-73.

Callahan, Daniel. "Living with the New Biology." *Center Magazine*, 5. No. 4 (1972), 4-12.

Casper, Barry M., and Paul D. Wellstone. "The Science Court on Trial in Minnesota." *The Hastings Center Report*, 8, No. 4 (1978), 5-7.

Cassell, Eric J. "Dying in a Technological Society," *Hastings Center Studies*, 2, No. 2 (1974), 31-36.

Cowan, Ruth S. "The Industrial Revolution in the Home: Household Technology and Social Change in the 20th Century." *Technology and Culture*, 17, No. 1 (1976), 1-23.

Crawley, Edward F. "Designing the Space Colony." *Technology Review*, 79, No. 8 (1977), 44-50.

Culliton, Barbara J. "The Clone Ranger." Rev. of *In His Image: The Cloning of Man*, by David Rorvik. *Columbia Journalism Review*, 17, No. 2 (1978), 58-62.

Easlea, Brian. "Who Needs The Liberation of Nature?" *Science Studies*, 4, No. 1 (1974), 77-92.

Eulau, Heinz. "Technology and the Fear of the Politics of Civility." *Journal of Politics*, 35, No. 2 (1973), 367-385.

Fekete, John. "McLuhanacy: Counter-revolution in Cultural Theory." *Telos*, No. 15 (1973), pp. 75-123.

Fowles, J. "The Improbability of Space Colonies." *Technological Forecasting and Social Change*, 12, No. 4 (1978), 365-78.

George, Francois. "Forgetting Lenin." *Telos*, No. 18 (1973-74), pp. 53-88.

Geymorat, L. "Neutrality Is Impossible." *Scientia* (Milan), 107, Nos. 9-10 (1972), 759-763.

Grange, Joseph. "Magic, Technology and Being." *Religious Humanism*, 8, No. 2 (1974), 88-91.

Jouvenel, Bertrand de. "The Political Consequences of the Rise of Science." *Bulletin of the Atomic Scientists*, 19, No. 10 (Dec. 1963), 2-8.

Kates, Robert W. "Human Issues in Human Rights." *Science*, 11 August 1978, pp. 502-506.

Kolata, Gina Bari, "In Vitro Fertilization: Is it Safe

and Repeatable?" *Science,* 25 August 1978, pp. 698-699.

La Porte, Todd R. "Nuclear Waste: Increasing Scale and Sociopolitical Impacts." *Science,* 7 July 1978, pp. 22-28.

Lewin, Leonard C. "Bioethical Questions." *Harper's,* August 1978, pp. 21-29.

Linden, Stanton J. "Francis Bacon and Alchemy: The Reformation of Vulcan." *Journal of the History of Ideas,* 35, No. 4 (1974), 547-560.

McGinn, Thomas. "Ecology and Ethics." *International Philosophical Quarterly,* 14, No. 2 (1974), 149-160.

Mazur, Allan, and Beverlie Conant. "Controversy Over a Local Nuclear Waste Repository." *Social Studies of Science,* 8, No. 2 (1978), 235-243.

Medawar, Peter. "What Is Human About Man in His Technology." *Smithsonian,* 4, No. 2 (1973), 22-28.

Modell, Michael. "Sustaining Life in a Space Colony." *Technology Review,* 79, No. 8 (1977), 36-43.

Morison, Robert S. "Misgivings about Life-Extending Technologies." *Daedalus,* 107, No. 2 (Spring 1978), 211-226.

Nelson, Daniel. "Taylorism and the Workers at Bethlehem Steel, 1898-1901." *Pennsylvania Magazine of History and Biography,* 101, No. 4 (Oct. 1977), 487-505.

Pimentel, David. "Realities of a Pesticide Ban." *Environment,* 15, No. 2 (1973), pp. 18-31.

Pursell, Carroll W. Jr., ed. "Two Hundred Years of American Technology." *Technology and Culture,* Special Issue, 20, No. 1 (1979), 1-2.

Rama Rao, P. S. S. "The Structure of a Non-Violent Society: An Analysis of Gandhian Thought," *Jounral of Thought,* 9, No. 1 (1974), pp. 39-46.

Rappaport, R. A. "The Flow of Energy in an Agricultural Society." *Scientific American,* 224, No. 3 (1971), 116-133.

Sagasti, Francisco R. "Underdevelopment, Science and Technology: The Point of View of the Underdeveloped Countries." *Science Studies,* 3, No. 1 (1973), 47-59.

Sinclair, George. "A Call for a Philosophy of Engineering." *Technology and Culture,* 18, No. 4 (1977), 685-689.

Sinsheimer, Robert L. "Humanism and Science." *Leonardo,* 10, No. 1 (1977), 59-62.

Weingart, Jerome Martin. "The 'Helios Strategy': An Heretical View of the Potential Role of Solar Energy in the Future of A Small Planet." *Technological Forecasting and Social Change.* 12, No. 4 (1978), 273-315.

Winner, Langdon. "On Criticizing Technology." *Public Policy,* 20, No. 1 (1972), 35-59.